W9-AWX-001

Basic Electronic Communication

ROY BLAKE

Niagara College of Applied Arts & Technology

WEST PUBLISHING COMPANY

Minneapolis/St. Paul   New York   Los Angeles   San Francisco

Text design: Sylvia Dovner, Technical Texts
Art: Illustrious, Inc.
Copyedit: Caroline Jumper
Composition: G & S Typesetters
Production, printing, and binding: West Publishing Company

West's Commitment to the Environment

In 1906, West Publishing Company began recycling materials left over from the production of books. This began a tradition of efficient and responsible use of resources. Today, up to 95 percent of our legal books and 70 percent of our college and school texts are printed on recycled, acid-free stock. West also recycles nearly 22 million pounds of scrap paper annually—the equivalent of 181,717 trees. Since the 1960s, West has devised ways to capture and recycle waste inks, solvents, oils, and vapors created in the printing process. We also recycle plastics of all kinds, wood, glass, corrugated cardboard, and batteries, and have eliminated the use of styrofoam book packaging. We at West are proud of the longevity and the scope of our commitment to the environment.

Library of Congress Cataloging-in-Publication Data
Blake, Roy.
    Basic electronic communication / Roy Blake.
        p.   cm.
    Includes index.
    ISBN 0-314-01200-1
    1. Telecommunication.   2. Electronics.   I. Title.
TK5101.B555   1993
621.382—dc20   92-35006
∞

To my wife Penny and my son Adam,
who never (well, hardly ever) complained about
the time I spent writing and editing.
I hope the result justifies their patience.

Contents

* Denotes optional sections

Preface

Intended for use in two-, three-, and four-year programs in electronics technology, *Basic Electronic Communication* accommodates a variety of course structures and curriculums. The book can be used for a single-term course (covering Chapters 1–9 plus the instructor's choices of later topics) or for a two-term course or two-course sequence (with Chapters 1–9 as the first course and 10–16 as the second).

The early chapters give students a strong foundation in analog communications, stressing amplitude and frequency modulation and their variations. Material on antennas, transmission lines, and propagation is also presented. Later chapters cover specialty subjects such as microwaves, fiber optics, digital and data communications, and television.

At colleges that offer separate courses on the topics covered in later chapters, *Basic Electronic Communication* will serve as a valuable introduction and reference, but it will be especially useful for programs that do not offer separate courses on these specialty subjects. Since basic material is reviewed when it is put to use in more specialized chapters, instructors can tailor the specialty topics covered—and the order in which they are covered—to best fit their course requirements.

Basic Electronic Communication introduces major topics through a block-diagram approach, using just enough mathematics to make the material clear, and progresses to discussions of the representative sections of most interest. It balances a systems orientation with sufficient detail about "real world" applications and procedures to equip students to do useful "real world" work. They are encouraged to think about systems in ways that will continue to serve them well as technology continues to evolve.

Each chapter opens with a list of objectives to which students can refer while reading the material. Each section covers essential theory before discussing one or more representative systems. Examples use actual equipment, complete with photographs and manufacturer's specifications. To further motivate students (and to anchor lab work into the main course), many chapters also include sections on troubleshooting applications, test equipment, and measurement techniques. Special interest boxes present additional links to everyday life and add historical perspective. More specialized optional sections (marked with asterisks in the table of contents) can be included or omitted from chapters at the instructor's discretion. Content is reinforced by a chapter summary and a list of important terms and equations. Each chapter concludes with numerous questions and problems, with the answers to odd-numbered problems provided at the back of the book.

Although an introductory chapter reviews several essential topics from prior coursework, students using this book should already be familiar with the basics of analog electronics (including amplifiers and oscillators) and have a basic understanding of solid-state devices. Their mathematical backgrounds should include algebra, basic trigonometry, and logarithms. Calculus is not required, although appendices are included to allow students with calculus backgrounds to look further into the mathematics behind frequency modulation and Fourier analyses.

Devices such as phase-locked loops, and Class C and other tuned amplifiers, which students may not have encountered in earlier courses, are covered as introductory topics. Tubes are introduced and covered briefly where needed: power amplifier tubes for transmitters, and cathode-ray tubes for television and video displays. An appendix is included for students who are not already familiar with decibel notation. Mixing and modulation, and techniques for making measurements at radio frequencies, are covered as new topics. No prior knowledge is assumed about essentials such as frequency-domain analysis or Fourier analysis.

Students will find that the book is written in a clear and readable style. Examination of the work of communications pioneers lends a historical framework to the field, and the text also offers numerous glimpses into future possibilities. Readers will find that this text makes the field of communications come alive.

An instructor's manual is available with complete solutions to all problems. A set of transparency masters and a laboratory manual complete the instructor's package.

Acknowledgments

This book could not have been written without the help of many people. Thanks especially to Tom Tucker, formerly an editor at West, who first suggested the book and got me started, and to Chris Conty, who took over from Tom and made many helpful suggestions as the work progressed. Laura Evans, who coordinated the production of the book, and Caroline Jumper, the copy editor who corrected my mistakes, also deserve special thanks. The artists at Illustrious had the job of deciphering my crude sketches and making legible drawings from them, and the folks at G & S Typesetters converted my marked-up typescript (which I was amused to find typesetters call "foul matter") into the neat and tidy book you see before you.

One of the ironies of publishing is that of all these people, Tom and Chris are the only ones I have met personally. I'm in Welland, Ontario, Chris is in Chicago, the production department of West is in Minnesota, the copy editor and artists are in California, and the typesetters are in Texas. Nonetheless all these people have become valued colleagues, and I thank them all.

In addition, many people reviewed this book, and made valuable suggestions. The book would be different, and not as good, without them. A list of reviewers follows.

Richard Anthony, Cuyahoga Community College (OH)
David Baldyga, University of Hartford
Thomas Bingham, St. Louis Community College
George Borchers, ITT Technical Institute (UT)
Ray Burns, Red River Community College (Manitoba)
Foster Chin, Tulsa Junior College—Northeast
Elaine Cooney, Purdue University—Indianapolis
G. J. Gerard, Greater New Haven State Technical College
Robert Hills, Erie Community College (NY)
Richard Honeycutt, Davidson County Community College (NC)
Bruce Koller, Diablo Valley College (CA)
Ronald Mackie, DeVry Institute of Technology (Ontario)
James E. McKay, DeVry Institute of Technology (GA)
Susan Meardon, Wake Technical Community College (NC)
Victor Michael, Pennsylvania College of Technology
Gary Mullett, Springfield Technical Community College (MA)
Mark Oliver, Monroe Community College (NY)

James Pearson, DeVry Institute of Technology (TX)
James Predko, Lansing Community College
Vicki Price, DeVry Institute of Technology (OH)
Alan Shapiro, Northern Virginia Community College
Larry Smith, Vincennes University (IN)
Jim Stewart, DeVry Institute of Technology (NJ)
Ronald Worley, San Diego Mesa Community College
David Zimny, Southern College of Technology (GA)

1 Introduction to Communication Systems

Objectives

After studying this chapter, you should be able to:

1. Describe the essential elements of a communication system.
2. Explain the need for modulation in communication systems.
3. Distinguish between baseband, carrier, and modulated signals, and give examples of each.
4. Write the equation for a modulated signal and use it to list and explain the various types of continuous-wave modulation.
5. Describe time-division and frequency-division multiplexing.
6. Explain the relationship between channel bandwidth, baseband bandwidth, and transmission time.
7. List the requirements for distortionless transmission, and describe some of the possible deviations from this ideal.
8. Use frequency-domain representations of signals, and convert simple signals between time and frequency domains.
9. Use a table of Fourier series to find the frequency-domain representations of common waveforms.
10. Describe several types of noise, and calculate the noise power and voltage for thermal noise.
11. Calculate signal-to-noise ratio, noise figure, and noise temperature for single and cascaded stages.
12. Use the spectrum analyzer for frequency, power, and signal-to-noise ratio measurements.

1.1 INTRODUCTION

Communication was one of the first applications of electrical technology. Today, in the age of fiber optics and satellite television, facsimile machines and cellular telephones, communication systems remain at the leading edge of electronics. Probably no other branch of electronics has as profound an effect on people's lives.

This book will introduce you to the study of electronic communication systems. After a brief survey of the history of communications, we will begin, in this chapter, with a consideration of the basic elements that are common to any such system: a **transmitter**, a **receiver**, and a **communication channel**. We will also begin the discussions of signals and of noise that will continue throughout the book.

Chapter 2 will review some of the fundamentals that you will need for later chapters, and introduce you to some of the circuits that are common with radio-frequency systems, and with which you may not be familiar. After that we will proceed, in succeeding chapters, to investigate the various ways of implementing each of the three basic elements.

We begin with analog systems. The reader who is impatient to get on to digital systems should realize that many of those also require analog technology to function. The modern technologist needs to have at least a basic understanding of both analog and digital systems before specializing in one or the other.

It is often said that we are living in the information age. Communication technology is absolutely vital to the generation, storage, and transmission of this information.

1.1.1 The Past, Present, and Future of Communications

Practical electrical communication began in 1837 with Samuel Morse's telegraph system. It was not the first system to use electricity to send messages, but it was the first to be commercially successful. Though not electronic, it had all the essential elements of the communication systems to be studied in this book. There was a transmitter, consisting of a telegraph key and a battery, to convert information into an electrical signal that could be sent along wires. There was a receiver, called a *sounder,* to convert the electrical signal into sound that could be perceived by the operator. A variant of this system used marks on a paper tape. There was also a transmission channel, consisting of wire on poles. Later, beginning in 1866, telegraph cables were run under water. By 1898, there were twelve transatlantic cables in operation.

Voice communication by electrical means began with the invention of the telephone by Alexander Graham Bell in 1876. Since that time, telephone systems have steadily expanded and have been linked together in complex ways. It is now possible to talk to almost anywhere in the world from nearly anywhere else.

The electronic content of telephony has also increased. In its early days, the telephone system did not involve electronics at all. Eventually, the vacuum tube, and later the transistor, allowed the use of amplifiers to increase the distance over which signals could be sent. Radio links were installed to allow communication where cables were impractical. Currently, there is a change underway to digital electronics, in both signal transmission and switching systems.

Radio is an important medium of communication. The theoretical framework was constructed by James Clerk Maxwell in 1865, and was verified experimentally by Heinrich Rudolph Hertz in 1887. Practical radiotelegraph systems were in use by the turn of the century, mainly for ship-to-shore and ship-to-ship service. The first transatlantic communication by radio was accomplished by Guglielmo Marconi in 1901.

Early radio transmitters used spark gaps to generate radiation, and were not well suited to voice transmission. By 1906, some transmitters were using specially designed high-frequency alternators, and one of these was used experimentally to transmit voice. The one-kilowatt transmitter was modulated by connecting a microphone in series with the antenna. The microphone was allegedly water-cooled. Regular radio broadcasting did not begin until 1920, by which time transmitters and receivers used vacuum-tube technology. The diode tube had been invented by Sir John Ambrose Fleming in 1904, and the triode, which could work as an amplifier, by Lee De Forest in 1906.

By the late 1920s, radio broadcasting was commonplace and experiments with television were under way. The United States and several European countries had experimental television services before World War II, and after the war it became a practical worldwide reality.

Satellite communication is a relatively new form of radio. The first artificial satellite, Sputnik I, was launched by the Soviet Union in 1957, but practical, reliable communication via satellite really began with the launch of Intelsat I in 1965. This satellite was still partly experimental; the first modern commercial geostationary satellite, Anik A-1, was launched by Canada in 1972.

Modern communication systems have benefited from the invention of the transistor in 1948 and of the integrated circuit in 1958. Modern equipment makes extensive use of both analog and digital solid-state technology, often in the same piece of equipment. The current cellular telephone system, for instance, uses analog FM radio, coupled with an elaborate digital control system that would be completely impractical without microprocessors, in the phones themselves and in the fixed stations in the network.

The coming years are likely to bring more exciting developments. Optical fiber is already a very important communication medium, and its use is bound to increase. The trend toward digital communication, even for such "analog" sources as voice and video, will continue.

It is impossible to predict accurately what will be developed in the future. Cellular telephone and facsimile are both improvements on systems that had been in use for many years and were regarded as not particularly exciting, yet they have both grown very quickly in the last several years. About the only thing that is certain about the future of communication technology is that it will be interesting.

Throughout this book an attempt will be made to keep the reader in touch with the past, with current developments, and, sometimes, with some possibilities for the future. This is done partly to make things more interesting, but also with the idea that a technologist may benefit from a knowledge of how a particular technology was developed, what the constraints were (whether technical, economic, or political), what is being done with it now, and what is proposed for the future.

1.2 ELEMENTS OF A COMMUNICATION SYSTEM

Any communication system moves information from a source to a destination through a **channel**. Figure 1.1 illustrates this very simple idea, which is at the heart of everything that will be discussed in this book. The information from the source will generally not be in a form that can travel through the channel, so a device called a *transmitter* will be employed at one end, and a *receiver* at the other.

1.2.1 The Source

The source or information signal can be analog or digital. Common examples are analog audio and video signals, and digital data. Sources are often described in terms of the frequency range that they occupy. Telephone-quality analog voice signals, for instance, contain frequencies from about 300 Hz to 3 kHz, while analog high-fidelity music needs a frequency range of approximately 20 Hz to 20 kHz.

Video requires a much larger frequency range than audio. An analog video signal of television-broadcast quality needs a frequency range from dc to about 4.2 MHz.

Digital sources can be derived from audio or video signals, or may consist of data (alphanumeric characters, for example). Digital signals can have almost any bandwidth depending on the number of bits transmitted per second, and the method used to convert binary ones and zeros into electrical signals.

1.2.2 The Channel

A communication channel can be almost anything: a pair of conductors or an optical fiber, for example. Much of this book deals with radio communication, where free space serves as the channel.

Sometimes a channel can carry the information signal directly. For example, an audio signal may be carried directly by a twisted-pair telephone cable. On the other hand, a radio link through free space cannot be used directly for voice signals. An antenna of enormous length would be required, and it would not be possible to transmit more than one signal without interference. Such situations require the use of a carrier signal whose frequency is such that it will travel, or propagate, through the channel. This carrier wave will be altered, or **modulated**, by the information signal, in such a way that the information can be recovered at the destination. When a carrier is

FIGURE 1.1

Elements of a
Communication System

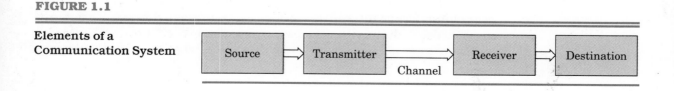

used, the information signal is also known as the *modulating signal*. Since the carrier frequency is generally much higher than that of the information signal, the frequency spectrum of the information signal is often referred to as the **baseband**. Thus, the three terms *information signal*, *modulating signal*, and *baseband* are equivalent in communication schemes involving modulated carriers.

1.2.3 Types of Modulation

All systems of modulation are variations on a small number of possibilities. A carrier is generated at a frequency much higher than the highest baseband frequency. Usually, the carrier is a sine wave. The instantaneous amplitude of the baseband signal is used to vary some parameter of the carrier.

A general equation for a sine-wave carrier is:

$$e(t) = E_c \sin(\omega_c t + \theta) \tag{1.1}$$

where $e(t)$ = instantaneous voltage as a function of time
E_c = peak voltage
ω_c = frequency in radians per second
t = time in seconds
θ = phase angle in radians

Radians per second are used in the mathematics concerning modulation to make the equations simpler. Of course, frequency is usually given in hertz, rather than in radians per second, when practical devices are being discussed. It is easy to convert between the two systems by recalling from basic ac theory that $\omega = 2\pi f$. Any reader for whom Equation (1.1) is unclear should refer to the review of ac theory in Appendix A for a refresher.

In modulation, the parameters that can be changed are amplitude E_c, frequency ω_c, and phase θ. Combinations are also possible; for example, many schemes for transmitting digital information use both amplitude and phase modulation.

The modulation is done at the transmitter. An inverse process, called *demodulation* or *detection*, takes place at the receiver, to restore the original baseband signal.

1.2.4 Signal Bandwidth

An unmodulated sine-wave carrier would exist at only one frequency and so would have zero bandwidth. However, a modulated signal is no longer a single sine wave, and it will therefore occupy a greater bandwidth. Exactly how much bandwidth is needed depends on the baseband frequency range (or data rate, in the case of digital communication) and the modulation scheme in use. There is a general rule that relates bandwidth and information capacity. Hartley's Law states that the amount of information that can be transmitted in a given time is proportional to bandwidth for a given modulation scheme.

$$I = ktB \tag{1.2}$$

where I = amount of information to be sent
        k = a constant that depends on the type of modulation
        t = time available
        B = channel bandwidth

Some modulation schemes use bandwidth more efficiently than others. The bandwidth of each type of modulated signal will be examined in detail in later chapters.

1.2.5   Frequency-Division Multiplexing

One of the benefits of using modulated carriers, even with channels that are capable of carrying baseband signals, is that several carriers can be used, at different frequencies. Each can be separately modulated with a different information signal, and filters at the receiver can separate the signals and demodulate whichever one is required.

Multiplexing is the term used in communications for the combining of two or more information signals. When the available frequency range is divided among the signals, the process is known as *frequency-division multiplexing* (FDM).

Radio and television broadcasting are everyday examples of FDM, where the available spectrum is divided among many signals. There are limitations to how many signals can be crowded into a given frequency range, due to the fact that each requires a certain bandwidth. For example, a television channel occupies a bandwidth of 6 MHz. Figure 1.2 shows how FDM applies to the VHF television band.

1.2.6   Time-Division Multiplexing

An alternative method of using a single communication channel to send many signals is to use *time-division multiplexing* (TDM). Instead of dividing the available bandwidth of the channel among many signals, the entire bandwidth is used for each signal, but only for a small part of the time. A glance at Equation (1.2) will confirm that time and bandwidth are equivalent in terms of information capacity. A nonelectronic example is the division of the total available time on a television channel among the various programs transmitted. Each program uses the whole bandwidth of the channel, but only for part of the time. Examples of TDM in electronic communication are not as common in everyday experience as FDM, but TDM is used extensively, especially with digital communication. The digital telephone system is a good example.

It is certainly possible to combine FDM and TDM. For example, the avail-

FIGURE 1.2

Frequency-Division Multiplexing in the VHF Television Band

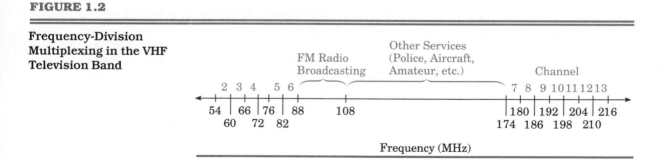

Frequency (MHz)

able bandwidth of a communication satellite is divided among a number of transmitter–receiver combinations called *transponders*. This is an example of FDM. A single transponder can be used to carry a large number of digital signals using TDM.

1.2.7 Frequency Bands

Hertz's early experiments were in the laboratory at frequencies in the range of approximately 50 to 500 MHz. When others, such as Marconi, tried to apply his results to practical communication, they initially found that results were much better at lower frequencies. Little was known about radio propagation at the time, or about antenna design, for that matter. It is now known that frequencies ranging from a few kilohertz to many gigahertz have their uses in radio communication systems.

In the meantime, a system of labeling frequencies came into use, with those most commonly used in the early days, from about 300 kHz to 3 MHz, being called *medium frequencies* (MF). Names were assigned to each order of magnitude of frequency, both upward and downward. The names persist to this day. Thus we have, besides medium frequencies, *low frequencies* (LF) from 30 to 300 kHz, and *very low frequencies* (VLF) from 3 to 30 kHz. Going the other way, we have *high frequencies* (HF) from 3 to 30 MHz, *very high frequencies* (VHF) from 30 to 300 MHz, and so on. Figure 1.3 shows the entire usable spectrum, with the appropriate labels.

FIGURE 1.3

The Radio Frequency Spectrum

FREQUENCY	WAVELENGTH	SAMPLE USES
300 GHz	1 mm	
Extremely High Frequencies (EHF)	Millimeter Waves	
30 GHz	1 cm	
Super High Frequencies (SHF)		Radar
3 GHz	10 cm Microwaves	Communications Satellites
Ultra High Frequencies (UHF)		Microwave Ovens / Cellular Telephones
300 MHz	1 m	
Very High Frequencies (VHF)		TV Broadcast / FM Broadcast
30 MHz	10 m	
High Frequencies (HF)	Short Waves	Shortwave Broadcast / Commercial
3 MHz	100 m	
Medium Frequencies (MF)	Medium Waves	AM Broadcast
300 kHz	1 km	
Low Frequencies (LF)	Long Waves	Navigation
30 kHz	10 km	Submarine Communications
Very Low Frequencies (VLF)		
3 kHz	100 km	
Voice Frequencies (VF)		
300 Hz	1000 km	Audio
Extremely Low Frequencies (ELF)		
30 Hz	10,000 km	Power Transmission

Radio waves can also be described according to wavelength, which is the distance a wave travels in one period. There is a general equation that relates frequency to wavelength for any wave:

$$v = f\lambda \tag{1.3}$$

where v = velocity of the wave in meters per second
        f = frequency of the wave in hertz
        λ = wavelength in meters

For a radio wave in free space, the velocity is the same as that of light, which is approximately 3×10^8 meters per second. The usual symbol for this quantity is c. Equation (1.3) then becomes

$$c = f\lambda \tag{1.4}$$

where c = speed of light, 3×10^8 m/s
        f = frequency in hertz
        λ = wavelength in meters

It can be seen from Equation (1.4) that a frequency of 300 MHz corresponds to a wavelength of 1 m, and that wavelength is inversely proportional to frequency. Low-frequency signals are sometimes called *long wave*, high frequencies correspond to *short wave*, and so on. The reader is no doubt familiar with the term *microwave* to describe signals in the gigahertz range.

EXAMPLE 1.1

Calculate the wavelength in free space corresponding to a frequency of:

(a) 1 MHz (AM radio broadcast band)

(b) 27 MHz (CB radio band)

(c) 4 GHz (used for satellite television)

Solution

Rearranging Equation (1.4),

$$c = f\lambda$$

$$\lambda = \frac{c}{f}$$

(a) $$\lambda = \frac{3 \times 10^8 \text{ m/s}}{1 \times 10^6 \text{ Hz}}$$
      $$= 300 \text{ m}$$

(b) $$\lambda = \frac{3 \times 10^8 \text{ m/s}}{27 \times 10^6 \text{ Hz}}$$
      $$= 11.1 \text{ m}$$

(c) $$\lambda = \frac{3 \times 10^8 \text{ m/s}}{4 \times 10^9 \text{ Hz}}$$
      $$= 0.075 \text{ m}$$
      $$= 7.5 \text{ cm}$$

1.2.8 Distortionless Transmission

The receiver should restore the baseband signal to exactly as it originated from the source. Of course, there will be a time delay due to the distance over which communication takes place, and there will likely be a change in amplitude as well. Neither effect is likely to cause problems, though there are exceptions. The time delay involved in communication via geostationary satellite can be a nuisance in telephone communication. Even though radio waves propagate at the speed of light, the great distance (about 70 thousand kilometers) over which the signal must travel causes a quarter-second delay.

Any other changes represent **distortion** that has a corrupting effect on the signal. There are many possible types, some of which are listed below. Not all of these may be immediately clear. Much of the rest of this book will be devoted to explaining them in more detail. Some possible types of distortion are:

- harmonic distortion: harmonics (multiples) of some of the baseband components are added to the original signal.
- intermodulation distortion: additional frequency components generated by combining (mixing) the frequency components in the original signal. Mixing will be described in Chapter 2.
- nonlinear frequency response: some baseband components are amplified more than others.
- nonlinear phase response: phase shift between components of the signal.
- noise: both transmitter and receiver add **noise**. The channel is also noisy. This noise adds to the signal and masks it. Noise will be discussed later in this chapter.
- interference: where more than one signal uses the same transmission medium, the signals may interact with each other.

One of the advantages of digital communication is the ability to regenerate a signal that has been corrupted by noise and distortion, provided that it is still identifiable as representing a one or a zero. In analog systems, however, noise and distortion tend to accumulate. There are cases in which distortion can be removed at a later point. If the frequency response of a channel is not flat but is known, for instance, equalization in the form of filters can be used to compensate. However, harmonic distortion, intermodulation, and noise, once present, are impossible to remove completely from an analog signal. A certain amount of immunity can be built into digital schemes, but excessive noise and distortion levels will be reflected in increased error rates.

1.3   TIME AND FREQUENCY DOMAINS

The reader will already be familiar with the **time-domain** representation of signals. An ordinary oscilloscope display, showing amplitude on one scale and time on the other, is a good example.

Signals can also be described in the **frequency domain**. In a frequency-domain representation, amplitude or power is shown on one axis and frequency is displayed on the other. A **spectrum analyzer** gives a frequency-domain representation of signals.

FIGURE 1.4

Sine Wave in Time and
Frequency Domains

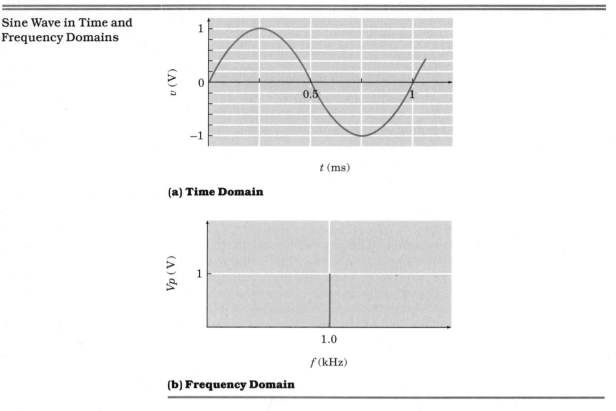

(a) Time Domain

(b) Frequency Domain

Any signal can be represented either way. For example, a 1 kHz sine wave is shown in both ways in Figure 1.4. The time-domain representation should need no explanation. (If it does, see Appendix A for a review of ac theory.) As for the frequency domain, a sine wave has energy only at its fundamental frequency, so it can be shown as a straight line at that frequency.

Notice that our way of representing the signal in the frequency domain fails to show one thing: the phase of the signal. The signal in Figure 1.4(b) could just as easily be a cosine wave as a sine wave.

Frequency-domain representations are very useful in the study of communication systems. For instance, the bandwidth of a modulated signal generally has some fairly simple relationship to that of the baseband signal. This bandwidth can easily be found if the baseband signal can be represented in the frequency domain. As we proceed, many other examples will occur where the ability to work with signals in the frequency domain will be required.

1.3.1 Fourier Series

Any well-behaved periodic waveform can be represented as a series of sine and/or cosine waves at multiples of its fundamental frequency, plus (sometimes) a dc offset. This very useful (and perhaps rather surprising) fact, was discovered in 1822 by Joseph Fourier, a French mathematician, in the course of research on heat conduction. Not all signals used in communications are strictly periodic, but they are often close enough for practical purposes.

Fourier's discovery, applied to a time-varying signal, can be expressed mathematically as follows:

$$f(t) = \frac{A_0}{2} + A_1 \cos \omega t + B_1 \sin \omega t + A_2 \cos 2\omega t$$

$$+ B_2 \sin 2\omega t + A_3 \cos 3\omega t + B_3 \sin 3\omega t + \cdots \tag{1.5}$$

where $f(t)$ = any well-behaved function of time as described above. For our purposes, $f(t)$ will generally be either a voltage $v(t)$ or a current $i(t)$.

A_n and B_n = real-number coefficients; that is, they can be positive, negative, or zero.

ω = radian frequency of the fundamental.

The radian frequency can be found from the time-domain representation of the signal by finding the period (that is, the time T after which the whole signal repeats exactly), and using the equations:

$$f = \frac{1}{T} \tag{1.6}$$

and

$$\omega = 2\pi f \tag{1.7}$$

The simplest ac signal is a sinusoid. The frequency-domain representation of a sine wave has already been described, and is shown in Figure 1.4(b) for a sine wave of voltage with a period of 1 ms and a peak amplitude of 1 V. A look at Equation (1.5) shows that, for this signal, all the coefficients are zero except for B_1, which has a value of 1 V. The equation becomes

$$v(t) = \sin (2000\pi) \text{ V}$$

which is certainly no surprise.

The examples presented below use equations from the table of Fourier series for common repetitive waveforms in Appendix B. For readers who are interested, the mathematics needed to find the series for any periodic waveform is also included in Appendix B.

EXAMPLE 1.2

Find and sketch the Fourier series corresponding to the square wave in Figure 1.5(a).

Solution

A square wave is another signal with a simple Fourier representation, though not quite as simple as that for a sine wave. For the signal shown in Figure 1.5(a), the frequency is 1 kHz, as before, and the peak voltage is 1 V.

According to the Fourier series for odd-function square waves given in Item 3(a) of Appendix B, Table B.1, this signal has components at an infinite number of frequencies: all odd multiples of the fundamental frequency of 1 kHz. However, the amplitude decreases with frequency, so that the third harmonic has an amplitude one-third that of the fundamental, the fifth harmonic an amplitude of one-fifth the fundamental, and so on. Mathematically, a square wave of voltage with a rising edge at $t = 0$ and no dc offset can be expressed as follows:

$$v(t) = \frac{4V}{\pi}\left(\sin \omega t + \frac{1}{3}\sin 3\omega t + \frac{1}{5}\sin 5\omega t + \cdots\right) \tag{1.8}$$

where V = peak amplitude of the square wave
          ω = radian frequency of the square wave
          t = time in seconds

For this example, Equation (1.8) becomes:

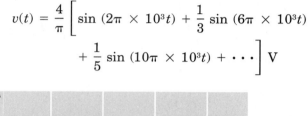

$$v(t) = \frac{4}{\pi}\left[\sin\left(2\pi \times 10^3 t\right) + \frac{1}{3}\sin\left(6\pi \times 10^3 t\right)\right.$$
$$\left. + \frac{1}{5}\sin\left(10\pi \times 10^3 t\right) + \cdots\right]\text{V}$$

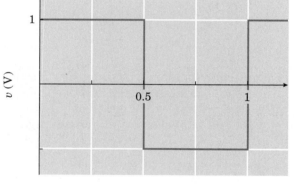

(a) Time Domain

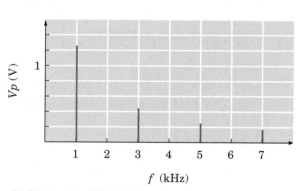

(b) Frequency Domain

FIGURE 1.5

The equation shows that the signal has frequency components at odd multiples of 1 kHz, that is, at 1 kHz, 3 kHz, 5 kHz, and so on. The 1 kHz component has a peak amplitude of

$$V_1 = \frac{4}{\pi} = 1.27 \text{ V}$$

The component at three times the fundamental frequency (3 kHz) has an amplitude one-third that of the fundamental, that at 5 kHz has an amplitude one-fifth that of the fundamental, and so on.

$$V_3 = \frac{4}{3\pi} = 0.424 \text{ V}$$

$$V_5 = \frac{4}{5\pi} = 0.255 \text{ V}$$

$$V_7 = \frac{4}{7\pi} = 0.182 \text{ V}$$

The result for the first four components is sketched in Figure 1.5(b). Theoretically, an infinite number of components would be required to describe the square wave completely, but, as the frequency increases, the amplitude of the components decreases rapidly.

The representations in Figures 1.5(a) and 1.5(b) are not two different signals, but merely two different ways of looking at the same signal. This can be shown graphically by adding the instantaneous values of several of the sine waves in the frequency-domain representation. If enough of these components are included, the result begins to look like the square wave in the time-domain representation. Figure 1.6 shows the results for two, four, and ten components. It was created by simply taking the instantaneous values of all the components at the same time and adding them algebraically. This was done for a number of time values, resulting in the graphs in Figure 1.6. Doing these calculations by hand would be simple but rather tedious, so a computer was used to make the calculations and plot the graphs. A perfectly accurate representation of the square wave would require an infinite number of components, but, as can be seen from the figure, ten components give a very good representation. This is because the amplitudes of higher-frequency components of the signal are very small.

It is possible to go back and forth at will between time and frequency domains, but it should be apparent that information about the relative phases of the frequency components in the Fourier representation of the signal is required to reconstruct the time-domain representation. The Fourier equations do have this information, but the sketch in Figure 1.5(b) does not. If the phase relationships between frequency components are changed in a communication system, the signal will be distorted in the time domain.

Figure 1.7 illustrates this point. The same ten coefficients were used as in Figure 1.6(c), but this time the waveforms alternated between sine and cosine: sine for the fundamental, cosine for the third harmonic, sine for the fifth, and so on. The result is a waveform that looks the same on the frequency-domain sketch of Figure 1.5(b), but very different in the time domain.

FIGURE 1.6

Construction of a
Square Wave from
Fourier Components

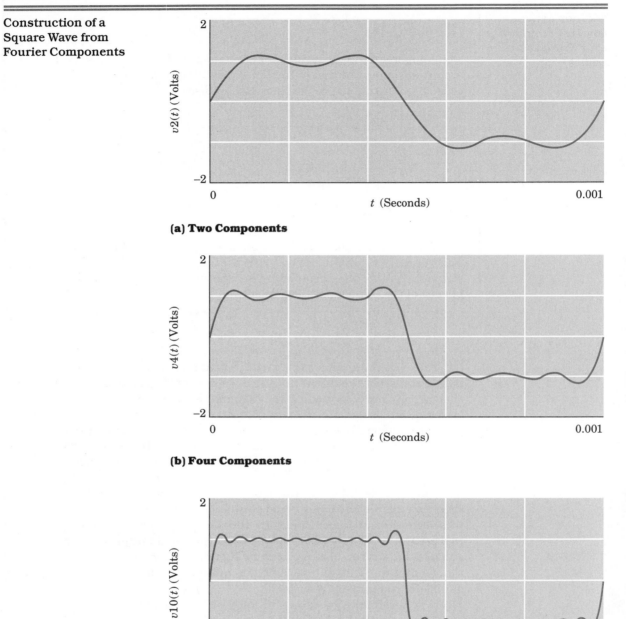

(a) Two Components

(b) Four Components

(c) Ten Components

FIGURE 1.7

Addition of Square-Wave
Fourier Components
with Wrong Phase
Angles

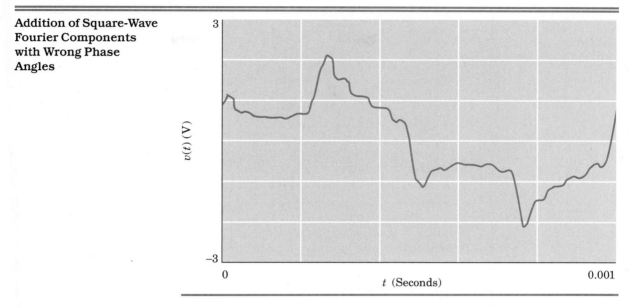

EXAMPLE 1.3

Find the Fourier series for the signal in Figure 1.8(a).

Solution

The positive-going sawtooth wave of Figure 1.8(a) has a Fourier series with a dc term and components at all multiples of the fundamental frequency. The general equation for such a wave is, from Appendix B:

$$v(t) = \frac{A}{2} - \left(\frac{A}{\pi}\right)\left(\sin \omega t + \frac{1}{2}\sin 2\omega t + \frac{1}{3}\sin 3\omega t + \cdots\right) \qquad (1.9)$$

The first (dc) term is simply the average value of the signal.

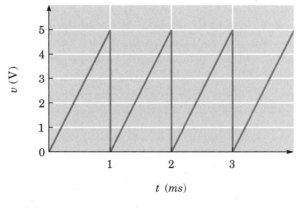

(a) Time Domain

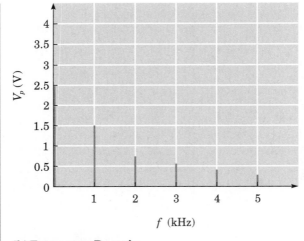

(b) Frequency Domain

FIGURE 1.8

For the signal in Figure 1.8, with a frequency of 1 kHz and a peak amplitude of 5 V, Equation (1.9) becomes:

$$v(t) = 2.5 - 1.59 \,[\sin (2\pi \times 10^3 t) + 0.5 \sin (4\pi \times 10^3 t)$$
$$+ \, 0.33 \sin (6\pi \times 10^3 t) + \cdots] \text{ V}$$

1.3.2 Effect of Filtering on Signals

As we have seen, many signals have a bandwidth that is theoretically infinite. Limiting the frequency response of a channel will remove some of the frequency components of these signals, causing the time-domain representation to be distorted. An uneven frequency response will emphasize some components at the expense of others, again causing distortion. Nonlinear phase shift will also affect the time-domain representation. For instance, shifting the phase angles of some of the frequency components in the square-wave representation of Figure 1.7 changed the signal to something other than a square wave.

However, Figure 1.6 shows that, while infinite bandwidth may theoretically be required, for practical purposes quite a good representation of a square wave may be obtained with a band-limited signal. This is true in general: the wider the bandwidth the better, but acceptable results may be obtained with a band-limited signal. This is welcome news, because practical systems always have finite bandwidth.

1.3.3 Bandwidth Limitations in Practical Systems

Although real communication systems have a solid theoretical base, they also involve many practical considerations. There is always a trade-off between fidelity to the original signal (that is, the absence of any distortion of its waveform) and such factors as bandwidth and cost. Increasing the bandwidth often increases the cost of a communication system. Not only is the hardware likely to be more expensive, but bandwidth itself may be in short supply. In radio communication, for instance, spectral allocations are assigned by national regulatory bodies in keeping with international agreements. This task is handled in the United States by the Federal Communications Commission (FCC) and in Canada by Communications Canada (formerly called the Department of Communication, or DOC). The allocations are constantly being reviewed, and there never seems to be enough room in the usable spectrum to accommodate all potential users.

In communication over cables, whether by electrical or optical means, the total bandwidth of a given cable is fixed by the technology employed. The more bandwidth used by each signal, the fewer signals can be carried by the cable.

Consequently, some services are constrained by cost and/or government regulation to use less than optimal bandwidth. For example, telephone systems use about 3 kHz of baseband bandwidth for voice. This is obviously not distortionless transmission: the difference between a voice heard "live," in the same room, and the same voice heard over the telephone is obvious. High-fidelity audio needs at least 15 kHz of baseband bandwidth. On the other hand, the main goal of telephone communication is the understanding of speech, and the bandwidth used is sufficient for this. Using a larger bandwidth would contribute little or nothing to intelligibility, and would greatly increase the cost of nearly every part of the system.

1.4   NOISE AND COMMUNICATIONS

Noise in communication systems originates both in the channel and in the communication equipment. Noise consists of undesired, usually random, variations that interfere with the desired signals and inhibit communication. It cannot be avoided completely, but its effects can be reduced by various means, such as reducing the signal bandwidth, increasing the transmitter power, and using low-noise amplifiers for weak signals.

It is helpful to divide noise into two types: *internal noise*, which originates within the communication equipment; and *external noise*, which is a property of the channel. (Most of the time in this book, the channel will be a radio link.)

1.4.1   External Noise

Assuming that the channel is a radio link, there are many possible sources of noise. The reader is no doubt familiar with the "static" that afflicts AM radio broadcasting during thunderstorms. Interference from automobile ignition systems is another common problem. Domestic and industrial electrical equipment, from light dimmers to vacuum cleaners to computers, can also cause objectionable noise. Not so obvious, but of equal concern to the communications specialist, is noise produced by the sun and other stars.

Man-Made Noise.   Noise is generated by equipment that produces sparks. Examples include automobile engines and electric motors with brushes. Any fast-risetime voltage or current can also generate interference, even without arcing. Light dimmers and computers are in this category.

Noise of this type has a broad frequency spectrum, but its energy is not equally distributed over the frequency range. This type of interference is generally more severe at lower frequencies, but the exact frequency distribution depends upon the source itself and any conductors to which it is connected.

Computers, for instance, may produce strong signals at multiples and sub-multiples of their clock frequency, and little energy elsewhere.

Man-made noise can propagate through space or along power lines. It is usually easier to control it at the source than at the receiver. A typical solution for a computer, for instance, involves shielding and grounding the case and all connecting cables, and installing a low-pass filter on the power line where it enters the enclosure.

Atmospheric Noise. Atmospheric noise is often called *static* because lightning, which is a static-electricity discharge, is a principal source of atmospheric noise. This type of disturbance can propagate for long distances through space. Lightning has most of its energy at relatively low frequencies, up to several megahertz.

Obviously nothing can be done to reduce atmospheric noise at the source. However, circuits are available to reduce its effect by taking advantage of the fact that this noise has a very high peak-to-average power ratio; that is, the noise occurs in short intense bursts with relatively long periods of time between bursts. It is often possible to improve communication by simply disabling the receiver for the duration of the burst. This technique is called *noise blanking*.

Space Noise. The sun is a powerful source of radiation over a wide range of frequencies, including the radio-frequency spectrum. Other stars also radiate noise called *cosmic*, *stellar*, or *sky noise*. Its intensity as received on earth is naturally much less than for solar noise, because of the greater distance.

Solar noise can be a serious problem with satellite reception, which becomes impossible when the satellite is in a line between the antenna and the sun. Space noise is more important at higher frequencies (VHF and above) because atmospheric noise dominates at lower frequencies and because most of the space noise at lower frequencies is absorbed by the upper atmosphere.

1.4.2 Internal Noise

Noise is generated in all electronic equipment. Both passive components (like resistors and cables) and active devices (like diodes, transistors, and tubes) can be noise sources. Several types of noise will be examined in this section, beginning with the most important, thermal noise.

Thermal Noise. Thermal noise is produced by the random motion of electrons in a conductor due to heat. The term "noise" is often used alone to refer to this type of noise, which is found everywhere in electronic circuitry.

The power density of thermal noise is constant with frequency, from zero to frequencies well above those used in electronic circuits. That is, there is equal power in every hertz of bandwidth. Thermal noise is thus an equal mixture of noise of all frequencies. It is sometimes called *white noise*, by analogy with white light, which is an equal mixture of all colors.

The noise power available from a conductor is a function of its temperature, as shown by the equation:

$$P_N = kTB \tag{1.10}$$

where P_N = noise power in watts

k = Boltzmann's constant, 1.38×10^{-23} joules/kelvin (J/K)

T = absolute temperature in kelvins (K). This can be found by adding 273 to the Celsius temperature.

B = noise power bandwidth in hertz

This equation is based on the assumption that power transfer is maximum. That is, the source (the resistance generating the noise) and the load (the amplifier or other device that receives the noise) are assumed to be matched in impedance.

Equation (1.10) shows that noise power is directly proportional to bandwidth. The bandwidth in question is the bandwidth over which the noise is observed. For an ideal bandpass filter, the bandwidth would be equal to the filter bandwidth. With practical filters, the noise power bandwidth will usually be somewhat greater than the 3 dB bandwidth.

The power referred to above is *average* power. The instantaneous voltage and power are random, but over time the voltage will average out to zero, as it does for an ordinary ac signal, and the power will average to the value given by Equation (1.10). Figure 1.9 shows this effect.

FIGURE 1.9

Noise Voltage and Power

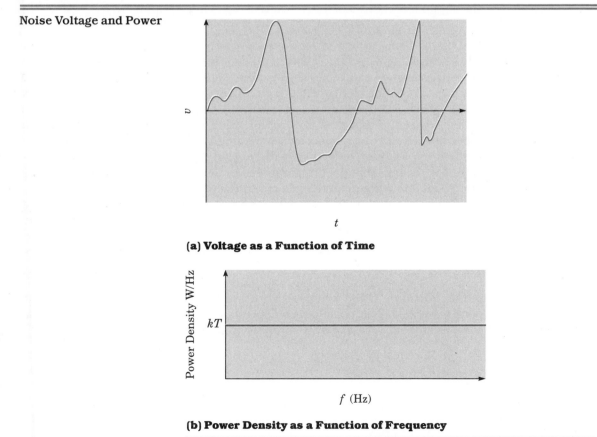

(a) Voltage as a Function of Time

(b) Power Density as a Function of Frequency

EXAMPLE 1.4

A receiver has a noise power bandwidth of 10 kHz. A resistor that matches the receiver input impedance is connected across its antenna terminals. What is the noise power contributed by that resistor in the receiver bandwidth, if the resistor has a temperature of 27°C?

Solution

First, convert the temperature to kelvins.

$$
\begin{aligned}
T(\text{K}) &= T(\text{°C}) + 273 \\
&= 27 + 273 \\
&= 300 \text{ K}
\end{aligned}
$$

According to Equation (1.10), a conductor at a temperature of 300 K would contribute a noise power, in that 10 kHz bandwidth, of:

$$
\begin{aligned}
P_N &= kTB \\
&= (1.38 \times 10^{-23} \text{ J/K}) (300 \text{ K}) (10 \times 10^3 \text{ Hz}) \\
&= 4.14 \times 10^{-17} \text{ W}
\end{aligned}
$$

This is obviously not a great deal of power, but it can be significant at the signal levels dealt with in sensitive receiving equipment.

Thermal noise power exists in all conductors and resistors at any temperature above absolute zero. The only way to reduce it is to decrease the temperature or the bandwidth of a circuit, or both. Amplifiers used with very low-level signals are often cooled artificially to reduce noise. The technique is called *cryogenics* and may involve, for example, cooling the first stage of a receiver for radio astronomy by immersing it in liquid nitrogen. The other method, *bandwidth reduction*, will be referred to many times throughout this book. Using a bandwidth greater than required for a given application is simply an invitation to problems with noise.

Thermal noise, as such, does not depend on the type of material involved or the amount of current passing through it. However, some materials and devices may also produce other types of noise that do depend on current. Carbon-composition resistors are in this category, for example, and so are semiconductor junctions.

Noise Voltage. Often we are more interested in the *noise voltage* than the power involved. The noise power depends only on bandwidth and temperature, as stated above. Power in a resistive circuit is given by the equation:

$$
P = \frac{V^2}{R} \tag{1.11}
$$

From this, we see that:

$$
\begin{aligned}
V^2 &= PR \\
V &= \sqrt{PR}
\end{aligned} \tag{1.12}
$$

FIGURE 1.10

Thermal Noise Voltage

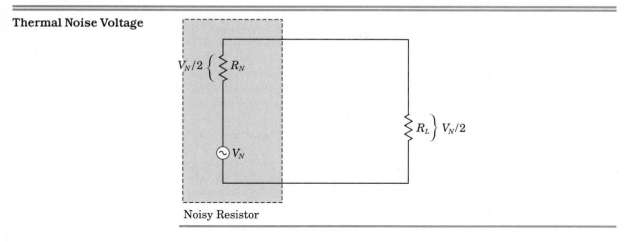

Noisy Resistor

Equation (1.12) shows that the noise voltage in a circuit must depend on the resistances involved, as well as temperature and bandwidth.

Figure 1.10 shows a resistor that serves as a noise source, connected across another resistor, R_L, that will be considered as a load. The noise voltage is represented as a voltage source V_N, in series with a noiseless resistance R_N. We know from Equation (1.10) that the noise power supplied to the load resistor, assuming a matched load, is:

$$P_N = kTB$$

In this situation, one-half the noise voltage appears across the load, and the rest is across the resistor that generates the noise.

The root-mean-square (RMS) noise voltage across the load is given by:

$$V_L = \sqrt{PR}$$
$$= \sqrt{kTBR}$$

An equal noise voltage will appear across the source resistance, so the noise source shown in Figure 1.10 must have double this voltage, or

$$V_N = 2\sqrt{kTBR}$$
$$= \sqrt{4kTBR} \qquad\qquad (1.13)$$

EXAMPLE 1.5

A 300 Ω resistor is connected across the 300 Ω antenna input of a television receiver. The bandwidth of the receiver is 6 MHz and the resistor is at room temperature of 293 K (20°C or 68°F). Find the noise power and noise voltage applied to the receiver input.

Solution

The noise power is given by Equation (1.10):

$$P_N = kTB$$
$$= (1.38 \times 10^{-23} \text{ J/K}) (293 \text{ K}) (6 \times 10^6 \text{ Hz})$$
$$= 24.2 \times 10^{-15} \text{ W}$$
$$= 24.2 \text{ fW}$$

The noise voltage is given by Equation (1.13):

$$V_N = \sqrt{4kTBR}$$
$$= \sqrt{4 (1.38 \times 10^{-23} \text{ J/K}) (293 \text{ K}) (6 \times 10^6 \text{ Hz}) (300 \ \Omega)}$$
$$= 5.4 \times 10^{-6} \text{ V}$$
$$= 5.4 \ \mu\text{V}$$

Of course, only one-half this voltage appears across the antenna terminals; the other half appears across the source resistance. Therefore the actual noise voltage at the input is 2.7 μV.

Shot Noise. This type of noise has a power spectrum that resembles that for thermal noise, having equal energy in every hertz of bandwidth, at frequencies from dc into the gigahertz region. However, the mechanism that creates it is different. Shot noise is due to random variations in current flow in active devices such as tubes, transistors, and semiconductor diodes. These variations are caused by the fact that current is a flow of carriers (electrons or holes), each of which carries a finite amount of charge. Current can thus be considered as a series of pulses, each consisting of the charge carried by one electron.

The name "shot noise" describes the electrons arriving randomly at the anode of a vacuum tube, like individual pellets of shot from a shotgun.

One would expect that the resulting noise power would be proportional to the device current, and this is true both for vacuum tubes and for semiconductor junction devices.

Shot noise is usually represented by a current source. The noise current for either a vacuum or a junction diode is given by the equation:

$$I_N = \sqrt{2qI_0B} \tag{1.14}$$

where I_N = RMS noise current, in amperes
  q = magnitude of the charge on an electron, equal to 1.6×10^{-19} coulomb
  I_0 = dc bias current in the device, in amperes
  B = bandwidth over which the noise is observed, in hertz

This equation is valid for frequencies much less than the reciprocal of the carrier *transit time* for the device. The transit time is the time a charge carrier spends in the device. Depending on the device, Equation (1.14) may be valid for frequencies up to a few megahertz or several gigahertz.

Since shot noise closely resembles thermal noise in its random amplitude

FIGURE 1.11

Diode Noise Generator

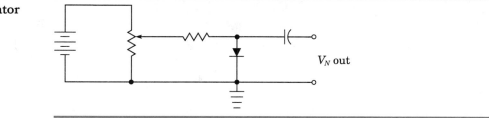

V_N out

and flat spectrum, it can be used as a substitute for thermal noise whenever a known level of noise is required. Such a noise source can be very useful, particularly when conducting measurements on receivers. A calibrated, variable noise generator can easily be constructed using a diode with adjustable bias current. Figure 1.11 shows a simplified circuit for such a generator.

EXAMPLE 1.6

A diode noise generator is required to produce 10 μV of noise in a receiver with an input impedance of 75 Ω, resistive, and a noise power bandwidth of 200 kHz. (These values are typical of FM broadcast receivers.) What must the current through the diode be?

Solution

First, convert the noise voltage to current, using Ohm's Law.

$$I_N = \frac{V_N}{R}$$

$$= \frac{10 \ \mu V}{75 \ \Omega}$$

$$= 0.133 \ \mu A$$

Next, rearrange Equation (1.14) and solve for I_0.

$$I_N = \sqrt{2qI_0B}$$

$$I_N^2 = 2qI_0B$$

$$I_0 = \frac{I_N^2}{2qB}$$

$$= \frac{(0.133 \times 10^{-6} \ A)^2}{2 \ (1.6 \times 10^{-19} \ C) \ (200 \times 10^3 \ Hz)}$$

$$= 0.276 \ A \quad \text{or} \quad 276 \ mA$$

Partition Noise. Partition noise is similar to shot noise in its spectrum and mechanism of generation, but it occurs only in devices where a single current separates into two or more paths. An example of such a device is a bipolar junction transistor, where the emitter current is the sum of the collector and base currents. As the charge carriers divide into one stream or the other, a random element in the currents is produced. A similar effect can occur in vacuum tubes.

The amount of partition noise depends greatly on the characteristics of the particular device, so no equation for calculating it will be given here. The same is true of shot noise in devices with three or more terminals. Device manufacturers provide noise figure information on their data sheets when a device is intended for use in circuits where signal levels are low and noise is important.

Partition noise is not a problem in field-effect transistors, where the gate current is negligible.

Excess Noise. Excess noise is also called *flicker noise* or $1/f$ noise (because the noise power varies inversely with frequency). Sometimes it is called *pink noise*, because there is proportionately more energy at the low-frequency end of the spectrum than with white noise, just as pink light has a higher proportion of red (the low-frequency end of the visible spectrum) than does white light. Excess noise is found in tubes, but is a more serious problem in semiconductors and in carbon resistors. It is not fully understood, but it is believed to be caused by variations in carrier density. Excess noise is rarely a problem in communication circuits, because it declines with increasing frequency, and is usually insignificant above approximately 1 kHz.

You may meet the term "pink noise" again. It refers to any noise that has equal power per octave, rather than per hertz. Pink noise is often used for testing and setting up audio systems.

High-Frequency Effects: Transit-Time Noise. Many junction devices produce more noise at frequencies approaching their cutoff frequencies. This high-frequency noise occurs when the time taken by charge carriers to cross a junction is comparable to the period of the signal. Some of the carriers may diffuse back across the junction, causing a fluctuating current that constitutes noise. Since most devices are used well below the frequency at which transit-time effects are significant, this type of noise is not usually important. The exception is for some microwave devices.

1.4.3 Addition of Noise from Different Sources

All of the noise sources we have looked at have random waveforms. The amplitude at a particular time cannot be predicted, though the average voltage is zero and the RMS voltage or current and average power are predictable (and *not* zero). Signals like this are said to be uncorrelated; that is, they have no definable phase relationship. In such a case, voltages and currents do not add directly, but the total voltage (series circuits) or current (parallel circuits) can be found by taking the square root of the sum of the squares of the individual voltages or currents.

Mathematically, we can say, for voltage sources in series:

$$V_{Nt} = \sqrt{V_{N1}^2 + V_{N2}^2 + V_{N3}^2 + \cdots}$$

(1.15)

and similarly, for current sources in parallel:

$$I_{Nt} = \sqrt{I_{N1}^2 + I_{N2}^2 + I_{N3}^2 + \cdots}$$

(1.16)

EXAMPLE 1.7

The circuit of Figure 1.12 shows two resistors in series, at two different temperatures. Find the total noise voltage and noise power produced at the load, over a bandwidth of 100 kHz.

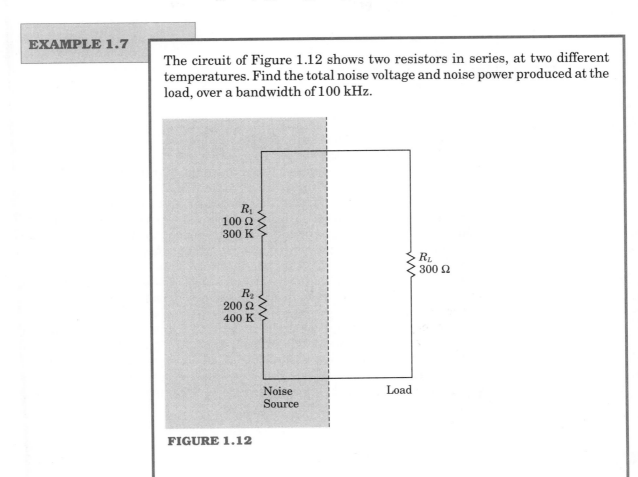

R_1
100 Ω
300 K

R_2
200 Ω
400 K

R_L
300 Ω

Noise
Source

Load

FIGURE 1.12

Solution

The open-circuit noise voltage can be found from Equation (1.15).

$$V_{Nt} = \sqrt{V_{R_1}^2 + V_{R_2}^2}$$

$$= \sqrt{(\sqrt{4kTBR_1})^2 + (\sqrt{4kTBR_2})^2}$$

$$= \sqrt{4kT_1BR_1 + 4kT_2BR_2}$$

$$= \sqrt{4kB(T_1R_1 + T_2R_2)}$$

$$= \sqrt{4\,(1.38 \times 10^{-23}\text{ J/K})\,(100 \times 10^3\text{ Hz})\,[(300\text{ K})\,(100\ \Omega) + (400\text{ K})\,(200\ \Omega)]}$$

$$= 779\text{ nV}$$

This is, of course, the open-circuit noise voltage for the resistor combination. Since in this case the load is equal in value to the sum of the resistors, one-half of this voltage, or 390 nV, will appear across the load. The easiest way to find the load power is from Equation (1.11):

$$P = \frac{V^2}{R}$$

$$= \frac{(390 \text{ nV})^2}{300 \ \Omega}$$

$$= 0.506 \times 10^{-15} \text{ W}$$

$$= 0.506 \text{ fW}$$

1.4.4 Signal-to-Noise Ratio

The main reason for studying and calculating noise power or voltage is the effect that noise has on the desired signal. In an analog system, noise makes the signal unpleasant to watch or listen to, and in extreme cases, difficult to understand. In digital systems, noise increases the error rate. In either case, it is not really the amount of noise that concerns us, but rather the amount of noise compared to the level of the desired signal. That is, it is the *ratio* of signal to noise power that is important, rather than the noise power alone. This **signal-to-noise** ratio, abbreviated *S/N* and usually expressed in decibels, is one of the most important specifications of any communication system. Typical values of *S/N* range from about 10 dB for barely intelligible speech to 90 dB or more for compact-disc audio systems.

Note that the ratio given above is a power, not a voltage ratio. Sometimes calculations result in a voltage ratio that must be converted to a power ratio in order to make meaningful comparisons. Of course, the value in decibels will be the same either way, provided that both signal and noise are measured in the same circuit, so that the impedances are the same. The two different equations below are used for power ratios and voltage ratios, respectively:

$$S/N \text{ (dB)} = 10 \log \frac{P_S}{P_N} \qquad (1.17)$$

$$S/N \text{ (dB)} = 20 \log \frac{V_S}{V_N} \qquad (1.18)$$

At this point, the reader who is not familiar with decibel notation should refer to Appendix E. Decibels are used constantly in communication theory and practice, and it is important to understand their use.

Although the signal-to-noise ratio is a fundamental characteristic of any communication system, it is often difficult to measure. For instance, it may be possible to measure the noise power by turning off the signal, but it is not possible to turn off the noise in order to measure the signal power alone. Consequently, a variant of *S/N*, called *(S+N)/N*, is often found in receiver specifications. This stands for the ratio of signal-plus-noise power to noise power alone.

There are other variants. For instance, there is the question of what to do about distortion. Though it is not a random effect like thermal noise, the effect on the intelligibility of a signal is rather similar. Consequently, it might be well to include it with the noise when measuring. This leads to the expression SINAD which stands for $(S+N+D)/(N+D)$, or signal-plus-noise and distortion, divided by noise and distortion. SINAD is usually used instead of S/N in specifications for FM receivers.

Both $(S+N)/N$ and SINAD are power ratios, and they are almost always expressed in decibels. Details on the measurement of $(S+N)/N$ will be provided in Chapter 5, and SINAD will be described in Chapter 9.

EXAMPLE 1.8

A receiver produces a noise power of 200 mW with no signal. The output level increases to 5 W when a signal is applied. Calculate $(S+N)/N$ as a power ratio and in decibels.

Solution

The power ratio is

$$(S+N)/N = \frac{5 \text{ W}}{0.2 \text{ W}}$$
$$= 25$$

In decibels, this is

$$(S+N)/N \text{ (dB)} = 10 \log 25$$
$$= 14 \text{ dB}$$

1.4.5 Noise Figure

Since thermal noise is produced by all conductors, and active devices add their own noise as described above, it follows that any stage in a communication system will add noise. An amplifier, for instance, will amplify equally both the signal and the noise at its input, but will, in addition, add some noise. It follows that the signal-to-noise ratio at the output will be lower than at the input.

Noise figure (abbreviated NF or just F) is a figure of merit, indicating how much a component, stage, or series of stages degrades the signal-to-noise ratio of a system. The noise figure is, by definition:

$$NF = \frac{(S/N)_i}{(S/N)_o} \tag{1.19}$$

where $(S/N)_i$ = input signal-to-noise power ratio (not in dB)

$(S/N)_o$ = output signal-to-noise power ratio (not in dB)

Very often both S/N and NF are expressed in decibels, in which case we have:

$$NF \text{ (dB)} = (S/N)_i \text{ (dB)} - (S/N)_o \text{ (dB)} \tag{1.20}$$

Since we are dealing with power ratios, it should be obvious that:

$$NF \text{ (dB)} = 10 \log NF \text{ (ratio)} \qquad (1.21)$$

Noise figure is occasionally called *noise factor*. Sometimes the term *noise ratio (NR)* is used for the simple ratio of Equation (1.19), with the term "noise figure" being reserved for the decibel form given by Equation (1.20).

EXAMPLE 1.9

The signal power at the input to an amplifier is 100 μW and the noise power is 1 μW. At the output, the signal power is 1 W and the noise power is 30 mW. What is the amplifier noise figure, as a ratio?

Solution

$$(S/N)_i = \frac{100 \ \mu W}{1 \ \mu W} = 100$$

$$(S/N)_o = \frac{1 \ W}{0.03 \ W} = 33.3$$

$$NF \text{ (ratio)} = \frac{100}{33.5} = 3$$

EXAMPLE 1.10

The signal at the input of an amplifier has an *S/N* of 42 dB. If the amplifier has a noise figure of 6 dB, what is the *S/N* at the output (in decibels)?

Solution

The *S/N* at the output can be found by rearranging Equation (1.20):

$$NF \text{ (dB)} = (S/N)_i \text{ (dB)} - (S/N)_o \text{ (dB)}$$
$$(S/N)_o \text{ (dB)} = (S/N)_i \text{ (dB)} - NF \text{ (dB)}$$
$$= 42 \text{ dB} - 6 \text{ dB}$$
$$= 36 \text{ dB}$$

1.4.6 Equivalent Noise Temperature

Equivalent **noise temperature** is another way of specifying the noise performance of a device. Noise temperature has nothing to do with the actual operating temperature of the circuit. Rather, as shown in Figure 1.13, it is the absolute temperature of a resistor that, connected to the input of a noiseless amplifier of the same gain, would produce the same noise at the output as the device under discussion. Figure 1.13(a) shows a real amplifier which, of course, produces noise at its output, even with no input signal. Figure

FIGURE 1.13

Equivalent Noise Resistance

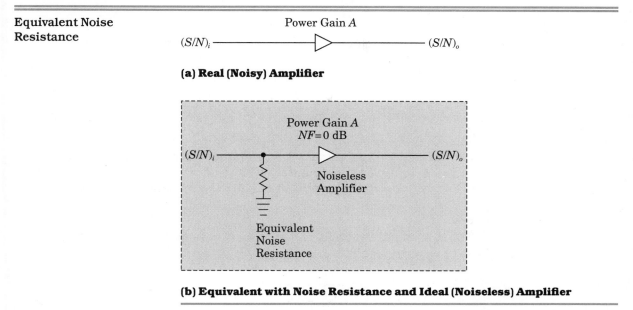

(a) Real (Noisy) Amplifier

(b) Equivalent with Noise Resistance and Ideal (Noiseless) Amplifier

1.13(b) shows a hypothetical noiseless amplifier with a noisy resistor at its input. The combination produces the same amount of a noise at its output as does the real amplifier of Figure 1.13(a).

From Equation (1.19), the noise figure is given by:

$$NF = \frac{(S/N)_i}{(S/N)_o}$$

$$= \frac{(S_i/N_i)}{(S_o/N_o)}$$

$$= \frac{S_i N_o}{N_i S_o}$$

$$= \frac{S_i}{S_o}\frac{N_o}{N_i} \tag{1.22}$$

where S_i = signal power at input

N_i = noise power at input

S_o = signal power at output

N_o = noise power at output

If we are going to refer all the amplifier noise to the input, then this noise, as well as the noise from the signal source, will be amplified by the gain of the amplifier. Let the amplifier power gain be A. Then:

$$A = \frac{S_o}{S_i} \tag{1.23}$$

Substituting this into Equation (1.22), we get

$$NF = \frac{N_o}{N_i A} \tag{1.24}$$

or

$$N_o = (NF)N_i A \tag{1.25}$$

Thus the total noise at the input is $(NF)N_i$.

Assuming that the input source noise N_i is thermal noise, it has the expression, first shown in Equation (1.10):

$$N_i = kTB$$

The equivalent input noise generated by the amplifier must be

$$
\begin{aligned}
N_{eq} &= (NF)N_i - N_i \\
&= (NF)kTB - kTB \\
&= (NF - 1)kTB \tag{1.26}
\end{aligned}
$$

If we suppose that this noise is generated in a (fictitious) resistor at temperature T_{eq}, and if we further suppose that the actual source is at a reference temperature T_0 of 290 K, then we get:

$$
\begin{aligned}
kT_{eq}B &= (NF - 1)kT_0B \\
T_{eq} &= (NF - 1)T_0 \\
&= 290(NF - 1) \tag{1.27}
\end{aligned}
$$

This may seem a trivial result, because it turns out that T_{eq} can be found directly from the noise figure, and thus contains no new information about the amplifier. However, T_{eq} is very useful when we look at microwave receivers connected to antennas by transmission lines. The antenna has a noise temperature due largely to space noise intercepted by the antenna. The transmission line and the receiver will also have noise temperatures. It is possible to get an equivalent noise temperature for the whole system by simply adding the noise temperatures of the antenna, transmission line, and receiver. From this, we can calculate the system noise figure merely by rearranging Equation (1.27):

$$
\begin{aligned}
T_{eq} &= 290(NF - 1) \\
NF - 1 &= \frac{T_{eq}}{290} \\
NF &= \frac{T_{eq}}{290} + 1 \tag{1.28}
\end{aligned}
$$

Equivalent noise temperatures of low-noise amplifiers are quite low, often less than 100 K. Note that this does not mean the amplifier operates at this temperature. It is quite common for an amplifier operating at an actual temperature of 300 K to have an equivalent noise temperature of 100 K.

EXAMPLE 1.11

An amplifier has a noise figure of 2 dB. What is its equivalent noise temperature?

Solution

First, we need the noise figure expressed as a ratio. This can be found from Equation (1.21):

$$NF \text{ (dB)} = 10 \log NF \text{ (ratio)}$$

From which we get:

$$NF \text{ (ratio)} = \text{antilog } \frac{NF \text{ (dB)}}{10}$$
$$= \text{antilog } 0.2$$
$$= 1.585$$

Now we can use Equation (1.27) to find the noise temperature:

$$T_{eq} = 290(NF - 1)$$
$$= 290(1.585 - 1)$$
$$= 169.6 \text{ K}$$

1.4.7 Noise Figure and Noise Temperature for Cascaded Amplifiers

When two or more stages are connected in *cascade*, as in a receiver, the noise figure of the first stage is the most important in determining the noise performance of the entire system. This is because noise generated in the first stage will be amplified in all succeeding stages. Noise produced in later stages will be amplified less, and noise generated in the last stage will be amplified least of all.

While the first stage is the most important noise contributor, the other stages cannot be neglected. It is possible to derive an equation that relates the total noise figure to the gain and noise figure of each stage. In what follows, the power gains and noise figures are expressed as ratios, not in decibels.

We will start with a two-stage amplifier like the one shown in Figure 1.14. The first stage has a gain A_1 and a noise figure NF_1, and the second-stage gain and noise figure are A_2 and NF_2 respectively.

Let the noise power input to the first stage be

$$N_{i1} = kTB$$

FIGURE 1.14

Noise Figure for a Two-
Stage Amplifier

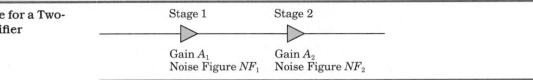

Then, according to Equation (1.25), the noise at the output is:

$$N_{o1} = NF_1 N_{i1} A_1$$
$$= NF_1 kTBA_1 \tag{1.29}$$

This noise appears at the input of the second stage. It is amplified by the second stage and appears at the output multiplied by A_2. However, the second stage also contributes noise, which can be expressed, according to Equation (1.26), as:

$$N_{eq2} = (NF_2 - 1)(kTB) \tag{1.30}$$

This noise also appears at the output, amplified by A_2. Therefore the total noise power at the output of the two-stage system is:

$$N_{o2} = N_{o1} A_2 + N_{eq2} A_2$$
$$= NF_1 A_1 kTBA_2 + (NF_2 - 1)kTBA_2$$
$$= kTBA_2(NF_1 A_1 + NF_1 - 1) \tag{1.31}$$

The noise figure for the whole system (NF_T) is, according to Equation (1.24):

$$NF_T = \frac{N_{o2}}{N_{i1} A_T}$$

Substituting the value for N_{o2} found in Equation (1.31), and making use of the fact that the total gain A_T is the product of the individual stage gains, we get

$$NF_T = \frac{kTBA_2(NF_1 A_1 + NF_2 - 1)}{kTBA_1 A_2}$$
$$= \frac{NF_1 A_1 + NF_2 - 1}{A_1}$$
$$= NF_1 + \frac{NF_2 - 1}{A_1} \tag{1.32}$$

This equation, known as *Friis' formula*, can be generalized to any number of stages:

$$NF_T = NF_1 + \frac{NF_2 - 1}{A_1} + \frac{NF_3 - 1}{A_1 A_2} + \frac{NF_4 - 1}{A_1 A_2 A_3} + \cdots \qquad (1.33)$$

This expression confirms our intuitive feeling by showing that the contribution of each stage is divided by the product of the gains of all the preceding stages. Thus the effect of the first stage is usually dominant. Note once again that the noise figures given in this section are ratios, not decibel values, and that the gains are power gains.

EXAMPLE 1.12

A three-stage amplifier has stages with the following specifications:

Stage	Power Gain	Noise Figure
1	10	2
2	25	4
3	30	5

Calculate the power gain, noise figure, and noise temperature for the entire amplifier, assuming matched conditions.

Solution

The power gain is simply the product of the individual gains.

$$A_T = 10 \times 25 \times 30 = 7500$$

The noise figure can be found from Equation (1.33):

$$NF_T = NF_1 + \frac{NF_2 - 1}{A_1} + \frac{NF_3 - 1}{A_1 A_2}$$

$$= 2 + \frac{4 - 1}{10} + \frac{5 - 1}{10 \times 25}$$

$$= 2.316$$

If answers are required in decibels, they can easily be found:

$$\text{Gain (dB)} = 10 \log 7500 = 38.8 \text{ dB}$$

$$NF \text{ (dB)} = 10 \log 2.316 = 3.65 \text{ dB}$$

The noise temperature can be found from Equation (1.27):

$$T_{eq} = 290(NF - 1)$$

$$= 290(2.316 - 1)$$

$$= 382 \text{ K}$$

1.5 TEST EQUIPMENT: THE SPECTRUM ANALYZER

In this chapter, we have discovered that there are two general ways of looking at signals: the time domain and the frequency domain. The ordinary oscilloscope is a convenient and versatile way of observing signals in the time domain, providing as it does a graph of voltage with respect to time. It would be logical to ask whether there is an equivalent instrument for the frequency domain: that is, something which could display a graph of voltage (or power) with respect to frequency.

The answer to the question is Yes: observation of signals in the frequency domain is the function of **spectrum analyzers**. These use one of two basic techniques. The first is to use a number of fixed-tuned bandpass filters, spaced throughout the frequency range of interest. The display indicates the output level from each filter, often with a bar graph consisting of light-emitting diodes (LEDs). The more filters and the narrower the range, the better the resolution that can be obtained. This method is quite practical for audio frequencies, and in fact a crude version, with five to twenty filters, is often found as part of a stereo system. More elaborate systems are often used in professional acoustics and sound-reinforcement work. Devices of this type are called *real-time spectrum analyzers*, as all of the filters are active all the time. Figure 1.15 shows a block diagram of a real-time spectrum analyzer.

When radio-frequency (RF) measurements are required, the limitations of real-time spectrum analyzers become very apparent. It is often necessary to observe signals over a frequency range with a width of megahertz or even gigahertz. Obviously, the number of filters required would be too large to be practical. For RF applications, it would be more logical to use only one, a tunable filter, and sweep it gradually across the frequency range of interest. This sweeping of the filter could coincide with the sweep of an electron beam across the face of a cathode-ray tube (CRT), and the vertical position of the beam could be made a function of the amplitude of the filter output. This is the essense of a *swept-frequency spectrum analyzer* like the one shown in Figure 1.16.

In practice, narrow-band filters that are rapidly tunable over a wide range with constant bandwidth are difficult to build. It is easier to build a

FIGURE 1.15

Real-Time Spectrum Analyzer

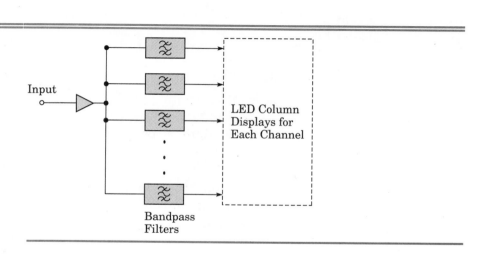

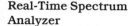

FIGURE 1.16

**Swept-Frequency
Spectrum Analyzer**

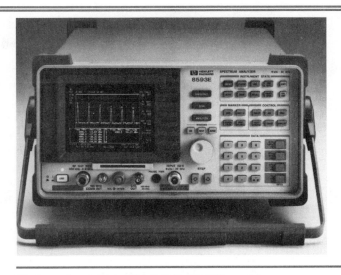

Photograph courtesy of Hewlett-Packard Company.

single, fixed-frequency filter with adjustable bandwidth, and sweep the signal through the filter. Figure 1.17 shows how this can be done. The incoming signal is applied to a mixer, along with a swept-frequency signal generated by a local oscillator in the analyzer. Mixers will be described in the next chapter; for now, it is sufficient to know that a mixer will produce outputs at the sum and the difference of the two frequencies that are applied to its input.

The filter can be arranged to be at either the sum or the difference frequency (usually the difference). As the oscillator frequency varies, the part of the spectrum that passes through the filter changes. The oscillator is a voltage-controlled oscillator (VCO) whose frequency is controlled by a sawtooth-wave generator that also provides the horizontal sweep signal for the CRT.

The filter output is an ac signal that must be rectified and amplified before it is applied to the vertical deflection plates of the CRT. If the amplification is linear, the vertical position of the trace will be proportional to the sig-

FIGURE 1.17

**Simplified Block
Diagram for a Swept-
Frequency Spectrum
Analyzer**

Input — Attenuator — Mixer — IF Filter — IF Amp — Envelope Detector — Vertical Amp

Bandwidth Control

Reference Level Control

VCO Center Frequency Control

Span Control

Sawtooth Timebase Generator

Horizontal Amp

CRT

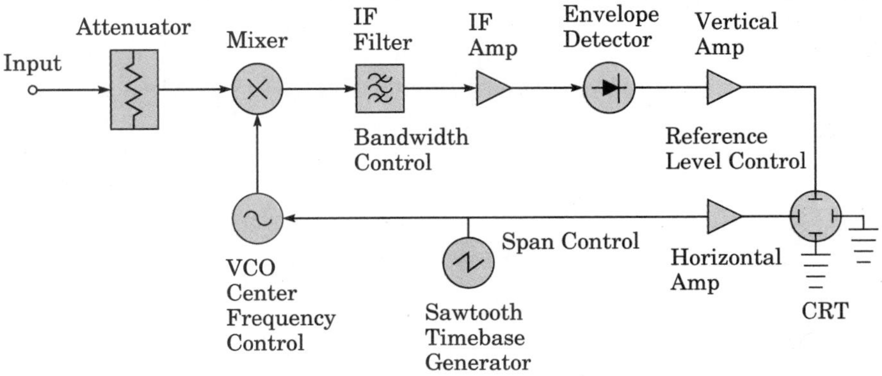

nal voltage amplitude at a given frequency. It is more common, however, to use logarithmic amplification, so that the display can be calibrated in decibels, with a reference level in decibels referenced to one milliwatt (dBm).

1.5.1 Using a Spectrum Analyzer

There are four main controls on a typical swept-frequency spectrum analyzer, and several less important ones. The major controls are shown on the block diagram of Figure 1.17. Their functions are as follows.

The *frequency control* sets either the center frequency of the sweep or the frequency at which it begins, called the start frequency. This is the frequency corresponding to the left edge of the display. Some analyzers provide a switch so that the display can show either the start or the center frequency.

The *span control* adjusts the range of frequencies on the display. There are two methods of specifying the span. Sometimes it is calibrated in kilohertz or megahertz per division. Other models give the span as the width of the whole display, which is normally ten divisions. Dividing this number by ten will give the span per division.

The frequency of a displayed signal can be found by adjusting the *tuning control* until the signal is in the center of the display (assuming the start/center control, if present, is at center). The frequency of a signal not at the center can be found by counting divisions and multiplying by the span per division:

$$f_{sig} = f_{center} + D_{sig}S \tag{1.34}$$

where f_{sig} = signal frequency in MHz
f_{center} = setting of the tuning control
D_{sig} = number of divisions by which the signal is to the right of the center of the display (This number is negative if the signal is left of center.)
S = span setting in MHz/division

The bandwidth control adjusts the filter bandwidth at the 3 dB points. Since the smaller the bandwidth the better the potential resolution of the analyzer, it might seem logical to build the filter with as narrow a bandwidth as possible, and use it that way at all times. However, sweeping a signal rapidly through a high-Q circuit (such as a narrow bandpass filter) gives rise to transients, which will reduce the accuracy of amplitude readings. If the frequency span being swept is small, a narrow bandwidth is necessary so that good resolution is obtained, but when a wide span is in use, narrow bandwidth is unnecessary. This is fortunate, for the sweep will naturally be faster, in terms of hertz per second, for the wider span. For wide spans, then, a wide bandwidth is appropriate, with narrow bandwidth being reserved for narrow span. Very often, the bandwidth and span controls are locked together, so that the correct filter bandwidth for a given span will be selected automatically. Provision is made for unlocking the span and bandwidth controls on the rare occasions when this is desirable.

It is important to realize that, as viewed on a spectrum analyzer, a sine wave, or a component of a more complex signal, will not always appear as a

FIGURE 1.18

Spectrum Analyzer Display

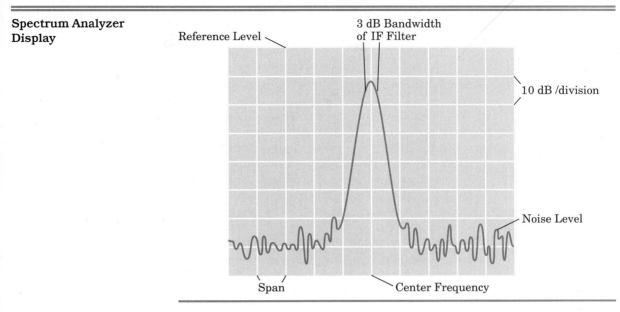

single vertical line. If the span is sufficiently narrow that the filter bandwidth is a significant portion of a division, the signal will appear to have a finite width. Such a display is shown in Figure 1.18. What is seen in the display is the width and shape of the filter passband, and not that of the signal.

The fourth major adjustment is the *reference level control*. Spectrum analyzers are set up so that amplitudes are measured with respect to a reference level at the top of the display (not the bottom, as one might assume). This level is set by the reference level control, which is normally calibrated in dBm.

There will also be an adjustable *attenuator* at the input to reduce strong signals to a level that can be handled by the input mixer. The attenuator control is normally coupled with the reference level setting, so that changing the attenuation automatically alters the displayed reference level.

Closely associated with the reference level control, there is often a switch that sets the *vertical scale factor*. Typically it has three settings: 10 dB/division, 1 dB/division, and linear. In the first position, one division vertically represents a 10 dB power difference at the input. The second position gives an expanded scale, with 1 dB per division. With either of these settings, the power of a signal can be measured in either of two ways: by setting the signal to the top of the display with the reference level controls, and then reading the reference level from the controls; or by counting divisions down from the top of the screen to the signal, multiplying by 1 or 10 dB according to the switch setting, and subtracting the result from the reference level setting:

$$P_{sig} = P_{ref} - F_s D_{sig} \tag{1.35}$$

where P_{sig} = signal power in dBm
P_{ref} = reference level in dBm

F_s = scale factor in dB/division

D_{sig} = number of divisions down from the top of the screen to the
      signal

The linear position is a little more complicated. The reference level is at
the top of the screen, as before. However, this time the screen is a linear voltage scale. Zero voltage is at the bottom, and the voltage corresponding to the
reference level is at the top. This can easily be calculated from the power in
dBm and the spectrum analyzer impedance, or you can use the nomograph
in Figure 1.19. Please note, however, that this nomograph is valid only if the
analyzer has an input impedance of 50 Ω, the most common value. Some analyzers allow the reference level to be set directly in volts.

Once the voltage that corresponds to the reference level is known, the
signal voltage can be found. If the reference level has been adjusted so that
the signal is just at the top of the screen, then the signal voltage is that corresponding to the reference level. Otherwise the signal voltage can be found
by using proportion. The signal voltage will be:

$$V_{sig} = \frac{V_{ref}U_{sig}}{N_{div}} \tag{1.36}$$

where V_{sig} = signal voltage in RMS volts

V_{ref} = RMS voltage corresponding to the reference level

FIGURE 1.19   Conversion Chart, dBm to Voltage (for 50 Ω)

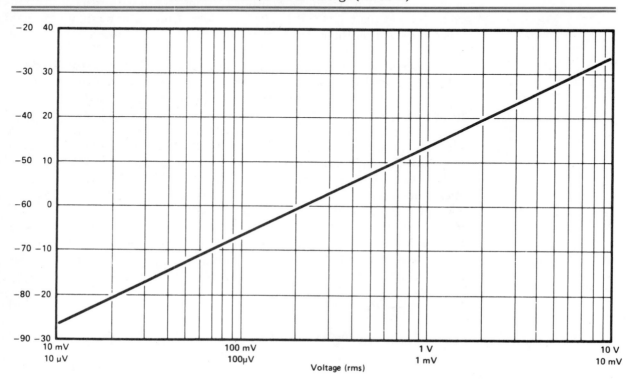

Courtesy Hewlett-Packard Company.

FIGURE 1.20

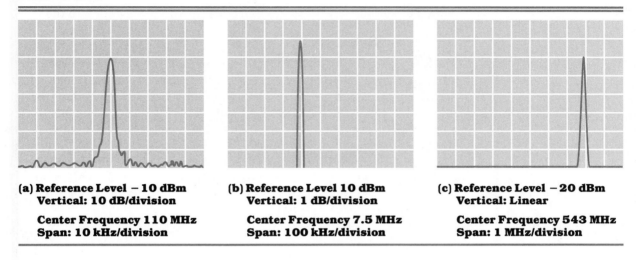

(a) **Reference Level − 10 dBm**
   Vertical: 10 dB/division

   Center Frequency 110 MHz
   Span: 10 kHz/division

(b) **Reference Level 10 dBm**
   Vertical: 1 dB/division

   Center Frequency 7.5 MHz
   Span: 100 kHz/division

(c) **Reference Level − 20 dBm**
   Vertical: Linear

   Center Frequency 543 MHz
   Span: 1 MHz/division

U_{sig} = position of the signal in divisions up from the bottom of the screen

N_{div} = number of divisions vertically on the display (usually 8)

The linear scale is not used very often with spectrum analyzers, but it can be handy on occasion.

EXAMPLE 1.13

For each of the spectrum-analyzer displays in Figure 1.20, find the frequency, power level (in dBm and watts), and voltage level of each of the signals. The input impedance of the analyzer is 50 Ω.

Solution

(a) The signal is in the center of the display horizontally, so its frequency is displayed on the analyzer's frequency readout, assuming that it is set to read the center frequency. In this example it is, so the signal frequency is 110 MHz. The signal peak is two divisions, at 10 dB/division, below the reference level of − 10 dBm, so its level is − 30 dBm. The equivalent power can be found from:

$$P \text{ (dBm)} = 10 \log \frac{P}{1 \text{ mW}}$$

$$P \text{ (mW)} = \text{antilog} \frac{\text{dBm}}{10}$$

$$= \text{antilog} \frac{-30}{10}$$

$$= 1 \times 10^{-3} \text{ mW}$$

$$= 1 \text{ } \mu\text{W}$$

The voltage can be found from the nomograph in Figure 1.19, or it can be found more accurately from:

$$P = \frac{V^2}{R}$$
$$V = \sqrt{PR}$$
$$= \sqrt{1 \ \mu V \times 50 \ \Omega}$$
$$= 7.07 \ mV$$

(b) The signal is one division to the left of center, at 100 kHz/division. The frequency is 100 kHz less than the reference frequency of 7.5 MHz, so

$$f = 7.5 \ MHz \ - \ 0.1 \ MHz \ = \ 7.4 \ MHz$$

As for the amplitude, this time the scale is 1 dB/division and the signal is one division down from the reference level, so the signal has a power level of

$$P \ (dBm) = 10 \ dBm \ - \ 1 \ dB \ = \ 9 \ dBm$$

This can be converted to watts and volts in the same way as for Part (a):

$$P \ (mW) = antilog \ \frac{dBm}{10}$$
$$= antilog \ \frac{9}{10}$$
$$= 7.94 \ mW$$
$$V = \sqrt{PR}$$
$$= \sqrt{7.94 \ mW \times 50 \ \Omega}$$
$$= 630 \ mV$$

(c) The signal is three divisions to the right of the center reference frequency of 543 MHz, at 1 MHz/division. Therefore, the frequency is:

$$f = 543 \ MHz \ + \ 3 \ \times \ 1 \ MHz$$
$$= 546 \ MHz$$

It is easier to find the signal voltage before the power, because this time the amplitude scale is linear. To arrive at the signal voltage, we need the voltage that corresponds to the power level at the top of the screen. Either from the graph or analytically, we can easily find that the reference level of -20 dBm corresponds to a power of 10 μW, which represents 22.4 mV. The level on the display is $\frac{6}{8}$ of this, because the scale is linear with 8 divisions representing the reference level voltage. Therefore the signal level is:

$$V = 22.4 \text{ mV} \times \frac{6}{8} = 16.8 \text{ mV}$$

Now it is easy to find the power. In watts, it is:

$$P = \frac{V^2}{R}$$

$$= \frac{(16.8 \text{ mV})^2}{50 \ \Omega}$$

$$= 5.64 \ \mu\text{W}$$

Now we find the power in dBm:

$$P \text{ (dBm)} = 10 \log \frac{P}{1 \text{ mW}}$$

$$= 10 \log \frac{5.64 \ \mu\text{W}}{1 \text{ mW}}$$

$$= -22.5 \text{ dBm}$$

1.5.2   Other Spectrum Analyzer Controls

Though the four controls described in the previous section are the most important in understanding the operation of a spectrum analyzer, practical analyzers generally have additional controls that can improve the accuracy of the display or the ease of use of the equipment. Some of these functions and the controls that regulate them will be described in this section.

Video Filter. This is simply an adjustable low-pass filter applied to the input of the vertical channel of the display. Used sparingly, it can help reduce the effect of noise. Noise, being a random and constantly changing voltage, can be expected to move the trace in the vertical direction more rapidly than does the signal. It should then be possible to average out some of the noise with minimal effect on the signal. This is accomplished by the video filter, which works quite well if the sweep speed is low. Overuse of the video filter will, however, reduce the amplitude of the signal in the display. If the signal is changing rapidly, as is the case, for example, for a signal with voice modulation, it will not be possible to use the video filter for accurate measurements. Figure 1.21 shows the video filter in use.

Another use for the video filter is in measuring the noise level itself. This will be discussed later in this section.

Baseline Clipper. Typically, the bottom division or so of a spectrum analyzer display will consist of noise. This part of the display can be blanked with the baseline clipper, giving the display a neater appearance, especially when photographed. The baseline clipper does not actually reduce the level of the noise, however, and it should not be confused with the video filter.

FIGURE 1.21

Effect of Video Filter

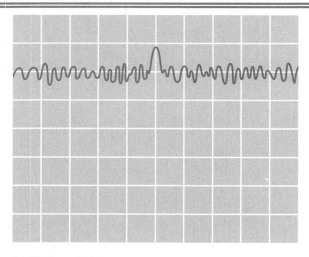

(a) Without Filter

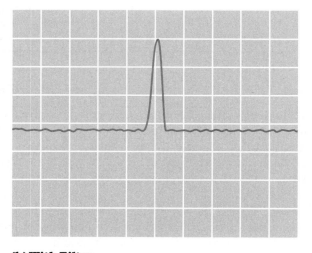

(b) With Filter

Display Controls. The display portion of a spectrum analyzer resembles that of an oscilloscope. The horizontal sweep and VCO frequency sweep must be synchronized. Usually, the sweep rate is determined automatically by the setting of the span control, so that it will not be too fast for the filter in use. Occasionally, it may still be too fast, as shown by distortion of the trace. In this case, it can be slowed down by adjusting the sweep speed manually.

Otherwise, the display will have the usual oscilloscope controls: intensity, focus, horizontal position, and vertical position. The setting of the latter two will affect the accuracy of the frequency and power scales respectively, and they should be adjusted during calibration of the analyzer and then left alone.

Sweeps using very narrow bandwidths will require a very low horizontal sweep speed, making a storage display desirable. Storage is useful as well for comparing displays or examining signals that appear only briefly. Storage can be either analog or digital, as with an oscilloscope. Digital storage has the advantage that digital averaging can be accomplished over several sweeps. This will reduce the prominence of random noise without changing the signal display, provided that the signal remains constant while the averaging is taking place. Some digital-storage spectrum analyzers have additional useful features, such as the ability to store waveforms on memory cards, the ability to interface with a computer, and adjustable markers to make power and frequency readout easier.

1.5.3 Noise Measurements with a Spectrum Analyzer

Almost any time a spectrum analyzer is used on the 10 dB/division setting, a fringe of noise, sometimes called *grass* because of its appearance, will be visible at the bottom of the screen. The noise level will vary with the bandwidth setting of the analyzer: the wider the bandwidth, the higher the noise level. This is because noise power is proportional to bandwidth.

When attempting to measure noise with a spectrum analyzer, remember that the noise power bandwidth of the analyzer filter is not necessarily equal to the 3 dB bandwidth shown on the bandwidth control. Nor is it necessarily equal to the noise power bandwidth of the system being measured.

To understand these problems, consider first a very simple case. Assume that the noise produced in the spectrum analyzer is negligible in comparison with external noise, and that the noise power bandwidth of the spectrum analyzer is identical with that of the system under test. Then the noise power in the system bandwidth, in dBm, could simply be read from the screen of the analyzer. If a signal-to-noise power ratio were required, it could be obtained, directly in dB, simply by displaying the signal and the noise at the same time, and counting the divisions between them. Figure 1.22 shows the type of display that would be observed.

For this idealized situation, the noise power in the given bandwidth is -50 dBm, and the signal power is -20 dBm, for a signal-to-noise power ratio of 30 dB.

In the real world, the noise power bandwidth of a spectrum analyzer may not be the same as the 3 dB bandwidth marked on the front panel. Typically it will be about 20% greater. Also, the detector used in the analyzer will not usually be accurate with noise signals. The logarithmic amplifier, used to give a decibel scale, will also reduce accuracy by amplifying noise peaks less than the rest of the noise signal. Typically, the detector and amplifier characteristics can be compensated for by adding a correction factor of approximately 2.5 dB to noise levels measured on the analyzer. The video filter should be on so that noise signals will be averaged on the display.

When finding the signal-to-noise ratio for a system, it is important to know the system bandwidth. Since most forms of noise have a power that varies directly with bandwidth, the more bandwidth the system has, the more noise will be let through. Ideally, the noise power bandwidth of the analyzer should equal the system bandwidth, but often this is not possible. In that case, a correction factor can be applied.

FIGURE 1.22

Spectrum Analyzer
Noise Measurements

Reference Level −20 dBm
Vertical: 10 dB/division

Bandwidth: 10 kHz

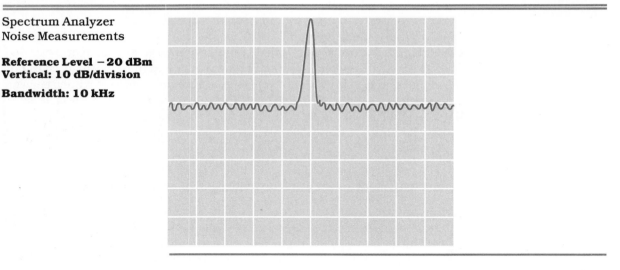

The correction can be made, without reference to any "cookbook" formulas, by the following method:

1. Find the noise level in dBm and calculate the noise power bandwidth of the spectrum analyzer.
2. Convert the noise power from dBm to watts.
3. Divide the noise power in watts by the bandwidth to arrive at noise power in watts per hertz.
4. Multiply the noise power in watts per hertz by the bandwidth of the system under test.
5. Convert back to dBm.
6. Subtract the noise level in dBm from the signal level in dBm to arrive at the signal-to-noise level in decibels.

This sequence is simple and logical but somewhat tedious. Instead, a condensed version can be used to find the noise in the system bandwidth:

$$N_{SB} = N_{AB} + 10 \log \frac{SB}{ANB} \qquad (1.37)$$

where N_{SB} = noise level in dBm in the system bandwidth
   N_{AB} = noise level in dBm measured on the screen of the spectrum analyzer, plus any required correction factor for the amplifier and detector
   SB = bandwidth of the system being investigated
   ANB = noise bandwidth of the spectrum analyzer which, as mentioned above, is usually somewhat larger than the bandwidth marked on the front panel

EXAMPLE 1.14

An FM radio station has a signal level of -50 dBm at the spectrum analyzer, and the noise measures -90 dBm with the analyzer set for a 100 kHz bandwidth. What is the signal-to-noise ratio, assuming an FM receiver has a bandwidth of 150 kHz?

Solution

First we add the required correction to the noise reading for our analyzer. We get this from the manufacturer's specifications, and find it to be 2.5 dB. This gives us a noise level of -87.5 dBm.

Next, we normalize the bandwidth. Again from the manufacturer's specifications, we find we have to increase the 3 dB bandwidth by 20% to get the noise bandwidth. This gives us a noise bandwidth of 120 kHz to use in Equation (1.37).

Substituting into that equation, we find:

$$N_{SB} = N_{AB} + 10 \log \frac{SB}{ANB}$$

$$= -87.5 + 10 \log \frac{150}{120}$$

$$= -87.5 + 0.969$$

$$= -86.5 \text{ dBm}$$

Now we can find the signal-to-noise ratio as:

$$S/N = -50 - (-86.5) = 36.5 \text{ dB}$$

One other problem can occur when measuring noise. All electronic equipment, including the spectrum analyzer itself, produces noise. If the noise level of the analyzer is not a good deal lower than that of the system under test, a low-noise preamplifier will have to be used with the analyzer.

Spectrum analyzers have a great many other uses in the analysis of communication systems. We shall return to the spectrum analyzer many times in the course of this book.

SUMMARY OF CHAPTER 1

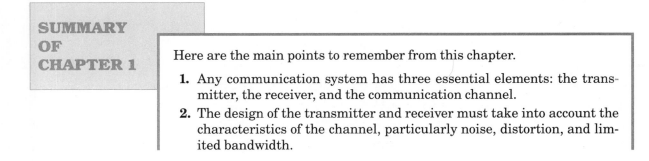

Here are the main points to remember from this chapter.

1. Any communication system has three essential elements: the transmitter, the receiver, and the communication channel.
2. The design of the transmitter and receiver must take into account the characteristics of the channel, particularly noise, distortion, and limited bandwidth.

3. The use of modulation is necessary with many types of communication channels. This involves a carrier waveform, some characteristic of which is changed in accordance with the amplitude of a lower-frequency signal known as the baseband, information, or modulating signal.

4. The characteristics of a carrier that can be modulated are its amplitude, frequency, and phase.

5. Time-division and frequency-division multiplexing are two systems for sharing a channel among several information signals.

6. For a given modulation scheme, the amount of information that can be transmitted is proportional to the time taken and the channel bandwidth employed.

7. Signal transmission is said to be distortionless if the information signal at the receiver output is identical to that at the transmitter input, except for time delay and change in amplitude. Any other change is known as distortion.

8. Signals can be represented in either the time or frequency domain. Fourier series can be used to find the frequency-domain representation for a periodic signal. Signals can be observed in the frequency domain with the help of a spectrum analyzer.

9. Noise is present in all communication systems and has a degrading effect on the signal. There are many types of noise, but the most common is thermal noise, which is a characteristic of all electronic equipment.

IMPORTANT EQUATIONS

Note: Each chapter will contain a brief list of the most commonly used equations, for handy reference. These equations must not be used blindly: you must understand how and where to use them. Refer to the text where necessary. These are not the only equations you will need to solve the problems in each chapter.

$$e(t) = E_c \sin(\omega_c t + \theta) \tag{1.1}$$

$$v = f\lambda \tag{1.3}$$

$$f(t) = A_0/2 + A_1 \cos \omega t + B_1 \sin \omega t + A_2 \cos 2\omega t + B_2 \sin 2\omega t$$
$$\quad + A_3 \cos 3\omega t + B_3 \sin 3\omega t + \cdots \tag{1.5}$$

$$f = 1/T \tag{1.6}$$

$$\omega = 2\pi f \tag{1.7}$$

$$P_N = kTB \tag{1.10}$$

$$V_N = \sqrt{4kTBR} \tag{1.13}$$

$$I_N = \sqrt{2qI_0 B} \tag{1.14}$$

$$V_{Nt} = \sqrt{V_{N1}^2 + V_{N2}^2 + V_{N3}^2 + \cdots} \tag{1.15}$$

$$S/N \text{ (dB)} = 10 \log(P_S/P_N) \tag{1.17}$$

$$S/N \text{ (dB)} = 20 \log (V_S / V_N) \tag{1.18}$$

$$NF = \frac{(S/N)_i}{(S/N)_o} \tag{1.19}$$

$$NF \text{ (dB)} = (S/N)_i \text{ (dB)} - (S/N)_o \text{ (dB)} \tag{1.20}$$

$$T_{eq} = 290(NF - 1) \tag{1.27}$$

$$NF_T = NF_1 + \frac{NF_2 - 1}{A_1} + \frac{NF_3 - 1}{A_1 A_2} + \frac{NF_4 - 1}{A_1 A_2 A_3} + \cdots \tag{1.33}$$

GLOSSARY

baseband the band of frequencies occupied by an information signal before it modulates the carrier

carrier a signal that can be modulated by an information signal

channel a path for the transmission of signals

continuous waves a system in which the transmission is continuous, as distinct from pulsed systems. The term is also used sometimes to represent on-off keying (switching) of an otherwise unmodulated carrier.

distortion any undesirable change in an information signal

Fourier series a way of representing periodic functions as a series of sinusoids

frequency domain a representation of a signal's power or amplitude as a function of frequency

modulation the process by which some characteristic of a carrier is varied by an information signal

multiplexing the transmission of more than one information signal over a single channel

noise any undesired disturbance superimposed on a signal that obscures its information content

noise figure ratio of input to output signal-to-noise ratio for a device

noise temperature equivalent temperature of a passive system having the same noise-power output as a given system

receiver device to extract the information signal from the signal propagating along a channel

signal-to-noise ratio (S/N) ratio of signal to noise power at a given point in a system

spectrum analyzer device for displaying signals in the frequency domain

time domain representation of a signal's amplitude as a function of time

transmission transfer of an information signal from one location to another

transmitter device that converts an information signal into a form suitable for propagation along a channel

QUESTIONS

1. Identify each of the following frequencies as to band.

    (a) 10 MHz (used for standard time and frequency broadcasts)
    (b) 2.45 GHz (used for microwave ovens)
    (c) 100 kHz (used for the LORAN navigation system for ships and aircraft)
    (d) 4 GHz (used for satellite television)
    (e) 880 MHz (used for cellular telephones)

2. Suppose that a voice frequency of 400 Hz is transmitted on an AM radio station operating at 1020 kHz. Which of these frequencies is

    (a) the information frequency?
    (b) the carrier frequency?
    (c) the baseband frequency?
    (d) the modulating frequency?

3. Name the parameters of a sine-wave carrier that can be modulated.

4. Explain briefly the concept of frequency-division multiplexing and give one practical example of its use.

5. Explain briefly the concept of time-division multiplexing and give one practical example of its use.

6. Describe what is meant by each of the following types of distortion:

    (a) harmonic distortion
    (b) nonlinear frequency response
    (c) nonlinear phase response

7. **(a)** What is the theoretical bandwidth of a 1 kHz square wave?
   (b) Suggest a bandwidth that would give reasonably good results in a practical transmission system.
   (c) Suppose that a 1 kHz square wave were transmitted through a channel that could not pass frequencies above 2 kHz. What would be the shape of the wave after passing through the channel?

8. Describe the way in which each of the following types of noise is generated:

    (a) man-made
    (b) atmospheric
    (c) space
    (d) thermal
    (e) shot
    (f) partition
    (g) excess
    (h) transit-time

9. What is the difference between white and pink noise?

10. What is meant by the noise power bandwidth of a system?

11. Why are amplifiers that must work with very weak signals sometimes cooled to extremely low temperatures, and what is this process called?

12. What is meant by the signal-to-noise ratio, and why is it important in communication systems?

13. State two ratios that are similar to the signal-to-noise ratio and are easier to measure.

14. Explain how noise figure relates to signal-to-noise ratio.

15. What is the relationship between noise figure and noise temperature?

16. Explain why the first stage in an amplifier is the most important in determining the noise figure for the entire amplifier.

17. Name the two basic types of spectrum analyzer, and briefly describe how each works.

18. Name the four most important controls on a swept-frequency spectrum analyzer, and describe the function of each.

19. What is the function of the video filter in a spectrum analyzer? How does a video filter differ from a baseline clipper?

20. What happens to the noise level displayed on a spectrum analyzer as its bandwidth is increased, and why?

PROBLEMS

Section 1.3

21. Sketch the spectrum for the half-wave rectified signal in Figure 1.23, showing harmonics up to the fifth. Show voltage and frequency scales, and indicate whether your voltage scale shows peak or RMS voltage.

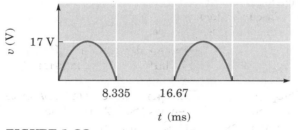

FIGURE 1.23

22. Sketch the frequency spectrum for the triangle wave shown in Figure 1.24, for harmonics up to the fifth. Show voltage and frequency scales.

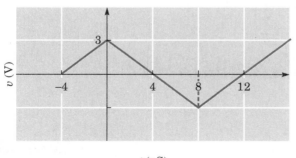

FIGURE 1.24

23. A 1 kHz square wave passes through each of three communication channels whose bandwidths are given below. Sketch the output in the time domain for each case.

(a) 0 to 10 kHz
(b) 2 kHz to 4 kHz
(c) 0 to 4 kHz

Section 1.4

24. A 50 Ω resistor operates at room temperature (21°C). How much noise power does it provide to a matched load over the bandwidth of

    (a) a CB radio channel (10 kHz)?
    (b) a TV channel (6 MHz)?

    Express your answers in both watts and dBm.

25. What would be the noise voltage generated for each of the conditions in Problem 24?

26. Calculate the noise current for a diode with a bias current of 15 mA, observed over a 25 kHz bandwidth.

27. The signal-to-noise ratio at the input to an amplifier is 30 dB and at the output it is 27.3 dB.

    (a) What is the noise figure of the amplifier?
    (b) What is its noise temperature?

28. (a) The signal voltage at the input of an amplifier is 100 μV, and the noise voltage is 2 μV. What is the signal-to-noise ratio in decibels?
    (b) The signal-to-noise ratio at the output of the same amplifier is 30 dB. What is the noise figure of the amplifier?

29. (a) A receiver has a noise temperature of 100 K. What is its noise figure in decibels?
    (b) A competing company has a receiver with a noise temperature of 90 K. Assuming its other specifications are equal, is this receiver better or worse than the one in Part (a)? Explain.

30. Suppose the noise power at the input to a receiver is 1 nW in the bandwidth of interest. What would be the required signal power for a signal-to-noise ratio of 25 dB?

31. A three-stage amplifier has power gain and noise figure (as ratios, not in decibels) for each stage as follows:

Stage	Power Gain	Noise Figure
1	10	3
2	20	4
3	30	5

    Calculate the total gain and noise figure, and convert both to decibels.

Section 1.5

32. The display in Figure 1.25 shows the fundamental of a sine-wave oscillator and its second harmonic, as seen on a spectrum analyzer. The analyzer has an input impedance of 50 Ω Find:

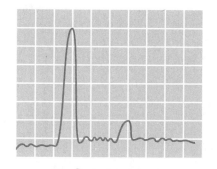

**Reference Level 10 dBm
Vertical: 10 dB/division**

Center Frequency 100 MHz

Span: 20 MHz/division
FIGURE 1.25

(a) the oscillator frequency
(b) the amplitude of the signal, in dBm and in volts
(c) the amount in decibels by which the second harmonic is less than the fundamental (usually expressed as "x dB down")

33. Find the frequency, power (both in dBm and in watts), and voltage for each of the signals shown in Figure 1.26.

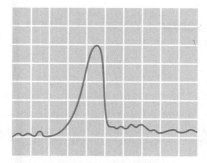

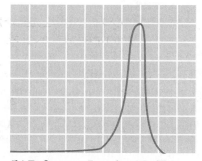

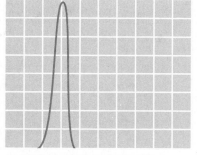

**(a) Reference Level − 30 dBm
Vertical: 10 dB/division**

**Center Frequency 872 MHz
Span: 10 MHz/division**

FIGURE 1.26

**(b) Reference Level − 18 dBm
Vertical: 1 dB/division**

**Center Frequency 79 MHz
Span: 500 kHz/division**

**(c) Reference Level 12 dBm
Vertical: Linear**

**Center Frequency 270 kHz
Span: 5 kHz/division**

34. From the spectrum analyzer display in Figure 1.27, calculate the system signal-to-noise ratio, assuming that the analyzer requires the same correction factors as described on page 43, and that the system bandwidth is 300 kHz.

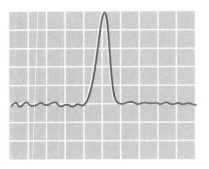

Reference Level −35 dBm
Vertical: 10 dB/division

Center Frequency 100.1 MHz
Bandwidth: 100 kHz

FIGURE 1.27

Comprehensive

35. The four stages of an amplifier have gains and noise figures (in decibels) as follows:

Stage	Power Gain	Noise Figure
1	12 dB	2 dB
2	15 dB	4 dB
3	20 dB	6 dB
4	17 dB	7 dB

Calculate the overall noise figure in decibels.

36. A three-stage amplifier is to have an overall noise temperature no greater than 70 K. The overall gain of the amplifier is to be at least 45 dB. The amplifier is to be built by adding a low-noise first stage to a two-stage amplifier that already exists, with gain and noise figure for each stage as shown below. (The stage numbers refer to their location in the new amplifier.)

Stage	Power Gain	Noise Figure
2	20 dB	3 dB
3	15 dB	6 dB

   (a) What is the minimum gain (in decibels) that the first stage can have?
   (b) Assuming the gain you calculated in Part (a), calculate the maximum noise figure (in decibels) that the first stage can have.
   (c) Suppose the gain of the first stage could be increased by 3 dB without affecting its noise figure. What would be the effect on the noise temperature of the complete amplifier?

37. The $(S+N)/N$ ratio at the output of an amplifier can be measured by measuring the output voltage with and without an input signal. When this is done for a

certain amplifier, it is found that the output is 2 V with the input signal switched on and 15 mV with it switched off. Calculate the $(S+N)/N$ ratio in decibels.

38. The frequency response of an amplifier can be measured by applying white noise to the input and observing the output on a spectrum analyzer. When this was done for a certain amplifier, the display shown in Figure 1.28 was obtained. Find the upper and lower cutoff frequencies (at 3 dB down) and the bandwidth of the amplifier.

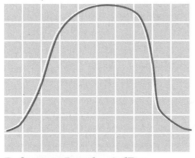

Reference Level − 4 dBm
Vertical: 1 dB/division

Center Frequency 210 MHz
Span: 10 MHz/division

FIGURE 1.28

39. Use calculus to verify the equations given in Appendix B for the Fourier series of a square wave and a sawtooth wave.

2

Radio-Frequency Circuits

Objectives

After studying this chapter, you should be able to:

1. Explain the differences in operation and construction between circuits that operate at low frequencies and those that operate at radio frequencies.
2. Describe the characteristics of Class A, B, and C amplifiers, and decide which type is the most suitable for a given application.
3. Analyze radio-frequency amplifier circuits, both narrowband and broadband, and choose the correct configuration for a given application.
4. Describe, draw circuits for, and analyze the most common types of radio-frequency oscillator circuits, and discuss their relative stability.
5. Explain the operation of varactor-tuned voltage-controlled oscillators, and calculate the variation of frequency with tuning voltage.
6. Describe the operation of crystal-controlled oscillators, and explain their advantages and disadvantages compared with *LC* oscillators. Perform frequency-stability calculations for crystal oscillators.
7. Describe the function of a mixer and analyze several circuits, explaining how and where they are used, and calculating output frequencies.

2.1 INTRODUCTION

In the previous chapter, we looked at some of the signals found in a communication system. In particular, we saw the need to modulate a carrier signal using an information (baseband) signal. The carrier must be much higher in frequency than the baseband signal. Carrier frequencies can be as low as a few kilohertz, but are typically much higher: megahertz or hundreds of megahertz. Microwave communications use carrier frequencies in the gigahertz range.

You are probably familiar, by this time, with amplifier and oscillator circuits that operate at audio frequencies. In this chapter, we will explore some of the differences in design and construction that are necessary so that these circuits will work at radio frequencies. We will also look at some techniques used in radio-frequency (RF) circuits, that are impossible or impractical to implement at lower frequencies. In addition, we will look at devices, such as frequency multipliers and mixers, that allow the frequency of a signal to be changed.

The purpose of this chapter is to give some insight into RF circuit design, and to provide the reader with some electronic "building blocks" that can be used in later chapters as we look into the design and construction of practical transmitters and receivers.

2.2 HIGH-FREQUENCY EFFECTS

When you first began to study electronics, you probably divided the frequency spectrum into two parts: ac and dc. A capacitor, for instance, would be considered an open circuit for dc and a short circuit for ac. This simplifying assumption works well in arriving at a general understanding of an audio amplifier circuit, and in calculating its gain at midband frequencies.

Later on, you found it necessary to be more careful in considering the effects of frequency. The bypass capacitor that looked like a short circuit at 1 kHz was no longer so simple at 20 MHz. In order to find the low-frequency response of a simple audio amplifier, it became necessary to consider capacitive reactance. Similarly, other capacitances had to be taken into account to calculate the high-frequency response. Some were actual components in the circuit, and some were incidental parts of components, for example, junction capacitances in transistors.

Perhaps the amplifier had a transformer or other inductive element in it. Again, a first approximation might simplify an inductor as representing an open circuit for ac and a short circuit for dc. A transformer would be represented as ideal. Further investigation, especially at the extremes of the amplifier's frequency range, might lead to the necessity of considering not only inductive reactance but also losses in the iron core of the inductor or transformer.

As we extend our study of electronic circuits to higher frequencies, we have to be more careful to include reactive effects, not only those that are included deliberately as circuit elements, but also the "stray" reactances in components and even within and between wires and circuit board traces. As we get still higher in frequency, into the UHF range, we find that conventional devices and construction methods become inefficient, and innovative approaches to circuit design become important. At microwave frequencies, many circuits seem to bear very little physical resemblance to those used at lower frequencies.

Microwave-circuit design will be discussed in a later chapter. For the present, we shall look at more conventional circuitry, operating in the range between approximately 300 kHz and 300 MHz, that is, in the MF, HF, and VHF frequency bands.

2.2.1 The Effect of Frequency on Device Characteristics

All electronic devices, whether active or passive, have capacitances and inductances that are not included intentionally, but are an inevitable result of the design and construction of the component. A capacitor, for instance, will exhibit inductance and resistance as well as capacitance. It can be represented by the equivalent circuit in Figure 2.1(a). The series inductive component L_S is mainly due to the leads. The resistive component can be divided into two parts, a small series component R_S due to lead resistance, and a large parallel resistance R_P representing dielectric losses.

As the frequency increases, so does the inductive reactance. Meanwhile, the capacitive reactance of the component decreases with increasing frequency. Eventually, a point will be reached where the two reactances are equal and the capacitor becomes a series-resonant circuit. This point is called the **self-resonant frequency.** Above this point, the magnitude of the inductive reactance becomes greater than that of the capacitive reactance, and our so-called capacitor behaves like an inductor.

Similar effects occur in transistors. Consider an ordinary bipolar transistor such as the 2N3904, for instance. Each of the two junctions has capacitance, which can be represented by capacitors drawn in between base and emitter and between collector and base, as shown in Figure 2.1(b). The size of these capacitors will depend partly on the physical structure of the transistor, and partly on its operating point. As frequency increases, the capacitive

FIGURE 2.1

High-Frequency Effects

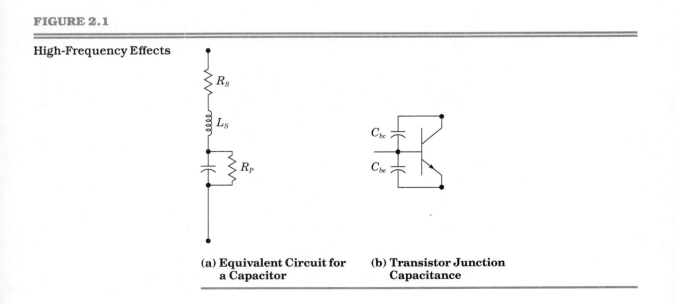

(a) Equivalent Circuit for   (b) Transistor Junction
    a Capacitor   Capacitance

reactances will decrease until the performance of the transistor is degraded. The base-to-collector capacitance, for instance, will cause feedback from output to input in an ordinary common-emitter amplifier circuit. The feedback can lower the gain of the amplifier or cause it to become unstable.

As the frequency increases into the gigahertz range, transit-time effects also become important. The transit time is the time it takes a charge carrier to cross a device. In an NPN transistor, it is the time taken for electrons to cross the base; in a PNP transistor, it is holes that exhibit transit time. In general, free electrons move more quickly than holes, so NPN transistors are preferred to PNP for high-frequency operation.

Transit time can be reduced by making devices physically small, but this causes problems with heat dissipation and breakdown voltage. It is difficult to remove large quantities of heat from a small area, and of course the dielectric strength of insulators is proportional to their thickness. There are limits to how small devices can be made, particularly if considerable power must be dissipated. For this reason, some devices include transit time as part of their design. Rather than simply making transit time as small as possible, for instance, it might be designed to be one or a number of periods at the operating frequency. Techniques such as these are used mainly in the microwave region of the spectrum, and are discussed in Chapter 13.

2.2.2 Lumped and Distributed Constants

At low frequencies, we generally assume that, for instance, capacitors have capacitance, resistors have resistance, and short sections of good conductors (for example, the traces on a circuit board) have neither. We saw in the previous section that this assumption is really a simplification that becomes less accurate as the frequency increases. For instance, in addition to resistance, a circuit board trace has a small amount of inductance. There will also be capacitance between this trace and every other trace on the board.

A little thought will show that the inductance and capacitance related to this trace cannot be shown exactly by a single capacitor and a single inductor. In fact, no finite number of "lumps" of capacitance and inductance will do, since they are both distributed along the entire length of the trace. We can often approximate these distributed constants by a few lumped constants with reasonable accuracy, but it is well to remember what is really happening, because some problems can only be solved using distributed constants. Figure 2.2 gives an idea of how an ordinary circuit board trace might look when analyzed this way. Of course, there is really no accurate way to sketch

FIGURE 2.2

A Circuit-Board Trace Using Distributed Constants

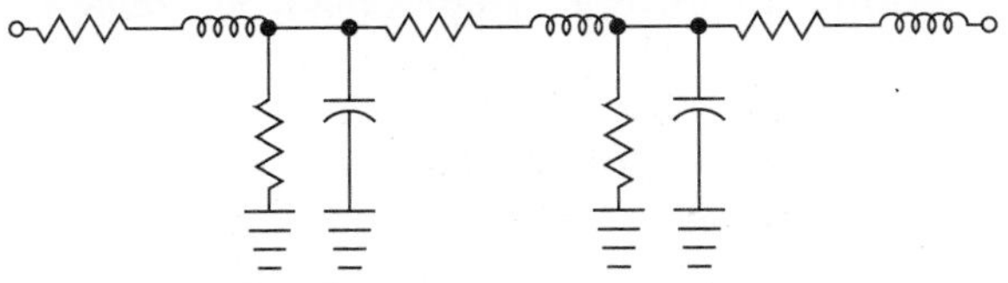

this trace, since the number of capacitors, resistors, and inductors should be infinite.

At frequencies in the UHF range and up, conductors even a few centimeters in length can no longer be ignored, or have only their lumped capacitance and inductance taken into account. They must be analyzed as transmission lines. This process includes the distributed constants that are really there all the time but can usually be ignored at lower frequencies. At microwave frequencies, almost all conductors must be analyzed in this way. Transmission lines will be covered in detail in Chapter 10.

2.2.3 High-Frequency Construction Techniques

It is possible to design circuitry to reduce the effect of "stray" capacitance and inductance resulting from the wiring and circuit board traces themselves. In general, keeping wires and traces short reduces inductance, and keeping them well separated reduces capacitance between them. Inductive coupling can be reduced by keeping conductors and inductors that are in close proximity at right angles to each other. The use of toroidal cores for inductors and transformers also helps to reduce stray magnetic fields.

There may seem to be contradictions here. How do we keep components far away from each other while simultaneously keeping connections short, for instance? Obviously, compromise is necessary.

Another way to reduce interactions between components is to use *shielding*. Coupling by way of electric fields can be reduced by shielding sensitive circuits with any good conductor, such as copper or aluminum. Ideally, the shielding should form a complete enclosure and be connected to an earth ground, but even a piece of aluminum foil glued to the inside of a plastic cabinet and connected to the circuit common point will sometimes provide adequate shielding. This "cheap and dirty" technique is often found in consumer electronics. When double-sided circuit boards are used, most of the copper is often left on one side and connected to ground. This **ground plane** can provide useful shielding. Figure 2.3 shows a typical VHF circuit, in this case a cable-television converter, that uses the techniques outlined here.

Generally, conductors in RF circuits are kept as short and as far away from others as possible. The neat right-angle circuit-board layouts common at low frequencies are often avoided in favor of more direct routing. It may not be as pretty, but it can avoid problems of inductive or capacitive coupling.

Anyone attempting to service RF equipment should remember that in high-frequency circuits the placement of components and wiring can be critical. Moving one component lead may require a complete realignment. In fact, sometimes a length of solid insulated hookup wire (a few centimeters long) will be seen, connected at only one end. This is called a **gimmick**, and represents a small capacitance to ground that is adjusted during circuit alignment by bending the wire slightly in one direction or another.

In order to prevent RF currents from traveling from one part of the circuit to another, careful **bypassing** is necessary. For instance, RF energy may travel from one stage to another that shares the same power supply. At first glance, this may seem unlikely, since power supplies typically contain large electrolytic capacitors that should have very low reactance at radio frequencies. Thus it would be expected that the power supply would look like a short circuit at high frequencies.

FIGURE 2.3

VHF Circuit Layout

Unfortunately, this ignores the inductance in the leads from the circuit to the power supply, and in the electrolytic capacitors themselves. To prevent energy from traveling from one circuit to another via this route, it will be necessary to provide small capacitors to ground, right at the power connection to each stage. Small capacitors are actually better than large electrolytics for this application, because they have less inductance. In difficult cases, either an inductance or a resistance can be added in the lead from the power supply, to further discourage the transfer of RF energy. Using an inductor reduces the dc voltage drop, of course, but it must be chosen with care because inductors, like capacitors, can exhibit self-resonance. Figure 2.4(a) shows a typical decoupling circuit between an amplifier stage and a power supply. If the power leads pass through a shielded enclosure, a feed-through capacitor like the one pictured in Figure 2.4(b) can be used to maintain the shielding.

2.3 RADIO-FREQUENCY AMPLIFIERS

Amplifiers for RF signals can be distinguished from their audio counterparts in several important ways. Wide bandwidth may or may not be required. Where it is not, gain can be increased and distortion reduced with the use of tuned circuits. Depending on the type of signal to be amplified, linearity of output with respect to input amplitude may or may not be required. If linearity is not necessary, efficiency can be improved by operating amplifiers in Class C, which will be described shortly. Impedance matching is likely to be more important than at lower frequencies, because of the possibility of trouble caused by signal reflections.

FIGURE 2.4

Decoupling

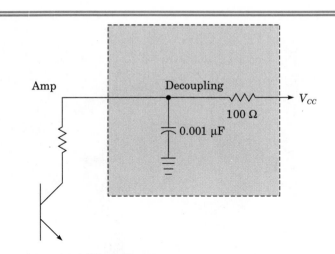

(a) Decoupling Circuit

(b) Feed-Through Capacitor

2.3.1 Narrowband Amplifiers

Often the signals in an RF communications system are restricted to a relatively narrow range of frequencies. In such circumstances it is unnecessary and, in fact, undesirable to use an amplifier with a wide bandwidth. Doing so invites problems with noise and interference. Consequently, many of the amplifiers found in both receivers and transmitters incorporate filters to restrict their bandwidth. In many cases these filters also increase the gain of the amplifier.

The simplest form of bandpass filter is, of course, a resonant circuit, and these are very common in RF amplifiers. Consider the bipolar common-emitter amplifiers shown in Figure 2.5, for instance. Figure 2.5(a) shows a conventional amplifier using RC coupling. As you will recall, this amplifier is generally biased so that the emitter voltage is about 10% of V_{CC}. The collector resistor will drop another 40% or so, leaving the voltage between collector

FIGURE 2.5

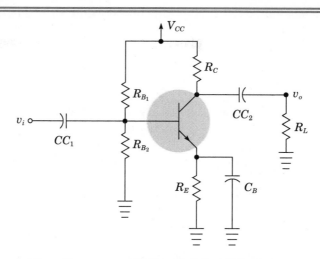

(a) Low-Frequency *RC*-Coupled Amplifier

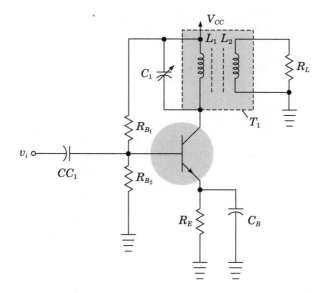

(b) High-Frequency Transformer-Coupled Tuned Amplifier

and emitter, V_{CE}, at about one-half the supply voltage. The voltage gain is given approximately by

$$A_V = \frac{-(R_C \parallel R_L)}{r_e} \tag{2.1}$$

where A_V = voltage gain as a ratio: $A_V = v_o/v_i$

$R_C \parallel R_L$ = parallel combination of the collector resistance and the load resistance

r_e = ac emitter resistance of the transistor

Do not confuse r_e, a transistor parameter, with R_E, which is part of the external circuit. The value of r_e depends on the bias current. It is given, very approximately, by

$$r_e = \frac{26 \text{ mV}}{I_E}$$

where r_e = emitter resistance in ohms
          I_E = dc emitter current in amperes

The tuned amplifier in Figure 2.5(b) is similar but not identical. For instance, the bias point is different. In the absence of a collector resistor, the dc collector voltage will obviously be equal to the supply voltage, less the very small drop due to the dc resistance of the transformer primary. The gain is still equal to the ratio of collector-circuit impedance to emitter-circuit impedance, but the collector-circuit impedance is now very much a function of frequency.

Once again, it is necessary to recall some basic electrical theory. (See Appendix A if you need a review of resonance.) The collector tuned circuit will be parallel resonant at a frequency that is given approximately by

$$f_o = \frac{1}{2\pi\sqrt{L_1 C_1}} \qquad\qquad (2.2)$$

where f_o = resonant frequency in hertz
          L_1 = primary inductance in henrys
          C_1 = primary capacitance in farads

Equation (2.2) is reasonably accurate as long as the loaded quality factor (Q) of the tuned circuit is greater than about ten. At resonance, the impedance of the tuned circuit is resistive and its magnitude is at a maximum. On either side of resonance, the impedance is lower in magnitude and reactive: inductive below resonance, and capacitive above.

The impedance of the tuned circuit depends on its Q, which depends in turn on the load R_L, as transformed into the primary circuit by the transformer, as well as on the losses in the inductor itself. For simplicity, let us assume at first that T_1 is an ideal transformer, bearing in mind that this is probably not a valid assumption for an RF amplifier. The exercise will be useful in demonstrating some of the differences between tuned and untuned amplifiers, as long as we bear in mind that the actual gain and bandwidth found by this method will be approximate.

Assuming that T_1 is an ideal transformer, it is easy to find the equivalent value of R_L, transformed to the primary circuit. For an ideal transformer, the impedance ratio is simply the square of the turns ratio, so we have

$$R_L' = R_L \left(\frac{N_1}{N_2}\right)^2 \qquad\qquad (2.3)$$

where R_L = actual load resistance connected to the secondary
          R_L' = equivalent load resistance transformed into the primary

N_1 = number of turns in the primary winding
N_2 = number of turns in the secondary winding

Assuming no losses in the tuned circuit itself, and a high value of shunt resistance due to the transistor, the gain of the stage at resonance, measured from the input to the transformer primary (that is, from base to collector of the transistor), would be

$$A'_{V_o} = \frac{R'_L}{r_e}$$

$$= \frac{R_L N_1^2}{r_e N_2^2} \tag{2.4}$$

Equation (2.4) does not include the voltage step-up or step-down effect of the transformer turns ratio. This must be included to find the gain of the complete amplifier, from base to load.

$$A_{V_o} = \frac{A'_{V_o} N_2}{N_1}$$

$$= \frac{R_L N_1^2}{r_e N_2^2} \times \frac{N_2}{N_1}$$

$$= \frac{R_L N_1}{r_e N_2} \tag{2.5}$$

The bandwidth of this circuit will depend on the loaded Q of the tuned circuit. The load impedance, transformed into the primary, will act as a resistance in parallel with the tuned circuit. Ignoring any losses in the resonant circuit or in the transistor, we can find the equivalent Q of the tuned circuit from

$$Q = \frac{R'_L}{X_L} = Z \text{ Tank}$$

$$= \frac{R'_L}{\omega_o L} = Z \text{ Tank} \tag{2.6}$$

where Q = equivalent Q at resonance
R'_L = load resistance transformed into the primary
X_L = reactance of the inductor at resonance
ω_o = radian frequency at resonance

Once the Q has been found, the bandwidth at 3 dB down from the resonant-frequency gain is given very simply by

$$B = \frac{f_o}{Q} \tag{2.7}$$

where B = bandwidth

   f_o = resonant frequency

   Q = equivalent Q at resonance, as given by Equation (2.6)

The reason the gain goes down away from resonance is that the impedance of the resonant circuit is greatest at resonance and drops rapidly at higher and lower frequencies. As can be seen from Equation (2.6), the loaded Q depends on the load resistance and the transformer turns ratio. The more heavily loaded the amplifier, the lower its Q and the wider its bandwidth.

An example may clarify this analysis, while also reviewing the process by which the emitter current may be found in a typical bipolar amplifier.

EXAMPLE 2.1

An RF amplifier has the circuit shown in Figure 2.6. Find:

(a) the operating frequency

(b) the gain at the operating frequency

(c) the bandwidth

Assume the transformer is ideal (no losses, and coupling coefficient equal to 1) and the loading effect of the transistor is negligible.

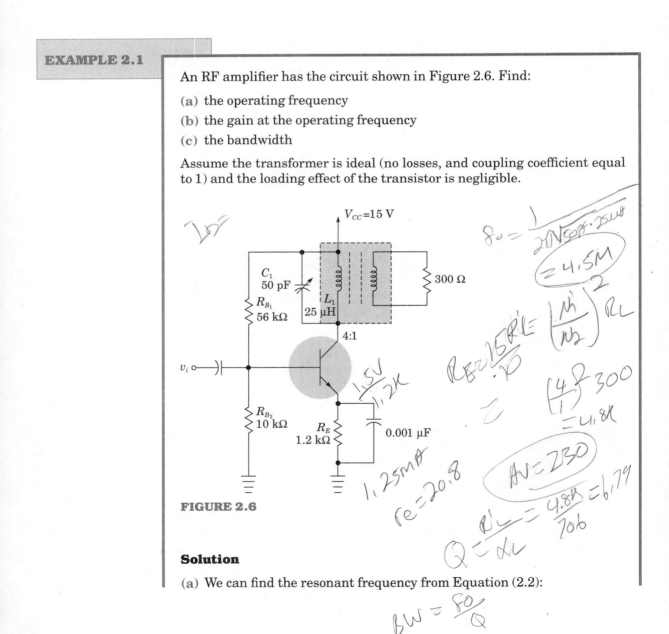

FIGURE 2.6

Solution

(a) We can find the resonant frequency from Equation (2.2):

$$f_o = \frac{1}{2\pi\sqrt{L_1 C_1}}$$

$$= \frac{1}{2\pi\sqrt{(25 \times 10^{-6})(50 \times 10^{-12})}}$$

$$= 4.50 \text{ MHz}$$

(b) First, it is necessary to find the dc emitter current. To do this, we need to find the emitter voltage. We recall that, for an NPN transistor in the active region, the emitter voltage will be approximately 0.7 V less than the base voltage.

We can find the base voltage from the voltage divider bias circuit. Assuming the base current is negligible compared to the current through R_{B_1} and R_{B_2} (which will be the case in a well-designed circuit), then

$$V_B = \frac{V_{CC} R_{B_2}}{R_{B_1} + R_{B_2}}$$

$$= \frac{15 \text{ V} \times 10}{56 + 10}$$

$$= 2.27 \text{ V}$$

Now we can find the emitter voltage:

$$V_E = V_B - 0.7 \text{ V}$$

$$= 2.27 \text{ V} - 0.7 \text{ V}$$

$$= 1.57 \text{ V}$$

The emitter current is easily found from Ohm's Law:

$$I_E = \frac{V_E}{R_E}$$

$$= \frac{1.57 \text{ V}}{1200 \ \Omega}$$

$$= 1.308 \times 10^{-3} \text{ A}$$

$$= 1.308 \text{ mA}$$

Next we can find r_e, which is needed for the gain calculation.

$$r_e = \frac{26 \text{ mV}}{I_E}$$

$$= \frac{26 \text{ mV}}{1.308 \text{ mA}}$$

$$= 19.9 \ \Omega$$

Let us now turn to the ac analysis. The gain, from Equation (2.5), is

$$A_{V_o} = \frac{R_L N_1}{r_e N_2}$$

$$= \frac{300 \ \Omega \ \times \ 4}{19.9 \ \Omega}$$

$$= 60.3$$

(c) To find the bandwidth, it will be necessary to transform the load resistance into the primary circuit, using Equation (2.3).

$$R_L' = R_L \left(\frac{N_1}{N_2}\right)^2$$

$$= 300 \ \Omega \ \times \ 4^2$$

$$= 4.8 \ k\Omega$$

The bandwidth can be found from Equation (2.7).

$$B = \frac{f_o}{Q}$$

The value for f_o was found in Part (a); it remains to find Q. From Equation (2.6),

$$Q_o = \frac{R_L'}{\omega_o L}$$

$$= \frac{4.8 \ \times \ 10^3}{2\pi(4.5 \ \times \ 10^6)(25 \ \times \ 10^{-6})}$$

$$= 6.79$$

Now, the bandwidth is easy to find using Equation (2.7).

$$B = \frac{f_o}{Q}$$

$$= \frac{4.5 \ \text{MHz}}{6.79}$$

$$= 663 \ \text{kHz}$$

There are some problems with this simplified approach. The first has to do with the transformer. While the ideal transformer is a useful approximation to a real transformer at audio frequencies, it is not always close to reality in RF circuits. With air-core transformers, the coefficient of coupling will be much lower than one, sometimes as low as 0.01 or so. For loosely coupled air-core transformers, it may be possible simply to ignore the load impedance when calculating voltage gain and bandwidth. This also means that the Q is likely to be larger, and the bandwidth smaller, than predicted using the assumption of an ideal transformer.

The other problems have to do with the transistor. If used in a circuit exactly like that of Figure 2.5(b), the output impedance of the transistor is likely to be low enough to reduce the Q of the tuned circuit to an unacceptably low value. The transistor output impedance can be represented as a resistance from collector to emitter that is effectively connected across the tuned circuit. We ignored the output impedance in Example 2.1, but for a bipolar transistor operating at radio frequencies, it may be only a few thousand ohms, and therefore it will have to be considered. Consequently, the transistor is usually connected across only part of the coil, as in Figure 2.7. The transformer primary acts as an autotransformer, increasing the effective impedance of the transistor and increasing the Q of the circuit. This more practical circuit also shows the use of a decoupling network. Consisting of R_D and C_D, this network connects the top end of the coil to ground for RF, and isolates it from the power supply. Note that, while the transistor is connected across only part of L_1, the capacitor C_1 tunes the whole inductor. Remember that the top end of the coil is effectively connected to ground for ac, and this will be clear.

The capacitance between collector and base of the transistor is also likely to cause trouble. By feeding back some of the output signal to the input, it reduces the gain of the circuit at high frequencies. In fact, for the common-emitter circuit shown in Figure 2.7, the effect of this capacitance, often called the *Miller effect*, is the same as if a much larger capacitance had been connected across the input. This **Miller capacitance** has a value of

FIGURE 2.7

Practical Common-
Emitter Amplifier with
Tapped Primary

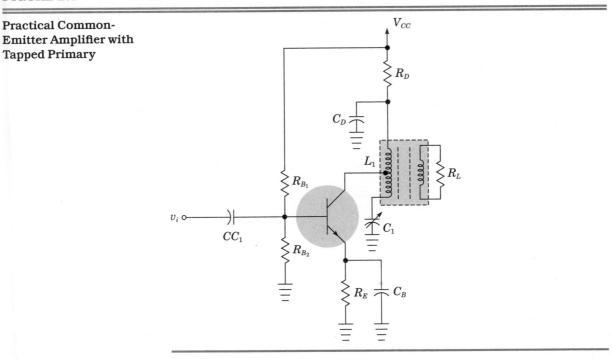

$$C_m = C_{bc}(A_{V_0} + 1) \tag{2.8}$$

where C_m = Miller capacitance across the input
         C_{bc} = capacitance between base and collector
         A_{V_0} = gain ignoring the Miller effect

The effect of this can be reduced somewhat by transformer-coupling the input as well as the output, and tuning the secondary of the input transformer. The Miller capacitance then becomes part of the capacitance necessary to tune the input circuit to resonance. Such a circuit is shown in Figure 2.8. The input is tapped down on the transformer secondary to reduce the loading effect of the transistor and its biasing circuit on the Q of the tuned circuit. C_1 tunes the input tuned circuit, and in this circuit C_2 is the output tuning capacitor.

The practice of tuning both the input and output of an amplifier can lead to instability. This can be corrected by neutralization, discussed later in this chapter.

Another way to avoid the problem of the Miller effect is to use a common-base amplifier, as shown in Figure 2.9. This circuit shows transformer coupling at both input and output. Common-base amplifiers are rare in low-frequency applications, but quite common at radio frequencies. The capacitance between collector and base appears across the output tuned circuit, where it simply reduces the external capacitance needed to achieve resonance. This will greatly extend the useful frequency range of the transistor.

FIGURE 2.8

Narrowband RF
Amplifier with Tuned
Input and Output

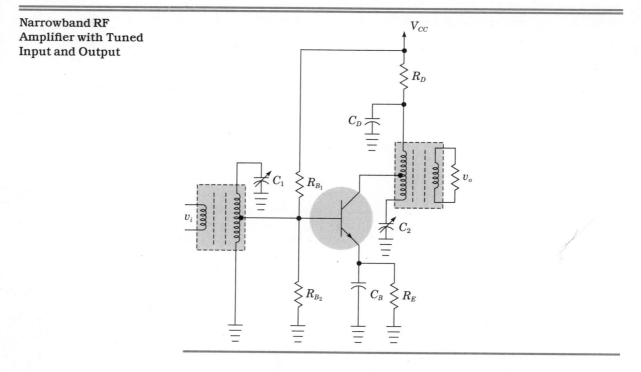

FIGURE 2.9

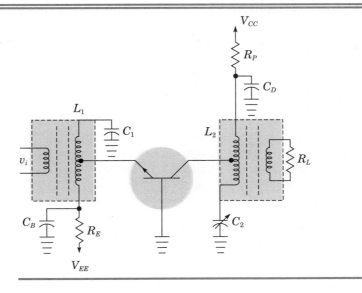

Common-Base RF Amplifier

In Figure 2.9, the input tuned circuit consists of L_1 and C_1, and the output tuned circuit is L_2 and C_2. Emitter bias is shown, but voltage-divider bias is also suitable and does not require a negative supply voltage.

The loading effect of the transistor impedances on tuned circuits can be greatly reduced by using field-effect, rather than bipolar, transistors. Field-effect transistors (FETs) are famous for their very high input impedance, of course, but their output impedance is also much higher than for bipolar transistors.

2.3.2 Wideband Amplifiers

Not all amplifiers used in communications have tightly restricted bandwidth. For instance, the amplifiers used for the baseband part of the system are usually *wideband* (also called *broadband*). You are no doubt already familiar with audio amplifiers. Those used for baseband video are similar, though their bandwidth is larger, about 4.2 MHz for broadcast television signals.

In this section, we will consider amplifiers designed for higher-frequency operation, where the response is required to extend over a relatively wide range of frequencies. For instance, an amplifier for a cable-television system might be required to amplify frequencies from 50 MHz to approximately 400 MHz. This is a fairly difficult requirement, especially if the system is expected to be *flat*, that is, to have equal gain across the entire bandwidth. (An actual cable-television amplifier would be designed for more gain at higher frequencies to compensate for greater loss in the cable.)

In most cases, it is not required that a broadband RF amplifier have response down to dc or even to audio frequencies. In fact, this is likely to be undesirable. We would not want our hypothetical cable-television amplifier to amplify 60 Hz hum, for example. In fact, we would prefer it to ignore AM radio broadcasts at about 1 MHz and CB radio transmitters at 27 MHz as well. Broadband amplifiers, then, generally incorporate some form of filtering so that the frequency response, while broad, is restricted to the range of interest.

Wideband RF amplifiers, like their narrowband counterparts, typically use transformer coupling. This technique was once popular for audio amplifiers as well, but the size, weight, and, especially, high cost of audio transformers has led to their virtual elimination from audio circuitry. For RF amplifiers they retain some advantages, however. Transformers are very convenient for impedance matching, which is likely to be more important in RF than in audio designs. The isolation between input and output is useful in helping to keep spurious signals at frequencies greatly different from the desired signal frequency from propagating through the system. Transformer coupling also makes it easy to couple balanced inputs or loads to the amplifier. Balanced lines have equal impedance from each conductor to ground. They are often used with antennas; ordinary television twin-lead is an example of a balanced line that should be transformer-coupled to the first stage in a television receiver.

The transformers used in wideband amplifiers need careful design to avoid self-resonance and to maintain relatively constant gain across the frequency range of interest. In particular, the reactance of the windings must be large compared to the impedances that are connected to them. Toroidal transformers with ferrite cores are common. In the case of an amplifier that must have a very wide bandwidth, the ferrite, having much higher permeability than air, allows for sufficient reactance at the low end of the frequency range. At higher frequencies, the ferrite is likely to be less effective, but the reactance of the winding will be sufficient, since reactance increases with frequency.

Figure 2.10 is the circuit of a broadband RF amplifier using a bipolar transistor. Notice that it is the same as Figure 2.5(b), except that C_1, the tuning capacitor, is gone, and the input as well as the output is transformer-coupled. If the transformer can be considered ideal (not always the case, as explained earlier), then the gain is the same as was found, in Equation (2.5), for the tuned amplifier.

$$A_{V_o} = \frac{R_L N_1}{r_e N_2}$$

where A_{V_o} = gain from the base of the transistor to the output

The difference is that the gain for this amplifier remains relatively constant over a wide frequency range. If the input is also transformer-coupled, the gain must of course be multiplied by the turns ratio of the input transformer.

It is also possible to use conventional resistance-capacitance coupling between broadband amplifier stages. Such circuits can be analyzed in the same way as their low-frequency counterparts.

Whatever design is used, if an amplifier operates over a sufficiently wide frequency range, it will tend to have higher gain at the low end of its range than at the high end, because the gain of the transistor falls off with increasing frequency. Negative feedback is often used to overcome this, with the amount of feedback reducing as the frequency increases. Another way is to

FIGURE 2.10

Broadband RF Amplifier

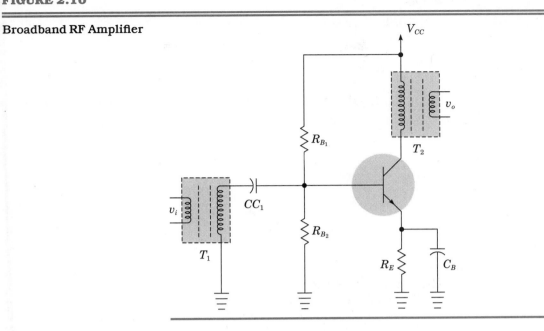

use a high-pass filter at the input, so that, as the frequency goes down, the amount of input signal reaching the amplifier is reduced.

An amplifier has the circuit of Figure 2.11. Calculate its voltage gain A_V, assuming ideal transformers and ignoring any loading by the transistor.

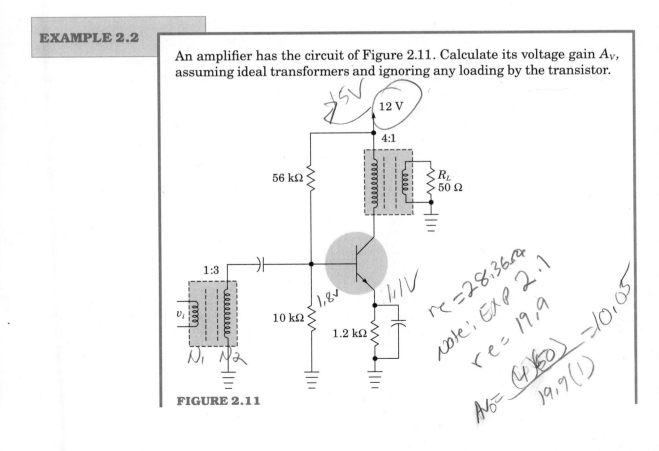

FIGURE 2.11

Solution

Since the dc bias circuit is the same as for Example 2.1, the value of r_e will be as found there.

$$r_e = 19.9 \ \Omega$$

Realizing that this is a somewhat idealized case, we can use Equation (2.5) to find the gain from base to output. Let us call that gain A'_{V_o}. Then

$$A_{V_o} = \frac{R_L N_1}{r_e N_2}$$
$$= \frac{50 \times 4}{19.9}$$
$$= 10.05$$

This must now be multiplied by the turns ratio N_2/N_1 for the input transformer.

$$A_V = 10.05 \times 3$$
$$= 30.15$$

2.3.3 Amplifier Classes

Amplifiers are classified according to the portion of the input cycle during which the active device conducts current. This is called the *conduction angle* and is expressed in degrees. The active device may be a bipolar transistor, FET, or tube; for the time being we will assume it is a bipolar transistor. For a push-pull amplifier, the conduction angle for one of the two amplifying devices is used.

Single-ended audio amplifiers are generally operated in Class A, where the transistor conducts current at all times, for a conduction angle of 360°. This is the only way to achieve linear operation, where the output is a reasonably faithful copy of the input, except for amplitude. The small-signal RF amplifiers described earlier in this chapter were also biased for Class A operation.

Push-pull amplifiers can be linear if at least one of the two transistors is conducting at all times. In a Class B amplifier, each transistor is biased at cutoff, so that each conducts for 180° of the input cycle. This results in greater efficiency than Class A, but the distortion is larger due to nonlinearity of the transistors near cutoff.

Most audio power amplifiers use Class AB, which is a compromise between Class A and Class B, in terms of both distortion and efficiency. Each transistor conducts for slightly more than 180° of the input cycle. This reduces the "crossover" distortion that occurs as the input signal passes through zero, where both transistors are near cutoff.

Figure 2.12 shows a simple Class B amplifier for RF operation. As with the Class A amplifiers shown earlier, this one uses transformer coupling

FIGURE 2.12

Class B RF Amplifier

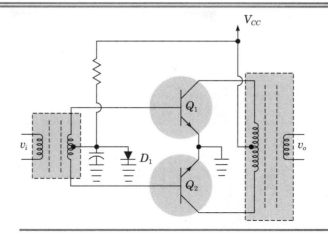

which, while rare in modern audio amplifiers, is still the most common way to design amplifiers for high-frequency operation. The example shown is broadband, but the circuit could just as easily be tuned.

Both transistors are biased near cutoff by the voltage drop across D_1. When the top end of the input transformer secondary goes positive, Q_1 turns on and Q_2 is cut off. The reverse happens on the other half of the input cycle; thus each transistor conducts for half the cycle.

Audio signals, and also some radio-frequency signals, have complex waveforms and require linear amplification to avoid distortion. Suppose, however, that we have to amplify an RF signal that we know is a sine wave. It might be possible to do this with an amplifier that created a great deal of distortion, provided that we had some means to remove that distortion after amplification, and to restore the signal to its original sinusoidal shape. This extra trouble would be worthwhile if the resulting amplifier were more efficient than either Class A or Class B amplification.

Such an amplifier class exists: it is called Class C. In it the active device conducts for less than 180° of the input cycle. The amplifier can be single-ended or push-pull; either way, it is apparent that the output for a sinusoidal input will resemble a series of pulses more than it does the original signal. These current pulses can be converted back into sine waves by an output tuned circuit. The resulting gain in efficiency can be quite worthwhile, as can be seen in Table 2.1.

Class C amplifiers achieve their maximum efficiency when the amplifying device almost saturates at peaks of the input cycle. It is, of course, cut off

TABLE 2.1

Review of Amplifier
Classes

Class	A	B	C
Conduction angle	360°	180°	<180°
Maximum efficiency	50%	78.5%	100%
Likely practical efficiency	25%	60%	75%

FIGURE 2.13

Class C Amplifier

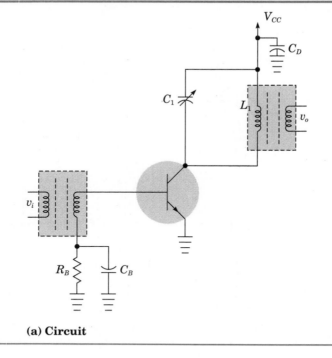

(a) Circuit

for most of the cycle. This type of operation minimizes power dissipation in the transistor, which is zero when the transistor is cut off, low when it is saturated, and much higher when it is in the normal, linear operating range. Since Class C amplifiers are biased beyond cutoff, they would be expected to have zero power dissipation with no input. This is indeed the case, provided the bias is independent of the input signal.

There remains the problem of distortion. Figure 2.13(a), a simplified circuit for a Class C amplifier, shows how the distortion is kept to a reasonable level. The process can be explained using either the time or the frequency domain. In the time domain, the output tuned circuit is excited, once per cycle, by a pulse of collector current. This keeps oscillations going in the resonant circuit. They are damped oscillations, of course, but the Q of the circuit will be high enough to ensure that the amount of damping that takes place in one cycle is negligible, and the output will be a reasonably accurate sine wave. The frequency-domain explanation is even simpler: the resonant circuit constitutes a bandpass filter which passes the fundamental frequency and attenuates harmonics and other spurious signals. In many cases, especially where the amplifier is the final stage of a transmitter, there will be additional filtering after the amplifier to further reduce the output of harmonics.

From the foregoing description, it might seem that Class C amplifiers would have to be narrowband. Many of them are, but this is not a requirement, as long as the output is connected to a low-pass filter that will attenuate all the harmonics that are generated.

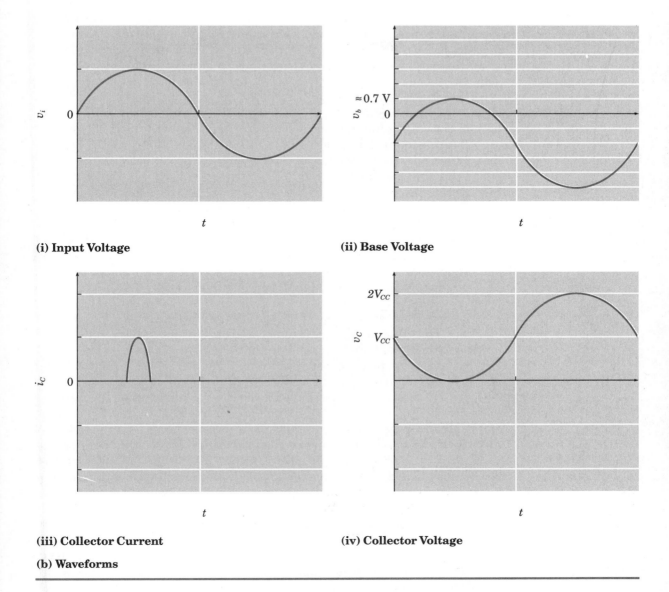

(i) Input Voltage

(ii) Base Voltage

(iii) Collector Current

(iv) Collector Voltage

(b) Waveforms

Let us look at the circuit of Figure 2.13(a) a little more closely. We begin with the bias circuit, which is unconventional. Far from being biased in the middle of the linear operating range, as for Class A, a Class C amplifier must be biased beyond cutoff. For a bipolar transistor, that can mean no base bias at all, since a base-to-emitter voltage of about 0.7 V is needed for conduction. With no signal at its input, the transistor will have no base current, no collector current, and will dissipate no power.

If a signal with a peak voltage of at least 0.7 V is applied to the input of the amplifier, the transistor will turn on during positive peaks. It might seem at first glance that an input signal larger than this would turn on the transistor for more and more of the cycle, so that with large signals the amplifier would approach Class B operation, but this is not the case. The base-emitter junction rectifies the input signal, charging C_B to a negative dc level that increases with the amplitude of the input signal. This self-biasing circuit will allow the amplifier to operate in Class C with a fairly large range of input signals.

The collector circuit can be considered next. At the peak of each input cycle, the transistor turns on almost completely. This effectively connects the bottom end of the tuned circuit to ground. Current flows through the coil L_1.

When the input voltage decreases a little, the transistor turns off. For the rest of the cycle, the transistor represents an open circuit between the lower end of the tuned circuit and ground. The current will continue to flow in the coil, decreasing gradually until the stored energy in the inductor has been transferred to capacitor C_1, which becomes charged. The process then reverses, and oscillation takes place. Energy is lost in the resistance of the inductor and capacitor, and of course energy is transferred to the load, so the amplitude of the oscillations would gradually be reduced to zero, except for one thing: Once each cycle, the transistor turns on, another current pulse is injected, and, because of this, oscillations continue indefinitely at the same level.

Figure 2.13(b) shows some of the waveforms associated with the Class C amplifier. Part (i) shows the input signal, and Part (ii) shows the actual signal applied to the base. Note the bias level that is generated by the signal itself. Part (iii) shows the pulses of collector current, and Part (iv) the collector voltage. It should not be surprising that this reaches a peak of almost $2V_{CC}$. The peak voltage across the inductor must be nearly V_{CC}, since, when the transistor is conducting, the top end of the coil is connected to V_{CC} and the bottom end is almost at ground potential ("almost" because, even when saturated, there will be a small voltage across the transistor). At the other peak of the cycle, the inductor voltage will have the same magnitude but opposite polarity, and will add to V_{CC} to make the peak collector voltage nearly equal to $2V_{CC}$.

This description implies that the output tuned circuit must be tuned fairly closely to the operating frequency of the amplifier, and that is indeed the case. Since the transistor must swing between cutoff and something close to saturation for Class C operation, it is also implicit in the design that this amplifier will be nonlinear; that is, doubling the amplitude of the input signal will not double the output. These two limitations restrict the use of Class C amplifiers to RF signals, and, in fact, only some RF applications can make use of this circuit.

Class C amplification can be used with either field-effect or bipolar transistors. It is also quite commonly used with vacuum tubes in large transmitters.

2.3.4 Neutralization

Device and stray capacitance tend to reduce gain and cause instability as frequency increases. Care must be taken to separate inputs and outputs to avoid feedback, but often a transistor or tube will itself introduce sufficient feedback to cause oscillations to take place. Sometimes this type of feedback can be cancelled by a process called **neutralization**.

FIGURE 2.14

Neutralized RF
Amplifier

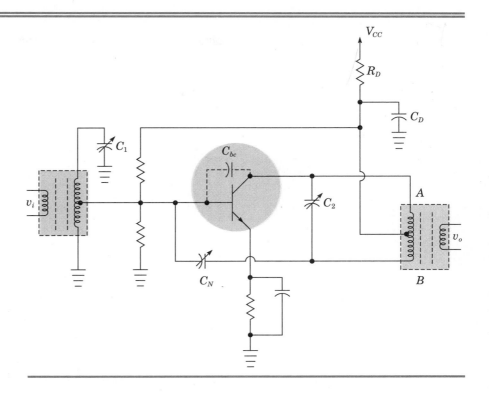

Neutralization is accomplished by deliberately feeding back a portion of
the output signal to the input in such a way that it has the same amplitude
as the unwanted feedback, but with opposite phase. Since the device capacitances vary from component to component, careful adjustment is necessary.

Figure 2.14 shows one type of neutralization. The basic circuit is that of
Figure 2.8, a narrowband RF amplifier with tuned input and output. As mentioned earlier, such an amplifier has a tendency to oscillate. This occurs because of the feedback through the base-collector capacitance of the transistor.
The method used in Figure 2.14 to provide feedback for neutralization is to
rearrange the primary of the output transformer so that the center is connected to V_{CC}. The power supply is, of course, at ground potential for ac (especially when the decoupling network consisting of R_D and C_D is used). The
transistor is still connected across only part of the output transformer, to reduce its loading effect as before, and there is no difference as far as the operation of the circuit is concerned.

The reason for rearranging the transformer connections is that now the
top and bottom ends of the winding have opposite polarities with respect to
ground, for the ac signal. The voltages may or may not be equal in magnitude:
there is no requirement that the transformer be tapped in the center as
shown. The internal capacitance C_{bc}, shown dotted, feeds a signal from the
top end of the coil, point A, to the transistor base. A small variable capacitor,
C_N, is adjusted to feed a signal of equal magnitude but opposite polarity from
the bottom of the coil, point B, to the base. When C_N is adjusted correctly, the
signals will cancel and the amplifier will be stable.

FIGURE 2.15 **Frequency Multiplier**

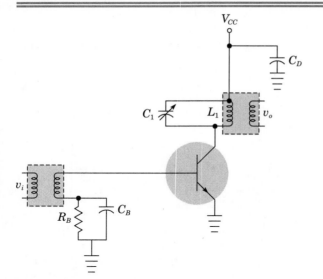

(a) Circuit

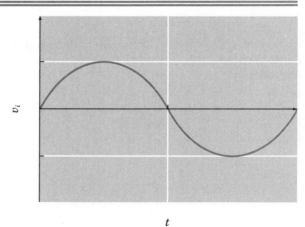

Input Signal

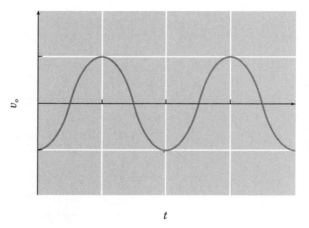

Output Signal

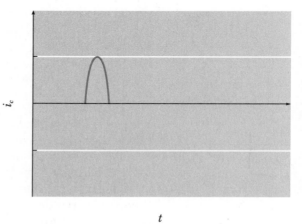

Collector Current

(b) Doubler Waveforms

FIGURE 2.16

Frequency
Multiplication

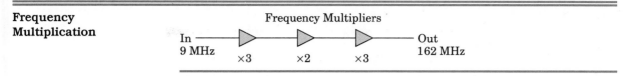

2.4   FREQUENCY MULTIPLIERS

The output circuit of a Class C amplifier can be tuned to a harmonic of the input signal. The amplifier will still operate, though with reduced efficiency. For instance, the pulses of collector current will sustain oscillations in an output tuned circuit that is tuned to twice the input frequency, by providing energy to the circuit during alternate cycles. Another way to look at the process is to recognize that the collector-current pulse in a Class C amplifier is rich in harmonics. Any one of these harmonics can be chosen as the output by tuning the output bandpass filter to the appropriate frequency. This harmonic operation can be very useful when a signal is required at a higher frequency than can be conveniently generated. Figure 2.15 shows a circuit for a *frequency doubler*, with input, output, and collector-current waveforms.

Because **frequency multipliers** operate at lower efficiencies than straight-through amplifiers, they are used at low power levels. Most multipliers operate at the second or third harmonic of the input frequency, and are known as **doublers** or **triplers** respectively. These are more efficient than multipliers operating at higher-order harmonics.

Multipliers can be used in cascade if greater multiplication is required. Figure 2.16 shows one way to get a multiplication of 18 times. This is a common arrangement in VHF transmitters, though it is not used in new designs as much as formerly.

2.5   RADIO-FREQUENCY OSCILLATORS

RF oscillators do not differ in principle from those used at lower frequencies, but the practical circuits are quite different. Where low-frequency oscillators usually use *RC* circuits in the frequency-determining section, *LC* circuits are more common at radio frequencies. In addition, many RF oscillators are crystal controlled.

Any amplifier can be made to oscillate if a portion of the output is fed back to the input in such a way that the following criteria, known as the *Barkhausen criteria*, are satisfied:

1. The gain around the loop must be equal to one. (If initially greater than one, it will become equal to one when oscillations start, due to some process such as transistor saturation; otherwise the output voltage would continue to increase without limit.)
2. The phase shift around the loop must total either 0° or some integer multiple of 360°, at the operating frequency (and not at other frequencies).

Taken together, these statements simply mean that, at the operating frequency, an input signal will be amplified then fed back in phase and with

FIGURE 2.17

Generalized Oscillator

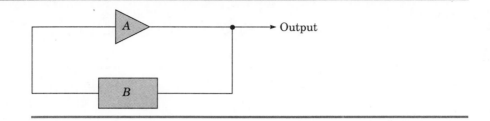

sufficient amplitude that it will maintain its value at the output without any further input. The initial signal needed to start the process can be noise or a transient caused by switching on the power to the oscillator circuit.

Figure 2.17 shows a "generic" oscillator consisting of an amplifier with gain A, and a feedback network with "gain" B. B will usually be less than 1, and only part of the output will be fed back to the input. B is often called the *feedback fraction*. According to the Barkhausen criteria, when oscillations are in progress, we will have:

$$AB = 1 \tag{2.9}$$

The phase shift around the loop will be zero or some multiple of 360°. This means that if the amplifier is an inverting amplifier, B must also invert the signal phase.

The foregoing must be true at only one frequency, so B has to have some form of frequency dependence. Either its phase shift, or its amplitude response, or both, must vary with frequency. There are many types of networks that meet this requirement, but one of the simplest, and the one most commonly used in radio-frequency oscillators, is the LC resonant circuit. Both series- and parallel-resonant circuits have amplitude and phase responses that are functions of frequency, and both have applications in oscillators.

2.5.1 LC Oscillators

It is now time to look at some practical oscillator configurations. We will begin with oscillators whose frequency is controlled by a resonant circuit using inductance and capacitance. Crystal-controlled oscillators will be described in Section 2.5.3.

All the circuits shown in this section have been in use for many years, and have been implemented with vacuum tubes, bipolar and field-effect transistors, and integrated circuits. In each case, we will begin by showing the amplifier as a simple gain block. All of these oscillator types can be implemented with either inverting or noninverting amplifiers, and will be shown both ways.

Hartley Oscillator. This oscillator type can be recognized by its use of a tapped inductor, part of a resonant circuit, to provide feedback.

Figure 2.18(a) shows a Hartley oscillator using a noninverting amplifier and Figure 2.18(b) shows the same oscillator using an inverting amplifier. In

FIGURE 2.18

Hartley Oscillators

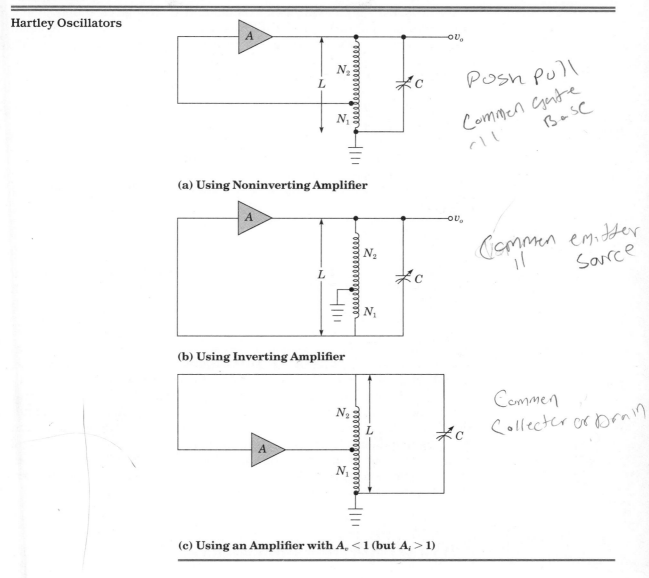

(a) Using Noninverting Amplifier

(b) Using Inverting Amplifier

(c) Using an Amplifier with $A_v < 1$ (but $A_i > 1$)

either case, the resonant frequency is that of the tuned circuit, including the whole inductor, that is

$$f_o = \frac{1}{2\pi\sqrt{LC}}$$ (2.10)

where f_o = frequency of oscillation in hertz
  L = inductance of the whole coil in henrys
  C = capacitance across the coil in farads

The feedback fraction is easy to find if the assumption is made that the inductor is a single coil with unity coupling. It is then given simply by the turns ratio of input turns to output turns. Thus there are two different equations, depending on whether the circuit uses an inverting or noninverting amplifier.

For the noninverting circuit of Figure 2.18(a),

$$B = \frac{N_1}{N_1 + N_2} \tag{2.11}$$

This circuit can be implemented with transistors connected in the common-base or common-gate configuration.

In the oscillator using the inverting amplifier, as in Figure 2.18(b),

$$B = \frac{-N_1}{N_2} \tag{2.12}$$

The negative sign indicates that the feedback has the opposite polarity to that of the output signal. This is required for an oscillator using an inverting amplifier, such as would be provided by a transistor connected for common-emitter or common-source operation.

Both of these circuits assume that the amplifier has a voltage gain greater than one. It might seem that this is a necessary condition for oscillation, but this is not the case. AB can be equal to one if A is less than one, provided that B is greater than one. The real amplifier requirement for oscillation is power gain, not voltage gain. The tapped coil of the Hartley oscillator can be used as an autotransformer to give a voltage feedback fraction greater than one. Of course, there will be a requirement for current gain from the amplifier in this case, a requirement that can be filled by using a common-collector or common-drain amplifier. Figure 2.18(c) demonstrates this configuration, in which the operating frequency remains as before, but the feedback fraction is now

$$B = \frac{N_1 + N_2}{N_1} \tag{2.13}$$

EXAMPLE 2.3

Two practical Hartley oscillators are shown in Figure 2.19. The first, in Figure 2.19(a), uses a junction field-effect transistor (JFET) in the common-source configuration, and that in Figure 2.19(b) uses the same transistor, connected as a common-drain amplifier (source follower). Determine whether each will oscillate, and if so, calculate the operating frequency.

Solution

Turning first to the common-source oscillator, we should make sure that the feedback has the correct polarity. The amplifier is a variation of the

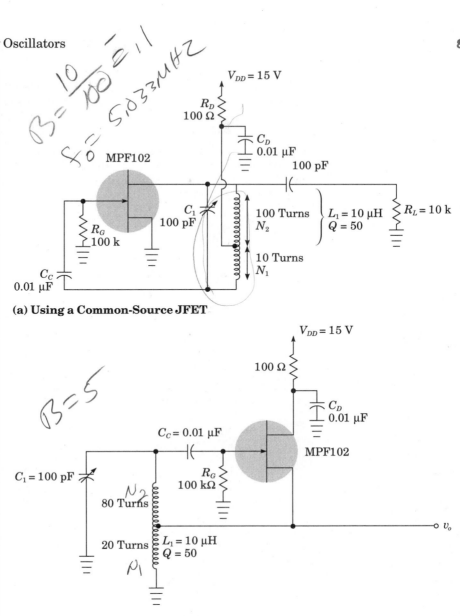

(a) Using a Common-Source JFET

(b) Using a Common-Drain JFET
FIGURE 2.19

inverting amplifier of Figure 2.18(b). Therefore the feedback should also be inverting, with the coil tap connected to ground. Actually it is connected to the power supply, V_{DD}, but, of course, that is an ac ground. The circuit appears to be workable.

In order to determine whether oscillations will take place, it is necessary to find both A and B and determine whether AB is at least equal to one. If it is more than one, it will be reduced by saturation effects once the oscillator starts. In fact, it would be good to have extra gain, since gain calculations are always approximate, and an oscillator with a loop gain even slightly less than one will not work at all. The *gate-leak* bias circuit helps with this. In a manner similar to that noted with the Class C amplifier, the bias is created by rectification of the signal voltage in the gate diode, which produces a small current through R_G and leaves the gate with a negative charge that varies with the strength of the signal. Since for a JFET, mutual conductance, and hence gain, reduces with increasing gate voltage, there is a form of negative feedback here that reduces the gain of the transistor as the signal voltage increases, until $AB = 1$.

Assuming that the circuit does oscillate, the resonant frequency is given by Equation (2.10).

$$f_o = \frac{1}{2\pi\sqrt{LC}}$$

$$= \frac{1}{2\pi\sqrt{(10 \times 10^{-6})(100 \times 10^{-12})}}$$

$$= 5.03 \text{ MHz}$$

The feedback fraction given in Equation (2.12) ignores any loading by the amplifier input, but that should not be a problem here because of the very high input impedance of the FET. For all practical purposes, the input loading consists only of the 100 kΩ resistor, R_G, which is large enough to ignore. The feedback fraction will then be

$$B = \frac{-N_1}{N_2}$$

$$= \frac{-10}{100}$$

$$= -0.1$$

To determine the minimum amplifier gain for oscillation, we start with Equation (2.9).

$$AB = 1$$

$$A = \frac{1}{B}$$

$$= \frac{1}{-0.1}$$

$$= -10$$

As usual, the negative sign denotes a phase inversion.

Finding the gain of the amplifier portion of the oscillator is a little more complicated. The gain that is required is that of the amplifier loaded by the feedback circuit. Since the gate circuit has a very high impedance, its contribution can be ignored, and the load resistance applied to the FET is the impedance of the tuned circuit at resonance in parallel with the load resistance. Note, by the way, that the effect of adding a load, that is, of extracting power from the oscillator, is to reduce the loop gain. If the circuit is too heavily loaded, it will stop oscillating.

For an FET in the common-source configuration, the voltage gain is given approximately by

$$A_V = -g_m r_D \tag{2.14}$$

where A_V = voltage gain (dimensionless)
    g_m = mutual conductance of the FET in siemens

and, in this circuit,

$$r_D = R_R \parallel R_L \tag{2.15}$$

Here, R_R is the impedance of the tuned circuit at resonance. (Actually, it will not be quite the whole impedance, since the coil tap is connected to ground, but the approximation is close enough for our purpose.) That will depend on its Q. In this case, the Q is 50. From basic electrical theory, we recall that

$$
\begin{aligned}
R_R &= QX_L \\
    &= Q \times 2\pi f_o L \\
    &= 50 \times 2\pi(5.03 \times 10^6)(10 \times 10^{-6}) \\
    &= 15.8 \text{ k}\Omega
\end{aligned}
$$

From Equation (2.15) we get

$$
\begin{aligned}
r_D &= R_R \parallel R_L \\
    &= 15.8 \text{ k}\Omega \parallel 10 \text{ k}\Omega \\
    &= \frac{15.8 \times 10}{15.8 + 10} \text{ k}\Omega \\
    &= 6.12 \text{ k}\Omega
\end{aligned}
$$

Now we need a value for the mutual conductance in order to find the gain. From the data sheet, the minimum value for g_m for the MPF102 can be found as 2000 µS, so we have, from Equation (2.14),

$$
\begin{aligned}
A_V &= -g_m r_D \\
    &= (-2000 \times 10^{-6})(6.12 \times 10^3) \\
    &= -12.2
\end{aligned}
$$

This is greater than required, so the circuit should oscillate.

The second circuit, in Figure 2.19(b), uses the same transistor, but in the common-drain configuration. It will have a voltage gain slightly less than one, but will still have reasonable current and power gain, and so the circuit can still be made to oscillate. In addition, the circuit is somewhat simpler, and a little more convenient. For instance, one side of C_1 is grounded. This can make its installation simpler, since variable capacitors generally have their movable section, called the *rotor,* connected to the case and the control shaft. In addition, the relatively low output impedance of this circuit reduces the effect of the load on the circuit.

The operating frequency will be the same as for the circuit in Figure 2.19(a), but the feedback fraction is different. Again neglecting any circuit loading effects, the feedback can be calculated using Equation (2.13).

$$B = \frac{N_1 + N_2}{N_1}$$

$$= \frac{80 + 20}{20}$$

$$= 5$$

Since the amplifier gain will be only slightly less than one, the feed-back fraction is more than sufficient to ensure oscillation.

Colpitts Oscillator. The Colpitts oscillator uses a capacitive voltage divider instead of a tapped inductor to provide feedback. Once again, the configuration of the feedback network depends on whether the amplifier is noninverting, as in Figure 2.20(a), or inverting, as in Figure 2.20(b).

The operating frequency is determined by the inductor and the series combination of C_1 and C_2.

$$f_o = \frac{1}{2\pi\sqrt{LC_T}} \tag{2.16}$$

FIGURE 2.20

Colpitts Oscillators

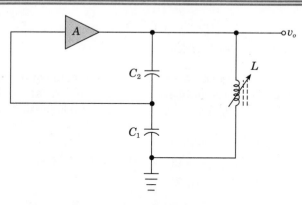

(a) Using Noninverting Amplifier

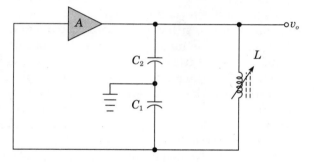

(b) Using Inverting Amplifier

where

$$C_T = \frac{C_1 C_2}{C_1 + C_2} \tag{2.17}$$

The feedback fraction is given by the ratio of reactances between output and input circuits. This is, of course, the reciprocal of the ratio of capacitances, because

$$X_C = \frac{1}{2\pi f C} \tag{2.18}$$

From this it follows that the ratio of the reactances of two capacitors C_a and C_b is

$$\begin{aligned}
\frac{X_{C_a}}{X_{C_b}} &= \frac{\dfrac{1}{2\pi f C_a}}{\dfrac{1}{2\pi f C_b}} \\[2mm]
&= \frac{2\pi f C_b}{2\pi f C_a} \\[2mm]
&= \frac{C_b}{C_a} \tag{2.19}
\end{aligned}$$

For the noninverting version of the oscillator, shown in Figure 2.20(a), the output is across the series combination of C_1 and C_2, which corresponds to C_T in Equation (2.16), and the input is the voltage across C_1. It is easy to see that the feedback fraction is

$$\begin{aligned}
B &= \frac{X_{C_1}}{X_{C_T}} \\[2mm]
&= \frac{C_T}{C_1} \\[2mm]
&= \frac{\dfrac{C_1 C_2}{C_1 + C_2}}{C_1} \\[2mm]
&= \frac{C_2}{C_1 + C_2} \tag{2.20}
\end{aligned}$$

The feedback fraction is even easier to determine for the inverting circuit of Figure 2.20(b). Since the output is applied across C_2 and the input is taken across C_1, the feedback fraction is

$$\begin{aligned}
B &= -\frac{X_{C_1}}{X_{C_2}} \\[2mm]
&= -\frac{C_2}{C_1} \tag{2.21}
\end{aligned}$$

As before, the negative sign indicates that the feedback signal is 180° out of phase with the output.

Since changing either C_1 or C_2, in order to tune the oscillator, will change the feedback fraction, it is quite common to use a variable inductor for tuning instead.

Like the Hartley, the Colpitts oscillator can be configured for an amplifier with power gain but no voltage gain. The derivation of the circuit and the feedback fraction is left as an exercise for the reader.

EXAMPLE 2.4

Determine the feedback fractions and operating frequencies for the oscillators whose circuits are shown in Figure 2.21.

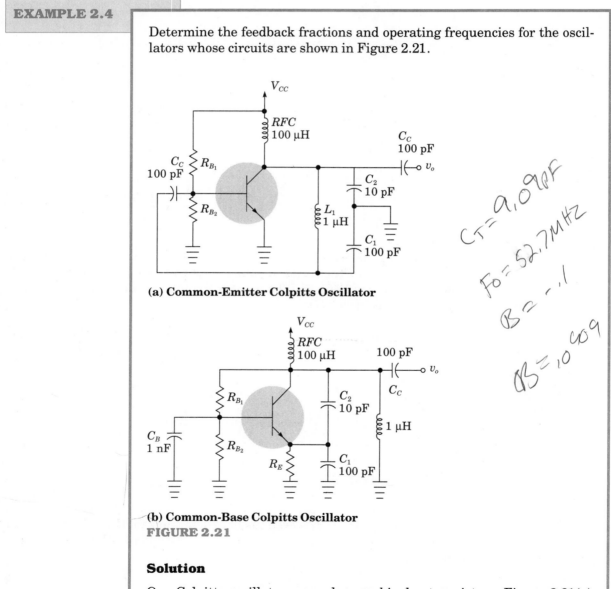

(a) **Common-Emitter Colpitts Oscillator**

(b) **Common-Base Colpitts Oscillator**

FIGURE 2.21

Solution

Our Colpitts oscillator examples use bipolar transistors. Figure 2.21(a) uses a common-emitter circuit, while the transistor is connected common-base in Figure 2.21(b). These two examples use inverting and noninverting amplifiers respectively, both with voltage gain greater than one. The use of a common-collector circuit is also possible.

A detailed analysis of these oscillators is a little more complex than for the FET circuits previously described, because of the lower impedances involved, which load the feedback circuit more heavily. However, approximate results, which are generally all that are required in practical situations, can be achieved very simply. In Figure 2.21(a), C_C is a coupling capacitor that prevents a dc short circuit from occurring between collector and base through the coil. The radio-frequency choke RFC takes the place of a collector resistor and keeps the ac at the collector from being short-circuited by the power supply. A collector resistor can be used, but the choke, because of its lower dc resistance, increases the output voltage and improves the efficiency of the circuit. R_{B_1} and R_{B_2} are bias resistors, of course. This leaves the frequency of the oscillator to be determined by L_1, C_1, and C_2. The effective capacitance for determining the frequency of operation is given by Equation (2.17):

$$C_T = \frac{C_1 C_2}{C_1 + C_2}$$

$$= \frac{10 \times 100}{10 + 100} \text{ pF}$$

$$= 9.09 \text{ pF}$$

The operating frequency can be found from Equation (2.16):

$$f_o = \frac{1}{2\pi\sqrt{LC_T}}$$

$$= \frac{1}{2\pi\sqrt{(1 \times 10^{-6})(9.09 \times 10^{-12})}}$$

$$= 52.8 \text{ MHz}$$

The feedback fraction is given approximately by Equation (2.21).

$$B = -\frac{C_2}{C_1}$$

$$= -\frac{10}{100}$$

$$= -0.1$$

For the common-base circuit, the operating frequency will be the same but the feedback fraction will be different. Of course, the sign will be positive because the amplifier is noninverting, but the magnitude is also slightly different. From Equation (2.20),

$$B = \frac{C_2}{C_1 + C_2}$$

$$= \frac{10}{100 + 10}$$

$$= 0.0909$$

Clapp Oscillator. The Clapp oscillator is a variation of the Colpitts circuit, designed to swamp device capacitances for greater stability. In the oscillators of Figure 2.22, the frequency of oscillation is determined by the inductor and the series combination of C_1, C_2, and C_3. In practice, the total capacitance is determined almost entirely by C_3, which is chosen to be much smaller than either C_1 or C_2. The total effective capacitance of the three capacitors in series is given by

$$C_T = \frac{1}{\dfrac{1}{C_1} + \dfrac{1}{C_2} + \dfrac{1}{C_3}} \tag{2.22}$$

After finding C_T from Equation (2.22), the operating frequency can easily be found from Equation (2.16), as for the Colpitts oscillator.

The feedback fraction is found in the same way as for the Colpitts oscillator.

One circuit will serve to illustrate the Clapp oscillator, since it is so similar to the Colpitts.

FIGURE 2.22

Clapp Oscillators

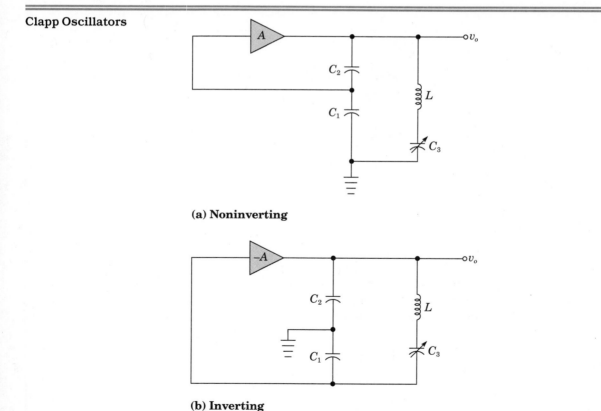

(a) Noninverting

(b) Inverting

EXAMPLE 2.5

Calculate the feedback fraction and oscillating frequency for the circuit in Figure 2.23.

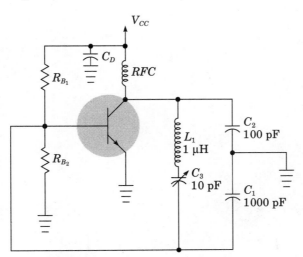

FIGURE 2.23

Solution

Figure 2.23 is the same as Figure 2.21(a) except for the addition of C_3 in series with the coil, and the increase in the values of C_1 and C_2. Any transistor capacitances will appear across C_1 and C_2, where they will have little effect on the frequency.

Before calculating the oscillating frequency, it is necessary to calculate the effective total capacitance from Equation (2.22).

$$C_T = \frac{1}{\dfrac{1}{C_1} + \dfrac{1}{C_2} + \dfrac{1}{C_3}}$$

$$= \frac{1}{\dfrac{1}{100} + \dfrac{1}{1000} + \dfrac{1}{10}} \text{ pF}$$

$$= 9.01 \text{ pF}$$

Note the relatively small effect of C_1 and C_2.

Now the operating frequency can be found in the usual way, from Equation (2.16).

$$f_o = \frac{1}{2\pi\sqrt{LC_T}}$$

$$= \frac{1}{2\pi\sqrt{(1 \times 10^{-6})(9.01 \times 10^{-12})}}$$

$$= 53.02 \text{ MHz}$$

FIGURE 2.24

Variable Capacitors and Inductors

(a)

2.5.2 Varactor-Tuned Oscillators

The frequency of an LC oscillator can be changed by varying, or tuning, either the inductive or the capacitive element in a tuned circuit. Inductors are typically tuned by moving a ferrite core into or out of the coil; this is known as *slug tuning*. Variable capacitors usually have two sets of plates which can be interleaved to a greater or lesser extent. Figure 2.24 shows typical examples of each.

Mechanical tuning tends to be awkward for a number of reasons. The components are bulky, expensive, and subject to accidental detuning, in the presence of vibration for instance. Variable capacitors and inductors are mechanical devices that have to be moved physically. This makes remote or automatic frequency control rather cumbersome.

Varactors are a more convenient substitute for variable capacitors in many applications. Essentially, a varactor is a reverse-biased silicon diode. As the reverse voltage increases, so does the width of the diode's depletion layer. As a result, the junction capacitance decreases. If this junction capacitance is made part of a resonant circuit, that circuit can be tuned simply by

(b)

varying the dc voltage on the varactor. This can be done in many ways, and is well adapted to remote or automatic control. The resulting circuit is often called a voltage-controlled oscillator (VCO).

It is, of course, necessary to separate the dc control voltage from the ac signal voltages. This is quite straightforward: Figure 2.25 shows one way to do it. The noninverting Clapp oscillator of Figure 2.22(a) has been adapted for use as a VCO by using a varactor for C_3. Resistor R prevents the RF in the circuit from being short-circuited by the circuit that provides the tuning voltage, and the extra capacitor, C_4, keeps the dc control voltage out of the rest of the circuit. C_4 is made much larger than C_3 so that its reactance will be negligible and C_3 will still control the operating frequency.

The variation of capacitance with voltage is not linear for a varactor. It is given approximately by

$$C = \frac{C_o}{\sqrt{1 + 2V}} \qquad\qquad (2.23)$$

FIGURE 2.25

Varactor-Tuned
Oscillator

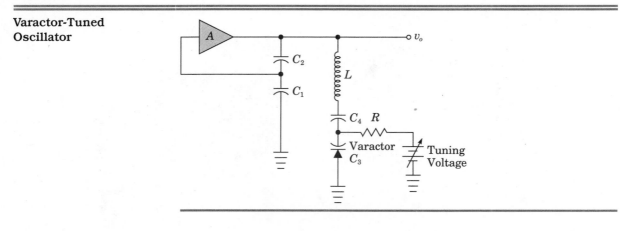

where C = capacitance at reverse voltage V
       C_o = capacitance with no reverse voltage

From this equation it can be seen that for relatively large reverse-bias voltages, the capacitance is approximately inversely proportional to the square
root of the applied voltage. A number of varactors are available with maximum capacitances varying from a few picofarads to more than 100 pF. From
Equation (2.23) it can be seen that the minimum capacitance will be limited
by the breakdown voltage of the diode and, of course, by the tuning voltage
available. In practice, a variation of about 5:1 in capacitance is quite practical. The magnitude of control voltage required for this can be found by letting
$C_o / C = 5$ in Equation (2.23).

$$C = \frac{C_o}{\sqrt{1 + 2V}}$$

$$\sqrt{1 + 2V} = \frac{C_o}{C}$$

$$\sqrt{1 + 2V} = 5$$

$$1 + 2V = 25$$

$$2V = 24$$

$$V = 12 \text{ V}$$

EXAMPLE 2.6

A varactor has a maximum capacitance of 80 pF and is used in a tuned
circuit with a 100 μH inductor.

(a) Find the resonant frequency with no tuning voltage applied.

(b) Find the tuning voltage necessary for the circuit to resonate at double
    the frequency found in Part (a).

Solution

(a) The maximum capacitance of the varactor occurs for zero bias voltage, so the capacitance in this case will be 80 pF. The resonant frequency can be calculated from Equation (2.10).

$$f_o = \frac{1}{2\pi\sqrt{LC}}$$

$$= \frac{1}{2\pi\sqrt{(100 \times 10^{-6})(80 \times 10^{-12})}}$$

$$= 1.78 \text{ MHz}$$

(b) From Equation (2.10) it is apparent that the resonant frequency is inversely proportional to the square root of capacitance, so that doubling the frequency will require reducing the capacitance by a factor of four. Alternatively we can use Equation (2.10) directly.

$$f_o = \frac{1}{2\pi\sqrt{LC}}$$

$$f_o^2 = \frac{1}{4\pi^2 LC}$$

$$C = \frac{1}{4\pi^2 f_o^2 L}$$

$$= \frac{1}{4\pi^2(2 \times 1.78 \times 10^6)^2(100 \times 10^{-6})}$$

$$= 20 \times 10^{-12} \text{ F}$$

$$= 20 \text{ pF}$$

Now we can find the required tuning voltage from Equation (2.23).

$$C = \frac{C_o}{\sqrt{1 + 2V}}$$

$$\sqrt{1 + 2V} = \frac{C_o}{C}$$

$$1 + 2V = \left(\frac{C_o}{C}\right)^2$$

$$V = \frac{\left(\frac{C_o}{C}\right)^2 - 1}{2}$$

$$= \frac{\left(\frac{80}{20}\right)^2 - 1}{2}$$

$$= 7.5 \text{ V}$$

2.5.3 Crystal-Controlled Oscillators

The frequency stability of any *LC* oscillator depends on that of its resonant circuit, including any stray or device reactances that may be present. Even with careful design, *LC* oscillators are subject to frequency change from such diverse sources as voltage variations, changes in load impedance, temperature changes, and mechanical vibration. All these problems can be reduced, but only at great expense.

Crystal oscillators achieve greater stability by using a small slab of quartz as a mechanical resonator, in place of an *LC* tuned circuit. Quartz is a **piezoelectric** material: that is, deforming it mechanically will cause the crystal to generate a voltage, and applying a voltage to the crystal will cause it to deform. Like any rigid body, the crystal slab has a mechanical resonant frequency. If it is pulsed with voltage at that frequency, it will vibrate. From the outside, the crystal will appear as an electrical resonant circuit.

Figure 2.26 is a photograph showing several typical crystals, some with the packaging removed to show the crystal slab. The actual quartz crystals from which RF crystals are made are shown in the background. Figure 2.27(a) is the schematic symbol, and Figure 2.27(b) shows an equivalent circuit for a crystal. Note that a crystal does not actually have this circuit. It is just a slab of quartz with electrodes. This circuit simply models its behavior, as seen from outside. The only part of the equivalent circuit that resembles the actual device is C_P, the parallel capacitance, which actually is the capacitance of the holder and the crystal itself. The series inductor and capacitor are electrical analogs of the mechanical properties of the crystal.

FIGURE 2.26

Quartz Crystals

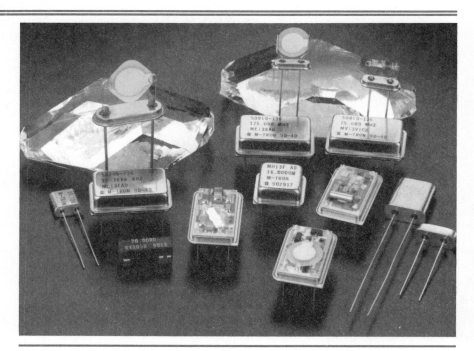

M-Tron Industries, Inc.

FIGURE 2.27

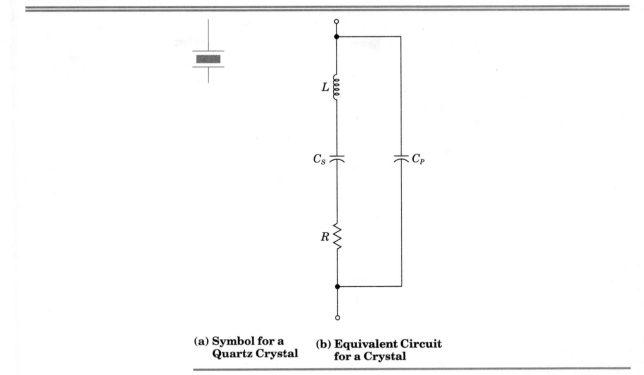

**(a) Symbol for a
   Quartz Crystal**

**(b) Equivalent Circuit
   for a Crystal**

In electrical terms, the inductor in the equivalent circuit is very large (on the order of henrys), while both capacitors are very small (picofarads or less), with the parallel capacitance much larger than the series. The resistance is small, and the Q of crystals is very high (in the range from 10^4 to 10^7). From the equivalent circuit, it is apparent that the crystal will actually have not one, but two, resonant frequencies: one series and one parallel. Figure 2.28 shows graphically how the reactance varies with frequency. The two reso-

FIGURE 2.28

Variation of Crystal
Reactance with
Frequency

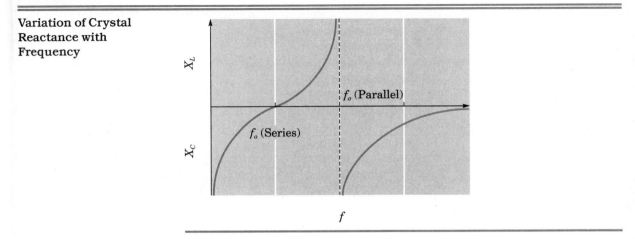

nances will vary by an amount that is on the order of 1% of the operating frequency, and it is necessary to know, when ordering a crystal, at which of these frequencies the circuit operates. For parallel resonance, it is also necessary to have a constant, known circuit capacitance across the crystal. It is also possible to operate the crystal at a *harmonic*, or *overtone*, of the fundamental frequency. This is necessary when frequencies above approximately 20 MHz are required.

Crystal oscillators offer great stability at the price of fixed-frequency operation. The operating frequency is typically accurate to $\pm 0.005\%$, and the stability can be much better if a temperature-controlled "oven" is used to keep the crystal at a constant temperature a little higher than the ambient temperature. Crystals are not immune to temperature changes, having temperature coefficients that can be either positive or negative depending on the way the crystal is cut, and that can be as high as approximately 100 parts per million (ppm) per degree Celsius. The frequency is given by

$$f_T = f_o + k f_o (T - T_o) \qquad\qquad (2.24)$$

where f_T = operating frequency at temperature T
 f_o = operating frequency at reference temperature T_o
 k = temperature coefficient per degree

EXAMPLE 2.7

A portable radio transmitter has to operate at temperatures from $-5°C$ to $+35°C$. If its signal is derived from a crystal oscillator with a temperature coefficient of $+10$ ppm/degree C, and it transmits at exactly 146 MHz at 20°C, find the transmitting frequency at the two extremes of the operating temperature range.

Solution

From Equation (2.24), the frequency at a temperature of 35°C will be

$$f_T = f_o + k f_o (T - T_o)$$
$$f_{max} = 146 \text{ MHz} + (146 \text{ MHz})(10 \times 10^{-6})(35 - 20)$$
$$= 146.0219 \text{ MHz}$$

By similar reasoning, at the lower temperature limit, the operating frequency will be

$$f_{min} = 146 \text{ MHz} + (146 \text{ MHz})(10 \times 10^{-6})(-5 - 20)$$
$$= 145.9635 \text{ MHz}$$

In other words, varying the temperature over a total range of 40°C caused the transmitter frequency to change by $(21.9 + 36.5) = 58.4$ kHz. This would not usually be acceptable in a practical transmitter, and some form of frequency compensation would be needed.

FIGURE 2.29

**Crystal Oscillator
Circuits**

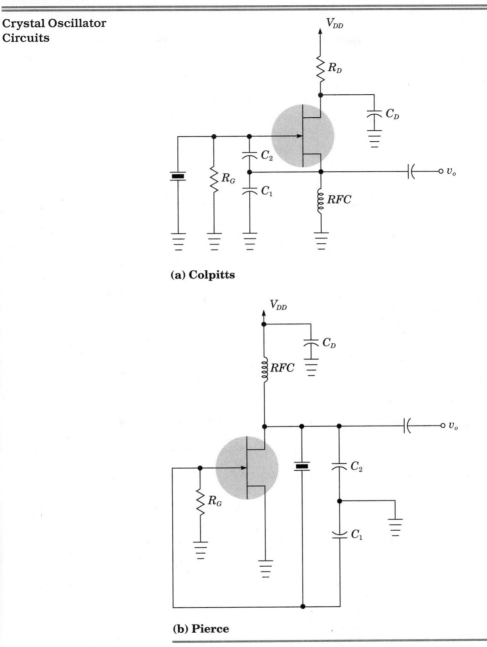

(a) Colpitts

(b) Pierce

Figure 2.29 shows two typical circuits for crystal oscillators. The Colpitts circuit in Figure 2.29(a) uses the parallel-resonance mode of the crystal, with capacitors C_1 and C_2 providing feedback. R_D and C_D **decouple** the drain from the power supply. The bias is generated from the signal itself using the gate-leak technique previously described for FET oscillators. The circuit in Figure 2.29(b), known as the *Pierce*, uses the crystal in place of the inductor in a series-resonant circuit consisting of C_1, C_2, and the crystal. It operates in the inductive region of the reactance curve of Figure 2.28, close to but slightly above the series resonant frequency.

Crystal oscillators are often supplied in convenient modules, including temperature compensation circuitry. In addition to crystals, typical modules are shown in Figure 2.26.

The frequency of a crystal oscillator can be adjusted slightly by placing a variable capacitance in series or in parallel with the crystal, depending on the type of circuit. An oscillator of this sort is known as a *variable crystal oscillator* (VXO). If the capacitor is actually a varactor, the frequency can be adjusted by a thermistor, for temperature compensation.

2.6 MIXERS

A **mixer**, as used in communications, is a nonlinear circuit that combines two signals in such a way as to produce the sum and difference of the two input frequencies at the output. Sometimes the original input frequencies, and perhaps other frequencies as well, are also present. Mixers provide a way of moving a signal, complete with any modulation that may be present, from one frequency to another.

It is important to distinguish mixing from *linear summing*, which will produce only the two input signal frequencies. Unfortunately the term *mixer* is often used in audio electronics to designate a linear summer. The "mixer" that regulates the sound in a recording studio is really a good example of a summer. Figure 2.30 illustrates the difference between mixing and summing.

Any nonlinear device can operate as a mixer. Familiar examples include a semiconductor diode and a Class C amplifier. In general, a nonlinear device produces a signal at its output that can be represented by a power series, for example

$$v_o = Av_i + Bv_i^2 + Cv_i^3 + \cdots \tag{2.25}$$

where v_o = instantaneous output voltage
v_i = instantaneous input voltage
$A, B, C, \cdots$ = constants

With a single input frequency, the output will contain all the harmonics as well as the fundamental. As the order of the harmonics increases, the magnitude decreases. This is the principle of the frequency multiplier, and also the cause of harmonic distortion.

When the input contains two different frequencies, we get cross products that represent sum and difference frequencies. If the two frequencies are f_1 and f_2, we get $mf_1 \pm nf_2$, where m and n are integers. Usually the most important are $f_1 + f_2$ and $f_1 - f_2$, where f_1 is assumed to be higher than f_2. When the cross products are not wanted, they are called *intermodulation distortion*.

FIGURE 2.30   Output Spectra for Mixer and Summer

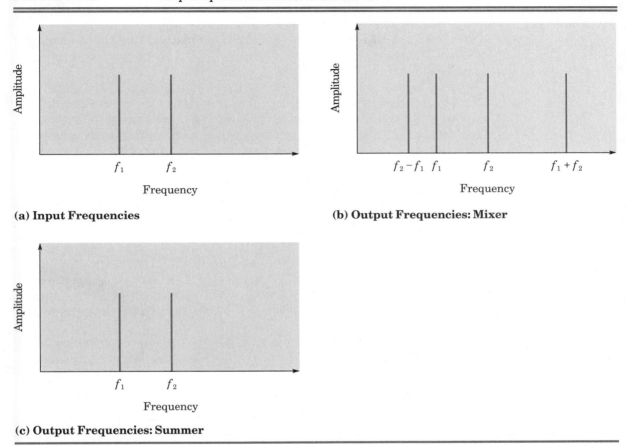

(a) Input Frequencies

(b) Output Frequencies: Mixer

(c) Output Frequencies: Summer

2.6.1 Square-Law Mixers

The *square-law mixer* is the simplest to understand mathematically, and it closely models the actual performance of mixers using FETs. Its output is given by a simplified form of Equation (2.25):

$$v_o = Av_i + Bv_i^2 \tag{2.26}$$

Let us apply two signals to this circuit, summing them at the input. The signals will be sine waves of different frequencies. For convenience, we can let each signal have an amplitude of 1 V peak. Then

$$v_i = \sin \omega_1 t + \sin \omega_2 t \tag{2.27}$$

and

$$v_o = Av_i + Bv_i^2$$
$$= A(\sin \omega_1 t + \sin \omega_2 t) + B(\sin \omega_1 t + \sin \omega_2 t)^2$$
$$= A \sin \omega_1 t + A \sin \omega_2 t + B \sin^2 \omega_1 t + B \sin^2 \omega_2 t \qquad (2.28)$$
$$+ 2B \sin \omega_1 t \sin \omega_2 t$$

The first two terms in Equation (2.28) are simply the input signals multiplied by A, which is a gain factor. The next two terms involve the squares of the input signals, which are signals at twice the input frequency (plus a dc component). This can be seen from the trigonometric identity

$$\sin^2 A = \frac{1}{2} - \frac{1}{2} \cos 2A$$

This means that the third term becomes

$$\frac{B}{2} - \frac{B}{2} \cos 2\omega_1 t$$

and the fourth term is

$$\frac{B}{2} - \frac{B}{2} \cos 2\omega_2 t$$

The final term is the interesting one. We can use the trigonometric identity

$$\sin A \sin B = \frac{1}{2} [\cos (A - B) - \cos (A + B)] \qquad (2.29)$$

to expand this term.

$$2B \sin \omega_1 t \sin \omega_2 t = \frac{2B}{2} [\cos (\omega_1 - \omega_2)t - \cos (\omega_1 + \omega_2)t]$$
$$= B [\cos (\omega_1 - \omega_2)t - \cos (\omega_1 + \omega_2)t]$$

As predicted, there are output signals at the sum and difference of the two input frequencies, in addition to the input frequencies themselves and their second harmonics. In a practical application, either the sum or the difference frequency would be used, and the others would be removed by filtering.

EXAMPLE 2.8

Sine-wave signals with frequencies of 10 MHz and 11 MHz are applied to a square-law mixer. What frequencies appear at the output?

Solution

Let f_1 = 11 MHz and f_2 = 10 MHz. Then the output frequencies are as follows:

$$f_1 = 11 \text{ MHz} \qquad 2f_1 = 22 \text{ MHz} \qquad f_1 + f_2 = 21 \text{ MHz}$$
$$f_2 = 10 \text{ MHz} \qquad 2f_2 = 20 \text{ MHz} \qquad f_1 - f_2 = 1 \text{ MHz}$$

2.6.2 Diode Mixers

There are two senses in which we can say that a diode is a nonlinear device. First, consider an ideal diode, which has zero reverse current for any reverse voltage, and zero forward voltage for any forward current. This voltage-current relationship is shown in Figure 2.31(a). There is a very obvious non-linearity at zero voltage.

Now consider a real diode with its forward voltage-current characteristics. A V-I curve for a typical silicon signal diode is shown in Figure 2.31(b). It is obviously not a straight line; that is, a real diode is nonlinear even when operated in the forward-bias region. Of course, the larger the forward current, the more the diode approaches linearity.

Diode **mixers** can use either type of nonlinearity. In the first type, one of the signals is strong enough to switch the diode between the reverse-biased and forward-biased states. The diode ring balanced mixer, to be described later, operates in this way. There is also another possibility, where the diode operates with a small forward bias, in the "knee" of the curve of Figure 2.31(b). The input signal is small, so that the diode stays in this highly nonlinear region of its characteristic curve. Figure 2.32 shows a circuit for a mixer of this type. Such mixers are not used very often because of their poor noise figure.

FIGURE 2.31   Diode Curves

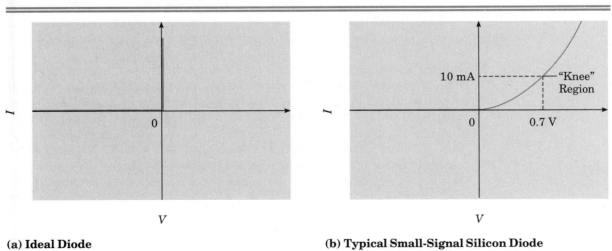

(a) Ideal Diode   **(b) Typical Small-Signal Silicon Diode**

FIGURE 2.32

Diode Mixer

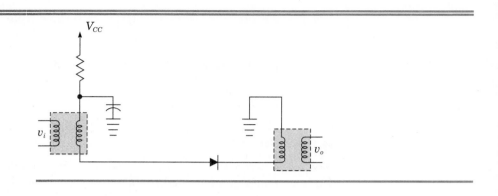

2.6.3 Transistor Mixers

All transistors, whether bipolar or field-effect types, have nonlinearities. In fact, much of the work in an introductory electronics course usually involves biasing these devices in such a way as to reduce the nonlinearity to a minimum. The success of these efforts is then measured in terms of percent distortion.

If a mixer is required, on the other hand, nonlinearities are desirable. Either a bipolar transistor or an FET can be operated in such a way that the input causes the device to enter nonlinear regions. The FET is especially convenient for use as a mixer because the parabolic shape of its transconductance curve gives it approximately a square-law response. The bipolar transistor will produce more spurious frequencies at its output, frequencies that will probably have to be removed by filtering. However, both types of transistor are in common use in mixer circuits.

Figure 2.33 includes circuits for three typical transistor mixers. Figure 2.33(a) uses a bipolar transistor, and Figure 2.33(b) uses a JFET. The circuit of Figure 2.33(c) uses a dual-gate metal oxide semiconductor FET (MOSFET) with one of the two signals to be mixed applied to each gate.

The bipolar-transistor mixer of Figure 2.33(a) resembles a conventional tuned RF amplifier, except that it has two inputs, one at the base and one at the emitter. One of the two signals to be mixed will be applied to each input, with the stronger of the two going to the emitter. The output may be tuned either to the sum or the difference of the two input signal frequencies, whichever is required. In this circuit, the resonant circuits L_1–C_1 and L_2–C_2 are tuned to f_1 and the required output frequency respectively. R_{B_1}, R_{B_2}, and R_E form the usual voltage-divider bias circuit, C_3 is a coupling capacitor that keeps L_1 from short-circuiting the base bias, and C_D and R_D form a decoupling network that keeps RF out of the power supply line. It is also possible to apply f_2, as well as f_1, to the base, summing them at this point.

The JFET circuit of Figure 2.33(b) illustrates the technique of summing the two input signals at a single input terminal. Both f_1 and f_2 are applied to the gate. The tuned circuits L_1–C_1 and L_2–C_2 have the same functions as before, and so do C_D and R_D. C_4 couples the f_2 input into the circuit, C_5 bypasses the source resistor, and R_S provides self-bias. In some JFET mixers one of the input signals is applied to the source.

Dual-gate MOSFETs are very easy to use as mixers, as one input signal can be applied to each gate. This is demonstrated in Figure 2.33(c). Dual-gate MOSFETs make excellent mixers. They have better dynamic range than mixers using bipolar transistors; that is, they can operate satisfactorily over a wider range of signal amplitudes. They also produce fewer unwanted output frequencies.

2.6.4 Balanced Mixers

A **balanced mixer** is one in which the input frequencies do not appear at the output. Ideally, the only frequencies that are produced are the sum and difference of the input frequencies.

A multiplier circuit, where the output amplitude is proportional to the product of two input signals, can be used as a balanced mixer. This is easy to show mathematically. Let the multiplier have the equation

$$v_o = Av_{i_1}v_{i_2} \qquad\qquad (2.30)$$

where v_o $\qquad\qquad$ = the instantaneous output voltage

$\qquad$ v_{i_1} and v_{i_2} = the instantaneous voltages applied at two input terminals

$\qquad$ A $\qquad\qquad$ = a constant

Let the input consist of two sine waves of different frequencies, as before. For simplicity, assume each signal has a peak amplitude of 1 V.

$$v_{i_1} = \sin \omega_1 t$$

$$v_{i_2} = \sin \omega_2 t$$

Then the output will be

$$v_o = Av_{i_1}v_{i_2}$$

$$= A \sin \omega_1 t \, \sin \omega_2 t$$

We can use the same trigonometric identity, Equation (2.29), that was used earlier with the square-law mixer.

$$\sin A \sin B = \frac{1}{2} [\cos (A - B) - \cos (A + B)]$$

to show that the output is

$$v_o = \frac{A}{2} [\cos (\omega_1 - \omega_2)t - \cos (\omega_1 + \omega_2)t]$$

As predicted, this mixer produces only the sum and difference of the input frequencies. This makes it an excellent choice for many applications. In fact, the usual block-diagram representation of a mixer is the same as for a multiplier (see Figure 2.34).

FIGURE 2.33

Transistor Mixers

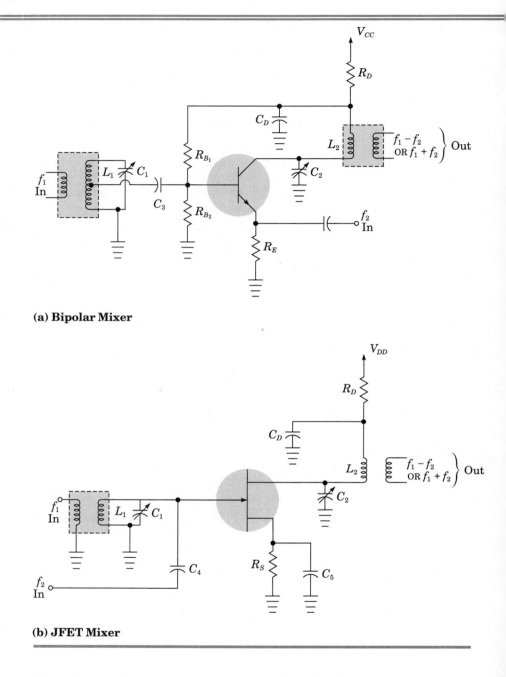

(a) Bipolar Mixer

(b) JFET Mixer

Though general-purpose integrated-circuit multipliers exist, there are a number of balanced mixers designed especially for communications use. One of these is the 1496, whose data sheet is shown in Figure 2.35. These devices are capable of operating over a very wide frequency range, and are extremely well balanced; that is, the input signal frequencies are reduced to very low levels.

It is also possible to build a balanced mixer using discrete components. Figure 2.36 shows a traditional circuit using four diodes in a ring configura-

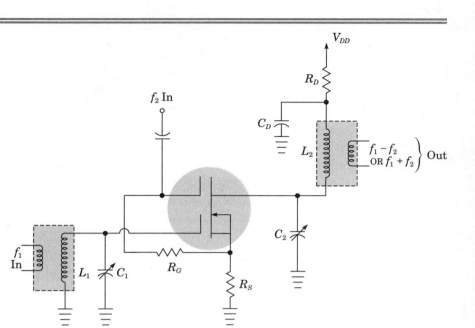

(c) Dual-Gate MOSFET Mixer

tion. For best results, the diodes should be carefully matched. The input signals should not have the same amplitude: the input at f_2 should have an amplitude large enough to turn the diodes on completely, and the other input signal should be of much lower voltage.

In order to understand the operation of this mixer, first consider an instant when the f_2 signal causes the secondary of transformer T_3 to have the polarity shown in Figure 2.36, that is, the left side is positive. This will cause diodes D_1 and D_2 to turn on, and D_3 and D_4 to turn off. D_1 and D_2 will then connect the much weaker signal at the secondary of T_1 directly to the primary of T_2 without any polarity change.

FIGURE 2.34

Mixer Symbol

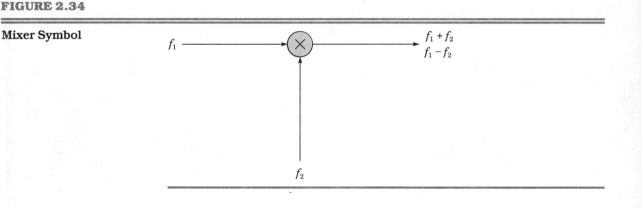

FIGURE 2.35　　　　　Integrated-Circuit Balanced Mixer

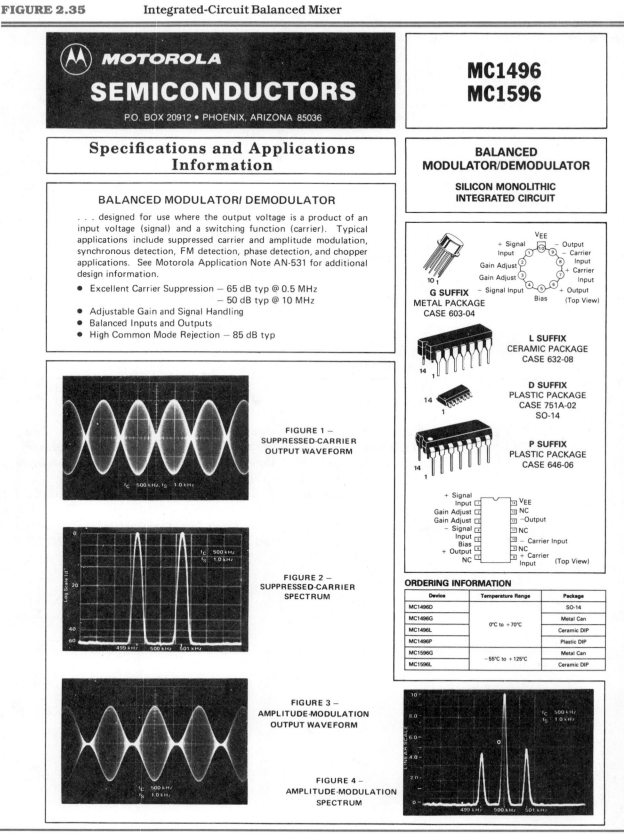

MOTOROLA
SEMICONDUCTORS
P.O. BOX 20912 • PHOENIX, ARIZONA 85036

MC1496
MC1596

Specifications and Applications Information

BALANCED MODULATOR/ DEMODULATOR

. . . designed for use where the output voltage is a product of an input voltage (signal) and a switching function (carrier). Typical applications include suppressed carrier and amplitude modulation, synchronous detection, FM detection, phase detection, and chopper applications. See Motorola Application Note AN-531 for additional design information.

- Excellent Carrier Suppression — 65 dB typ @ 0.5 MHz
  — 50 dB typ @ 10 MHz
- Adjustable Gain and Signal Handling
- Balanced Inputs and Outputs
- High Common Mode Rejection — 85 dB typ

BALANCED MODULATOR/DEMODULATOR

SILICON MONOLITHIC INTEGRATED CIRCUIT

G SUFFIX
METAL PACKAGE
CASE 603-04

L SUFFIX
CERAMIC PACKAGE
CASE 632-08

D SUFFIX
PLASTIC PACKAGE
CASE 751A-02
SO-14

P SUFFIX
PLASTIC PACKAGE
CASE 646-06

FIGURE 1 —
SUPPRESSED-CARRIER
OUTPUT WAVEFORM

FIGURE 2 —
SUPPRESSED-CARRIER
SPECTRUM

FIGURE 3 —
AMPLITUDE-MODULATION
OUTPUT WAVEFORM

FIGURE 4 —
AMPLITUDE-MODULATION
SPECTRUM

ORDERING INFORMATION

Device	Temperature Range	Package
MC1496D		SO-14
MC1496G	0°C to +70°C	Metal Can
MC1496L		Ceramic DIP
MC1496P		Plastic DIP
MC1596G	−55°C to +125°C	Metal Can
MC1596L		Ceramic DIP

Courtesy Motorola, Inc.

MAXIMUM RATINGS* ($T_A = +25°C$ unless otherwise noted)

Rating	Symbol	Value	Unit
Applied Voltage ($V_6 - V_7$, $V_8 - V_1$, $V_9 - V_7$, $V_9 - V_8$, $V_7 - V_4$, $V_7 - V_1$, $V_8 - V_4$, $V_6 - V_8$, $V_2 - V_5$, $V_3 - V_5$)	ΔV	30	Vdc
Differential Input Signal	$V_7 - V_8$ $V_4 - V_1$	$+5.0$ $\pm(5 + I_5 R_e)$	Vdc
Maximum Bias Current	I_5	10	mA
Thermal Resistance, Junction to Air Ceramic Dual In-Line Package Plastic Dual In-Line Package Metal Package	$R_{\theta JA}$	 100 100 160	°C/W
Operating Temperature Range MC1496 MC1596	T_A	 0 to $+70$ -55 to $+125$	°C
Storage Temperature Range	T_{stg}	-65 to $+150$	°C

ELECTRICAL CHARACTERISTICS* ($V_{CC} = +12$ Vdc, $V_{EE} = -8.0$ Vdc, $I_5 = 1.0$ mAdc, $R_L = 3.9$ kΩ, $R_e = 1.0$ kΩ,
$T_A = +25°C$ unless otherwise noted) (All input and output characteristics are single-ended unless otherwise noted.)

Characteristic	Fig.	Note	Symbol	MC1596 Min	MC1596 Typ	MC1596 Max	MC1496 Min	MC1496 Typ	MC1496 Max	Unit
Carrier Feedthrough $V_C = 60$ mV(rms) sine wave and $\quad f_C = 1.0$ kHz offset adjusted to zero $\quad f_C = 10$ MHz $V_C = 300$ mVp-p square wave: offset adjusted to zero $\quad f_C = 1.0$ kHz offset not adjusted $\quad f_C = 1.0$ kHz	5	1	V_{CFT}	 — — — —	 40 140 0.04 20	 — — 0.2 100	 — — — —	 40 140 0.04 20	 — — 0.4 200	μV(rms) mV(rms)
Carrier Suppression $f_S = 10$ kHz, 300 mV(rms) $f_C = 500$ kHz, 60 mV(rms) sine wave $f_C = 10$ MHz, 60 mV(rms) sine wave	5	2	V_{CS}	 50 —	 65 50	 — —	 40 —	 65 50	 — —	dB k
Transadmittance Bandwidth (Magnitude) ($R_L = 50$ ohms) Carrier Input Port, $V_C = 60$ mV(rms) sine wave $f_S = 1.0$ kHz, 300 mV(rms) sine wave Signal Input Port, $V_S = 300$ mV(rms) sine wave $|V_C| = 0.5$ Vdc	8	8	BW_{3dB}	 — —	 300 80	 — —	 — —	 300 80	 — —	MHz
Signal Gain $V_S = 100$ mV(rms), $f = 1.0$ kHz; $|V_C| = 0.5$ Vdc	10	3	A_{VS}	2.5	3.5	—	2.5	3.5	—	V/V
Single-Ended Input Impedance, Signal Port, $f = 5.0$ MHz Parallel Input Resistance Parallel Input Capacitance	6	—	r_{ip} c_{ip}	 — —	 200 2.0	 — —	 — —	 200 2.0	 — —	 kΩ pF
Single-Ended Output Impedance, $f = 10$ MHz Parallel Output Resistance Parallel Output Capacitance	6	—	r_{op} c_{oo}	 — —	 40 5.0	 — —	 — —	 40 5.0	 — —	 kΩ pF
Input Bias Current $I_{bS} = \dfrac{I_1 + I_4}{2}$; $I_{bC} = \dfrac{I_7 + I_8}{2}$	7	—	I_{bS} I_{bC}	 — —	 12 12	 25 25	 — —	 12 12	 30 30	μA
Input Offset Current $I_{ioS} = I_1 - I_4$; $I_{ioC} = I_7 - I_8$	7	—	$|I_{ioS}|$ $|I_{ioC}|$	 — —	 0.7 0.7	 5.0 5.0	 — —	 0.7 0.7	 7.0 7.0	μA
Average Temperature Coefficient of Input Offset Current ($T_A = -55°C$ to $+125°C$)	7	—	$|TC_{io}|$	—	2.0	—	—	2.0	—	nA/°C
Output Offset Current ($I_6 - I_9$)	7	—	$|I_{oo}|$	—	14	50	—	14	80	μA
Average Temperature Coefficient of Output Offset Current ($T_A = -55°C$ to $+125°C$)	7	—	$|TC_{oo}|$	—	90	—	—	90	—	nA/°C
Common-Mode Input Swing, Signal Port, $f_S = 1.0$ kHz	9	4	CMV	—	5.0	—	—	5.0	—	Vp-p
Common-Mode Gain, Signal Port, $f_S = 1.0$ kHz, $|V_C| = 0.5$ Vdc	9	—	ACM	—	-85	—	—	-85	—	dB
Common-Mode Quiescent Output Voltage (Pin 6 or Pin 9)	10	—	V_{out}	—	8.0	—	—	8.0	—	Vp-p
Differential Output Voltage Swing Capability	10	—	V_{out}	—	8.0	—	—	8.0	—	Vp-p
Power Supply Current $I_6 + I_9$ I_{10}	7	6	I_{CC} I_{EE}	 — —	 2.0 3.0	 3.0 4.0	 — —	 2.0 3.0	 4.0 5.0	mAdc
DC Power Dissipation	7	5	P_D	—	33	—	—	33	—	mW

* Pin number references pertain to this device when packaged in a metal can. To ascertain the corresponding pin numbers for plastic or
ceramic packaged devices refer to the first page of this specification sheet.

(continued)

FIGURE 2.35   *(Continued)*

GENERAL OPERATING INFORMATION *

Note 1 — Carrier Feedthrough

Carrier feedthrough is defined as the output voltage at carrier frequency with only the carrier applied (signal voltage = 0).

Carrier null is achieved by balancing the currents in the differential amplifier by means of a bias trim potentiometer (R_1 of Figure 5).

Note 2 — Carrier Suppression

Carrier suppression is defined as the ratio of each sideband output to carrier output for the carrier and signal voltage levels specified.

Carrier suppression is very dependent on carrier input level, as shown in Figure 22. A low value of the carrier does not fully switch the upper switching devices, and results in lower signal gain, hence lower carrier suppression. A higher than optimum carrier level results in unnecessary device and circuit carrier feedthrough, which again degenerates the suppression figure. The MC1596 has been characterized with a 60 mV(rms) sinewave carrier input signal. This level provides optimum carrier suppression at carrier frequencies in the vicinity of 500 kHz, and is generally recommended for balanced modulator applications.

Carrier feedthrough is independent of signal level, V_S. Thus carrier suppression can be maximized by operating with large signal levels. However, a linear operating mode must be maintained in the signal-input transistor pair — or harmonics of the modulating signal will be generated and appear in the device output as spurious sidebands of the suppressed carrier. This requirement places an upper limit on input-signal amplitude (see Note 3 and Figure 20). Note also that an optimum carrier level is recommended in Figure 22 for good carrier suppression and minimum spurious sideband generation.

At higher frequencies circuit layout is very important in order to minimize carrier feedthrough. Shielding may be necessary in order to prevent capacitive coupling between the carrier input leads and the output leads.

Note 3 — Signal Gain and Maximum Input Level

Signal gain (single-ended) at low frequencies is defined as the voltage gain,

$$A_{VS} = \frac{V_o}{V_S} = \frac{R_L}{R_e + 2r_e} \text{ where } r_e = \frac{26 \text{ mV}}{I_5 \text{ (mA)}}$$

A constant dc potential is applied to the carrier input terminals to fully switch two of the upper transistors "on" and two transistors "off" ($V_C = 0.5$ Vdc). This in effect forms a cascode differential amplifier.

Linear operation requires that the signal input be below a critical value determined by R_E and the bias current I_5

$$V_S \leq I_5 R_E \text{ (Volts peak)}$$

Note that in the test circuit of Figure 10, V_S corresponds to a maximum value of 1 volt peak.

Note 4 — Common-Mode Swing

The common-mode swing is the voltage which may be applied to both bases of the signal differential amplifier, without saturating the current sources or without saturating the differential amplifier itself by swinging it into the upper switching devices. This swing is variable depending on the particular circuit and biasing conditions chosen (see Note 6).

Note 5 — Power Dissipation

Power dissipation, P_D, within the integrated circuit package should be calculated as the summation of the voltage-current products at each port, i.e. assuming $V_9 = V_6$, $I_5 = I_6 = I_9$ and ignoring base current, $P_D = 2 I_5 (V_6 - V_{10}) + I_5 (V_5 - V_{10})$ where subscripts refer to pin numbers.

Note 6 — Design Equations

The following is a partial list of design equations needed to operate the circuit with other supply voltages and input conditions. See Note 3 for R_e equation.

A. Operating Current

The internal bias currents are set by the conditions at pin 5. Assume:

$$I_5 = I_6 = I_9$$

$$I_B \ll I_C \text{ for all transistors}$$

then:

$$R_5 = \frac{V^- - \phi}{I_5} - 500 \ \Omega \quad \text{where: } R_5 \text{ is the resistor between pin 5 and ground}$$
$$\phi = 0.75 \text{ V at } T_A = +25^\circ C$$

The MC1596 has been characterized for the condition $I_5 = 1.0$ mA and is the generally recommended value.

B. Common-Mode Quiescent Output Voltage

$$V_6 = V_9 = V^+ - I_5 R_L$$

Note 7 — Biasing

The MC1596 requires three dc bias voltage levels which must be set externally. Guidelines for setting up these three levels include maintaining at least 2 volts collector-base bias on all transistors while not exceeding the voltages given in the absolute maximum rating table;

$$30 \text{ Vdc} \geq [(V_6, V_9) - (V_7, V_8)] \geq 2 \text{ Vdc}$$

$$30 \text{ Vdc} \geq [(V_7, V_8) - (V_1, V_4)] \geq 2.7 \text{ Vdc}$$

$$30 \text{ Vdc} \geq [(V_1, V_4) - (V_5)] \geq 2.7 \text{ Vdc}$$

The foregoing conditions are based on the following approximations:

$$V_6 = V_9, \quad V_7 = V_8, \quad V_1 = V_4$$

Bias currents flowing into pins 1, 4, 7, and 8 are transistor base currents and can normally be neglected if external bias dividers are designed to carry 1.0 mA or more.

Note 8 — Transadmittance Bandwidth

Carrier transadmittance bandwidth is the 3-dB bandwidth of the device forward transadmittance as defined by:

$$y_{21C} = \frac{i_o \text{ (each sideband)}}{v_s \text{ (signal)}} \Big|_{V_o = 0}$$

Signal transadmittance bandwidth is the 3-dB bandwidth of the device forward transadmittance as defined by:

$$y_{21S} = \frac{i_o \text{ (signal)}}{v_s \text{ (signal)}} \Big|_{V_C = 0.5 \text{ Vdc}, V_o = 0}$$

*Pin number references pertain to this device when packaged in a metal can. To ascertain the corresponding pin numbers for plastic or ceramic packaged devices refer to the first page of this specification sheet.

Note 9 — Coupling and Bypass Capacitors C_1 and C_2

Capacitors C_1 and C_2 (Figure 5) should be selected for a reactance of less than 5.0 ohms at the carrier frequency.

Note 10 — Output Signal, V_o

The output signal is taken from pins 6 and 9, either balanced or single-ended. Figure 12 shows the output levels of each of the two output sidebands resulting from variations in both the carrier and modulating signal inputs with a single-ended output connection.

Note 11 — Negative Supply, V_{EE}

V_{EE} should be dc only. The insertion of an RF choke in series with V_{EE} can enhance the stability of the internal current sources.

Note 12 — Signal Port Stability

Under certain values of driving source impedance, oscillation may occur. In this event, an RC suppression network should be connected directly to each input using short leads. This will reduce the Q of the source-tuned circuits that cause the oscillation.

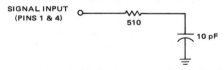

An alternate method for low-frequency applications is to insert a 1 k-ohm resistor in series with the inputs, pins 1 and 4. In this case input current drift may cause serious degradation of carrier suppression.

TEST CIRCUITS

FIGURE 5 — CARRIER REJECTION AND SUPPRESSION

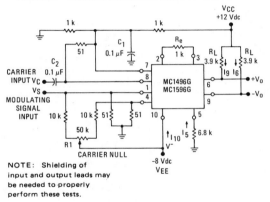

NOTE: Shielding of input and output leads may be needed to properly perform these tests.

FIGURE 6 — INPUT-OUTPUT IMPEDANCE

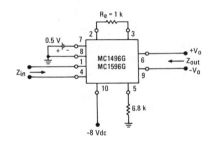

FIGURE 7 — BIAS AND OFFSET CURRENTS

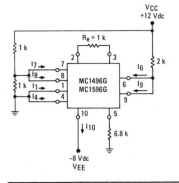

FIGURE 8 — TRANSCONDUCTANCE BANDWIDTH

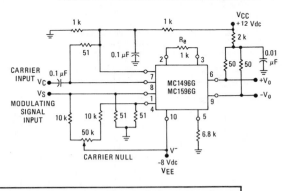

NOTE: Pin number references pertain to this device when packaged in a metal can. To ascertain the corresponding pin numbers for plastic or ceramic packaged devices refer to the first page of this specification sheet.

(continued)

FIGURE 2.35   *(Continued)*

TEST CIRCUITS (continued)

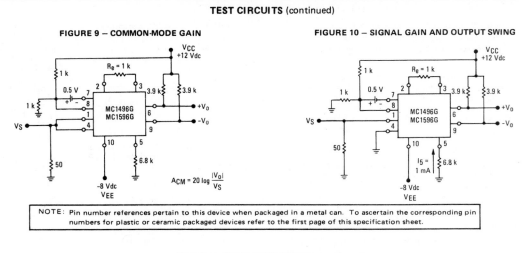

FIGURE 9 – COMMON-MODE GAIN

FIGURE 10 – SIGNAL GAIN AND OUTPUT SWING

$$A_{CM} = 20 \log \frac{|V_0|}{V_S}$$

NOTE: Pin number references pertain to this device when packaged in a metal can. To ascertain the corresponding pin numbers for plastic or ceramic packaged devices refer to the first page of this specification sheet.

TYPICAL CHARACTERISTICS (continued)

Typical characteristics were obtained with circuit shown in Figure 5, f_C = 500 kHz (sine wave), V_C = 60 mV(rms), f_S = 1 kHz, V_S = 300 mV(rms), T_A = +25°C unless otherwise noted.

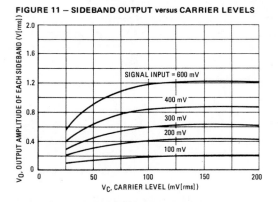

FIGURE 11 – SIDEBAND OUTPUT versus CARRIER LEVELS

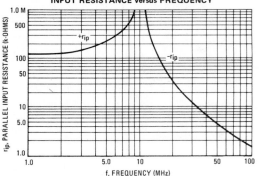

FIGURE 12 – SIGNAL-PORT PARALLEL-EQUIVALENT INPUT RESISTANCE versus FREQUENCY

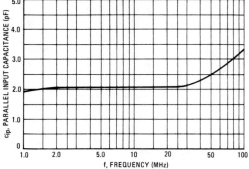

FIGURE 13 – SIGNAL-PORT PARALLEL-EQUIVALENT INPUT CAPACITANCE versus FREQUENCY

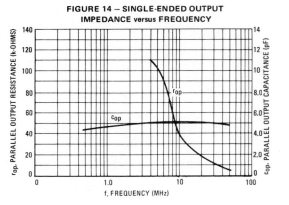

FIGURE 14 – SINGLE-ENDED OUTPUT IMPEDANCE versus FREQUENCY

TYPICAL CHARACTERISTICS (continued)

Typical characteristics were obtained with circuit shown in Figure 5, f_C = 500 kHz (sine wave),
V_C = 60 mV(rms), f_S = 1 kHz, V_S = 300 mV(rms), T_A = +25°C unless otherwise noted.

FIGURE 15 – SIDEBAND AND SIGNAL PORT
TRANSADMITTANCES versus FREQUENCY

FIGURE 16 – CARRIER SUPPRESSION
versus TEMPERATURE

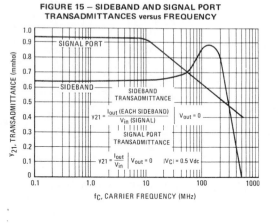

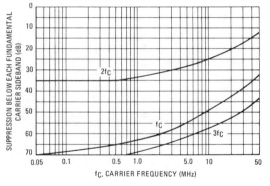

FIGURE 17 – SIGNAL-PORT FREQUENCY RESPONSE

FIGURE 18 – CARRIER SUPPRESSION versus FREQUENCY

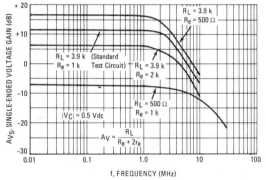

FIGURE 19 – CARRIER FEEDTHROUGH versus FREQUENCY

FIGURE 20 – SIDEBAND HARMONIC SUPPRESSION
versus INPUT SIGNAL LEVEL

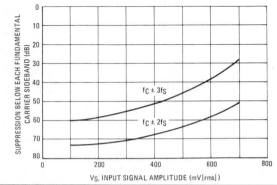

(continued)

FIGURE 2.35   (*Continued*)

TYPICAL CHARACTERISTICS (continued)

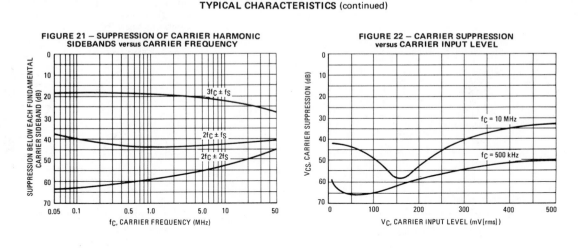

FIGURE 21 – SUPPRESSION OF CARRIER HARMONIC SIDEBANDS versus CARRIER FREQUENCY

FIGURE 22 – CARRIER SUPPRESSION versus CARRIER INPUT LEVEL

OPERATIONS INFORMATION

The MC1596/MC1496, a monolithic balanced modulator circuit, is shown in Figure 23.

This circuit consists of an upper quad differential amplifier driven by a standard differential amplifier with dual current sources. The output collectors are cross-coupled so that full-wave balanced multiplication of the two input voltages occurs. That is, the output signal is a constant times the product of the two input signals.

Mathematical analysis of linear ac signal multiplication indicates that the output spectrum will consist of only the sum and difference of the two input frequencies. Thus, the device may be used as a balanced modulator, doubly balanced mixer, product detector, frequency doubler, and other applications requiring these particular output signal characteristics.

The lower differential amplifier has its emitters connected to the package pins so that an external emitter resistance may be used. Also, external load resistors are employed at the device output.

Signal Levels

The upper quad differential amplifier may be operated either in a linear or a saturated mode. The lower differential amplifier is operated in a linear mode for most applications.

For low-level operation at both input ports, the output signal will contain sum and difference frequency components and have an amplitude which is a function of the product of the input signal amplitudes.

For high-level operation at the carrier input port and linear operation at the modulating signal port, the output signal will contain sum and difference frequency components of the modulating signal frequency and the fundamental and odd harmonics of the carrier frequency. The output amplitude will be a constant times the modulating signal amplitude. Any amplitude variations in the carrier signal will not appear in the output.

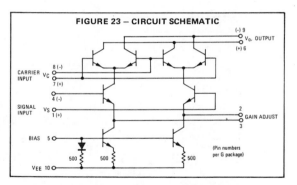

FIGURE 23 – CIRCUIT SCHEMATIC

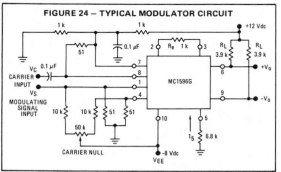

FIGURE 24 – TYPICAL MODULATOR CIRCUIT

NOTE: Pin number references pertain to this device when packaged in a metal can. To ascertain the corresponding pin numbers for plastic or ceramic packaged devices refer to the first page of this specification sheet.

OPERATIONS INFORMATION (continued)

The linear signal handling capabilities of a differential amplifier are well defined. With no emitter degeneration, the maximum input voltage for linear operation is approximately 25 mV peak. Since the upper differential amplifier has its emitters internally connected, this voltage applies to the carrier input port for all conditions.

Since the lower differential amplifier has provisions for an external emitter resistance, its linear signal handling range may be adjusted by the user. The maximum input voltage for linear operation may be approximated from the following expression:

$$V = \left(I_5\right)\left(R_E\right) \text{volts peak.}$$

This expression may be used to compute the minimum value of R_E for a given input voltage amplitude.

FIGURE 25 – TABLE 1
VOLTAGE GAIN AND OUTPUT FREQUENCIES

Carrier Input Signal (V_C)	Approximate Voltage Gain	Output Signal Frequency(s)
Low-level dc	$\dfrac{R_L \, V_C}{2(R_E + 2r_e)\left(\frac{KT}{q}\right)}$	f_M
High-level dc	$\dfrac{R_L}{R_E + 2r_e}$	f_M
Low-level ac	$\dfrac{R_L \, V_C(rms)}{2\sqrt{2}\left(\frac{KT}{q}\right)(R_E + 2r_e)}$	$f_C \pm f_M$
High-level ac	$\dfrac{0.637 \, R_L}{R_E + 2r_e}$	$f_C \pm f_M, \; 3f_C \pm f_M,$ $5f_C \pm f_M, \; \ldots$

The gain from the modulating signal input port to the output is the MC1596/MC1496 gain parameter which is most often of interest to the designer. This gain has significance only when the lower differential amplifier is operated in a linear mode, but this includes most applications of the device.

As previously mentioned, the upper quad differential amplifier may be operated either in a linear or a saturated mode. Approximate gain expressions have been developed for the MC1596/MC1496 for a low-level modulating signal input and the following carrier input conditions:

1) Low-level dc
2) High-level dc
3) Low-level ac
4) High-level ac

These gains are summarized in Table 1, along with the frequency components contained in the output signal.

NOTES:
1. Low-level Modulating Signal, V_M, assumed in all cases. V_C is Carrier Input Voltage.
2. When the output signal contains multiple frequencies, the gain expression given is for the output amplitude of each of the two desired outputs, $f_C + f_M$ and $f_C - f_M$.
3. All gain expressions are for a single-ended output. For a differential output connection, multiply each expression by two.
4. R_L = Load resistance.
5. R_E = Emitter resistance between pins 2 and 3.
6. r_e = Transistor dynamic emitter resistance, at +25°C;

$$r_e \approx \frac{26 \text{ mV}}{I_5 \text{ (mA)}}$$

7. K = Boltzmann's Constant, T = temperature in degrees Kelvin, q = the charge on an electron.

$$\frac{KT}{q} \approx 26 \text{ mV at room temperature}$$

APPLICATIONS INFORMATION

Double sideband suppressed carrier modulation is the basic application of the MC1596/MC1496. The suggested circuit for this application is shown on the front page of this data sheet.

In some applications, it may be necessary to operate the MC1596/MC1496 with a single dc supply voltage instead of dual supplies. Figure 26 shows a balanced modulator designed for operation with a single +12 Vdc supply. Performance of this circuit is similar to that of the dual supply modulator.

AM Modulator

The circuit shown in Figure 27 may be used as an amplitude modulator with a minor modification.

All that is required to shift from suppressed carrier to AM operation is to adjust the carrier null potentiometer for the proper amount of carrier insertion in the output signal.

However, the suppressed carrier null circuitry as shown in Figure 27 does not have sufficient adjustment range. Therefore, the modulator may be modified for AM operation by changing two resistor values in the null circuit as shown in Figure 28.

Product Detector

The MC1596/MC1496 makes an excellent SSB product detector (see Figure 29).

This product detector has a sensitivity of 3.0 microvolts and a dynamic range of 90 dB when operating at an intermediate frequency of 9 MHz.

The detector is broadband for the entire high frequency range. For operation at very low intermediate frequencies down to 50 kHz the 0.1 μF capacitors on pins 7 and 8 should be increased to 1.0 μF. Also, the output filter at pin 9 can be tailored to a specific intermediate frequency and audio amplifier input impedance.

As in all applications of the MC1596/MC1496, the emitter resistance between pins 2 and 3 may be increased or decreased to adjust circuit gain, sensitivity, and dynamic range.

This circuit may also be used as an AM detector by introducing carrier signal at the carrier input and an AM signal at the SSB input.

The carrier signal may be derived from the intermediate frequency signal or generated locally. The carrier signal may be introduced with or without modulation, provided its level is sufficiently high to saturate the upper quad differential amplifier. If the carrier signal is modulated, a 300 mV(rms) input level is recommended.

(continued)

FIGURE 2.35   (*Continued*)

APPLICATIONS INFORMATION (continued)

Doubly Balanced Mixer

The MC1596/MC1496 may be used as a doubly balanced mixer with either broadband or tuned narrow band input and output networks.

The local oscillator signal is introduced at the carrier input port with a recommended amplitude of 100 mV(rms).

Figure 30 shows a mixer with a broadband input and a tuned output.

Frequency Doubler

The MC1596/MC1496 will operate as a frequency doubler by introducing the same frequency at both input ports.

Figures 31 and 32 show a broadband frequency doubler and a tuned output very high frequency (VHF) doubler, respectively.

Phase Detection and FM Detection

The MC1596/MC1496 will function as a phase detector. High-level input signals are introduced at both inputs. When both inputs are at the same frequency the MC1596/MC1496 will deliver an output which is a function of the phase difference between the two input signals.

An FM detector may be constructed by using the phase detector principle. A tuned circuit is added at one of the inputs to cause the two input signals to vary in phase as a function of frequency. The MC1596/MC1496 will then provide an output which is a function of the input signal frequency.

NOTE: Pin number references pertain to this device when packaged in a metal can. To ascertain the corresponding pin numbers for plastic or ceramic packaged devices refer to the first page of this specification sheet.

TYPICAL APPLICATIONS

FIGURE 26 – BALANCED MODULATOR
(+12 Vdc SINGLE SUPPLY)

FIGURE 27 – BALANCED MODULATOR-DEMODULATOR

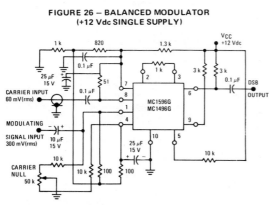
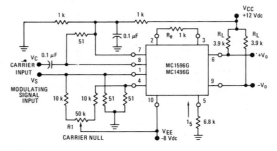

FIGURE 28 – AM MODULATOR CIRCUIT

FIGURE 29 – PRODUCT DETECTOR
(+12 Vdc SINGLE SUPPLY)

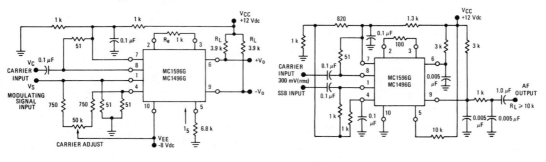

TYPICAL APPLICATIONS (continued)

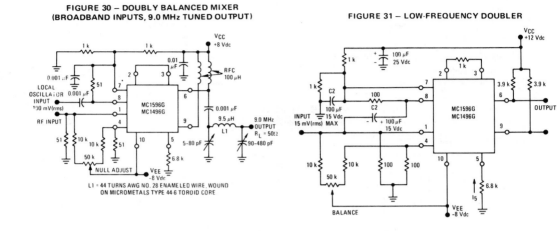

FIGURE 30 — DOUBLY BALANCED MIXER
(BROADBAND INPUTS, 9.0 MHz TUNED OUTPUT)

FIGURE 31 — LOW-FREQUENCY DOUBLER

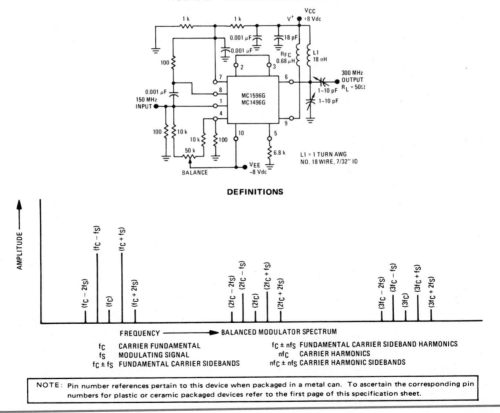

FIGURE 32 — 150 to 300 MHz DOUBLER

DEFINITIONS

f_C	CARRIER FUNDAMENTAL
f_S	MODULATING SIGNAL
$f_C \pm f_S$	FUNDAMENTAL CARRIER SIDEBANDS

$f_C \pm nf_S$	FUNDAMENTAL CARRIER SIDEBAND HARMONICS
nf_C	CARRIER HARMONICS
$nf_C \pm nf_S$	CARRIER HARMONIC SIDEBANDS

NOTE: Pin number references pertain to this device when packaged in a metal can. To ascertain the corresponding pin numbers for plastic or ceramic packaged devices refer to the first page of this specification sheet.

(*continued*)

FIGURE 2.35   (*Continued*)

THERMAL INFORMATION

The maximum power consumption an integrated circuit can tolerate at a given operating ambient temperature, can be found from the equation:

$$P_{D(T_A)} = \frac{T_{J(max)} - T_A}{R_{\theta JA}(Typ)} \geq V_I\,I_S - V_O\,I_O$$

Where: $P_{D(T_A)}$ = Power Dissipation allowable at a given operating ambient temperature.

$T_{J(max)}$ = Maximum Operating Junction Temperature as listed in the Maximum Ratings Section

T_A = Maximum Desired Operating Ambient Temperature

$R_{\theta JA}(Typ)$ = Typical Thermal Resistance Junction to Ambient

I_S = Total Supply Current

OUTLINE DIMENSIONS

G SUFFIX
METAL PACKAGE
CASE 603-04

NOTE:
LEADS WITHIN 0.18 mm (0.007) RADIUS OF TRUE POSITION AT SEATING PLANE AT MAXIMUM MATERIAL CONDITION.

DIM	MILLIMETERS		INCHES	
	MIN	MAX	MIN	MAX
A	8.51	9.39	0.335	0.370
B	7.75	8.51	0.305	0.335
C	4.19	4.70	0.165	0.185
D	0.407	0.533	0.016	0.021
E	—	1.02	—	0.040
F	0.406	0.483	0.016	0.019
G	5.84 BSC		0.230 BSC	
H	0.712	0.864	0.028	0.034
J	0.737	1.14	0.029	0.045
K	12.70	—	0.500	—
L	6.35	12.70	0.250	0.500
M	36° BSC		36° BSC	
P	—	1.27	—	0.050
Q	3.56	4.06	0.140	0.160
R	0.254	1.02	0.010	0.040

L SUFFIX
CERAMIC PACKAGE
CASE 632-08

NOTES:
1. DIMENSIONING AND TOLERANCING PER ANSI Y14.5M, 1982.
2. CONTROLLING DIMENSION: INCH.
3. DIMENSION L TO CENTER OF LEAD WHEN FORMED PARALLEL.
4. DIM F MAY NARROW TO 0.76 (0.030) WHERE THE LEAD ENTERS THE CERAMIC BODY.

⌖ | 0.25 (0.010) Ⓜ | T | A Ⓢ

⌖ | 0.25 (0.010) Ⓜ | T | B Ⓢ

DIM	MILLIMETERS		INCHES	
	MIN	MAX	MIN	MAX
A	19.05	19.94	0.750	0.785
B	6.23	7.11	0.245	0.280
C	3.94	5.08	0.155	0.200
D	0.39	0.50	0.015	0.020
F	1.40	1.65	0.055	0.065
G	2.54 BSC		0.100 BSC	
J	0.21	0.38	0.008	0.015
K	3.18	4.31	0.125	0.170
L	7.62 BSC		0.300 BSC	
M	0°	15°	0°	15°
N	0.51	1.01	0.020	0.040

OUTLINE DIMENSIONS

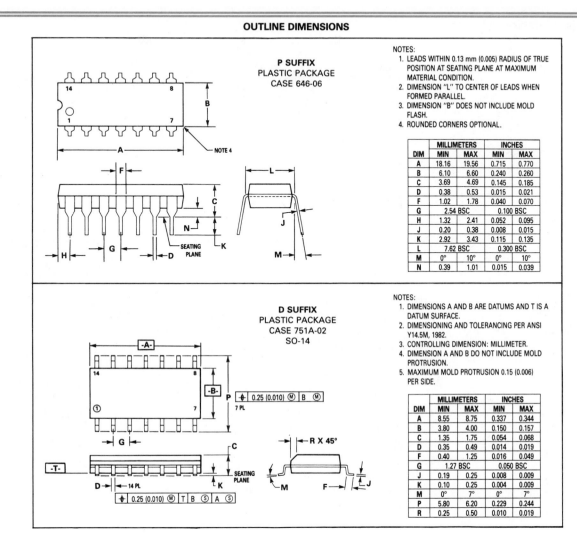

P SUFFIX
PLASTIC PACKAGE
CASE 646-06

NOTES:
1. LEADS WITHIN 0.13 mm (0.005) RADIUS OF TRUE POSITION AT SEATING PLANE AT MAXIMUM MATERIAL CONDITION.
2. DIMENSION "L" TO CENTER OF LEADS WHEN FORMED PARALLEL.
3. DIMENSION "B" DOES NOT INCLUDE MOLD FLASH.
4. ROUNDED CORNERS OPTIONAL.

DIM	MILLIMETERS		INCHES	
	MIN	MAX	MIN	MAX
A	18.16	19.56	0.715	0.770
B	6.10	6.60	0.240	0.260
C	3.69	4.69	0.145	0.185
D	0.38	0.53	0.015	0.021
F	1.02	1.78	0.040	0.070
G	2.54 BSC		0.100 BSC	
H	1.32	2.41	0.052	0.095
J	0.20	0.38	0.008	0.015
K	2.92	3.43	0.115	0.135
L	7.62 BSC		0.300 BSC	
M	0°	10°	0°	10°
N	0.39	1.01	0.015	0.039

D SUFFIX
PLASTIC PACKAGE
CASE 751A-02
SO-14

NOTES:
1. DIMENSIONS A AND B ARE DATUMS AND T IS A DATUM SURFACE.
2. DIMENSIONING AND TOLERANCING PER ANSI Y14.5M, 1982.
3. CONTROLLING DIMENSION: MILLIMETER.
4. DIMENSION A AND B DO NOT INCLUDE MOLD PROTRUSION.
5. MAXIMUM MOLD PROTRUSION 0.15 (0.006) PER SIDE.

⊕ 0.25 (0.010) Ⓜ B Ⓜ
7 PL

⊕ 0.25 (0.010) Ⓜ T B Ⓢ A Ⓢ

DIM	MILLIMETERS		INCHES	
	MIN	MAX	MIN	MAX
A	8.55	8.75	0.337	0.344
B	3.80	4.00	0.150	0.157
C	1.35	1.75	0.054	0.068
D	0.35	0.49	0.014	0.019
F	0.40	1.25	0.016	0.049
G	1.27 BSC		0.050 BSC	
J	0.19	0.25	0.008	0.009
K	0.10	0.25	0.004	0.009
M	0°	7°	0°	7°
P	5.80	6.20	0.229	0.244
R	0.25	0.50	0.010	0.019

FIGURE 2.36

Double-Balanced Diode Mixer

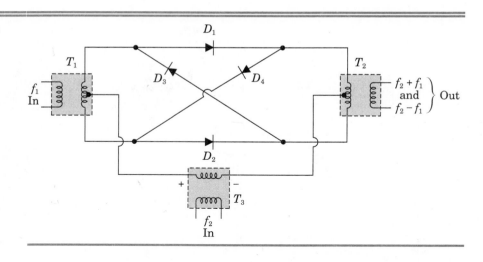

Now consider what happens when the polarity of the secondary of T_3 reverses. D_1 and D_2 will now be off, and D_3 and D_4 will be on. The signal from T_1 will still be connected to T_2, but this time the polarity will be reversed.

From this description, it can be seen that the circuit multiplies the signal at f_1 by a factor of $+1$ or -1, alternating at the f_2 rate. This is equivalent to multiplying the signal at f_1 by a square wave with frequency f_2 and amplitude 1 V peak. The result will be the sum and difference frequencies described earlier, plus higher-frequency components at $3f_2 \pm f_1$, $5f_2 \pm f_1$, and so on. Generally, f_2 is the higher frequency, and these spurious signals will be at such high frequencies compared to the desired frequency that they will be very easy to filter out.

Balanced mixers are also called *balanced modulators*, and will be seen again under that name when we discuss the generation of suppressed-carrier signals in Chapter 6.

SUMMARY OF CHAPTER 2

Here are the main points to remember from this chapter.

1. As the frequency increases, the effects of distributed capacitance and inductance become more significant.

2. At VHF frequencies and above, the transit time of charge carriers within a device may become a substantial part of the period of the signal. This means that transit time must be taken into account in the design of high-frequency devices and circuits.

3. When the length of a conductor is an appreciable part of a wavelength, it is necessary to use transmission-line techniques in circuit analysis.

4. Interactions between components in high-frequency circuits can often be avoided by the use of shielding and bypassing.

5. Many RF amplifiers are designed for a relatively narrow range of frequencies to avoid problems with noise and spurious signals. These are called *narrowband amplifiers*. *Wideband amplifiers* are also used.

6. Depending on the application, an RF amplifier may operate in Class A, B, or C. Class A has the least distortion, but is very inefficient. Class C is the most efficient, but its extreme nonlinearity makes it unsuitable for many signals.

7. A frequency multiplier can be created by tuning the output circuit of a Class C amplifier to a multiple of the input frequency.

8. Any oscillator requires a feedback loop with unity gain and zero phase shift around the loop, at the operating frequency.

9. Most variable-frequency RF oscillators use a resonant circuit for the feedback network. Crystal-controlled oscillators, in which a quartz crystal is used as the feedback element, are also common.

10. When two signals at different frequencies are applied to a nonlinear circuit, signals at the sum and difference of the frequencies are produced. A circuit designed to accomplish this is called a *mixer*.

11. A balanced mixer is designed to cancel out the actual input frequencies, producing only the sum and difference of the input signal frequencies.

IMPORTANT EQUATIONS

Each chapter has a brief list of the most commonly used equations, for handy reference. These equations must not be used blindly: you must understand how and where to use them. Refer to the text for this. These are not the only equations you will need to solve the problems in this chapter.

$$A_V = \frac{-(R_C \parallel R_L)}{r_e} \tag{2.1}$$

$$f_o = \frac{1}{2\pi\sqrt{L_1 C_1}} \tag{2.2}$$

$$R_L' = R_L \left(\frac{N_1}{N_2}\right)^2 \tag{2.3}$$

$$A_{V_o} = \frac{R_L N_1}{r_e N_2} \tag{2.5}$$

$$Q = \frac{R_L'}{\omega_o L} \tag{2.6}$$

$$B = \frac{f_o}{Q} \tag{2.7}$$

$$C_m = C_{bc}(A_{V_0} + 1) \tag{2.8}$$

$$AB = 1 \tag{2.9}$$

$$A_V = -g_m r_D \tag{2.14}$$

$$C = \frac{C_o}{\sqrt{1 + 2V}} \qquad\qquad (2.23)$$

$$f_T = f_o + k f_o (T - T_o) \qquad\qquad (2.24)$$

$$\sin A \sin B = \frac{1}{2}\left[\cos(A - B) - \cos(A + B)\right] \qquad\qquad (2.29)$$

GLOSSARY

balanced mixer   a mixer in which the input frequencies are cancelled, so that they are not present at the output

bypassing   removal of an unwanted signal by providing a low-impedance path to ground

crystal   a small slab of quartz with attached electrodes; used as a resonant circuit

decoupling   prevention of the undesired passage of signals between circuits

doubler   a frequency multiplier whose output frequency is twice that of the input signal

frequency multiplier   a circuit whose output frequency is a small integer multiple of the input signal frequency

gimmick   a small length of wire, connected at only one end and used as a capacitance to ground

ground plane   an artificial ground, often consisting of an area of foil left on one side of a circuit board

Miller capacitance   an equivalent capacitance across the input of an amplifier that represents the effect of feedback combined with internal capacitance in the amplifying device

mixer   a nonlinear circuit designed to generate sum and difference frequencies when two or more frequencies are present at its input(s)

multiplier   a circuit whose output is proportional to the product of the instantaneous amplitudes of two input signals

neutralization   a means of avoiding instability in amplifiers by using negative feedback

piezoelectric effect   an effect that occurs with some materials, such as quartz and some ceramics, whereby a voltage is produced across the material when it is deformed. The converse is also true: applying a voltage to the material will cause mechanical deformation.

self-resonance   resonance of a single component, either an inductor or capacitor, due to the presence of stray capacitance or inductance respectively

tripler   a frequency multiplier whose output frequency is three times that of the input signal

varactor   a reverse-biased diode used as a voltage-variable capacitor

QUESTIONS

1. What is the difference between the behavior of a resistor at low and high frequencies?
2. Explain what is meant by "self-resonance."
3. Distinguish between lumped and distributed constants, and explain why both concepts are useful.
4. Explain how a circuit board can be made to incorporate shielding. Gimmick
5. Why are bypass capacitors necessary at power-supply connections?
6. Why are narrowband amplifiers often preferred for RF applications?
7. Give one example of an RF application where a wideband amplifier would be preferred.
8. State the characteristics of Class A, AB, B, and C amplifiers, in terms of conduction angle, efficiency, and distortion level.
9. Why are Class C amplifiers unsuitable for use as audio amplifiers?
10. What is neutralization, and why is it sometimes needed in RF amplifiers?
11. How does a frequency multiplier differ from an ordinary RF amplifier?
12. State the Barkhausen criteria and explain them in your own words.
13. Draw the circuit for each LC oscillator type listed in this chapter and show the feedback path.
14. What advantages does varactor tuning have compared with variable capacitors or inductors?
15. Explain why a quartz crystal has two resonant frequencies. Name the two types of resonance, and state which one occurs at the lower frequency.
16. Give two advantages of crystal-controlled over LC oscillators. Do crystal-controlled oscillators have any disadvantages?
17. What is the function of a mixer? Distinguish a mixer from a summer.
18. List all the frequency components produced when two sine waves of different frequencies are applied to a square-law mixer.
19. List all the frequency components produced when two sine waves of different frequencies are applied to a balanced mixer.
20. Compared to a bipolar transistor, what advantages does a dual-gate MOSFET have when used as a mixer?

PROBLEMS

Section 2.3

21. The narrowband RF amplifier of Figure 2.37 has a signal at the input with a frequency of 12 MHz and an amplitude of 2.5 mV RMS. Calculate:
    (a) the value to which C_1 should be adjusted for best performance at the signal frequency
    (b) the maximum gain for this amplifier (assuming no transistor loading and an ideal transformer)
    (c) the output signal voltage with the given input signal

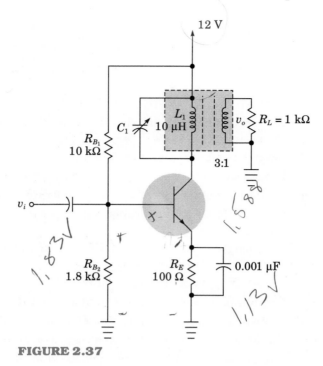

FIGURE 2.37

22. The wideband RF amplifier of Figure 2.38 is to have an output signal of 100 mV
    amplitude. What must be the input amplitude, assuming ideal transformers and
    no transistor loading?

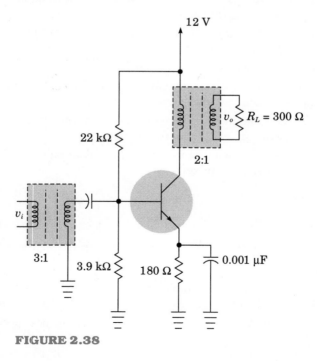

FIGURE 2.38

23. The circuit of Figure 2.39 represents an RF amplifier.

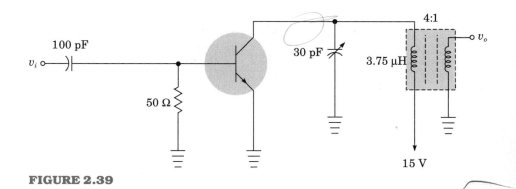

FIGURE 2.39

(a) What class of operation is it designed for? How can you tell?
(b) What is the collector current with no input signal?
(c) Sketch the input voltage and collector current waveforms with an input signal at 15 MHz. Sketch them one above the other using the same time scale. Current and voltage scales are not required.
(d) Assuming an ideal transformer, what would be the maximum load voltage?

24. An amplifier is required to deliver 100 W to a load. Calculate the power it would draw from the power supply if it were to operate:

(a) Class A with an efficiency of 30%
(b) Class B with an efficiency of 55%
(c) Class C with an efficiency of 80%

25. A transistor has a power dissipation rating of 25 W. Assume that the transistor is the only element that dissipates power in the circuit. Calculate the power an amplifier using this transistor could deliver to the load if it operates:

(a) Class A with an efficiency of 30%
(b) Class C with an efficiency of 80%

Section 2.4

26. A transmitter uses a frequency multiplication circuit consisting of two triplers and three doublers to get a frequency of 540 MHz from a crystal oscillator. What frequency should the crystal have?

27. The circuit of Figure 2.40 represents a frequency multiplier. If the input frequency is 10 MHz and the circuit operates as a tripler, calculate the value of L_1.

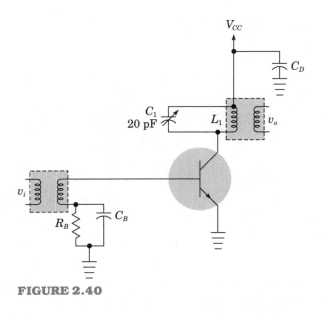

FIGURE 2.40

Section 2.5

28. The circuit of Figure 2.41 shows an oscillator.

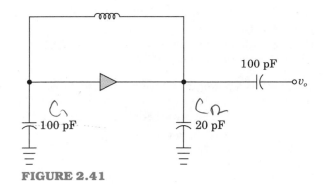

FIGURE 2.41

   (a) Name the type of oscillator.
   (b) Should the amplifier be inverting or noninverting? Explain why.
   (c) What is the minimum gain the amplifier could have, for the circuit to oscillate?
   (d) What value should L have, if the operating frequency is to be 5 MHz?

29. Suppose L were replaced by a quartz crystal in the circuit of Figure 2.41.

   (a) What is the name of this type of oscillator?
   (b) Sketch the curve of reactance as a function of frequency for a quartz crystal, and indicate where on the curve this oscillator would work.

30. (a) Calculate the operating frequency for the Clapp oscillator shown in Figure 2.42.

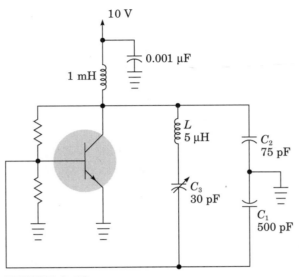

10 V

0.001 μF

1 mH

L
5 μH

C_2
75 pF

C_3
30 pF

C_1
500 pF

FIGURE 2.42

(b) Suppose that transistor loading doubles the value of C_1. By what percentage does this change the operating frequency?

31. A varactor-tuned oscillator uses an inductance of 25 μH with a varactor having a maximum capacitance of 45 pF in the frequency-determining circuit. Calculate the range of frequencies for this oscillator, with a tuning voltage that varies from 1 V to 10 V.

32. A 10 MHz crystal oscillator has a temperature coefficient of −5 ppm/degree C. It operates over a frequency range from +10°C to +30°C, and has been set to exactly 10 MHz at 20°C.

    (a) What are its output frequencies at 10° and 30°?
    (b) What is its percent accuracy over the operating temperature range?
    (c) Suppose the output of this oscillator is applied to a doubler. Recalculate the range of output frequencies and percent accuracy.

33. A quartz watch is guaranteed accurate to 15 seconds per month. Assuming a month has 30 days, calculate the accuracy of the crystal oscillator in the watch in parts per million.

Section 2.6

34. Signals at 12 MHz and 9 MHz are applied to the input of a mixer. What frequencies will be found at the output if the mixer is:

    (a) a square-law type?
    (b) a multiplier?

Comprehensive

35. The narrowband RF amplifier of Figure 2.43 has a bandwidth of 70 kHz and a voltage gain of 100 at a center frequency of 10 MHz when unloaded. What will be the gain and bandwidth when it is loaded by being connected to a 50 Ω antenna?

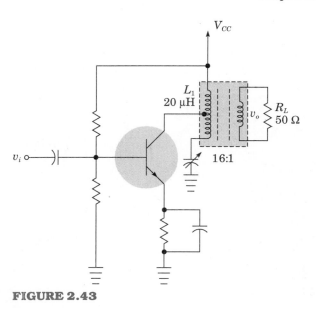

FIGURE 2.43

36. Draw a circuit for a Colpitts oscillator that will operate with a noninverting amplifier with a voltage gain of less than 1. Design it with a feedback fraction of 2, and an operating frequency of 16 MHz. The inductor should have a value of 5 μH.

37. Refer to Figure 1.17, the simplified block diagram for the swept-frequency spectrum analyzer discussed in Chapter 1. Let the bandpass filter have a center frequency of 20 MHz, and calculate the range of oscillator frequencies that will be necessary to sweep the input frequency range of 100 to 200 MHz. Assume the oscillator frequency is higher than the input signal frequency.

3

Amplitude Modulation

Objectives

After reading this chapter, studying the examples, and solving the problems, you should be able to:

1. Write the time-domain equation for an AM signal, and describe how the equation relates to the signal itself.
2. Sketch an AM signal in both the time and frequency domains.
3. Define the modulation index, calculate it, and measure it using either an oscilloscope or a spectrum analyzer.
4. Describe the effects of overmodulation, and explain why it must be avoided.
5. Sketch, and explain the operation of, simple AM modulators and demodulators.
6. Calculate the bandwidth of an AM signal, and explain why bandwidth is an important factor in a communications system.
7. Calculate power and voltage for an AM signal, and for each of its components.

3.1 INTRODUCTION

In Chapter 1, we mentioned the several ways in which an information signal can modulate a carrier wave, in order to produce a higher-frequency signal that carries the information. The most straightforward of these, and historically the first to be used, is **amplitude modulation** (AM). AM has the advantage of being usable with very simple modulators and demodulators. It does have some disadvantages, including poor performance in the presence of noise and inefficient use of transmitter power. However, its simplicity, and the fact that it was the first system to become established, have ensured its continued popularity. Applications include broadcasting in the medium- and high-frequency bands, aircraft communications in the VHF frequency range, and CB radio, among others.

The basic technique of amplitude modulation can also be modified to serve as the basis for a variety of more sophisticated schemes, which are found in applications as diverse as television broadcasting and long-distance telephony. Thus, it is essential to understand the process of amplitude modulation in some detail, both for its own sake and as a foundation for further study.

3.2 FULL-CARRIER AM: AN OVERVIEW

This section will provide you with a brief glimpse of the entire AM system. The description is deliberately oversimplified in an attempt to give you a feeling for what is happening. In order to do anything very useful with AM, it will be necessary to understand the process in more detail, of course. That will be the subject of the rest of this chapter and of Chapters 4 and 5.

3.2.1 The Signal

An AM signal can be produced by using the instantaneous amplitude of the information signal (the baseband or modulating signal) to vary the peak amplitude of a higher-frequency signal. Figure 3.1(a) shows a 1 kHz sine wave, which can be combined with the 10 kHz signal shown in Figure 3.1(b) to produce the AM signal of Figure 3.1(c). If the peaks of the individual waveforms of the modulated signal are joined, an **envelope** results that resembles the original modulating signal. It repeats at the modulating frequency, and the shape of each "half" (positive or negative) is the same as that of the modulating signal.

The higher-frequency signal that is combined with an information signal to produce the modulated waveform is called the *carrier*. Figure 3.1(c) shows a case where there are only 10 cycles of the carrier for each cycle of the modulating signal. In practice, the ratio between carrier frequency and modulating frequency is usually much greater. For instance, an AM broadcasting station might have a carrier frequency of 1 MHz and a modulating frequency on the order of 1 kHz. A waveform like this is shown in Figure 3.2. Since there are 1000 cycles of the carrier for each cycle of the envelope, the individual RF cycles are not visible, and only the envelope can be seen.

Note that amplitude modulation is not the simple linear addition of the two signals. That would produce the waveforms shown in Figure 3.3. Figure

FIGURE 3.1   **Amplitude Modulation**

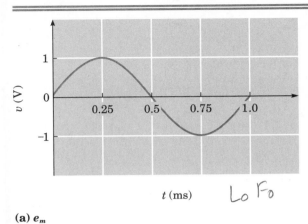

Lo Fo

(a) e_m

Hfo   Carrier

(b) e_c

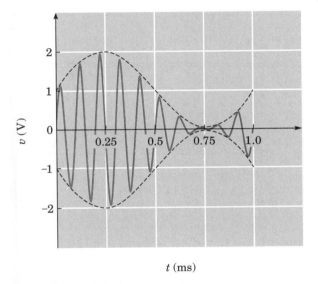

t (ms)

(c) 100% Amplitude Modulation

3.3(a) shows a low-frequency signal, Figure 3.3(b) shows a higher-frequency signal, and Figure 3.3(c) shows the result of the addition of the two signals.

Amplitude modulation is essentially a nonlinear process. As with any nonlinear interaction between signals, sum and difference frequencies are produced that, in the case of amplitude modulation, contain the information to be transmitted. Another interesting thing about AM is that, even though we seem to be varying the amplitude of the carrier, and in fact this is what is implied by the term "amplitude modulation," a look at the frequency domain shows that the signal component at the carrier frequency survives intact, with the same amplitude and frequency as before! This mystery is easily unravelled with the aid of a little mathematics, as we will see shortly; for the moment, just remember that AM is a bit of a misnomer, since the amplitude of the carrier remains constant. The amplitude of the entire signal does change with modulation, however, as is very clearly shown in Figure 3.1.

FIGURE 3.2

Envelope of an
AM Signal

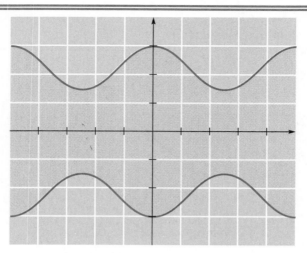

Vertical: 25 mV/division

Horizontal: 200 µs/division

FIGURE 3.3 Linear Addition of Two Signals

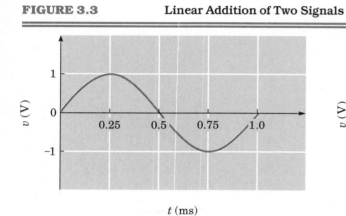

(a) e_m

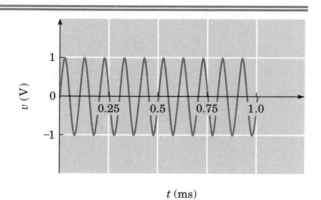

(b) e_c

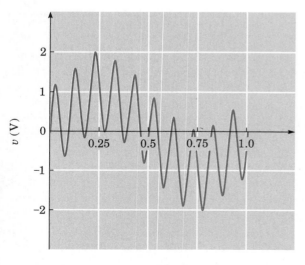

(c) $e_c + e_m$

FIGURE 3.4

AM and Linear Addition in the Frequency Domain

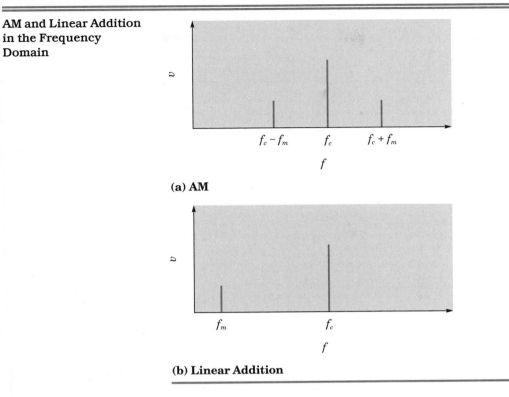

(a) AM

(b) Linear Addition

Figure 3.4(a) shows frequency-domain representations for amplitude modulation, and the linear addition of two signals is shown in Figure 3.4(b). The AM signal has no component at the modulating frequency: all the information is transmitted at frequencies near that of the carrier. By contrast, linear addition has accomplished nothing: the frequency-domain sketch shows that the information and carrier signals remain separate, each at its original frequency.

3.2.2 A Simple Modulator Circuit

There are many ways to achieve the result shown in Figure 3.1. Perhaps the easiest circuit to understand is that shown in Figure 3.5. The RF carrier is amplified in a Class C stage. The modulating signal is applied to the primary of a transformer, the secondary of which is in series with the supply voltage to the amplifier. This is a simplified version of the type of circuit used in many AM transmitters. Transmitters will be studied in Chapter 4.

You will recall from Chapter 2 that the peak output voltage of a tuned Class C amplifier like that in Figure 3.5 is approximately twice the collector supply voltage. In this circuit, that voltage, and hence the output of the modulated stage, varies with the instantaneous value of the modulating signal. As the level of the audio signal increases, so does the effect on the RF signal. When the modulating voltage, at the modulating transformer secondary, has a peak amplitude equal to V_{CC}, the collector supply voltage (measured at point A in the figure) will vary from zero to $2V_{CC}$, and the amplitude

FIGURE 3.5

Simplified AM Modulator

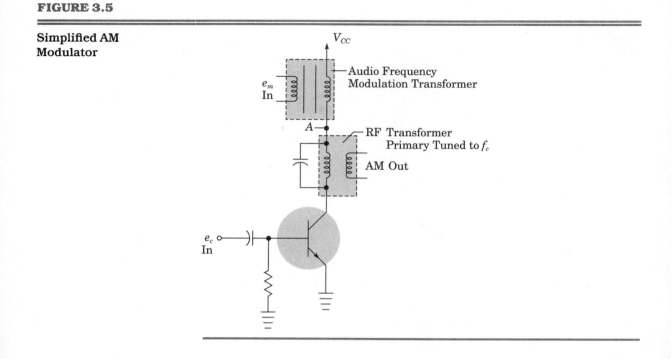

of the output waveform will vary from zero to twice its value without modulation, as in Figure 3.1(c). This is called *100% modulation*.

3.2.3 Demodulator Circuits

As explained in Chapter 1, a complete communications system requires both that baseband information be used to modulate a carrier at one end of the chain (the transmitter), and that the original information be recovered (demodulated) at the other end (the receiver). Demodulation can be done very simply with full-carrier AM. As shown in Figure 3.6, the original signal can be recovered by first rectifying the modulated signal, then low-pass filtering the result. This is called an **envelope detector**. The extreme simplicity (and hence low cost) of AM demodulation is a major reason for its continued popularity.

An ideal diode would provide a faithful representation of the envelope, but a real diode will spend some time in the *square-law* portion of its curve, where output is not directly proportional to input. Using a large input signal will minimize this type of distortion. Distortionless operation is not possible even so, since for 100% modulation the level of the envelope will reach zero once each cycle of the modulating waveform. There is also the probability that the low-pass filter will remove some of the modulation as well as the carrier-frequency component. Diode detectors commonly have distortion on the order of a few percent. Better detectors are available, but are found more often in the suppressed-carrier variations of AM, to be discussed in Chapter 6, than in the basic system.

In a practical receiver, means will be provided to isolate the desired signal from interference, and to provide gain both before and after demodulation. Receiver design will be considered in Chapter 5.

FIGURE 3.6

Simple AM
Demodulator

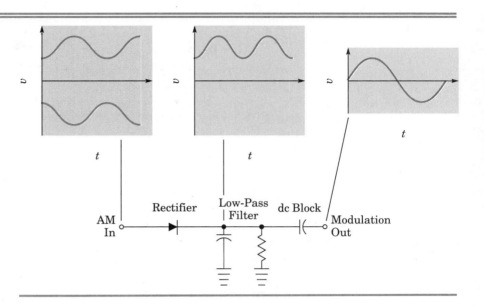

3.3 TIME-DOMAIN ANALYSIS

Now that the general idea of AM has been described, it is time to examine
the system in greater detail. We will look at the modulated signal in both the
time and frequency domains, as each method emphasizes some of the important characteristics of AM. The time domain is probably more familiar, so let
us begin here.

Amplitude modulation is created by using the instantaneous modulating
signal voltage to vary the amplitude of the modulated signal. The carrier is
almost always a sine wave. The modulating signal can be a sine wave, but is
more often an arbitrary waveform, such as an audio signal. However, an
analysis of sine-wave modulation is very useful, since Fourier analysis often
allows complex signals to be expressed as a series of sinusoids.

We can express the above relationship by means of an equation:

$$v(t) = (E_c + e_m) \sin \omega_c t \qquad (3.1)$$

where $v(t)$ = instantaneous amplitude of the modulated signal in volts
       E_c = peak amplitude of the carrier in volts
       e_m = instantaneous amplitude of the modulating signal in volts
       ω_c = radian frequency of the carrier
       t = time in seconds

The addition of E_c and e_m is algebraic. That is, the peak amplitude of the
modulated signal is both increased and decreased by the modulation.

Once again, let us note the difference between amplitude modulation
and simple addition. Just adding the carrier and modulating signals would
give the equation

$$v'(t) = E_c \sin \omega_c t + e_m \qquad (3.2)$$

which is definitely not the same as Equation (3.1). The summing of two wave-forms simply adds their instantaneous values at all times. Amplitude modulation, on the other hand, involves the addition of the instantaneous baseband amplitude to the peak carrier amplitude.

If the modulating (baseband) signal is a sine wave, Equation (3.1) has the following form:

$$v(t) = (E_c + E_m \sin \omega_m t) \sin \omega_c t \qquad (3.3)$$

where E_m = peak amplitude of the modulating signal in volts
          ω_m = radian frequency of the modulating signal

and the other variables are as defined for Equation (3.1).

EXAMPLE 3.1

A carrier wave with an RMS voltage of 2 V and a frequency of 1.5 MHz is modulated by a sine wave with a frequency of 500 Hz and amplitude of 1 V RMS. Write the equation for the resulting signal.

Solution

First, note that Equation (3.3) requires peak voltages and radian frequencies. We can easily get these as follows:

$$E_c = \sqrt{2} \times 2 \text{ V}$$
$$= 2.83 \text{ V}$$
$$E_m = \sqrt{2} \times 1 \text{ V}$$
$$= 1.41 \text{ V}$$

The equation also requires radian frequencies:

$$\omega_c = 2\pi \times 1.5 \times 10^6$$
$$= 9.42 \times 10^6 \text{ rad/s}$$
$$\omega_m = 2\pi \times 500$$
$$= 3.14 \times 10^3 \text{ rad/s}$$

So the equation is

$$v(t) = (E_c + E_m \sin \omega_m t) \sin \omega_c t$$
$$= [2.83 + 1.41 \sin (3.14 \times 10^3 t)] \sin (9.42 \times 10^6 t) \text{ V}$$

3.3.1 The Modulation Index

The amount by which the signal amplitude is changed in modulation depends on the ratio between the amplitudes of the modulating signal and the carrier. For convenience, this ratio is defined as the **modulation index** m. It can be expressed mathematically as

$$m = \frac{E_m}{E_c} \qquad (3.4)$$

Modulation can also be expressed as a percentage, with percent modulation found by multiplying m by 100. For example, $m = 0.5$ corresponds to 50% modulation.

Substituting m into Equation (3.3) gives:

$$v(t) = E_c(1 + m \sin \omega_m t) \sin \omega_c t \qquad (3.5)$$

EXAMPLE 3.2

Calculate m for the signal of Example 3.1 and write the equation for this signal in the form of Equation (3.5).

Solution

To avoid an accumulation of round-off errors we should go back to the original voltage values to find m.

$$m = \frac{E_m}{E_c}$$

$$= \frac{1}{2}$$

$$= 0.5$$

It is all right to use the RMS values for calculating this ratio, as the factors of $\sqrt{2}$, if used to find the peak voltages, will cancel.

Now we can rewrite the equation:

$$v(t) = E_c(1 + m \sin \omega_m t) \sin \omega_c t$$
$$= 2.83[1 + 0.5 \sin (3.14 \times 10^3 t)] \sin (9.42 \times 10^6 t)$$

It is worthwhile to examine what happens to Equation (3.5), and to the modulated waveform, as m varies. To start with, when $m = 0$, $E_m = 0$ and we have the original, unmodulated carrier. As m varies between 0 and 1, the changes due to modulation become more pronounced. Resultant waveforms for several values of m are shown in Figure 3.7. Note especially the result for $m = 1$ or 100%. Under these conditions, the peak signal voltage will vary between zero and twice the unmodulated carrier amplitude.

3.3.2 Overmodulation

When the modulation index is greater than 1, the signal is said to be **overmodulated**. There is nothing in Equation (3.3) that would seem to prevent E_m from being greater than E_c, that is, m greater than 1. There are practical difficulties, however. Figure 3.8(a) shows the result of simply substituting $m = 2$ into Equation (3.5). As you can see, the envelope no longer resembles

FIGURE 3.7   AM Envelopes for Various Values of *m*

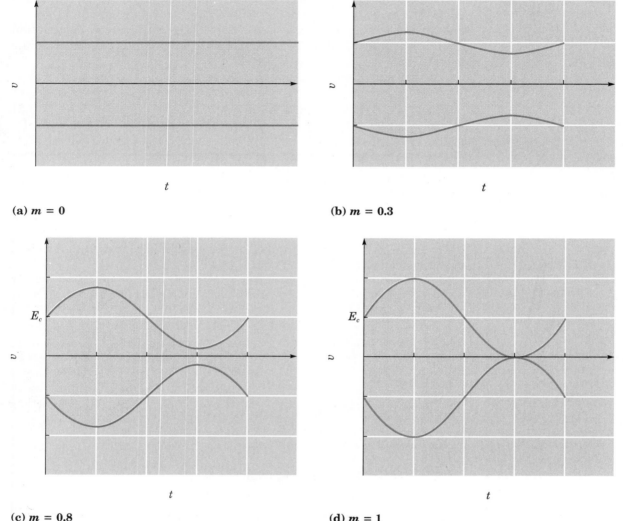

(a) *m* = 0   **(b)** *m* = 0.3

(c) *m* = 0.8   **(d)** *m* = 1

the modulating signal. Thus the type of demodulator described earlier no longer gives undistorted results, and the signal is no longer a full-carrier AM signal. Therefore, *m* must be kept less than or equal to 1.

Whenever we work with mathematical models, we must remember to keep checking against physical reality. This situation is a good example. It is possible to build a circuit that does produce an output that agrees with Equation (3.5), for *m* greater than 1. But suppose we try to operate the circuit of Figure 3.5 under these conditions. We find that for *m* = 1, the modulating voltage, at the secondary of the modulation transformer, has a peak value equal to V_{CC}, which reduces the collector supply voltage to zero during negative peaks of the modulating signal. Then, for *m* greater than 1, there will be

FIGURE 3.8   Overmodulation

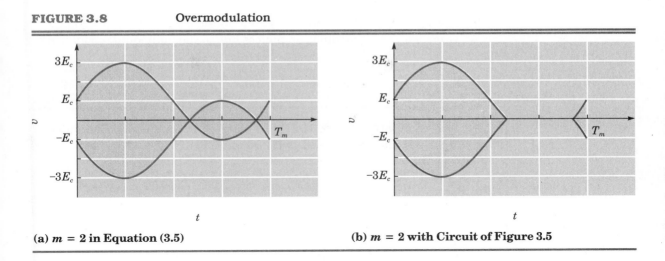

(a) $m = 2$ in Equation (3.5)   (b) $m = 2$ with Circuit of Figure 3.5

times when the transistor actually has a negative collector supply voltage. It obviously will not operate under these conditions. Thus, for $m = 2$, the circuit will produce the output shown in Figure 3.8(b). This is different from that given by Equation (3.5), but it also has the characteristic that the modulation envelope is no longer an accurate representation of the modulating signal. In fact, the sharp "corners" on the waveform as the output goes to zero on negative modulation peaks, if subjected to Fourier analysis, would be found to represent high-frequency components added to the original baseband signal. This type of overmodulation will create side frequencies further from the carrier than would otherwise be the case. These spurious frequencies are known as **splatter**, and cause the modulated signal to have increased bandwidth.

From the foregoing, we can draw the conclusion that for full-carrier AM, m must be in the range from 0 to 1. Overmodulation creates distortion in the demodulated signal, and may result in the signal occupying a larger bandwidth than normal. Since spectrum space is tightly controlled by law, overmodulation of an AM transmitter is actually illegal, and means must be provided to prevent it.

3.3.3 Modulation Index for Multiple Modulating Frequencies

Practical AM systems are seldom used to transmit sine waves, of course. The information signal is more likely to be a voice signal, which contains many frequencies. Though a typical audio signal is not strictly periodic, it is close enough to periodic that we can use the idea of Fourier series to consider it as a series of sine waves of different frequencies.

When there are two or more sine waves of different, uncorrelated frequencies (that is, frequencies that are not multiples of each other) modulating a single carrier, m is calculated by using the equation

$$m_T = \sqrt{m_1^2 + m_2^2 + \cdots}$$ (3.6)

where m_T = total resultant modulation index

$m_1, m_2, \ldots$ = modulation indices due to the individual modulating components

EXAMPLE 3.3

Find the modulation index if a 10 V carrier is amplitude modulated by three different frequencies, with amplitudes of 1 V, 2 V, and 3 V respectively.

Solution

The three separate modulation indices are:

$$m_1 = \frac{1}{10} = 0.1 \qquad m_2 = \frac{2}{10} = 0.2 \qquad m_3 = \frac{3}{10} = 0.3$$

$$\begin{aligned} m_T &= \sqrt{m_1^2 + m_2^2 + m_3^2} \\ &= \sqrt{0.1^2 + 0.2^2 + 0.3^2} \\ &= 0.374 \end{aligned}$$

3.3.4 Measurement of Modulation

If we let E_m and E_c be the peak modulation and carrier voltages respectively, then we can see, either by using Equations (3.3) and (3.5) or by inspecting Figure 3.9, that the maximum envelope voltage is simply

$$E_{max} = E_c + E_m$$

or

$$E_{max} = E_c(1 + m) \tag{3.7}$$

and the minimum envelope voltage is

$$E_{min} = E_c - E_m$$

or

$$E_{min} = E_c(1 - m) \tag{3.8}$$

Note, by the way, that this agrees with the conclusions expressed earlier: for $m = 0$, the peak voltage is E_c, and for $m = 1$, the envelope voltage ranges from $2E_c$ to zero.

Applying a little algebra to the above expressions, it is easy to show that

$$m = \frac{E_{max} - E_{min}}{E_{max} + E_{min}} \tag{3.9}$$

FIGURE 3.9

Voltage Relationships in an AM Signal

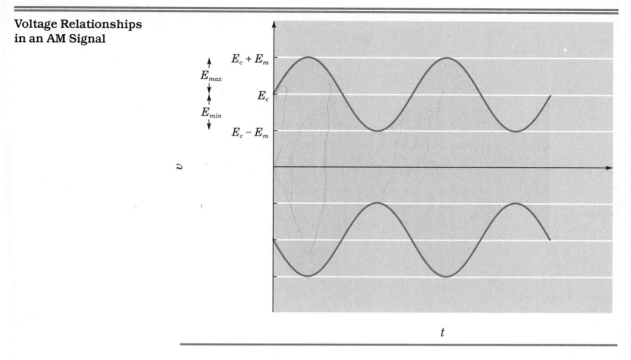

Of course, doubling both E_{max} and E_{min} will have no effect on this equation, so it is quite easy to find m by displaying the envelope on an oscilloscope and measuring the maximum and minimum peak-to-peak values for the envelope voltage. Other time-domain methods for measuring the amplitude modulation index will be described along with AM transmitters, in the next chapter.

EXAMPLE 3.4

Calculate the modulation index for the waveform shown in Figure 3.2.

Solution

It is easiest to use peak-to-peak values with an oscilloscope. From the figure we see that:

$$E_{max} = 150 \text{ mV} \qquad E_{min} = 80 \text{ mV} \qquad m = \frac{E_{max} - E_{min}}{E_{max} + E_{min}}$$

$$= \frac{150 - 80}{150 + 80}$$

$$= 0.304 \quad \text{or} \quad 30.4\%$$

▬▬▬▬ 3.4 AM IN THE FREQUENCY DOMAIN

So far, we have looked at the AM signal exclusively in the time domain, that is, as it can be seen on an ordinary oscilloscope. In order to find out more about this signal, however, it is necessary to consider its spectral makeup. We could use Fourier methods to do this, but for a simple AM waveform it is easier, and just as valid, to use trigonometry.

To start, we should observe carefully that although both the carrier and the modulating signal may be sine waves, the modulated AM waveform is *not* a sine wave. This can be seen from a simple examination of the waveform of Figure 3.1(c). It is important to remember that the modulated waveform is not a sine wave when, for instance, trying to find RMS from peak voltages. The usual formulas, so laboriously learned in fundamentals courses, do not apply here!

If an AM signal is not a sine wave, then what is it? We already have a mathematical expression, given by Equation (3.5):

$$v(t) = E_c(1 + m \sin \omega_m t) \sin \omega_c t$$

Expanding it and using a trigonometric identity will prove useful. Expanding gives

$$v(t) = E_c \sin \omega_c t + mE_c \sin \omega_m t \sin \omega_c t$$

The first term is just the carrier. The second can be expanded using two trigonometric identities:

$$\sin A \sin B = \frac{1}{2} [\cos (A - B) - \cos (A + B)]$$

and

$$\cos A = \cos (-A)$$

to give

$$v(t) = E_c \sin \omega_c t + \frac{mE_c}{2} [\cos (\omega_c - \omega_m)t - \cos (\omega_c + \omega_m)t]$$

which can be separated into three distinct terms:

$$v(t) = E_c \sin \omega_c t + \frac{mE_c}{2} \cos (\omega_c - \omega_m)t - \frac{mE_c}{2} \cos (\omega_c + \omega_m)t \qquad (3.10)$$

We now have, besides the original carrier, two additional sinusoidal waves, one above the carrier frequency and one below. When the complete signal is sketched in the frequency domain, as in Figure 3.10, we see the carrier and two additional frequencies, one to each side. These are called, logically enough, **side frequencies**. The separation of each side frequency from the carrier is equal to the modulating frequency, and the relative amplitude of the side frequency, compared with the carrier, is proportional to m,

FIGURE 3.10

AM in the Frequency
Domain

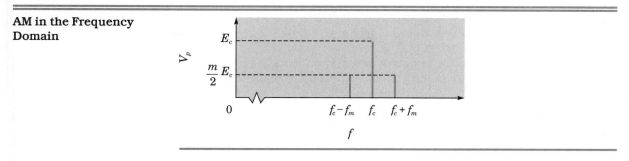

becoming half the carrier voltage for $m = 1$.

In a realistic situation there will generally be more than one set of side frequencies, because there will be more than one modulating frequency. Each modulating frequency will produce two side frequencies. Those above the carrier can be grouped into a band of frequencies called the **upper sideband**. Similarly, there will be a **lower sideband**, which looks like a mirror image of the upper, reflected in the carrier.

From now on, we will generally use the term *sideband*, rather than *side frequency*, even for the case of single-tone modulation, because it is more general and more commonly used in practice.

Mathematically, we have:

$$f_{usb} = f_c + f_m \tag{3.11}$$

$$f_{lsb} = f_c - f_m \tag{3.12}$$

$$E_{lsb} = E_{usb} = \frac{mE_c}{2} \tag{3.13}$$

where f_{usb} = frequency of the upper sideband
  f_{lsb} = frequency of the lower sideband
  E_{usb} = peak voltage of the upper-sideband component
  E_{lsb} = peak voltage of the lower-sideband component

EXAMPLE 3.5

(a) A 1 MHz carrier with an amplitude of 1 V peak is modulated by a 1 kHz signal with $m = 0.5$. Sketch the voltage spectrum.

(b) An additional 2 kHz signal modulates the carrier with $m = 0.2$. Sketch the voltage spectrum.

Solution

(a) The frequency scale is easy. There will be three frequency components. The carrier will be at:

$$f_c = 1 \text{ MHz}$$

The upper sideband will be at:

$$f_{usb} = f_c + f_m$$
$$= 1 \text{ MHz} + 1 \text{ kHz}$$
$$= 1.001 \text{ MHz}$$

The lower sideband will be at:

$$f_{lsb} = f_c - f_m$$
$$= 1 \text{ MHz} - 1 \text{ kHz}$$
$$= 0.999 \text{ MHz}$$

Next, we have to determine the amplitudes of the three components. The carrier is unchanged with modulation, so it remains at 1 V peak. The two sidebands will have the same peak voltage:

$$E_{lsb} = E_{usb} = \frac{mE_c}{2}$$
$$= \frac{0.5 \times 1}{2}$$
$$= 0.25 \text{ V}$$

Figure 3.11(a) shows the solution.

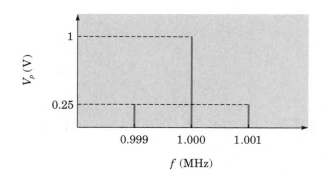

(a) $f_c = 1 \text{ MHz}$ $f_m = 1 \text{ kHz}$ $m = 0.5$ $E_c = 1 \text{ V}$

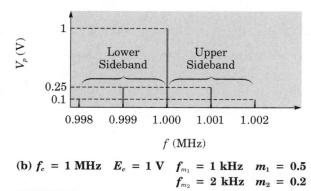

(b) $f_c = 1 \text{ MHz}$ $E_c = 1 \text{ V}$ $f_{m_1} = 1 \text{ kHz}$ $m_1 = 0.5$
$f_{m_2} = 2 \text{ kHz}$ $m_2 = 0.2$

FIGURE 3.11

(b) The addition of another modulating signal at a different frequency simply adds another set of side frequencies. It does not change anything that was done for Part (a). The new frequency components will be at 1.002 and 0.998 MHz, and their amplitude will be 0.1 V. The result is shown in Figure 3.11(b).

3.4.1 Bandwidth of an AM Signal

Signal bandwidth is one of the most important characteristics of any modulation scheme. In general, a narrow bandwidth is desirable. In any situation where spectrum space is limited, a narrow bandwidth will allow more signals to be transmitted simultaneously than will a wider bandwidth. It also allows a narrower bandwidth to be used in the receiver. Since ordinary thermal noise is evenly distributed over the frequency domain, using narrower bandwidth in receivers will include less noise, thereby increasing the signal-to-noise ratio. (There is one major exception to the general rule that reducing the bandwidth improves the signal-to-noise ratio. The exception is for wideband frequency modulation, which will be described in Chapter 8.) However, the receiver must have a wide enough bandwidth to pass the complete signal including all the sidebands, or distortion will result. Consequently, we will have to calculate the signal bandwidth for each of the modulation schemes we consider.

A glance at Equation (3.10) and Figure 3.10 will show that this calculation is very easy for AM. The signal extends from the lower side frequency, which is at the carrier frequency less the modulation frequency, to the upper side frequency, at the carrier frequency plus the modulation frequency. The difference between these is simply twice the modulation frequency.

If we have a complex modulating signal, with more than one modulating frequency, as in Figure 3.11(b), the bandwidth will be twice the *highest* modulating frequency. For telephone-quality voice, for instance, a bandwidth of about 6 kHz would suffice, while a video signal with a 4 MHz maximum baseband frequency would need 8 MHz of bandwidth, if transmitted in this way. (Since a television channel is only 6 MHz wide, we can surmise, correctly, that television must actually be transmitted by a more complex modulation scheme that uses less bandwidth.)

Mathematically, the relationship is:

$$B = 2F_m \tag{3.14}$$

where B = bandwidth in hertz
F_m = highest modulating frequency in hertz

EXAMPLE 3.6

CB radio channels are 10 kHz apart. What is the maximum modulation frequency that can be used if a signal is to remain entirely within its assigned channel?

Solution

From Equation (3.14) we have

$$B = 2F_m$$

or

$$F_m = \frac{B}{2}$$
$$= \frac{10 \text{ kHz}}{2}$$
$$= 5 \text{ kHz}$$

By the way, many people assume that because AM radio broadcast channels are also assigned at 10 kHz intervals, there is a similar limitation on the audio frequency response for broadcasting. This is not the case: AM broadcast transmitters typically have an audio frequency response extending to about 10 kHz, giving a theoretical signal bandwidth of 20 kHz. This works because adjacent channels are not assigned in the same locality, so some overlap is possible. For instance, if your city has a station at 1000 kHz, there will not be one at 990 or 1010 kHz. In order to reduce interference from distant stations, many AM receivers do have narrow bandwidth, and consequently limited audio frequency response.

3.5 POWER RELATIONSHIPS

Power is important in any communications scheme, because the crucial signal-to-noise ratio at the receiver depends as much on the signal power being large as the noise power being small. The power that is most important, however, is not the total signal power, but only that portion that is used to transmit information. Since the carrier in an AM signal remains unchanged with modulation, it contains no information. Its only function is to aid in demodulating the signal at the receiver. This makes AM inherently wasteful of power, compared with some other modulation schemes to be described later.

3.5.1 Power in an AM Signal

The easiest way to look at the power in an AM signal is to use the frequency domain. We can find the power in each frequency component, then add to get total power. We will assume that the signal appears across a resistance R, so that reactive volt-amperes can be ignored. We will also assume that the power required is average power. (See Appendix A for a review of the various ways of describing power in ac signals.)

Suppose that the modulating signal is a sine wave. Then the AM signal consists of three sinusoids, the carrier and two side frequencies (usually called sidebands), as shown in Figure 3.10.

The power in the carrier is easy to calculate, since the carrier by itself is a sine wave. The carrier is given by the equation

$$e_c = E_c \sin \omega_c t$$

where e_c = instantaneous carrier voltage
E_c = peak carrier voltage

Since E_c is the peak carrier voltage, the power P_c developed when this signal appears across a resistance R is simply

$$P_c = \frac{(E_c/\sqrt{2})^2}{R}$$

$$= \frac{E_c^2}{2R} \text{ W}$$

The next step is to find the power in each sideband. The two frequency components have the same amplitude, so they will have equal power. Assuming sine-wave modulation, each sideband is a cosine wave whose peak voltage is given by Equation (3.13).

$$E_{lsb} = E_{usb} = \frac{mE_c}{2}$$

Since the carrier and both sidebands are part of the same signal, the sidebands will appear across the same resistance R as the carrier. The two sidebands will have equal power. Looking at the lower sideband,

$$P_{lsb} = \frac{E_{lsb}^2}{2R}$$

$$= \frac{(mE_c/2)^2}{2R}$$

$$= \frac{m^2 E_c^2}{4 \times 2R}$$

$$= \frac{m^2}{4} \times \frac{E_c^2}{2R}$$

$$= \frac{m^2}{4} P_c \qquad\qquad (3.15)$$

The same reasoning will show that

$$P_{usb} = \frac{m^2}{4} P_c$$

Since the two sidebands have equal power, the total sideband power is given by

$$P_{sb} = \frac{m^2}{2} P_c \qquad\qquad (3.16)$$

The total power P_t in the whole signal is just the sum of the power in the carrier and the sidebands, so it is

$$P_t = P_c + \left(\frac{m^2}{2}\right)P_c$$

or

$$P_t = P_c\left(1 + \frac{m^2}{2}\right) \tag{3.17}$$

These latest equations tell us several useful things:

1. The total power in an AM signal increases with modulation, reaching a value 50% greater than that of the unmodulated carrier for 100% modulation. As we shall see, this has implications for transmitter design.
2. The extra power with modulation goes into the sidebands: the carrier power does not change with modulation.
3. The useful power, that is, the power that carries information, is rather small, being a maximum of one third of the total signal power for 100% modulation, and much less at lower modulation indices. For this reason, AM transmission is more efficient when the modulation index is as close to 1 as practicable.

EXAMPLE 3.7

An AM broadcast transmitter has a carrier power output of 50 kW. What would be the total power produced with 80% modulation?

Solution

$$P_t = P_c\left(1 + \frac{m^2}{2}\right)$$

$$= (50 \text{ kW})\left(1 + \frac{0.8^2}{2}\right)$$

$$= 66 \text{ kW}$$

3.5.2 Measuring the Modulation Index in the Frequency Domain

Since the ratio between sideband and carrier power is a simple function of m, it is quite possible to measure the modulation index by observing the spectrum of an AM signal. The only slight complication is that spectrum analyzers generally display power ratios in decibels. The power ratio between sideband and carrier power can easily be found from the relation:

$$\frac{P_{lsb}}{P_c} = \text{antilog } (dB/10) \tag{3.18}$$

where P_c = carrier power

P_{lsb} = power in one sideband (The lower has been chosen for the example, but of course the upper sideband has the same power.)

dB = difference between sideband and carrier signals, measured in decibels (This number will be negative.)

Once the ratio between carrier and sideband power has been found, it is easy to find the modulation index from Equation (3.15):

$$P_{lsb} = \frac{m^2}{4} P_c$$

$$m^2 = \frac{4P_{lsb}}{P_c}$$

$$m = 2\sqrt{\frac{P_{lsb}}{P_c}} \tag{3.19}$$

Figure 3.12 is a handy nomograph for finding m directly from the difference in decibels between sideband and carrier powers.

FIGURE 3.12

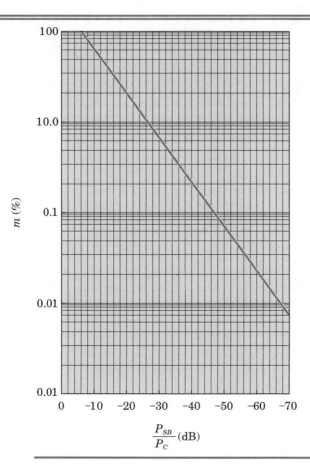

Courtesy Hewlett-Packard.

Although the time-domain measurement described earlier is simpler and uses less-expensive equipment, frequency-domain measurement enables much smaller values of m to be found. A modulation level of 5%, for instance, would be almost invisible on an oscilloscope, but it is quite obvious, and easy to measure, on a spectrum analyzer. The spectrum analyzer also allows the contribution from different modulating frequencies to be observed and calculated separately.

EXAMPLE 3.8

Calculate the modulation frequency f_m and modulation index m for the spectrum analyzer display shown in Figure 3.13.

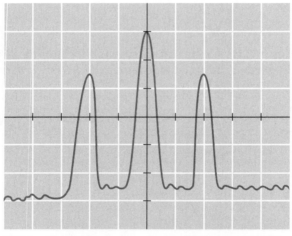

Ref. Level: 10 dBm   Center Frequency: 10 MHz
Vert: 10 dB/division   Span: 5 kHz/division
FIGURE 3.13

Solution

First let us find f_m. The difference between the carrier and either sideband is 2 divisions at 5 kHz/division, or 10 kHz. So f_m is 10 kHz.

Next, we need to find the modulation index. The two sidebands have the same power, so we can use either. Let us use the lower. The spectrum analyzer is set for 10 dB/division, and the sideband is 1.5 divisions, or 15 dB, below the carrier. This corresponds to a power ratio, as given in Equation (3.17), of

$$\frac{P_{lsb}}{P_c} = \text{antilog}\left(\frac{-15}{10}\right)$$

$$= 0.0316$$

From Equation (3.19),

$$m = 2\sqrt{\frac{P_{lsb}}{P_c}}$$
$$= 2\sqrt{0.0316}$$
$$= 0.356$$

We could have used the nomograph instead, with the same result. Try it.

3.5.3 RMS and Peak Voltages and Currents

The well-known power equation

$$P = \frac{V^2}{R}$$

is valid for any kind of waveform, provided that P represents average power, and V represents RMS voltage. To find the RMS voltage when the power is known, rearrange this equation to get

$$V = \sqrt{PR} \tag{3.20}$$

The total power in an AM waveform was found earlier. It is given by Equation (3.17) as

$$P_t = P_c\left(1 + \frac{m^2}{2}\right)$$

Substituting this into Equation (3.20) gives

$$V = \sqrt{P_c\left(1 + \frac{m^2}{2}\right)R} \tag{3.21}$$

Similarly, from the fundamental equation

$$P = I^2R$$

we can easily find the RMS current:

$$I = \sqrt{\frac{P}{R}} \tag{3.22}$$

For a modulated signal, the power required is the total power, giving the equation

$$I = \sqrt{\frac{P_c(1 + m^2/2)}{R}} \tag{3.23}$$

EXAMPLE 3.9

Find the RMS voltage and current at the output of a transmitter that has a carrier power of 500 W into a load impedance of 50 Ω:

(a) without modulation

(b) with 70% modulation

Solution

(a) Without modulation, the signal is a sine wave and the RMS voltage is given by Equation (3.20):

$$V = \sqrt{PR}$$
$$= \sqrt{(500 \text{ W})(50 \text{ } \Omega)}$$
$$= 158 \text{ V}$$

The current is found from Equation (3.22):

$$I = \sqrt{\frac{P}{R}}$$
$$= \sqrt{\frac{500 \text{ W}}{50 \text{ } \Omega}}$$
$$= 3.16 \text{ A}$$

(b) With modulation, the RMS voltage and current are found in the same way, except that the total power now includes the power in the sidebands. For voltage, we can use Equation (3.21):

$$V = \sqrt{P_c\left(1 + \frac{m^2}{2}\right)R}$$
$$= \sqrt{(500 \text{ W})\left(1 + \frac{0.7^2}{2}\right)(50 \text{ } \Omega)}$$
$$= 176 \text{ V}$$

Similarly the RMS current is given by Equation (3.23):

$$I = \sqrt{\frac{P_c(1 + m^2/2)}{R}}$$
$$= \sqrt{\frac{(500 \text{ W})(1 + 0.7^2/2)}{50 \text{ } \Omega}}$$
$$= 12.45 \text{ A}$$

It is not necessary to remember Equations (3.21) and (3.23). The only thing to remember is that, to find RMS voltage or current, you use total average power. This is true for an AM signal just as for any other signal.

Sometimes an RF ammeter in the transmission line from transmitter to antenna is used to measure modulation index for an AM station. It is easy to derive an expression to show how this can be done. Once again, it is not necessary to memorize this equation; the derivation is presented as one more example of the many things that can be learned about an AM signal once its basic structure is understood. From Equation (3.23) we know that:

$$I = \sqrt{\frac{P_c(1 + m^2/2)}{R}}$$

We can rearrange this equation to express m in terms of the current:

$$1 + \frac{m^2}{2} = \frac{I^2 R}{P_c}$$

$$\frac{m^2}{2} = \frac{I^2 R}{P_c} - 1$$

$$m = \sqrt{2\left(\frac{I^2 R}{P_c} - 1\right)}$$

In this equation, of course, I is the current with modulation, as measured by the RF ammeter. The ammeter will not measure P_c directly, but we do know that

$$P_c = I_0^2 R$$

where I_0 = current without modulation

If we substitute this into our equation for m, we get

$$m = \sqrt{2\left(\frac{I^2 R}{I_0^2 R} - 1\right)}$$

R cancels out, so we find

$$m = \sqrt{2\left(\frac{I^2}{I_0^2} - 1\right)} \tag{3.24}$$

An RF ammeter in the transmission line from transmitter to antenna measures 5 A without modulation and 5.5 A with modulation. What is the modulation index?

Solution

From Equation (3.24), we know that:

$$m = \sqrt{2\left(\frac{I^2}{I_0^2} - 1\right)}$$

$$= \sqrt{2\left(\frac{5.5^2}{5^2} - 1\right)}$$

$$= 0.648$$

Now, how about peak values for voltage and current? We may be tempted to use the formula

$$V_p = \sqrt{2}V_{RMS}$$

but we had better resist the temptation, because that equation is valid only for sine waves. Instead, let us go back to Equation (3.7), which shows us that the peak voltage is

$$E_{max} = E_c(1 + m)$$

Thus, we can find the peak voltage of an AM signal by first finding the peak carrier voltage and the modulation index. We are more likely to know the carrier power than the carrier voltage, but since the carrier is a sine wave we should have no difficulty.

EXAMPLE 3.11

Find the peak voltage at the output of a CB transmitter with 4 W carrier power output, operating into a 50 Ω resistive load, with $m = 0.8$.

Solution

From the carrier power, we can get the carrier voltage. The RMS voltage will be given by Equation (3.20):

$$V_c = \sqrt{P_c R}$$

but we want the peak carrier voltage which, since the carrier is a sine wave, will be the RMS voltage multiplied by $\sqrt{2}$, or

$$E_c = \sqrt{2P_c R}$$

$$= \sqrt{2(4 \text{ W})(50 \text{ }\Omega)}$$

$$= 20 \text{ V}$$

Now, the peak voltage for the signal is, from Equation (3.7),

$$E_{max} = E_c(1 + m)$$
$$= 20(1 + 0.8)$$
$$= 36 \text{ V}$$

3.6 EXAMPLE: AM ANALYSIS

The following example covers most of the topics discussed in this chapter. It serves to demonstrate the wealth of information that can be obtained about an AM signal, given only a few parameters. Please note that all the information can be derived quite simply, given a knowledge of the shape of an AM wave and the equation that defines it. It is certainly not necessary to memorize all the equations given in this chapter; most of them can be derived on the spot, as needed. This method is used in the example that follows, in an attempt to illustrate the thought processes that could be used in examining an AM signal.

EXAMPLE 3.12

An AM transmitter has a carrier power of 1 kW at a carrier frequency of 10 MHz. It operates into an antenna with an impedance of 50 Ω, resistive. It is modulated with $m = 0.5$, by a sine wave with a frequency of 1 kHz.

(a) Find:
   (i)   the signal bandwidth
   (ii) the total signal power
   (iii) the RMS voltage of the signal
   (iv) the peak signal voltage

(b) sketch the signal in the time domain

(c) sketch the signal in the frequency domain

Solution

(a) (i) From Equation (3.14),

$$B = 2F_m = 2 \times 1 \text{ kHz} = 2 \text{ kHz}$$

(ii) From Equation (3.17),

$$P_t = P_c\left(1 + \frac{m^2}{2}\right)$$
$$= (1 \text{ kW})\left(1 + \frac{0.5^2}{2}\right)$$
$$= 1.125 \text{ kW}$$

(iii) For any signal, whether sine wave or not, the following equation is valid:

$$P = \frac{V^2}{R}$$

where $P =$ average power in watts
        $V =$ RMS voltage in volts
        $R =$ the resistance across which the signal appears, in ohms

Since we know the power and the resistance, we can find the RMS voltage:

$$V = \sqrt{PR}$$
$$= \sqrt{(1.125 \times 10^3 \text{ W})(50 \text{ }\Omega)}$$
$$= 237 \text{ V}$$

(iv) Be careful here. The temptation is to use the formula

$$V_p = \sqrt{2}V_{RMS}$$

but that is valid only for a sine wave, and an AM equation is NOT a sine wave, as we know. We will have to go back to our time-domain representation of an AM signal. We note that the peak voltage is simply

$$V_p = E_c(1 + m)$$

We already know m, so if we can find E_c, the problem is solved. Now, E_c is just the peak voltage of the unmodulated carrier. We know the carrier power is 1 kW, and the carrier is a sine wave, so the RMS voltage for the carrier alone is

$$V = \sqrt{PR}$$
$$= \sqrt{1000 \text{ W} \times 50 \text{ }\Omega}$$
$$= 223.6 \text{ V}$$

Now, since the carrier is a sine wave, its peak voltage is

$$E_c = 223.6 \text{ V} \times \sqrt{2} = 316 \text{ V}$$

We're almost done. The peak voltage for the whole signal will be

$$V_p = (316 \text{ V})(1 + 0.5)$$
$$= 474 \text{ V}$$

Please observe that this is *not* the same as would be found from the sine wave formula. Try it and see.

(b) We will not be able to see both the individual RF waveforms and the envelope on one sketch. Of the two, the envelope will be much more useful. We already know the general shape of an AM envelope. All we really have to do is find suitable time and voltage values for the axes.

The period of the envelope will be the same as that of the modulating signal. For a modulating frequency of 1 kHz, the period will of course be $T = 1/f = 1$ ms. That takes care of the time scale.

Now we need some voltages. The maximum voltage for the envelope was found in Part (a)(iv) to be $E_c(1 + m) = 474$ V, and the minimum voltage will be

$$E_{min} = E_c(1 - m)$$
$$= 316(1 - 0.5)$$
$$= 158 \text{ V}$$

Finally, we can sketch the signal in the time domain, as shown in Figure 3.14(a).

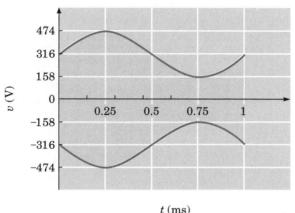

(a) Time Domain
FIGURE 3.14

(c) Again, we already know the general form of this signal in the frequency domain. (See Figure 3.10.) We need a frequency scale for the horizontal axis, and either a voltage or a power scale for the vertical axis. Let's do it both ways.

First, the frequency scale. The carrier frequency is given as 10 MHz. The upper sideband will be at

$$f_{usb} = f_c + f_m$$
$$= 10 \text{ MHz} + 1 \text{ kHz}$$
$$= 10.001 \text{ MHz}$$

Similarly, the lower sideband is at

$$f_{lsb} = f_c - f_m$$
$$= 10 \text{ MHz} - 1 \text{ kHz}$$
$$= 9.999 \text{ MHz}$$

Now for the vertical axis. Let us do power first, since this is likely to be more useful. (Spectrum analyzers, for instance, are generally calibrated in terms of power.)

The carrier power is easy: it is given as 1 kW in the statement of the problem. As for the power in each sideband, we can easily find it from Equation (3.15):

$$P_{lsb} = \frac{m^2}{4}\, P_c$$

$$= \left(\frac{0.5^2}{4}\right)1000$$

$$= 62.5 \text{ W}$$

Now we can sketch the spectrum shown in Figure 3.14(b).

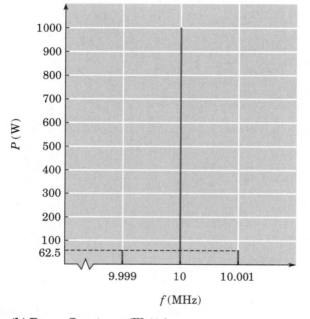

(b) Power Spectrum (Watts)
FIGURE 3.14   (*Continued*)

Actually, most spectrum analyzers indicate power in dBm rather than watts, so it might be instructive to sketch this spectrum with the vertical axis in dBm. (You are probably familiar with decibels, dBm, dBW, and dBf from earlier work; if not, refer to Appendix E.) We can find the carrier power in dBm from:

$$P_c \text{ (dBm)} = 10 \log \frac{P}{1 \text{ mW}}$$

$$= 10 \log \frac{1000 \text{ W}}{0.001 \text{ W}}$$

$$= 60 \text{ dBm}$$

Similarly, the power in the lower sideband is

$$P_{lsb} \text{ (dBm)} = 10 \log \frac{62.5 \text{ W}}{0.001 \text{ W}}$$

$$= 48 \text{ dBm}$$

The reading for the upper sideband will, of course, be the same. The result is shown in Figure 3.14(c).

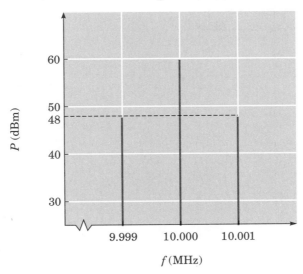

(c) Power Spectrum (dBm)
FIGURE 3.14   (*Continued*)

You will notice that the decibel plot, being logarithmic, is more convenient to use when observing the carrier and sideband signals at the same time. With a linear scale, the sideband power is "lost" at the bottom of the plot.

In passing, a word of caution is in order. While it is quite proper to add the carrier power to that in the sidebands to arrive at the total signal power of $1000 + 62.5 + 62.5 = 1125$ W, it is *not* all right to add the powers in dBm, because of the logarithmic nature of that unit. The total signal power 1125 W corresponds to a total power of

$$P_t \text{ (dBm)} = 10 \log \frac{1125 \text{ W}}{0.001 \text{ W}}$$

$$= 60.51 \text{ dBm}$$

which is a great deal different from the number you would get from adding up the dBm values of the three components.

Lastly, we could plot the signal in the frequency domain in terms of voltages, either peak or RMS. Let us use RMS. We can find the RMS voltage of each of the three sinusoidal components from:

$$V = \sqrt{PR}$$

For the carrier this gives:

$$V_c = \sqrt{(1000 \text{ W})(50 \text{ }\Omega)}$$

$$= 224 \text{ V}$$

and for each sideband:

$$V_{sb} = \sqrt{(62.5 \text{ W})(50 \text{ }\Omega)}$$

$$= 55.9 \text{ V}$$

These results are shown in Figure 3.14(d).

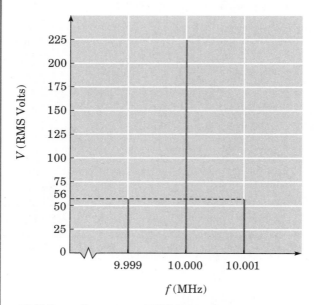

(d) Voltage Spectrum (RMS Volts)
FIGURE 3.14   (*Continued*)

Another way to get the same result would be to use the value for E_c, the peak carrier voltage, found in Part a(iv). Since the carrier is a sine wave, we can find the RMS voltage V_c, by using:

$$V_{RMS} = \frac{V_p}{\sqrt{2}}$$

$$V_c = \frac{316 \text{ V}}{\sqrt{2}}$$

$$= 223 \text{ V}$$

(The difference is due to the fact that E_c was already rounded to three significant digits.)

Another way to find the sideband voltages is to look at Equation (3.13), which shows that the peak voltage for each sideband is

$$E_{sb} = \frac{mE_c}{2}$$

$$= \frac{0.5 \times 316 \text{ V}}{2}$$

$$= 79 \text{ V}$$

so the RMS voltage for each sideband is

$$V_{sb} = \frac{79 \text{ V}}{\sqrt{2}}$$

$$= 55.9 \text{ V}$$

**Improved Quality
for AM**

There is no inherent reason why AM radio has to have poor audio quality. Much of the problem rests with receivers whose bandwidth is too narrow to include all the sidebands of the broadcast signal. Many AM receivers, including those included with high-fidelity audio systems, have an audio response that extends only to about 4 kHz.

In an attempt to compensate for this, most AM broadcasters boost the higher audio frequencies before modulation. This is called pre-emphasis. The problem is that until now there has been no standard for the amount of boost used.

Recent decisions by the FCC in the United States have standardized the amount of pre-emphasis and have also determined that stations may transmit an audio signal up to 10 kHz in frequency (20 kHz total bandwidth). The AM band has now been extended to 1700 kHz. At the same time, the Electronics Industry Association in the United States has developed a voluntary standard for receivers requiring reasonably flat audio response to at least 7.5 kHz, with an option for selectable reduced bandwidth to avoid interference.

It will be interesting to see whether these improvements, plus the standardizing of AM stereo to Motorola's C-Quam™ system (discussed in Section 3.7), will reverse the decline in popularity of AM compared to FM broadcasting.

It should be seen from this example that there is usually more than one way to solve a problem dealing with complex signals. There may also be some wrong ways. Many of these pitfalls can be avoided by keeping one eye on physical reality. For instance, if we had made the common mistake of adding sideband and carrier powers, given in dBm, we would have come up with $60 + 48 + 48 = 156$ dBm. A quick conversion to watts would give

$$P \text{ (mW)} = \text{antilog (dBm/10)}$$
$$= 3.98 \times 10^{15} \text{ mW}$$
$$= 3.98 \times 10^{12} \text{ W}$$

This is about 4 trillion watts, which is, of course, absurd. You should always proceed carefully, and check each answer against what you know of physical reality. For instance, you know that the total power of any AM signal will be somewhere between the carrier power (for $m = 0$) and a value 50% greater (for $m = 1$). Any value outside this range is cause for rechecking your work. Similarly, the peak voltage of an AM envelope will be somewhere between the peak carrier voltage and a value twice as large.

3.7 QUADRATURE AM AND AM STEREO

It is possible to send two separate information signals using amplitude modulation at one carrier frequency. This can be accomplished by generating two carriers, both at the same frequency but separated in phase by 90°. Then, each is modulated by a separate information signal and the two resulting

FIGURE 3.15

Quadrature AM

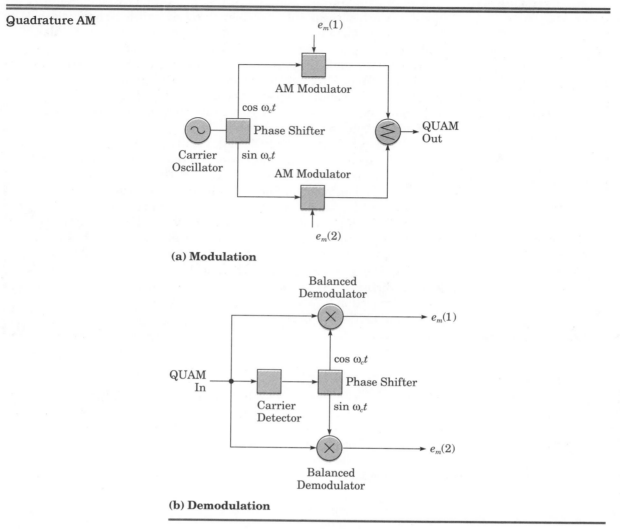

(a) Modulation

(b) Demodulation

signals are summed. Because of the 90° phase shift involved, the scheme is called **quadrature AM** (QUAM or QAM). Figure 3.15(a) shows how it can be implemented.

Recovery of the two information signals requires **synchronous detection** using two balanced demodulators. These must be supplied with reference carriers having exactly the same frequency and phase as the original carriers. Otherwise the output from each detector will be some combination of the two baseband signals. Figure 3.15(b) gives the general idea. Balanced demodulators are essentially the same as the balanced mixers described in Chapter 2. Their operation as detectors will be covered in detail in Chapter 6, in connection with single-sideband systems.

Quadrature AM is a much more complex scheme than ordinary AM. Until recently, its main applications have been in data transmission and in color television. The main reason for including a brief introduction to it here is that a variation of QUAM is also used for AM stereo broadcasting. The scheme used is called C-Quam™ and was developed by Motorola, Inc.

Quadrature AM could be used to transmit stereo signals, without any increase in bandwidth, by modulating each carrier with one channel (left or right). Unfortunately, ordinary quadrature AM would not be very suitable for stereo broadcasts because it would not be compatible with the envelope detectors used in ordinary AM radios. For compatibility, such a detector should produce a monaural (mono) signal consisting of the sum of the two channels. Using the system just described would result in distorted signals in an ordinary (mono) radio. Mathematically, the signal produced by an ideal envelope detector would be

$$v(t) = \sqrt{[1 + v_L(t) + v_R(t)]^2 + [v_L(t) - v_R(t)]^2}$$

where $v_L(t)$ and $v_R(t)$ represent the left and right signal voltages, respectively. It is easy to see that this represents a sum $(v_L + v_R)$ signal (plus a dc component) only when $v_L - v_R = 0$, that is, when the original signal is monaural. Otherwise, there will terms of the form $v_L v_R$, which represent intermodulation distortion.

The C-Quam™ system solves this problem by predistorting the QUAM signal so that the envelope of the resulting signal represents the sum of the two stereo channels (referred to as the L+R signal). Thus a mono receiver, with a conventional envelope detector, will receive an L+R signal. A stereo receiver will have the necessary circuitry to convert the C-Quam™ signal to an ordinary QUAM signal and extract the L-R information.

In addition to the predistorted QUAM signal, a stereo AM signal has a 25 Hz pilot carrier that indicates the presence of stereo to the receiver. Single-chip C-Quam™ decoders are available that sense the 25 Hz carrier, and automatically switch to stereo operation in the presence of a stereo signal.

SUMMARY OF CHAPTER 3

Here are the main points to remember from this chapter.

1. Amplitude modulation is the simplest form of modulation.
2. In the time domain, the process of amplitude modulation creates a signal with an envelope that closely resembles the original information signal.
3. In the frequency domain, an AM signal consists of the carrier, which is unchanged from its unmodulated state, and two sidebands. The total bandwidth of the signal is twice the maximum modulating frequency.
4. Amplitude modulation can be generated by using the modulating signal to vary the supply voltage to a Class C amplifier.
5. An AM signal can be demodulated by an envelope detector, which consists of a diode followed by a low-pass filter.
6. The peak voltage of an AM signal varies with the modulation index,

becoming twice that of the unmodulated carrier for the maximum modulation index of 1.

7. The power in an AM signal increases with modulation. The extra power goes into the sidebands. At maximum modulation, the total power is 50% greater than the power in the unmodulated carrier.

8. The modulation index can easily be measured using either an oscilloscope or a spectrum analyzer. The oscilloscope method requires less-expensive equipment but the spectrum analyzer can measure much lower modulation indices.

9. When calculating power and voltage values for an AM signal, it is important to take into account the actual makeup of the signal, and avoid reliance on formulas derived for sine waves.

IMPORTANT EQUATIONS

$$v(t) = (E_c + e_m) \sin \omega_c t \tag{3.1}$$

$$v(t) = (E_c + E_m \sin \omega_m t) \sin \omega_c t \tag{3.3}$$

$$m = \frac{E_m}{E_c} \tag{3.4}$$

$$v(t) = E_c(1 + m \sin \omega_m t) \sin \omega_c t \tag{3.5}$$

$$m_T = \sqrt{m_1^2 + m_2^2 + \cdots} \tag{3.6}$$

$$E_{max} = E_c(1 + m) \tag{3.7}$$

$$E_{min} = E_c(1 - m) \tag{3.8}$$

$$m = \frac{E_{max} - E_{min}}{E_{max} + E_{min}} \tag{3.9}$$

$$v(t) = E_c \sin \omega_c t + \frac{mE_c}{2} \cos (\omega_c - \omega_m)t$$
$$- \frac{mE_c}{2} \cos (\omega_c + \omega_m)t \tag{3.10}$$

$$f_{usb} = f_c + f_m \tag{3.11}$$

$$f_{lsb} = f_c - f_m \tag{3.12}$$

$$E_{lsb} = E_{usb} = \frac{mE_c}{2} \tag{3.13}$$

$$B = 2F_m \tag{3.14}$$

$$P_{lsb} = \frac{m^2}{4} P_c \tag{3.15}$$

$$P_{sb} = \frac{m^2}{2} P_c \tag{3.16}$$

$$P_t = P_c\left(1 + \frac{m^2}{2}\right) \tag{3.17}$$

$$m = 2\sqrt{\frac{P_{lsb}}{P_c}} \tag{3.19}$$

$$V = \sqrt{P_c\left(1 + \frac{m^2}{2}\right)R} \tag{3.21}$$

$$I = \sqrt{\frac{P_c(1 + m^2/2)}{R}} \tag{3.23}$$

$$m = \sqrt{2\left(\frac{I^2}{I_0^2} - 1\right)} \tag{3.24}$$

GLOSSARY

amplitude modulation (AM) a modulation scheme in which the amplitude of a high-frequency signal is varied in accordance with the instantaneous amplitude of an information signal

demodulator circuit that restores the original information signal when supplied with a modulated signal

envelope curve produced by joining the tips of the individual RF cycles of a modulated wave

envelope detector demodulator that works by extracting the information signal from the envelope of an AM signal

modulation index measure of the extent of modulation of a signal

modulator circuit that produces a modulated wave when supplied with a carrier wave and an information signal

overmodulation modulation to a greater depth than allowed. (For AM, this means a modulation index greater than one.)

quadrature AM transmission of two separate information signals using two amplitude-modulated carriers at the same frequency but differing in phase by 90°

side frequency a signal component in a modulated signal, at a frequency different from that of the carrier

sideband all of the side frequencies to one side of the carrier frequency

splatter colloquial term used to describe additional side frequencies produced by overmodulation or distortion in an AM system

synchronous detection system of demodulation that makes use of the original carrier frequency and phase

QUESTIONS

1. What is meant by the *envelope* of an AM waveform, and what is its significance?

2. Although amplitude modulation certainly involves changing the amplitude of the

signal, it is not true to say that the amplitude of the carrier is modulated. Explain this statement.

3. Why is it desirable to have the modulation index of an AM signal as large as possible, without overmodulating?

4. Describe what happens when a typical AM modulator is overmodulated, and explain why overmodulation is undesirable.

5. Explain the difference between amplitude modulation and linear addition of the carrier and information signals.

6. How does the bandwidth of an AM signal relate to the information signal?

7. Describe three different ways in which the modulation index of an AM signal can be measured.

8. By how much does the power in an AM signal increase with modulation, compared to the power of the unmodulated carrier?

9. Draw a simple circuit for an AM modulator, and explain briefly how it works.

10. Draw a simple circuit for an AM demodulator, and explain briefly how it works.

11. Suppose a modulation system could be devised that was similar to AM but with half the bandwidth. Would this be an improvement? Explain.

PROBLEMS

Section 3.3

12. An AM signal has the equation:

$$v(t) = [15 + 4 \sin (44 \times 10^3 t)] \sin (46.5 \times 10^6 t) \text{ V}$$

(a) Find the carrier frequency.

(b) Find the frequency of the modulating signal.

(c) Find the value of m.

(d) Find the peak voltage of the unmodulated carrier.

(e) Sketch the signal in the time domain, showing voltage and time scales.

13. An AM signal has a carrier frequency of 3 MHz and an amplitude of 5 V peak. It is modulated by a sine wave with a frequency of 500 Hz and a peak voltage of 2 V. Write the equation for this signal and calculate the modulation index.

14. An AM signal consists of a 10 MHz carrier modulated by a 5 kHz sine wave. It has a maximum positive envelope voltage of 12 V and a minimum of 4 V.

(a) Find the peak voltage of the unmodulated carrier.

(b) Find the modulation index and percent.

(c) Sketch the envelope.

(d) Write the equation for the signal voltage as a function of time.

15. An AM transmitter is modulated by two sine waves, at 1 kHz and 2.5 kHz, with a modulation due to each of 25% and 50% respectively. What is the effective modulation index?

16. For the AM signal sketched in Figure 3.16, calculate:

(a) the modulation index

(b) the peak carrier voltage

(c) the peak modulating-signal voltage

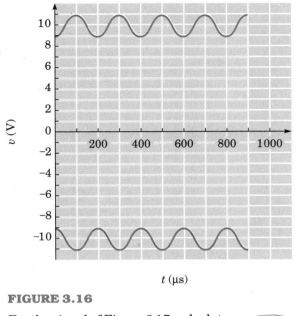

FIGURE 3.16

17. For the signal of Figure 3.17, calculate:

    (a) the index of modulation
    (b) the RMS voltage of the carrier without modulation

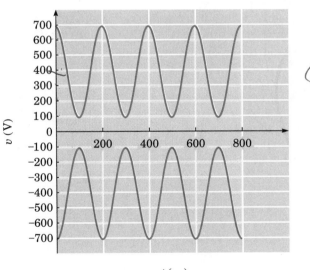

FIGURE 3.17

Section 3.4

18. An audio system requires a frequency response from 50 Hz to 15 kHz for high fidelity. If this signal were transmitted using AM, what bandwidth would it require?

19. A transmitter operates with a carrier frequency of 7.2 MHz. It is amplitude modulated by two tones with frequencies of 1500 and 3000 Hz. What frequencies are produced at its output?

20. Sketch the signal whose equation is given in Problem 12, in the frequency domain, showing frequency and voltage scales.

21. An AM signal has the following characteristics:

$$f_c = 150 \text{ MHz} \qquad f_m = 3 \text{ kHz} \qquad E_c = 50 \text{ V} \qquad E_m = 40 \text{ V}$$

For this signal, find:

(a) the modulation index
(b) the bandwidth
(c) the peak voltage of the upper side frequency

Section 3.5

22. An AM signal, observed on a spectrum analyzer, shows a carrier at +12 dbm, with each of the sidebands 8 dB below the carrier. Calculate:

(a) the carrier power in milliwatts
(b) the modulation index

23. An AM transmitter supplies 10 kW of carrier power to a 50 Ω load. It operates at a carrier frequency of 1.2 MHz, and is 80% modulated by a 3 kHz sine wave.

(a) Sketch the signal in the frequency domain, with frequency and power scales. Show the power in dBW. (If necessary, refer to the discussion of dBW in Appendix E.)
(b) Calculate the total average power in the signal, in watts and dBW.
(c) Calculate the RMS voltage of the signal.
(d) Calculate the peak voltage of the signal.

24. A transmitter with a carrier power of 10 W at a frequency of 25 MHz operates into a 50 Ω load. It is modulated at 60% by a 2 kHz sine wave.

(a) Sketch the signal in the frequency domain. Show power and frequency scales. The power scale should be in dBm.
(b) What is the total signal power?
(c) What is the RMS voltage of the signal?

Comprehensive

25. The signal of Figure 3.18 is the output of an AM transmitter with a carrier frequency of 12 MHz and a carrier power of 150 W.

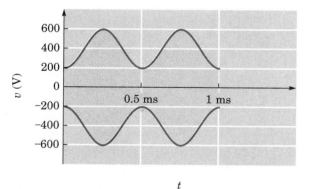

FIGURE 3.18

(a) Calculate the percent modulation.
(b) Calculate the RMS voltage of the carrier and use it to find the transmitter's load resistance.

(c) Sketch the signal in the frequency domain, with power in dBm on the vertical axis.

26. An AM transmitter with a carrier power of 20 kW is connected to a 50 Ω antenna. In order to design a lightning arrestor for the antenna, it is necessary to know the maximum instantaneous voltage that will appear at the antenna terminals. Calculate this voltage, assuming 100% modulation for the transmitter.

27. The spectrum analyzer display in Figure 3.19 represents the output of an AM transmitter. The analyzer has an input impedance of 50 Ω and is connected to the transmitter output through a 60 dB attenuator. Sketch the envelope of the signal in the time domain, as it would appear at the transmitter output terminals. Be sure to show both time and voltage scales.

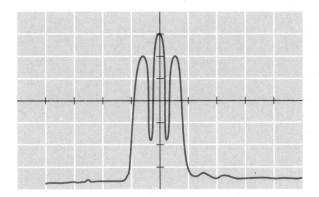

Reference Level: −10 dBm
Vertical: 10 dB/division

Center Frequency: 21.200 MHz
Span: 5 kHz/division
FIGURE 3.19

28. An AM CB transmitter with a carrier frequency of 27.005 MHz is connected as shown in Figure 3.20(a). The oscilloscope display is shown in Figure 3.20(b). Sketch the same signal as it would appear on the spectrum analyzer.

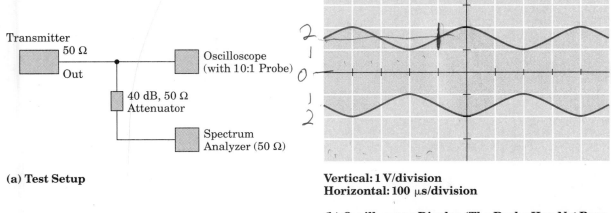

(a) Test Setup

Vertical: 1 V/division
Horizontal: 100 μs/division

(b) Oscilloscope Display (The Probe Has *Not* Been Taken into Account.)

FIGURE 3.20

4

AM Transmitters

Objectives

After studying this chapter, you should be able to:

1. Discuss the requirements and specifications of AM transmitters, and determine whether a given transmitter is suitable for a particular application.
2. Draw block diagrams for several types of transmitters, and explain their operation.
3. Analyze the operation of transmitter circuits.
4. Perform measurements on AM transmitters.

4.1 INTRODUCTION

Chapter 3 dealt with the theory behind amplitude modulation, and showed some simple modulator and demodulator circuits. This chapter covers the design of AM transmitters, using the theory from Chapter 3 and many of the circuit elements, such as oscillators and amplifiers, of Chapter 2. AM transmitter designs range from unlicensed toys developing 100 mW or less, to transmitters used in international shortwave radio broadcasting, with powers in the megawatt range. Representative AM transmitters are illustrated in Figure 4.1. Though not a particularly efficient mode of radio communication, AM has remained popular, largely because of the simplicity of receiver design associated with the mode.

Later in this book, transmitters for other communication systems, such as single-sideband and FM, will be described. Many of the principles, and a good number of the circuit features, in these transmitters resemble those in AM transmitters, so much of the material in this chapter will be useful beyond its application to AM.

4.2 TRANSMITTER REQUIREMENTS

Before looking at actual circuits, it would be well to determine what a transmitter has to do. We already know the characteristics of the AM signal that it must generate. The signal must be generated with sufficient power, at the right frequency, with reasonable efficiency, and be coupled into an antenna. The demodulated signal must be a reasonably faithful copy of the original modulating signal.

Early Transmitters

The very first transmitters used a spark gap connected directly to an antenna. The only "tuned circuit" was the antenna itself. Needless to say, the bandwidth of such a transmitter was rather wide. It resembled more the interference from automobile ignition than any radio signal transmitted today. In fact, the consensus in the earliest days of radio, up to about the turn of the century, was that only one signal could be received in a given area. In other words, frequency-division multiplexing was unknown.

The next step was to connect one or more tuned circuits between the spark gap and the antenna. The original name for this idea was "syntony." This greatly reduced the bandwidth and made it possible to transmit more than one signal in the same area, at different frequencies.

Spark-gap transmitters became more and more complex, involving multiple gaps, rotating gaps, ac supplies, and elaborate tuning systems, until they were eventually superseded by the development of vacuum tubes.

FIGURE 4.1

AM Transmitters

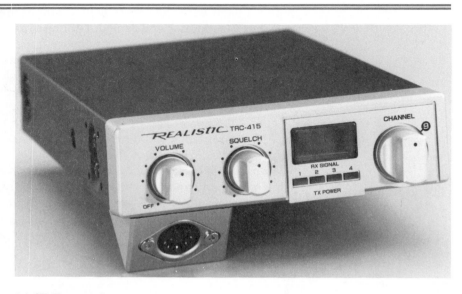

(a) CB Transceiver

(b) Broadcast Transmitter

(a) Courtesy of Radio Shack, a division of Tandy Corporation. (b) Photo provided by
Continental Electronics Corporation, Dallas, Texas, 1992.

4.2.1 Frequency Accuracy and Stability

The accuracy and stability of the transmitter frequency are essentially fixed by the oscillator. The exact requirements vary with the use to which the transmitter is put, and are set by government regulatory bodies: the Federal Communications Commission in the United States and Communications Canada in Canada, for example. Depending on the application, frequency accuracy and stability may be specified in hertz or as a percentage of the operating frequency. It is easy to convert between the two methods, as the following example shows.

EXAMPLE 4.1

A crystal oscillator is accurate within 0.005%. How far off frequency could its output be at 27 MHz?

Solution

The frequency could be out by 0.005% of 27 MHz, which is

$$27 \times 10^6 \text{ Hz} \times \frac{0.005}{100} = 1350 \text{ Hz}$$
$$= 1.35 \text{ kHz}$$

4.2.2 Frequency Agility

Frequency agility refers to the ability to change operating frequency rapidly, without extensive retuning. In a broadcast transmitter, this is not a requirement, since such stations rarely change frequency. When they do, it is a major, time-consuming operation, involving extensive changes to the antenna system as well as the retuning of the transmitter.

With other services, for example CB radio, the situation is different. Rapid retuning to any of the 40 available channels is essential to any modern CB transceiver. In addition to a frequency synthesizer for setting the actual transmitting frequency, such transmitters are required to use broadband techniques throughout so that frequency changes can be made instantly with no retuning.

4.2.3 Spectral Purity

All transmitters produce spurious signals. That is, they emit signals at frequencies other than those of the carrier and the sidebands required for the modulation scheme in use. Spurious signals are often harmonics of the operating frequency, or of the carrier oscillator if it operates at a different frequency. Any amplifier will produce harmonic distortion. Class C amplifiers, very common in AM transmitters, produce a large amount of harmonic energy. All frequencies except the assigned transmitting frequency must be filtered out to avoid interference with other transmissions.

The filtering of harmonics can never be perfect, of course, but in modern, well-designed transmitters it is very effective. As an example, Figure 4.2 shows the spectrum produced by a typical CB transmitter. Note the funda-

FIGURE 4.2

Output of a CB Transmitter in the Frequency Domain

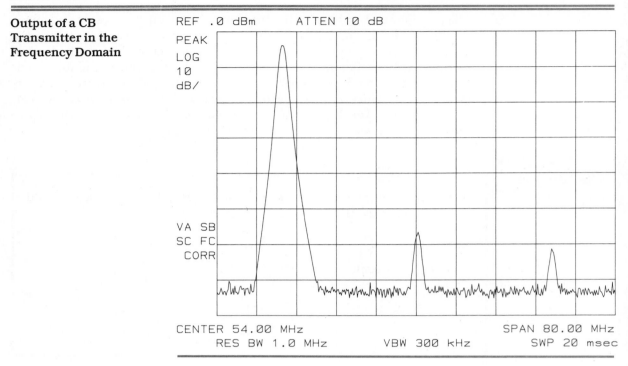

REF .0 dBm   ATTEN 10 dB

PEAK
LOG
10
dB/

VA SB
SC FC
CORR

CENTER 54.00 MHz   SPAN 80.00 MHz
      RES BW 1.0 MHz   VBW 300 kHz   SWP 20 msec

mental at about 27 MHz and the harmonics at twice and three times the fundamental frequency. How far down from the fundamental are these harmonic emissions?

4.2.4 Power Output

As pointed out in Chapter 3, there are a number of ways to measure the power in an AM signal. AM transmitters are generally rated in terms of carrier power output. Sometimes dc power input to the RF power amplifier stage has been used, because it is easier to measure. However, different amplifier efficiencies result in a considerable difference in output power for the same power input, so output power is a more useful rating.

When testing communications transmitters, the technologist should be aware of the rated **duty cycle** of the transmitter. Many transmitters designed for two-way voice communication are not rated for continuous operation at full power, since it is assumed that the operator will normally be talking for less than half the time, and probably only for a few seconds to a minute at a time. This is easy to forget when testing, and the result can be overheated transistors or other components. Broadcast transmitters, of course, are rated for continuous duty, 24 hours a day at full power.

4.2.5 Efficiency

Transmitter efficiency is important for two reasons. The most obvious one is *energy conservation*. This is especially important where very large power levels are involved, as in broadcasting, or, at the other extreme of the power-level range, where hand-held operation from batteries is required. Another reason for achieving high efficiency becomes apparent when we consider what happens to the power that enters the transmitter from the power supply but does not exist via the antenna: it is converted into heat in the transmitter, and this heat must be dissipated. Large amounts of heat require large components, heat sinks, fans, and—in the case of some high-powered transmitters—even water cooling. All of these add to the cost of the equipment.

When discussing efficiency, it is important to distinguish between the efficiency of an individual stage and that of the entire transmitter. Knowing the efficiency of an amplifier stage is useful in designing cooling systems and sizing power supplies. In calculating energy costs, on the other hand, what is important is the **overall efficiency** of the system. This is the ratio of output power to power input from the primary power source, whether it be the ac power line or a battery. Overall efficiency is reduced by factors such as tube-heater power and losses in the power supply.

4.2.6 Modulation Fidelity

As mentioned in Chapter 1, an ideal communications system allows the original information signal to be recovered exactly, except for a time delay. Any distortion introduced at the transmitter is likely to remain; in most cases it will not be possible to remove it at the receiver. It might be expected, then, that an AM transmitter would be capable of modulating any baseband frequency onto the carrier, at any modulation level from 0% to 100%, to preserve the information signal as much as possible. In practice, however, the baseband spectrum often must be restricted in order to keep the transmitted bandwidth (which is twice the highest frequency in the baseband) within legal limits.

In addition, some form of **compression**, where low-level baseband signals are amplified more than high-level signals, is often used to keep the modulation percentage high. This distorts the original signal by reducing the **dynamic range**, which is the ratio between the levels of the loudest and the quietest passages in the audio signal. The result is an improved signal-to-noise ratio at the receiver, at the expense of some distortion.

The effects of compression on dynamic range can be removed by applying an equal and opposite expansion at the receiver. Such expansion would involve giving more gain to signals at higher levels. Compression-expansion combinations are quite common in communications systems, but are not normally used with full-carrier AM.

AM communications transmitters tend to use a relatively simple **automatic-level-control (ALC) circuit**, which (as far as possible) keeps the modulation at a level approaching, but never exceeding, 100%. There is a great deal of dynamic range reduction, but this does not affect the intelligibility of speech. Broadcast transmitters, on the other hand, are often used with quite elaborate processors, applying varying amounts of compression to different frequency ranges. See Figure 4.3 for a photograph of one of these devices. They are set up very carefully to achieve the best compromise between natural-sounding music and good signal-to-noise ratio (which also means increased transmitter range).

FIGURE 4.3

**Audio Processor for AM
Broadcasting**

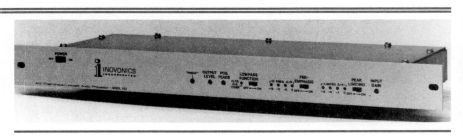

Courtesy Inovonics, Inc.

Other kinds of distortion, such as harmonic and intermodulation distortion, also have to be kept within reasonable limits. As would be expected, low distortion levels are more important in the broadcast service than the mobile-radio service.

4.3 TYPICAL TRANSMITTER TOPOLOGIES

AM transmitters can take several different forms, as shown in block diagram form in Figure 4.4. In Figure 4.4(a), an oscillator is modulated and connected directly to an antenna. This is not a practical circuit for several reasons. Oscillators are susceptible to frequency changes due to variation of any of the operating conditions of the active device used. Amplitude modulation represents such a change and, when applied to an oscillator, is likely to result in frequency modulation as well. In addition, changes in load impedance (such as would be created by ice on the antenna, for example) will also result in frequency changes. Finally, oscillators should run at very low power levels to reduce heating effects that can also result in frequency changes. Even with crystal oscillators, low power levels are necessary; otherwise, the crystal will be damaged. In spite of all these disadvantages, "transmitters" with this topology are built—but only as toys. The "wireless microphones" that allow children to talk over an AM radio are sometimes this simple.

A more practical transmitter is shown in Figure 4.4(b). A Class C amplifier stage has been added to provide isolation and power gain. The amplifier, rather than the oscillator, is modulated.

Most practical transmitters use more than one stage between oscillator and antenna, as shown in Figure 4.4(c). There are two extra stages shown. The buffer isolates the oscillator from any load changes caused by modulation of the power amplifier, and the **driver** supplies the power needed at the input to the power amplifier.

If an AM transmitter is to operate in the HF or VHF range, or higher in frequency, one or more frequency multipliers will probably be used between the oscillator and the driver, as shown in Figure 4.4(d). Operation at a relatively low frequency allows for a more stable oscillator design.

All of the designs shown so far have the RF power stage modulated. This allows all of the RF amplifying stages to operate Class C for maximum effi-

FIGURE 4.4   **AM Transmitter Topologies**

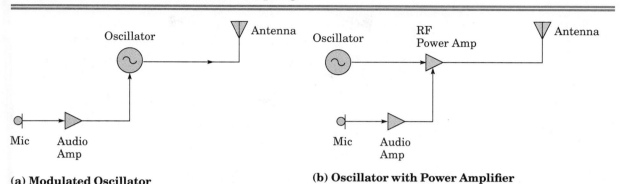

(a) Modulated Oscillator   **(b) Oscillator with Power Amplifier**

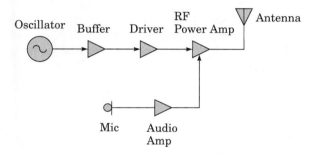

(c) Transmitter with Buffer and Driver Stages

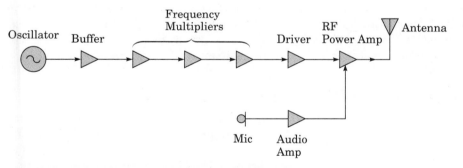

(d) VHF Transmitter with Frequency Multiplication

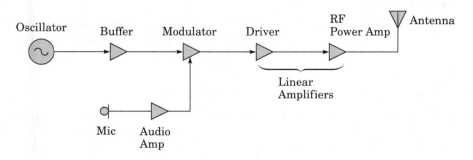

(e) Transmitter with Low-Level Modulation and Linear Amplifier

ciency. However, it is possible to apply the modulation to an earlier stage, as demonstrated in Figure 4.4(e). The drawback is that all amplifier stages following the modulator must be linear (Class A or AB). This is so wasteful of power that this topology is seen only in very low-power transmitters or in those designed mainly for another modulation scheme. Many transmitters that are designed for single-sideband suppressed-carrier operation fit this description. Single-sideband transmitters will be studied in Chapter 6.

In the balance of this chapter, it will be assumed that the modulation is applied to the RF power amplifier stage. Sometimes the driver is modulated as well, to reduce distortion.

4.3.1 Low-Level and High-Level Modulation

By definition, **high-level modulation** is accomplished in the output circuit of the RF power amplifier stage. Usually this will be the collector of a transistor or the anode of a tube. Anything else is **low-level modulation**, and is much less efficient. On the other hand, it requires a great deal less power from the modulating amplifier. The configuration shown in Figure 4.4(e), where the modulated stage is followed by a linear amplifier, is an example of low-level modulation, but so is a transmitter in which the modulation is accomplished in the base or grid circuit of the power amplifier.

4.4 TRANSMITTER STAGES

All of the stages in a transmitter except the power amplifier, and possibly the driver, operate at low power levels and are quite similar to the circuits examined in Chapter 2. Often this part of the transmitter, exclusive of the power-handling stages, is called the **exciter**.

4.4.1 The Oscillator Stage

Good stability in modern transmitter design generally requires a crystal-controlled oscillator. Where variable-frequency operation is required the usual practice is to use a frequency synthesizer locked to a crystal-controlled master oscillator. (See Chapter 7 for details of frequency synthesizers.) It is possible to use a **variable-frequency oscillator (VFO)** in applications where frequency agility is important and exact frequencies are not too important (as in the military and amateur services), but here, too, crystal-controlled synthesizers have taken over in virtually all new designs over the last few years.

4.4.2 The Buffer and Multiplier Stages

Whatever type of oscillator or synthesizer is used to generate the operating frequency, it must be isolated from any changes in load impedance in order to maintain good stability. The **buffer** stage accomplishes this. In modern designs it is likely to be a wideband amplifier to minimize adjustments when changing frequency. The buffer stage operates at low power, where efficiency is less important than spectral purity, so it is likely to operate Class A.

Multiplier stages, on the other hand, are required to generate harmonics as part of their design. They will operate Class C to ensure this, with tuned output circuits designed to pass the desired frequency and attenuate **spurious signals**.

4.4.3 The Driver Stage

Depending on the output power of the transmitter, its power-amplifier stage may require considerable power at its input. For instance, if the power amplifier of a 10·kW transmitter has a power gain of 20 dB, it will need an input power of 100 W. This calls for a power amplifier, probably operating Class C, to drive the final stage. This stage may be referred to as the **intermediate power amplifier** (IPA).

4.4.4 The Power Amplifier/Modulator

In small transmitters, a single transistor operating in Class C is likely to be used in the final amplifier, with collector modulation accomplished by means of a push-pull Class AB audio amplifier. For powers much above 100 W, it is necessary to use more than one output transistor. Simply putting transistors in parallel will not do, since the internal capacitances will add to unacceptable levels. Accordingly, it is necessary to use several complete amplifiers whose output power is then combined before being delivered to the antenna.

An alternative, still much used for high-power transmitters, is to use vacuum tubes in the final amplifier, modulating amplifier, and sometimes the driver stage as well. Vacuum tubes, while obsolete in receivers and low-powered transmitters, are still used in high-powered transmitters (in the kilowatt range and up), where they are often more cost-effective than transistors. (Vacuum tubes will be discussed in more detail in Section 4.9.)

Figure 4.5 shows a circuit for the modulated power amplifier of a low-powered AM transmitter. This circuit is similar to the AM modulator introduced in Chapter 3, which was in turn based on the Class C amplifier described in Chapter 2. It is now time to have a closer look at it.

First, the essential idea is that the modulating signal is used to vary the collector supply voltage. The most straightforward—though certainly not the only—way of doing this is to use a transformer. This also provides a means of transforming the audio amplifier's optimum load impedance to the impedance of the modulated stage. Ideally, for 100% modulation, the voltage at point A, the transistor side of the modulation transformer secondary, will vary between zero and twice the supply voltage, V_{CC}.

The power supply voltage does not change with modulation (assuming a properly designed supply), and neither does the average collector current. In fact, a milliammeter in the collector supply lead should not show any change in reading, since the variations with modulation are too rapid for the meter to follow. From this it appears that the power provided directly by the power supply to the final amplifier does not change with modulation.

Basic AM theory tells us that modulation will increase the power output of the amplifier by 50% for 100% modulation. Therefore, the power input will have to increase by the same percentage. Where does the extra power come from, if not directly from the supply? There is only one other possible source,

FIGURE 4.5

Modulated Class C Amplifier

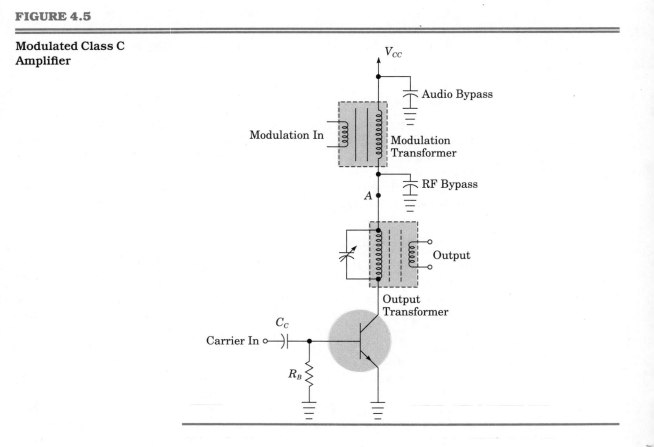

and that is the modulating amplifier, in this case assumed to be an audio amplifier. It is apparent that this amplifier will have to provide 50% of the dc carrier power input to the final amplifier. That is,

$$P_a = 0.5P_i \tag{4.1}$$

where P_a = power required from the audio amplifier
        P_i = dc power input to the final amplifier

EXAMPLE 4.2

A transmitter has a carrier power output of 10 W at an efficiency of 70%. How much power must be supplied by the modulating amplifier for 100% modulation?

Solution

The dc power input to the amplifier can be found from

$$eff = \frac{P_o}{P_i}$$

$$P_i = \frac{P_o}{eff}$$

$$= \frac{10}{0.7}$$

$$= 14.3 \text{ W}$$

The audio power required is

$$P_a = 0.5 P_i$$

$$= 7.14 \text{ W}$$

The impedance looking from the transformer secondary into the modulated stage is easy to calculate. For 100% modulation, we know from Equation (4.1) that

$$P_a = 0.5 P_i$$

P_i is just the dc power from the supply, so

$$P_i = V_{CC} I_c \qquad\qquad (4.2)$$

where V_{CC} = power supply voltage

   I_c = average collector current

Therefore,

$$P_a = 0.5 V_{CC} I_c \qquad\qquad (4.3)$$

For 100% modulation, we also know that the voltage from the modulating amplifier, when added to the supply voltage, must be capable of reducing it to zero on one peak and doubling it on the other. In other words, at the transformer secondary

$$V_{a(pk)} = V_{CC} \qquad\qquad (4.4)$$

where V_a = audio voltage measured across the modulation transformer
         secondary

Assuming the modulating signal is a sine wave,

$$V_{a(RMS)} = \frac{V_{a(pk)}}{\sqrt{2}}$$

$$= \frac{V_{CC}}{\sqrt{2}} \qquad\qquad (4.5)$$

Now, letting Z_a be the impedance seen looking into the power amplifier from the modulation transformer secondary, and using the power equation for RMS voltage and impedance:

$$P = \frac{V^2}{Z}$$

which can be rearranged to get

$$Z = \frac{V^2}{P}$$

$$Z_a = \frac{V^2_{a(RMS)}}{P_a}$$

$$= \frac{\left(\dfrac{V_{CC}}{\sqrt{2}}\right)^2}{0.5 V_{CC} I_c}$$

$$= \frac{0.5 V^2_{CC}}{0.5 V_{CC} I_c}$$

$$= \frac{V_{CC}}{I_c} \tag{4.6}$$

EXAMPLE 4.3

A transmitter operates from a 12 V supply, with a collector current of 2 A. The modulation transformer has a turns ratio of 4 : 1. What is the load impedance seen by the audio amplifier?

Solution

The impedance at the transformer secondary is, from Equation (4.6),

$$Z_a = \frac{V_{CC}}{I_c}$$

$$= \frac{12\ \text{V}}{2\ \text{A}}$$

$$= 6\ \Omega$$

This impedance is multiplied by the square of the transformer turns ratio. Letting the impedance looking into the primary be Z_p, we have:

$$Z_p = Z_a \left(\frac{N_1}{N_2}\right)^2$$

$$= (6\ \Omega)(4^2)$$

$$= 96\ \Omega$$

As noted previously, the collector voltage in an unmodulated Class C amplifier can vary between almost zero (the "almost" is due to the fact that there will be a small voltage, typically less than 1 V, across the transistor when it is saturated) to twice the supply voltage. With 100% modulation, the supply voltage can double, so the collector voltage now reaches a maximum of

$$V_{c(max)} = 4V_{CC} \qquad\qquad (4.7)$$

This fact must be borne in mind when choosing transistors and other components, such as capacitors, in the output circuit. For instance, for the amplifier in Example 4.2 above, the collector voltage can reach 48 V, even though the circuit operates from a 12 V supply.

Another factor to consider is power dissipation. Though the theoretical maximum efficiency of a Class C amplifier is 100%, the actual efficiency is unlikely to be much greater than 75%. Most of the remaining power is dissipated in the transistor, though some will be converted to heat in the output circuit and other components. Though the efficiency is likely to remain about the same with modulation, the power dissipation goes up.

EXAMPLE 4.4

A collector-modulated Class C amplifier has a carrier output power P_c of 100 W and an efficiency of 70%. Calculate the input power and the transistor power dissipation with 100% modulation.

Solution

Assume all the power dissipation occurs in the transistor. This will give a conservative answer, providing some safety margin. The output power with 100% modulation is

$$P_o = 1.5P_c$$
$$= 1.5 \times 100 \text{ W}$$
$$= 150 \text{ W}$$

The input power is

$$P_i = \frac{P_o}{eff}$$
$$= \frac{150 \text{ W}}{0.7}$$
$$= 214 \text{ W}$$

The power dissipated P_D is the difference between the input and the output power.

$$P_D = P_i - P_o$$
$$= 214 \text{ W} - 150 \text{ W}$$
$$= 64 \text{ W}$$

By the way, it was stated above that the supply voltage should not change with modulation. Sometimes it does: since the supply is required to supply up to 50% more power with modulation than without, a poorly regulated supply may provide a lower voltage when the transmitter is fully modulated. If this is the case, the carrier will have a lower amplitude with modulation than without. The phenomenon is called **carrier shift**, and is undesirable.

4.4.5 Audio Circuitry

Audio circuitry is required to amplify the very small signal from a microphone, on the order of 1 mV, to a sufficient level to modulate the transmitter. For a large transmitter of 50 kW or more, that can be a considerable amount of power.

In a communications transmitter, the audio circuitry will be straightforward. There will probably be some form of automatic level control to prevent overmodulation while keeping the modulation as close to 100% as possible. There will likely be some frequency-response shaping as well, to emphasize those frequencies in the voice range, from about 300 Hz to 3 kHz, that are most important for intelligibility. Lower frequencies contribute little to intelligibility and waste energy if transmitted, while higher frequencies waste bandwidth.

Most of the audio circuitry involved in broadcast radio is actually not in the transmitter. Studio equipment will amplify the tiny signals from microphones, phonograph cartridges, and tape heads, *mix* them (electrically, the process is really *summation*), and apply any required equalization. The transmitter must be located in an area where there is a considerable amount of land for the antenna system; the studio may be located in the same building, but is often at a more central location. In that case, the audio signal will be sent to the transmitter over a leased telephone line or a microwave radio link. The transmitter itself will receive a *line level* signal, often either +4 or +8 dBm into 600 Ω, and amplify that to the very considerable power levels needed by the modulator. The final audio stage will undoubtedly be push-pull, operating Class B, and will very often use vacuum tubes. (See Section 4.9).

4.5 OUTPUT IMPEDANCE MATCHING

Most practical transmitters are designed to operate into a 50 Ω resistive load, to match the characteristic impedance of the coaxial cable that is generally used to carry the transmitter power to the antenna. The antenna itself may not have this impedance, but, in that case, there will be matching circuitry at the antenna (or as close to it as practical). Often the match will not be exact, especially where the system must operate over a range of frequencies, and many transmitters have some adjustment latitude.

The transmitter output circuitry must be designed to transform the standard load resistance at the output terminal to whatever is required by the active device or devices. Here, the requirements vary dramatically, depending on the type of amplifier. Transistors usually require a load impedance less than 50 Ω, while tubes are suited for much higher impedances.

The simplest way to find the appropriate required load impedance for a Class C amplifier is to note that, for an unmodulated carrier, the collector voltage varies between approximately zero and $2V_{CC}$. The peak value of the output voltage is then

$$V_{o(pk)} = V_{CC}$$

$$V_{o(RMS)} = \frac{V_{CC}}{\sqrt{2}}$$

Assuming that the amplifier is being designed for a specific value of power output at a given supply voltage, we can use the usual power equation for RMS voltage and a resistive load:

$$P = \frac{V^2}{R}$$

Since we know P and V and require R, we rearrange this to read

$$R = \frac{V^2}{P}$$

Now, substitute the output voltage and power into this equation to get

$$R_L = \frac{\left(\dfrac{V_{CC}}{\sqrt{2}}\right)^2}{P_c}$$

$$= \frac{V_{CC}^2}{2P_c} \tag{4.8}$$

where R_L = load resistance seen at the collector (or **plate**, for a tube)
 P_c = carrier power output, without modulation
 V_{CC} = collector (or plate) supply voltage

EXAMPLE 4.5

An AM transmitter is required to produce 10 W of carrier power when operating from a 15 V supply. What is the required load impedance as seen from the collector?

Solution

From Equation (4.8),

$$R_L = \frac{V_{CC}^2}{2P_c}$$

$$= \frac{15^2}{2 \times 10}$$

$$= 11.25 \ \Omega$$

FIGURE 4.6

Narrowband Output
Circuit

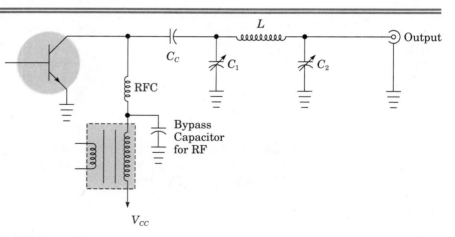

(a) Pi Network

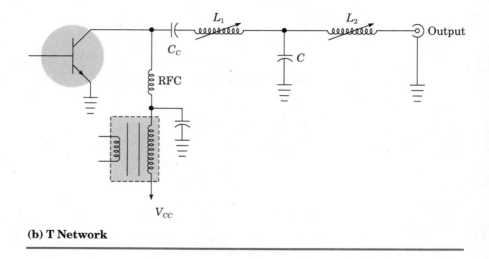

(b) T Network

For many years, a popular output circuit has been the *pi network*, shown in Figure 4.6(a). Its name stems from the resemblance of its schematic diagram to the shape of the Greek letter π. The pi network can be used to transform impedances either up or down, but is best suited to active devices that require a fairly high load impedance, such as tubes. With devices that require a very low load impedance, like bipolar transistors with more than a few watts of output, the design can be accomplished in theory, but some of the required component values are likely to be too small to be practicable.

High-power transistor amplifiers, with their requirement for small values of load impedance, can use a variety of narrowband coupling circuits. One common network is shown in Figure 4.6(b). For obvious reasons, it is called a *T network*. There are equations and charts available to aid in the design of these circuits, but an easier way is to make use of a computer program. There are many such programs; one handy one is called APPCAD,™ and is available free from Hewlett-Packard.

To be practical, both of these circuits need an extra capacitor (labeled C_c in the figures) to keep the dc voltage on the collector or plate away from the antenna. Some antenna systems represent a short circuit for dc. An RF choke is used to allow power to reach the output stage.

Besides providing impedance transformation, the pi and T networks also act as low-pass filters, aiding in the reduction of harmonic levels. Very often, they are followed by additional low-pass filtering to further attenuate harmonics and other spurious signals.

Any transmitter should operate into an impedance close to its rated load at all times. Severe mismatches, such as an open or short circuit at the load, can cause final-amplifier voltages or currents to be much higher than normal. Destruction of output transistors or other components is quite likely under these circumstances. Many solid-state transmitters have circuits to sense these conditions and automatically protect the final amplifier, either by reducing power or shutting it off completely. However, not all transmitters are so equipped, and in any case it is poor practice to tempt fate. Whether on the bench or in the field, a transmitter should always operate into a matched load. When conducting tests, a noninductive resistor of the correct value (usually 50 Ω), and capable of dissipating the transmitter's rated power, is used. Such a resistor is called a **dummy load**.

4.6   AN AM CITIZENS' BAND TRANSMITTER

A CB radio transmitter is always found as part of a transceiver, that is, a combination receiver and transmitter. Compared with a separate transmitter and receiver, transceivers are compact, convenient to install and operate, and economical due to the use of some of the same components for both the transmit and receive functions.

Figure 4.7 is the complete block diagram of a typical CB transceiver using amplitude modulation. This is the same transceiver that is shown in the photograph of Figure 4.1. Figure 4.8 shows only those stages that are used in transmitting. Some of these, such as the oscillator and the audio amplifier, are also used for receiving. (The operation of the receiver section will be described in Chapter 5.)

The oscillator is actually a frequency synthesizer, to maintain crystal-controlled frequency accuracy and stability. It operates at half the output frequency, so a doubler is the next stage after the oscillator. This is followed by two more gain stages before the RF power amplifier.

The audio circuitry consists of a microphone preamplifier, followed by an integrated-circuit amplifier, which also provides the power to the loudspeaker when the transceiver is in the receive mode. There is an automatic level control, which is designed to keep the modulation index as high as possible without permitting overmodulation. Because of this, there is no necessity for the operator to use an audio gain control while transmitting, or to monitor the modulation percentage.

The complete circuit of the transceiver is shown in Figure 4.9. If it seems

FIGURE 4.7   **Block Diagram of CB Transceiver**

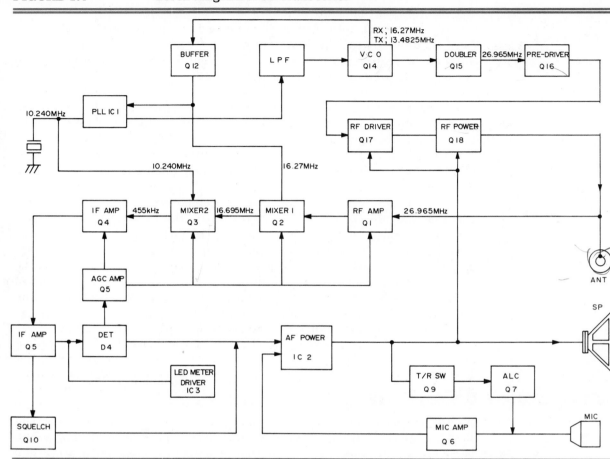

Courtesy of Radio Shack, a division of Tandy Corporation.

a bit intimidating at first, here are a few hints to put some order into it. First, locate the antenna and microphone connections. Now work your way back from the antenna through the RF chain, comparing transistor numbers on the circuit with those on the block diagram. Similarly, start at the microphone and work your way through the audio circuitry. Of course, the fact that this is a transceiver makes the circuit a little more complicated.

Some of the circuit details should be familiar by now. Note the tuned transformer-coupled output on Q_{15}, the frequency doubler. Note also the fact that the collector is tapped down on the primary winding, so as not to reduce the Q of the tuned circuit too much. (This technique was discussed in Chapter 2.)

The output circuit for the final amplifier is very similar to the T network described in the previous section, but with an extra LC section. Although this is essentially a narrowband amplifier chain, it is found that the bandwidth is sufficient to cover the entire CB frequency range from 26.965 to

FIGURE 4.8

Transmitter Section of CB Transceiver

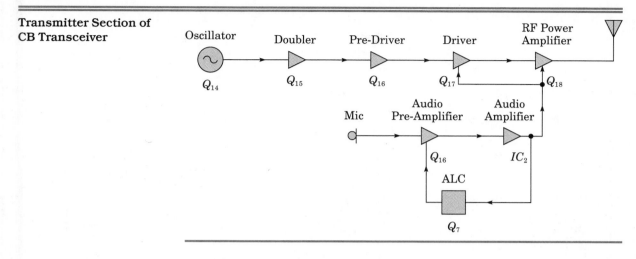

27.405 MHz when the transmitter is tuned for maximum output on Channel 18 (27.175 MHz), near the center of the band.

The audio circuitry is also interesting. The output of IC_2 modulates both the driver and the power amplifier via T_1, which is an autotransformer rather than the conventional transformer previously discussed. Modulation of the driver as well as the power amplifier is very common with transistor modulators, because otherwise the saturation voltage of the power-amplifier transistor prevents 100% modulation from being achieved. A sample of the output signal from the modulation transformer goes to the ALC circuit, which adjusts the gain of the microphone preamplifier Q_6 to keep the audio output relatively constant.

4.7 MODERN AM TRANSMITTER DESIGN

AM transmitters have been built since the invention of the vacuum tube, and, in some ways, it might be said that their design has changed little since then. Recently, however, some fresh approaches have been tried. Of course, the best approach technically might be to abandon full-carrier AM altogether, in favor of some of the more efficient systems to be described shortly. For historical and economic reasons, however, that is not likely to happen in the near future.

High-power AM transmitters are necessarily large and expensive, because of the amount of power to be handled by the power amplifier and modulating amplifier, the high-voltage power supplies required for the tubes, and the need for a modulation transformer that is capable of handling very large amounts of audio-frequency power. Recent efforts at improving AM transmitters have involved the development of high-power solid-state power amplifiers and the use of pulse-duration modulation and switching amplifiers in the modulation process. A representative example of each technique will be presented.

FIGURE 4.9

Schematic Diagram of CB Transceiver

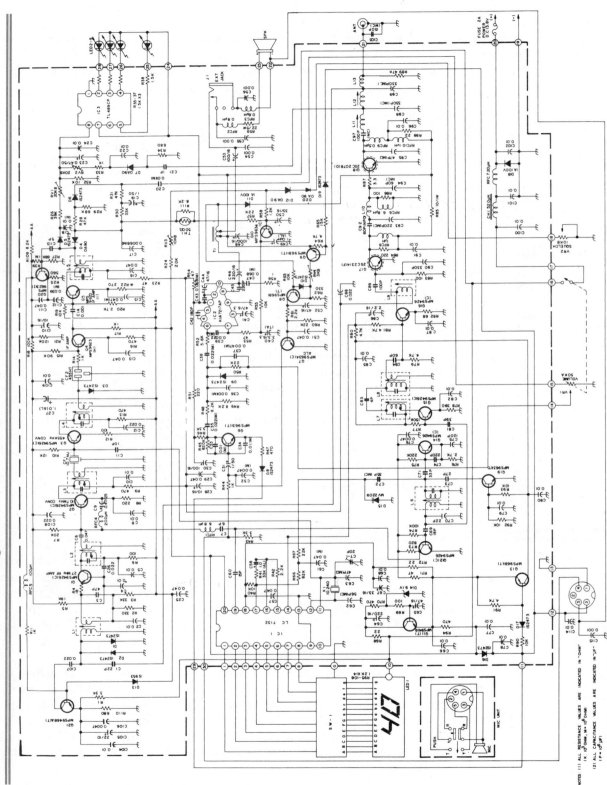

Courtesy of Radio Shack, a division of Tandy Corporation.

4.7.1 Solid-State Radio-Frequency Power Amplifiers

Transistors have been slow to take over in transmitter power amplifiers because of the greater simplicity of using one or two large tubes instead of many separate transistor amplifiers and an elaborate power combiner. The use of solid-state components can provide greater reliability and higher efficiency, however, as well as reducing the physical size of the transmitter by about 50%. The fact that many output modules are used allows operation at only slightly reduced power in the event of the failure of an output transistor, and also allows instantaneous and efficient switching of power levels. Quite often this is required of broadcast transmitters, as many station licenses require reduced power at night to avoid interference. Signals in the standard AM broadcast band tend to travel farther at night than in the daytime. See Chapter 11 for an explanation of this phenomenon.

Figure 4.10 shows the block diagram of a typical solid-state transmitter. Both the RF and audio power amplifiers are made up of a number of solid-state modules, typically six to twelve. The outputs of the RF modules must be combined before being applied to the antenna.

4.7.2 Pulse Duration-Modulators

The traditional Class B (or AB) modulator described earlier becomes very expensive when large power levels are involved. Class C, while more efficient, cannot be used for audio because of the extreme amount of distortion that would be generated. There is another way to amplify audio linearly and yet efficiently, however. Sometimes it is called Class D or *switching* amplification, but essentially it is pulse-duration modulation.

FIGURE 4.10

**High-Power Solid-State
AM Transmitter**

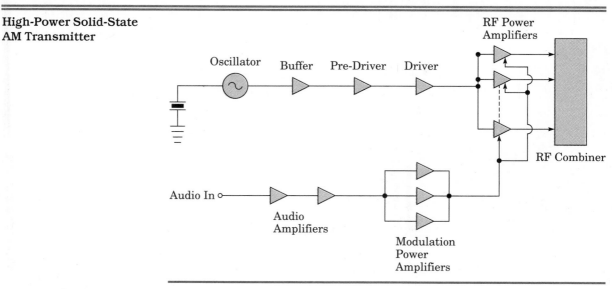

Suppose that the audio signal of Figure 4.11(a) is sampled at frequent intervals. (In theory, it must be sampled at least twice per cycle, but in practice the rate must be somewhat higher.) The result will be the discrete analog samples shown in Figure 4.11(b). Then suppose that an electronic switch is turned on for a length of time that is proportional to the amplitude of the sample. The result is *pulse-duration modulation* (PDM), also called *pulse-width modulation*, as shown in Figure 4.11(c). The pulses are produced by switching, so the theoretical efficiency so far can be 100%. (Since an ideal

FIGURE 4.11   **Pulse-Duration Modulation** , *Pulse width modulation*

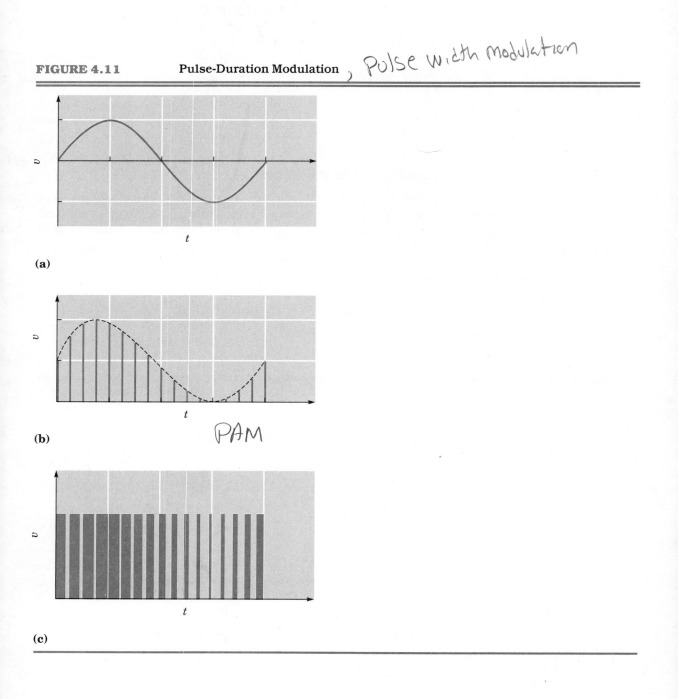

(a)

(b)   *PAM*

(c)

FIGURE 4.12

**AM Transmitter
Using Pulse-Duration
Modulation**

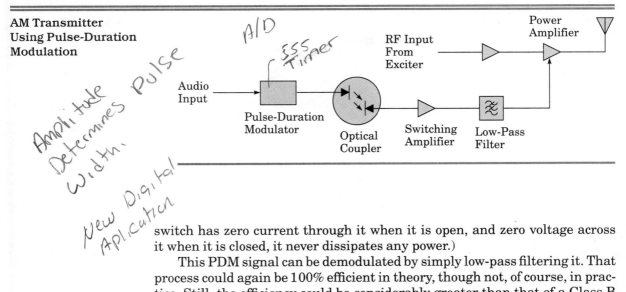

[handwritten annotations: A/D; 555 Timer; Amplitude Determines Pulse Width; New Digital Aplication]

switch has zero current through it when it is open, and zero voltage across it when it is closed, it never dissipates any power.)

This PDM signal can be demodulated by simply low-pass filtering it. That process could again be 100% efficient in theory, though not, of course, in practice. Still, the efficiency could be considerably greater than that of a Class B amplifier.

This configuration has been used occasionally in the relatively low-powered audio amplifiers found in home and commercial sound systems, but it is relatively complicated and has not become very popular. In large transmitters, however, the gain in efficiency is worth almost any amount of complication. Figure 4.12 shows one possible implementation, in which a solid-state PDM modulator is followed by a tube-type switching amplifier, which is used to modulate a vacuum-tube RF power amplifier. Note the optical coupler, which isolates the solid-state circuitry from the high voltages in the vacuum-tube part of the transmitter.

4.7.3 Transmitters for AM Stereo

When AM stereo was introduced, there were several incompatible systems, and adoption was slow. Now that the Motorola C-Quam™ system has become the standard, there is likely to be more interest in the mode.

A major concern of broadcasters who are considering improvements must always be cost. In particular, replacing a transmitter, especially the high-power portion, represents a major capital investment. Luckily, AM stereo can be implemented without replacing the transmitter power amplifier and modulator. The way this is done is by separating the quadrature AM signal into a phase and an amplitude component. The amplitude component modulates the signal in the usual way; in fact, with the C-Quam™ system, the amplitude component is simply the monaural left-plus-right signal. Phase modulation must be applied to one of the transmitter's early stages. Phase modulation will be studied in Chapter 8, but we can note at this time that since phase modulation has no effect on the amplitude of the RF signal, a phase-modulated signal can be amplified by a Class C amplifier. Thus only the low-power exciter circuitry of the transmitter must be modified. Figure 4.13 is a simplified block diagram of an AM stereo transmitter.

FIGURE 4.13

Simplified Block Diagram of AM Stereo Transmitter

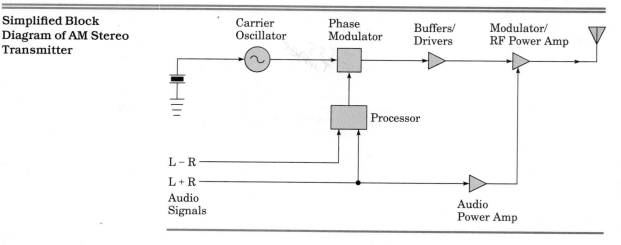

4.8 AM TRANSMITTER MEASUREMENTS AND TROUBLESHOOTING

Many of the measurements that must be made in transmitter circuitry are quite ordinary, involving the measurement of dc or low-frequency ac quantities. No special techniques are required, but a few precautions are in order.

First, there are safety requirements. Technologists who are used to solid-state circuits may be intimidated by the voltage levels (in the kilovolts) found in tube-type transmitters. They should be. These voltages often come from supplies capable of providing several amperes of current, and mishaps are very likely to be fatal. High-voltage compartments will be interlocked, so that the power is removed when they are opened, and the temptation to defeat the interlocks must be resisted. The beginner should follow the manual exactly, and should work with a more experienced person.

There is also the possibility of RF burns. The output from a multikilowatt transmitter is very obviously dangerous, but even a small transmitter is capable of inflicting painful burns if the antenna leads are touched during operation. The author found this out the hard way many years ago with a 30 W transmitter! Any operating transmitter must be treated with respect and care.

Speaking of RF, strong radio-frequency fields can cause problems when seemingly unrelated measurements are made. All electronic test equipment must be properly grounded and shielded to avoid interference.

Transmitters should always be tested off-the-air whenever possible, to avoid creating interference. Suitable loads exist for this, and are described in the next section.

4.8.1 Measurement of Radio-Frequency Power

There are many situations where it is required to measure the power in an RF circuit. Conventional wattmeters, such as are used at power-line frequencies, are useless because of the reactance of their coils. In this section, we will look at a few of the many ways in which power can be measured at high frequencies.

FIGURE 4.14

Transmitter Dummy Load

MFJ Enterprises, Inc., P. O. Box 494, Mississippi State, MS 39762.

It is possible to measure power by measuring the voltage across a known resistance, and this can certainly be done using an oscilloscope or a specially designed high-frequency ac rectifier-type voltmeter. The problem here is to be sure that the resistance is just that: resistive, and free from inductive or capacitive reactance. Ordinary carbon resistors can be used for low power at frequencies into the VHF range, but wire-wound resistors have far too much inductance. Noninductive resistors can be made for RF use; an example is shown in the photograph of Figure 4.14. This one can dissipate a kilowatt. Resistors of this sort are often called *dummy loads*, as they can be connected to a transmitter instead of an antenna for testing.

One way of measuring true power is to measure its heating effect. This has the advantages of ignoring any reactive power and of not requiring any connections of test equipment to high-frequency or high-voltage points. It is a common technique to measure the power of large transmitters operating into dummy loads. Perhaps surprisingly, a variation of the same idea can be used to measure quite small amounts (milliwatts) of power at microwave frequencies.

Figure 4.15 illustrates one way to measure power by a **calorimeter** technique. The dummy load heats the circulating water. If the rate of flow is known and controlled, the amount of heat given to the water can be calculated from the rise in temperature between input and output. Notice that the only actual measurements involved are of temperature.

When the transmitter load is a real antenna, power measurement becomes more complicated. The antenna may look like something approaching a pure resistance, but this is certainly not always the case. Antenna impedance will be discussed in detail later in the book; for now, let us just remark that it varies with frequency, and may very well have a reactive component that can be inductive at one frequency and capacitive at another. Furthermore, if the load is not exactly matched to the transmission line that feeds it, some power will be reflected from the load, and measurements of voltage or current at the transmitter end of the line can be very deceptive. This is due

FIGURE 4.15

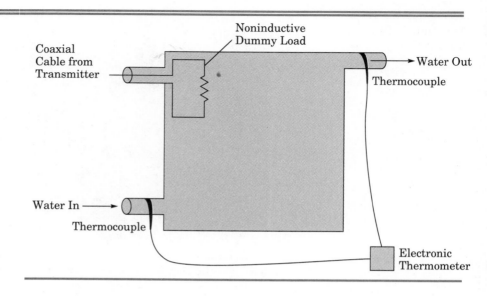

**Calorimeter Technique
for RF Power
Measurement**

to *standing waves*, which will be described in the discussion of transmission lines in Chapter 10. Broadcast stations use an RF ammeter at the antenna, which will give an accurate indication of carrier power provided that the resistive component of the antenna impedance is known.

There are meters available that have **directional couplers** which allow them to distinguish between power flowing from the transmitter toward the antenna and reflected power from the antenna. Subtracting the two readings gives the actual power delivered to the antenna. Figure 4.16 is a photograph of one such meter. The plug-in elements allow for the measurement of a wide variety of power levels at different frequencies. The direction of the power flow to be measured is given by the arrow on the plug-in element, and can be changed by simply rotating the element. The meter movement is a dc microammeter, and can easily be removed from the box for remote mounting.

It should be noted that these meters measure carrier power, and do not show the increase in power output with amplitude modulation. This is because the internal circuits sense rectified average voltage, which does not change with modulation even though average power does.

4.8.2 Measurement of Modulation Percentage

The use of an envelope display on an oscilloscope to measure AM modulation depth was discussed in Chapter 3. That technique works quite well in a laboratory situation where the modulating signal is a sine wave with constant amplitude and frequency, but it is inconvenient in actual operation where the baseband signal is more likely to be voice or music. With a random signal, the oscilloscope's horizontal sweep will not be triggered consistently. Actual overmodulation can still be seen if the display is observed carefully, most obviously as brief flashes of light at the *x* axis when overmodulation occurs in the negative direction. Another drawback to this system is that it is difficult to tell from an envelope pattern whether the modulator is distorting the sig-

FIGURE 4.16

Bird Thruline™
Wattmeter

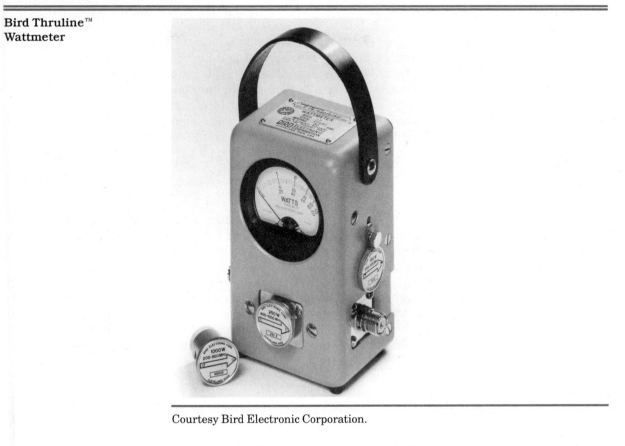

Courtesy Bird Electronic Corporation.

nal. With a constant sine wave, deviation from the sinusoidal shape may be noticeable if it is severe, but there is little hope of seeing distortion with normal, rapidly varying program material.

The setup shown in Figure 4.17(a) solves some of these problems. This time the oscilloscope is set to x–y mode. The vertical input samples the mod-

FIGURE 4.17   **Trapezoidal Method of AM Modulation Monitoring**

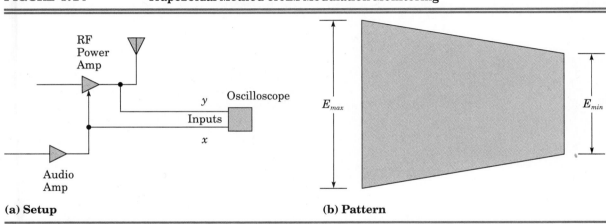

(a) Setup   **(b) Pattern**

ulated waveform as before. The horizontal input, however, is connected to the output of the modulating amplifier; that is, it receives the baseband signal just before modulation. This creates a **trapezoidal pattern** on the screen, as shown in Figure 4.17(b). This is easier to observe than the envelope pattern, because the top and bottom edges of the trapezoid are (or should be) straight lines, regardless of the shape of the modulating waveform. This is because both the vertical and horizontal amplitudes change in proportion to the amplitude of the modulating signal.

The calculation of the modulation index from this pattern is the same as for the envelope pattern. That is, as shown in Chapter 3,

$$m = \frac{E_{max} - E_{min}}{E_{max} + E_{min}} \tag{4.9}$$

Figure 4.18 illustrates the appearance of some representative values of m on the trapezoidal display. Note that for $m = 0$ the trapezoid becomes a vertical line, while for $m = 1$ it is actually a triangle.

Detecting overmodulation is easy with this setup. In the positive direction, it will simply result in the long side of the trapezoid being too long, which may not be obvious unless the oscilloscope has been properly calibrated in advance with an unmodulated carrier. In the negative direction, however, the results will be very obvious, as in Figure 4.18(d). The horizontal line extending to one side of the trapezoid (actually a triangle) indicates that

FIGURE 4.18   **Trapezoidal Patterns**

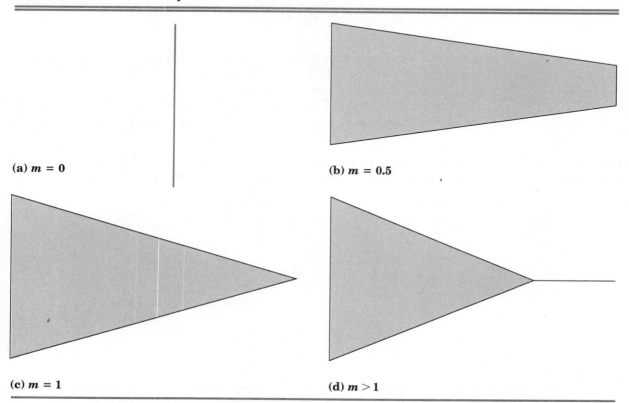

(a) $m = 0$

(b) $m = 0.5$

(c) $m = 1$

(d) $m > 1$

there is no RF output for part of the baseband cycle. By the way, it is *negative* overmodulation that is most important, because it produces spurious signals and severe distortion of the envelope. Slight overmodulation in the positive direction does not create these problems and is, in fact, sometimes employed deliberately.

EXAMPLE 4.6

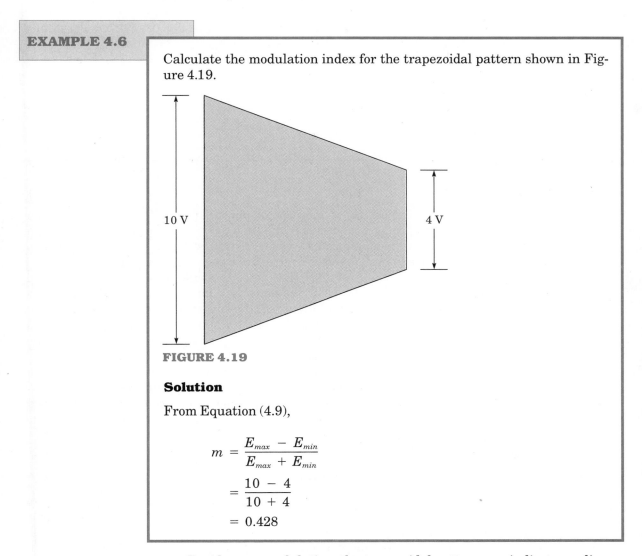

Calculate the modulation index for the trapezoidal pattern shown in Figure 4.19.

10 V

4 V

FIGURE 4.19

Solution

From Equation (4.9),

$$m = \frac{E_{max} - E_{min}}{E_{max} + E_{min}}$$

$$= \frac{10 - 4}{10 + 4}$$

$$= 0.428$$

Besides overmodulation, the trapezoidal pattern can indicate nonlinearity in the modulation stage. If the peak output signal level varies exactly in accordance with the instantaneous amplitude of the modulating signal, the top and bottom of the trapezoid will be straight lines, regardless of the waveshape of the modulating signal. Conversely, any deviation from straightness represents distortion in the modulated stage. Obviously it is easier to see deviations from a straight line than from a sinusoidal shape, and much easier than trying to notice deviations from the unknown waveform of program material. If the audio output, as heard on a receiver, is distorted, but the trapezoidal pattern is normal, then the fault must be in the audio portion of the transmitter and not in the modulator.

FIGURE 4.20

**Modulation Monitor
(Simplified)**

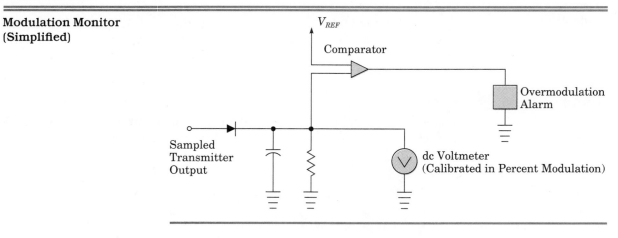

While very useful, oscilloscope displays require interpretation. For continuous use, it is more practical to have a device that reads percent modulation on a meter, and indicates overmodulation with a visible or audible alarm. One way to accomplish this is to demodulate the signal and measure the peak amplitude of the output. The detector will have to be calibrated, of course, and this can easily be done with an oscilloscope. After that, the meter will measure modulation directly. A comparator can be used with a calibrated reference voltage to operate an alarm whenever the modulation reaches a preset level. A highly simplified circuit for this type of modulation meter is shown in Figure 4.20.

If the calibrated detector is incorporated into a receiver, the modulation check can easily be done remotely, and the audio output from the receiver can be used to get a qualitative indication of signal quality. Commercial units to accomplish this are available.

4.9 VACUUM-TUBE APPLICATIONS

A vacuum tube consists of an evacuated glass or metal envelope containing two or more elements. Tubes are classified as diodes, triodes, tetrodes, pentodes, etc., according to the number of elements.

The simplest tube is the diode, consisting of a hot cathode which emits electrons by **thermionic emission**, and an anode, often called the *plate*. The cathode may have a separate heater, or it may be heated by passing a current through it. Generally, small tubes have a separate heater to minimize the modulating effect on emission of the ac used to operate the heater. Large transmitting tubes use a combined heater-cathode for efficiency.

In operation, some electrons receive sufficient thermal energy to leave the cathode. If the anode is more positive than the cathode, electrons will flow through the vacuum from cathode to anode, causing a conventional current flow in the opposite direction. There will be no current with reversed polarity, as the anode is not capable of electron emission.

FIGURE 4.21

Diode Tube

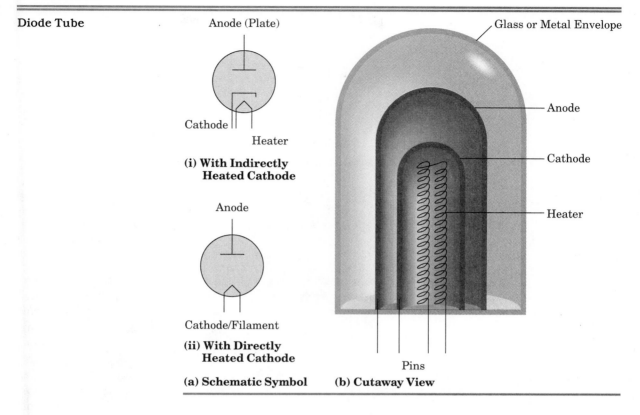

(i) **With Indirectly Heated Cathode**

(ii) **With Directly Heated Cathode**

(a) **Schematic Symbol**   (b) **Cutaway View**

Thus, the diode tube has a circuit function similar to that of a semiconductor diode, and can be used in power supplies to rectify ac, for instance.
Diode tubes are not as efficient as semiconductor diodes, however, and have
now been almost entirely superseded. Figure 4.21(a) shows the schematic
symbol for a diode. It is drawn as it is for clarity and simplicity. In fact, the
structure is generally concentric, with the cathode surrounding the heater
and close to it, and the anode forming another concentric cylinder around the
cathode. Figure 4.21(b) is a cutaway view of a small diode tube.

4.9.1 Triodes

Adding a structure of fine wire called a **control grid** in the space between
cathode and anode allows the amount of current flowing between anode and
cathode to be varied by changing the voltage on the grid. The grid is biased
at a negative potential with respect to the cathode, and therefore no current
flows in the grid circuit. (Class C amplifiers are an exception, where the grid
goes positive, and grid current flows, during a part of each cycle.) If the grid
is sufficiently negative, all electrons reaching it will be repelled back to the
vicinity of the cathode. No plate current will flow, and the tube is said to be in
cutoff. Reducing the negative grid voltage will allow more current to flow in
the anode-cathode circuit.

FIGURE 4.22

Triode Tubes

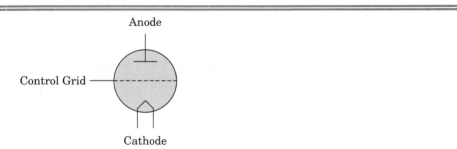

(a) Schematic Symbol (with Directly Heated Cathode)

(b) Transmitting Tubes

Photograph courtesy of Varian Power Grid Tube Products.

Figure 4.22(a) shows the schematic symbol for a triode. Figure 4.22(b) is a photograph of typical transmitting tubes.

4.9.2 Tetrodes and Pentodes

To reduce the capacitance between control grid and anode, which degrades high-frequency performance, an extra grid called the **screen grid** may be installed. The screen is grounded for the signal frequency, providing isolation, but is operated at a dc potential between that of the cathode and the anode. This allows it to aid in accelerating electrons from the cathode. Since the screen is positive with respect to the cathode, there will be a small screen current, but most of the electrons emitted by the cathode will pass through to reach the anode. The schematic symbol for a tetrode is shown in Figure 4.23(a).

FIGURE 4.23 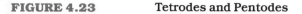 **Tetrodes and Pentodes**

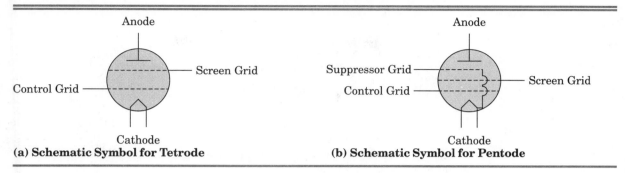

(a) Schematic Symbol for Tetrode

(b) Schematic Symbol for Pentode

Further isolation can be achieved by inserting a **suppressor grid** between screen and anode. This grid is usually at the same potential as the cathode, and is often connected internally to it. It is therefore strongly negative with respect to the anode. In addition to further reducing interelectrode capacitance, it serves to repel any secondary electrons emitted by the anode, back to it. **Secondary emission** occurs when electrons strike the anode with sufficient energy to dislodge other electrons from it. This can be a problem with tetrodes due to the velocity with which electrons strike the anode. Figure 4.23(b) shows the symbol for a pentode.

Some tetrodes, called *beam power tetrodes*, are designed in such a way that the electrons form narrow beams on their way from cathode to anode. These tubes allow large currents at relatively low voltages, and also make the use of a suppressor grid unnecessary.

4.9.3 Vacuum-Tube Specifications

A vacuum tube is a voltage-controlled current source, as is an FET. As with FETs, the most important ac specification is mutual conductance g_m, followed by plate resistance r_p, which is the equivalent of drain resistance r_d for an FET.

Mutual conductance is defined as

$$g_m = \frac{\Delta i_p}{\Delta v_g} \quad \text{with } V_p \text{ constant} \tag{4.10}$$

where g_m = mutual conductance in siemens

Δi_p = a small change in plate current in amperes

Δv_g = a small change in grid-to-cathode voltage in volts

V_p = plate-to-cathode voltage in volts

Mutual conductance, also called *transconductance*, should be familiar from FETs. It has essentially the same meaning with tubes; that is, it is the ratio of a small change in output current to the small change in input voltage that causes it.

The *plate resistance* of a tube is the ac resistance of the plate circuit, and is defined as

$$r_p = \frac{\Delta v_p}{\Delta i_p} \quad \text{with } V_g \text{ constant}$$

where r_p = plate resistance in ohms

Δv_p = a small change in plate voltage in volts

V_g = grid-to-cathode voltage in volts

and the other quantities are as defined above. (If this reminds you of the drain resistance r_d of an FET, it should.)

Another common vacuum-tube specification is the amplification factor μ, which is defined as follows:

$$\mu = \frac{\Delta v_p}{\Delta v_g} \quad \text{with } I_p \text{ constant} \tag{4.11}$$

where I_p = plate current in amperes

That is, the amplification factor describes the maximum variation in plate-to-cathode voltage for a small change in grid voltage.

The three specifications given above are not independent. They are related by the equation

$$\mu = g_m r_p \tag{4.12}$$

(which should also remind you of FET theory).

Tube amplifiers can be operated in Class A, B, or C, just like transistors, and with similar results in terms of distortion and efficiency. Actual circuits will differ considerably, of course, because of different bias requirements and the much higher voltage and impedance levels typical of tubes. Since tubes are used almost exclusively as power amplifiers in modern equipment, it would be worthwhile to look at two representative examples of vacuum-tube power amplifiers: a Class C modulator and a Class B audio amplifier.

4.9.4 Vacuum-Tube Modulators

As already mentioned, vacuum tubes are often used in the driver and power amplifier stages of a large AM transmitter. These amplifiers will operate in Class C for best efficiency. Figure 4.24 shows a typical circuit for a Class C plate-modulated power amplifier using a tube.

The RF input signal from the driver is transformer-coupled to the grid of the tube by T_2, which is tuned by C_1 to the operating frequency. Note the neutralizing capacitor C_N, which prevents oscillation due to the feedback via the plate-to-grid capacitance. C_N would probably be unnecessary with a tetrode or pentode, because of the reduced internal plate-to-grid capacitance.

Bias is supplied via the negative supply V_{GG}. For Class C, the tube will be biased at a voltage greater than cutoff, so the tube conducts only on input peaks. These peaks will actually drive the grid positive, so that a small amount of grid current flows.

FIGURE 4.24

Vacuum-Tube
Modulator

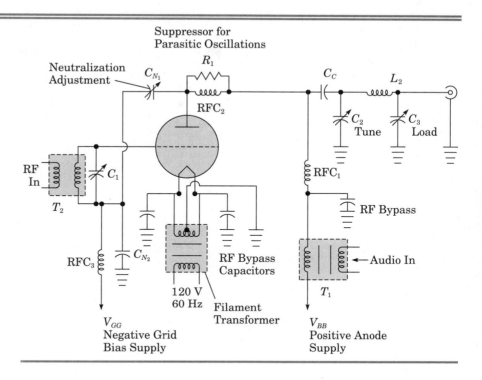

This tube, like most transmitting tubes, has a directly heated cathode. Note the way power is provided to this, from a separate power-transformer winding. Since the cathode in this circuit is at ground potential, no special insulation is required in the transformer.

The plate circuit uses a pi network to match the relatively high impedance at the anode to a 50 Ω load impedance. Modulation is provided in the usual way, by using a transformer to vary the anode voltage.

4.9.5 Vacuum-Tube Audio Amplifiers

There is a need for very high audio powers in a large AM transmitter. Recall from Chapter 3 that high-level modulation requires the modulating amplifier to provide a power equal to one-half of the input power to the final amplifier. Since no real amplifier has an efficiency of 100%, this means that the modulating amplifier will have to supply a power greater than half the output power of the transmitter: probably at least 60% of that power. This would mean an audio power output of 6 kW for the modulator of a rather ordinary 10 kW transmitter. Such large amounts of power can be produced by solid-state amplifiers using a large number of output transistors, but tubes are still commonly used. Of course, modulating amplifiers must be linear. Figure 4.25 shows a circuit for a push-pull Class B amplifier using tubes. This amplifier could be used to supply the audio power for the circuit of Figure 4.24.

FIGURE 4.25

Tube-Type Class B
Audio Amplifier

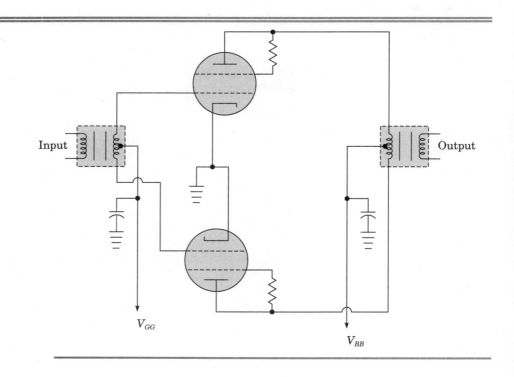

For variety, this circuit uses tetrodes with indirectly heated cathodes. The heater connections are not shown. This stage is transformer-coupled at both input and output. A phase-splitter circuit could be used at the input instead of a transformer.

SUMMARY OF CHAPTER 4

Here are the main points to remember from this chapter.

1. An AM transmitter is required to generate a signal at the correct carrier frequency, with the right power level, and with a modulation envelope that is an accurate reflection of the original information signal.

2. It is important in the design of any transmitter to minimize the presence of spurious signals at the output.

3. The efficiency of a transmitter is an important specification, particularly in view of the large power levels often used.

4. A typical AM transmitter has a crystal-controlled or frequency-synthesized oscillator followed by several stages of amplification. Usually the final amplifier stage, called the power amplifier, is modulated. This allows all stages to operate Class C.

5. High-level modulation is used in most AM transmitters for greater efficiency. It involves modulation at the output element of the output stage of the transmitter. Any other type of modulation is low-level modulation.

6. For conventional high-level modulation, the modulating amplifier must have a power output equal to one-half the dc power input to the transmitter's output stage, in order to achieve 100% modulation.

7. Transmitters require an impedance-matching circuit between the output of the power amplifier and the antenna. A properly designed matching circuit can also act as a filter to prevent spurious frequencies from being radiated.

8. In an effort to improve the overall efficiency of AM transmitters, designers have used innovative techniques such as pulse-width modulation in modulator circuits.

9. Techniques for the measurement of transmitter output power include calorimeter wattmeters, directional couplers, and RF ammeters, among others.

10. Oscilloscopes can be used for modulation monitoring, but in a practical situation a calibrated detector is easier to use.

11. Though solid-state devices are used in most modern transmitter circuits, high-powered amplifier stages still make use of vacuum tubes.

IMPORTANT EQUATIONS

$$P_a = 0.5P_i \tag{4.1}$$

$$P_a = 0.5V_{CC}I_c \tag{4.3}$$

$$V_{a(pk)} = V_{CC} \tag{4.4}$$

$$Z_a = \frac{V_{CC}}{I_c} \tag{4.6}$$

$$V_{c(max)} = 4V_{CC} \tag{4.7}$$

$$R_L = \frac{V_{CC}^2}{2P_c} \tag{4.8}$$

$$m = \frac{E_{max} - E_{min}}{E_{max} + E_{min}} \tag{4.9}$$

$$g_m = \frac{\Delta i_p}{\Delta v_g} \quad \text{with } V_p \text{ constant} \tag{4.10}$$

$$\mu = \frac{\Delta v_p}{\Delta v_g} \quad \text{with } I_p \text{ constant} \tag{4.11}$$

$$\mu = g_m r_p \tag{4.12}$$

GLOSSARY

automatic-level-control (ALC) circuit   a circuit for keeping the amplitude of a signal within prescribed limits

buffer   an amplifier stage used to isolate two stages. Often used to prevent changes in load impedance from affecting the frequency of an oscillator

calorimeter wattmeter   device for measuring power by sensing a change in temperature

carrier shift   change in amplitude of the carrier component of an AM signal with modulation

compression   system that provides more gain for low-level signals than for higher-level signals

control grid   the element in a vacuum tube that is used to vary the flow of electrons through the tube

directional coupler   device which allows a signal moving along a transmission line in one direction to be measured

driver   amplifier which supplies the required input power for a power amplifier

dummy load   a noninductive power resistor used to simulate an antenna or loudspeaker when testing a transmitter or audio power amplifier respectively

duty cycle   ratio of time in use to total time for an electronic system

dynamic range   ratio between largest and smallest signal present at a point in a system; usually expressed in decibels

exciter   the stages of a transmitter that operate at low power levels

frequency agility   ability of a transmitter to tune rapidly from one operating frequency to another

high-level modulation   amplitude modulation of the output element of the output stage of a transmitter

intermediate power amplifier (IPA)   the driver stage of a large transmitter

low-level modulation   modulation of a transmitter at any point before the output element of the output stage

overall efficiency   ratio of the power output of a device such as a transmitter to the total power required from the primary power source

plate   the anode of an electron tube

screen grid   an element of a vacuum tube used to shield the control grid from the anode

secondary emission   emission of electrons from the anode of a vacuum tube due to the energy with which other electrons strike it

spectral purity   absence of spurious signals in the output of a transmitter

spurious signal   any emission from a transmitter other than the carrier and the sidebands required by the modulation scheme in use (in suppressed-carrier systems the carrier is also a spurious signal)

suppressor grid   an element in a vacuum tube designed to reduce the effects of secondary emission

thermionic emission   emission of electrons from the heated cathode of a vacuum tube due to thermal energy

trapezoidal pattern   a means of measuring amplitude modulation using an oscilloscope

variable-frequency oscillator (VFO)   an oscillator whose frequency can be changed easily and quickly, by adjusting a variable capacitor or inductor, or by changing the bias voltage on a varactor

QUESTIONS

1. What is frequency agility, and under what circumstances is it desirable?
2. Why is it necessary to suppress the emission of harmonics and other spurious signals by a transmitter?
3. What is meant by the overall efficiency of a transmitter?
4. Why is compression used with many AM transmitters?
5. Why is it poor practice to apply amplitude modulation directly to an oscillator?
6. What is the most common way of rating the power output of an AM transmitter?
7. What is meant by the duty cycle of a transmitter?
8. Is it possible to use Class C amplifiers to amplify an AM signal? Explain.
9. What advantages does the use of a frequency synthesizer for the oscillator stage of a transmitter have over:
   (a) a crystal-controlled oscillator?
   (b) a VFO?
10. It is possible to apply a modulating signal to the control grid of a tube or the base of a transistor, used as the power amplifier (output) stage of a transmitter. State whether this procedure is considered to be high- or low-level modulation, and explain.
11. Explain the function of each of the following stages in a transmitter: buffer; multiplier; driver.
12. What is carrier shift, and how is it caused?
13. Draw the schematic diagram for, and explain the operation of, a triode tube.
14. What is the function of the screen grid in a tetrode, and how is it connected?
15. What is the function of the suppressor grid in a pentode, and how is it connected?
16. Why is it undesirable to connect a large number of transistors in parallel in an RF power amplifier to achieve a greater power-dissipation capability?
17. What is meant by a switching audio amplifier?
18. Explain the operation of a calorimeter-type power meter.
19. What is a dummy load, and how is it useful?
20. What advantages does the trapezoidal method of modulation measurement have over the envelope method?

PROBLEMS

Section 4.2

21. The oscillator of a CB transmitter is guaranteed accurate to $\pm 0.005\%$. What are the maximum and minimum frequencies at which it could actually be transmitting, if it is set to transmit on channel 20, with a nominal carrier frequency of 27.205 MHz?

22. A CB transmitter is rated to supply 4 W of carrier power to a 50 Ω load, while operating from a power supply that provides 13.8 V. The nominal supply current is 1 A without modulation and 1.8 A with 80% modulation. Calculate the overall efficiency of this transmitter, with and without modulation.

23. The audio frequency response of an AM transmitter, measured from microphone input to the secondary of the modulation transformer, is 3 dB down at 3 kHz from its level at 1 kHz. If a 1 kHz signal modulates the transmitter to 90%, what will be the modulation percentage due to a 3 kHz signal with the same level at the input?

Section 4.3

24. Draw a block diagram for an AM transmitter using high-level modulation, with an oscillator, buffer, driver, and power amplifier. Indicate the probable class of operation of each amplifier stage.

25. Draw a block diagram for an AM transmitter with an oscillator, buffer, pre-driver, driver, and power amplifier. Modulation is applied to the pre-driver stage. Indicate the probable class of operation of each amplifier stage.

Section 4.4

26. The RF power amplifier stage of a transmitter has an output of 50 kW and a gain of 15 dB. How much power must be supplied to this stage by the previous stage?

27. The power amplifier of an AM transmitter has an output carrier power of 25 W and an efficiency of 70%, and is collector modulated. How much audio power will have to be supplied to this stage for 100% modulation?

28. If the transmitter in the previous question operates from a 24 V supply, what will be the impedance seen looking into the power amplifier from the modulation transformer secondary?

29. A transistor RF power amplifier, operating Class C, is designed to produce 30 W output with a supply voltage of 50 V.
    (a) If the efficiency of the stage is 70%, what is the average collector current?
    (b) Assuming high-level modulation, what is the impedance seen by the modulation transformer secondary?
    (c) What power output would be required from the audio stages for 100% modulation of the amplifier?
    (d) What is the maximum voltage that appears between collector and emitter of the transistor?

Section 4.5

30. Calculate a suitable load impedance for the amplifier in Problem 29. Sketch a suitable matching network to allow this amplifier to drive a 50 Ω load (component values not required).

Section 4.7

31. A transmitter uses twelve modules in its solid-state output stage. Calculate the power reduction in decibels that would occur if one module fails.

32. What would be the minimum allowable sampling rate for a pulse-duration modulator if it is required to handle a baseband frequency range from 50 Hz to 10 kHz?

Section 4.8

33. An in-line wattmeter, connected in the transmission line between a transmitter and its antenna, reads 50 W in the forward direction and 30 W in the reverse direction. How much output power is actually being produced?

34. Calculate the percent modulation for each of the trapezoidal patterns in Figure 4.26.

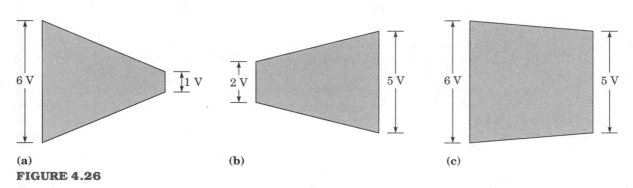

(a)   **(b)**   **(c)**

FIGURE 4.26

35. The circuit of Figure 4.27 can be used to indicate overmodulation.

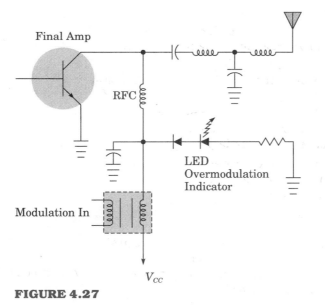

FIGURE 4.27

(a) Explain how it works.
(b) Why will it not show very small amounts of overmodulation?

Section 4.9

36. A transmitter power amplifier uses a tube. It has an average plate current of 2.5 A at a supply voltage of 10 kV. The carrier power output into the antenna is 20 kW.

(a) What is the efficiency of this stage (neglecting heater power)?
(b) How much audio power must be provided to plate-modulate this stage?
(c) What impedance is seen by the secondary of the modulation transformer?
(d) This transformer uses a pi network for coupling to the antenna. Would 20 kV be a sufficiently high voltage rating for the coupling capacitor between the anode and the pi network? Explain.

37. A certain tube has a plate impedance of 100 kΩ and an amplification factor of 50 at its operating point. Calculate its transconductance.

Comprehensive

38. A transmitter operates into a 50 Ω resistive load. The RMS voltage measured at the output is 250 V without modulation and 300 V with modulation, using a true-RMS reading meter. Find:

    (a) the power with modulation
    (b) the power without modulation
    (c) the modulation index
    (d) the peak voltage with modulation
    (e) the overall efficiency with modulation if the transmitter draws 3 kW from the ac line when modulated
    (f) the amount of power that will be drawn from the ac line when the transmitter is unmodulated, if the efficiency is the same as in Part (e)

39. Use a computer program such as APPCAD™ to solve the following problems. In each case, try several configurations, and choose one with reasonable component values.

    (a) Match a 10 ohm source to a 50 ohm load at 1 MHz.
    (b) Match a 1000 ohm source to a 50 ohm load at 10 MHz.

AM Receivers

Objectives

After studying this chapter, you should be able to:

1. Describe the basic superheterodyne system, and explain why it is the preferred design for most receivers.
2. Distinguish between single- and multiple-conversion receivers, and decide which type would be more suitable for a given application.
3. Choose suitable intermediate frequencies, and calculate image rejection for a receiver.
4. Explain the requirements for each stage in a receiver, and suggest suitable types of circuits to fulfill the requirements.
5. Test and troubleshoot receivers.
6. Analyze specifications for receivers, and use them to determine suitability for a given application.
7. Analyze a circuit diagram for a receiver or transceiver to find the function of each stage.

5.1 INTRODUCTION

In Chapter 3, we described the AM signal and discussed a simple demodulator that would recover the original baseband signal. It was pointed out that a practical receiver would also need means of amplifying the signal and of limiting the bandwidth to reduce noise and interference. The time has now come to see how this can be done.

Most of the emphasis in this chapter is on receivers for AM signals, because that is the only modulation scheme that has been described so far. However, the principles introduced here will be used again and again, with all types of modulation. In the chapters that follow, we will only have to look at the modifications that are necessary for different applications.

There are two important specifications that are fundamental to all receivers. **Sensitivity** is a measure of the signal strength required to achieve a given signal-to-noise ratio, and **selectivity** is the ability to reject unwanted signals at frequencies different from that of the desired signal. An understanding of these concepts will help you to follow the discussion of receiver types that follows. The exact mathematical definitions and the techniques for measuring these parameters differ depending on the application, and will be

---

Early Receivers

The very first receiver, used by Heinrich R. Hertz for his experiments in 1887, was a loop of wire with a small spark gap in the center. This was obviously not sensitive enough to be useful outside the laboratory. The first practical receiving device was the *coherer*, first used by Edouard Branly (1844–1940) in 1890. It consisted of a tube of metal filings through which the RF signal was passed, along with a dc current. The signal caused the filings to adhere together, or "cohere," reducing the resistance. This change in resistance increased the dc current, which activated a telegraph sounder. Unfortunately, the particles remained stuck together after the RF signal was removed, and the tube had to be tapped periodically to ascertain whether the signal was still present. Nonetheless, the coherer could be used to detect radiotelegraph signals. An improved version of the coherer was used for Guglielmo Marconi's first marine radio installations around the turn of the century. It was connected across a tuned circuit, and had an automatic "tapper" to separate or "decohere" the metal filings between the dots and dashes of the Morse code.

The coherer could not demodulate AM signals, and was quickly replaced by solid-state detectors made of galena, a semiconductor, and a thin wire "cat's whisker" that was carefully adjusted to touch a sensitive spot on the crystal, forming, in effect, a point-contact diode. Vacuum tubes soon replaced these early "crystal" detectors, since the tube could provide gain. It may be the only time in history that a vacuum tube replaced a solid-state device. One can only wonder what would have happened if, instead of devoting all its resources to the new vacuum tube, the electronics industry had developed the potential inherent in the early and unreliable "cat's whisker" detector. Perhaps the transistor would have been invented thirty years sooner.

described later in this chapter, along with several other important specifications.

5.2   RECEIVER TYPES

Almost all modern receiver designs use the **superheterodyne** principle, which will be described shortly. However, in order to recognize its advantages, we should look first at some more straightforward methods.

The simplest conceivable receiver would be an envelope detector (of the sort described in Chapter 3) connected directly to an antenna, as in Figure 5.1(a). Any AM signal arriving at the antenna would be demodulated, and the detector output would be connected to sensitive headphones. Since this is a passive circuit, only strong signals received by a good antenna could be heard at all. In addition, this receiver would have no ability to discriminate against unwanted signals and noise, and so would receive all local AM stations at once. Obviously the results would not be at all satisfactory.

This receiver could be improved by adding a tuned circuit at the input, as shown in Figure 5.1(b). This would provide some selectivity; that is, the receiver could be tuned to a particular station. Signals at the resonant frequency of the tuned circuit would be passed to the detector, and those at other frequencies would be attenuated. However, there is still no gain. Receivers of this type are often called *crystal radios* ("crystal" being an early term for a semiconductor diode). They can still be found as toys.

FIGURE 5.1   **Simple Receivers**

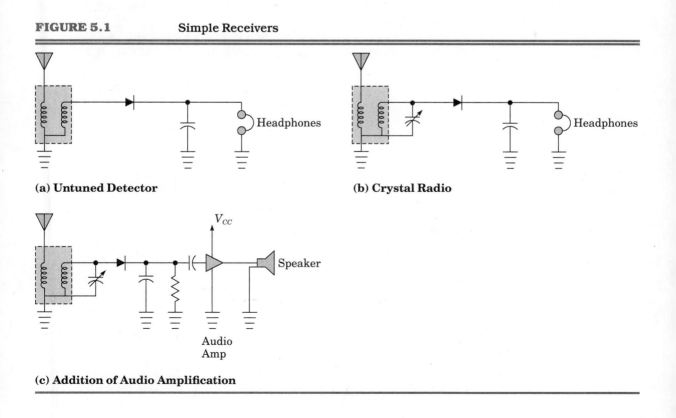

(a) Untuned Detector   **(b) Crystal Radio**

(c) Addition of Audio Amplification

The addition of an audio amplifier, as shown in Figure 5.1(c), could provide enough output power to operate a speaker. However, the selectivity will remain poor because of the single tuned circuit, and the receiver will not be sensitive enough to receive weak signals. This is because a diode detector needs a relatively large input voltage to operate efficiently, with low noise and distortion. More sensitive detectors can be devised, but a better solution is to provide gain before the detector.

5.2.1 The Tuned-Radio-Frequency Receiver

Figure 5.2 shows a block diagram for a **tuned-radio-frequency (TRF) receiver**. Several RF amplifiers, each tuned to the signal frequency, provide gain and selectivity before the detector. An audio amplifier after the detector supplies the necessary power amplification to drive the speaker.

The problems with this receiver are with the RF stages. To achieve satisfactory gain and selectivity, several stages will probably be needed. All of their tuned circuits must tune together to the same frequency, or **track** very closely, which tends to cause problems both electrical and mechanical. Having several high-gain tuned amplifier stages in close physical proximity is likely to lead to oscillation, since it is difficult to prevent feedback. In addition, some means must be provided to tune all the circuits simultaneously. The usual way, when TRF receivers were popular, was to use several variable capacitors connected together mechanically, either *ganged* (mounted on the same shaft) or linked by belts or gears. Component tolerances will cause the circuits to tune to slightly different frequencies when their capacitors are at the same position, unless the frequencies are corrected with small adjustable capacitors called *trimmers* and *padders*. Even with these adjustments, the tracking will never be perfect.

Another problem arises from the fact that the bandwidth of a tuned circuit does not remain constant as its resonant frequency is changed. At first glance this may seem surprising, since for an inductor,

$$Q = \frac{X_L}{R} \tag{5.1}$$

FIGURE 5.2

TRF Receiver

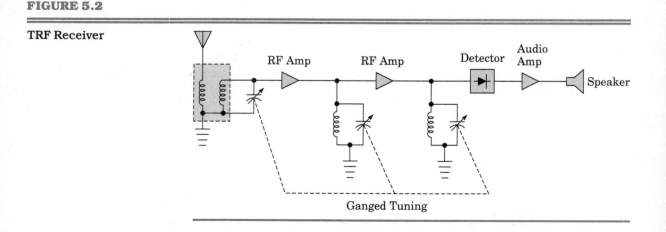

Ganged Tuning

and X_L increases directly with frequency. The required Q to achieve a given bandwidth also varies directly with frequency, as can be seen from the equation

$$B = \frac{f_o}{Q} \tag{5.2}$$

However, in practice the resistance of a coil will also increase with frequency. This is caused by the **skin effect**. At higher frequencies, internal magnetic fields in the wire cause the current to flow mainly in the region near the surface of the conductor. This decreases the effective cross-sectional area of the conductor, increasing its resistance. The resistance varies with the square root of frequency. Therefore, the bandwidth of a tuned circuit increases approximately with the square root of frequency.

This means that a receiver with the correct selectivity at the low end of its tuning range will have too wide a bandwidth at the high end. An example will make this clear.

EXAMPLE 5.1

A tuned circuit tunes the AM radio broadcast band (from 540 to 1700 kHz). If its bandwidth is 10 kHz at 540 kHz, what is it at 1700 kHz?

Solution

The bandwidth varies with the square root of frequency. Therefore, at the high end the bandwidth will increase to:

$$B = 10 \text{ kHz} \times \sqrt{\frac{1700}{540}}$$

$$= 17.7 \text{ kHz}$$

Assuming that the bandwidth is correct at the lower end, there may well be interference from adjacent stations at the top end.

For high signal frequencies and narrow bandwidths, it may not be possible to obtain satisfactory results with a reasonable number of RF stages, because of limitations on the Q of conventional tuned circuits. Circuits with higher Q, such as **crystal filters**, are not practical because of the difficulty in operating them over a wide range of frequencies. In addition, most active devices show a reduction of gain with increasing frequency.

The TRF system is now used only for simple, fixed-frequency receivers, where most of its disadvantages do not apply.

5.2.2 The Superheterodyne Receiver

The **superheterodyne receiver** or *superhet* was invented by Edwin H. Armstrong (1890–1954) in 1918, and is still almost universally used, in many variations. Figure 5.3 shows its basic layout.

FIGURE 5.3

Basic Superheterodyne
Receiver

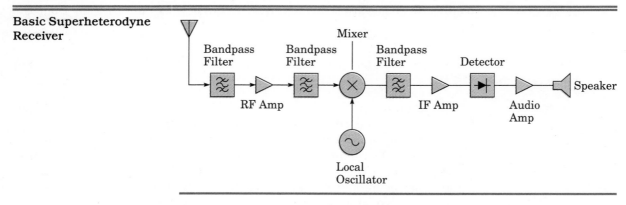

There may be one or more stages of RF amplification. The RF stage may be tuned, as in the TRF receiver, or may be broadbanded. This stage should have a good noise figure as, being the first stage in the receiver, it is largely responsible for the noise performance of the entire system. Low-cost receivers sometimes omit the RF amplifier, but do include some sort of input filter, such as a tuned circuit. The input filter and RF stage (or mixer, if there is no RF stage) are sometimes referred to as the **front end** of a receiver.

The next stage is a mixer. The signal frequency is mixed with a sine-wave signal generated by an associated stage called the **local oscillator**. A difference frequency is created, which is called the **intermediate frequency** (IF). The local oscillator is tunable, so that the IF is fixed, regardless of the signal frequency. The combination of mixer and local oscillator is known as a **converter**.

Figure 5.4 shows the signal and local oscillator frequencies for a typical AM broadcast receiver with an IF of 455 kHz. In Figure 5.4(a), the frequency of the desired signal is 740 kHz and the local oscillator is set to 1195 kHz. The difference frequency is 455 kHz, and this is passed to the next stage. The mixer will also produce a sum frequency of 1650 kHz, but this is easily re-

FIGURE 5.4   Signal and Local-Oscillator Frequencies

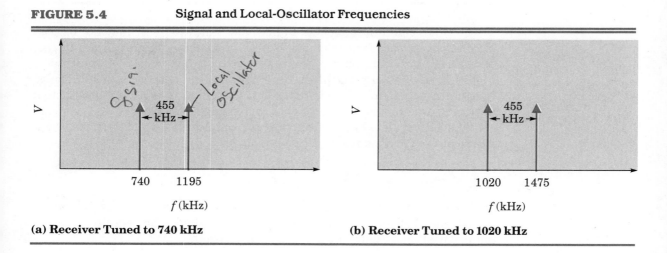

(a) Receiver Tuned to 740 kHz   (b) Receiver Tuned to 1020 kHz

moved by filtering. In Figure 5.4(b), the receiver is tuned to 1020 kHz. This involves retuning the local oscillator to 1475 kHz, so that the difference frequency is still 455 kHz.

Receivers with conventional variable-capacitor tuning usually use a two- or three-gang tuning capacitor. One section tunes the local oscillator and the other section or sections tune the mixer input circuit and the input circuit for the RF amplifier (if present).

The mixer is followed by the IF amplifier, which provides most of the receiver's gain and selectivity. Generally there are two or more IF stages, with selectivity provided either by resonant circuits or, in more advanced designs, a **crystal filter** or a **ceramic filter**. The use of a fixed IF greatly simplifies the problem of achieving adequate gain and selectivity.

The remainder of the receiver is straightforward, and resembles the TRF design. There is a detector to demodulate the signal, and an audio amplifier to increase the signal power to the level required to operate a loudspeaker.

The **automatic gain control (AGC)** adjusts the gain of the IF—and sometimes the RF—stages in response to the strength of the received signal, providing more gain for weak signals. This allows the receiver to cope with the very large variation in signal levels found in practice. This range may be more than 100 dB for a communications receiver.

The block diagram in Figure 5.3 represents a receiver for AM audio use, but the superheterodyne system is also used with other modulation schemes (for instance, FM) and other modulating signals (such as video). Once you are thoroughly familiar with the basic system, it will be easy to understand the variations.

5.2.3 Example: An AM Broadcast-Band Receiver

Figure 5.5(a) shows the schematic of a typical AM broadcast-band receiver, and Figure 5.5(b) shows its block diagram. Note the similarity to the generic block diagram of Figure 5.3.

For reasons of economy, the RF stage has been omitted. The received signal is coupled to the mixer by way of a single tuned circuit. In fact, the inductive element of this tuned circuit is actually the receiving antenna, which is a ferrite *loopstick* (that is, a coil wound around a ferrite rod). An example is shown in the photograph of Figure 5.6(a).

The mixer and local oscillator use a single transistor, in a configuration called an **autodyne converter**. This is done for economy: it would be better from the point of view of stability and the reduction of **spurious responses** to use a separate local oscillator.

The IF is 455 kHz, and **high-side injection** is used. This means that the local oscillator is tuned so that it generates a frequency that is always 455 kHz higher than the incoming signal frequency. This is accomplished by using two variable capacitors, one for the input circuit and one for the local oscillator, mounted on a single shaft. Such a capacitor is shown in Figure 5.6(b). The local oscillator section has fewer plates and lower capacitance than the mixer section. Adjustments must be provided to allow reasonably accurate tracking across the tuning range, but this is not nearly as critical as for a TRF receiver, since the input tuned circuit is quite broad, and slight mistuning of that circuit will have little effect. The actual selection of the desired station is done by the local oscillator tuning.

FIGURE 5.5   **AM Broadcast Receiver**

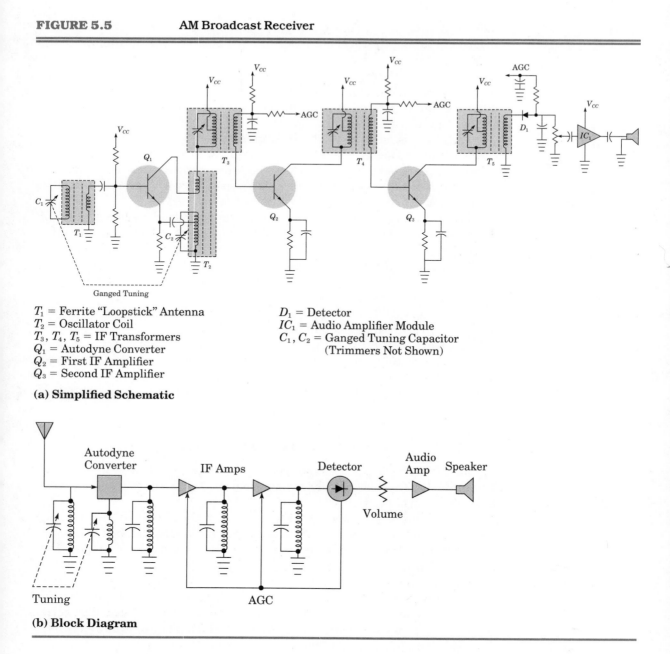

T_1 = Ferrite "Loopstick" Antenna
T_2 = Oscillator Coil
T_3, T_4, T_5 = IF Transformers
Q_1 = Autodyne Converter
Q_2 = First IF Amplifier
Q_3 = Second IF Amplifier

D_1 = Detector
IC_1 = Audio Amplifier Module
C_1, C_2 = Ganged Tuning Capacitor
          (Trimmers Not Shown)

(a) Simplified Schematic

(b) Block Diagram

The mixer will, of course, produce the sum and difference of the local oscillator and signal frequencies. The sum is not used in this application; the difference of 455 kHz is amplified by the IF stages. Here, conventional tuned circuits are used to give the desired bandwidth, which, for a small AM radio, will be about 7 to 10 kHz.

Low-side injection—where the local oscillator frequency is lower than that of the received signal—could have been used, but it would require the oscillator to have a wider tuning range, in percentage terms, which would

FIGURE 5.6

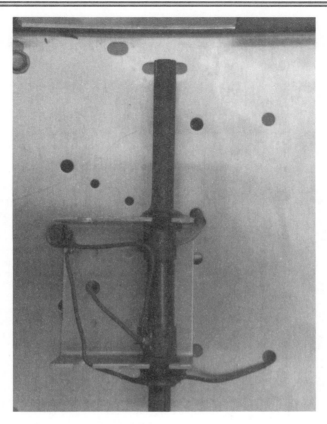

(a) Ferrite "Loopstick" Antenna

(b) Dual-Gang Variable Capacitor

FIGURE 5.7   Data Sheet for the LM1863

National Semiconductor

LM1863 AM Radio System for Electronically Tuned Radios

General Description

The LM1863 is a high performance AM radio system intended primarily for electronically tuned radios. Important to this
application is an on-chip stop detector circuit which allows
for a user adjustable signal level threshold and center frequency stop window. The IC uses a low phase noise, level-
controlled local oscillator.

Low phase noise is important for AM stereo which detects
phase noise as noise in the L-R channel. A buffered output
for the local oscillator allows the IC to directly drive a phase
locked loop synthesizer. The IC uses a RF AGC detector to
gain reduce an external RF stage thereby preventing overload by strong signals. An improved noise floor and lower
THD are achieved through gain reduction of the IF stage.
Fast AGC settling time, which is important for accurate stop
detection, and excellent THD performance are achieved
with the use of a two pole AGC system. Low tweet radiation

and sufficient gain are provided to allow the IC to also be
used in conjunction with a loopstick antenna.

Features

- Low supply current
- Level-controlled, low phase noise local oscillator
- Buffered local oscillator output
- Stop circuitry with adjustable stop threshold and adjust-
  able stop window
- Open collector stop output
- Excellent THD and stop time performance
- Large amount of recovered audio
- RF AGC with open collector output
- Meter output
- Compatible with AM stereo

Block Diagram

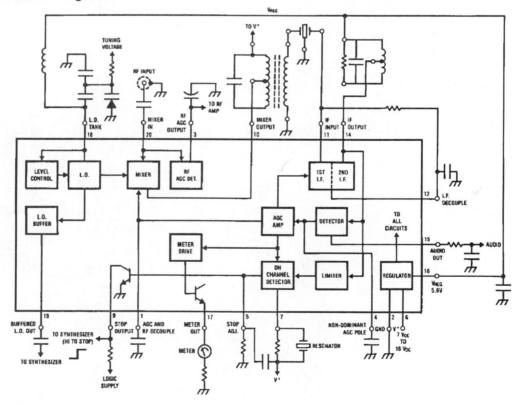

Courtesy National Semiconductor.

make the mechanical design of the tuning system more difficult. An example will make this clear. For high-side injection, the oscillator must tune from:

$$f_l = 540 \text{ kHz} + 455 \text{ kHz} = 995 \text{ kHz}$$

to:

$$f_h = 1600 \text{ kHz} + 455 \text{ kHz} = 2055 \text{ kHz}$$

The ratio of highest to lowest frequency is:

$$\frac{f_h}{f_l} = \frac{2055}{995} = 2.065$$

With low-side injection, on the other hand, the oscillator would have to tune from 85 to 1145 kHz, a ratio of 13.47. Larger capacitance and inductance values would also require the components to be physically larger. Nonetheless, low-side injection is sometimes used for receivers operating at higher frequencies, as will be seen later in this chapter.

The AGC in this receiver is quite simple. Low-pass filtering the detector output (with a time constant of about a second) swamps the variations due to modulation and leaves a dc voltage proportional to the carrier level. This voltage is used to alter the bias on the IF amplifier transistors, which in turn changes the gain of the IF stages. The details of this and other types of AGC will be covered in Section 5.5.

In addition to AGC, there is a manual volume control, which varies the level of signal provided from the detector to the audio amplifier.

Practically all of the active circuitry for a radio such as this can be incorporated into a custom integrated circuit (IC). Figure 5.7 shows a typical example, the National Semiconductor LM1863. The circuit shown needs only an audio amplifier, which could be another IC, to complete it.

5.3 RECEIVER CHARACTERISTICS

At the beginning of our discussion of receivers, we looked at two important receiver parameters: *sensitivity* and *selectivity*. It is now time to look more closely at these, and to introduce some others.

5.3.1 Sensitivity

The transmitted signal may have a power level, at the transmitting antenna, ranging from milliwatts to hundreds of kilowatts. However, the losses in the path from transmitter to receiver are so great that the power of the received signal is often measured in dBf (that is, decibels relative to one femtowatt: $1 \text{ fW} = 1 \times 10^{-15} \text{ W}$). A great deal of amplification is needed to achieve a useful power output. To operate an ordinary loudspeaker, for instance, requires a power on the order of 1 W, or 150 dB greater than 1 fW.

Because received signals are often quite weak, noise added by the receiver itself can be a problem. Diode detectors are inherently noisy and also operate best with fairly large signals (hundreds of millivolts), so some of the amplification must take place before demodulation.

The ability to receive weak signals with an acceptable signal-to-noise ratio (S/N) is called *sensitivity*. It is expressed in terms of the voltage or power at the antenna terminals necessary to achieve a specified signal-to-noise ratio, or some more easily measured equivalent. One common specification for AM receivers is the signal strength required for a 10 dB signal-plus-noise to noise [$(S+N)/N$] ratio, at a specified output power level. Procedures for measuring sensitivity will be given later in this chapter.

5.3.2 Selectivity

In addition to noise generated within the receiver, there will be noise coming in with the signal, as well as interfering signals with frequencies different from that of the desired signal. All of these problems can be reduced by limiting the receiver bandwidth to that of the signal, including all its sidebands. When interference is severe, an even smaller bandwidth can be used in an AM receiver, at the expense of reducing the response to high frequencies in the original modulating signal. The ability to discriminate against interfering signals is known as *selectivity*.

Selectivity can be expressed in various ways. The bandwidth of the receiver at two different levels of attenuation can be specified. The bandwidth at the points where the signal is 3 or 6 dB down is helpful in determining whether all the sidebands of the desired signal will be passed without attenuation. To indicate the receiver's effectiveness in rejecting interference, a bandwidth for much greater attenuation, for example 60 dB, should also be given.

The frequency-response curve for an ideal IF filter would have a square shape, with no difference between its bandwidths at 6 dB and 60 dB down. The closer the two bandwidths are, the better the design. The ratio between these bandwidths is called the **shape factor**. That is,

$$SF = \frac{B_{-60\ \text{dB}}}{B_{-6\ \text{dB}}}$$ (5.3)

where SF = shape factor

   $B_{-60\ \text{dB}}$ = bandwidth at 60 dB down from maximum

   $B_{-6\ \text{dB}}$ = bandwidth at 6 dB down from maximum

The shape factor should be as close to one as possible. The following example shows the effect of changing the shape factor on the rejection of interfering signals.

EXAMPLE 5.2

Calculate the shape factors for the two IF response curves shown in Figure 5.8, and calculate the amount by which the interfering signal shown would be attenuated in each case.

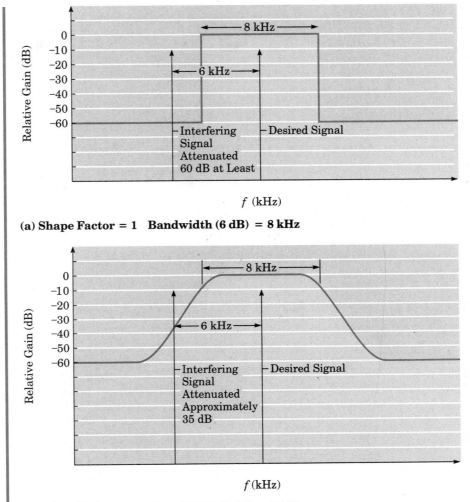

(a) Shape Factor = 1   Bandwidth (6 dB) = 8 kHz

(b) Shape Factor = 2   Bandwidth (6 dB) = 8 kHz
FIGURE 5.8

Solution

Figure 5.8(a) shows an ideal filter. Since the -6 dB and -60 dB band-widths are equal, $SF = 1$. The interfering signal is attenuated by 60 dB.

In Figure 5.8(b), the -6 dB bandwidth is 8 kHz and the -60 dB band-width is 16 kHz, so $SF = 2$. The interfering signal is attenuated approximately 35 dB compared to the desired signal.

Adjacent channel rejection is another way of specifying selectivity that is commonly used with channelized systems, such as CB radio. It is defined as the number of decibels by which an **adjacent channel** signal must be stronger than the desired signal for the same receiver output.

Alternate channel rejection is also used in systems, such as FM broadcasting, where stations in the same locality are not assigned to adjacent channels. The **alternate channel** is two channels removed from the desired

FIGURE 5.9

Adjacent and Alternate
Channels

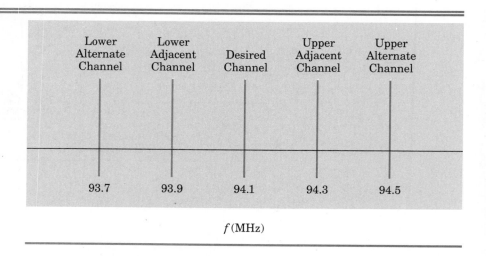

f (MHz)

one. It is also known as the *second adjacent channel*. Figure 5.9 shows the relation between adjacent and alternate channels. For example, if 94.1 MHz were assigned to a station, the adjacent channel at 94.3 MHz would not be assigned in the same community, but the alternate channel, at 94.5 MHz, might be. Any signal at 94.3 MHz, then, will be relatively weak, and a receiver's ability to reject strong adjacent channel signals will be less important than its ability to reject the alternate channel (in this case, the one at 94.5 MHz).

5.3.3 Distortion

In addition to good sensitivity and selectivity, an ideal receiver would reproduce the original modulation exactly. A real receiver, however, will subject the signal to several types of *distortion*. They are the same types encountered in other analog systems: harmonic and intermodulation distortion, uneven frequency response, and phase distortion.

Harmonic distortion occurs when the frequencies generated are multiples of those in the original modulating signal. Envelope detectors produce significant harmonic distortion because they operate part of the time in the square-law portion of the diode curve.

Intermodulation takes place when frequency components in the original signal mix in a nonlinear device, creating sum and difference frequencies. A type of intermodulation peculiar to receivers consists of mixing between the desired signal and an interfering one, which is outside the IF passband of the receiver but within the passband of the RF stage, and is therefore present in the mixer. This can result in interference from a strong local station that is not at all close in frequency to the desired signal.

EXAMPLE 5.3

A receiver is tuned to a station with a frequency of 1 MHz. A strong signal with a frequency of 2 MHz is also present at the amplifier. Explain how intermodulation between these two signals could cause interference.

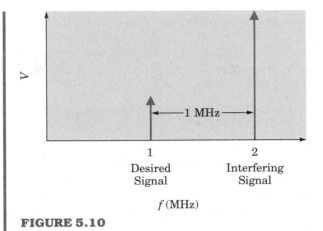

FIGURE 5.10

Solution

See Figure 5.10. Assume that the tuned circuit at the mixer input has insufficient attenuation at 2 MHz to block the interfering signal completely. (Remember that the interfering signal may be much stronger than the desired signal, and therefore may still have similar strength to the desired signal, even after being attenuated by the input tuned circuit.) The two signals could mix in the mixer stage to give a difference frequency of 1 MHz, at the same frequency as the original signal. This interfering signal will then pass through the rest of the receiver in the same way as the desired signal.

In addition, any nonlinearities in the detector or any of the amplification stages can cause intermodulation between components of the original baseband signal.

The frequency response of an AM receiver is related to its IF bandwidth. Restricting the bandwidth to less than the full width of the signal reduces the upper limit of response, since the sidebands corresponding to higher modulating frequencies are further away from the carrier. For example, a receiver bandwidth of 10 kHz will allow a modulation frequency of 5 kHz to be reproduced. Reducing the bandwidth to 6 kHz reduces the maximum audio frequency that can be reproduced to 3 kHz. In addition, any unevenness within the IF passband will affect the audio frequency response. Of course, as would be expected, the audio amplifier also has an effect on frequency response.

Phase distortion is slightly more difficult to understand than frequency response. Of course, the signal at the receiver output will not be in phase with the input to the transmitter: There will be some time delay, which can be translated into a phase shift that increases linearly with frequency. Phase distortion consists of irregular shifts in phase, and is quite a common occurrence when signals pass through filters.

The impact of distortion depends to a great extent on the application. For instance, phase distortion is unimportant for voice communications, of questionable importance for music (some audiophiles claim to be able to hear it), and fatal to some other applications, such as color television (where it results in incorrect colors) and certain types of data transmission. For voice communications, an audio bandwidth of more than about 3 kHz is simply a waste of the radio spectrum, while for high-fidelity music an audio bandwidth of at least 15 kHz is essential. In a CB radio at maximum output power, 10% harmonic distortion may be acceptable, while 1% might be considered excessive in a quality FM broadcast tuner.

5.3.4 Dynamic Range

As mentioned above, a receiver must operate over a considerable range of signal strengths. The response to weak signals is usually limited by noise generated within the receiver. On the other hand, signals that are too strong will overload one or more stages, causing unacceptable levels of distortion. The ratio between these two signal levels, expressed in decibels, is the *dynamic range* of the receiver.

The above description is actually a bit simplistic. A well-designed AGC system can easily vary the gain of a receiver by 100 dB or so. Thus, almost any receiver can cope with very wide variations of signal strength, provided only one signal is present at a time. This range of signal strengths is sometimes referred to as *dynamic range*, but should properly be called *AGC range*.

The hardest test of receiver dynamic range occurs when two signals with slightly different frequencies and very different power levels are applied to the antenna input simultaneously. This can cause overloading of the receiver input stage by the stronger signal, even though the receiver is not tuned to its frequency. The result can be **blocking**, also called *desensitization* or "*desense*," which is a reduction in sensitivity to the desired signal. Intermodulation between the two signals is also possible.

EXAMPLE 5.4

A receiver has a sensitivity of 0.5 μV and a blocking dynamic range of 70 dB. What is the strongest signal that can be present along with a 0.5 μV signal without blocking taking place?

Solution

Since both signal voltages are across the same impedance, the input impedance of the receiver, the general equation

$$\frac{P_1}{P_2} \text{ (dB)} = 20 \log \frac{V_1}{V_2} \tag{5.4}$$

can be used.

Here, we can let $V_2 = 0.5$ μV and

$$\frac{P_1}{P_2} \text{ (dB)} = 70$$

and solve for V_1 as follows. From Equation (5.4),

$$\frac{V_1}{V_2} = \text{antilog} \frac{P_1/P_2 \text{ (dB)}}{20}$$

$$V_1 = V_2 \text{ antilog} \frac{P_1/P_2 \text{ (dB)}}{20}$$

$$= (0.5 \text{ } \mu V) \text{ antilog} \frac{70}{20}$$

$$= 1581 \text{ } \mu V$$

$$= 1.58 \text{ mV}$$

There is yet another type of dynamic range, *audio dynamic range*, which is essentially the usable range of modulation depth with a given carrier level. Generally, a strong signal is specified for this measurement, which relates to the sound quality that can be expected with a good signal. This type of dynamic range specification is more common with FM than with AM receivers.

5.3.5 Spurious Responses

The superheterodyne receiver has many important advantages compared to simpler receivers, but it is not without its problems. In particular, it has a tendency to receive signals at frequencies to which it is not tuned, and sometimes to generate signals internally, which can interfere with reception. Careful design can reduce these *spurious responses* almost to insignificance, but they will still be present.

Image Frequencies. The intermediate-frequency signal in a superheterodyne receiver is the difference between the received signal and local oscillator frequencies. It does not matter whether the local oscillator is higher in frequency than the received signal (high-side injection) or lower (low-side injection).

It follows that in any receiver there will be two frequencies, one below and one above the local oscillator frequency, that will mix with it to produce a signal at the intermediate frequency f_{IF}. One of these will be the frequency f_{sig} to which the radio is tuned; the other is called the **image frequency** f_{image}. The image is an equal distance from the local oscillator frequency f_{LO} on the other side of it from the signal. This can be shown mathematically as follows. Assuming high-side injection,

$$f_{IF} = f_{LO} - f_{sig}$$

so

$$f_{LO} = f_{IF} + f_{sig} \tag{5.5}$$

[handwritten: f_{LO} = freq. Local oscillator]

For the image,

$$f_{IF} = f_{image} - f_{LO}$$

therefore

$$f_{LO} = f_{image} - f_{IF} \tag{5.6}$$

Combining Equations (5.5) and (5.6),

$$f_{IF} + f_{sig} = f_{image} - f_{IF}$$
$$f_{image} = f_{sig} + 2f_{IF} \tag{5.7}$$

In a similar way, it can be shown that for low-side injection

$$f_{image} = f_{sig} - 2f_{IF} \tag{5.8}$$

Figure 5.11 shows the similarity with reflection in a mirror. Think of the local oscillator signal as the mirror, with the desired signal and the image equidistant from it, on opposite sides.

An image must be *rejected* prior to mixing: Once it has entered the IF chain, the image will be indistinguishable from the desired signal, and impossible to filter out. Image rejection is accomplished by tuned circuits or other filters before the mixer. Using a higher IF will improve image rejection by placing the image further away in frequency from the desired signal, where it can more easily be removed by tuned circuits before the mixer.

Image rejection is defined as the ratio of voltage gain at the input frequency to which the receiver is tuned to gain at the image frequency. Image rejection IR is usually expressed in decibels. For a single tuned circuit:

$$IR = \frac{A_{sig}}{A_{image}} = \sqrt{1 + Q^2 x^2} \tag{5.9}$$

where Q = Q of the tuned circuit

A_{image} = voltage gain at image frequency

A_{sig} = voltage gain at signal frequency

x $= \dfrac{f_{image}}{f_{sig}} - \dfrac{f_{sig}}{f_{image}}$

Since we are dealing with voltage ratios across the same impedance, it is easy to convert to decibels:

FIGURE 5.11   **Image Response**

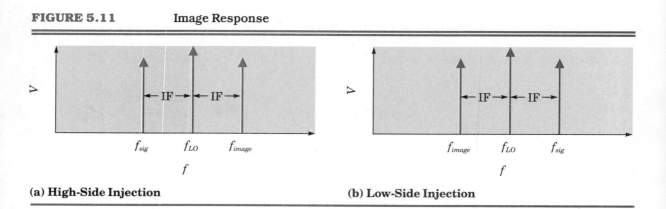

(a) High-Side Injection   **(b) Low-Side Injection**

$$IR \text{ (dB)} = 20 \log \frac{A_{sig}}{A_{image}} \tag{5.10}$$

For multiple uncoupled tuned circuits, multiply the image rejection ratios (or add, if they are given in decibels). Many receivers have a single RF amplifier stage with a tuned circuit at its input, followed by a mixer, also with a tuned input. These two resonant circuits are uncoupled, and the above equation applies.

EXAMPLE 5.5

The receiver of Figure 5.4 is tuned to a station at 590 kHz. *Page 218*

(a) Find the image frequency.

(b) Calculate the image rejection in decibels, assuming that the input filter consists of one tuned circuit with a Q of 40.

Solution

(a) This receiver uses high-side injection of the local oscillator. From Equation (5.7),

$$\begin{aligned} f_{image} &= f_{sig} + 2f_{IF} \\ &= 590 \text{ kHz} + 2(455 \text{ kHz}) \\ &= 1500 \text{ kHz} \end{aligned}$$

Since 1500 kHz is also within the AM broadcast band, a strong local signal at that frequency is a real possibility.

(b) The image rejection can be found using Equation (5.9). First, find

$$\begin{aligned} x &= \frac{f_{image}}{f_{sig}} - \frac{f_{sig}}{f_{image}} \\ &= \frac{1500 \text{ kHz}}{590 \text{ kHz}} - \frac{590 \text{ kHz}}{1500 \text{ kHz}} \\ &= 2.149 \end{aligned}$$

Now,

$$\begin{aligned} \frac{A_{sig}}{A_{image}} &= \sqrt{1 + Q^2 x^2} \\ &= \sqrt{1 + 40^2 \times 2.149^2} \\ &= 85.97 \end{aligned}$$

In decibels, the image rejection is

$$\begin{aligned} IR \text{ (dB)} &= 20 \log 85.97 \\ &= 38.7 \text{ dB} \end{aligned}$$

The use of a higher IF in Example 5.5 would have improved image rejection, but at the expense of requiring more elaborate filtering in the IF amplifier to achieve the desired bandwidth. Choice of IF is always a compromise. Over the years, 455 kHz has come to be a very common frequency for the IF in receivers designed for the standard AM broadcast band.

Image response is usually undesirable, but in certain cases, for example some of the "scanners" used to receive police and fire calls and the like, image reception is used deliberately to increase the frequency coverage of a receiver.

The term *double-spotting* is sometimes used to describe the same phenomenon as image response, but seen from a different point of view. For example, suppose there is a strong AM station at 1500 kHz. It may be possible to tune a receiver to a frequency lower than this by twice the IF, and receive the signal. For a typical AM radio with an IF of 455 kHz, this would require tuning it to 1500 kHz − 455 kHz × 2 = 590 kHz. Obviously, what has been done is to tune the receiver to a frequency such that this station is an image response. However, we could look at it the other way around and say that the station comes in at *two spots* on the dial, hence the term *double-spotting*.

Other Spurious Responses.   Besides images, superheterodyne receivers are subject to other problems. The local oscillator will have harmonics, for instance, and these can mix with incoming signals to produce spurious responses. The incoming signal may also have harmonics, created by distortion in the RF stage. In fact, it is possible for a receiver to respond to any frequency given by the equation

$$f_s = \left(\frac{m}{n}\right)f_{LO} \pm \frac{f_{IF}}{n} \tag{5.11}$$

where f_s = frequency of the spurious response

f_{LO} = local oscillator frequency

f_{IF} = intermediate frequency

m, n = any integers

EXAMPLE 5.6

An AM high-frequency receiver has an IF of 1.8 MHz using high-side injection. If it is tuned to a frequency of 10 MHz, calculate the frequencies that can cause an IF response, for values of m and n ranging up to 2.

Solution

First, we find f_{LO}:

$$f_{LO} = f_{sig} + f_{IF}$$
$$= 10 \text{ MHz} + 1.8 \text{ MHz}$$
$$= 11.8 \text{ MHz}$$

Now, the problem is easily solved using Equation (5.11) and a table of values. All frequencies in the table are in MHz.

m	n	$\left(\dfrac{m}{n}\right)f_{LO}$	$\dfrac{f_{IF}}{n}$	$\left(\dfrac{m}{n}\right)f_{LO} + \dfrac{f_{IF}}{n}$	$\left(\dfrac{m}{n}\right)f_{LO} - \dfrac{f_{IF}}{n}$
1	1	11.8	1.8	13.6	10.0
1	2	5.9	0.9	6.8	5.0
2	1	23.6	1.8	25.4	21.8
2	2	11.8	0.9	12.7	10.9

Rearranging the results in the last two columns in order of ascending frequency gives us the frequencies to which the receiver can respond. In MHz, they are:

5.0	12.7
6.8	13.6
10.0	21.8
10.9	25.4

In spite of the problems listed above, however, the superheterodyne remains the preferred arrangement for almost all receiving applications. It proves to be easier to improve the design to reduce its problems than to go to a different system.

5.4   RECEIVER CIRCUITS

Now that we have looked at the general structure of a typical superheterodyne receiver, it is time to consider the design of the various stages in a little more detail. We will start at the input and work our way through to the output. The assumption at this point is that the receiver will be used for AM. Modifications to allow it to work with different types of modulation will be described in later chapters.

5.4.1   The Radio-Frequency Amplifier

Inexpensive receivers often omit the RF amplifier stage, especially if they are designed for operation at low to medium frequencies, where atmospheric noise entering the receiver with the signal is likely to be more significant than noise generated within the receiver itself. That is why our example AM broadcast-band receiver did not have an RF amplifier. On the other hand, weak-signal reception at microwave frequencies requires very careful design of the first RF amplifier for the best possible noise figure. The low-noise amplifier (LNA) used with television satellite receivers is just an RF amplifier, using a gallium arsenide field-effect transistor (GaAsFET), and is located right at the antenna in order to provide gain before the attenuation of the antenna cable can degrade the signal-to-noise ratio.

Between these extremes, we find high-frequency communications receivers and FM broadcast receivers, for instance. These receivers generally have one RF amplifier stage, located in the receiver and not at the antenna. The stage may be tuned to the signal frequency, in which case it has to track the local oscillator tuning. This can be accomplished by adding a third section to the dual-gang tuning capacitor shown earlier in Figure 5.6(b).

On the other hand, an RF stage may use relatively broadband filters, so that the frequency band of interest is covered, but image and other spurious responses are excluded. Such designs require a fairly high IF, so that a useful tuning range can be covered with one setting of the input filter (without allowing both the desired signal and the image ever to be within the bandwidth of the filter at the same time). Communications receivers, for instance, often divide their coverage into "bands" 1 MHz wide, and retune the input filter only when switching between bands. The IF must then be considerably higher than 1 MHz. This system is especially popular in modern designs, which usually use a frequency synthesizer to generate the local oscillator signal, and thus do not provide a convenient mechanical way for the RF amplifier to track the local oscillator tuning.

The RF stage is a Class A amplifier. It should have a good noise figure and a wide dynamic range. AGC can be used, but designers often prefer not to apply it to the RF stage, because any alteration of the stage gain from the optimum level will degrade the noise figure. To prevent overloading, a switch is sometimes provided to remove the RF stage from the signal path for strong signals; such signals are applied directly to the mixer after going through the input filter. To prevent very strong signals from overloading the mixer, another position on the same switch may add a few decibels of attenuation. Of course, both removing the RF amplifier and adding attenuation have adverse effects on the receiver's noise performance, but for very strong signals it is less important to maintain the noise figure than to avoid overloading.

Figure 5.12 shows some typical RF amplifier circuits. Figure 5.12(a) shows a bipolar amplifier in a common-base configuration. This is quite common in RF amplifiers, as it gives better stability and a higher cutoff frequency than a common-emitter circuit. The circuit in Figure 5.12(a) is a narrowband amplifier. The variable capacitors must track the receiver's local oscillator tuning.

FIGURE 5.12

RF Amplifiers

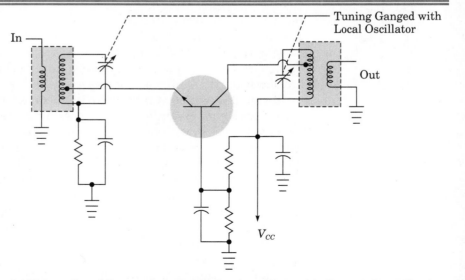

(a) Narrowband Common-Base Bipolar Amplifier with Conventional Tuning

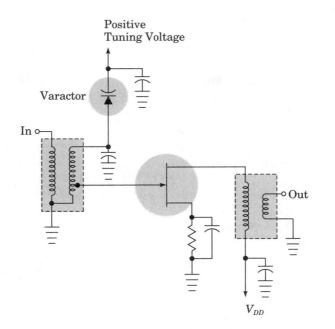

(b) Common-Source FET Amplifier with Varactor Tuning

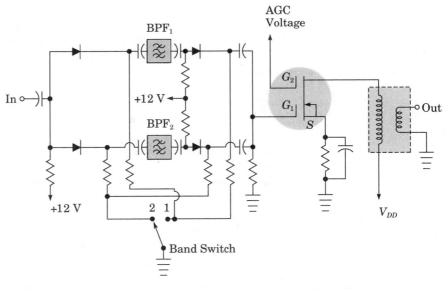

(c) RF Amplifier Using Dual-Gate MOSFET and Diode Switching

Figure 5.12(b) shows an amplifier using an FET. FETs are more common than bipolar transistors for RF amplifiers in modern receiver designs, for two main reasons. Their very high input impedance provides less loading to the input tuned circuit, improving its Q and allowing better image rejection. Perhaps more important, FETs exhibit lower levels of third-order intermodulation distortion than bipolar transistors. This reduces the possibility of interference between signals that are close enough in frequency to pass through the input filter together, and improves the receiver's dynamic range. Figure 5.12(a) is also a narrowband amplifier, but this one is tuned electronically by a varactor diode. The dc tuning voltage that controls the amplifier tuning would be used for the local oscillator tuning as well.

In the event that AGC is required, using a dual-gate MOSFET provides a handy extra element to which the AGC voltage can be applied. This is shown in Figure 5.12(c), which also shows the use of bandpass filters (BPF$_1$ and BPF$_2$) in place of the input tuned circuit.

This broadband RF amplifier does not require tracking, but bandswitching is needed. In Figure 5.12(c), it is accomplished by diode switching. This method takes advantage of the fact that a forward-biased diode has a low impedance to ac signals, while the impedance of a reverse-biased diode is very high. Diode switching avoids some problems that arise with conventional bandswitches. Because leads in RF circuits must be short to reduce stray inductance and capacitance, the switched circuits must be grouped around the switch. When more than one stage is switched, this can easily result in unwanted coupling between stages. In any case, it leads to an awkward and expensive mechanical design. With diode switching, the diode can be placed right next to the component being switched. Since the control voltage is dc, the switch can be located anywhere. In fact, the switching can be done by remote control, or even by a computer.

5.4.2 The Mixer/Converter

Any of the mixer circuits described in Chapter 2 will work. Diode mixers are generally rejected as too noisy and too lossy, except in the simplest receivers. For example, some simple devices used to detect police radar signals have used a diode mixer connected directly to the antenna. Either bipolar transistors or FETs can be used as mixers, but the latter are preferred because they create fewer intermodulation distortion components. The problem here, as in the RF amplifier, is not so much distortion of the modulating signal, as the creation of spurious responses due to interactions between desired and interfering signals.

The mixer and local oscillator can be combined for economy. The combination is called an **autodyne converter** or a *self-excited mixer*, and is very common in simple AM broadcast receivers like the one discussed above. However, better designs use a separate local oscillator.

It is extremely important that the local oscillator be stable, as any frequency change will result in the receiver drifting away from the station to which it is tuned. *LC* oscillators must be carefully designed for stability, following the methods described in Chapter 2. This is expensive, of course, and some designs have used various types of *automatic frequency control* (AFC), which reduce drift by feeding an error signal back to the local oscillator. FM broadcast receivers and television receivers have made extensive use of AFC (often called AFT, for *automatic fine tuning*, when used in television receivers), and the methods used will be discussed along with those types of receivers. The higher the local oscillator frequency the worse the drift, in kilohertz, for a given percentage frequency change; thus AFC is more commonly seen at VHF and up than at lower frequencies.

Another approach to local oscillator stability is to use crystal control. The local oscillator can be a simple crystal oscillator, with a switch to change crystals for different channels. This becomes unwieldy with more than a few channels, so crystal-controlled frequency synthesizers have become very popular in new designs. The use of a synthesizer also allows remote control, direct frequency or channel number entry, and other similar conveniences. AFC is not necessary with a properly designed synthesizer. (See Chapter 7 for a discussion of synthesizers.)

Oscillator *spectral purity* is also important. Any frequency components at the oscillator output have the possibility of mixing with incoming signals and creating undesirable spurious responses. Noise generated by the local oscillator will degrade the noise performance of the receiver, and should be minimized. Some designs use a bandpass filter to remove spurious signal components and noise from the local oscillator signal before it is applied to the mixer.

5.4.3 The Intermediate-Frequency Amplifier

This is basically a linear, fixed-tuned amplifier. The IF amplifier of an AM receiver must be Class A to avoid distorting the signal envelope.

The IF amplifier accounts for most of the receiver's gain and selectivity. The classical method to provide this is to use several stages, to give sufficient gain, coupled by tuned transformers that provide the selectivity. These transformers can be single-tuned. In that case, either the primary or secondary winding will be part of a *resonant circuit*. Alternatively, both windings can be included in resonant circuits, creating a double-tuned transformer. In either case, the capacitors that are needed for resonance are generally combined with the inductors in a shielded enclosure. Tuning is done with a nonmetallic screwdriver or "alignment tool." Figure 5.13(a) is a photograph of two of these IF transformers.

The choice of transformer depends on the frequency, bandwidth, and shape factor required. When narrow bandwidth is required, single- or double-tuned transformers with loose coupling may be used. For single-tuned transformers, tighter coupling means more gain but broader bandwidth, as shown in Figure 5.13(b).

The ideal IF passband has a flat top and steep sides. This is often achieved by using overcoupled double-tuned transformers. *Overcoupling* means k greater than the critical coupling factor k_c given by

$$k_c = \frac{1}{\sqrt{Q_p Q_s}} \tag{5.12}$$

where Q_p, Q_s = primary and secondary Q, respectively

As a rule of thumb, a good compromise between steep skirts and flat passband is given by using the optimum coupling factor

$$k_{opt} = 1.5 k_c \tag{5.13}$$

The effect of varying k for an overcoupled double-tuned transformer is illustrated in Figure 5.13(c). The bandwidth for a double-tuned amplifier with $k = k_{opt} = 1.5 k_c$ is given approximately by:

$$B = k f_o \tag{5.14}$$

FIGURE 5.13   **IF Transformers**

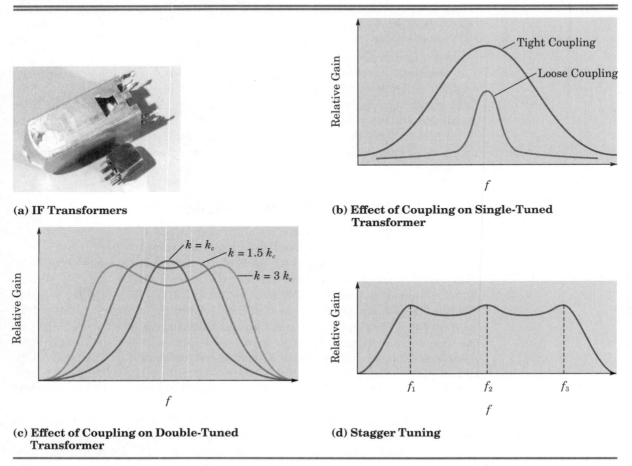

(a) IF Transformers

(b) Effect of Coupling on Single-Tuned Transformer

(c) Effect of Coupling on Double-Tuned Transformer

(d) Stagger Tuning

EXAMPLE 5.7

An IF transformer operates at 455 kHz. The primary circuit has a Q of 40 and the secondary has a Q of 30. Find:

(a) the critical coupling factor
(b) the optimum coupling factor
(c) the bandwidth using the optimum coupling factor

Solution

(a) The critical coupling factor is given by Equation (5.12):

$$k_c = \frac{1}{\sqrt{Q_p Q_s}}$$

$$= \frac{1}{\sqrt{40 \times 30}}$$

$$= 0.0289$$

(b) From Equation (5.13),

$$k_{opt} = 1.5k_c$$
$$= 1.5 \times 0.0289$$
$$= 0.0433$$

(c) The bandwidth is given by Equation (5.14):

$$B = kf_o$$
$$= 0.0433 \times 455 \text{ kHz}$$
$$= 19.7 \text{ kHz}$$

When more than one IF transformer is used, as is practically always the case, their responses are multiplied (or added, if expressed in decibels). If all are tuned to the same frequency, this will cause a reduction in bandwidth. If a wide passband with steep sides is required, two or more transformers may be *stagger tuned*, (that is, tuned to slightly different frequencies), as shown in Figure 5.13(d). Three single-tuned transformers are tuned to three slightly different frequencies, f_1, f_2, and f_3. The result is a relatively broad passband with three peaks.

The broadcast receiver we looked at earlier (Figure 5.5) has a conventional IF amplifier. There are two stages of amplification, using three single-tuned transformers.

In modern receivers, other means of increasing selectivity (for example, crystal and ceramic filters) are popular. These will be discussed later. The filter is often placed at the beginning of the IF chain, right after the mixer. The IF amplifier itself, in a modern design, is more likely to use specially designed integrated circuits than discrete transistors. Two common ICs are the CA3028A and the MC1590G. Specifications and typical applications for these are shown in Figures 5.14 and 5.15, respectively.

5.4.4 The Detector

The envelope (peak) detector introduced in Chapter 2 is the one most commonly used for full-carrier AM. Other more elaborate detectors are possible, but are more commonly used for AM variations, such as single-sideband AM, and will be described later.

An *envelope detector* is essentially a rectifier followed by an RC network, as in Figure 5.16(a). Figure 5.16(b) shows a time-domain representation of its operation. The modulated signal, at the IF frequency, is applied at point A. Sketch (i) shows the envelope for a 1 kHz sinusoidal modulating signal. The diode removes half of the envelope. The result is shown in sketch (ii). Please note, however, that this sketch ignores the low-pass filter that is connected to point B.

The capacitor C_1 charges to the peak value of the RF waveform, and is slowly discharged by the resistor. If the RC_1 time constant is much longer than the period of the RF waveform, and also much shorter than that of the modulating signal, the output will be a reasonably faithful reproduction of the modulating signal. There will be an additional dc component, which can easily be removed by a blocking capacitor, C_2.

FIGURE 5.14   Data Sheet for CA3028A

Features

- **Controlled for Input Offset Voltage, Input Offset Current and Input Bias Current (CA3028 Series Only)**

- **Balanced Differential Amplifier Configuration with Controlled Constant-Current Source**

- **Single-Ended and Dual-Ended Operation**

Applications

- **RF and IF Amplifiers (Differential or Cascode)**

- **DC, Audio and Sense Amplifiers**

- **Converter in the Commercial FM Band**

- **Oscillator**

- **Mixer**

- **Limiter**

- **Companion Application Note, ICAN 5337 "Application of the CA3028 Integrated Circuit Amplifier in the HF and VHF Ranges." This note covers characteristics of different operating modes, noise performance, mixer, limiter, and amplifier design considerations.**

Description

The CA3028A and CA3028B are differential/cascode amplifiers designed for use in communications and industrial equipment operating at frequencies from DC to 120MHz.

The CA3028B is like the CA3028A but is capable of premium performance particularly in critical DC and differential amplifier applications requiring tight controls for input offset voltage, input offset current, and input bias current.

The CA3053 is similar to the CA3028A and CA3028B but is recommended for IF amplifier applications.

The CA3028A, CA3028B, and CA3053 are available in 8-lead packages as shown below. When ordering these devices, it is important to add the appropriate suffix letter to the device.

Package/Lead Options

SMALL OUTLINE (150 MIL)	STRAIGHT LEAD TO-5	DUAL-IN-LINE FORMED-LEAD TO-5	DUAL-IN-LINE PLASTIC (MINI-DIP)
CA3028AM	CA3028A*	CA3028AS	CA3028AE
CA3028BM	CA3028B*	CA3028BS	CA3028BE
CA3053M	CA3053*	CA3053S	CA3053E

*Most types in a straight-lead TO-5 package carry a "T" suffix. This one does not. Order type number as shown.

Schematic Diagram

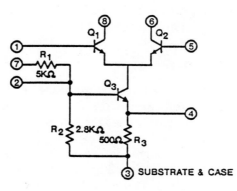

CA3028A, CA3028B AND CA3053

CAUTION: These devices are sensitive to electrostatic discharge. Proper I.C. handling procedures should be followed.

Courtesy Harris Corporation.

FIGURE 5.15   Data Sheet for MC15906

RF/IF/AUDIO AMPLIFIER

. . . an integrated circuit featuring wide-range AGC for use in RF/IF amplifiers and audio amplifiers over the temperature range, −55 to +125°C. See Motorola Application Note AN513 for design details.

- High Power Gain — 50 dB Typ at 10 MHz
  45 dB Typ at 60 MHz
  35 dB Typ at 100 MHz
- Wide-Range AGC — 60 dB min, dc to 60 MHz
- Low Reverse Transfer Admittance — <10 μmhos Typ at 60 MHz
- 6.0 to 15-Volt Operation, Single-Polarity Power Supply

WIDEBAND AMPLIFIER WITH AGC

SILICON MONOLITHIC INTEGRATED CIRCUIT

PIN CONNECTIONS

**G SUFFIX
METAL PACKAGE
CASE 601
TO-99**

Case Ground 8
Inv. Input 1
V_{CC} 7
AGC Input 2
Output (−) 6
Non-Inv. Input 3
Output (+) 5
Substrate Ground 4

MAXIMUM RATINGS (T_A = +25°C unless otherwise noted)

Rating	Symbol*	Value	Unit
Power Supply Voltage	V_{CC}	+18	Vdc
Output Supply	V_O	+18	Vdc
AGC Supply	$V_{2(AGC)}$	V_{CC}	Vdc
Differential Input Voltage	V_I	5.0	Vdc
Operating Temperature Range	T_A	−55 to +125	°C
Storage Temperature Range	T_{stg}	−65 to +150	°C
Junction Temperature	T_J	+175	°C

ADMITTANCE PARAMETERS (V_{CC} = +12 Vdc, T_A = +25°C)

Parameter	Symbol	f = MHz Typ 30	f = MHz Typ 60	Unit
Single-Ended Input Admittance	g_{11}	0.4	0.6	mmhos
	b_{11}	1.2	−3.0	
Single-Ended Output Admittance	g_{22}	0.05	0.1	mmho
	b_{22}	0.50	1.0	
Forward Transfer Admittance (Pin 1 to Pin 5)	$|Y_{21}|$	175	150	mmhos
	θ_{21} (Polar)	−30	−105	degrees
Reverse Transfer Admittance*	g_{12}	−0	−0	μmhos
	b_{12}	−5.0	−10	

*The value of Reverse Transfer Admittance includes the feedback admittance of the test circuit used in the measurement. The total feedback capacitance (including test circuit) is 0.025 pF and is a more practical value for design calculations than the internal feedback of the device alone. (See Figure 10.)

REPRESENTATIVE CIRCUIT SCHEMATIC

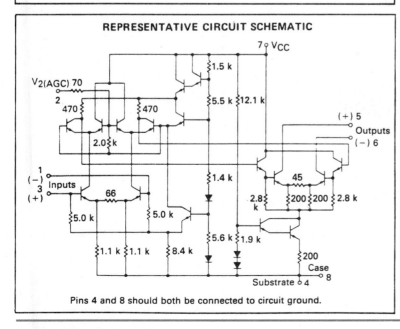

Pins 4 and 8 should both be connected to circuit ground.

SCATTERING PARAMETERS (V_{CC} = +12 Vdc, T_A = +25°C, Z_O = 50 Ω)

Parameter	Symbol	f = MHz Typ 30	f = MHz Typ 60	Unit
Input Reflection Coefficient	$|S_{11}|$	0.95	0.93	–
	θ_{11}	−7.3	−16	degrees
Output Reflection Coefficient	$|S_{22}|$	0.99	0.98	–
	θ_{22}	−3.0	−5.5	degrees
Forward Transmission Coefficient	$|S_{21}|$	16.8	14.7	–
	θ_{21}	128	64.3	degrees
Reverse Transmission Coefficient	S_{12}	0.00048	0.00092	–
	θ_{12}	84.9	79.2	degrees

FIGURE 5.16   Envelope Detector

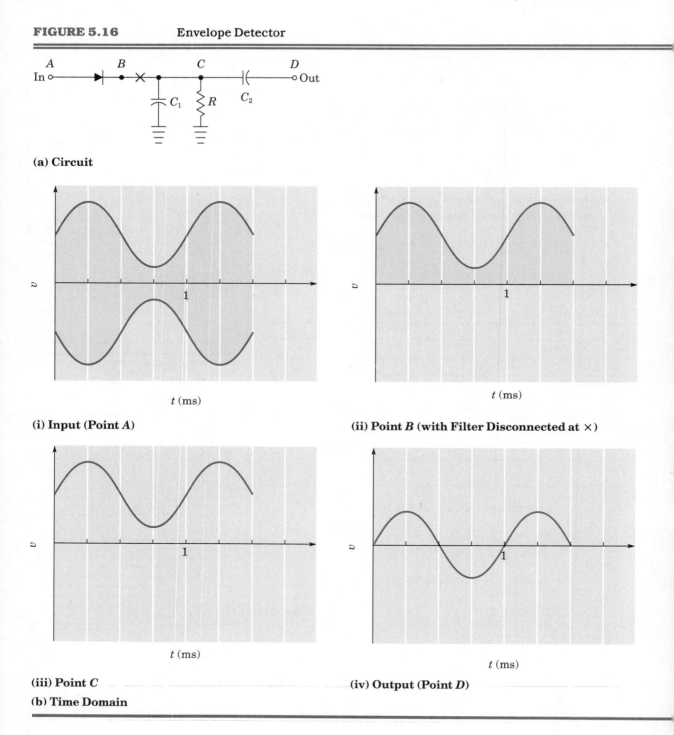

(a) Circuit

(i) Input (Point A)

(ii) Point B (with Filter Disconnected at ×)

(iii) Point C

(iv) Output (Point D)

(b) Time Domain

      In the frequency domain, an envelope detector can be considered as a diode mixer that mixes the carrier and sideband frequencies to give sum and difference frequencies, followed by a low-pass filter to reject the sum, as well as the original carrier and sideband frequencies. The difference is the recovered modulating signal. Figure 5.16(c) shows this interpretation for an IF of 1 MHz and a modulating frequency of 1 kHz.

      For best results, the signal voltage should be sufficient to fully turn on

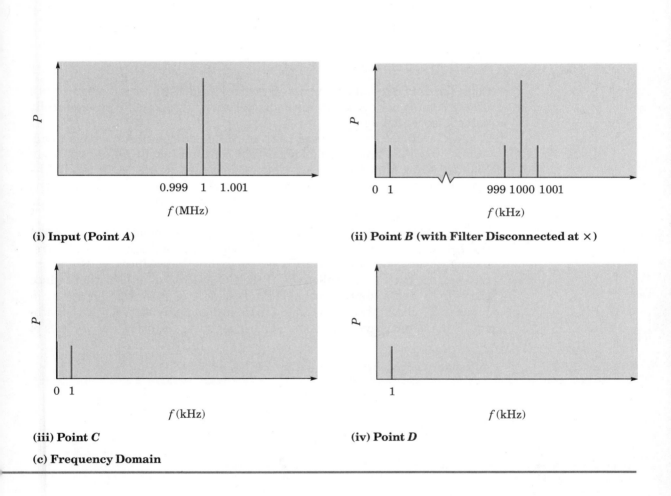

(i) Input (Point A)

(ii) Point B (with Filter Disconnected at ×)

(iii) Point C

(iv) Point D

(c) Frequency Domain

the diode, even at those moments when the envelope has lowest amplitude. Germanium diodes are often used for detectors because of their low turn-on voltage of about 300 mV. Even so, it can be seen that, as modulation approaches 100%, there will inevitably be distortion due to the nonlinearity of the diode.

The low-pass filter is usually a single RC section. Its time constant must be much greater than the period corresponding to the IF to filter out the carrier component, and much shorter than the period of the modulation to avoid distorting the envelope. This is more a theoretical than a practical problem, since the IF of the receiver is usually several orders of magnitude greater than the frequency of the modulating signal. As an example, consider a receiver for the AM broadcast band. Assuming a carrier frequency of 1 MHz modulated by an audio frequency of 1 kHz, the carrier T_c and modulating signal T_m would have periods of 1 μs and 1 ms respectively. A good value for the detector time constant T_{det} would be approximately the geometric mean between the two periods. That is,

$$
\begin{aligned}
T_{det} &= \sqrt{T_c T_m} \\
&= \sqrt{(1 \times 10^{-3} \text{ s})(1 \times 10^{-6} \text{ s})} \\
&= 31.6 \times 10^{-6} \text{ s} \\
&= 31.6 \text{ μs}
\end{aligned}
$$

5.4.5 Automatic Gain Control

Some form of gain control is necessary before the detector to reduce gain with strong signals and prevent overloading. This is usually done automatically using a feedback circuit.

AGC voltage can be derived from an AM diode detector by using an additional low-pass filter with a longer time constant (about 1 s), as shown in Figure 5.17(a). Sometimes, a separate diode is used for the AGC circuit. The resulting dc voltage will vary with carrier amplitude and is used to adjust the bias on transistors in IF and sometimes RF amplifiers. Its polarity can be reversed by simply reversing the diode.

Figure 5.18 shows approximately how the gain of a bipolar transistor varies with collector current. Starting at the current I_c, either increasing or reducing the current will reduce the gain. Circuits that reduce the gain by increasing the current are called *forward AGC systems*, and those that reduce the gain by decreasing the current are *reverse AGC systems*. The circuit in Figure 5.17 uses reverse AGC: an increase in signal strength causes the cathode of the detector diode to become more negative, resulting in the bases of

FIGURE 5.17   Automatic Gain Control

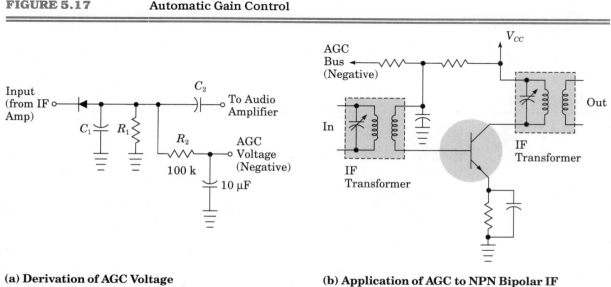

(a) Derivation of AGC Voltage

(b) Application of AGC to NPN Bipolar IF Amplifier

FIGURE 5.18

Variation of Transistor
Gain with Collector
Current

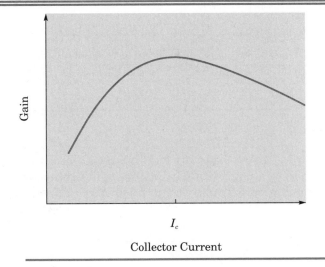

Collector Current

the controlled transistors becoming less positive. Since the transistors are NPN, the bias is reduced with increasing signal strength.

A representative IF amplifier stage, with AGC applied, is shown in Figure 5.17(b). The amplifier circuit uses conventional voltage-divider bias, except that the negative end of the voltage divider is connected to the AGC bus instead of to ground. As the signal becomes stronger at the detector, the AGC bus becomes more negative, and the voltage at the transistor base becomes lower. This reduces the transistor emitter current, lowering the gain of the stage.

Many communications receivers allow for a manual override of the AGC, using a control misleadingly labeled "RF Gain," although it is usually the gain of the IF stages that is being adjusted.

Many modern IF amplifiers use integrated circuits, in which case there will be a gain-control terminal to which an AGC voltage can be applied.

Delayed AGC. Ordinary AGC reduces gain even for quite weak signals. More advanced AGC systems operate only when the signal reaches a threshold value, set sufficiently low to avoid amplifier overload. Thus the "delay" in the name is really based on the assumption that the signal gradually increases in amplitude over time. The effect of delayed AGC is to increase sensitivity.

5.5 RECEIVER VARIATIONS

Previous sections have covered receiver basics. Next, we will look at some extra features that are found in some, but not all, receivers.

5.5.1 Crystal and Ceramic Filters

The use of quartz crystals as frequency-determining components in oscillators was described in Chapter 2. It should be apparent that the properties of

FIGURE 5.19

Crystal Filter

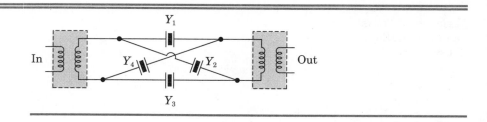

a crystal that make it useful in an oscillator, namely very high Q and excellent stability, could also be used to advantage in the design of bandpass filters. Using only one crystal (instead of a tuned circuit) would actually make the bandwidth too narrow for most purposes, but several crystals, tuned to slightly different frequencies, can be used in a *crystal lattice* to make a bandpass filter with a bandwidth of several kilohertz and an excellent shape factor. Figure 5.19 shows the schematic diagram for a typical *crystal filter*.

Some ceramics exhibit a piezoelectric effect similar to that of quartz, and can also be used as resonators and filters. The Q of *ceramic filters* is lower than for quartz, and the bandwidth is wider. They are quite useful for wideband signals, such as broadcast FM.

Crystal and ceramic filters have the additional advantage, compared with conventional IF amplifier circuitry, of requiring no adjustment. IF amplifiers with multiple tuned transformers require a fairly elaborate process of alignment to set all the tuned circuits to the correct frequencies. Due to slight differences between different components, this must be done by hand *after* the receiver is constructed. This labor-intensive process is not required with crystal and ceramic filters, nor with the mechanical and surface-acoustic-wave filters (to be described next).

It is quite possible to include more than one crystal filter in a receiver, for use with signals having different bandwidths. The correct filter can be selected with an ordinary switch, or—more elegantly—diode switching can be used.

5.5.2 Mechanical and Surface-Acoustic-Wave Filters

An older technique, no longer much used, is to use mechanical resonators (consisting of discs and rods) as elements in a bandpass filter. Transducers at the input and output convert electrical energy to mechanical, kinetic energy and back again. These **mechanical filters** work very well, but only at frequencies up to about 500 kHz.

A modern filter type based on similar principles is the **surface-acoustic-wave** (SAW) filter. Again, transducers are used, and electrical energy is converted into mechanical vibrations. This time, however, the vibrations are transverse waves on the surface of a piezoelectric substrate. The transducers are simply electrodes on the surface. By careful shaping of the substrate and the transducers, a wide variety of responses can be recreated at frequencies well into the megahertz range. For instance, television receivers, with an IF of 45.75 MHz for the picture signal, often use SAW filters. The problem in a television receiver is not to achieve a narrow bandwidth, since the signal is almost 6 MHz wide, but to produce a *carefully shaped* (not flat) response. The SAW filter is very well suited to this application.

5.5.3 Double Conversion

The choice of IF usually involves a compromise. It should be high for image rejection, but satisfactory gain and selectivity are more easily obtained at lower frequencies. "High" and "low" are relative, of course, and over the years standard designs have evolved that achieve reasonable compromises. For instance, 455 kHz gives fair image rejection at AM broadcast frequencies (on the order of 1 MHz), and the relatively narrow bandwidth of most AM broadcast receivers (about 10 kHz) can easily be obtained at that frequency. FM broadcasting, with a signal frequency 100 times as high, requires a higher IF for reasonable image rejection, but, luckily, the receiver bandwidth need not be as narrow; in fact, about 150 to 200 kHz is needed. A frequency of 10.7 MHz has become very common for the IF in such receivers.

Suppose, however, that the signal frequency is high and the required bandwidth is narrow. Such a condition can be found in high-frequency and VHF communications receivers. There may be no satisfactory compromise for an IF that provides both excellent image rejection and satisfactory selectivity. There are two solutions to this problem. One is to use a high IF for image rejection, and to achieve narrow bandwidth by using more exotic filters than the simple tuned circuits referred to above. Crystal filters can be used well into the megahertz range, and so they allow the construction of excellent single-conversion receivers for narrow-bandwidth signals into the VHF range.

Another approach, which actually is often combined with the first, is to use double conversion. (See Figure 5.20 for a block diagram.) The first mixer is conventional, using a tunable local oscillator and converting all incoming signals to a first IF, which will be at a relatively high frequency for excellent image rejection. In fact, in some designs, the first IF is actually higher than the incoming signal frequency, and the sum, rather than the difference, of the signal and local oscillator frequencies is used. This type of mixing is called *up-conversion*.

FIGURE 5.20   **Double-Conversion Receiver**

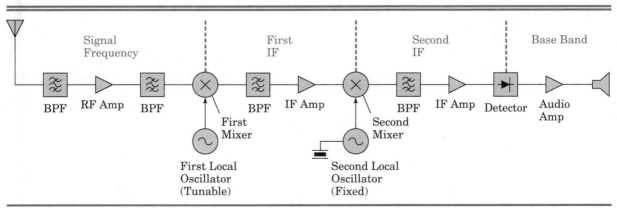

The first mixer is followed by a relatively wideband filter and perhaps some gain. Then the signal enters another mixer whose local oscillator operates at a fixed, crystal-controlled frequency. The signal is converted to a lower second IF, where most of the gain and selectivity are provided. The filter in the first IF stage is there just to avoid image responses in the second mixer; it is the second IF filter that sets the bandwidth of the receiver. Double-conversion receivers require careful design because the extra mixer and local oscillator can give rise to additional spurious signals.

Of course, there is no reason that the number of conversions has to be limited to two. Some communications receivers use triple and even quadruple conversion.

It is also possible to make the first local oscillator fixed and the second variable in frequency. This is often done in systems where the first conversion takes place in a separate unit, often called, logically, a *converter*. The *block converters* used in some cable and satellite television systems (to be described in Chapter 14) are examples of this method, which has the advantage of requiring no adjustments on the converter unit.

EXAMPLE 5.8

For the receiver whose block diagram is shown in Figure 5.21, find the intermediate frequencies, and state whether each local oscillator uses high-side or low-side injection.

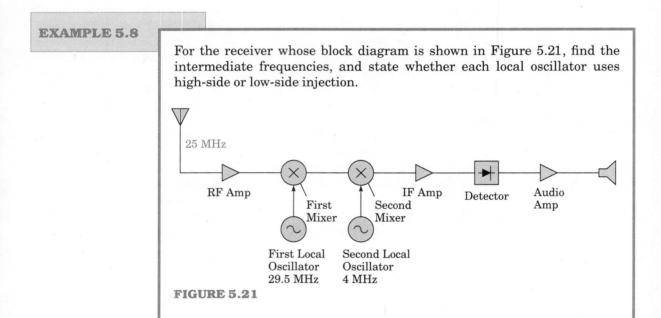

FIGURE 5.21

Solution

The signal input frequency is 25 MHz and the first local oscillator frequency is 29.5 MHz. Therefore the first local oscillator uses high-side injection, and the first IF is

$$f_{IF_1} = f_{LO_1} - f_{sig}$$
$$= 29.5 \text{ MHz} - 25 \text{ MHz}$$
$$= 4.5 \text{ MHz}$$

The second local oscillator operates at 4 MHz. Therefore it uses low-side injection, and the second IF is

$$f_{IF_2} = f_{IF_1} - f_{LO_2}$$
$$= 4.5 \text{ MHz} - 4 \text{ MHz}$$
$$= 0.5 \text{ MHz}$$
$$= 500 \text{ kHz}$$

5.6 COMMUNICATIONS RECEIVERS

Of course, all receivers are used for communication, but the term *communications receiver* is used mainly for general-purpose receivers that cover a relatively wide range of frequencies. For instance, a typical receiver might cover the range from 100 kHz to 30 MHz. It would include low-frequency navigation signals, the AM standard broadcast band, and the many services that use the high-frequency range, including shortwave broadcasting, military and commercial radioteletype, amateur radio, citizens' band, and others. Other similar receivers are available for the VHF and UHF frequency ranges.

Communications receivers generally divide their frequency coverage into several bands. They are usually equipped to receive several types of modulation, including AM. In addition, they will have some of the special features listed below, which are included to make reception easier under difficult conditions.

5.6.1 Squelch

A **squelch** circuit disables the receiver audio in the absence of a signal. It can be implemented using the AGC voltage. When the voltage is very low, the audio amplifier is biased off. Squelch is very convenient for two-way mobile radio, as it eliminates channel noise in the absence of a carrier.

More elaborate squelch systems, which will shut off the audio unless a particular station is being received, are also possible. These respond to a transmitted signal, which can vary from a simple low-frequency (subaudible) tone to a complex digital code sent at the beginning of each transmission. This type of squelch is useful in mobile radio systems, as it enables a central dispatcher to address the mobile units individually or in any desired combination.

5.6.2 Noise Limiters and Blankers

Impulse noise, unlike thermal noise, can be removed to some extent from a signal in the receiver. This type of noise derives from sources such as lightning and automobile ignition systems, and is characterized by short pulses with relatively large amplitude and fast rise times.

The simplest way to deal with these noise pulses is simply to use a diode *limiter*, or *clipper*, in the audio section of the receiver. The limiting threshold

FIGURE 5.22   **Noise Limiters for AM Receivers**

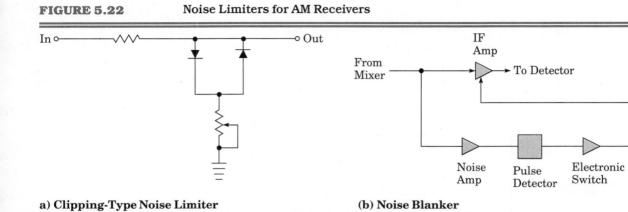

a) **Clipping-Type Noise Limiter**   (b) **Noise Blanker**

should be set just above the peak level of the audio signal. Noise pulses, then, will be limited to about the same peak amplitude as the signal itself. Since the noise pulses are generally of large amplitude and short duration, clipping them to the same peak level as the rest of the signal removes much of their energy, rendering them less disturbing. Figure 5.22(a) shows the circuit of a simple clipper that will remove peaks greater than the diode threshold voltage.

A better system is to *blank* noise pulses by muting the receiver output for the duration of the pulse. A block diagram of such a system is shown in Figure 5.22(b). It works by switching off the IF stage, rather than the audio stages. This is more effective because it avoids any broadening of the noise pulse by the IF bandpass filter.

Both of these systems sense noise pulses by their large amplitude and fast rise time. Neither is effective in removing thermal noise, which is generally at a level below that of the signal.

5.6.3 Notch Filters

One of the most annoying types of interference occurs when two stations have almost the same carrier frequency. If an interfering signal has a carrier frequency 1 kHz above that of the desired signal, for example, the two carriers will mix in the detector to produce a difference frequency signal at 1 kHz, which will cause a very audible whistle.

Some communications receivers incorporate a tunable band-reject (*notch*) filter in the IF stage. This filter can be tuned to the frequency of the interfering carrier, greatly attenuating it and avoiding the audible whistle. Of course, this solution is not perfect: the notch filter will not attenuate the sidebands of the interfering signal. However, as shown in Chapter 3, these sidebands have an amplitude that is much smaller than that of the interfering carrier, so their potential for interference is less. The notch filter will also attenuate some of the sidebands of the desired signal, causing distortion, but—if the filter is narrow—the distortion created will be less objectionable than the interference eliminated.

5.6.4 The S-Meter

The **S-meter** is designed to indicate signal strength for comparison and as an aid to tuning the receiver, and sometimes, for adjusting the antenna. It is usually calibrated in units from 1 to 9, and then in "decibels above S-9." According to one standard, S-9 represents 50 μV at the antenna terminals, and each S-number represents 6 dB. However, there is really very little standardization, and S-meter readings are almost useless, except as an aid to tuning.

The AGC line provides a handy source for a voltage that varies with signal strength, and is often used as a source for the S-meter circuit.

EXAMPLE 5.9

An S-meter as described above reads S-6. Calculate the signal strength at the receiver input.

Solution

Since each S-number represents 6 dB, the signal will be 18 dB less than 50 μV. Therefore, using the equation

$$dB = 20 \log \frac{V_1}{V_2}$$

we can let $dB = 18$. We know that

$$V_1 = 50 \ \mu V$$

so we can solve for V_2:

$$\frac{V_1}{V_2} = \text{antilog} \ \frac{dB}{20}$$

$$V_2 = \frac{V_1}{\text{antilog} \ (dB/20)}$$

$$= \frac{50 \ \mu V}{\text{antilog} \ (18/20)}$$

$$= 6.29 \ \mu V$$

5.6.5 Example: A Modern Communications Receiver

Figure 5.23(a) is a photograph of a typical communications receiver of modern design, the ICOM IC-R71A. The specifications for this receiver are given in Figure 5.23(b), and the block diagram can be found in Figure 5.23(c). You will note from the specifications that this receiver can receive several other "modes" (types of modulation) besides AM. We will return to this type of receiver in later chapters as we investigate the other modulation schemes. For the present, however, we are equipped to understand most of the receiver's characteristics, since they are the same for all modes.

FIGURE 5.23   The ICOM IC-R71A

(a) Communications Receiver

IC-R71A/E/D: SUPERIOR SHORTWAVE RECEIVER.

Put the world at your fingertips with the advanced IC-R71A/E/D, Icom's popular all-purpose communications receiver. Listen to both domestic and overseas broadcasts as well as maritime, aeronautical, news agency, government, CB (Citizen's Band) and amateur radio messages. The easy-to-use receiver is ideal for anyone wanting to listen to worldwide communications: professional and amateur radio operators, SWLs (Short Wave Listeners) and BCLs (Broadcasting Listeners). No previous shortwave operating experience is necessary.

ALL MODE AND GENERAL COVERAGE

The IC-R71A/E/D is a high performer from 100 kHz to 30 MHz continuously*[1] in SSB (LSB, USB), CW, AM, FM*[2] and RTTY modes. These modes are suitable for receiving most long wave, medium wave and shortwave stations.

*[1] Refer to frequency coverage in specifications.
*[2] An optional IC-EX257 FM UNIT is required.

SUPERIOR RF CIRCUITS

Front end circuits in the IC-R71A/E/D incorporate Icom's DFM (Direct Feed Mixer) which ensures higher intermodulation and cross modulation rejections. The IC-R71A/E/D has a wide 100 dB dynamic range and utilizes a well-designed triple up-conversion scheme which minimizes image and spurious responses. These circuits make it possible to receive signals from DX (long distance) or weak signal stations.

*FM mode : double conversion

NOTCH FILTER SYSTEM

Icom's notch filter can be tuned to eliminate interfering signals. This system ensures you clear and comfortable listening.

DIRECT FREQUENCY ENTRY

Direct frequency entry from the KEYBOARD allows complete programming versatility. This is especially convenient when the frequency is already known as with a broadcast station.

(b) Specifications

32 MEMORY CHANNELS

The IC-R71A/E/D is equipped with 32 fully tunable memory channels. The operating frequency and mode can be stored in any of 32 memory channels. Memory channel selection using the MEMORY-CH selector allows you to quickly recall all memory channels. In addition, the frequencies in the memory channels can be easily changed via the TUNING CONTROL.

REMOTE CONTROL

Utilizing the optional RC-11 WIRELESS REMOTE CONTROLLER, you can remotely control the IC-R71A/E/D's numerous functions such as power ON/OFF, audio level, receive frequencies, and memory channels.

SCAN FUNCTIONS

Three convenient scan functions are available in the IC-R71A/E/D:
- Programmed scan repeatedly scans between two programmable scan edges.
- Memory channel scan repeatedly scans all memory channels in succession.
- Selected mode memory scan repeatedly scans memory channels with the same mode.

OPTIONAL VOICE SYNTHESIZER UNIT

The optional EX-310 VOICE SYNTHESIZER UNIT announces frequency readings in English.

OPTIONAL FILTERS

Icom offers you the optional FL-32A and FL-63A CW AND RTTY NARROW FILTERS for clear CW and RTTY reception.
- FL-32A : 9.0106 MHz, 500 Hz/−6 dB
- FL-63A : 9.0106 MHz, 250 Hz/−6 dB

CI-V SYSTEM

Utilizing an optional CT-17 CI-V LEVEL CONVERTER with a UX-14 CI-IV/CI-V CONVERTER, many control functions can be carried out from a personal computer equipped with an RS-232C socket.

ADDITIONAL FUNCTIONS

The IC-R71A/E/D has the following additional functions:
- Selectable FAST, SLOW and OFF AGC control
- Level and wide/narrow pulse width selectable noise blanker
- All mode squelch
- Receive tone control

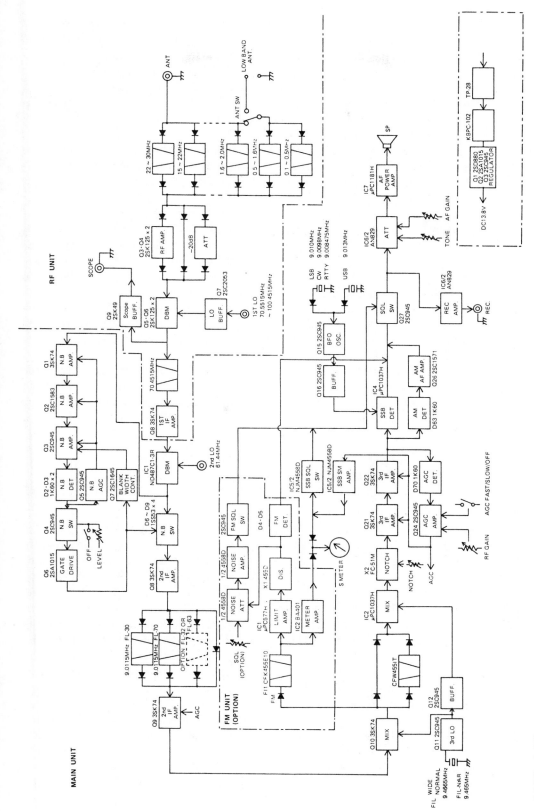

(c) Block Diagram

Courtesy ICOM America, Inc.

The specifications are typical for good modern receivers. The sensitivity is specified in terms of a 10 dB $(S+N)/N$ ratio, as is usual for AM receivers. The specified sensitivity of 0.5 µV is good, but not outstanding. The selectivity for AM is given in terms of the bandwidth at 6 dB and 50 dB down from the center frequency. The 6 dB bandwidth for AM is 6 kHz maximum, and can be reduced. This clearly indicates that the receiver is designed with communications—and not entertainment—in mind, as the bandwidth is too narrow for good music reproduction.

The receiver is actually quadruple conversion, though it has only three different intermediate frequencies. The fourth conversion is used for a variable bandwidth feature. The high first IF of 70.4515 MHz is largely responsible for the excellent image-rejection specification of 60 dB.

The receiver front end uses fixed-tuned bandpass filters, which are selected by diode switching. Instead of applying AGC to the RF amplifier (Q_3 and Q_4), this receiver has a switch that allows the operator to select the amplifier, to send the signal straight through, bypassing the RF stage, or to select a 20 dB attenuator for very strong signals. Diode switching is used to accomplish this selection.

The first mixer is a double-balanced type (Q_5 and Q_6). The mixer uses high-side injection of the local oscillator, which is actually a frequency synthesizer. The first IF signal passes through a bandpass filter and one stage of amplification (Q_8).

The second mixer (IC_1) is also a double-balanced type. Its local oscillator operates at a fixed frequency of 61.44 MHz, producing a second IF whose frequency is the difference between the first IF and the second local oscillator frequency. This second IF is:

$$f_{IF_2} = f_{IF_1} - f_{LO_2}$$
$$= 70.4515 \text{ MHz} - 61.44 \text{ MHz}$$
$$= 9.0115 \text{ MHz}$$

Note that the second local oscillator uses low-side injection.

The next stage is the noise blanker switch (D_8 and D_9). By accomplishing noise blanking at this point, the receiver designers have avoided the broadening of noise pulses that could occur in the narrow IF filters that follow.

There is provision for a choice of three filters, selected by diode switching, at the second IF, though the third is optional. This provides a selection of bandwidths for different modulation types. For AM, a bandwidth of 6 kHz is selected. The filters are located between the two stages of amplification at the second IF (Q_8 and Q_9).

The next part of the receiver is rather unusual. The signal is mixed down, in Q_{10}, to a third IF of 455 kHz. It passes through a crystal filter and is then immediately mixed back up, in IC_2, to 9.0115 MHz. The same local oscillator is used for both mixers, so the fourth IF will be exactly the same frequency as the third. However, this oscillator is a variable crystal oscillator (VXO) which can be tuned over about a 4 kHz range. This allows the position of the incoming signal to be moved within the crystal filter passband, which can be a useful way of avoiding interference. There is no amplification at the 455 kHz IF.

The last IF section, in addition to two more gain stages (Q_{21} and Q_{22}), has a tunable notch filter that, as described earlier, can be used to filter out annoying interfering carriers.

The rest of the receiver is simple enough. There is an AM detector, a squelch circuit, and audio amplifiers. (There are SSB and FM detectors as well, but we will discuss them in later chapters.) An S-meter is also provided.

As can be seen, this is a very complex receiver. However, a careful approach, beginning with the block diagram, can help to unravel some of its mysteries.

5.7   TRANSCEIVERS

A transceiver is essentially just a transmitter and a receiver in the same box. As well as being convenient, transceivers allow certain economies to be made.

Most two-way radio schemes do not require *full-duplex operation*, which allows the operator to talk and listen simultaneously, as on an ordinary telephone. *Half-duplex* communication, where the station transmits and receives alternately, is more common. Since the transmitter and receiver are never used at the same time, it is possible to use some of the same circuitry for both transmitting and receiving.

Audio circuitry is one example. The transmitter needs an audio amplifier to boost the microphone output to sufficient power to modulate the transmitter. The receiver also needs an audio amplifier, to amplify the detector output to an adequate level to drive a loudspeaker. With suitable switching, the same amplifier can be used for both purposes.

5.7.1   Example: A Citizens' Band Transceiver

Figure 5.24 is the block diagram of a typical CB transceiver. The frequencies shown are those that are present for an operating frequency of 26.965 MHz (channel 1). We looked at the transmitter section of this unit in Chapter 4. Now, let us look at the receiver section, and see how certain economies have been made by using some stages for both transmitting and receiving.

A look at the schematic, Figure 5.25, shows that the actual circuitry is quite conventional. In particular, the receiver and transmitter both use discrete, bipolar technology, except for the frequency synthesizer and the audio power amplifier, which employ integrated circuits.

In Chapter 3, we found that the transmitter section has five stages: oscillator (which is a frequency synthesizer), frequency doubler, and three amplifier stages. The driver and power amplifier are both modulated. When transmitting, the frequency synthesizer operates at half the transmitting frequency (13.4825 MHz in this example).

The frequency synthesizer and audio power amplifier are also used in the receiver. The transceiver thus has economies compared with a separate receiver and transmitter.

The receiver is double conversion. It has a single RF stage, followed by a mixer that converts the incoming frequency to a 10.695 MHz first IF. This is high enough to achieve the specified image rejection of at least 60 dB. This

FIGURE 5.24 Block Diagram of CB Transceiver

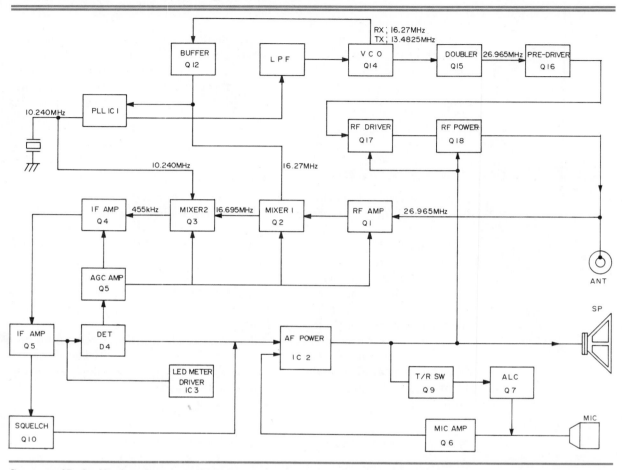

Courtesy of Radio Shack, a division of Tandy Corporation.

particular choice of frequency also allows the use of components, such as IF transformers and a ceramic filter, designed for the 10.7 MHz IF used in FM broadcast receivers.

The synthesizer that sets the transmitting frequency also serves as the first local oscillator for the receiver: it switches frequencies between transmit and receive. Note that the first mixer uses low-side injection. This requires it to operate at the receive frequency less 10.695 MHz (16.270 MHz in this example).

The synthesizer is based on a 10.240 MHz crystal oscillator, which also provides a fixed local oscillator signal for the receiver's second mixer. That mixer also uses low-side injection. The difference between the first IF of 10.695 MHz and 10.240 MHz is 455 kHz, which is the second IF. You will recall that this is the usual IF for AM broadcast receivers. Again, it offers the designer a choice of available, low-cost components.

The rest of the receiver is conventional, with two stages of IF amplifica-

FIGURE 5.25

Schematic Diagram of CB Transceiver

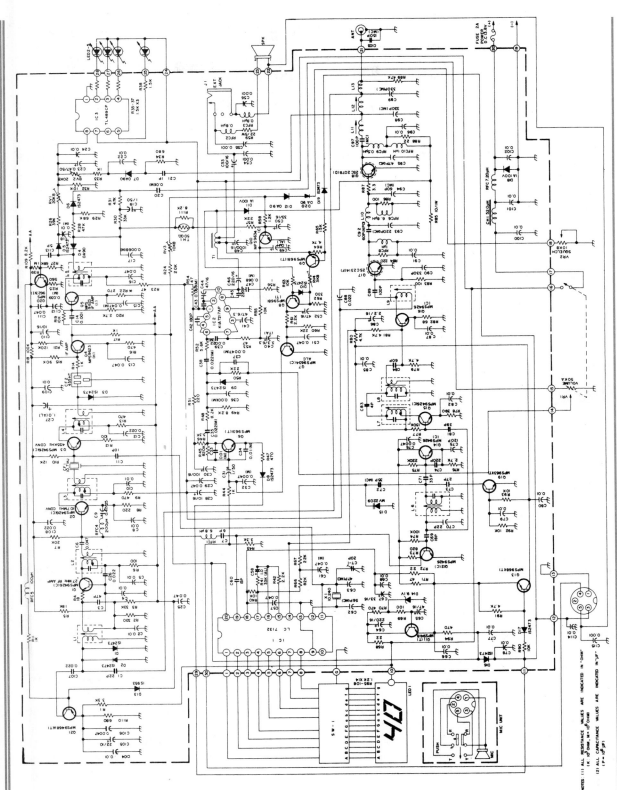

NOTES (1) ALL RESISTANCE VALUES ARE INDICATED IN "OHM"
1K = 10³ OHM, M = 10⁶ OHM

(2) ALL CAPACITANCE VALUES ARE INDICATED IN "µF"
(P = ×10⁻⁶ µF)

Courtesy of Radio Shack, a division of Tandy Corporation.

tion, an envelope detector, and an audio amplifier. Squelch and S-meter circuits are provided.

Looking more closely at the schematic diagram, we see that Q_1, the RF amplifier, is in the common-base configuration. The pair of diodes at its input (D_1 and D_2) are to protect Q_1 from damage while the transceiver is transmitting. D_1 and D_2 limit the voltage at the receiver input to approximately 700 mV peak.

The receiver RF stage is fixed-tuned, so no tracking with the local oscillator is necessary. The output of the first mixer Q_2 is coupled by a single-tuned transformer L_3 to a 10.7 MHz ceramic filter CF_1, and then directly to the second mixer. There is no amplifier stage at the first IF. The second mixer Q_3 sends its output through another single-tuned transformer L_4 to another ceramic filter CF_2, which operates at 455 kHz. Two stages of IF amplification follow.

The detector is a conventional envelope type, using diode D_4. D_5 provides a voltage proportional to carrier strength for the S-meter, which is a row of LEDs driven by IC_3. The detected signal is low-pass filtered and amplified by Q_{20} before being used as an AGC voltage to control the bias on the RF, mixer, and IF stages. The audio is amplified by an integrated amplifier IC_2, and routed to the speaker.

5.8   AM RECEIVER TEST PROCEDURES

At the beginning of this chapter, we looked at some of the major criteria for receiver quality, such as sensitivity and selectivity. In the meantime, we have found that there are other important receiver characteristics, some of which (like image rejection) are unique to the superheterodyne system. We will now consider how some of these characteristics can be measured.

5.8.1   Sensitivity

Any specification of *sensitivity* requires some mention of noise level to have much meaning. Signal-to-noise ratio (S/N) is difficult to measure directly because of the difficulty of completely separating the two quantities. Usually, signal-plus-noise to noise [$(S+N)/N$], or SINAD (signal-plus-noise-and-distortion to noise-and-distortion) are measured instead. Manufacturers usually use the $(S+N)/N$ method, to be described here, to specify the sensitivity of AM receivers. (The SINAD method is more commonly used with FM receivers, so it will be studied in Chapter 9.)

AM receiver sensitivity is usually specified as the minimum signal level, with 30% modulation, that will give a 10 dB $(S+N)/N$ ratio with at least 500 mW of audio output. Measuring it requires a calibrated RF signal generator and an audio voltmeter. Figure 5.26 shows a typical test setup. The receiver audio gain, and RF gain if present, should be at maximum, and any squelch circuit must be off. The RF generator output impedance should be the same as the receiver input impedance. If it is not, an impedance-matching network must be used, and any losses in the network must be taken into account. All covers should be on the receiver, and all cables should

FIGURE 5.26

Test Setup for
Measurement of
Receiver Sensitivity

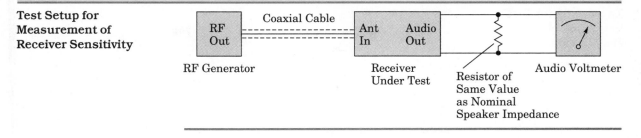

be shielded, to prevent the entry of noise other than that generated in the receiver.

To measure $(S+N)/N$, first measure the receiver audio output with the signal modulated to a depth of 30% by a 1 kHz tone. This measurement will include the signal plus the noise. Then switch the modulation off at the generator, leaving the carrier on. The remaining output from the receiver is the noise. The carrier is left on so that the receiver AGC level will not change. Otherwise, the AGC is likely to increase the receiver gain, and hence its noise output, when the carrier is turned off.

If both of the above measurements are taken in decibels with respect to the same reference (in dBV, for example), then simple subtraction will yield $(S+N)/N$ in decibels. If the result is greater than 10 dB, the RF generator carrier level should be reduced, and the measurements repeated. If the ratio is less than 10 dB, the generator level must be increased.

When the 10 dB ratio is reached, the audio output level should be checked to make sure it is at least 500 mW, or as specified by the manufacturer. This is rarely a problem. The power can, of course, easily be calculated from the voltage across a known load resistance. A loudspeaker does not make a very satisfactory load, as its impedance may vary widely with frequency, and may not be at all close to its nominal value. Besides, even 500 mW of a 1 kHz tone is annoyingly loud! A noninductive resistor, of the same value as the nominal speaker impedance, and used in place of the speaker, is better.

Assuming that at least the minimum audio output power is achieved, the carrier level necessary to achieve 10 dB $(S+N)/N$ is the sensitivity; otherwise, the carrier level is increased until the specified audio output is obtained, and that carrier level is the sensitivity.

5.8.2 Selectivity

There are as many ways to measure selectivity as there are to specify it. Two representative methods will be described; others can be found in manufacturers' service manuals.

A straightforward method using only one RF generator is available. The test setup is the same as for sensitivity measurements (see Figure 5.26). The generator is tuned to the same frequency as the receiver, and set for the level that provides a 10 dB $(S+N)/N$ reading, just as for sensitivity measurements. Then either the receiver or the generator is detuned, and the genera-

FIGURE 5.27

Two-Generator Test
Setup for Selectivity
Measurements

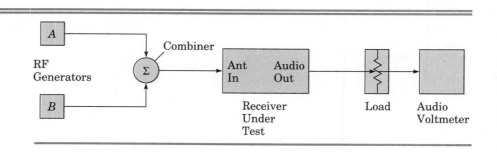

tor level increased until the 10 dB $(S+N)/N$ ratio is restored. The ratio, in decibels, between the two generator levels, is the amount by which the off-frequency signal is attenuated. With this method, it helps to switch off or otherwise disable the AGC, so that the receiver gain will remain constant.

This method is easy to understand and to perform, but its results may not predict very accurately the receiver's performance under conditions of actual interference. This is because a strong interfering signal may overload the receiver's front end, causing **blocking**. Figure 5.27 shows a test setup for selectivity measurement using two RF generators that approximates real-word conditions more closely. The attenuation of the combiner network must be taken into account in the measurements. Generator A is set to the receiver's frequency, and is modulated to a depth of 30% by a 1 kHz tone. Generator B is set to the adjacent channel, and is modulated 30% by a 400 Hz tone. With generator B's output off, generator A is adjusted to give an $(S+N)/N$ reading of 10 dB, as for a sensitivity measurement. Then generator B is switched on, and its carrier level is increased until the $(S+N)/N$ ratio drops to 6 dB. The ratio between the output power levels of the two generators, measured in decibels, is the adjacent channel selectivity.

This second method gives reasonably good results, as it approximates actual operating conditions where the adjacent channel signal interferes with the reception of a signal on the channel to which the receiver is tuned.

5.8.3 Intermediate Frequency and Image Rejection

These can be measured in the same way as selectivity: instead of being on the adjacent channel, the interfering signal is inserted at the intermediate frequency, or at an image frequency. Generally the first method described above, using one generator, is used.

5.8.4 Other Receiver Tests

While sensitivity and selectivity are probably the most important, they are certainly not the only receiver characteristics that can be measured. For instance, audio frequency response and distortion characteristics are often specified, and can be measured in almost the same way as for audio amplifiers. Of course, a modulated RF signal will be required, and it will be necessary to use a strong signal for distortion measurements, so that the distortion analyzer will not misinterpret noise as distortion. Many other specifications supplied with receivers are similarly easy to interpret. For instance, the squelch-circuit range can easily be measured by observing the strength of the signal required to *open the squelch* (turn on the audio) at each end of the squelch control's rotation. For these and other tests it is advisable to consult the manufacturer's service information.

From the foregoing, it should be obvious that it is only possible to compare specifications for different receivers if they have been obtained in the same way.

5.8.5 Troubleshooting

The tests above will identify some receiver problems such as low sensitivity and poor selectivity, and will allow for the comparison of receivers. Assuming that a receiver met its specifications when new, any deviation from specified performance represents a fault which can be found and corrected. A reasonably detailed knowledge of the particular receiver is required to correct subtle faults like poor sensitivity. Often this indicates a need for realignment of the tuned circuits. Information on alignment can be found in the receiver service manual.

Much of the time, receiver faults are not subtle enough to require detailed testing. For instance, the unit may not work at all, or its output may be so distorted as to be unintelligible. Again, the receiver may work, but be so insensitive that it requires an input signal measured in millivolts rather than microvolts.

The power supply should always be suspected when a radio is completely dead. Assuming all is in order there, cases like this can often be solved by using either of two time-honored and rather straightforward techniques known as *signal injection* and *signal tracing*.

In signal tracing, the object is to follow the signal through the receiver and see where it disappears or becomes severely distorted. Beginning at the antenna input, a sensitive detector can be moved through the receiver, checking the signal at the input and output of each stage. The detector can be an oscilloscope or a sensitive amplifier with a speaker. For RF and IF signals, a *demodulator probe* can be used. This is simply a diode detector that provides a demodulated audio signal to the amplifier or oscilloscope.

Signal injection is similar, except that a signal is applied to the device under test, starting at a point close to the output, and working back towards the antenna. The receiver's loudspeaker can be used as a monitoring device. Once again, the point where the signal disappears indicates the defective stage. Of course, the signal must be appropriate, in both amplitude and frequency, to the point in the circuit to which it is applied.

Once the problem has been localized to a single stage, other techniques must be used. In-circuit voltage measurements can be used to form an opinion as to the defective component. Then, individual components can be tested. Bear in mind that power-supply problems can sometimes cause the failure of only one stage (where a regulated supply is used for a critical circuit, for example).

Signal tracing and signal injection each have their difficulties. There may be a problem finding a device sensitive enough to detect a signal in the RF stage of a receiver, for instance, while injecting such a signal is no problem: an ordinary RF generator will do. It should be coupled through a capacitor to the circuit, to avoid any problems with dc voltages that may be present.

Signal injection requires a variety of signals. For an ordinary AM receiver, AM signals must be available at the carrier and the intermediate frequencies. A baseband audio signal must also be available, and the correct signal must be applied at each test point. Care must also be taken not to overload sensitive circuits too greatly (to the point where damage could occur).

An objection might be made that in modern receivers much of the circuitry is likely to be on one integrated circuit, and it makes very little difference which part of the chip is defective. That is certainly true; once a defect is localized to an integrated circuit, the whole chip must be replaced. Before doing so, however, and especially if the IC is soldered in place, it is a good idea to make sure that the chip is receiving all the correct supply voltages and signals.

Most AM broadcast receivers, except those incorporated in high-fidelity tuners, are so inexpensive that it is not cost-effective to repair them. A radio that sells new for twenty dollars is not worth very much of a technician's time. Nevertheless, the basic techniques of signal tracing and signal injection are applicable to many more complex and expensive systems, and a simple AM radio is a good place to learn these techniques.

EXAMPLE 5.10

Find suitable test points for signal injection or tracing in the circuit of Figure 5.5.

Solution

This is a typical AM broadcast-band receiver. Suitable test points are at the input and output of each stage. Some easily identifiable points in the mixer and IF stages are the transistor bases and collectors. The tap of the volume control is easy to find, and will provide access to the audio-amplifier input.

SUMMARY OF CHAPTER 5

Here are the main points to remember from this chapter.

1. A receiver must separate the desired signal from other signals and noise, and then demodulate the signal. A considerable amount of gain is also necessary in a practical receiver.

2. By far the most common receiver type is the superheterodyne, which uses a mixer/local oscillator combination to transfer all incoming signal frequencies to a common IF.

3. Better-quality receivers, particularly at the higher frequencies, use at least one stage of RF amplification before the mixer. The RF amplifier is principally responsible for setting the noise figure for the receiver.

4. Superheterodyne receivers can receive signals at other frequencies than that to which the receiver is tuned. The most important such signal is called the image frequency.

5. Images and other spurious responses must be rejected before the mixer. This requires bandpass filtering in the receiver before the mixer.

6. The IF must be high enough to provide good image rejection, but low enough to allow the required selectivity to be obtained with the type of filter in use.

7. The stability of a superheterodyne receiver depends directly on that of the local oscillator. Changing the local oscillator frequency tunes the receiver.

8. The IF amplifier is principally responsible for the selectivity of the receiver, and also provides most of the pre-detection gain.

9. Most AM receivers use envelope detectors. These have the advantage of great simplicity and the disadvantage of relatively high levels of distortion.

10. Receivers require some form of AGC to compensate for the very great range in signal strength that appears at the antenna.

11. The most important specifications for a receiver are sensitivity and selectivity. *Sensitivity* refers to the signal strength required for a satisfactory signal-to-noise ratio, and *selectivity* refers to the ability of the receiver to reject interference and out-of-channel noise.

12. Signal injection and signal tracing are useful troubleshooting techniques for receivers and many other types of electronic systems.

IMPORTANT EQUATIONS

$$Q = \frac{X_L}{R} \tag{5.1}$$

$$B = \frac{f_o}{Q} \tag{5.2}$$

$$SF = \frac{B_{-60 \text{ dB}}}{B_{-6 \text{ dB}}} \tag{5.3}$$

$$f_{image} \text{ (high-side injection)} = f_{sig} + 2\,f_{IF} \tag{5.7}$$

$$f_{image} \text{ (low-side injection)} = f_{sig} - 2f_{IF} \tag{5.8}$$

$$IR = \frac{A_{sig}}{A_{image}} = \sqrt{1 + Q^2 x^2} \tag{5.9}$$

where $x = \dfrac{f_{image}}{f_{sig}} - \dfrac{f_{sig}}{f_{image}}$

$$IR \text{ (dB)} = 20 \, \log \frac{A_{sig}}{A_{image}} \tag{5.10}$$

$$f_s = \left(\frac{m}{n}\right) f_{LO} \pm \left(\frac{f_{IF}}{n}\right) \tag{5.11}$$

$$k_c = \frac{1}{\sqrt{Q_p Q_s}} \tag{5.12}$$

$$k_{opt} = 1.5 k_c \tag{5.13}$$

$$B = k f_o \tag{5.14}$$

GLOSSARY

adjacent channel   the communications channel immediately above or below the desired channel in frequency

alternate channel   the communications channel next beyond the adjacent channel

autodyne converter   a combined mixer and local oscillator using the same transistor or tube for both

automatic gain control (AGC)   a circuit to adjust the gain of a system in accordance with the input signal strength

blocking   reduction of gain for a weak signal due to a strong signal close to it in frequency

ceramic filter   a bandpass filter using piezoelectric ceramic elements

converter   the combination of a mixer and a local oscillator that is used to move a signal from one frequency to another

crystal filter   a bandpass filter using piezoelectric quartz elements

front end   the first stage of a receiver

high-side injection   application to a mixer of a signal from a local oscillator that operates at a frequency above that of the incoming signal

image frequency   in a frequency converter, a second input frequency that will produce the same output frequency

intermediate frequency (IF)   a frequency to which a signal is shifted as an intermediate step in reception or transmission

local oscillator   an oscillator used in conjunction with a mixer to shift a signal to a different frequency

low-side injection   application to a mixer of a signal from a local oscillator that operates at a frequency below that of the incoming signal

mechanical filter   a bandpass filter that makes use of mechanical resonators

S-meter   a meter on a receiver that indicates the strength of the received signal

surface-acoustic-wave (SAW) filter   a filter that uses acoustic waves on the surface of a substrate to achieve the desired response

selectivity   the ability of a receiver to reject signals of frequencies other than that to which the receiver is tuned

sensitivity   the ability of a receiver to receive weak signals with a satisfactory signal-to-noise ratio

shape factor   for a bandpass filter, the ratio between the bandwidths for two specified amounts of attenuation

skin effect   reduction in effective cross-sectional area of a conductor with increasing frequency

spurious response   reception of signals at frequencies other than that to which a receiver is tuned

squelch   system which disables the output of a receiver in the absence of a suitable signal

superheterodyne receiver   receiver in which the signal is moved, using a mixer, to an intermediate frequency before demodulation

tracking   adjustment of two or more tuned circuits so that they can be tuned simultaneously with one adjustment

tuned-radio-frequency (TRF) receiver   receiver in which the signal is amplified at its original frequency before demodulation

QUESTIONS

1. Compare the TRF and superheterodyne receiver types. Which is better and why?
2. Why does the resistance of an inductor increase with frequency?
3. Explain the purpose and operation of trimmer and padder capacitors.
4. Distinguish between low- and high-side injection of the local oscillator signal.
5. Explain how image-frequency signals are received in a superheterodyne receiver. How may these signals be rejected?
6. Why do some superheterodyne receivers use an RF stage while others do not?
7. What are the main characteristics of a well-designed RF stage?
8. What is an autodyne converter?
9. Why is the stability of the local oscillator important?
10. In addition to images, what spurious responses are possible with superheterodyne receivers, and how are they caused?
11. What is AGC, and why is it required in a practical receiver?
12. Why do designers of communications receivers often use a switchable RF stage and a switchable input attenuator, instead of applying AGC to the RF stage?
13. Describe four types of filters that can be used in the IF stages of a receiver, and give an application for each.
14. State what is meant by the shape factor of a filter, and explain why a small value for the shape factor is better for the IF filter of a receiver.
15. What causes the relatively high distortion that is a characteristic of envelope detectors?
16. What is the advantage of delayed AGC in a communications receiver?
17. What advantage is gained by using double conversion in a receiver? Are there any disadvantages?
18. Describe a system for reducing the effect in a receiver of noise due to lightning. Will the same system work for thermal noise? Explain.

19. Describe one technique for measuring the sensitivity of a receiver.

20. Describe two techniques for measuring the selectivity of a receiver.

21. List and describe two troubleshooting techniques that are useful with receivers.

PROBLEMS

Section 5.2

22. A tuned circuit has a Q of 60 at 5 MHz. Find its bandwidth at 5 MHz and 20 MHz.

23. A superheterodyne receiver is tuned to a frequency of 5 MHz when the local oscillator frequency is 6.65 MHz.

    (a) What is the IF?
    (b) Which type of injection is in use?

Section 5.3

24. One receiver has a sensitivity of 1 μV and another a sensitivity of 10 dBf, under the same measurement conditions. Both receivers have an input impedance of 50 Ω. Which receiver is more sensitive?

25. A receiver has a sensitivity of 0.3 μV. The same receiver can handle a signal level of 75 mV without overloading. What is its AGC range in decibels?

26. The receiver of Problem 25 has a blocking dynamic range of 80 dB. If the desired signal has a level of 10 μV, what is the maximum signal level that can be tolerated within the receiver passband?

27. A receiver uses low-side injection for the local oscillator, and an IF of 1750 kHz. If the local oscillator is operating at 15.750 MHz,

    (a) to what frequency is the receiver tuned?
    (b) what is the image frequency?

28. An AM broadcast receiver with high-side injection and an IF of 455 kHz is tuned to a station at 910 kHz.

    (a) What is the local oscillator frequency?
    (b) What is the image frequency?
    (c) Find four other spurious frequencies that could be picked up by this receiver.

29. A receiver's IF filter has a shape factor of 2.5 and a bandwidth, at the 6 dB down points, of 6 kHz. What is its bandwidth at 60 dB down?

30. What is the shape factor of the filter sketched in Figure 5.28?

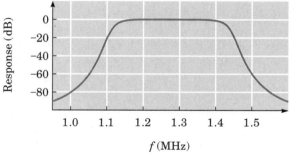

FIGURE 5.28

Section 5.4

31. A receiver has two uncoupled tuned circuits before the mixer, each with a Q of 75. The signal frequency is 100.1 MHz, and the IF is 10.7 MHz. The local oscillator uses high-side injection.

    (a) Calculate the image rejection ratio in decibels.
    (b) To show the advantage of a high IF, recalculate the image rejection, this time assuming an IF of 455 kHz.

32. An AM broadcast receiver tunes from 530 to 1700 kHz. The IF is 455 kHz and the local oscillator uses high-side injection. The local oscillator uses a variable capacitor with a maximum value of 365 pF. Calculate:

    (a) the range of frequencies that must be generated by the local oscillator
    (b) the value of inductor needed so that the local oscillator will tune the receiver to the lowest frequency on the band when the capacitor is at maximum
    (c) the required minimum value for the variable capacitor so that the receiver will be tuned to the highest frequency on the band when the capacitor is set at minimum

33. A double-tuned IF transformer has $k = 1.5k_c$. The primary Q is 50 and the secondary Q is 40. Calculate the bandwidth of the transformer at a frequency of 10.7 MHz.

Section 5.5

34. The block diagram of Figure 5.29 shows a double-conversion receiver.

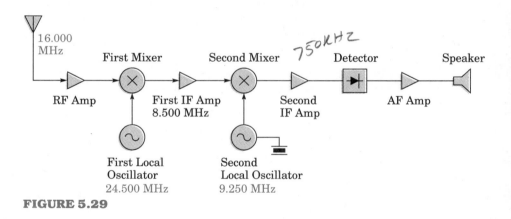

FIGURE 5.29

    (a) Does the first mixer use low- or high-side injection?
    (b) What is the second IF?
    (c) Suppose that the input signal frequency were changed to 17.000 MHz. What would the frequencies of the two local oscillators be now?

Section 5.6

35. A receiver is equipped with an S-meter as described in the text. What would it read for the following signal levels?

    (a) 12.5 μV
    (b) 100 μV
    (c) 2 mV

Section 5.7

36. Refer to the schematic diagram for the transceiver described in Section 5.7, and determine how the unit is switched from receive to transmit; that is, electrically, what happens when the push-to-talk button is pressed?

Section 5.8

37. What frequency and type of signal would be applied to each of the following points during signal injection with a typical AM broadcast-band receiver?
    (a) audio-amplifier input
    (b) detector input
    (c) mixer output
    (d) mixer input

Comprehensive

38. Suppose a receiver has the following gain structure (and that all stages are operating at maximum gain):

    RF amp: 12 dB gain   detector: 3 dB loss
    mixer: 1 dB loss   audio amp: 35 dB gain
    IF amp: 76 dB gain

    (a) What is the total power gain of the receiver from antenna to speaker?
    (b) What would be the minimum signal required at the antenna in order to get a power of 0.5 W into the speaker? Express your answer both in watts and in microvolts, assuming a 50 Ω input impedance.
    (c) Now suppose a 100 mV signal is applied to the antenna. Calculate the output power. Is this a reasonable answer? Explain. What would actually happen in a real receiver?

6 Suppressed-Carrier AM Systems

Objectives

After studying this chapter, you should be able to:

1. Explain the advantages of suppressed-carrier, and of single-sideband, transmission over standard AM.

2. Calculate the improvements in signal-to-noise ratio that result from the use of suppressed-carrier and single-sideband techniques.

3. Describe the operation of a balanced modulator and sketch the output in both time and frequency domains, for typical input signals.

4. Describe the filter method of single-sideband generation, and calculate suitable filter and oscillator parameters to produce a given sideband at a given carrier frequency.

5. Describe the operation of the phasing method of single-sideband generation, and draw a block diagram of a suitable arrangement to produce a given sideband.

6. Describe the operation of communications receivers that are capable of single-sideband reception, and show how they differ from receivers designed only for full-carrier AM.

7. Trace the signal path for transmit and receive operation, in a block diagram of a single-sideband transceiver, and determine which components are used for both.

8. Perform tests on single-sideband transmitters to measure peak envelope power and the suppression of harmonics and other spurious signals.

9. Describe several practical communication systems that use suppressed-carrier and/or single-sideband techniques.

6.1 INTRODUCTION

Although full-carrier AM is simple, it is not a particularly efficient form of modulation in terms of bandwidth or signal-to-noise ratio. We have seen that the transmission bandwidth is twice the highest modulating frequency, because there are two sidebands containing the same information. We have also noticed that two-thirds or more of the transmitted power is found in the carrier, which contains no information and merely serves as an aid to demodulation.

Over the years many variations on the basic AM system have been used, and some of these have become quite popular for specific applications. This chapter looks at a number of these schemes from both the theoretical and practical points of view.

6.2 IMPROVING AM PERFORMANCE

There are two main ways to improve the performance of an AM system: The **power efficiency** can be increased by partly or completely suppressing the carrier, and the bandwidth can be reduced by partly or completely removing one of the two sidebands. Let us look at each of these possibilities in turn.

6.2.1 Efficiency Improvement by Carrier Suppression

As stated above, at least two-thirds of the power in an ordinary double-sideband (DSB) AM signal is in the carrier. The proportion is given by the following equation, which was derived in Chapter 3:

$$P_t = P_c\left(1 + \frac{m^2}{2}\right) \qquad (6.1)$$

where P_t = total signal power

$\quad\quad P_c$ = carrier power

$\quad\quad m$ = modulation index

From this, we see that

$$P_c = \frac{P_t}{1 + m^2/2}$$

For $m = 1$, we get

$$P_c = \frac{P_t}{1.5}$$
$$= 0.667 P_t$$

For lower values of the modulation index, the carrier has much more of the total power. For instance, if $m = 0.5$,

$$P_c = \frac{P_t}{1 + m^2/2}$$

$$= \frac{P_t}{1 + 0.5^2/2}$$

$$= \frac{P_t}{1.125}$$

$$= 0.889 P_t$$

Another way of looking at these figures is to notice that the proportion of the total power that is in the sidebands is 33% and 11%, respectively, for the two examples above.

Removing the carrier before power amplification takes place would allow all of the transmitter power to be devoted to the sidebands, resulting in a substantial increase in sideband power. Removing the carrier from a fully modulated AM signal would change the power available for the sidebands from one-third of the total to all of it. The power increase in the sidebands would be the total available power divided by the power in the sidebands with full carrier:

$$A_P = \frac{P_t}{(1/3)P_t}$$

$$= 3$$

where A_P = power advantage gained by suppressing the carrier

This could easily be expressed in decibels:

$$A_P \text{ (dB)} = 10 \log A_P$$

$$= 4.77 \text{ dB} \tag{6.2}$$

The power advantage of almost 5 dB given by Equation (6.2) is a minimum value. A practical AM system will operate at less than 100% modulation most of the time.

Figure 6.1 shows the effect of removing the carrier from a fully modulated AM signal, in both the frequency and time domains. Figure 6.1(a) shows a 1 MHz carrier 100% modulated by a 1 kHz sine wave and applied to a 50 Ω load. The carrier power is 1 W (30 dBm). Each sideband has one-quarter the carrier power, or 24 dBm. In Figure 6.1(b), we have the result of using the same total signal power (1.5 W) to produce a *double-sideband suppressed-carrier* (DSBSC) signal. Since there is no carrier, each sideband has half the total power: 0.75 W (28.8 dBm).

Obviously, the envelope of the signal is no longer a faithful representation of the modulating signal. In fact, it is merely the sum of the upper and lower sideband signals. When these two sine waves (one at 0.999 MHz and the other at 1.001 MHz) are added, there is reinforcement when the two signals are in phase and cancellation when they are out of phase. The result is an envelope with a frequency equal to the difference between the frequencies

FIGURE 6.1   **DSBSC in Time and Frequency Domains**

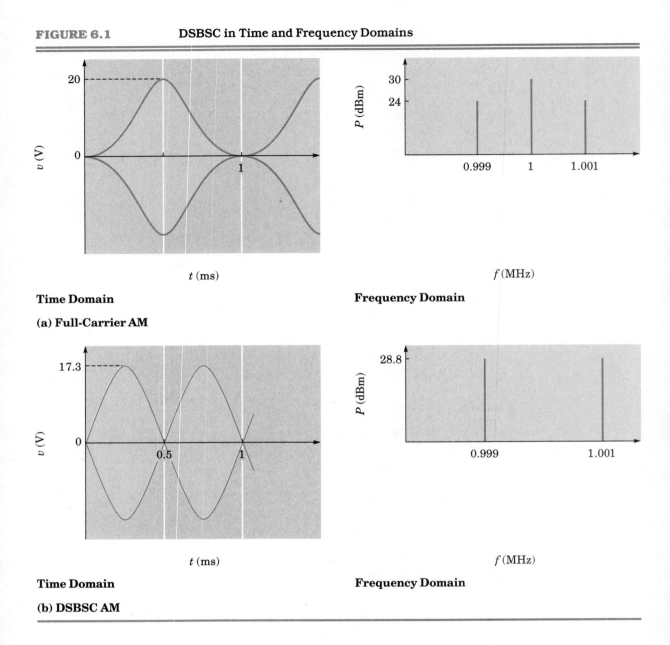

Time Domain   **Frequency Domain**

(a) Full-Carrier AM

Time Domain   **Frequency Domain**

(b) DSBSC AM

of the two sidebands; that is, the envelope frequency is twice the modulating frequency.

Demodulation cannot be accomplished by the method used for full-carrier AM. An envelope detector would produce a distorted signal at twice the baseband frequency. To demodulate this signal, it would be necessary to reinsert the missing carrier at the receiver.

The peak amplitude of this signal can be found as follows. Each sideband has a power of 0.75 W. In a 50 Ω load, the RMS voltage corresponding to one sideband can be found from

$$P = \frac{V_{RMS}^2}{R}$$

$$V_{RMS} = \sqrt{PR}$$
$$= \sqrt{0.75 \text{ W} \times 50 \text{ }\Omega}$$
$$= 6.12 \text{ V}$$

We need the peak voltage. Since one individual sideband is a sine wave, this is given by

$$V_p \text{ (for one sideband)} = \sqrt{2}V_{RMS}$$
$$= \sqrt{2} \times 6.12 \text{ V}$$
$$= 8.66 \text{ V}$$

When the two signals are in phase, the peak envelope voltage will be the sum of the individual peak voltages, or

$$V_p \text{ (for the whole signal)} = 8.66 \text{ V} + 8.66 \text{ V}$$
$$= 17.3 \text{ V}$$

EXAMPLE 6.1

A spectrum analyzer connected to a DSBSC signal has the display shown in Figure 6.2.

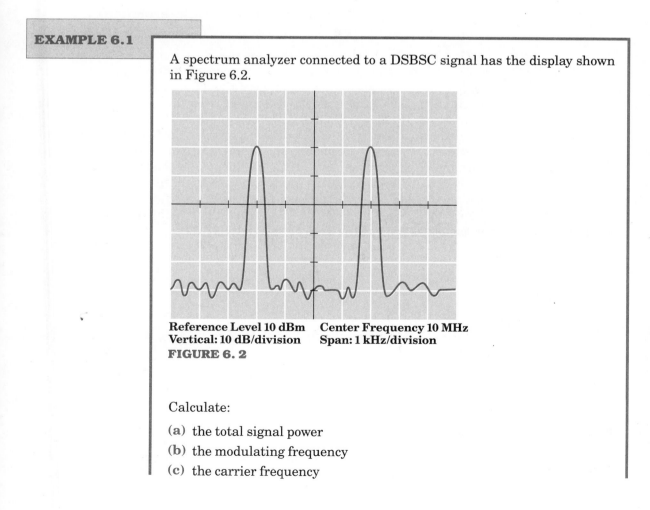

Reference Level 10 dBm   **Center Frequency 10 MHz**
Vertical: 10 dB/division   **Span: 1 kHz/division**
FIGURE 6. 2

Calculate:

(a) the total signal power
(b) the modulating frequency
(c) the carrier frequency

Solution

(a) Each of the two sidebands is two divisions (20 dB) below the reference level of 10 dBm. Therefore each sideband has a power of -10 dBm (100 μW). The total power is twice this, or 200 μW.

(b) The two sidebands are separated by four divisions at 1 kHz per division, or 4 kHz total. The separation between the sidebands is twice the modulating frequency, which must therefore be 2 kHz.

(c) Even though this signal has no carrier, it still has a carrier frequency. This is the frequency the carrier had before it was suppressed. The carrier frequency is midway between the two sidebands, so it is 10 MHz.

DSBSC AM is not often found on its own as a modulation scheme. It is used as the basis for generating **single-sideband suppressed-carrier** (SSBSC, or just SSB) signals, as discussed in the next section, and it is also found as a component in some rather complex multiplexed signals. Examples of these are color television and stereo FM signals. These will be described later in this book.

6.2.2 Bandwidth Reduction Using Single-Sideband Transmission

In Chapter 3, it was discovered that the two sidebands of an AM signal are mirror images of each other, since one consists of the sum of the carrier and modulation frequencies, and the other of the difference. Thus one sideband is redundant, assuming the carrier frequency is known, and it should not be necessary to transmit both in order to communicate.

Removing one sideband will obviously reduce the bandwidth by at least a factor of two. Since the modulating signal rarely extends right down to dc, the bandwidth improvement will usually be greater than two.

Figure 6.3 illustrates this effect. The baseband, shown in Figure 6.3(a), is a voice signal extending over a frequency range from 300 Hz to 3 kHz. Figure 6.3(b) shows this signal transmitted by DSBSC AM with a carrier frequency of 1 MHz. The bandwidth will be

$$
\begin{aligned}
B &= 2\, f_{m(max)} \\
&= 2 \times 3 \text{ kHz} \\
&= 6 \text{ kHz}
\end{aligned}
$$

With SSB transmission, as shown in Figure 6.3(c), the bandwidth of one sideband is

$$
\begin{aligned}
B &= f_{m(max)} - f_{m(min)} \\
&= 3 \text{ kHz} - 0.3 \text{ kHz} \\
&= 2.7 \text{ kHz}
\end{aligned}
$$

This bandwidth reduction has two benefits. Perhaps the more obvious one is that the signal takes up less spectrum. This allows twice as many signals to be transmitted in a given spectrum allotment.

FIGURE 6.3

DSBSC and SSB
Transmission of a Voice
Signal

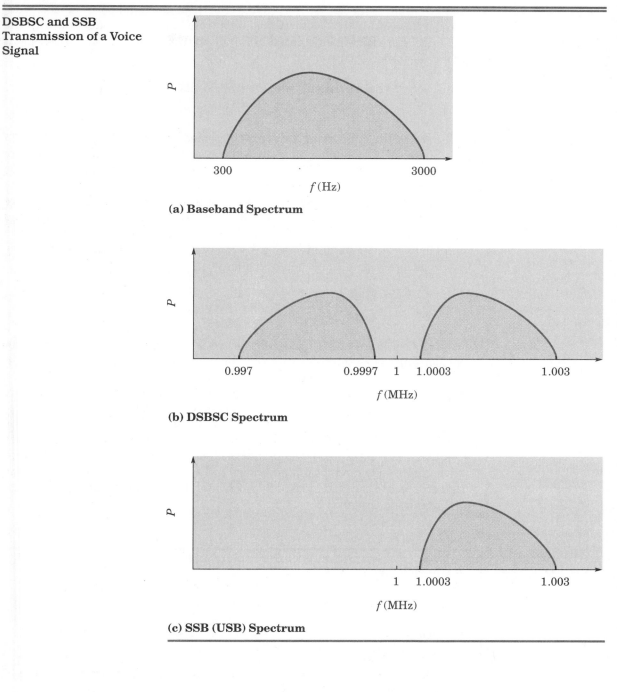

(a) Baseband Spectrum

(b) DSBSC Spectrum

(c) SSB (USB) Spectrum

At least as important, however, is the improvement in the signal-to-noise ratio that can be achieved by reducing bandwidth. If the bandwidth of the transmitted signal is reduced by 50%, the receiver bandwidth can be reduced by an equivalent amount. Since noise power is proportional to bandwidth, reducing the receiver bandwidth by one-half eliminates one-half the noise. Assuming the signal power remains constant, this represents a 3 dB improvement in signal-to-noise ratio.

The signal-to-noise improvement from bandwidth reduction is in addition to that achieved by increasing the transmitted power in the sidebands. Combining the 3 dB improvement from bandwidth reduction with the 4.77 dB improvement calculated in Equation (6.2) gives a total improvement in signal-to-noise ratio of

$$S/N \text{ improvement} = 4.77 \text{ dB} + 3 \text{ dB}$$
$$= 7.77 \text{ dB}$$

compared with full-carrier AM with 100% modulation.

Since the two sidebands of an AM signal contain identical information, either upper-sideband (USB) or lower-sideband (LSB) transmission can be used, and both are found in practice.

SSB transmissions are even more unlike full-carrier AM than are DSBSC signals. In fact, if we look at a single-tone SSB signal, we will find that there is no envelope at all. For instance, consider a USB signal where a 1 MHz carrier is modulated by a 1 kHz baseband signal. The USB will be simply a sinusoid at a frequency given by

$$f_{usb} = f_c + f_m$$
$$= 1.001 \text{ MHz}$$

Another form of SSB modulation, which is often used for transmitter measurements, is the **two-tone test** signal. For an example, see Figure 6.4. A 1 MHz carrier is modulated by two baseband frequencies, 1 kHz and 3 kHz, producing the AM signal shown in Figure 6.4(a). The carrier and the LSB are suppressed, leaving the two USB components, at 1.001 and 1.003 MHz. This is quite obvious in the frequency domain, as shown in Figure 6.4(b). In the time domain, illustrated in Figure 6.4(c), the summation of the two frequency components creates an envelope that is identical to that of the DSBSC signal with single-tone modulation, as shown in Figure 6.1. As before, an envelope detector would give a distorted 2 kHz signal, having no relation to the original baseband signal.

EXAMPLE 6.2

A transmitter generates an LSB signal with a carrier frequency of 8 MHz. What frequencies will appear at the output with a two-tone modulating signal with frequencies of 2 kHz and 3.5 kHz?

Solution

Since the signal is LSB, the components are found by subtracting the modulating frequency from the carrier frequency. Therefore the output frequencies are

$$8 \text{ MHz} - 2 \text{ kHz} = 7.998 \text{ MHz}$$

and

$$8 \text{ MHz} - 3.5 \text{ kHz} = 7.9965 \text{ MHz}$$

FIGURE 6.4

Two-Tone Modulation

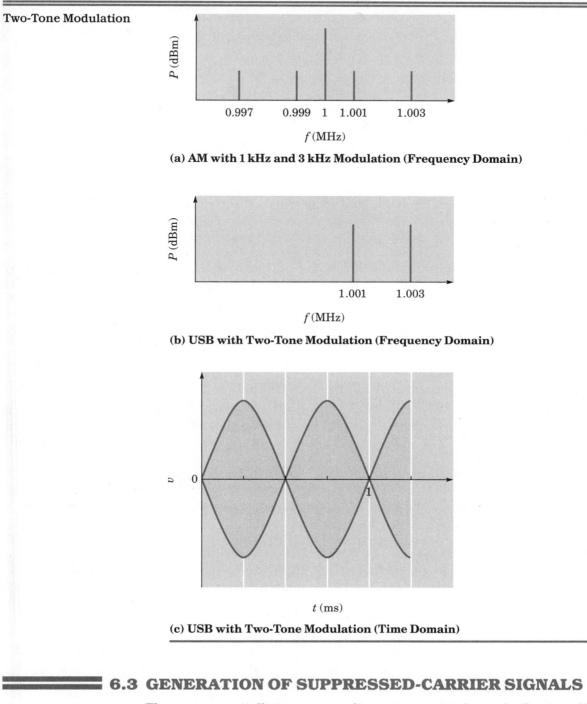

(a) AM with 1 kHz and 3 kHz Modulation (Frequency Domain)

(b) USB with Two-Tone Modulation (Frequency Domain)

(c) USB with Two-Tone Modulation (Time Domain)

6.3 GENERATION OF SUPPRESSED-CARRIER SIGNALS

There are essentially two ways to eliminate unwanted signals: One is to filter them out, and the other is to remove them by phase cancellation. In SSB systems, the usual method is to cancel out the carrier using a balanced modulator, producing a DSBSC signal, and then to remove the unwanted sideband by filtering. However, it is possible to remove both the carrier and the unwanted sideband by phase cancellation.

6.3.1 Balanced Modulators for Double-Sideband Suppressed-Carrier Generation

Balanced mixers were introduced in Chapter 2. A balanced modulator is the same thing used for a different purpose. In fact, the type MC1496 IC shown in Chapter 2 is widely used as both a balanced mixer and a balanced modulator.

Essentially, mixing and amplitude modulation are the same process. Both amplitude modulators and ordinary (unbalanced) mixers are provided with two input frequencies. They produce sum and difference signals at their output, as well as the two original signals. In a transmitter modulator, the two inputs are the baseband and the carrier signals. The sum and difference signals are, of course, the USB and LSB. The carrier signal feeds through the modulator to the output. The baseband input does not feed through in a practical circuit, because of the choice of output components. In the usual case, the baseband is audio and is completely rejected by the output tuned circuits, which are tuned to the carrier frequency several orders of magnitude higher.

In a receiver, on the other hand, the two inputs to the mixer are the incoming signal and the local oscillator signal. Usually the difference will be used by the IF amplifier, and the other outputs will be rejected by filters. The only differences between a mixer and a modulator, then, are practical considerations: modulators generally operate at higher power levels, and use one RF and one audio-frequency signal, while receiver mixers operate at low power and make use of RF signals for all inputs and outputs.

Figure 6.5 should help to clarify this. Figure 6.5(a) shows a receiver mixer and its output, and Figure 6.5(b) demonstrates the similarity with a transmitter modulator.

When discussing balanced mixers, we noted that they produce only the sum and difference of the two input signal frequencies that are applied. The input frequencies themselves are cancelled, or *balanced*, in the circuitry. Applying the same idea to an amplitude modulator, we would end up with a signal that had upper and lower sidebands, but no carrier; in other words, a DSBSC signal. In fact, that is how it is done. The baseband and carrier signals are applied to the inputs of a balanced modulator, and the output is DSBSC, as shown in Figure 6.5(c).

Any of the balanced mixer configurations described in Chapter 2 can be employed, but, in modern designs, specialized ICs, (such as the MC1496) are generally used. In theory, there is no reason why balanced modulation could not be accomplished at high power, but for practical reasons—which will very shortly become apparent—the DSBSC signal is always generated at low level, at an early stage in the transmitter.

Mathematically, we showed previously that an ideal balanced modulator is equivalent to a multiplier. Figure 6.6 shows such a modulator fed with sine-wave carrier and baseband signals. The result (ignoring amplitude factors) is simply

$$v = \sin \omega_m t \, \sin \omega_c t$$
$$= 0.5[\cos (\omega_c - \omega_m)t - \cos (\omega_c + \omega_m)t] \tag{6.3}$$

This is indeed a DSBSC signal, with two components representing the upper and lower sidebands.

FIGURE 6.5   **Mixing and Modulation**

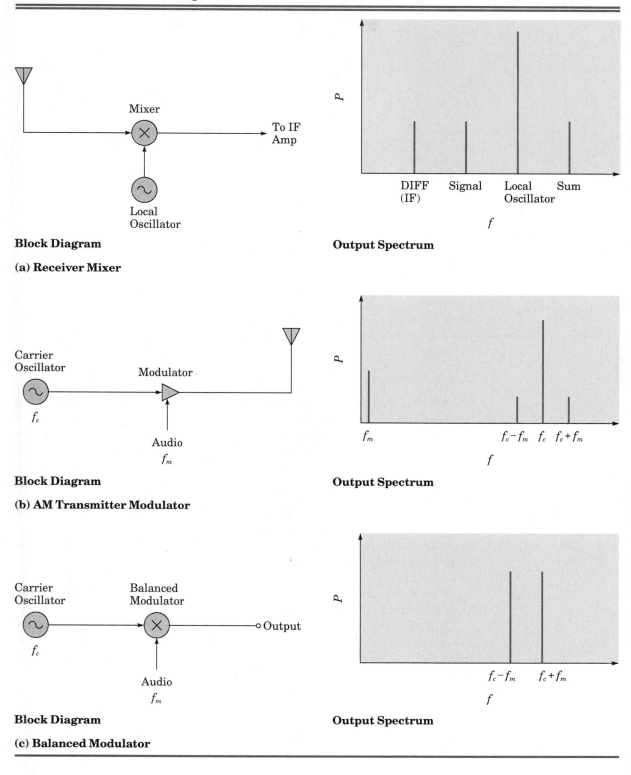

Block Diagram   **Output Spectrum**

(a) Receiver Mixer

Block Diagram   **Output Spectrum**

(b) AM Transmitter Modulator

Block Diagram   **Output Spectrum**

(c) Balanced Modulator

FIGURE 6.6   **DSBSC Generation**

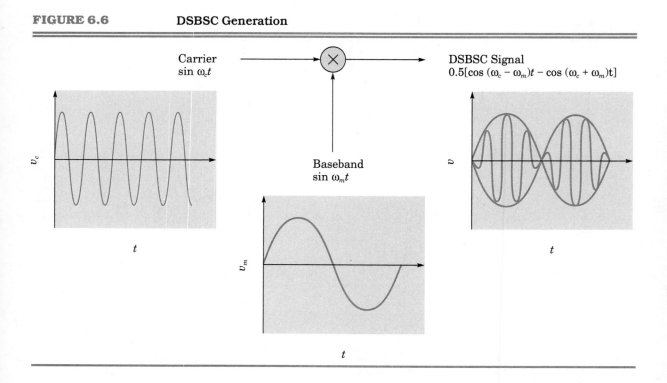

6.3.2 Generating Single-Sideband Signals: Filter Method

The obvious way to eliminate the unwanted sideband, once the carrier has been suppressed, is to use a bandpass filter. You might wonder why we do not simply filter out the carrier as well. The problem is that in an AM signal the carrier is both very large in amplitude and very near the wanted sideband. There would be practical difficulties in removing the carrier while leaving the sideband untouched. It is more convenient to null the carrier with a balanced modulator, and then remove one of the sidebands by filtering.

Figure 6.7 shows the filter method of SSB generation. The block diagram in Figure 6.7(a) is the same for either LSB or USB: which sideband is passed depends on the position of the filter passband relative to the signal. If the carrier frequency is at or near the low end of the filter passband, the USB will pass through the filter and the LSB will be suppressed. On the other hand, with the carrier frequency at or near the high end of the passband, the filter will pass the LSB and suppress the USB.

The sidebands are shown as triangles, not because that is the shape of the spectrum, but in order to emphasize that they are *mirror images* of each other and to make clear which sideband is present when a single-sideband signal is shown.

Even with the carrier suppressed, the cutoff of the filter must be quite sharp. The lowest baseband frequency transmitted is likely to be about 300 Hz in a communications-quality voice signal. Thus there is a gap of 600 Hz between the two sidebands. The attenuation of the filter must change from zero to at least 60 dB within this span.

The generation of SSB requires a high-order filter with high Q. It can be done with LC filters, but only at low frequencies, and it is not an economical proposition. *Mechanical filters* have also been used.

FIGURE 6.7

SSB Generation by Filter Method

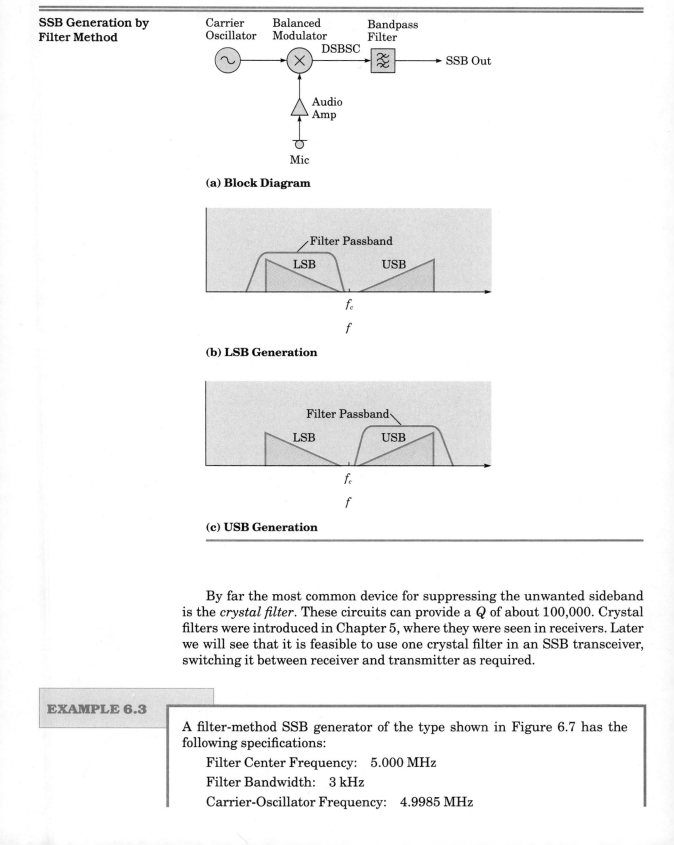

(a) Block Diagram

(b) LSB Generation

(c) USB Generation

By far the most common device for suppressing the unwanted sideband is the *crystal filter*. These circuits can provide a Q of about 100,000. Crystal filters were introduced in Chapter 5, where they were seen in receivers. Later we will see that it is feasible to use one crystal filter in an SSB transceiver, switching it between receiver and transmitter as required.

EXAMPLE 6.3

A filter-method SSB generator of the type shown in Figure 6.7 has the following specifications:

   Filter Center Frequency: 5.000 MHz

   Filter Bandwidth: 3 kHz

   Carrier-Oscillator Frequency: 4.9985 MHz

(a) Which sideband will be passed by the filter?

(b) What frequency should the carrier oscillator have if it is required to generate the other sideband?

Solution

(a) Since the carrier frequency is at the low end of the filter passband, the USB will be passed.

(b) To generate the LSB, the carrier frequency should be moved to the high end of the filter passband, at 5.0015 MHz.

6.3.3 Generating Single-Sideband Signals: Phasing Method

The SSB signal can be generated without filters by using two balanced modulators and two sets of phase-shift networks. Figure 6.8 shows how. The essence of the system is that the two balanced modulators are fed signals, both baseband and carrier, that are 90° apart in phase. This can be accomplished with one 90° phase-shift network for the baseband and one for the carrier frequency, but it is more convenient to use two 45° networks, one leading and one lagging, for each signal, as shown in the figure.

Both of the balanced modulators will produce DSBSC signals, but they will not be identical. Modulator A will produce a signal that is given by the following equation (ignoring amplitude factors):

FIGURE 6.8

SSB Generation by
Phasing Method

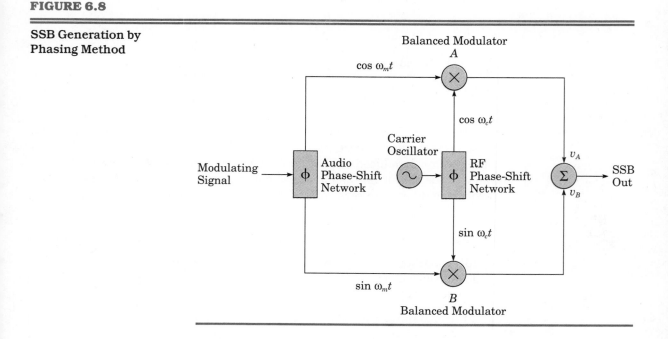

$$v_A = \cos \omega_m t \cos \omega_c t$$
$$= 0.5[\cos (\omega_c - \omega_m)t + \cos (\omega_c + \omega_m)t] \qquad (6.4)$$

Modulator B produces a different version of the DSBSC signal.

$$v_B = \sin \omega_m t \sin \omega_c t$$
$$= 0.5[\cos (\omega_c - \omega_m)t - \cos (\omega_c + \omega_m)t] \qquad (6.5)$$

The summer shown can either add or subtract these two signals. If it adds, we get the output signal

$$v_o = v_A + v_B = \cos (\omega_c - \omega_m)t$$

which is an LSB signal. Subtracting gives

$$v_o = v_A - v_B = \cos (\omega_c + \omega_m)t$$

This is upper sideband. Switching sidebands is very easy with this method.

This system, called the *phasing method* of SSB generation, is not as common as the filter method. One reason is that an accurate 45° or 90° phase shift across the whole audio range is difficult to achieve (although it certainly can be done). The other is that an SSB transmitter is usually found as part of a transceiver. In a transceiver, the same filter can be used in the transmitter section for generating SSB and in the receiver as an IF bandpass filter.

The phasing method does have advantages over the filter method. First and most obvious is the fact that no filter is required, which can reduce the cost. Another advantage is the fact that the phasing method can be used at any carrier frequency, or at a number of carrier frequencies. The filter method, on the other hand, requires the SSB signal to be generated at the same carrier frequency at all times, since the filter is fixed-tuned. This makes adjusting the transmitter frequency more complicated.

There is a third method of SSB generation, sometimes called the *Weaver method*, but usually just referred to as the *third method*. It uses four balanced modulators, but has the advantage of needing only single-frequency phase-shift networks. It is rather complex and seldom seen, so it will not be described in more detail here.

6.4 SINGLE-SIDEBAND TRANSMITTERS

When the filter method is used, the SSB signal must be generated at a single carrier frequency, since a crystal filter cannot be retuned. If the transmitter is to operate at more than one frequency, the output frequency can be changed by using one or more mixers.

In the past, SSB was generated at low frequencies (100 to 500 kHz) because of the problem of designing a high-frequency filter with sufficient Q. The signal frequency was then raised by two or more up-conversions, using mixers and local oscillators. Modern practice is to use crystal filters and generate SSB at higher frequencies (9 to 11 MHz is common). One stage of mixing is generally all that is required for a high-frequency transmitter.

FIGURE 6.9   **An SSB Transmitter Using the Filter Method**

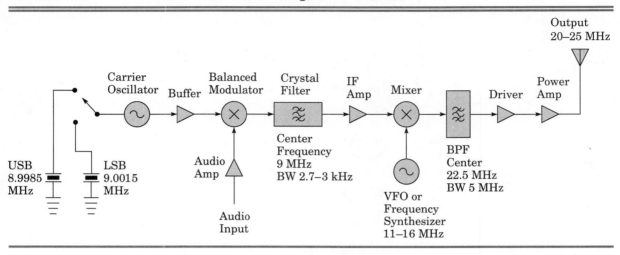

The block diagram of a typical SSB transmitter is shown in Figure 6.9. This transmitter uses a carrier oscillator and a balanced modulator to generate DSBSC, followed by a crystal filter to remove the unwanted sideband. By changing the frequency of the carrier oscillator, either sideband can be generated using the same crystal filter. This is less expensive than using two crystal filters. The transmitter in the figure is designed to transmit either sideband, over a frequency range of approximately 20 to 25 MHz, using a single crystal filter with a center frequency of 9 MHz. Once the SSB signal has been generated, it is mixed with a local oscillator signal. The sum of the two input frequencies is chosen at the output of the mixer, and amplified by the driver and power amplifier.

Filter-method SSB transmitters require that the carrier frequency be positioned near one edge or the other of the filter passband. Figure 6.10 shows one possibility. Here, the center frequency of the filter is 9.0000 MHz, and it has a bandwidth of 3 kHz. For simplicity, let us assume that the filter is ideal, so that it passes everything in its passband with no attenuation, and blocks all other frequencies completely. In that case, when the carrier frequency is at the low end of the passband (8.9985 MHz), all of the USB, and none of the LSB, will be passed. The same filter can be used to generate an LSB signal by simply placing the carrier frequency at the top end of the passband (9.0015 MHz).

Note that the term *carrier frequency* does not imply that the carrier signal actually exists. By this point in the transmitter, the carrier has already been suppressed.

A real crystal filter will not have an infinite slope between passband and stopband. Consequently, placing the carrier frequency exactly at the filter's cutoff frequency will allow some of the unwanted sideband to pass through the filter. Therefore, it is more usual to place the carrier frequency very slightly outside the passband. This will provide more attenuation of the unwanted sideband. It will also attenuate that part of the desired sideband that

FIGURE 6.10

Use of a Single Crystal Filter for Both USB and LSB Signals

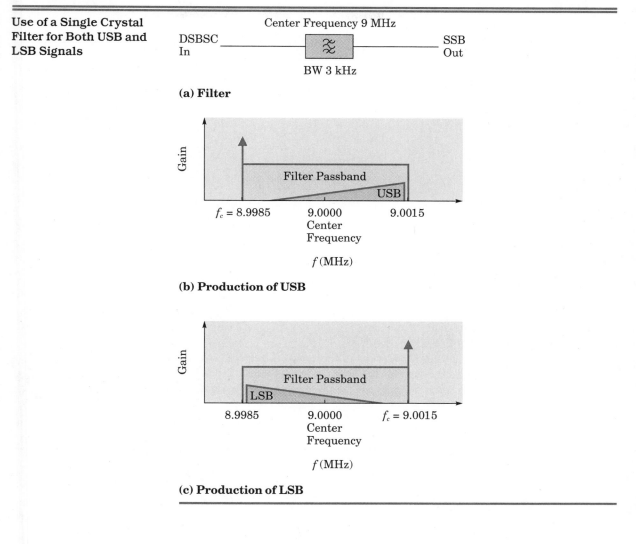

Center Frequency 9 MHz

DSBSC In ———————— [≈] ———————— SSB Out

BW 3 kHz

(a) Filter

Gain

Filter Passband

USB

$f_c = 8.9985$ 9.0000 Center Frequency   9.0015

f (MHz)

(b) Production of USB

Gain

Filter Passband

LSB

8.9985   9.0000 Center Frequency $f_c = 9.0015$

f (MHz)

(c) Production of LSB

is very close to the carrier in frequency. However, these sideband frequencies represent low-frequency components of the baseband signal. As long as baseband frequencies above 300 Hz are passed, there is no problem for voice communications. In our example, this modification could be made by leaving the carrier frequencies as they are and using a filter with a nominal bandwidth of, for example, 2.7 kHz. See Figure 6.11 for the results.

These filter considerations make it difficult to build a filter-method SSB transmitter that will handle very-low-frequency baseband components. This is one reason why SSB is not used for broadcasting, where music signals with frequencies as low as 50 Hz or so would have to be accommodated.

After filtering, the SSB signal is moved to the required operating frequency by means of a mixer and local oscillator combination. In this case, the transmitted frequency is the sum of the carrier-oscillator and local oscillator frequencies. That is,

FIGURE 6.11

SSB Generation Using a
Practical Crystal Filter

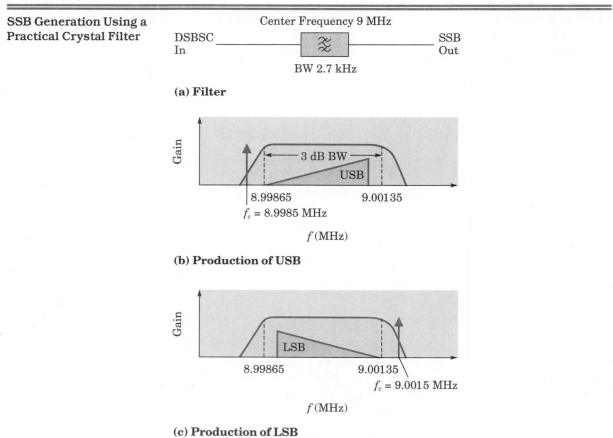

(a) Filter

(b) Production of USB

(c) Production of LSB

$$f_o = f_{CO} + f_{LO} \tag{6.6}$$

where f_o = carrier frequency of the transmitted signal
f_{CO} = carrier-oscillator frequency
f_{LO} = local-oscillator frequency

After mixing, the signal is amplified to the required power level. The power amplifier must be linear, in order to preserve the amplitude variations in the signal.

EXAMPLE 6.4

The SSB transmitter shown in Figure 6.9 is required to transmit a USB signal at a carrier frequency of 21.5 MHz. What must be the frequency of the local oscillator?

Solution

In this transmitter, the carrier-oscillator and local oscillator frequencies are summed to get the transmitting frequency. For USB operation with this transmitter, the carrier-oscillator frequency is 8.9985 MHz. We can easily find the local oscillator frequency from Equation (6.6).

$$f_o = f_{CO} + f_{LO}$$
$$f_{LO} = f_o - f_{CO}$$
$$= 21.5 \text{ MHz} - 8.9985 \text{ MHz}$$
$$= 12.5015 \text{ MHz}$$

Assuming that the baseband spectrum extends from zero to 3 kHz, the frequency spectra at various points in the transmitter will be as shown in Figure 6.12.

Each of the functional blocks in a typical SSB transmitter will now be analyzed in a little more detail.

FIGURE 6.12

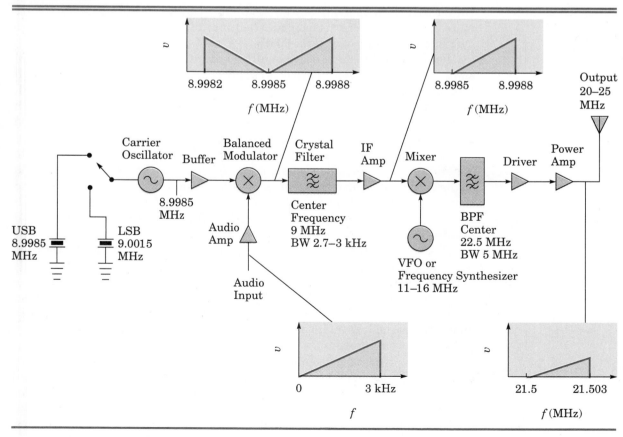

6.4.1 The Carrier Oscillator and Variable-Frequency Oscillator

From the above, we can see that the requirements for the carrier oscillator are quite simple. It must generate a stable signal at exactly two selectable frequencies, which must be accurate to within one or two hundred hertz. The obvious way to meet these requirements is to use a crystal oscillator with switched crystals, as shown in Figure 6.9. Often, a trimmer capacitor is used across the crystal to allow the frequency to be adjusted over a small range.

The local oscillator, on the other hand, will have to have good stability and accuracy over a range of frequencies. The exact range will depend on the use to which the transmitter will be put. A transmitter for the amateur radio service, for instance, may need to cover several segments of the high-frequency band.

Either a variable-frequency oscillator (VFO) or a frequency synthesizer can be used for the local oscillator. Modern designs generally use synthesizers, which can have better stability and accuracy of frequency readout. Where a VFO is used, particular attention must be paid to stability, because even a small amount of mistuning (as little as 100 Hz) can make SSB reception difficult or impossible.

6.4.2 The Balanced Modulator

Either a discrete or IC balanced modulator can be used, but the latter are more common in modern designs. When carefully adjusted, these can achieve carrier suppression of at least 60 dB.

SSB transmitters are occasionally required to transmit a full-carrier AM signal. One application where this capability is useful is CB radio. Although SSB CB transceivers are available, most operators still use full-carrier AM. Therefore, if the user of an SSB transceiver wishes to communicate with someone who is using an AM transceiver, it will be necessary to transmit a signal that is compatible with the envelope detector in the AM transceiver.

It is possible to deliberately unbalance a balanced modulator in order to generate full-carrier AM. This signal would still have one sideband removed in the crystal filter, in our example, but the result, consisting of one sideband plus carrier, could be detected by a simple envelope detector. Of course, full-carrier double-sideband AM could be transmitted by bypassing the crystal filter as well.

SSB transmitters are also used for nonvoice communications. Morse code is usually transmitted by switching an unmodulated carrier on and off, a technique usually called *continuous-wave* (CW) transmission. The same process is also sometimes called *on-off keying* (OOK). An SSB transmitter can easily be used for CW transmission by unbalancing the balanced modulator and switching off the audio amplifier.

Digital communications via high-frequency radio are usually accomplished by means of a technique called *frequency-shift keying* (FSK). Two transmitted frequencies are used, separated by a few hundred hertz. This can easily be accomplished with an SSB transmitter by shifting the frequency of an audio oscillator that is used to modulate the transmitter. The application of single-frequency modulation to the transmitter will result in a single output frequency that shifts with changes in the modulating frequency. (Digital communication is covered in more detail in Chapter 15.)

6.4.3 Mixing

Older SSB transmitter designs practically always require the SSB signal emerging from the filter to be raised in frequency, and often need two or more mixer stages to do so. In modern designs, it may be necessary to move the signal either up or down, or perhaps both, depending on the frequency of operation. For instance, an amateur radio transmitter that generates an SSB signal at 9 MHz will have to use up-conversion for the 14 MHz band and down-conversion for the 7 MHz band.

Mixers for SSB transmitters are straightforward and resemble their receiver counterparts, except that the power levels involved are somewhat higher. Obviously, for down-conversion, the difference between the signal from the filter and the local-oscillator signal must be used. The local-oscillator frequency may be either higher or lower than that of the carrier oscillator. For up-conversion, either the sum or the difference can be used, but if the difference is used, the local oscillator must use high-side injection; that is, it must be higher in frequency than the carrier oscillator. Figure 6.13 shows a few possibilities. In each case, a USB signal with a carrier frequency of 9 MHz is shown at the input to the mixer.

One thing that may not be immediately obvious is that some types of mixing will interchange sidebands; that is, a USB signal becomes LSB, and vice versa. This can be seen by referring to Figure 6.13(a). Suppose the incoming signal is a USB signal with a (suppressed) carrier frequency of 9 MHz, modulated with 1 kHz and 3 kHz tones. Then it has the spectrum of Figure 6.14(a). There are components at 9.001 and 9.003 MHz. The mixing process produces the difference between the local oscillator frequency and each of these two components. These difference-frequency signals will be found at

$$12 \text{ MHz} - 9.001 \text{ MHz} = 2.999 \text{ MHz}$$

and

$$12 \text{ MHz} - 9.003 \text{ MHz} = 2.997 \text{ MHz}$$

while the new carrier frequency is

$$12 \text{ MHz} - 9 \text{ MHz} = 3 \text{ MHz}$$

It is clear from Figure 6.14(b) that what has happened is that the USB signal has been changed into an LSB signal by the mixing process. This is based on simple arithmetic: subtracting a larger number gives a smaller result. The effect occurs only for subtractive mixing with high-side injection of the local oscillator: try it with the other examples in Figure 6.13 to verify this.

Exactly the same thing happens in a receiver mixer. It goes unnoticed in an AM receiver because the two sidebands of a DSB AM signal are mirror

FIGURE 6.13

**Examples of Frequency
Conversion in SSB
Transmitters**

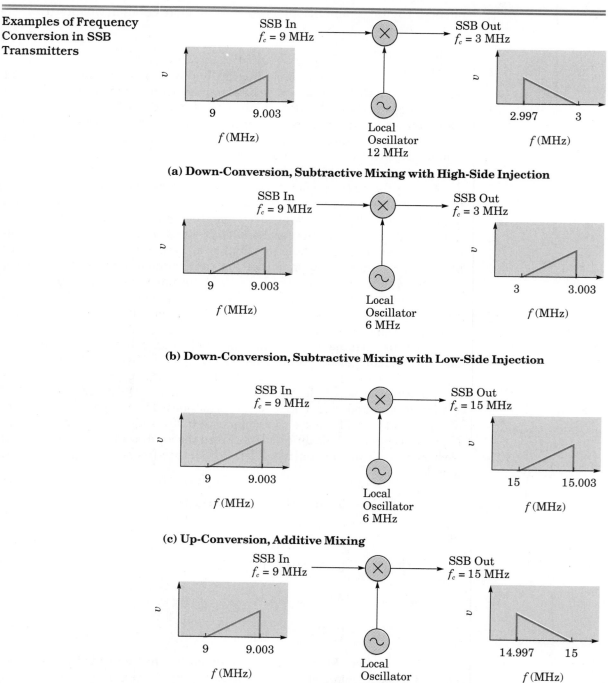

(a) Down-Conversion, Subtractive Mixing with High-Side Injection

(b) Down-Conversion, Subtractive Mixing with Low-Side Injection

(c) Up-Conversion, Additive Mixing

(d) Up-Conversion, Subtractive Mixing

FIGURE 6.14   **Effect of Subtractive Mixing with High-Side Injection on SSB Signals**

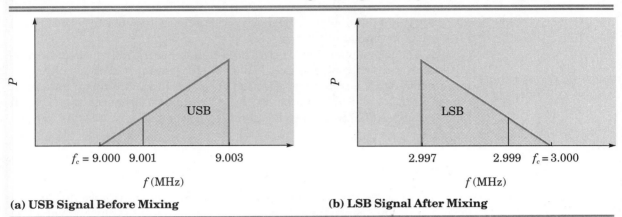

(a) USB Signal Before Mixing   **(b) LSB Signal After Mixing**

images, and interchanging them makes no difference. With SSB transmitters and receivers, this effect must be taken into account, however.

Note that changing from one sideband to the other by switching the frequency of the carrier oscillator also changes the carrier frequency at the output. This will require a change in the frequency readout for the transmitter.

EXAMPLE 6.5

Suppose that the transmitter in Example 6.4 is switched to LSB by simply switching carrier-oscillator crystals. What is the new carrier frequency of the transmitted signal?

Solution

The new carrier-oscillator frequency is 9.0015 MHz (see Figure 6.9). The local-oscillator frequency remains 12.5015 MHz, as found in Example 6.4. The new output carrier frequency can easily be found from Equation (6.6):

$$f_o = f_{CO} + f_{LO}$$
$$= 9.0015 \text{ MHz} + 12.5015 \text{ MHz}$$
$$= 21.503 \text{ MHz}$$

This is 3 kHz higher than before. In order to maintain an accurate frequency readout, the dial display would have to change when sidebands are changed, or the local-oscillator frequency would have to change automatically when sidebands are switched. Either method is quite easy when frequency synthesis is used for the local oscillator.

6.4.4 Power Amplification

The power-amplification stages of an SSB transmitter must be linear. SSB is a variation of AM, and any AM signal must be amplified in such a way that the amplitude variations are preserved; if not, the demodulated signal will be distorted. As we saw in Chapter 4, the need for linear amplification can be avoided in a full-carrier AM transmitter by modulating the RF power amplifier. This is not possible in a filter-type SSB transmitter, because the crystal filter must operate at low power.

The enormous power levels that are often found in AM broadcast transmitters are rare with SSB. Since SSB is usually used for point-to-point communication with reasonably effective receivers and receiving antennas, transmitter power levels can be much lower than for broadcasting. Typical high-frequency SSB transmitters operate at power levels from 100 W to a few kilowatts.

6.4.5 Single-Sideband Transmitter Specifications

Speaking of power, you might wonder how an SSB transmitter would be rated for power output. Since there is no carrier, carrier power is obviously useless as a specification. The actual transmitter output power varies greatly with modulation, from zero with no modulation to some maximum level with full modulation. That maximum is referred to as the **peak envelope power** (PEP). It is simply the power at modulation peaks, calculated using the RMS formula.

$$PEP = \frac{(V_p/\sqrt{2})^2}{R_L}$$

$$= \frac{V_p^2}{2R_L} \tag{6.7}$$

where PEP = peak envelope power in watts
     V_p = peak signal voltage in volts
     R_L = load resistance in ohms

Figure 6.15 illustrates this calculation. The signal shown represents an SSB signal with two-tone modulation, as previously discussed in Section

---

FIGURE 6.15

PEP Calculation

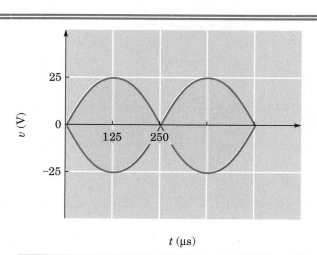

t (μs)

6.2.2. The peak voltage is 25 V, and let us assume that the load is 50 Ω, resistive.

Then the peak envelope power is

$$PEP = \frac{V_p^2}{2R_L}$$

$$= \frac{(25 \text{ V})^2}{2 \times 50 \text{ } \Omega}$$

$$= 6.25 \text{ W}$$

PEP is not the same as instantaneous peak power. In fact, it is one-half the maximum instantaneous power. This can easily be seen by recalling that instantaneous power is simply

$$P = \frac{v^2}{R}$$

where $v =$ instantaneous voltage across resistance R

The maximum instantaneous voltage is simply V_p, so the maximum instantaneous power is

$$P_{max} = \frac{V_p^2}{R_L} \tag{6.8}$$

which is just twice the PEP given by Equation (6.7).

It is also important to distinguish PEP from average power. For a well-defined test signal, a relation can be found between them. With single-tone modulation, the PEP and the average power will be equal, since the signal is a sine wave with the same peak voltage for every cycle. With the two-tone signal of Figure 6.15, the relationship is a little more complicated, but still easily found. There are a number of ways to do this, of which one is illustrated in the following example.

EXAMPLE 6.6

Find the average power of the signal illustrated in Figure 6.15.

Solution

It helps to remember that this waveform is created from the algebraic addition of two sine waves of equal amplitude, separated in frequency by a small amount (2 kHz, in this example). When the signals are exactly 180° out of phase, they will cancel completely, as shown at $t = 0$ and $t = 250$ μs. On the other hand, when they are exactly in phase, the peak amplitudes will add. This is shown in the figure at $t = 125$ μs. Since the two sine waves have equal amplitude, the peak amplitude of each must be one-half the peak voltage of the combined signal, or 12.5 V each.

The total average power in the signal will be the sum of the average powers of the two components. Each component, being a sine wave, will have a power equal to

$$P = \frac{V_{RMS}^2}{R_L}$$

$$= \frac{(V_p/\sqrt{2})^2}{R_L}$$

$$= \frac{V_p^2}{2R_L}$$

$$= \frac{(12.5 \text{ V})^2}{2 \times 50 \ \Omega}$$

$$= 1.56 \text{ W}$$

The total average signal power will be twice this, or 3.125 W. This is one-half the PEP of 6.25 W already found.

There is no simple relationship between PEP and average power for a random voice signal. In general the average power with voice modulation varies from PEP/4 to PEP/3.

The PEP rating for a transmitter will generally be limited by clipping, while the average power is limited by heat dissipation or power-supply capacity. The duty cycle of many SSB transmitters is less than 100% even for voice, since they are designed for two-way communication, where short transmissions alternate with longer idle periods. Consequently, care must be taken to avoid overheating when using or testing such transmitters with single-tone or two-tone signals.

Other important specifications for SSB transmitters relate to output-frequency stability and accuracy, and to spectral purity (that is, the absence of harmonics and other unwanted signals). The former are the same as for full-carrier AM transmitters. Spectral purity figures are a little more complex for SSB transmitters because the carrier and unwanted sideband are among the spurious signals that must be eliminated.

Table 6.1 shows some nominal specifications for a typical small SSB transmitter, in this case the transmitter part of a CB transceiver. This particular transmitter is also capable of operating in the full-carrier AM mode.

Note that the PEP rating is considerably greater than the AM carrier power. Also note that the carrier and unwanted-sideband suppression are not quite as good as the harmonic suppression. The unwanted-sideband suppres-

TABLE 6.1

SSB Transmitter Specifications		
	AM carrier power	3.8 W
	PEP, two-tone SSB	12 W
	Harmonic suppression	65 dB
	Carrier suppression, SSB	55 dB
	Unwanted sideband suppression at 2500 Hz	55 dB

sion is given at a point 2500 Hz from the carrier; it would undoubtedly be worse for lower modulating frequencies.

Measurement techniques for SSB transmitters will be discussed later in this chapter.

6.5   SINGLE-SIDEBAND RECEIVERS

The block diagram of a basic SSB receiver (see Figure 6.16) is similar to that for an AM receiver. There are some differences, however. We have already seen that SSB signals require a carrier to be reinserted at the receiver. This necessitates a different sort of detector from that of a receiver designed for full-carrier AM. Since SSB signals have much smaller bandwidth than full-carrier AM transmissions, we might also expect a narrower IF filter response. Finally, since tuning is very critical with SSB, an SSB receiver must have a very stable local oscillator. This is desirable in any case, but it is more critical with SSB than with full-carrier AM.

The example receiver in Figure 6.16 is designed to cover the same frequency range as the transmitter in Figure 6.9. In fact, the receiver is almost a mirror image of the transmitter.

The signal from the antenna, after RF amplification and filtering to reject image frequencies, is applied to a mixer. The one in the illustration uses low-side injection, but high-side injection works equally well. A crystal filter provides IF selectivity; other than that, the IF amplifier is conventional. At the **product detector**, a signal is injected at the carrier frequency of the incoming signal. The oscillator that produces this signal is usually called the **beat-frequency oscillator** (BFO), largely for historical reasons.

Of course, at this point in the receiver, the carrier frequency has been moved by the mixer from its original value to an IF of approximately 9 MHz. Mathematically, for any receiver using low-side injection, we have

$$f_{IF} = f_{sig} - f_{LO}$$

FIGURE 6.16

Block Diagram of SSB Receiver

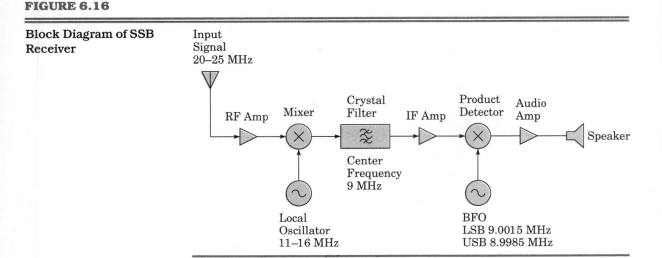

The BFO and CW Reception

Early communications receivers were usually designed for two types of signals: full-carrier AM and CW, the latter standing for *continuous waves*. The carrier was not actually transmitted continuously; rather, it was switched on and off in the dot-and-dash patterns of the Morse code.

The BFO was employed for CW reception. In that application, the BFO signal mixed with the transmitted carrier in the detector to produce an audible difference, or beat, frequency. CW reception remains quite possible with any receiver designed for SSB, though it can be optimized by using a narrower IF filter and a different BFO frequency. Though gradually falling into disuse, CW is still quite popular with radio amateurs, and sees limited use in marine and military communications. It is especially useful for emergency communication: in the presence of severe noise and interference, slow-speed CW is much more intelligible than voice.

where f_{IF} = carrier frequency in the IF amplifier
f_{sig} = carrier frequency of the input signal
f_{LO} = local-oscillator frequency

For this example, f_{IF} is equal to the BFO frequency f_{BFO}, so

$$f_{BFO} = f_{sig} - f_{LO} \tag{6.9}$$

Just as with the SSB transmitter studied earlier, the carrier is injected at one of two frequencies near either edge of the IF passband, depending on whether the received signal is USB or LSB. For the USB, the BFO frequency is near the lower edge of the passband, so that the USB signal passes through the filter. In the receiver, the purpose of the filter is not to reject the LSB of the received signal; that has already been done at the transmitter. Rather, the crystal filter rejects noise and interfering signals outside the frequency range occupied by the desired signal.

EXAMPLE 6.7

The receiver illustrated in Figure 6.16 is tuned to an LSB signal with a (suppressed) carrier frequency of 23 MHz. To what frequency should the local oscillator be tuned for best reception?

Solution

The mixer must shift the carrier frequency of the received signal to that of the BFO, which for the LSB is 9.0015 MHz. From Equation (6.9),

$$f_{BFO} = f_{sig} - f_{LO}$$
$$f_{LO} = f_{sig} - f_{BFO}$$
$$= 23.0000 \text{ MHz} - 9.0015 \text{ MHz}$$
$$= 13.9985 \text{ MHz}$$

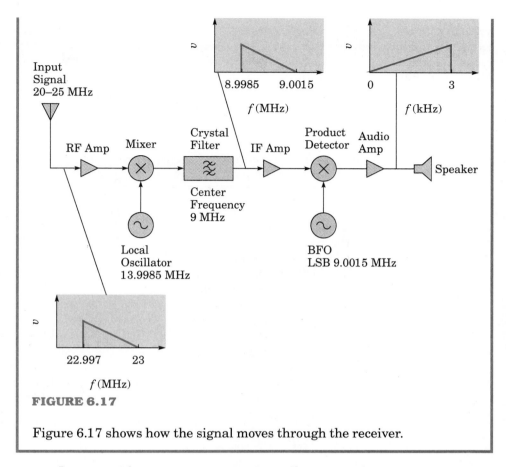

FIGURE 6.17

Figure 6.17 shows how the signal moves through the receiver.

Just as with transmitters, receivers that use subtractive mixing and high-side injection will switch sidebands in the mixing process; that is, a USB signal at the antenna will become LSB in the IF amplifier, and an LSB signal will be changed to USB.

6.5.1 Detectors for Suppressed-Carrier Signals

A diode detector used alone will not work for SSB or DSBSC because the envelope is different from that of AM. A locally generated carrier can be injected along with the SSB signal into a diode detector, but it is better to use an arrangement similar to that which generates suppressed-carrier signals (that is, a balanced modulator). When used in receivers, such a device is called a *product detector*. Any of the balanced modulator circuits previously studied will work, but most contemporary designs use ICs such as the MC1496.

It may not be obvious that SSB and DSBSC can be demodulated by essentially the same process that created them, but it is easy to show mathematically that this is the case. Consider the detector in Figure 6.18, and remember that balanced modulators and product detectors both are actually multiplier circuits; that is, they have a single output whose instantaneous amplitude is proportional to the product of the amplitudes of two input signals.

The two inputs to the detector consist of the signal to be demodulated and a locally generated carrier produced by the BFO. In modern receivers,

FIGURE 6.18

**Product Detector for
DSBSC and SSB**

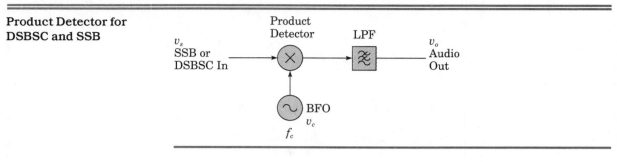

the BFO is crystal-controlled and switchable so that the carrier can be injected at the exact frequency of the (suppressed) signal carrier in the IF. This
will differ depending on whether the signal is USB or LSB. Since DSBSC is
seldom used by itself, communications receivers do not make special provision for it; a DSBSC signal would be treated as SSB, with one of its sidebands
removed in the IF before detection.

Consider an SSB signal with single-tone modulation. Let us choose USB.
(The LSB case is left as an exercise for the reader.)

Assuming that this signal was originally produced by modulating a sine-
wave carrier with radian frequency ω_c with another sine wave at ω_m, it will
have the following equation (neglecting amplitude factors which are not relevant to this discussion):

$$v_{sig} = -\cos(\omega_c + \omega_m)t \qquad (6.10)$$

This is simply the USB component of the DSBSC signal shown in Equation
(6.3).

Let the locally generated carrier be

$$e_c = \sin \omega_c t \qquad (6.11)$$

The product detector will multiply these two signals to produce the output

$$v_o = \sin \omega_c t [-\cos(\omega_c + \omega_m)t]$$
$$= -\sin \omega_c t \cos(\omega_c + \omega_m)t$$

Using the trigonometric identity

$$\sin a \cos b = 0.5[\sin(a - b) + \sin(a + b)]$$

we get

$$v_o = -0.5[\sin(-\omega_m t) + \sin(2\omega_c + \omega_m)t]$$
$$= 0.5 \sin \omega_m t - 0.5 \sin(2\omega_c + \omega_m)t \qquad (6.12)$$

This signal has two components. The first is simply the demodulated
baseband signal. The second component can be removed by passing the out-

put signal through a low-pass filter. This is quite easy because the unwanted component has a frequency more than twice the carrier frequency, and therefore is several orders of magnitude higher than the baseband frequency.

The same method can be used to show that the product detector will work for multiple modulation frequencies, whether the signal is USB, LSB, or DSBSC.

In the foregoing, it was assumed that the carrier inserted at the receiver was exactly the same as the original carrier in both frequency and phase. This is *not* a trivial matter. In most communications receivers, the phase relationship between the original and locally generated carriers will be random, and the two frequencies will not be exactly equal. This is because there is no transmitted carrier to act as a reference, and the receiver is generally tuned by ear for the most natural-sounding voice. The effect can be determined by repeating the mathematical analysis performed earlier for a BFO frequency not quite equal to ω_c. However, the effect on an SSB signal can be seen more easily by looking at Figure 6.19. Figure 6.19(a) shows an example of a USB signal with a (suppressed) carrier frequency of 1 MHz and two modulating signals of 1 kHz and 3 kHz. This results in an RF spectrum with components at 1001 and 1003 kHz.

Figure 6.19(b) shows this signal being detected correctly, with the BFO at exactly the carrier frequency of 1 MHz. The output, after low-pass filtering, consists of the difference signals that result from subtracting the BFO frequency from each of the transmitted components. This gives the original 1 kHz and 3 kHz baseband signals at the output.

Figure 6.19(c) shows what happens when the BFO signal is 100 Hz too low, at 999.9 kHz. The difference signals are now

$$1001 \text{ kHz} - 999.9 \text{ kHz} = 1.1 \text{ kHz}$$

and

$$1003 \text{ kHz} - 999.9 \text{ kHz} = 3.1 \text{ kHz}$$

That is, a BFO that is set 100 Hz low causes each component of the demodulated baseband signal to be 100 Hz higher than it should be. It is obvious that this will have a distorting effect on voice signals, but provided the difference is small (less than 100 Hz or so), the information signal will still be intelligible. It is not obvious from the figure, but the phase relationships between components of the baseband signal will also be disrupted. This is of no consequence for voice-grade communications. In fact, experts disagree as to whether phase coherence is important even for high-fidelity audio systems. In any case, as explained earlier, SSB is never used for high fidelity.

The situation is more critical with a DSBSC signal, since a BFO frequency offset will simultaneously raise one of the two difference frequencies between sideband and inserted carrier, while lowering the other. This will cause severe distortion. Phase relationships must also be preserved in many applications of DSBSC. In color television, for example, the hue of the color is determined by the phase angle of the color signal.

We can conclude, then, that a product detector with a BFO that is at almost the original carrier frequency is acceptable for SSB voice, but that DSBSC or phase-critical signals require a locally generated carrier that is

FIGURE 6.19   Effect of Receiver Mistuning on SSB Reception

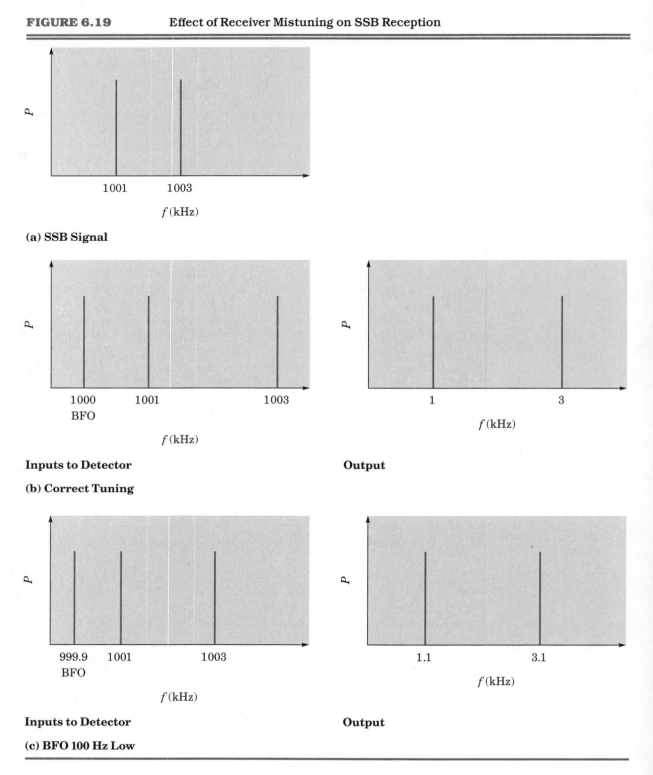

(a) SSB Signal

Inputs to Detector   **Output**

(b) Correct Tuning

Inputs to Detector   **Output**

(c) BFO 100 Hz Low

phase-locked to the original carrier. This is called *coherent detection*. There are several ways of accomplishing this, the simplest of which is to transmit a **pilot carrier** with reduced amplitude to which the receiver BFO can be locked. Other methods will be described at the appropriate times.

6.5.2 Beat-Frequency Oscillators

In our discussion of SSB transmitters using the filter method, it was found that the most cost-effective method was to use one crystal filter and switch the carrier-oscillator frequency so that the filter passed the desired sideband. For USB generation, the carrier frequency would be placed near the low end of the filter passband, and for LSB it would be placed near the high end. We also noted that, under certain conditions, the sidebands would be interchanged in the process of frequency changing by mixing.

Exactly the same factors operate in receivers; in fact, in a typical SSB transceiver, the same IF section (including the crystal filter) is used in both transmit and receive modes. The IF in the receiver can be seen as having almost the same functions as in the transmitter, but in reverse. The transmitter modulates the signal at the IF, filters it, then moves it to the operating frequency; the receiver takes in the signal at the operating frequency, moves it to the IF, filters it, then demodulates it.

From the above, it appears that the receiver BFO should operate at the same two frequencies as the transmitter carrier oscillator. The BFO will inject a carrier near one side or the other of the passband of the single IF filter: the top end for LSB and the bottom for USB, as seen in the IF amplifier. Of course, with subtractive mixing and high-side injection, the sideband at the antenna will be opposite to that in the IF amplifier. Crystal control is usual for the BFO, for greater stability.

6.5.3 Single-Sideband Receiver Intermediate Frequency Circuitry

The IF section of an SSB receiver is similar to that of an AM receiver, with minor differences. The bandwidth will be narrower, and a crystal filter will almost certainly be used. The AGC system will be different: since there is no carrier, the AGC will have to respond to the modulation level of the signal. The AGC will need to have a short attack time and a relatively long decay time; that is, it will reduce the IF gain quickly when the signal strength increases, to avoid overloading the receiver at the beginning of a transmission. The gain will increase slowly when the signal strength decreases. This will avoid an increase in the background noise level during short pauses between words and sentences, when the received signal level drops briefly to zero.

6.6 SINGLE-SIDEBAND TRANSCEIVERS

SSB is not currently used for broadcasting, though there has been some talk of using it in shortwave radio broadcasting. Most of the applications for SSB are in two-way half-duplex communication, where each station transmits and receives alternately. Combining the transmitter and receiver into a single piece of equipment (called a *transceiver*) provides greater convenience and a considerable reduction in cost.

Figure 6.20 is a simplified block diagram of a typical SSB transceiver. The dotted lines show the transmit signal path and the dashed lines show the received signal. The expensive crystal filter and frequency synthesizer are used for both transmitting and receiving. There is also a single crystal oscillator to provide the carrier signal for the balanced modulator when transmitting and to act as a BFO while receiving.

FIGURE 6.20 **Block Diagram of SSB Transceiver**

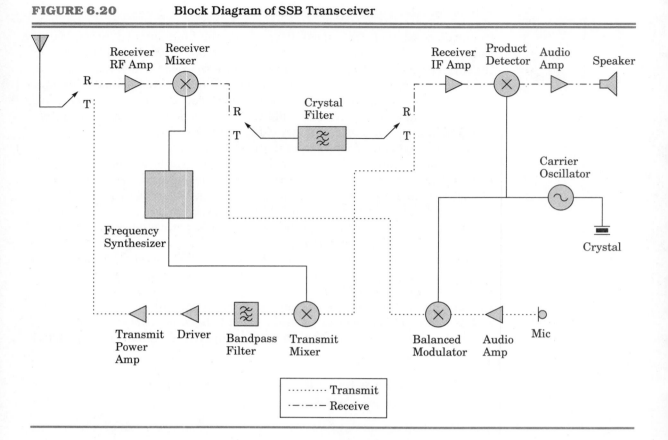

The transmitting process begins at the microphone. The audio signal is amplified and applied to a balanced modulator, along with the carrier-oscillator signal. The output of the balanced modulator is a DSBSC signal at the IF frequency. It is sent through a crystal filter to remove the unwanted sideband, and then to a mixer. The transmit mixer combines the SSB signal with a signal from the frequency synthesizer to produce an output signal at the operating frequency. This signal is then amplified to produce the required output power.

The received signal goes first to a stage of RF amplification, then to a mixer, where it is converted to the same IF as is used for transmitting. The crystal filter provides the receiver selectivity, and is followed by IF amplification. A product detector, with the carrier oscillator serving as BFO, recovers the modulation. The baseband signal is then amplified and sent to a loudspeaker.

EXAMPLE 6.8

An SSB transceiver has the block diagram of Figure 6.20. It is designed to operate at a frequency of approximately 30 MHz. When the unit is switched to receive, the synthesizer produces a frequency of 21.000 MHz. The crystal filter has a center frequency of 9.000 MHz and a bandwidth of 2.8 kHz.

(a) Determine which sideband will be received.

(b) What should the carrier frequency of the received signal be for proper reception?

(c) What frequency should the synthesizer produce when transmitting, in order that the same sideband, with the same carrier frequency, will be transmitted?

Solution

(a) First, notice that the receive local oscillator operates at 21 MHz, which is approximately 9 MHz lower than the received signal frequency. This means that the mixer is using subtractive mixing with low-side injection. Therefore there will be no sideband inversion; that is, the sideband that is present in the IF amplifier will be the same sideband that is received. Since the BFO is near the high end of the IF filter, the lower sideband will be received.

(b) The carrier frequency f_i at the antenna must satisfy the equation

$$f_i - f_{LO} = f_{IF}$$

because low-side injection and subtractive mixing are used. Therefore

$$
\begin{aligned}
f_i &= f_{IF} + f_{LO} \\
&= 9.0015 \text{ MHz} + 21.000 \text{ MHz} \\
&= 30.0015 \text{ MHz}
\end{aligned}
$$

The frequency of the carrier oscillator, rather than the filter center frequency, is used for f_{IF}, so that the carrier frequency will be moved to the same frequency as the BFO. This is a requirement for proper demodulation.

(c) The transmitted SSB signal is generated at the same IF as is used in the receiver, and must be raised to the transmitting frequency. Therefore the transmit mixer must be additive, so that

$$f_o = f_{IF} + f_{LO}$$

The same carrier oscillator can be used as before. This will generate an LSB signal with a carrier frequency of 9.0015 MHz. The output frequency must be 30.0015 MHz, as calculated above. Rearranging the above equation gives

$$
\begin{aligned}
f_{LO} &= f_o - f_{IF} \\
&= 30.0015 \text{ MHz} - 9.0015 \text{ MHz} \\
&= 21.000 \text{ MHz}
\end{aligned}
$$

This is the same frequency as used for transmitting. If the transmit and receive frequencies are the same, it is not even necessary to retune the synthesizer when switching between transmit and receive operation.

6.7 TEST EQUIPMENT AND PROCEDURES: TWO-TONE TESTS OF SINGLE-SIDEBAND TRANSMITTERS

Measurement of output power is somewhat more complicated with SSB than with AM transmitters. Obviously, carrier-power measurements are useless, since SSB transmitters are not intended to produce a carrier. We noted above that PEP (peak envelope power) is the usual way to measure the power output of SSB transmissions. This power is often limited by clipping in the transmitter output stage. On the other hand, such limitations as power-supply capacity and heat dissipation will tend to put limits on the average power output. The ratio between PEP and average power depends on the waveform of the modulating signal and the duty cycle of the transmitter.

Figure 6.21 shows the envelope of an SSB signal for several situations. Continuous modulation with a single tone (a sine wave), as shown in Figure 6.21(a), is the hardest test, since PEP and average power output are equal. As shown in Figure 6.21(b), voice signals have much higher peak-to-average power ratios, about 3:1 or 4:1, but of course this varies with the voice. It was

FIGURE 6.21   **SSB Signal Envelopes**

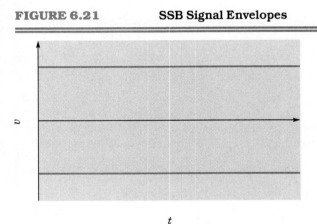

(a) Single-Tone Modulation

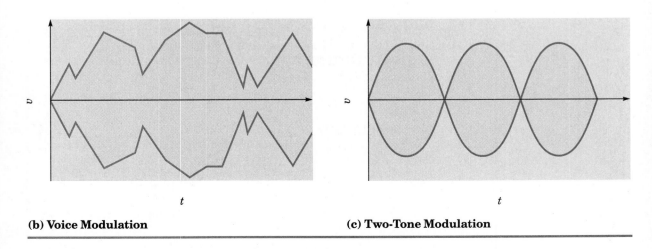

(b) Voice Modulation   **(c) Two-Tone Modulation**

shown previously that a transmitter modulated by a *two-tone test signal*, consisting of two sine waves of equal amplitude but different frequencies, as in Figure 6.21(c), has a PEP-to-average-power ratio of 2:1. This is a little more severe than voice operation.

Any of the modulating signals shown in Figure 6.21 can be continuous or intermittent. Two-way half-duplex communication, the commonest form of SSB service, usually represents a duty cycle of 50% or less; that is, the operator spends at least as much time listening as talking. A 50% duty cycle reduces transmitter cooling requirements for a given PEP rating. In fact, many transmitters have PEP ratings that vary according to the use to which the transmitter will be put.

PEP can easily be measured, using a two-tone test signal, by using the test setup sketched in Figure 6.22(a). Two audio generators are needed, and they should be well isolated from each other by a combining network. A resistive network like the one shown in Figure 6.22(b) can be used with generators having a 50 Ω output impedance. The frequencies of the audio generators should not be harmonically related.

The transmitter power output can be measured using an oscilloscope or a spectrum analyzer. Figure 6.23(a) shows normal displays for each type of instrument. The scope allows severe distortion, such as clipping, to be seen as a change in the shape of the waveform, while the spectrum analyzer will

FIGURE 6.22

Two-Tone Transmitter Tests

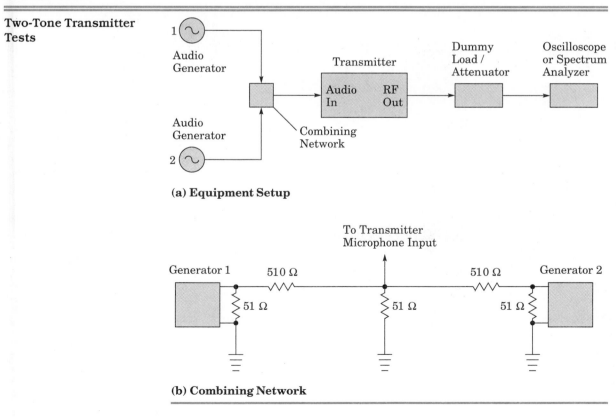

(a) Equipment Setup

(b) Combining Network

FIGURE 6.23   Two-Tone SSB Test Results

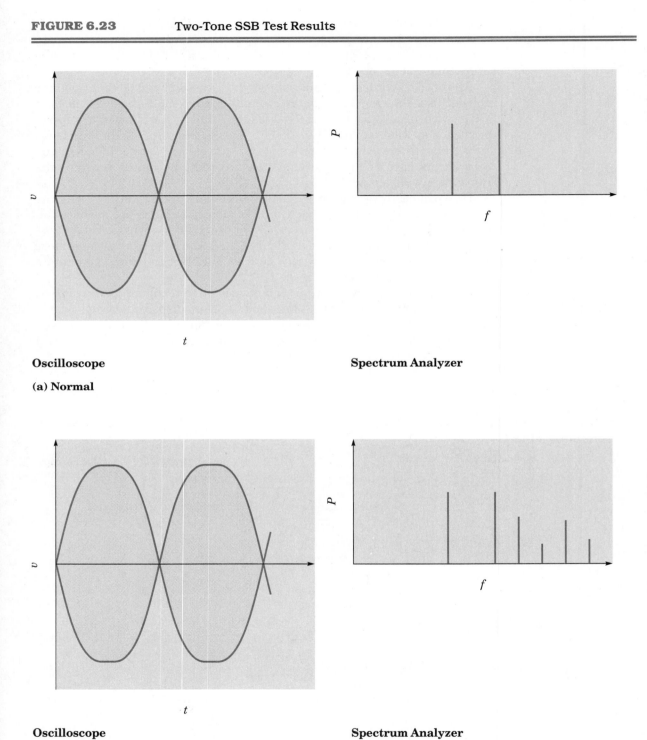

Oscilloscope   Spectrum Analyzer

(a) Normal

Oscilloscope   Spectrum Analyzer

(b) Clipped

FIGURE 6.24

Two-Tone Test of SSB Transmitter

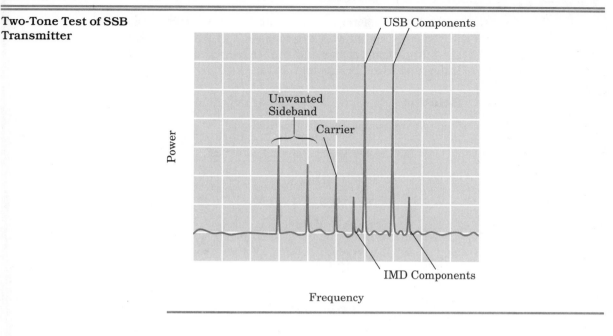

display spurious signals if distortion is present. See Figure 6.23(b) for examples of distorted signals.

Besides power output, operators of SSB transmitters are very concerned with carrier and unwanted-sideband suppression and intermodulation products. All of these are easily measured with a two-tone test signal and a spectrum analyzer. Figure 6.24 shows a typical spectrum-analyzer display. The carrier and the unwanted sideband can be clearly seen, even though they are very much weaker than the desired sideband. There will always be some carrier output, because no balanced modulator is ever perfectly balanced, and there will always be some unwanted-sideband energy because no bandpass filter has infinite rejection.

There are also spurious signals that result from the addition and subtraction of the desired frequency components and their harmonics. These are due to undesired nonlinearities in the amplifier stages in the transmitter, and can never be removed entirely. Most, but unfortunately not all, of the possible combinations will be so far removed in frequency from the main signal that they will not be transmitted. Two of these spurious signals are shown in the display of Figure 6.24. In order to measure the effectiveness of the suppression of unwanted frequency components, it is only necessary to note the number of decibels by which the level of each component is less than that of each of the desired signal components (which should be at the same level).

The most bothersome of these intermodulation distortion (IMD) products are the third-order products. If the two transmitted frequencies with a two-tone test signal are f_A and f_B, the third-order intermodulation products have frequencies of $(2f_A - f_B)$ and $(2f_B - f_A)$. Since f_A and f_B are very close together, these difference frequencies will be close in frequency to f_A and f_B, and will be in the passband of any filter that is applied to the transmitter output.

EXAMPLE 6.9

An SSB transmitter operating on the USB with a carrier frequency of 26.9 MHz is modulated with two tones at 1 kHz and 1.6 kHz, respectively. Find the output frequencies and the frequencies of the third-order intermodulation distortion products.

Solution

The two desired output frequencies are:

$$f_A = 26.9 \text{ MHz} + 1 \text{ kHz}$$
$$= 26.901 \text{ MHz}$$
$$f_B = 26.9 \text{ MHz} + 1.6 \text{ kHz}$$
$$= 26.9016 \text{ MHz}$$

The third-order IMD products are at:

$$2f_A - f_B = 2 \times 26.901 \text{ MHz} - 26.9016 \text{ MHz}$$
$$= 26.9004 \text{ MHz}$$
$$2f_B - f_A = 2 \times 26.9016 \text{ MHz} - 26.901 \text{ MHz}$$
$$= 26.9022 \text{ MHz}$$

Figure 6.24 shows how these distortion products relate to the transmitted signal. Both of them are within the range of voice frequencies that would be transmitted by this transmitter. Once generated, these distortion products would be impossible to remove by filtering.

6.8 OTHER SUPPRESSED-CARRIER SYSTEMS

SSB suppressed-carrier AM is used in its simplest form, as described above, for a variety of communications systems, including government, ship-to-ship and ship-to-shore, amateur and CB radio. There are many variations of the basic system with a multitude of other uses, ranging from long-distance telephony to television broadcasting. This section will provide a brief introduction to some of these applications.

6.8.1 Vestigial Sideband

It is difficult to transmit low-frequency baseband information using an SSB system, because of the difficulty of passing the wanted sideband while suppressing the unwanted one. This is not a serious problem for voice communications, where it is unnecessary to transmit frequencies below about 300 Hz. However, it would cause difficulties with music or video signals, both of which require the transmission of large amounts of low-frequency information. High-fidelity audio is not currently transmitted using SSB techniques, but video is. A variant of SSB called **vestigial sideband** (VSB) is the

**Single Sideband
in Amateur Radio**

Amateur radio operators have taken advantage of the benefits of SSB for many years. In fact, very few amateurs use full-carrier AM at present. Figure 6.25 shows a typical modern amateur radio transceiver for use in the high-frequency bands. Anyone with a communications receiver can listen to amateur radio signals. Tune in the vicinity of 3.8, 7.2, or 14.2 MHz to start. You will find that LSB is used on the lowest of these bands and USB on the others.

Amateur Radio SSB Transceiver
FIGURE 6.25
Courtesy Yaesu U.S.A.

Many portable radios with shortwave bands will receive these frequencies, but will not give intelligible results with SSB signals. By now you know why: they lack the essential BFO.

After listening for a while, many people decide they would like to transmit as well. This requires a license, of course, but getting one is not difficult. By the time you finish this book, you will know the required theory. It is also necessary to study the relevant government regulations and, for some classes of license, to learn the Morse code. More details can be obtained from a local amateur radio club (many colleges have them) or, in the United States, from

> American Radio Relay League
> Newington, CT 06111

In Canada, contact

> Canadian Radio Relay League
> Box 56
> Arva, ON N0M 1C0

Amateur radio is an enjoyable hobby and a good way to study practical communications.

FIGURE 6.26

Spectrum of TV Transmission on Channel 2

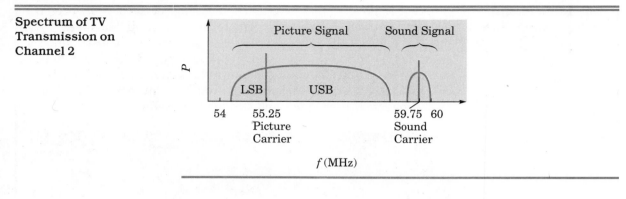

method used to transmit the video part of a television signal in ordinary terrestrial (not satellite) television broadcasting.

Figure 6.26 shows the simplified spectrum of a television signal on channel 2. The sound is transmitted using FM, and will be described later. The video is transmitted using the full USB and a partial LSB, filtered to reduce its bandwidth from 4.2 MHz to about 0.75 MHz. This allows the low-frequency components of the baseband signal to be transmitted on both sidebands, while only the USB carries the high-frequency information. The result is a compromise between DSB and SSB transmission in terms of efficiency and bandwidth, but it does allow the transmission of low-frequency information, right down to dc.

By the way, television video signals do not have the carrier suppressed. This allows for easier detection, using envelope detectors.

6.8.2 Independent Sideband

If only one sideband is required for one channel of baseband information, then why not use the other sideband for a second, totally different information signal? It can be done, giving a spectrum like that shown in Figure 6.27. The situation is similar to transmitting two SSB signals (one USB and one

FIGURE 6.27

ISB Spectrum

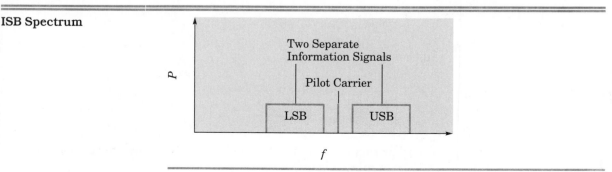

LSB) at the same time with the same carrier frequency. In an **independent sideband** (ISB) transmitter, however, the same carrier oscillator would be used for both, so that there would be no possibility of one transmitter drifting and causing interference to the other.

ISB is used mainly for military communications.

6.8.3 Pilot Carriers

When the carrier is completely suppressed in an SSB or DSB signal, tuning becomes difficult. It can be done by ear for voice signals, but that requires the attention of a skilled operator. If the carrier is transmitted, but at a level considerably below that needed for full-carrier AM, it becomes relatively easy to devise automatic tuning systems, where the carrier oscillator (BFO) in the receiver is synchronized to the transmitted carrier, using a phase-locked loop (PLL), for example. Of course, transmitting any carrier energy reduces the power efficiency of the system, but the carrier level need not be large, and the bandwidth reduction of SSB is still achieved. *Pilot carriers* also allow coherent demodulation, essential if phase information is to be preserved.

The systems described in Sections 6.8.4 and 6.8.5 both use pilot carriers.

6.8.4 Amplitude-Compandered Single-Sideband

The power in an SSB signal varies according to the level of the modulating signal. As the transmitted power level increases, so does the signal-to-noise ratio. Therefore, it would seem logical to keep the modulating signal relatively constant in amplitude, so that the transmitter power output can be kept close to maximum.

Unfortunately, real signal sources such as the human voice do not have constant power output. In order to increase efficiency, many communications transmitters employ signal processing that applies more gain to lower-level signals and less to stronger signals. This is called **compression**. Unfortunately, the use of very much compression causes the human voice to lose character, and with extreme amounts intelligibility suffers.

The solution is to undo the effects of compression on voice quality by using a similar but complementary process at the receiver. That is, stronger signals should be amplified more than weaker signals at the receiver. This is called **expansion**, and the use of a combination of compression at the transmitter and expansion at the receiver is called **companding**. For best results, of course, the compression and expansion should follow the same curve, or *law*. The result will be a received signal with improved signal-to-noise ratio but no other noticeable changes.

Amplitude-compandered SSB (ACSSB) is a standardized form of companding that is used in some mobile communications. A pilot carrier is transmitted with this system; the level of the information signal relative to that of the carrier provides the information the receiver needs to apply the correct amount of gain to the received signal.

6.8.5 Telephony Using Frequency-Division Multiplexing

Most of what has been said so far in this chapter assumes the presence of a radio link between an SSB transmitter and receiver. One of the most impor-

FIGURE 6.28

FDM Telephone System
Spectrum

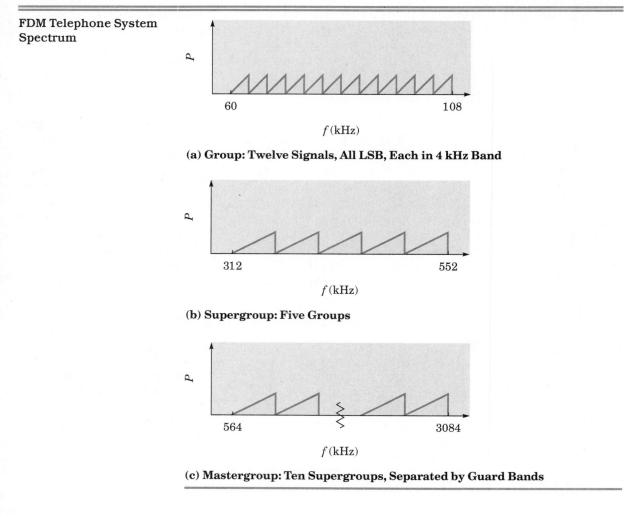

(a) Group: Twelve Signals, All LSB, Each in 4 kHz Band

(b) Supergroup: Five Groups

(c) Mastergroup: Ten Supergroups, Separated by Guard Bands

tant applications of SSB, however, often does not involve radio at all. SSB is used for long-distance telephone trunks over coaxial and twisted-pair cables, as well as with microwave radio links, both terrestrial and via satellite. Each voice channel occupies 4 kHz of bandwidth. Pilot carriers are transmitted to allow for automatic demodulation. The usable bandwidth of the transmission medium can then be filled with signals, each at a different carrier frequency. This is an example of frequency-division multiplexing (FDM).

The spectrum of signals on a typical coaxial cable carrying FDM telephone signals is shown in Figure 6.28. Twelve voice channels are combined into a **group**; five groups form a **supergroup**; and ten supergroups form a **mastergroup**. This mastergroup can handle a total of $12 \times 5 \times 10 = 600$ channels. Microwave links and large-capacity cables can carry even more channels.

FDM telephony is an analog technique that is gradually being replaced by digital systems using time-division multiplexing (TDM). However, the FDM system will remain in use for many years to come.

SUMMARY
OF
CHAPTER 6

Here are the main points to remember from this chapter.

1. Suppressing the carrier of an AM signal leads to greater efficiency with no loss of information content.

2. Removing one of the two sidebands of an AM signal reduces its bandwidth by approximately 50%, leading to an improved signal-to-noise ratio and allowing more signals to occupy a given frequency range.

3. A DSBSC AM signal can be generated by means of a balanced modulator, to which baseband and carrier-frequency signals are applied.

4. The most popular method for generating an SSB signal is to use a crystal filter to remove the unwanted sideband from a DSBSC signal. This is called the *filter method* of SSB generation.

5. A second way to generate an SSB signal is to use two balanced modulators to cancel out both the carrier and the unwanted sideband. This technique is known as the *phasing method* of SSB generation.

6. Once an SSB signal has been generated, it can be moved to the transmitting frequency by means of a mixer–local oscillator combination.

7. SSB and DSBSC signals, like all AM signals, must be amplified using linear amplifiers, in both transmitters and receivers.

8. In order to demodulate an SSB or DSBSC signal, it is necessary to reinsert the carrier at the receiver. The locally generated carrier frequency must be accurate to within approximately 100 Hz for intelligible voice reception.

9. Where half-duplex communication is required, considerable reductions in cost and complexity are possible by combining a transmitter and receiver into a transceiver. Generally, a single crystal filter, local oscillator, and frequency converter can be used for both transmission and reception.

10. Common variations of SSB include: the use of a pilot carrier for easier tuning; the transmission of a reduced-amplitude vestigial sideband (VSB) instead of completely suppressing one sideband, for better transmission of low-frequency information; and independent sideband transmission (ISB), which allows the transmission of two separate information signals, one on each sideband.

IMPORTANT EQUATIONS

$$\sin \omega_m t \, \sin \omega_c t = 0.5[\cos (\omega_c - \omega_m)t - \cos (\omega_c + \omega_m)t] \tag{6.3}$$

$$\cos \omega_m t \, \cos \omega_c t = 0.5[\cos (\omega_c - \omega_m)t + \cos (\omega_c + \omega_m)t] \tag{6.4}$$

$$PEP = \frac{V_p^2}{2R_L} \tag{6.7}$$

$$v_o \text{ (product detector)} = 0.5 \sin \omega_m t - 0.5 \sin (2\omega_c + \omega_m)t \tag{6.12}$$

GLOSSARY

beat-frequency oscillator (BFO)   an oscillator that reinserts the carrier signal in a single-sideband or continuous-wave receiver

compander   a combination of a compressor, used in the transmitter, and an expander, used in the receiver, designed to improve the signal-to-noise ratio of a communications system

compressor   an amplifier whose gain decreases with increasing amplitude of the input signal

expander   an amplifier whose gain increases with increasing amplitude of the input signal

group   in frequency-division multiplexed telephony, a combination of twelve voice channels

independent sideband (ISB)   an AM communications system in which each of the two sidebands carries a separate information signal

mastergroup   in frequency-division multiplexed telephony, a combination of ten supergroups

peak envelope power (PEP)   the power measured at modulation peaks in an AM or single-sideband signal

pilot carrier   a reduced-amplitude carrier signal transmitted to aid in synchronizing the receiver

power efficiency   a measure of how much of the transmitted power carries useful information

product detector   a balanced modulator used as a detector for suppressed-carrier signals

single-sideband (SSB)   any AM scheme where only one of the two sidebands is transmitted

supergroup   in frequency-division multiplexed telephony, a combination of five groups

suppressed-carrier signal   an AM signal in which the carrier-frequency component is eliminated, and only one or both sidebands are transmitted

two-tone test signal   a signal consisting of two audio frequencies, not harmonically related, used to test single-sideband transmitters

vestigial sideband (VSB)   an AM signal in which one sideband is transmitted in full, but only those components of the other sideband that correspond to relatively low modulation frequencies are transmitted

QUESTIONS

1. Give two advantages of SSB operation compared with full-carrier AM.
2. Does full-carrier AM have any advantages over SSB? Explain.
3. Why, and by how much, does suppressing the carrier improve the efficiency of an AM signal?
4. Why, and by how much, does eliminating one sideband improve the efficiency of an AM signal?

5. Why is the transmitted bandwidth of a typical SSB signal actually less than one-half that of a full-carrier AM signal transmitting the same information signal?

6. What is the audible effect of slight mistuning of an SSB receiver?

7. What is the audible effect of setting an SSB receiver to the wrong sideband?

8. Describe two methods of generating SSB signals.

9. Which method of SSB generation is the most popular, and why?

10. Why are Class C amplifiers unsuitable for use in SSB transmitter power amplifiers?

11. How is the proper output of an SSB transmitter specified?

12. What is a BFO, and why is it necessary in an SSB receiver?

13. Why is local-oscillator stability more important in SSB receivers than in those for full-carrier AM?

14. Suggest suitable bandwidths for the AM, SSB, and CW filters in a communications receiver.

15. Why is VSB, rather than DSB AM or SSB, used for television broadcasting?

16. What is ISB transmission and how is it used?

17. Explain compression and expansion, and show how they can be used to make a communications system more efficient.

18. What do the following circuits have in common: balanced mixer, balanced modulator, product detector?

19. Explain, using a block diagram, how a two-tone test of an SSB transmitter can be performed.

20. Explain how SSB techniques are used in FDM telephony.

PROBLEMS

Section 6.2

21. A 5 MHz carrier is modulated by a 5 kHz sine wave. Sketch the result in both the time and frequency domains for each of the following modulation types. Time and frequency scales are required but amplitude scales are not.

    (a) DSB full-carrier AM
    (b) DSBSC AM
    (c) SSBSC AM (USB)

22. If a transmitter power of 100 W is sufficient for reliable communication over a certain path using SSB, approximately what power level would be required using each of the following?

    (a) DSBSC
    (b) full-carrier AM

23. (a) An AM transmitter has a carrier power of 50 W at a carrier frequency of 12 MHz. It is modulated at 80% by a 1 kHz sine wave. How much power is contained in the sidebands?

    (b) Suppose the transmitter in Part (a) can also be used to transmit a USB signal with an average power level of 50 W. By how much (in decibels) will the signal-to-noise ratio be improved when the transmitter is used in this way, compared with the situation in Part (a)?

24. Sketch the output signal from each of the transmitters described in the previous

problem, in both the time and frequency domains. Assume the load impedance is 50 Ω

25. An SSB transmitter has a PEP of 250 W. It is modulated equally by two audio frequencies at 400 Hz and 900 Hz. The carrier frequency is 14.205 MHz and the transmitter generates the LSB. Sketch the output in the time and frequency domains, for a load impedance of 50 Ω.

Section 6.3

26. A carrier at 10 MHz and a sine-wave modulating signal at 2 kHz are applied to a balanced modulator. Sketch the output in the time and frequency domains showing time and frequency scales, respectively.

27. A filter-type SSB generator uses an ideal bandpass filter with a center frequency of 5.000 MHz and a bandwidth of 2.7 kHz. What frequency should be used for the carrier oscillator if the generator is to produce a USB signal with a baseband frequency response having a lower limit of 280 Hz?

28. A variation on a phasing-type SSB generator is shown in Figure 6.29.

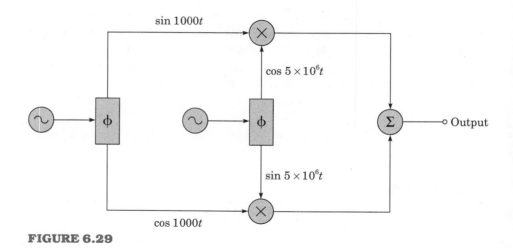

FIGURE 6.29

(a) What will be the carrier frequency?
(b) Which sideband will be produced?

Section 6.4

29. An SSB transmitter has the block diagram shown in Figure 6.30.

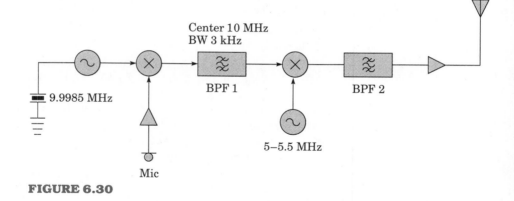

FIGURE 6.30

(a) Using a VFO which tunes from 5.0 to 5.5 MHz, the transmitter will operate over two frequency ranges which can be selected by a suitable choice of band-pass filter BPF2. What ranges are they?

(b) Which sideband would be produced at the output for each of the frequency ranges specified in Part (a)?

(c) The other sideband could be generated by changing the carrier-oscillator frequency. Choose a suitable value for this frequency.

30. The block diagram of an SSB transmitter is shown in Figure 6.31. The local oscillator frequency is higher than the frequency at which the SSB signal is generated, and the difference between the two frequencies is used at the output.

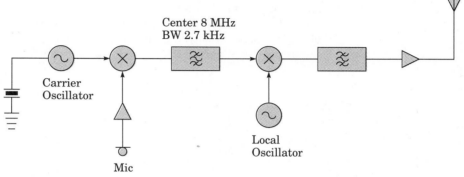

FIGURE 6.31

(a) Choose a suitable frequency for the carrier oscillator if the transmitter is to produce a USB signal.

(b) What should be the frequency of the local oscillator if the (suppressed) carrier frequency at the antenna is to be exactly 30 MHz?

(c) Suppose that the transmitter is modulated by a single sine-wave tone at 1 kHz. It is operating with a PEP of 100 W into a 50 Ω load. Sketch the output in the time and frequency domains, showing all appropriate scales.

31. Consider a full-carrier AM signal with $m = 1$. Show mathematically that the PEP of this signal is equal to four times the carrier power.

Section 6.5

32. The block diagram of an SSB receiver is shown in Figure 6.32.

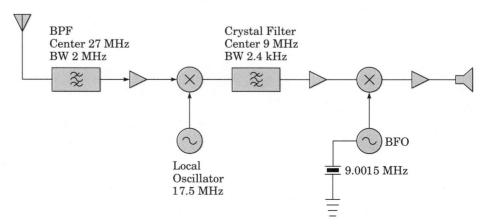

FIGURE 6.32

(a) Which sideband will be received with the values shown?

(b) What will be the carrier frequency of the received signal?

(c) What changes would have to be made so that this receiver will be set up to receive the other sideband at the same carrier frequency?

33. Sometimes an AM receiver is made capable of SSB reception by simply adding a BFO, with the IF bandwidth left as for DSB AM. What would be the effect of this on the signal-to-noise ratio, compared with a receiver with the proper IF filter?

34. A CW signal is being transmitted at 7.100 MHz. Explain how it can be received by an SSB receiver, and suggest suitable frequencies to which the receiver could be tuned. Does it matter whether the receiver is set to USB or LSB?

Section 6.6

35. A block diagram of an SSB transceiver is given in Figure 6.33. Redraw it to show the transmitter and receiver functions in two separate diagrams. Note that some blocks are used for both transmitting and receiving.

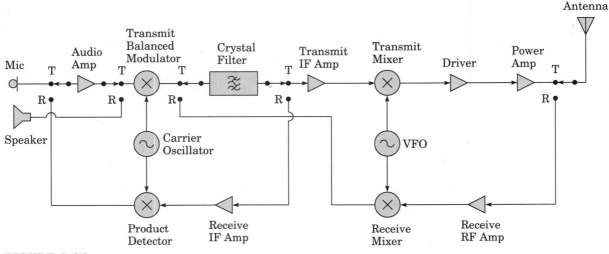

FIGURE 6.33

Section 6.7

36. The results of a two-tone test on an SSB transmitter with a carrier frequency of 17 MHz are shown in both the time and frequency domains in Figure 6.34. Use the information provided to find:

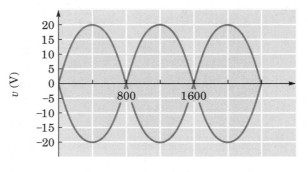

t (μs)

FIGURE 6.34

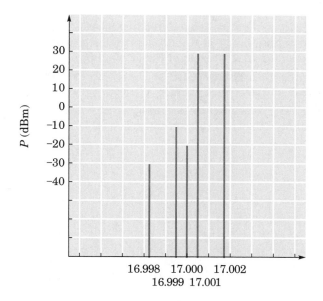

FIGURE 6.34 (*Continued*)

(a) the two tones
(b) whether the transmitter is USB or LSB
(c) the PEP (assume the transmitter has a 50 Ω load)
(d) the carrier suppression
(e) the unwanted-sideband suppression at each of the test frequencies

Comprehensive

37. SSB transmitters use Class AB power amplifiers while AM transmitters generally use more efficient Class C circuits. Suppose that 100 W total dc power is used to power the final stages of each of two transmitters as follows. (For the AM transmitter, the 100 W includes the power supplied to the modulating amplifier.)

   Transmitter 1: Full-carrier AM. The modulating amplifier is Class AB with 70% efficiency, the final amplifier is Class C with 85% efficiency. The modulation index is 75%.

   Transmitter 2: SSB. The RF power amplifier is Class AB with 65% efficiency.

   Which transmitter will produce a better signal-to-noise ratio at the receiver, and by how much?

38. An SSB transmitter transmits on the USB with a (suppressed) carrier frequency of 7.2 MHz. It is modulated by two tones, with frequencies 1 kHz and 2.5 kHz and equal amplitude. The transmitter PEP is 75 W into a 50 Ω load. The carrier and unwanted sideband are each suppressed by 60 dB.

   (a) Calculate the peak voltage across the load.
   (b) Calculate the average power into the load.
   (c) Sketch the signal envelope in the time domain, showing voltage and time scales.
   (d) Sketch the signal in the frequency domain, showing frequency and power scales.

39. An SSB transmitter is modulated by tones of 1, 2, and 3 kHz. A receiver is tuned to this signal, but it is slightly misadjusted: it is tuned 100 Hz too high. What frequencies will be present at the output of this receiver if the transmitter is:

   (a) transmitting on the USB?
   (b) transmitting on the LSB?

40. Using FDM, how many telephone-quality voice channels could be transmitted in the bandwidth of an ordinary television channel?

7

Frequency Synthesis

Objectives

After studying the material in this chapter, you should be able to:

1. Explain the operation of a phase-locked loop.
2. Calculate the loop gain for a phase-locked loop.
3. Define and calculate the lock and capture range for a phase-locked loop.
4. Explain and calculate lock time for a simple phase-locked loop.
5. Draw block diagrams for several types of frequency synthesizer, explain their operation, and calculate output frequencies.
6. Interpret the specifications for a frequency synthesizer.
7. Describe the operation of electronic counters, and use them to measure frequency and period.

7.1 INTRODUCTION

In Chapter 2, we found that free-running *LC* oscillators are easily tuned to different frequencies, because their operating frequency is usually determined by tuned circuits. For that reason, they are often referred to as variable-frequency oscillators (VFOs). Unfortunately, they also exhibit undesired frequency changes as a result of vibration, voltage or temperature changes, component aging, and so on. In addition, accurately setting these oscillators to a particular frequency is tricky, requiring precision-built variable capacitors or inductors and expensive dials using complicated arrangements of gears and pulleys.

Early Frequency Synthesizers

Before phase-locked-loop frequency synthesizers became practical, attempts were made to generate multiple frequencies from a smaller number of crystals by using mixers and frequency multiplication. A simplified example of a synthesizer that uses banks of switched crystals is shown in Figure 7.1. There are two banks with four crystals in each. (In practice, there would likely be more.) There are two oscillators, each of which can work with any crystal in its associated bank. The mixer combines the two oscillator output frequencies, and the switchable bandpass filter can select either of the two crystal frequencies, their sum or their difference. The range could be enhanced by using more mixers.

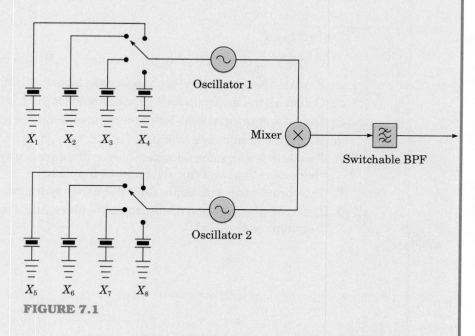

FIGURE 7.1

Systems like this were once popular in CB transceivers. Though the number of crystals required is lower than for a simple switched-crystal oscillator, it is still considerable, and there is also the problem of filtering out spurious frequencies generated by the mixers.

Crystal oscillators, on the other hand, have very good stability. With voltage regulation and temperature control or compensation, frequency drift can be reduced to a few parts per million over long time periods. The disadvantage, for many applications, is that the frequency of a crystal oscillator can be changed only a very small amount by adjusting series or parallel capacitors. This makes its use awkward or impossible for any application that requires continuous frequency variation, or even operation at more than a few discrete frequencies.

Most receivers and many transmitters require the frequency agility of the VFO coupled with the stability and accuracy of the crystal oscillator. For many years, it was customary to use VFOs for applications where tuning had to be continuous or where there were large numbers of frequencies in use, and to use crystal oscillators with switchable crystals when operation was required on only a relatively small number of different frequencies. In recent years, however, the phase-locked frequency synthesizer has become very popular. It is now the preferred method of frequency generation in most modern receivers and transmitters. In fact, it is often possible to save money as well as improve performance when a synthesizer is used instead of a VFO, because of reduced requirements for mechanical precision.

7.2 PHASE-LOCKED LOOPS

Before looking more closely at synthesizers, it is necessary to know something about the **phase-locked loop** (PLL), because it is the basis of practically all modern synthesizer design. Though the PLL was actually invented in 1932, it is only since 1970—when it was first produced on an IC—that it has been much more than a laboratory curiosity. A PLL synthesizer can certainly be constructed using tubes or discrete transistors, but so many devices would be needed that the technique is not practical.

Figure 7.2 shows the essentials of a simple PLL. The loop consists of a phase detector, a **voltage-controlled oscillator (VCO)**, and a low-pass filter (LPF) called the **loop filter**. An external reference signal is compared with the VCO signal in the **phase detector**. An error voltage is produced whose amplitude varies with the phase difference between the two signals. After filtering, this error signal is applied as a control voltage to the VCO.

The purpose of the PLL is to lock the VCO to the reference signal. That is, the two signals will have the same frequency. Since the frequency of the two signals is exactly the same, the phase angle between them will remain constant, hence the term *phase-locked*.

FIGURE 7.2

Phase-Locked Loop

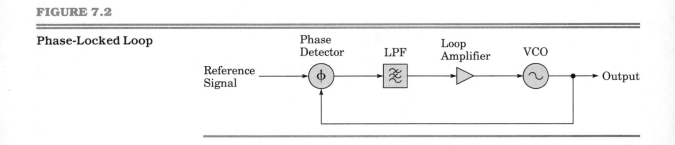

When the loop starts operating, the VCO will operate at its **free-running frequency** (that is, the frequency at which it operates when the control voltage is zero). This will probably not be the same as the reference frequency. The loop is said to be *unlocked*. The phase detector will generate a control voltage, which will cause the VCO frequency to change until it is exactly that of the external input signal. This is called the *acquisition of phase lock*. Once phase lock has been acquired, the loop will remain locked indefinitely. Any tendency of the VCO to drift in frequency will result in a change in the control voltage in the direction required to bring the loop back into a locked condition.

A practical loop will likely have some gain for the control signal. This is the purpose of the loop amplifier shown in Figure 7.2. There will also be some limits on how far apart the free-running VCO frequency and the reference frequency can be for lock to be acquired or maintained. Of course, it will also take a small but finite amount of time to achieve lock.

7.2.1 Phase-Locked Loop Operation

Suppose that the external reference frequency and the VCO frequency are initially very far apart. If the reference is gradually brought closer to that of the VCO, there will come a point where the VCO frequency will suddenly change to that of the external signal, and the loop will lock. The range over which the reference frequency can be varied and still achieve phase lock is called the **capture range**. This is an important PLL specification, as it determines how far apart the external and internal frequencies initially can be for the loop to achieve lock.

Suppose that lock has been achieved, so that the VCO is synchronized to the reference frequency. Now, we change the reference frequency, gradually moving it further from the free-running frequency of the VCO. For a while, the VCO will *track*; that is, its frequency will change to follow the external signal. Eventually, however, phase lock will be lost, and the VCO will return to its free-running frequency. The total frequency range within which lock, once achieved, can be maintained, is called the **lock range**.

Figure 7.3 should help to make all this clear. Please note the difference between capture range and lock range. The lock range is almost always larger than the capture range.

FIGURE 7.3

PLL Frequency
Specifications

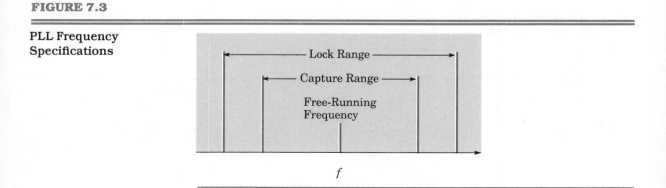

EXAMPLE 7.1

A phase-locked loop has a VCO with a free-running frequency of 12 MHz. As the frequency of the reference input is gradually raised from zero, the loop locks at 10 MHz and comes out of lock again at 16 MHz.

(a) Find the capture range and lock range.

(b) Suppose that the experiment were repeated, but this time the reference input began with a very high frequency and steadily moved downward. Predict the frequencies at which lock would be achieved and lost.

Solution

(a) The capture range is approximately twice the difference between the free-running frequency and the frequency at which lock is first achieved. For this example,

$$\text{Capture Range} = 2(12\,\text{MHz} - 10\,\text{MHz})$$
$$= 4\,\text{MHz}$$

The lock range is approximately twice the difference between the frequency where lock is lost and the free-running frequency. Here,

$$\text{Lock Range} = 2(16\,\text{MHz} - 12\,\text{MHz})$$
$$= 8\,\text{MHz}$$

(b) The PLL frequency response is (at least approximately) symmetrical; that is, the free-running frequency is in the center of the lock range and capture range. The frequency at which lock will be acquired, moving downward in frequency, is

$$12\,\text{MHz} + 2\,\text{MHz} = 14\,\text{MHz}$$

Lock will be lost on the way down at

$$12\,\text{MHz} - 4\,\text{MHz} = 8\,\text{MHz}$$

Figure 7.4 shows these relationships for this example.

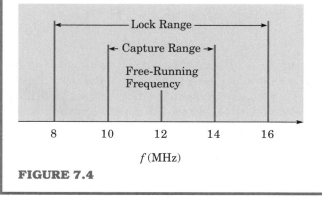

FIGURE 7.4

In order to understand how the results summarized above are achieved, it would be useful to look briefly at each of the components in turn. Great detail is not required, as the elements to be described (except sometimes for the loop filter) are usually found as parts of an integrated circuit. The technologist is not likely to have to design a PLL from scratch, but some knowledge of its operation will be useful when attempting to evaluate the performance of, or find faults in, a PLL synthesizer.

7.2.2 The Voltage-Controlled Oscillator

Let us begin with the VCO. Any type of oscillator could be used, but astable multivibrators are very common. The frequency-determining element must be voltage sensitive; a varactor diode fulfills this requirement.

The VCO must be constructed so that a control voltage of one polarity raises the frequency above its free-running value, while the opposite polarity lowers the frequency. Since a varactor must always be reverse-biased, a bias voltage in addition to the control voltage will have to be applied to the varactor.

The variation of frequency with voltage will not be perfectly linear, but it will be sufficiently close to linear (over a restricted frequency range) that a constant of proportionality can be derived.

$$k_f = \frac{\Delta f}{\Delta v} \tag{7.1}$$

where k_f = VCO constant of proportionality in hertz per volt
   Δf = a small change in frequency
   Δv = the change in control voltage required for the change in frequency Δf

EXAMPLE 7.2

A VCO has a free-running frequency of 12 MHz and

$$k_f = 50 \text{ kHz/V}$$

The variation of frequency with voltage is linear for control voltages between +3 and −3 V. Draw a graph showing the relationship between control voltage and frequency for this oscillator.

Solution

Since the VCO frequency response is linear, the graph will be a straight line. Therefore, it is only necessary to find two points. One is given: for a voltage of zero, the frequency will be exactly 12 MHz. The other point can be anywhere within the control range. If we use a voltage of 3 V, we find that the frequency is

$$f = 12 \text{ MHz} + 3 \text{ V} \times 50 \text{ kHz/V}$$
$$= 12.15 \text{ MHz}$$

The resulting graph is shown in Figure 7.5.

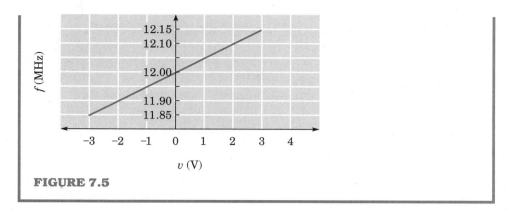

FIGURE 7.5

7.2.3 The Phase Detector

Moving around the loop, the next stage is the phase detector. Phase detectors are found in other communications circuits as well, so it is worthwhile to understand how they work in general. Phase detectors can be based on either analog or digital techniques. Analog circuits use mixers or multipliers; recall that a multiplier is a balanced mixer. The digital versions use exclusive-or gates or S-R flip-flops. In all cases, the result is the same: an output containing both ac and dc components, whose average value is proportional to the phase difference between two input signals.

A brief look at one form of phase detector will make this clear. The balanced mixer, or multiplier, has already been studied in Chapter 2 and again in Chapter 6. An ideal multiplier produces an output whose amplitude is proportional to the product of two input signals. In all three cases, we are referring to instantaneous amplitudes. The constant of proportionality is of no concern at the moment, so let it be equal to 1. To simplify the equations, let the amplitudes of the reference signal input v_{ref} and the VCO signal v_{VCO} be equal to 1 V peak. Let the two inputs be

$$v_{ref} = \sin(\omega_r t + \theta_r)$$

and

$$v_{VCO} = \sin(\omega_v t + \theta_v)$$

where ω_r and ω_v are the radian frequencies of the reference signal and the VCO respectively. The phase angles θ_r and θ_v are included because they will become significant later in the analysis of the phase-locked loop.

If these two sine waves of different frequencies are applied to the phase-detector inputs, the output will be their product, which is

$$
\begin{aligned}
v_o &= v_{ref}v_{VCO} \\
&= \sin(\omega_r t + \theta_r)\sin(\omega_v t + \theta_v) \\
&= 0.5\{\cos[(\omega_r - \omega_v)t + \theta_r - \theta_v] \\
&\quad - \cos[(\omega_r + \omega_v)t + \theta_r + \theta_v]\}
\end{aligned}
\tag{7.2}
$$

If the two frequencies are equal, as they are when the loop is locked, the first term will contain only a dc component that varies with the phase angle. This term is

$$\cos\,[(\omega_r - \omega_v)t + \theta_r - \theta_v]$$

(the constant 0.5 has been removed for simplicity). Since $\omega_r = \omega_v$, the first part is zero and the term reduces to

$$\cos\,(\theta_r - \theta_v)$$

The angle $(\theta_r - \theta_v)$ is, of course, the phase angle between the two inputs. Its cosine will be zero when the angle is 90°, positive for smaller angles, and negative for larger ones. The variation of output voltage with phase is not linear, of course, but it will be almost linear provided that the phase angle is close to 90°. In that case, a constant of proportionality for the phase detector can be defined, relating the output voltage to the change in the phase angle between the two inputs. The units will be volts per radian (V/rad). Sometimes this constant is called the *gain* of the phase detector.

$$k_p = \frac{v_o}{\phi} \tag{7.3}$$

where k_p = phase-detector gain in volts per radian

   v_o = output voltage in volts

   ϕ = phase angle between the two inputs, in radians (after subtracting $\pi/2$ radians, for the example above)

The second term in Equation (7.2) is a component at twice the reference frequency, which will be removed by the loop filter.

For unequal frequencies, which occur when the loop is unlocked, the situation is a little more complicated. At the sum of the reference and VCO frequencies, there will be a component which will be removed by the loop filter. In addition, there will be a component at the difference frequency. This will cause the VCO frequency to sweep until it becomes the same as that of the reference input, and lock will be acquired. Once lock is acquired, of course, the two frequencies are equal.

EXAMPLE 7.3

A phase detector has $k_p = 3$ V/rad. Its output is zero when the phase angle between the two input signals is 90°. What is the output voltage when the phase angle is 100°?

Solution

First, it is necessary to subtract the 90° offset, giving a net phase difference of 10°. Next, we convert the 10° into radians. The conversion formula is easy to forget, but almost as easy to derive. Remembering that a circle has 360° or 2π rad, we find that 1° is equivalent to $(2\pi/360) = 0.01745$ rad. Therefore, 10° are equivalent to 0.1745 rad, and the output voltage is

$$v_o = k_p\phi$$
$$= 3 \times 0.1745$$
$$= 0.524 \text{ V}$$

7.2.4 The Loop Filter

The ac components of the phase-detector output will be at the sum and difference of the two signal frequencies, but the low-pass filter will remove the sum component. This leaves the difference frequency and the dc component. When the loop is locked, the difference frequency is zero, and only the dc voltage remains, providing the control voltage required to keep the VCO at the correct frequency.

If the difference frequency is much higher than the loop filter's cutoff frequency, then the difference will also be filtered out, and the loop will not achieve lock. Thus the choice of loop filter has an important effect on the behavior of the loop in its unlocked state.

The other effect of the loop filter is to increase the speed with which the VCO follows changes in the reference frequency. The PLL is a negative feedback circuit, in which a difference between the reference and VCO frequencies eventually causes a change in the VCO control voltage in the direction that causes the VCO frequency to follow the reference frequency. All of this takes time. The loop filter combines with the VCO and the phase detector to produce a second-order system that, if carefully designed, will allow the VCO to follow changes in the reference frequency more quickly than it would without a loop filter. (For further details about this process, see Section 7.8.1 near the end of this chapter.)

7.2.5 The Amplifier

The lock range of the PLL is limited to the frequency range over which the VCO can be tuned with the voltage available from the phase detector. With most phase detectors (including the one described earlier), this allows a phase shift of 90° in either direction from the angle that produces a zero error signal. In our example, a zero error signal corresponded to a phase difference of 90° between the reference and VCO signals.

The frequency range over which lock can be maintained can be increased by inserting a dc-coupled amplifier into the loop. However, besides responding to dc, the amplifier must also have very good high-frequency response, with a cutoff frequency much larger than the loop bandwidth. Otherwise, it can introduce instability because of the phase shift that occurs in the neighborhood of its cutoff frequency.

As with any other amplifier, this one has a gain, which can be called A_v, and is simply the ratio of output to input voltage. As such, it is of course dimensionless.

7.2.6 Loop Gain

As in any feedback circuit, the loop gain can be found by multiplying the gains of the individual components around the loop. There are two differences between the PLL and the more familiar amplifier or oscillator situations. First, all of the signal is fed back in a PLL. Second, the individual gains are not all in the same units.

Let us proceed around the loop in the same order as before, multiplying these gains to find the loop gain and its units. Then it will be necessary to consider what this specification means in physical terms.

Since we are currently studying the loop in the locked condition, the low-pass filter will have no effect. We will assume it is lossless; if not, its only function will be to reduce the gain of the amplifier.

Our first step is to multiply all the stage gains together and see what develops. Let us call the product the loop gain k_L.

$$k_L = k_p A_v k_f$$

where k_p = phase-detector gain in volts per radian
        A_v = loop amplifier gain (dimensionless)
        k_f = VCO proportionality constant, in hertz per volt

The units for k_L will be

$$(\text{volts/radian}) \times (\text{hertz/volt}) = \text{hertz/radian}$$

The loop gain can also be specified in more fundamental units by defining a new gain k_v which is defined as k_L multiplied by 2π:

$$k_v = 2\pi k_L \tag{7.4}$$
$$= 2\pi k_p A_v k_f$$

Now, the units are

$$(\text{radians/second})/\text{radian} = \text{seconds}^{-1} \quad \text{or} \quad \text{hertz}$$

What this tells us, in terms of practical, physical reality, is the amount of frequency variation that is possible. The maximum phase difference that can be accommodated is 90° in each direction or 180° total, which corresponds to π radians. Thus, the lock range is the change in frequency which corresponds to π radians:

$$\text{Lock Range (Hz)} = \pi k_L \tag{7.5}$$
$$= \frac{k_v}{2}$$

Please note that this derivation contains the important assumption that the phase detector is linear. For frequencies well away from the VCO's free-running frequency, this is likely to be untrue, and the lock range found from Equation (7.5) will be only an approximation.

What does follow from this discussion, however, is the concept that high loop gain results in a wide lock range, and that anything that reduces the loop gain also reduces the lock range. We shall have occasion to make use of this finding very shortly.

EXAMPLE 7.4

A PLL has a phase detector that is linear over a total range of 180°. The phase detector has

$$k_p = 3 \text{ V/rad}$$

The phase detector is followed by an amplifier with a voltage gain of 10. The output of the amplifier controls a VCO with

$$k_f = 20 \text{ kHz/V}$$

Find the lock range.

Solution

$$
\begin{aligned}
\text{Lock Range} &= \pi k_L \\
&= \pi k_p A_v k_f \\
&= \pi \times 3 \text{ V/rad} \times 10 \times (20 \times 10^3 \text{ Hz/V}) \\
&= 1.88 \times 10^6 \text{ Hz} \\
&= 1.88 \text{ MHz}
\end{aligned}
$$

7.3 SIMPLE FREQUENCY SYNTHESIZERS

It may not be obvious how a PLL can fulfill our original goal of creating an oscillator with crystal-controlled stability and VFO agility, without using a great number of crystals. To achieve this goal, we need to add one more component to the loop: a programmable divider. See Figure 7.6 for a sketch of a **frequency synthesizer**, reduced to its simplest terms.

In this circuit, the phase detector still compares two frequencies and produces a control voltage that results in the two becoming locked together. However, while one of these is still an external reference frequency, which could be generated by a crystal oscillator, the other is no longer the VCO frequency itself. That frequency is divided by some integer number N, and then compared with the reference frequency. Using a programmable divider allows N to be varied. It is easy to see that, assuming the PLL is locked,

$$f_{ref} = \frac{f_{VCO}}{N}$$

so

FIGURE 7.6

A Simple Frequency Synthesizer

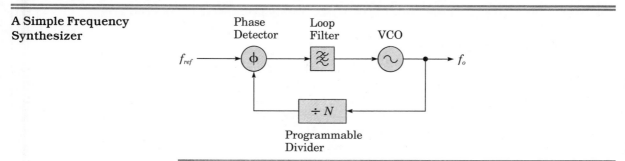

FIGURE 7.7

Frequency Synthesizer
with Divider for the
Reference Frequency

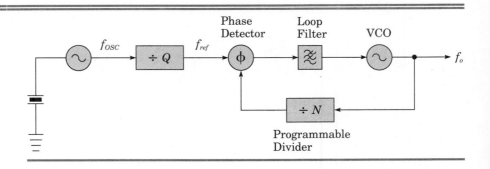

$$f_{VCO} = Nf_{ref}$$

Generally, the VCO generates the output frequency f_o. Then

$$f_o = Nf_{ref} \tag{7.6}$$

This means that a large number of different output frequencies can be generated, all locked to a single crystal-controlled reference frequency, simply by changing the **modulus** (the value of N). This will be done by changing the voltages on some of the pins of the divider chip. Therefore the technique lends itself very easily to computer control and/or remote control. Expensive variable capacitors and inductors are also eliminated. The resultant cost savings can actually make a frequency synthesizer cheaper than a conventional VFO, in spite of its greater complexity.

There is one problem. This circuit cannot generate just any frequency, but only those that are multiples of f_{ref}. For instance, if f_{ref} is 100 kHz, the circuit can, at least in theory, generate any multiple of 100 kHz. This would be satisfactory in an FM broadcast receiver, since FM broadcast channels are 200 kHz apart, but it would have its limitations in AM broadcasting, where the channel spacing is only 10 kHz.

The obvious solution is to reduce f_{ref}. Crystals with frequencies much below 100 kHz are impractical, but a fixed-modulus divider can be used to divide down the reference frequency, as shown in Figure 7.7.

EXAMPLE 7.5

Configure a simple PLL synthesizer using a 10 MHz crystal so that it will generate the AM broadcast frequencies from 540 to 1700 kHz.

Solution

Our synthesizer will have the block diagram of Figure 7.7. Since the channel spacing in AM broadcasting is 10 kHz and all channels are at integer multiples of 10 kHz, it would be logical to use this value for f_{ref}; in that case, each time N is incremented or decremented by 1, the output frequency will move to the adjacent channel.

Since a 10 MHz crystal is used, it will be necessary to divide it by a factor (shown as Q in Figure 7.7) to get 10 kHz. This factor can easily be found as follows.

$$Q = \frac{f_{osc}}{f_{ref}}$$

$$= \frac{10 \text{ MHz}}{10 \text{ kHz}}$$

$$= 1000$$

Next, we should specify the range of values of N that will be required. We have already seen that changing N by 1 changes channels. All we need to do then, is find N at each end of the tuning range. We can rearrange Equation (7.6)

$$f_o = Nf_{ref}$$

to get

$$N = \frac{f_o}{f_{ref}}$$

At the low end of the band, we have

$$N = \frac{540 \text{ kHz}}{10 \text{ kHz}}$$

$$= 54$$

At the high end,

$$N = \frac{1700 \text{ kHz}}{10 \text{ kHz}}$$

$$= 170$$

7.3.1 Example: A Synthesized CB Transceiver

Figure 7.8(a) is the block diagram of a frequency synthesizer for use in a CB transceiver like the one discussed in Chapters 4 and 5. The synthesizer uses a PLL IC. The IC includes a crystal oscillator, fixed and programmable dividers, and phase detector. The rest of the synthesizer consists of a VCO, a buffer amplifier, and a low-pass filter.

The synthesizer output is taken from the VCO. When transmitting, it produces a signal at one-half the operating frequency, which goes to a frequency-doubler stage that is part of the transmitter. When receiving, the VCO provides the local-oscillator signal for the first mixer of a double-conversion receiver. The 10.240 MHz crystal oscillator that is included in the PLL is used to generate the 2.5 kHz reference frequency and is also used directly to provide the local-oscillator signal for the second receiver mixer. Thus, all the frequencies required for a transmitter and a double-conversion

FIGURE 7.8

Use of Frequency
Synthesizer in a CB
Transceiver

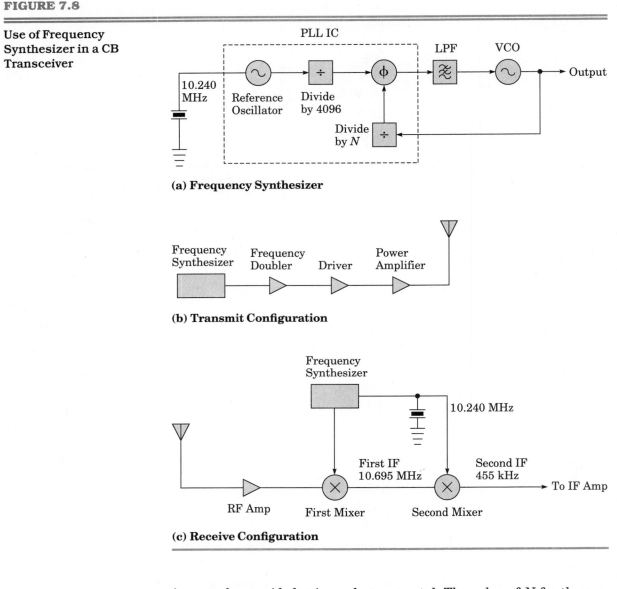

(a) Frequency Synthesizer

(b) Transmit Configuration

(c) Receive Configuration

receiver can be provided using only one crystal. The value of N for the programmable divider that is part of the IC can be set by a multipole switch.

It is quite easy to calculate the value of N for a given output frequency, as the next example shows.

EXAMPLE 7.6

Calculate the correct values of N in the synthesizer shown in Figure 7.8 for transmitting and receiving on CB Channel 10 (27.075 MHz).

Solution

For transmitting, the synthesizer must generate a signal at one-half the transmitting frequency (13,537.5 kHz). The value for N can easily be calculated by rearranging Equation (7.6):

$$f_o = N f_{ref}$$

$$N = \frac{f_o}{f_{ref}}$$

$$= \frac{13537.5 \text{ kHz}}{2.5 \text{ kHz}}$$

$$= 5415$$

For receiving, the first IF is at 10.695 MHz and the local oscillator uses low-side injection. Therefore, the synthesizer, which provides the local-oscillator signal, must operate at a frequency f_o where

$$f_o = f_i - f_{IF}$$

$$= 27.075 \text{ MHz} - 10.695 \text{ MHz}$$

$$= 16.380 \text{ MHz}$$

The value of N for receiving can now be found.

$$N = \frac{f_o}{f_{ref}}$$

$$= \frac{16380 \text{ kHz}}{2.5 \text{ kHz}}$$

$$= 6552$$

7.4 PRESCALING

There is a problem with the basic synthesizer when output frequencies in the VHF range and higher are required. Programmable dividers are simply not available at frequencies much above 100 MHz. With current technology, it is not possible to build a UHF synthesizer with the simple topology of Figure 7.7. The simplest way to get a synthesizer to work at frequencies beyond those at which programmable dividers operate is to add a fixed-modulus divider in front of the programmable one, as illustrated by Figure 7.9. This divider could employ *emitter-coupled logic* (ECL), a digital technology which

FIGURE 7.9

Frequency Synthesizer with Fixed Prescaler

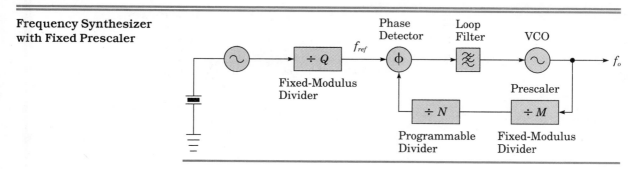

can be used at frequencies above 1 GHz. For even higher frequencies, discrete transistors using gallium arsenide could be used. This will work well into the microwave region.

There is one drawback to this idea. The VCO frequency is now divided by the fixed modulus M, then by the programmable modulus N, before it is compared with the reference frequency. Thus

$$f_{ref} = \frac{f_o}{MN}$$

or

$$f_o = MNf_{ref} \tag{7.7}$$

Since only N can be changed, the minimum amount by which the frequency can be changed is now Mf_{ref}. For example, if f_{ref} is 10 kHz, and a 10:1 **prescaler** is used, the minimum step by which the frequency can be changed is 100 kHz. We seem to have taken one step forward and one back.

A truly elegant solution to the problem, and one that is often used in synthesizers for VHF and UHF, is to use a *two-modulus prescaler*. This is a divider that can be programmed to divide by either of two consecutive integers (for instance, 10 and 11 or 15 and 16). We can let these integers be P and $(P + 1)$. The frequency limitations of such dividers are not nearly as severe as for fully programmable dividers: using ECL, they can work up to at least 1.2 GHz. In addition, they are much more flexible than single-modulus counters.

Figure 7.10 shows a synthesizer with a two-modulus prescaler. The main counter divides by N, as before, but the prescaler can divide by either P or $(P + 1)$. In addition, one more programmable counter is needed. Let it divide by M. The output of this counter switches the modulus of the two-modulus counter between P and $P + 1$. Remember that M, N, and P are all integers. M and N can be changed by writing to registers on the counters, but P cannot

FIGURE 7.10

Frequency Synthesizer with Two-Modulus Prescaler

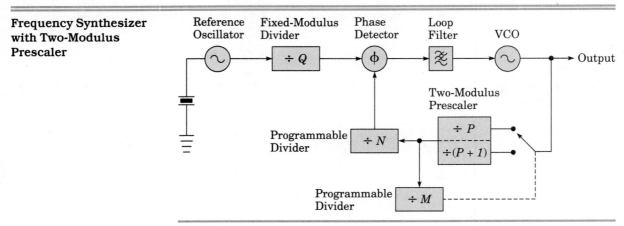

be changed. In addition, it is necessary to have N greater than M, as will shortly become obvious.

In order to understand the operation of this synthesizer, suppose that initially the prescaler is set to divide by $(P + 1)$ and the divide-by-M counter has just been reset. The two-modulus counter will switch to the P mode when the divide-by-M counter changes state (after it has received M transitions from the VCO). Then the two-modulus counter will divide by P until the divide-by-M counter changes state again. The trick here is that the M counter remains inactive until reset; that happens when the divide-by-N programmable counter reaches zero. That is, on reaching zero, the divide-by-N counter resets both itself and the divide-by-M counter, setting each to the value stored in its respective programming register.

Since initially the two-modulus counter divides by $(P + 1)$, it will take $(P + 1)$ transitions from the VCO to apply one count to the divide-by-M counter. Therefore, this counter will reach zero after $M(P + 1)$ transitions on the VCO output line. The same pulses that go to the divide-by-M counter are input to the divide-by-N counter. By the time the divide-by-M counter has reached zero, the divide-by-N counter has reached $(N - M)$. This value must be greater than zero, which accounts for the condition imposed above that N must be greater than M.

At this point, the prescaler switches to its other mode. The next P transitions from the VCO produce one transition at the prescaler output. This continues until the divide-by-N counter reaches zero. Since this counter begins this section of the process with a count of $(N - M)$, it must receive $(N - M)$ output transitions from the prescaler, which accounts for $P(N - M)$ state changes of the VCO output. At this point, both programmable counters reset to their programmed values, and the two-modulus counter switches back to its $(P + 1)$ mode.

If we look at the whole sequence, we see that the total number of pulses emanating from the VCO for one output pulse to the phase detector is

$$M(P + 1) + P(N - M) = MP + M + NP - MP$$
$$= M + NP$$

That is, the whole system including both the prescaler and the main counter has divided the VCO output by a factor of $(M + NP)$. Therefore

$$f_o = (M + NP)f_{ref} \tag{7.8}$$

The reader who has followed this rather convoluted argument may feel that this is not a very spectacular result from so many counters and so much head-scratching. It is actually worth all the trouble, however, as an example will show.

EXAMPLE 7.7

The synthesizer of Figure 7.10 has $P = 10$ and $f_{ref} = 10$ kHz. Find the minimum frequency step size and compare with that obtained using a fixed divide-by-10 prescaler.

Solution

With a fixed-modulus prescaler, the minimum frequency step would be

$$\text{Step Size} = Mf_{ref}$$
$$= 10 \times 10 \text{ kHz}$$
$$= 100 \text{ kHz}$$

To find the step size with the two-modulus system, let the main divider modulus N remain constant, and increase the modulus M to $(M + 1)$ to find how much the frequency changes. For the first case, the output frequency would be

$$f_o = (M + NP)f_{ref}$$
$$= (M + NP)10 \text{ kHz}$$

If we now leave N alone, but change M to $(M + 1)$, the new frequency is

$$f'_o = (M + 1 + NP)f_{ref}$$
$$= (M + 1 + NP)10 \text{ kHz}$$

The difference is

$$f'_o - f_o = (M + 1 + NP)10 \text{ kHz} - (M + NP)10 \text{ kHz}$$
$$= (M + 1 + NP - M - NP)10 \text{ kHz}$$
$$= 10 \text{ kHz}$$

This is the same step size as would have been obtained without prescaling.

Thus, the two-modulus prescaler achieves the benefit of prescaling, increased high-frequency capability, without the disadvantage of poorer resolution.

7.5 FREQUENCY TRANSLATION

All of the synthesizers shown so far are capable of generating very low frequencies, right down to f_{ref}. When such a synthesizer is used to generate a high frequency, a very large value of N is required. Dividing the VCO output by N divides the loop gain by the same amount, and very large values of N can cause instability.

In many practical applications, it is completely unnecessary for the synthesizer to operate at very low frequencies. For instance, an FM broadcast receiver must tune at 200 kHz intervals, but only between 88.1 and 107.9 MHz. One way to produce the local oscillator signal for such a receiver would be to generate it at a relatively low frequency and then raise it by mixing. Such a system is shown in Figure 7.11. The final output does not cover the FM band, but rather a range of frequencies 10.7 MHz higher, since it is a local oscillator signal that is being generated, and 10.7 MHz is the usual IF in FM

FIGURE 7.11

Synthesizer with
Frequency Shifting

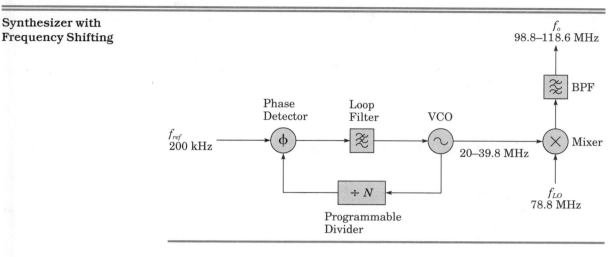

FIGURE 7.12

Synthesizer with Mixer
in the Loop

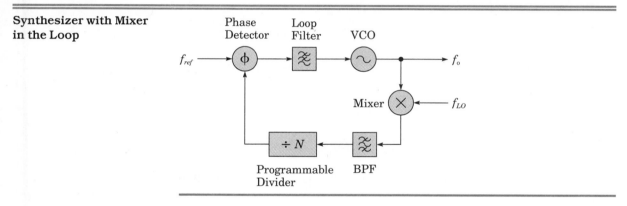

broadcast receivers. The VCO output varies from 20 MHz to 39.8 MHz in 200 kHz steps. It is mixed with a fixed-frequency oscillator at 78.8 MHz, and the sum of the two frequencies is used for the output, which varies from 98.8 to 118.6 MHz, as required. A bandpass filter is used to remove the difference component as well as the VCO and crystal oscillator frequencies from the mixer output.

Another possibility is to include the mixer and local oscillator within the loop, as shown in Figure 7.12. In this case, the VCO provides the output frequency directly, but its output is mixed down before being applied to the programmable divider.

EXAMPLE 7.8

A synthesizer of the type shown in Figure 7.12 has $f_{ref} = 20$ kHz, and a local oscillator operating at 10 MHz. Find the frequency range of the output as the value of N ranges from 10 to 100. Also find the minimum amount by which the frequency can be varied.

Solution

The frequency at the input to the divider must be Nf_{ref}, by the same logic that was used with the simple synthesizer. This is the output of a mixer that subtracts the fixed 10 MHz frequency from the VCO frequency. Other mixer outputs are removed by the bandpass filter.

Therefore the VCO frequency, which is also the output frequency, is

$$f_o = Nf_{ref} + f_{LO}$$

For $f_{LO} = 10$ MHz, $f_{ref} = 20$ kHz, and $N = 10$, we have

$$f_o = 10 \times 20 \text{ kHz} + 10 \text{ MHz}$$
$$= 10.2 \text{ MHz}$$

If N changes to 100, the output frequency will be

$$f_o = 100 \times 20 \text{ kHz} + 10 \text{ MHz}$$
$$= 12 \text{ MHz}$$

By now we know that the step size will be constant, so we can simply divide the total frequency range by the number of steps to find it.

$$\text{Step Size} = \frac{12 \text{ MHz} - 10.2 \text{ MHz}}{100 - 10}$$
$$= \frac{1.8 \text{ MHz}}{90}$$
$$= 20 \text{ kHz}$$

As you would expect, the step size is simply equal to f_{ref}.

It is not always necessary to use a second crystal oscillator for f_{LO}; with careful choice of frequencies it may be possible to use the local-oscillator frequency, suitably divided down, as the reference frequency. This technique is shown in Figure 7.13. In this case, the only additional stage required is the mixer itself.

FIGURE 7.13

Derivation of f_{LO} and f_{ref} from the Same Source

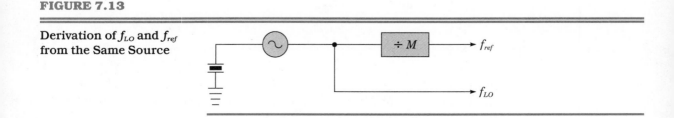

7.6   FREQUENCY SYNTHESIZER SPECIFICATIONS

As with other oscillators, the accuracy and stability of the operating frequency are important. There are some other specifications that are not characteristic of other oscillators, however.

7.6.1   Frequency Accuracy and Stability

Synthesizers were invented largely for the purpose of obtaining crystal-controlled stability and accuracy with the convenience and range of operation of a VFO. It is now apparent just how well the PLL synthesizer fulfills these expectations. When the loop is locked, the VCO will be an exact multiple of the reference frequency. Its frequency accuracy and stability will be the same as for the oscillator that generates the reference frequency. Since there is usually only one crystal oscillator in a synthesizer, it is quite practical where necessary to use a crystal oven to maintain a constant temperature and to regulate the supply voltage very precisely. Consequently, the stability of synthesizers ranges from very good to excellent, with drift of a part per million per month not impossible to achieve. Even low-cost synthesizers, like those in typical consumer products, will drift only a few parts per million.

Accuracy and stability are not the same thing, of course. A generator can be very stable, consistently generating the wrong frequency for months or years at a time! The accuracy of a crystal oscillator varies with the care with which the crystal is ground and the closeness with which the circuit approximates the load conditions for which the crystal was constructed. Low-cost crystals are generally specified to ±0.005% accuracy, but this can be improved greatly, either through the purchase of more accurate crystals, or by *pulling* the frequency slightly, as described in Chapter 2. For a PLL synthesizer, it is only necessary to perform this adjustment once. If one output frequency is correct, all will be.

Conventional VFOs rely for frequency accuracy on careful dial calibration and a variety of trimmer and padder capacitors that attempt to make the oscillator frequency *track* the dial calibrations. Anyone who has compared an

---

Crystal Calibrators

One indication of the great improvement in accuracy achieved by synthesizers is the virtual disappearance of crystal calibrators from communications receivers. These are, or were, crystal oscillators that operated at a low frequency, usually 100 kHz, but produced a waveform rich in harmonics. Coupled into the antenna circuit, the calibrator provided an audible signal every 100 kHz throughout the receiver tuning range. The idea was that the dial calibrations would be accurate enough so that the operator would know which harmonic was being received; the calibrator could then be used to calibrate the dial more precisely for the particular 100 kHz segment being used. Modern synthesized receivers have no need for crystal calibrators, as the crystal oscillator that controls the synthesizer is as accurate as the calibrator, and is, of course, operative at all frequencies.

ordinary dial-type broadcast radio or communications receiver with a modern synthesized receiver will have noticed the great difference in calibration accuracy. With the older radio, finding a particular frequency for the first time is often a matter of guesswork and trial-and-error tuning; in the modern receiver, it is only necessary to set the display to the correct frequency, and the station appears.

7.6.2 Resolution

Theoretically, frequency **resolution** is one area where a VFO could have an advantage over a synthesizer, since the tuning of a VFO is continuous. However, for many applications, such as with channelized systems, the steps inherent in synthesizer tuning are actually an advantage. For others, it is possible to reduce the step size to any required value (though not, of course, to zero). High-frequency communications receivers generally have 10 Hz resolution, and laboratory generators can be obtained with resolution of 1 Hz or even less.

For demanding applications like the reception of SSB signals, it is possible to achieve continuous tuning over a narrow range. A variable crystal oscillator (VXO) can be used for the reference-frequency generator, or for a local oscillator if the synthesizer uses mixing. The VXO tuning control is often called a *clarifier* because of the audible effect it has on SSB signals.

In any case, the theoretically infinite resolution of the VFO is not really achieved in practice. Resolution is limited by the mechanical construction of capacitor bearings, tuning dials, and so on. Any play in the bearings, or backlash in the dial mechanism, will reduce the precision with which the variable capacitor or inductor can be adjusted.

7.6.3 Phase Noise

All oscillators, in fact all electronic components and circuits, generate noise. PLL synthesizers, however, produce their own unique variety of noise. Known as **phase noise**, it arises from instability in the correction voltage from the phase detector, which causes random alterations in the phase of the output signal. In critical applications like receiver local oscillators, phase noise can degrade performance.

In very critical applications, the first mixer sometimes uses a fixed-frequency crystal oscillator, with the synthesizer being used for the second local oscillator of a double-conversion receiver. Figure 7.14 illustrates the idea. By the time the signal reaches the second mixer, it will be sufficiently strong to mask the effect of phase noise. Of course, the first IF must cover a wide frequency range, and the receiver's selectivity will be provided by the second IF.

7.6.4 Lock Time

A PLL synthesizer is useful only when the loop is locked. When the frequency setting is changed, the loop will be unlocked momentarily. Of course, assuming a workable design, lock will quickly be reacquired at the new frequency. In the meantime, however, there is a short time period during which the out-

FIGURE 7.14

Dual-Conversion Receiver Using Frequency Synthesis for Second Local Oscillator

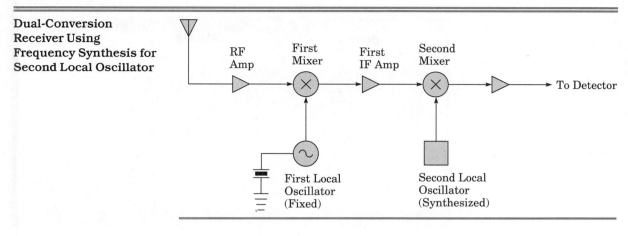

put frequency is indeterminate. In a transmitter, this could cause out-of-band operation. In a transceiver, the **lock time** can limit the rapidity with which transmit-receive switching can take place. Generally, the delay is only a few milliseconds and presents no serious problems. The designer must build in a delay, so that the transmitter does not turn on until the synthesizer has settled at the new frequency.

7.7 TEST EQUIPMENT AND PROCEDURES: FREQUENCY COUNTERS

The operating frequency of modern communications equipment using frequency synthesizers is so accurate that often the readout on a frequency display is taken for granted. It is occasionally necessary to calibrate synthesizers, however, and to check that the operating frequency is correct. In addition there are many other circumstances in which accurate frequency measurement is required.

One method of frequency measurement has already been discussed, in Chapter 1: the spectrum analyzer. Unfortunately, however, most spectrum analyzers are not very accurate when used to measure frequency. This is because their local oscillators are required to sweep over a considerable frequency range and are not locked to a crystal reference. In fact, the output frequency of even a low-cost synthesized transmitter, like a CB transmitter, is likely to be much more accurate than the analyzer used to measure it.

The most convenient and accurate method of everyday frequency measurement is the frequency counter, a typical example of which is pictured in Figure 7.15. Counters are available with a wide variety of frequency ranges, accuracy specifications, and precision. The counter supplements the oscilloscope and spectrum analyzer; it does not replace them. Though its frequency measurements can be much more accurate than either of the other instruments, it gives much less information about the signal. It tells nothing about harmonics, noise, and other distortions. In fact, these can lead the counter to produce erroneous readings.

FIGURE 7.15

Frequency Counter

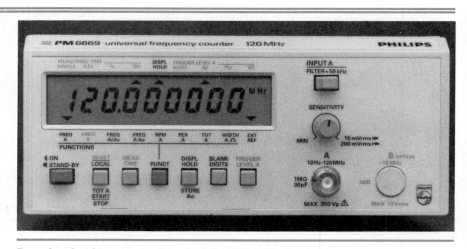

Reproduced with permission from the John Fluke Mfg. Co., Inc.

7.7.1 Frequency Counter Operation

Figure 7.16 shows the block diagram of a very basic frequency counter. Input cycles are counted during one part of the timebase cycle, called the *gate time*. The display is updated, and the counter reset, during the remainder of the cycle. The number of pulses that will be counted in one gate time is just the frequency multiplied by the gate time; that is,

$$N = f_i T_G \tag{7.9}$$

where N = counter reading
f_i = frequency of the input signal
T_G = gate time

FIGURE 7.16

Frequency Counter Block Diagram

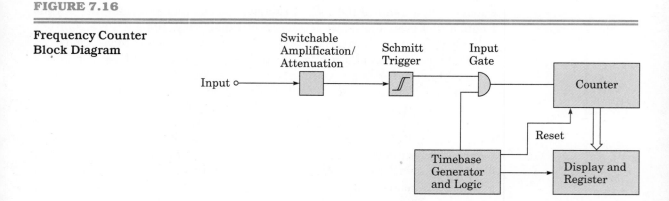

Since the counter counts only discrete events, N must be an integer. Simply ignore anything after the decimal point.

If the gate time is 1 s, then the counter will read the frequency directly in hertz, with 1 Hz resolution. Generally, the counter will have a register for the display so that it will show the previous count until the current one is complete. With a 1 s gate time, the reading will be updated approximately once each second.

With this type of counter, the resolution and the time needed to update the reading both depend on the timebase. With a 1 ms gate time, the count will take only 1 ms, but the readout will only be to 1 kHz precision. On the other hand, a 10 s gate time will give a resolution of 0.1 Hz, but will take 10 s to update the display. Since any frequency change during a count will not be recorded accurately, 20 s may have to elapse before a change in frequency can be determined.

EXAMPLE 7.9

A frequency of 2,345,678.9 Hz is applied to a counter with an eight-digit display. What will be the reading if the gate time is:

(a) 1 ms   (b) 1 s   (c) 10 s

Solution

(a) In 1 ms, the counter will count to the kilohertz so the display will read 00002345. This can also be found from Equation (7.9).

$$N = f_i T_G$$
$$= 2{,}345{,}678.9 \text{ Hz} \times 0.001 \text{ s}$$
$$= 2345.6789$$

Remember that the counter itself can only count integers, so anything after the decimal must be ignored. Most counters have circuitry to blank leading zeros. Also, with any counter, the least significant digit can be out by 1 either way, so readings of 2344 and 2346 are also possible.

(b) In 1 s, the counter can achieve 1 Hz resolution, so, subject to the notes above, the display would be 2345678.

(c) Here we will get 0.1 Hz resolution for a display of 23456789. All counters have some way of keeping track of units, whether an illuminated decimal point, units such as hertz and kilohertz appearing in the display, or both. In this case, a decimal point could be illuminated after the number 8 for a display in hertz, or after the 5 for a display calibrated in kilohertz.

The accuracy of a counter depends both on the precision resulting from the choice of timebase and on the accuracy of the crystal oscillator that provides the timebase. The better laboratory counters use crystal ovens and

FIGURE 7.17

Period Measurement Using a Counter

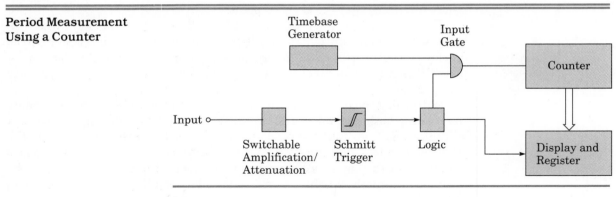

make provision for adjusting the timebase to bring it into line with a more accurate reference-frequency standard.

Attempting to measure very low frequencies accurately with this type of counter can be frustrating, because very long gate times are necessary for reasonable accuracy. An alternative method is to measure the period of the signal, from which the frequency can easily be calculated. This can be done by rearranging the basic counter slightly, as shown in Figure 7.17. This time, the instrument counts cycles of the internal timebase, which for this application has a high frequency. The input gate is opened and closed by the input signal, so that the display shows the number of cycles of the timebase oscillator that occur during the period of time corresponding to one cycle of the input. If, for instance, the timebase is 1 μs (corresponding to a frequency of 1 MHz), then the period of the input signal will be measured directly in microseconds. With this configuration, the time to update the reading is equal to the period of the signal to be measured. For more accurate readings, many counters allow several period measurements to be made and automatically averaged.

The counter shown in the photograph of Figure 7.15 is capable of both frequency and period measurements. With this counter, the operator would have to calculate the frequency, if required, from the period measurements, but more expensive counters are available that will do this automatically.

EXAMPLE 7.10

A frequency of exactly 60 Hz is measured using a counter with a gate time of 10 s, then its period is measured using a 1 μs timebase. For each measurement, compare the number of significant digits possible, and the time taken.

Solution

The readout with frequency measurement will be to a resolution of 0.1 Hz, for a count of 600 (three significant digits).

The period will be measured to the microsecond, giving a readout of 16666, which has five significant digits. To find the frequency, it would be necessary to take the reciprocal.

> The time taken is 10 s for the first measurement. The second takes one period of the waveform, or 16.7 ms. Of course, getting an accurate reading after a change of frequency would take two measurement times.

Since frequency counters use digital divider circuits, they suffer from some of the same frequency limitations as PLL synthesizers. However, the divider does not have to be programmable, so its frequency range can easily be extended into the hundreds of megahertz without major problems. For higher frequencies still, a fixed prescaler may be used. For frequencies in the microwave region, a common technique is to down-convert the input signal, using a mixer and an accurately calibrated, fixed local oscillator frequency. This produces a difference frequency that can be measured more easily than the original. Since the resulting counter is essentially a superheterodyne receiver, there is a possibility of erroneous readings due to image frequencies with this method.

7.7.2 Counter Specifications

Figure 7.18 shows the specifications for the Philips PM6669, a typical general-purpose counter. It covers a range of 10 Hz to 120 MHz, and can measure either frequency or period.

The *input specifications* portion of the specifications is very important. Though not as delicate as spectrum analyzers, counters can easily be damaged by overvoltage. This could happen if the counter is connected directly to a transmitter output, for instance. For this counter, the overload voltage decreases with increasing frequency. There is also a minimum input voltage below which the counter may not function reliably.

The input impedance is also essential information, especially for RF applications. This example has a relatively high input impedance, intended for *bridging*; that is, the counter is intended to be connected between a signal test point and ground, without affecting the operation of the circuit very much. This will not always work with RF circuits, because of the loading effect of the counter's input capacitance. Some counters are made with a 50 Ω *terminating* impedance, designed to match the output of RF generators and transmitters (though an attenuator will be required with transmitters), and others have switchable input impedance. A 50 Ω input is an option with this model.

The available accuracy with any counter depends on two things: the resolution of the display, and the accuracy of the timebase. The crystal oscillator in this or almost any other counter can be adjusted very accurately by comparison with a frequency standard. After that, it is the amount of frequency drift that matters. The specifications for this counter give both short-term and long-term stability information, as well as the effect of temperature variations. This counter does not have a crystal oven; even so, the variation is typically 10 ppm over a normal operating temperature range with the standard oscillator. A more accurate oscillator is available for this counter, which reduces the drift to about 0.2 ppm. Called by Philips an MTCXO (for mathematically temperature-compensated crystal oscillator), it achieves its greater accuracy by looking up pre-stored correction values for each temperature in a nonvolatile memory. The timebase accuracy and stability of this counter are

FIGURE 7.18   **Specifications for the PM6669 Counter**

Technical Specifications

Measuring Functions

Frequency A or B (optional)

Frequency Range:

Freq A: 10 Hz to 120 MHz
            (typically to 160 MHz)
Freq B: 70 MHz to 1.3 GHz
            (PM 9608B)
Mode: Reciprocal frequency measurement
LSD Unit Displayed:

$$\frac{2.5 \times 10^{-7} \times \text{FREQUENCY}}{\text{measuring time}}$$

Frequency A/A$_0$ (PM 6669 only)

A FREQ A measurement is divided by the constant A$_0$ before display. A$_0$ is read in frequency mode using the STORE button. At power-on A$_0$ is set to 1 (default).

Frequency A-A$_0$ (PM 6669 only)

A FREQ A measurement is subtracted by the constant A$_0$ before display. A$_0$ is read in frequency mode using the STORE button. At power-on A$_0$ is set to 0 (default).

RPM A (PM 6669 only)

A FREQ A measurement is multiplied by 60, and displayed as revolutions per minute (RPM).
Range: 6 RPM to 720×10^6 RPM

Period A (PM 6669 only)

Range: 8 ns to 2×10^8 s
Mode: Single period measurement (SINGLE) or period average measurement (at 0.2, 1 or 10s meas. times)
LSD Displayed:
SINGLE period measurement:
100 ns   (TIME <100s)

$$\frac{5 \times \text{PERIOD}}{10^9}\text{s} \qquad (\text{TIME} > 100\text{s})$$

Period average measurement:

$$\frac{2.5 \times 10^{-7} \times \text{PERIOD}}{\text{Measuring time}}$$

Totalize A (PM 6669 only)

Event counting is controlled by the START/STOP button. Sequential start-stop counts are accumulated.
Range: 0 to 1×10^{13} with indication of k or M (kilo-pulses or mega-pulses). The result is truncated to 9 digits.

Frequency range:
Sine-wave: 10 Hz to 12 MHz
Pulse: 0 Hz to 12 MHz
Pulse pair resolution: 80 ns

Width A (PM 6669 only)

A positive pulse width measurement is performed. Measuring time selection is not valid (always SINGLE)
Range: 100 ns to 2×10^8 s
LSD Displayed: 100 ns   (TIME < 100 s)

$$\frac{5 \times \text{WIDTH}}{10^9} \qquad (\text{TIME} \geq 100 \text{ s})$$

Input Specifications

Input A (PM 6662 only)

Frequency Range: 10 Hz to 120 MHz (typically 160 MHz)
Coupling: AC
Impedance: 1 MΩ//30 pF
Max Sensitivity:
Sinewave: 15 mVrms 10 Hz to 70 MHz
              30 mVrms 70 MHz to 120 MHz
              typ 60 mVrms at 160 MHz
Pulse: 50 mVpp 10 Hz to 120 MHz
Minimum Pulse Duration: 4 ns
Attenuation: X1, X3, X10, X30, X100 and X300. The Attenuation selector is labelled sensitivity: 15 mV, 50 mV, 150 mV, 500 mV, 1.5V, 5V
Auto Trigger Level: A fixed (+, 0 or –) trigger level offset is automatically applied to ensure proper triggering on any waveform and duty cycle
Maximum Voltage Without Damage: 350V (DC+ACpeak) DC to 440 Hz falling to 12Vrms at 1 MHz

Input A (PM 6669 only)

Frequency Range:
10 Hz to 120 MHz (typically up to 160 MHz with 30 mVrms input signal)
Sensitivity:
Sine: 10 mVrms, 10 Hz to 120 MHz
          (30 mVrms, 120 to 160 MHz typ)
Pulse: 30 mVpp 0 Hz to 120 MHz
Coupling: AC
Impedance: 1 MΩ//30 pF
Attenuation: Continuously variable in two ranges between x1 and x400
Filter: Switchable 50 kHz low pass noise filter with a suppression of 20 dB at 200 kHz.

Trigger Levels: 3 different levels for triggering on signals with various duty factors, and AUTO.

used for symmetrical input signals with a duty factor of 0.25 to 0.75.

used for input signals with duty factor <0.25.

used for input signals with duty factor >0.75.

AUTO Trigger Level: The counter automatically selects a suitable trigger level setting (not active in TOT-A measurements)
Input Signal
Repetition Rate: >100 Hz
Trigger Slopes (via GPIB only): + or –
Maximum Voltage Without Damage: 350V (DC+ACpeak) between dc and 440 Hz, falling to 11Vrms at 1 MHz.

Input B (Option PM 9608B)

Frequency Range: 70 MHz to 1.3 GHz
Coupling: AC
Operating Input Voltage Range:
10 mVrms to 12 Vrms; 70 to 900 MHz
15 mVrms to 12 Vrms; 900 to 1100 MHz
40 mVrms to 12 Vrms; 1100 to 1300 MHz
AM Tolerance: 94% at max 100 kHz modulation frequency. Minimum signal must exceed minimum operating input voltage requirement
Input Impedance: 50Ω nominal, VSWR <2:1
Max Voltage Without Damage: 12 Vrms; overload protection with pin diodes

External Reference Input

The external reference input is automatically selected when an external signal of 9.9...10.1 MHz is connected.
Input Frequency: 10 MHz ±0.1 MHz
Coupling: AC
Operating Input Voltage Range:
500 mVrms to 15 Vrms (sine)
Maximum Voltage Without Damage: 15 Vrms
Impedance: Approx 300Ω at 10 MHz

good enough that its accuracy is often limited by the resolution of the display itself.

7.7.3 Counter Resolution and Accuracy

Any counter has an inherent possible error of ±1 LSD; that is, the least-significant digit (the one on the right) can be in error by 1 either way. For instance, a reading of 123 could really represent 122 or 124. If timebase error is neglected, this fact will allow the accuracy of a counter in a given application to be calculated quite easily, in percent or parts per million. Once this calculation is made, it can easily be seen whether the uncertainty of the timebase frequency is significant. If so, the maximum timebase error can simply be added in.

EXAMPLE 7.11

A counter has a timebase accurate to 5 ppm. It has a six-digit display, and the gate time is set to 1 s. What is its accuracy in parts per million and percent if it is used to measure a frequency of:

(a) 100 Hz   (b) 800 kHz

Solution

(a) The last digit can be off by ±1. Since the timebase is 1 s, the counter will read directly in hertz, so the counter, neglecting timebase error, is accurate to ±1 Hz in 100 Hz, or 1%. This corresponds to 1×10^4 ppm, so the timebase error is insignificant by comparison and the accuracy is ±1%. By the way, the accuracy for this low-frequency measurement could be increased by using a longer gate time or by measuring period rather than frequency.

(b) The display can still be off by ±1 Hz, but this time the error is 1 part in 800×10^3, or 1.25 ppm, which corresponds to (1.25×10^{-4})%. Here, the timebase error of 5 ppm must be added, giving a total possible error of 6.25 ppm or (6.25×10^{-4})%.

Though the basic accuracy of most frequency counters is excellent, sometimes very large errors can occur. This is often due to noise or distortion, or the presence of modulation. The input circuitry of the counter has *hysteresis*; that is, the input voltage must cross a *window* or voltage range in order to trigger the counter. Noise or distortion can cause the number of transitions to increase, giving inaccurate readings. Two examples of this effect are shown in Figure 7.19. Some counters have adjustable hysteresis to improve their performance in some situations. In any case, it is wise to use counter readings with some care. In doubtful situations, an oscilloscope will often show what the problem is. Frequency measurements made with an oscilloscope are not very accurate (a few percent at best), but the scope can give a much better qualitative picture of the waveform, showing waveshape and the presence of noise and distortion. Thus, the oscilloscope and counter complement each other very well. In fact, one useful practice is to connect the channel 1 output (if available) from the oscilloscope to a frequency counter. The counter will

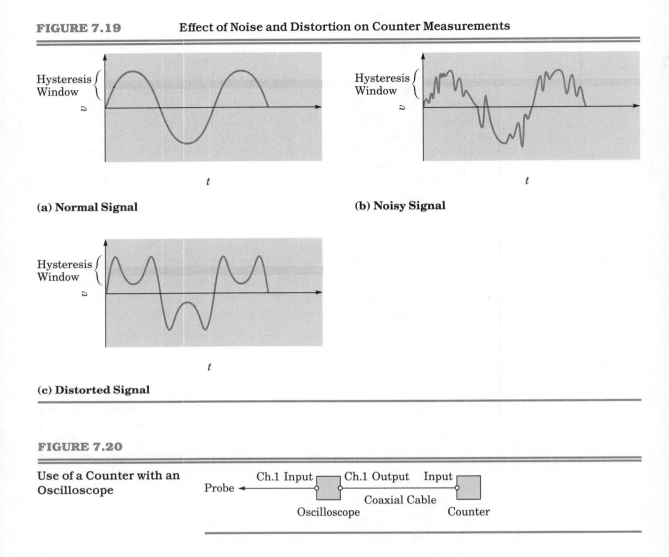

FIGURE 7.19   **Effect of Noise and Distortion on Counter Measurements**

(a) **Normal Signal**

(b) **Noisy Signal**

(c) **Distorted Signal**

FIGURE 7.20

Use of a Counter with an
Oscilloscope

then measure whatever signal is connected to the oscilloscope. The oscilloscope will, when adjusted to give a useful display, provide a signal of suitable
amplitude to the counter, and the circuit loading will be that of the scope
alone. Figure 7.20 shows this setup.

7.8 PHASE-LOCKED LOOP STABILITY AND FREQUENCY RESPONSE

Previously, it was mentioned that PLLs have limitations as to their capture
and lock range, and that acquiring lock took a finite, though usually small,
time. It was also suggested that, under certain conditions, a PLL—like any
other feedback circuit—could become unstable. In this section, we will look
into these characteristics of PLLs in more detail; in particular, we will investigate the effect of the loop filter on PLL operation.

7.8.1 Frequency Response of Loop Components

When the loop is locked, the VCO and phase detector must be capable of working at the operating frequency, and the amplifier and loop filter (if present) must be dc-coupled, so that they can pass on the dc error voltage from the phase detector to the control terminal of the VCO. When the loop is unlocked, however, the error signal will be ac and the bandwidths of the amplifier and loop filter become important. So does the frequency response of the VCO itself.

In the following discussion, the frequency referred to is that of the difference signal produced by the phase detector. This frequency can vary from zero, when the loop is locked, to the *pull-in range* of the loop. For example, if the loop is free-running at 10 MHz with a capture range of 2 MHz, the pull-in range is 1 MHz, and that is the maximum frequency that will occur in the loop amplifier and filter, and at the VCO control terminal.

Normally, the loop amplifier will have a cutoff frequency much higher than the highest frequency it is required to handle. This is necessary because any corner frequency, or *pole*, that occurs within the pull-in range will cause a phase shift that is likely to make the loop unstable.

The response of the phase detector, on the other hand, will always be that of an integrator. This results from the fact that phase is the integral of frequency. The VCO frequency varies directly with the control voltage. However, the phase detector responds to phase, not directly to frequency, and so its response is proportional to the integral of the VCO control voltage.

An *integrator* is a low-pass filter whose gain decreases with increasing frequency at a constant rate of 6 dB/octave or 20 dB/decade. Since in an uncompensated loop (one with no loop filter) the VCO is, for practical purposes, the only component whose frequency response is other than flat, the whole loop will have the same shape of Bode plot as the VCO (see Figure 7.21). The gain of the loop will be equal to k_v at a frequency of 1 rad/s, declining at 6 dB/octave, until it reaches 1 (0 dB) at a radian frequency of

$$\omega_o = k_v \qquad\qquad (7.10)$$

FIGURE 7.21

**Bode Plot of VCO
Frequency Response**

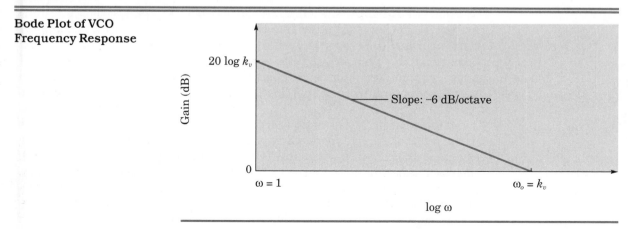

In hertz, this is

$$f_o = \frac{k_v}{2\pi}$$

$$= k_L \tag{7.11}$$

Remember that the frequency referred to on this plot is that of the control voltage applied to the VCO; that is, it is the difference between the current frequency of the VCO and the reference frequency to which it will (we hope) eventually become locked. From the graph, it is obvious that there will be some frequency difference at which the gain of the loop will be too small to achieve lock; that is, that there is a limit to the PLL pull-in range. This point will be the frequency at which the loop gain equals 0 dB. From the Bode plot, it appears that the pull-in range will be equal to k_L. In fact, it will be a little less than that, for a more detailed analysis would show that the loop gain at f_o is actually -3 dB.

The actual process of achieving lock will take some time. In fact, the uncompensated loop behaves like a first-order low-pass filter in terms of its response to a step change in the input frequency. The time constant of this filter is

$$\tau = \frac{1}{k_v} \tag{7.12}$$

where τ = loop time constant in seconds

k_v = loop gain in hertz (or seconds^{-1})

That is, the uncompensated PLL responds to frequency changes the same way a simple first-order RC filter responds to voltage changes. Accordingly, it will take a time of about 5τ to move to the new frequency.

From the above equation, it can also be seen that the higher the loop gain, the shorter the time constant and the more quickly the loop can respond to changes in the reference frequency. This explains a problem that arises in frequency synthesizers, when a very large value of N reduces the loop gain, causing the synthesizer to be slow to change frequency.

Now, what happens when a filter is added? The simplest loop filter is a first-order RC low-pass filter, like the one shown in Figure 7.22. The filter combines with the response of the VCO to produce a second-order system. Sometimes this type of loop filter is called *lag compensation* because of the phase lag that it introduces. Its effects are to reduce the loop bandwidth, and therefore the pull-in range, while also adding a phase lag that reduces the phase margin of the system.

If done carefully, lag compensation can cause the loop to follow changes in the reference signal more quickly. On the other hand, too much phase lag can result in *overshoot*, where the VCO frequency moves more than it should, then oscillates around the correct value before settling at the right frequency. In extreme cases, the loop can become unstable, in which case the VCO frequency will never settle at the correct value.

All second-order systems have a parameter called *damping factor*. This is found in many different systems, from loudspeakers to motor speed con-

FIGURE 7.22

First-Order *RC* Low-Pass Filter

trols, and it appears here as well. A damping factor of 1 or more will produce no overshoot but will produce a response that is slower than optimum. For values between 1 and 0.5 or so, overshoot will be small and the loop will settle quickly. Smaller values are undesirable because of the instability that results.

Lag compensation does reduce the capture range, compared to an uncompensated PLL. It might seem that the bandwidth for a PLL should be as large as possible. Like all electronic systems, however, PLLs are subject to noise, and therefore should have as much bandwidth as needed, and no more. The required bandwidth varies with the application.

The simple lag filter does not allow independent control of bandwidth once damping factor and loop gain are fixed. A circuit that does is the lead-lag compensator, shown in Figure 7.23. This can be used where narrow bandwidth is desirable, as in sensitive receivers that have to follow a slowly varying carrier frequency. The analysis of these systems will not be done here, but can be found in any book on control systems theory. For most practical purposes, "cookbook" formulas, like the one on the PLL data sheet in Figure 7.24, can be used.

7.8.2 Example: An Integrated-Circuit Phase-Locked Loop

Figure 7.24 is the data sheet for a commonly used PLL, the LM565. The VCO, phase detector, and loop amplifier are all on the chip. The VCO free-running frequency is set by an external resistor and capacitor. The loop filter is external, but design equations for both lag and lead-lag filters are provided. As we

FIGURE 7.23

Lead-Lag Compensator

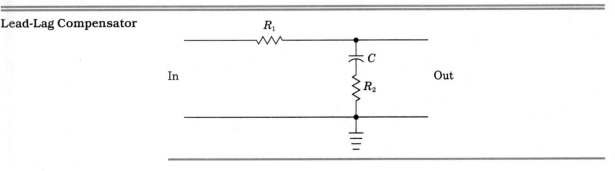

FIGURE 7.24   Data Sheet for the LM565

April 1987

LM565/LM565C Phase Locked Loop

General Description

The LM565 and LM565C are general purpose phase locked loops containing a stable, highly linear voltage controlled oscillator for low distortion FM demodulation, and a double balanced phase detector with good carrier suppression. The VCO frequency is set with an external resistor and capacitor, and a tuning range of 10:1 can be obtained with the same capacitor. The characteristics of the closed loop system—bandwidth, response speed, capture and pull in range—may be adjusted over a wide range with an external resistor and capacitor. The loop may be broken between the VCO and the phase detector for insertion of a digital frequency divider to obtain frequency multiplication.

The LM565H is specified for operation over the −55°C to +125°C military temperature range. The LM565CH and LM565CN are specified for operation over the 0°C to +70°C temperature range.

Features

■ 200 ppm/°C frequency stability of the VCO
■ Power supply range of ±5 to ±12 volts with 100 ppm/% typical
■ 0.2% linearity of demodulated output
■ Linear triangle wave with in phase zero crossings available
■ TTL and DTL compatible phase detector input and square wave output
■ Adjustable hold in range from ±1% to > ±60%

Applications

■ Data and tape synchronization
■ Modems
■ FSK demodulation
■ FM demodulation
■ Frequency synthesizer
■ Tone decoding
■ Frequency multiplication and division
■ SCA demodulators
■ Telemetry receivers
■ Signal regeneration
■ Coherent demodulators

Connection Diagrams

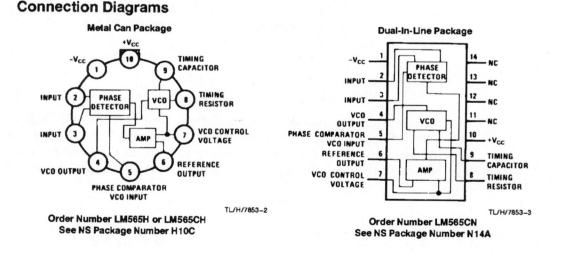

Metal Can Package

Order Number LM565H or LM565CH
See NS Package Number H10C

TL/H/7853–2

Dual-In-Line Package

Order Number LM565CN
See NS Package Number N14A

TL/H/7853–3

Absolute Maximum Ratings

If Military/Aerospace specified devices are required, please contact the National Semiconductor Sales Office/Distributors for availability and specifications.

Supply Voltage	$\pm 12V$
Power Dissipation (Note 1)	1400 mW
Differential Input Voltage	$\pm 1V$

Operating Temperature Range
LM565H	$-55°C$ to $+125°C$
LM565CH, LM565CN	$0°C$ to $+70°C$
Storage Temperature Range	$-65°C$ to $+150°C$
Lead Temperature (Soldering, 10 sec.)	$260°C$

Electrical Characteristics AC Test Circuit, $T_A = 25°C$, $V_{CC} = \pm 6V$

Parameter	Conditions	LM565 Min	LM565 Typ	LM565 Max	LM565C Min	LM565C Typ	LM565C Max	Units
Power Supply Current			8.0	12.5		8.0	12.5	mA
Input Impedance (Pins 2, 3)	$-4V < V_2, V_3 < 0V$	7	10			5		$k\Omega$
VCO Maximum Operating Frequency	$C_o = 2.7$ pF	300	500		250	500		kHz
VCO Free-Running Frequency	$C_o = 1.5$ nF $R_o = 20$ $k\Omega$ $f_o = 10$ kHz	-10	0	$+10$	-30	0	$+30$	%
Operating Frequency Temperature Coefficient			-100			-200		ppm/°C
Frequency Drift with Supply Voltage			0.1	1.0		0.2	1.5	%/V
Triangle Wave Output Voltage		2	2.4	3	2	2.4	3	V_{p-p}
Triangle Wave Output Linearity			0.2			0.5		%
Square Wave Output Level		4.7	5.4		4.7	5.4		V_{p-p}
Output Impedance (Pin 4)			5			5		$k\Omega$
Square Wave Duty Cycle		45	50	55	40	50	60	%
Square Wave Rise Time			20			20		ns
Square Wave Fall Time			50			50		ns
Output Current Sink (Pin 4)		0.6	1		0.6	1		mA
VCO Sensitivity	$f_o = 10$ kHz		6600			6600		Hz/V
Demodulated Output Voltage (Pin 7)	$\pm 10\%$ Frequency Deviation	250	300	400	200	300	450	mV_{p-p}
Total Harmonic Distortion	$\pm 10\%$ Frequency Deviation		0.2	0.75		0.2	1.5	%
Output Impedance (Pin 7)			3.5			3.5		$k\Omega$
DC Level (Pin 7)		4.25	4.5	4.75	4.0	4.5	5.0	V
Output Offset Voltage $\lvert V_7 - V_6 \rvert$			30	100		50	200	mV
Temperature Drift of $\lvert V_7 - V_6 \rvert$			500			500		μV/°C
AM Rejection		30	40			40		dB
Phase Detector Sensitivity K_D			.68			.68		V/radian

Note 1: The maximum junction temperature of the LM565 and LM565C is $+150°C$. For operation at elevated temperatures, devices in the TO-5 package must be derated based on a thermal resistance of $+150°C$/W junction to ambient or $+45°C$/W junction to case. Thermal resistance of the dual-in-line package is $+85°C$/W.

(*continued*)

FIGURE 7.24   (*Continued*)

Typical Performance Characteristics

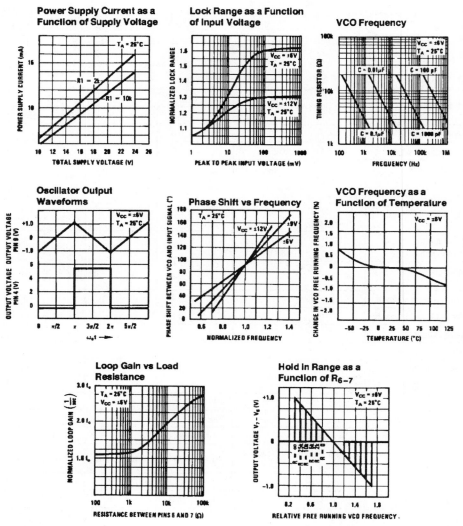

TL/H/7853–4

Schematic Diagram

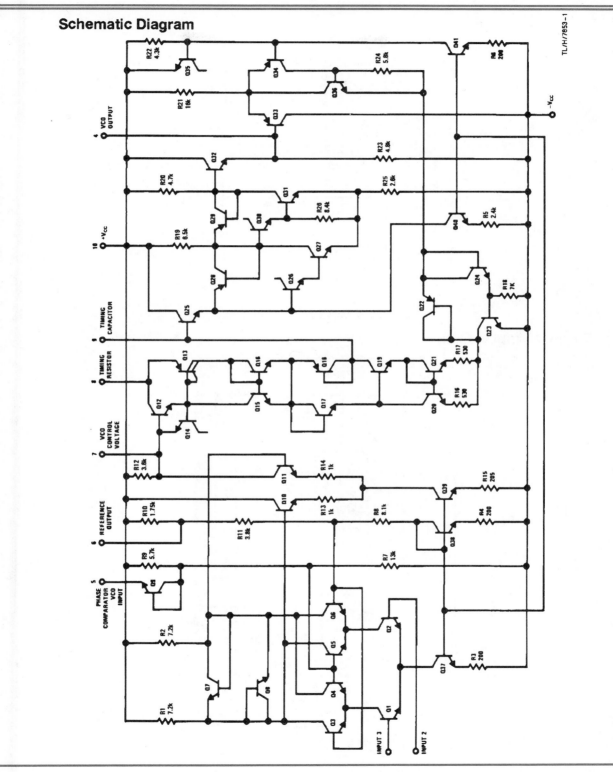

TL/H/7853–1

(continued)

FIGURE 7.24   (*Continued*)

AC Test Circuit

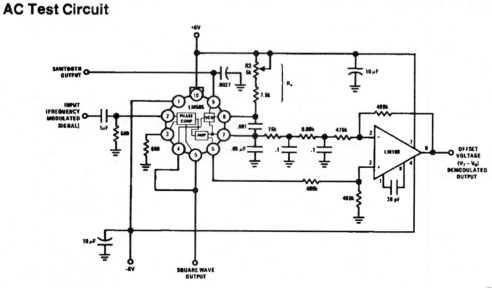

Note: S_1 open for output offset voltage ($V_7 - V_6$) measurement.

TL/H/7853–5

Typical Applications

2400 Hz Synchronous AM Demodulator

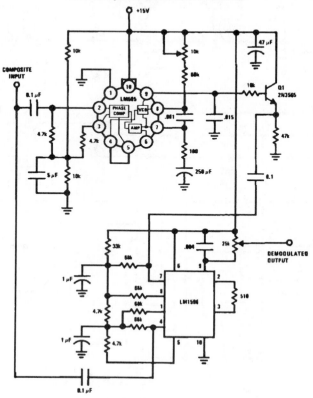

TL/H/7853–6

Typical Applications (Continued)

FSK Demodulator (2025–2225 cps)

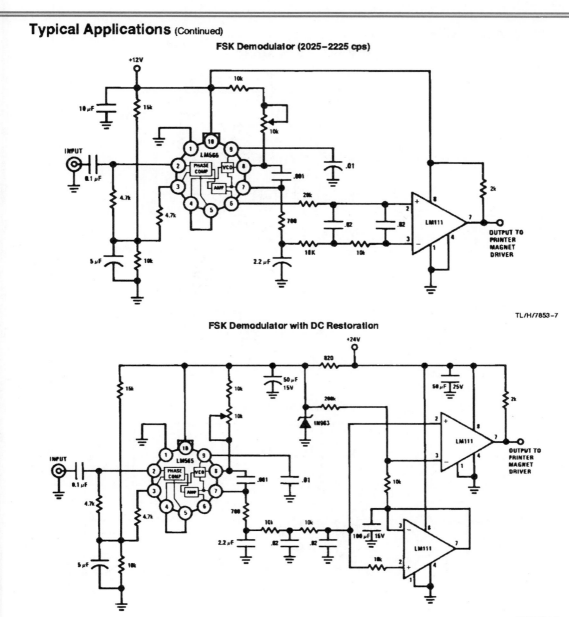

TL/H/7853–7

FSK Demodulator with DC Restoration

TL/H/7853–8

(continued)

FIGURE 7.24   (*Continued*)

Typical Applications (Continued)

Frequency Multiplier (×10)

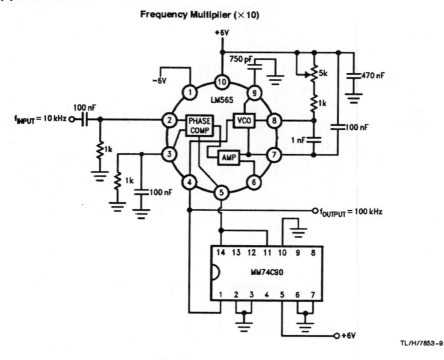

TL/H/7853–9

IRIG Channel 13 Demodulator

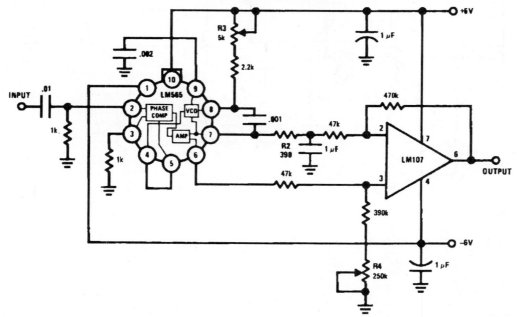

TL/H/7853–10

Applications Information

In designing with phase locked loops such as the LM565, the important parameters of interest are:

FREE RUNNING FREQUENCY

$$f_o \cong \frac{0.3}{R_o C_o}$$

LOOP GAIN: relates the amount of phase change between the input signal and the VCO signal for a shift in input signal frequency (assuming the loop remains in lock). In servo theory, this is called the "velocity error coefficient."

$$\text{Loop gain} = K_o K_D \left(\frac{1}{\text{sec}}\right)$$

K_o = oscillator sensitivity $\left(\frac{\text{radians/sec}}{\text{volt}}\right)$

K_D = phase detector sensitivity $\left(\frac{\text{volts}}{\text{radian}}\right)$

The loop gain of the LM565 is dependent on supply voltage, and may be found from:

$$K_o K_D = \frac{33.6\, f_o}{V_c}$$

f_o = VCO frequency in Hz
V_c = total supply voltage to circuit

Loop gain may be reduced by connecting a resistor between pins 6 and 7; this reduces the load impedance on the output amplifier and hence the loop gain.

HOLD IN RANGE: the range of frequencies that the loop will remain in lock after initially being locked.

$$f_H = \pm \frac{8\, f_o}{V_c}$$

f_o = free running frequency of VCO
V_c = total supply voltage to the circuit

THE LOOP FILTER

In almost all applications, it will be desirable to filter the signal at the output of the phase detector (pin 7); this filter may take one of two forms:

Simple Lag Filter

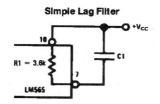

TL/H/7853–11

Lag-Lead Filter

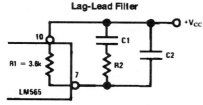

TL/H/7853–12

A simple lag filter may be used for wide closed loop bandwidth applications such as modulation following where the frequency deviation of the carrier is fairly high (greater than 10%), or where wideband modulating signals must be followed.

The natural bandwidth of the closed loop response may be found from:

$$f_n = \frac{1}{2\pi} \sqrt{\frac{K_o K_D}{R_1 C_1}}$$

Associated with this is a damping factor:

$$\delta = \frac{1}{2} \sqrt{\frac{1}{R_1 C_1 K_o K_D}}$$

For narrow band applications where a narrow noise bandwidth is desired, such as applications involving tracking a slowly varying carrier, a lead lag filter should be used. In general, if $1/R_1 C_1 < K_o K_D$, the damping factor for the loop becomes quite small resulting in large overshoot and possible instability in the transient response of the loop. In this case, the natural frequency of the loop may be found from

$$f_n = \frac{1}{2\pi} \sqrt{\frac{K_o K_D}{\tau_1 + \tau_2}}$$

$$\tau_1 + \tau_2 = (R_1 + R_2) C_1$$

R_2 is selected to produce a desired damping factor δ, usually between 0.5 and 1.0. The damping factor is found from the approximation:

$$\delta \approx \pi \tau_2 f_n$$

These two equations are plotted for convenience.

Filter Time Constant vs Natural Frequency

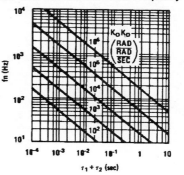

TL/H/7853–13

Damping Time Constant vs Natural Frequency

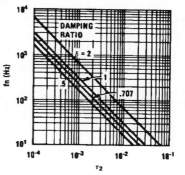

TL/H/7853–14

Capacitor C_2 should be much smaller than C_1 since its function is to provide filtering of carrier. In general $C_2 \leq 0.1\, C_1$.

would expect, the lock range is set by varying the loop gain, which can be adjusted by changing the supply voltage. It can also be lowered by adding an external resistor between pins 6 and 7. Again as we would expect, the capture range is set by the design of the loop filter; the data sheet refers to capture range as the "natural bandwidth of the closed loop response," which is another way of saying the same thing. An equation for damping factor is also provided.

Let us use this data sheet to design a phase-locked loop with the following specifications:

Free-Running Frequency: 100 kHz

Lock Range: 100 kHz

Capture Range: 20 kHz

We begin by setting the free-running frequency. From the data sheet,

$$f_o \approx \frac{1}{3.7R_oC_o}$$

The approximately-equal sign reminds us that we should provide some adjustment for R_o. We can set R_o and then find C_o by rearranging the equation:

$$C_o = \frac{1}{3.7R_of_o}$$

For linearity, R_o should be between 2 kΩ and 20 kΩ. Since the equation is approximate, a potentiometer will likely be used to set R_o. For greatest range of adjustment either way, we could let R_o be the geometric mean of the above range. Let

$$R_o = \sqrt{2 \text{ k}\Omega \times 20 \text{ k}\Omega}$$
$$= 6.3 \text{ k}\Omega$$

Then

$$C_o = \frac{1}{3.7R_of_o}$$
$$= \frac{1}{3.7(6.3 \times 10^3 \text{ }\Omega)(100 \times 10^3 \text{ Hz})}$$
$$= 429 \times 10^{-12} \text{ F}$$
$$= 429 \text{ pF}$$

Since there is plenty of possible variation in the resistance, a standard value of capacitance, such as 390 or 470 pF, would be used.

Next, let us set the lock range. Actually, the **hold-in range** f_H is given on the data sheet, but this is simply one-half the lock range. Therefore we want to set the hold-in range to 50 kHz. The simplest way (mathematically, at any rate) is to set the supply voltage using the equation

$$f_H = \frac{\pm 8f_o}{V_c} \tag{7.13}$$

Here, the plus-or-minus sign indicates only that it is the hold-in range, which can be to either side of the free-running frequency, that is being dealt with, so we can drop it in actual use. V_c is defined as the total supply voltage to the circuit; that is, since this circuit runs on a split supply, it is the sum of the positive and negative supply voltages. We should note, to check our work, the maximum supply voltage rating of ± 12 V.

Since f_H and f_o are known, we rearrange Equation (7.13) in terms of V_c:

$$V_c = \frac{8f_o}{f_H} \text{ V}$$

$$= \frac{8 \times 100}{50} \text{ V}$$

$$= 16 \text{ V}$$

This requires a ± 8 V supply, which is within the allowable range (from 5 V to 12 V).

Next, we have to set the loop bandwidth, which is approximately equivalent to the capture range. Let us use lag compensation. This takes the form of an RC filter, for which the resistor is part of the chip, and the capacitor is external. The data sheet provides the equation

$$f_n = \frac{1}{2\pi}\sqrt{\frac{K_o K_D}{R_1 C_1}} \tag{7.14}$$

where f_n = capture range

This is not as formidable as it looks: R_1 is an internal resistor with a value of 3.6 kΩ. The product $K_o K_D$ is the loop gain, and is found from a separate equation on the data sheet:

$$K_o K_D = \frac{33.6f_o}{V_c} \tag{7.15}$$

where f_o = VCO frequency in hertz
         V_c = total supply voltage

Since we already know both f_o and V_c, we will first find the value of the product $K_o K_D$ from Equation (7.15), then substitute into Equation (7.14).

$$K_o K_D = \frac{33.6f_o}{V_c}$$

$$= \frac{33.6(100 \times 10^3)}{16}$$

$$= 210 \times 10^3 \text{ s}^{-1}$$

Rearranging Equation (7.14) to make C_1 the unknown, we get

$$f_n^2 = \frac{1}{4\pi^2} \times \frac{K_o K_D}{R_1 C_1}$$

$$f_n^2 = \frac{K_o K_D}{4\pi^2 R_1 C_1}$$

$$C_1 = \frac{K_o K_D}{4\pi^2 R_1 f_n^2}$$

$$= \frac{210 \times 10^3}{4\pi^2(3.6 \times 10^3)(100 \times 10^3)^2}$$

$$= 148 \times 10^{-12} \text{ F}$$

$$= 148 \text{ pF}$$

At this point, it would be a good idea to calculate the damping factor to be sure that it is reasonable. It will not be possible to change it without changing either the loop gain or bandwidth, however. The data sheet gives an equation for calculating damping factor:

$$\delta = \frac{1}{2}\sqrt{\frac{1}{R_1 C_1 K_o K_D}}$$

$$= \frac{1}{2}\sqrt{\frac{1}{(3.6 \times 10^3)(148 \times 10^{-12})(210 \times 10^3)}}$$

$$= 1.49$$

Thus the circuit will be stable.

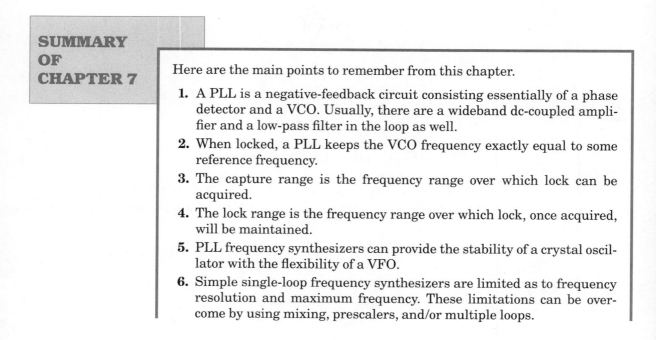

SUMMARY OF CHAPTER 7

Here are the main points to remember from this chapter.

1. A PLL is a negative-feedback circuit consisting essentially of a phase detector and a VCO. Usually, there are a wideband dc-coupled amplifier and a low-pass filter in the loop as well.

2. When locked, a PLL keeps the VCO frequency exactly equal to some reference frequency.

3. The capture range is the frequency range over which lock can be acquired.

4. The lock range is the frequency range over which lock, once acquired, will be maintained.

5. PLL frequency synthesizers can provide the stability of a crystal oscillator with the flexibility of a VFO.

6. Simple single-loop frequency synthesizers are limited as to frequency resolution and maximum frequency. These limitations can be overcome by using mixing, prescalers, and/or multiple loops.

7. Frequency and period can be measured accurately using electronic counters. The available accuracy depends both on the internal time-base accuracy and the time taken for the measurement.

IMPORTANT EQUATIONS

$$k_f = \frac{\Delta f}{\Delta v} \tag{7.1}$$

$$k_p = \frac{v_o}{\phi} \tag{7.3}$$

$$k_v = 2\pi k_L \tag{7.4}$$
$$= 2\pi k_p A_v k_f$$

$$\text{Lock Range (Hz)} = \pi k_L \tag{7.5}$$
$$= \frac{k_v}{2}$$

$$f_o = N f_{ref} \tag{7.6}$$
$$f_o = MN f_{ref} \tag{7.7}$$
$$f_o = (M + NP) f_{ref} \tag{7.8}$$
$$N = f_i T_G \tag{7.9}$$
$$\omega_o = k_v \tag{7.10}$$
$$f_o = \frac{k_v}{2\pi} \tag{7.11}$$
$$= k_L$$
$$\tau = \frac{1}{k_v} \tag{7.12}$$

GLOSSARY

capture range   the total frequency range over which a PLL can become locked to a signal

free-running frequency   the frequency at which a VCO operates when its control voltage is zero

frequency synthesizer   a device that can produce a large number of output frequencies from a smaller number of fixed-frequency oscillators

frequency translation   movement of a signal from one frequency to another using a mixer-oscillator combination

hold-in range   frequency range within which a PLL can remain locked, measured to one side of the free-running frequency. Equal to half the lock range

lock range   total range of frequencies within which a PLL, once locked, can remain locked

lock time   time taken for a PLL to achieve phase-lock, when the frequency shifts

loop filter   a low-pass filter included within a phase-locked loop

modulus   the number by which a digital divider chain divides

phase detector   device whose output voltage is a function of the phase difference between two input signals

phase noise   random fluctuations in the phase of a signal

phase-locked loop (PLL)   a device which locks the frequency of a VCO exactly to that of an input signal

pull-in range   the range of signal frequencies to which a PLL can become locked, measured to one side of the VCO free-running frequency, equal to one-half the capture range

resolution   in a frequency synthesizer, the smallest amount by which the output frequency can be changed

voltage-controlled oscillator (VCO)   an oscillator whose frequency can be controlled by changing an external control voltage

QUESTIONS

1. Sketch the block diagram of a PLL. Label all the components.
2. Why are low-pass filters found in most PLLs?
3. Why is it necessary that the loop amplifier, if present, be dc-coupled?
4. Sketch the block diagram of a basic PLL frequency synthesizer.
5. What is the effect of the divider in a basic synthesizer on the loop gain?
6. What is the smallest amount by which the frequency of a basic synthesizer can be changed?
7. Why are prescalers necessary when synthesizers are used at UHF frequencies?
8. Explain the operation of a two-modulus prescaler.
9. Why is a mixer often incorporated into a frequency synthesizer?
10. Under what conditions is it preferable to measure period rather than frequency with an electronic counter?
11. What is the function of a crystal oven?
12. Why is it sometimes desirable to restrict the capture range of a PLL?
13. What is phase noise, and how does it arise in a PLL?
14. How does the design of the loop filter affect the lock time of a PLL?
15. Draw the block diagram of a basic frequency counter, and explain its operation.
16. What is meant by instability in a PLL?
17. What is the requirement for the bandwidth of the loop amplifier in a PLL?
18. Why are synthesizers often lower in cost than VFOs when good stability is required?
19. Why does a VCO have a low-pass frequency response?
20. How are low-frequency reference signals generated in a synthesizer?

PROBLEMS

Section 7.2

21. A PLL has a free-running frequency of 10 MHz, with a capture range of 1 MHz, and a lock range of 2 MHz. Graph the VCO frequency as a function of the reference input frequency as the latter varies from 5 to 15 MHz.

22. A PLL has a phase detector with $k_p = 1$ V/rad, a loop amplifier with a gain of 10, and a VCO with $k_f = 10$ kHz/V. Calculate the lock range.

23. A PLL has the following characteristics:

    Free-Running Frequency: 3 MHz

    Lock Range: 200 kHz

    Capture Range: 50 kHz

    What is the highest input frequency that

    (a) will cause an unlocked loop to lock?
    (b) can be tracked by a locked loop?

Section 7.3

24. Figure 7.25 shows a simple frequency synthesizer.

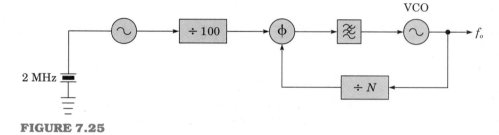

FIGURE 7.25

(a) What frequency range does it generate as N varies from 100 to 200?
(b) What is the smallest possible frequency step with this synthesizer?

25. Draw a block diagram of a simple frequency synthesizer that will generate frequencies from 1 to 10 MHz in 500 kHz steps. Find the range of values of N that will be needed for the programmable divider.

Section 7.4

26. A simple frequency synthesizer can function up to 50 MHz, with a frequency step size of 10 kHz. Calculate new values for these specifications, if a fixed 10:1 prescaler is used. The prescaler itself is capable of responding to frequencies up to 1 GHz.

27. Calculate the output frequency for the synthesizer shown in Figure 7.26, for $N = 50$ and $M = 10$.

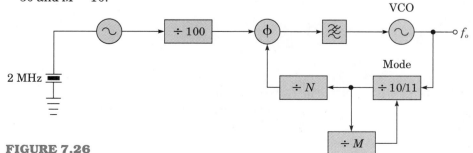

FIGURE 7.26

Section 7.5

28. Figure 7.27 shows a synthesizer with external frequency translation. Calculate the output frequency for $N = 50$.

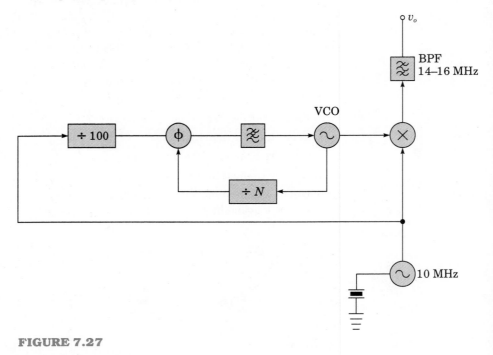

FIGURE 7.27

29. The synthesizer in Figure 7.28 has frequency translation within the loop. Calculate its output frequency for $N = 30$.

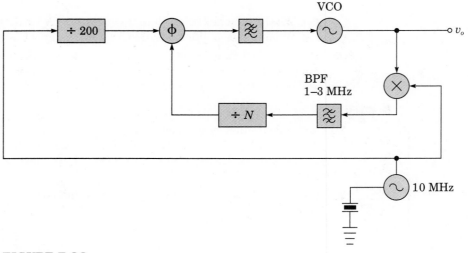

FIGURE 7.28

Section 7.6

30. The 10 MHz crystal oscillator in a PLL frequency synthesizer has been calibrated to a reference standard known to be accurate to 5 ppm. The oscillator is guaranteed to drift no more than 10 ppm per month.

(a) A year later, what are the maximum and minimum frequencies that the oscillator could be producing?

(b) Suppose this oscillator is used in a simple single-loop PLL to generate a frequency of 45 MHz. What are the maximum and minimum values that the output frequency could have, under the conditions of Part (a)?

31. A synthesized RF generator has a phase noise specification of -45 decibels below carrier output (dBC). If the generator is producing a CW signal of 100 μV, what is the maximum allowable noise output?

Section 7.7

32. Assuming that the timebase generator is perfectly accurate, calculate the possible percent error in the reading of a six-digit counter with a 1 s gate time, as it measures the frequency of each of the following signals: 10 Hz, 1 kHz, 455 kHz.

33. Repeat the previous problem, except that this time the counter is measuring period by counting cycles of a 100 kHz timebase over one period of the input signal.

34. What would the display show if the counter of Problem 32 were connected to a signal with a frequency of 3.456789 MHz? Would it be possible to use the counter under these conditions? If so, how?

Section 7.8

35. An LM565 PLL is configured with a lag filter, as shown in Figure 7.24. If V_c = 10 V, R_o = 5 kΩ, C_o = 180 pF, and C_1 = 100 pF, calculate:

(a) the free-running frequency

(b) the capture range

(c) the lock range

(d) the damping factor of the loop filter

Comprehensive

36. The output of the phase detector described in the text is proportional to cos θ, where θ is the phase angle between the two inputs. It was stated that cos θ is approximately proportional to $(θ - 90°)$ for θ close to 90°. Verify this statement using values for θ of 70°, 80°, 90°, 100°, and 110°. Does the proportionality hold for $θ = 150°$?

37. An AM broadcast radio receiver uses high-side injection and an IF of 455 kHz. Design a frequency synthesizer to serve as the local oscillator for this receiver.

38. Draw a block diagram for a synthesizer with a two-modulus divide-by-10/divide-by-11 prescaler, and choose suitable values so that it will generate frequencies in the range from 100 to 200 MHz, at 1 MHz intervals.

39. Design a synthesizer with frequency translation within the loop, to generate the frequencies between 144 and 148 MHz, at 10 kHz intervals.

40. Use the LM565 data sheet in Figure 7.24 to design a PLL with the following specifications:

Free-Running Frequency: 200 kHz

Capture Range: at least 50 kHz

Lock Range: at least 100 kHz

Angle Modulation

Objectives

After studying the material in this chapter, you should be able to:

1. Describe and explain the differences between amplitude and angle modulation schemes, and the advantages and disadvantages of each.
2. Describe and explain the differences between frequency and phase modulation, and show the relationship between the two.
3. Calculate bandwidth, sideband frequencies, carrier and sideband voltage and power levels, and modulation index for frequency- and phase-modulated signals.
4. Explain the capture effect and noise threshold level for FM signals, and calculate the signal-to-noise ratio for simple situations.
5. Relate deviation, bandwidth, and signal-to-noise improvement for FM systems.
6. Explain the use of pre-emphasis and de-emphasis in FM systems, and calculate component values for pre-emphasis and de-emphasis circuits.
7. Describe the system used for FM stereo broadcasting, and draw a diagram showing the spectrum of an FM stereo signal.
8. Perform measurements on FM signals using a spectrum analyzer.

8.1   INTRODUCTION

At the beginning of this book, we observed that there are only three parameters of a carrier wave that can be changed, or modulated, in order for it to carry information: amplitude, frequency, and phase. The last two are closely related, since frequency (expressed in radians per second) is the rate of change of phase angle (in radians). If either frequency or phase is changed in a modulation system, the other will change as well. Consequently, it is useful to group frequency and phase modulation together under the heading of **angle modulation**.

Both *frequency modulation* (FM) and *phase modulation* (PM) are widely used in communications systems. FM is more familiar in daily life, since it is used extensively for radio broadcasting. FM is also used for the sound signal in television, for two-way fixed and mobile radio systems, for satellite communications, and for cellular telephone systems, to name only a few of the more common applications.

While PM may be less familiar, it is used extensively in data communications. It is also used in some FM transmitters as an intermediate step in the generation of FM. FM and PM are closely related mathematically, and it is quite easy to change one to the other.

The most important advantage of FM or PM over AM is the possibility of a greatly improved signal-to-noise ratio. A penalty is paid for this in increased bandwidth: an FM signal may occupy several times as much bandwidth as that required for an AM signal. There may seem to be a contradiction here, as we found that, for AM, decreasing the bandwidth improved the signal-to-noise ratio. This seeming contradiction will be resolved shortly.

In our discussion of amplitude modulation, we found that the amplitude of the modulated signal varied in accordance with the instantaneous amplitude of the modulating signal. In FM, it is the *frequency* of the modulated

---

Historical Development of FM

FM was considered very early in the development of radio communications. At first, it was thought that FM might make a reduced transmission bandwidth possible compared with AM. This was refuted by experimental tests, and also mathematically by John Renshaw Carson (1887–1940) in 1922.

Carson failed to notice that FM does have an advantage over AM in terms of signal-to-noise ratio. Edwin Armstrong (yes, the same Armstrong who invented the superheterodyne receiver) did notice this, and in 1936 he proposed a practical FM system. FM broadcasting began in the United States in 1939, but suffered a setback in 1944 when its frequency allocation was abruptly shifted from 42–50 MHz to its present range of 88–108 MHz. FM broadcasting gradually became popular because of its noise and fidelity advantages over AM. By now, there are actually more FM than AM listeners.

Unfortunately, Armstrong did not benefit from the success of FM broadcasting. He spent the remainder of his life involved in lawsuits in an attempt to receive royalties from his inventions, and finally, in 1954, a broken man, he committed suicide.

signal that varies with the amplitude of the modulating signal. In PM, the *phase* varies directly with the modulating-signal amplitude. This is important to remember: in all types of modulation, it is the amplitude of the modulating signal that varies the carrier wave.

Unlike the case for AM, the amplitude and the power of an FM or PM signal do not change with modulation. Thus, an FM signal does not have an envelope that reproduces the modulation. This is actually an advantage: an FM receiver does not have to respond to amplitude variations, and this lets it ignore noise to some extent. Similarly, FM transmitters can use Class C amplifiers throughout, since amplitude linearity is not important. Modulation can be accomplished at low power levels.

8.2 FREQUENCY MODULATION

Figure 8.1 shows FM, with a square wave modulating a sine-wave carrier. Figure 8.1(a) shows the unmodulated carrier and the modulating signal. Figure 8.1(b) shows the modulated signal in the time domain, as it would appear on an oscilloscope. The amount of frequency change has been exaggerated for clarity. The amplitude remains as before, and the frequency changes can be seen in the changing times between zero-crossings for the waveforms.

The next two sections are interesting. Figure 8.1(c) shows how the signal frequency varies with time in accordance with the amplitude of the modulating signal. Figure 8.1(d) shows how the phase changes with time. For this figure, the phase angle of the unmodulated carrier is used as a reference. When the frequency is greater than the carrier frequency, the phase angle gradually moves ahead, and when the frequency is lower than the carrier frequency, the phase begins to lag.

One further inference can be drawn from Figure 8.1. While the unmodulated carrier is a sine wave, the modulated signal is not. A sine wave that changes frequency is not really a sine wave at all. In our study of AM, we found that changing the amplitude of a sine wave generated extra frequencies called *side frequencies* or *sidebands*. This happens in FM, too. In fact, for an FM signal, the number of sets of sidebands is theoretically infinite.

There are many ways to generate FM, some of which will be described in the next chapter. The simplest method is to use a voltage-controlled oscillator (VCO) to generate the carrier frequency, and to apply the modulating signal to the oscillator's control signal input, as indicated in Figure 8.2. As might be expected, this is a little too simple for most practical transmitters; in particular, the stability of a free-running VCO is not likely to be good enough. It does, however, give us a conceptual model to use for the time being.

8.2.1 Frequency Deviation

Assume that the carrier frequency is f_c. Modulation will cause the signal frequency to vary, or deviate, from its resting value. If the modulation system is properly designed, this deviation will be proportional to the amplitude of the modulating signal. Sometimes this is referred to as *linear modulation*, though no modulation process is really linear. FM can be called linear only in the sense that the graph relating instantaneous modulating-signal amplitude e_m to instantaneous frequency deviation Δf is a straight line. The slope

FIGURE 8.1   Frequency Modulation of a Sine-Wave Carrier by a Square Wave

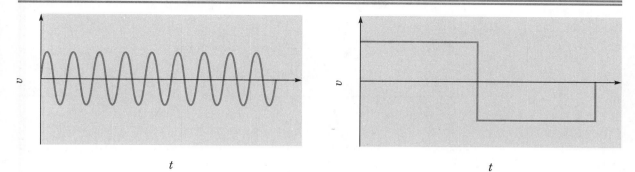

Carrier   **Modulation**

(a) Carrier and Modulating Signals Before Modulation

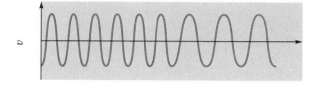

(b) Modulated Signal (Time Domain)

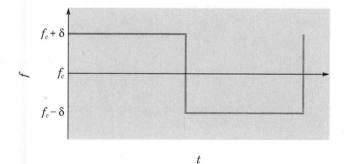

(c) Frequency as a Function of Time

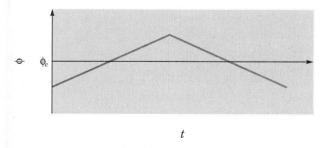

(d) Phase as a Function of Time

FIGURE 8.2

Simplified FM
Generator

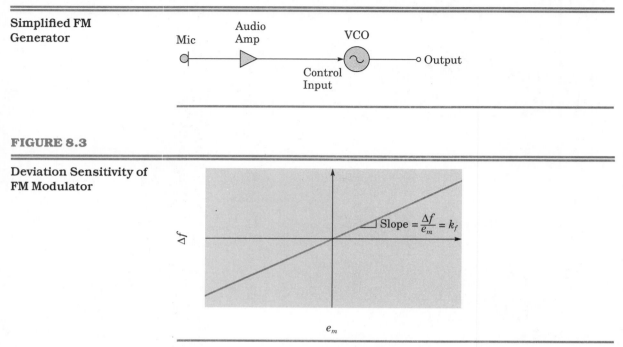

FIGURE 8.3

Deviation Sensitivity of
FM Modulator

of this line is the ratio $\Delta f / e_m$, and it represents the deviation sensitivity of
the modulator, with units of hertz per volt. Let us call this constant k_f. Then

$$k_f = \frac{\Delta f}{e_m} \tag{8.1}$$

Figure 8.3 demonstrates this.

Once again, remember that the **frequency deviation** is proportional to
the amplitude, not the frequency, of the modulating signal. The number of
times per second that the frequency varies from its lowest to its highest value
is, of course, equal to the modulating-signal frequency.

It is possible to write an equation for the signal frequency as a function
of time:

$$f_{sig}(t) = f_c + k_f e_m(t) \tag{8.2}$$

where $f_{sig}(t)$ = signal frequency as a function of time
       f_c = unmodulated carrier frequency
       k_f = modulator deviation constant
       $e_m(t)$ = modulating voltage as a function of time

EXAMPLE 8.1

An FM modulator has $k_f = 30$ kHz/V, and operates at a carrier frequency
of 175 MHz. Find the output frequency for an instantaneous value of the
modulating signal equal to:

(a) 150 mV   (b) −2 V

Solution

Equation (8.2) can be used for both parts of the question.

(a) $f_{sig} = (175 \times 10^6 \text{ Hz}) + (30 \times 10^3 \text{ Hz/V})(150 \times 10^{-3} \text{ V})$
     $= 175.0045 \times 10^6 \text{ Hz}$
     $= 175.0045 \text{ MHz}$

(b) $f_{sig} = (175 \times 10^6 \text{ Hz}) + (30 \times 10^3 \text{ Hz/V})(-2 \text{ V})$
     $= (175 \times 10^6 \text{ Hz}) - (30 \times 10^3 \text{ Hz/V})(2 \text{ V})$
     $= 174.94 \times 10^6 \text{ Hz}$
     $= 174.94 \text{ MHz}$

In our study of AM, we found it convenient to assume a sine wave for the modulating signal. This treatment can then be generalized to cover any periodic signal using Fourier techniques. If the modulating signal is a sine wave with the equation

$$e_m(t) = E_m \sin \omega_m t \qquad (8.3)$$

then Equation (8.2) becomes

$$f_{sig}(t) = f_c + k_f E_m \sin \omega_m t \qquad (8.4)$$

and the peak frequency deviation, each side of the carrier frequency, will be $k_f E_m$ Hz. Usually, the peak frequency deviation is given the symbol δ. Then

$$\delta = k_f E_m \qquad (8.5)$$

where δ = peak frequency deviation in hertz
       k_f = modulator sensitivity in hertz per volt
       E_m = peak value of the modulating signal in volts

EXAMPLE 8.2

The same FM modulator as in the previous example is modulated by a 3 V sine wave. Calculate the deviation.

Solution

Unless otherwise stated, ac voltages are assumed to be RMS. On the other hand, δ is a peak value. Therefore, the modulating voltage must be converted to a peak value before Equation (8.5) can be used.

$$E_m = 3\sqrt{2} \text{ V}$$
$$= 4.24 \text{ V}$$
$$\delta = k_f E_m$$
$$= 30 \text{ kHz/V} \times 4.24 \text{ V}$$
$$= 127.2 \text{ kHz}$$

Using frequency deviation, Equation (8.4) becomes

$$f_{sig}(t) = f_c + \delta \sin \omega_m t \qquad (8.6)$$

Another basic term for FM is the **frequency modulation index** m_f (not to be confused with f_m, which is the modulating frequency). By definition, for sine-wave modulation,

$$m_f = \frac{\delta}{f_m} \tag{8.7}$$

The reason for this rather peculiar definition, which includes not only the frequency deviation but also the modulating frequency, will become apparent very soon. Meanwhile, note one other peculiarity of m_f: unlike the amplitude modulation index, which cannot exceed one, there are no theoretical limits on m_f. It can exceed one, and often does.

EXAMPLE 8.3

An FM broadcast transmitter operates at its maximum deviation of 75 kHz. Find the modulation index for a sinusoidal modulating signal with a frequency of:

(a) 15 kHz   (b) 50 Hz

Solution

(a) $m_f = \dfrac{\delta}{f_m}$

$= \dfrac{75 \text{ kHz}}{15 \text{ kHz}}$

$= 5.00$

(b) $m_f = \dfrac{\delta}{f_m}$

$= \dfrac{75 \times 10^3 \text{ Hz}}{50 \text{ Hz}}$

$= 1500$

Substituting Equation (8.7) into Equation (8.6) gives

$$f_{sig}(t) = f_c + m_f f_m \sin \omega_m t \tag{8.8}$$

as an equation for the frequency of an FM signal with sine-wave modulation.

8.3 PHASE MODULATION

In phase modulation, it is the phase shift, rather than the frequency deviation, that is proportional to the instantaneous amplitude of the modulating signal. In a similar way to that for FM, we can define a constant for a phase modulator that relates change in phase angle to the amplitude of the modulating signal:

$$k_p = \frac{\phi}{e_m} \tag{8.9}$$

where k_p = phase modulator sensitivity in radians per volt
       ϕ = phase deviation in radians
       e_m = modulating-signal amplitude in volts

An equation can be written for the phase of a PM signal as a function of time, similar to Equation (8.2) for the frequency of an FM signal:

$$\theta(t) = \theta_c + k_p e_m(t) \tag{8.10}$$

Once again it would be useful to express this in terms of sine-wave modulation. If

$$e_m(t) = E_m \sin \omega_m t$$

then

$$\phi = k_p E_m \sin \omega_m t \tag{8.11}$$

and

$$\theta(t) = \theta_c + k_p E_m \sin \omega_m t$$

The peak phase deviation, in radians, is defined as m_p, the *phase modulation index*. Then

$$\theta(t) = \theta_c + m_p \sin \omega_m t \tag{8.12}$$

EXAMPLE 8.4

A phase modulator has k_p = 2 rad/V. What would be the RMS voltage of a sine wave that would cause a peak phase deviation of 60°?

Solution

Remembering that a circle has 360° or 2π rad, we see that

$$360° = 2\pi \text{ rad}$$

$$60° = \frac{2\pi \text{ rad} \times 60}{360}$$

$$= \frac{\pi}{3} \text{ rad}$$

The voltage to cause this deviation can be found from Equation (8.9):

$$k_p = \frac{\phi}{e_m}$$

$$e_m = \frac{\phi}{k_p}$$

$$= \frac{(\pi/3) \text{ rad}}{2 \text{ rad/V}}$$

$$= \frac{\pi}{6} \text{ V}$$

$$= 0.524 \text{ V}$$

This is peak voltage. We can find its RMS value in the usual way:

$$V_{RMS} = \frac{V_{\text{peak}}}{\sqrt{2}}$$

$$= \frac{0.524}{\sqrt{2}}$$

$$= 0.37 \text{ V}$$

8.4 THE RELATIONSHIP BETWEEN FREQUENCY MODULATION AND PHASE MODULATION

As mentioned earlier, either FM or PM will result in changes in both the frequency and phase of the modulated waveform. It has also been pointed out that frequency (in radians per second) is the rate of change of phase (in radians). That is, frequency is the derivative of phase. This leads to a relatively simple relationship between FM and PM that can make it easier to understand both, and to perform calculations with either. The results will be presented here, with enough discussion to make them plausible. (For readers familiar with calculus, a more mathematical treatment can be found in Appendix F.)

For any angle-modulated signal with sine-wave modulation, the modulation index m_p or m_f represents the peak phase deviation from the phase of the unmodulated carrier, in radians. This is obvious for PM, but not quite so obvious for FM. We know that

$$m_f = \frac{\delta}{f_m}$$

That is, the modulation index (which corresponds to peak phase deviation) is proportional to frequency deviation, and inversely proportional to modulating frequency. The first statement sounds very reasonable. Suppose that the frequency increases. Then the higher frequency represents a phase angle that changes more quickly than before. The greater the frequency change, the greater the increase in phase angle.

The second statement needs more explanation. Why should an increased modulating frequency result in a reduced change in phase angle? Consider a low modulating frequency, say 1 Hz, that causes a frequency deviation of 1 kHz. For simplicity, suppose that the modulating signal is a square wave. Then the signal frequency will increase to a point 1 kHz above the carrier frequency, stay there for one-half second, then decrease by 2 kHz to a point 1 kHz below the carrier frequency for the next one-half second. Figure 8.4(a) shows the way the instantaneous signal frequency varies.

When the signal frequency is higher than normal, the phase angle, with respect to that of the unmodulated carrier, steadily increases as the modulated wave gets further and further ahead of the unmodulated signal. This continues until the frequency lowers. At that point, the signal's phase angle starts to lag, letting the carrier phase angle catch up and then overtake the angle of the modulated signal.

FIGURE 8.4

**Relation Between
Modulating Frequency
and Modulation Index
for FM**

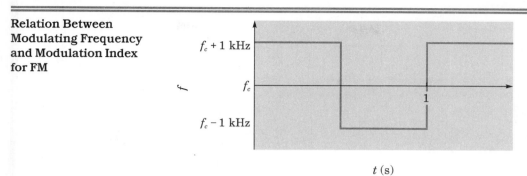

(a) Frequency Shift for 1 Hz Modulating Signal

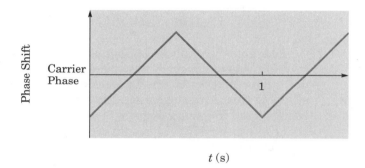

(b) Phase Shift for 1 Hz Modulating Signal

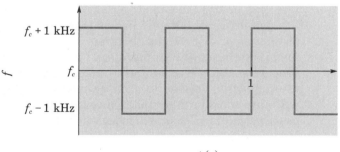

(c) Frequency Shift for 2 Hz Modulating Signal

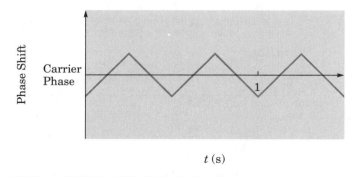

(d) Phase Shift for 2 Hz Modulating Signal

FIGURE 8.5   **Use of an Integrator to Convert PM to FM**

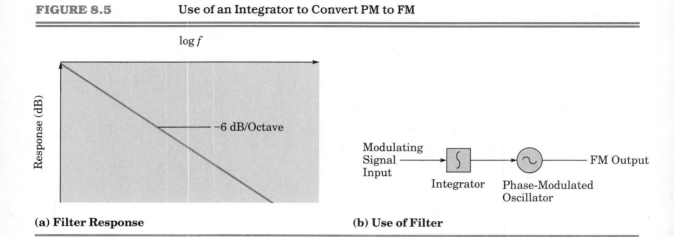

(a) Filter Response   (b) Use of Filter

The amount of phase change is proportional to the length of time the instantaneous frequency stays above the carrier frequency; that is, the phase deviation is proportional to the period of the modulating signal. Another way of saying this is that phase deviation is inversely proportional to the modulating-signal frequency. Figures 8.4(c) and (d) show the frequency and phase change, respectively, for a modulating signal with the same amplitude as before but twice the frequency. The phase deviation is only one-half as much as before.

This simple relationship between FM and PM suggests that it would be easy to convert one to the other, and this is true. For instance, a phase modulator can be used to generate FM. The baseband signal is passed through a low-pass filter with the frequency response shown in Figure 8.5(a). This type of filter is often referred to as an *integrator*. For equal amplitudes at the modulator input, the amplitude of the output from the filter will be inversely proportional to the modulating frequency. Figure 8.5(b) shows the low-pass filtered signal applied to a phase modulator. The phase-modulated output will have a modulation index inversely proportional to the modulating frequency; that is, it will be identical to FM!

EXAMPLE 8.5

An FM communications transmitter has a maximum frequency deviation of 5 kHz and a range of modulating frequencies from 300 Hz to 3 kHz. What is the maximum phase shift that it produces?

Solution

We now know that the peak phase shift in radians is equal to the frequency modulation index m_f. Since, by Equation (8.7),

$$m_f = \frac{\delta}{f_m}$$

m_f will be largest for the lowest possible value of f_m, in this case 300 Hz. In that case, the phase shift is

$$\phi_{max} = m_f$$

$$= \frac{\delta}{f_m}$$

$$= \frac{5000}{300}$$

$$= 16.7 \text{ rad}$$

EXAMPLE 8.6

A phase modulator has a sensitivity of $k_p = 3$ rad/V. How much frequency deviation does it produce with a sine-wave input of 2 V peak at a frequency of 1 kHz?

Solution

The maximum phase shift is easily found from Equation (8.11):

$$\phi = k_p E_m \sin \omega_m t$$

The maximum value of ϕ is m_p and it occurs for the peak modulating voltage.

$$m_p = \phi_{max}$$
$$= k_p E_m$$
$$= 3 \text{ rad/V} \times 2 \text{ V}$$
$$= 6 \text{ rad}$$

This has the same value as m_f, if the signal is considered as frequency modulation. From Equation (8.7),

$$m_f = \frac{\delta}{f_m}$$
$$\delta = m_f f_m$$
$$= 6 \times 1 \text{ kHz}$$
$$= 6 \text{ kHz}$$

8.5 THE ANGLE MODULATION SPECTRUM

Angle modulation produces an infinite number of sidebands, even for single-tone modulation. These sidebands are separated from the carrier by multiples of f_m, but their amplitude tends to decrease as their distance from the carrier frequency increases. Those sidebands with amplitude less than about 1% of the total signal voltage can usually be ignored, so that for practical purposes an angle-modulated signal can be considered to be band-limited. In most cases, though, its bandwidth is much larger than that of an AM signal.

TABLE 8.1

m	J_0	J_1	J_2	J_3	J_4	J_5	J_6	J_7	J_8	J_9	J_{10}	J_{11}	J_{12}	J_{13}	J_{14}	J_{15}	J_{16}	J_{17}	J_{18}	J_{19}	J_{20}
0	1.00																				
0.25	0.98	0.12																			
0.5	0.94	0.24	0.03																		
0.75	0.86	0.35	0.07	0.01																	
1	0.77	0.44	0.11	0.02																	
1.25	0.65	0.51	0.17	0.04	0.01																
1.5	0.51	0.56	0.23	0.06	0.01																
1.75	0.37	0.58	0.29	0.09	0.02																
2	0.22	0.58	0.35	0.13	0.03	0.01															
2.25	0.08	0.55	0.40	0.17	0.05	0.01															
2.4	0.00	0.52	0.43	0.20	0.06	0.02															
2.5	−0.05	0.50	0.45	0.22	0.07	0.02															
2.75	−0.16	0.43	0.47	0.26	0.10	0.03	0.01														
3	−0.26	0.34	0.49	0.31	0.13	0.04	0.01														
3.5	−0.38	0.14	0.46	0.39	0.20	0.08	0.03	0.01													
4	−0.40	−0.07	0.36	0.43	0.28	0.13	0.05	0.01													
4.5	−0.32	−0.23	0.22	0.42	0.35	0.20	0.08	0.03	0.01												
5	−0.18	−0.33	0.05	0.36	0.39	0.26	0.13	0.05	0.02												
5.5	0.00	−0.34	−0.12	0.26	0.40	0.32	0.19	0.09	0.03	0.01											
6	0.15	−0.28	−0.24	0.11	0.36	0.36	0.25	0.13	0.06	0.02	0.01										
6.5	0.26	−0.15	−0.31	−0.03	0.28	0.37	0.30	0.18	0.09	0.04	0.01										
7	0.30	−0.01	−0.30	−0.17	0.16	0.35	0.34	0.23	0.13	0.06	0.02	0.01									
7.5	0.27	0.14	−0.23	−0.26	0.02	0.28	0.35	0.28	0.17	0.09	0.04	0.01									
8	0.17	0.24	−0.11	−0.29	−0.11	0.19	0.34	0.32	0.22	0.13	0.06	0.03	0.01								
8.5	0.04	0.27	0.02	−0.26	−0.21	0.07	0.29	0.34	0.27	0.17	0.09	0.04	0.02	0.01							
8.65	0.00	0.27	0.06	−0.24	−0.23	0.03	0.27	0.34	0.28	0.18	0.10	0.05	0.02	0.01							
9	−0.09	0.25	0.14	−0.18	−0.27	−0.06	0.20	0.33	0.30	0.21	0.13	0.06	0.03	0.01							
10	−0.25	0.04	0.26	0.06	−0.22	−0.23	−0.01	0.22	0.32	0.29	0.21	0.12	0.06	0.03	0.01						
11	−0.17	−0.18	0.14	0.23	−0.01	−0.24	−0.20	0.02	0.23	0.31	0.28	0.20	0.12	0.06	0.03	0.01					
12	0.05	−0.22	−0.08	0.20	0.18	−0.07	−0.24	−0.17	0.04	0.23	0.30	0.27	0.20	0.12	0.07	0.03	0.01				
13	0.21	−0.07	−0.22	0.00	0.22	0.13	−0.12	−0.24	−0.14	0.07	0.23	0.29	0.26	0.19	0.12	0.07	0.03	0.01			
14	0.17	0.13	−0.15	−0.18	0.08	0.13	−0.23	−0.11	0.08	0.24	0.29	0.25	0.19	0.12	0.07	0.03	0.02	0.01			
15	−0.01	0.20	0.04	−0.19	−0.12	0.13	0.21	0.03	−0.17	−0.22	−0.09	0.10	0.24	0.28	0.25	0.18	0.12	0.07	0.03	0.02	0.01
16	−0.17	0.09	0.19	−0.04	−0.20	−0.06	0.17	0.18	−0.01	−0.19	−0.21	−0.07	0.11	0.24	0.27	0.24	0.18	0.11	0.07	0.03	0.02
17	−0.17	−0.10	0.16	0.14	−0.11	−0.19	0.00	0.19	0.15	−0.04	−0.20	−0.19	−0.05	0.12	0.24	0.27	0.23	0.17	0.11	0.07	0.04

8.5.1 Bessel Functions

Both FM and PM signals have similar equations. In fact, without a knowledge of the baseband signal, it would be impossible to tell them apart. For the modulation of a carrier with amplitude A and radian frequency ω_c by a single-frequency sinusoid, the equation is of the form

$$v(t) = A \sin (\omega_c t + m \sin \omega_m t) \tag{8.13}$$

The factor m can represent m_f for FM or m_p for PM.

   This equation cannot be simplified by ordinary trigonometry, as is the case for amplitude modulation. About the only useful information that can be gained by inspection is the fact that the signal amplitude remains constant regardless of the modulation index. This observation is significant, since it demonstrates one of the major differences between AM and FM or PM, but it provides no information about the sidebands.

   This signal can be expressed as a series of sinusoids by using Bessel functions of the first kind. Proving this is beyond the scope of this text, but it can be done. The Bessel functions themselves are rather tedious to evaluate numerically, but that, too, has been done. Some results are presented in Table 8.1 and Figure 8.6.

   The table and graph of Bessel functions represent normalized voltages for the various frequency components of an FM or PM signal; that is, the numbers in the tables will represent actual voltages if the unmodulated carrier has an amplitude of 1 V. Here J_0 represents the component at the carrier frequency. The variable J_1 represents each of the first set of sidebands, at frequencies of $(f_c + f_m)$ and $(f_c - f_m)$, and J_2 represents the amplitude of each of the second set of sidebands, which are separated from the carrier frequency by twice the modulating frequency, and so on. Figure 8.7

FIGURE 8.6

Bessel Functions

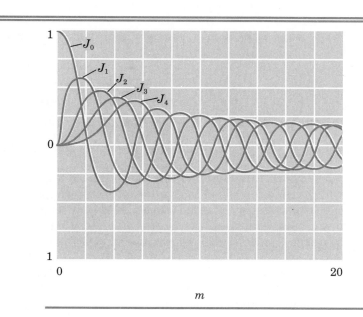

FIGURE 8.7   **FM in the Frequency Domain, for an Unmodulated Carrier Amplitude of 1 V**

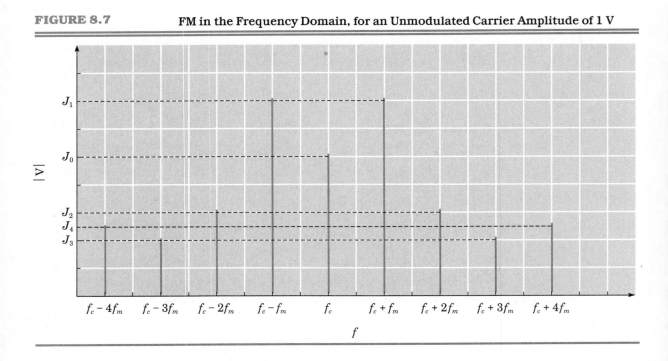

shows this on a frequency-domain plot. All of the Bessel terms should be multiplied by the voltage of the unmodulated carrier to find the actual sideband amplitudes. Of course, the Bessel coefficients are equally valid for peak or RMS voltages, but the user should be careful to keep track of which type of measurement is being used. There certainly will be a difference when sideband power calculations are involved.

When Bessel functions are used, the signal of Equation (8.13) becomes

$$
\begin{aligned}
v(t) &= A \sin (\omega_c t + m \sin \omega_m t) \\
&= A\{J_0(m) \sin \omega_c t \\
&\quad - J_1(m)[\sin (\omega_c - \omega_m)t - \sin (\omega_c + \omega_m)t] \\
&\quad + J_2(m)[\sin (\omega_c - 2\omega_m)t + \sin (\omega_c + 2\omega_m)t] \\
&\quad - J_3(m)[\sin (\omega_c - 3\omega_m)t + \sin (\omega_c + 3\omega_m)t] \\
&\quad + \cdots\}
\end{aligned}
\tag{8.14}
$$

With angle modulation, the total signal voltage and power do not change with modulation. Therefore, the appearance of power in the sidebands indicates that the power at the carrier frequency must be reduced below its unmodulated value, in the presence of modulation. In fact, the carrier-frequency component disappears for certain values of m (for example, 2.4 and 5.5).

This constant-power aspect of angle modulation can be demonstrated using the table of Bessel functions. For simplicity, normalization can be used. Let the unmodulated signal have a voltage of 1 V RMS across a resistance of

$1 \, \Omega$. Its power is, of course, 1 W. Now when modulation is applied, the carrier voltage will be reduced and sidebands will appear. From Table 8.1, J_0 will represent the RMS voltage at the carrier frequency, and the power at the carrier frequency will be

$$
\begin{aligned}
P_c &= \frac{V_c^2}{R} \\
    &= \frac{J_0^2}{1} \\
    &= J_0^2
\end{aligned}
$$

Similarly, the power in each of the first set of sidebands will be

$$
P_{SB_1} = J_1^2
$$

The combined power in the first set of sidebands will be twice as much as this, of course. The power in the whole signal will then be

$$
P_T = J_0^2 + 2J_1^2 + 2J_2^2 + \cdots \tag{8.15}
$$

If the series is carried on far enough, the result will be equal to 1 W, regardless of the modulation index.

For a signal with a power different from 1 W, it is only necessary to multiply the terms in Equation (8.15) by the total signal power.

The bandwidth of an FM or PM signal is to some extent a matter of definition. The Bessel series is infinite, but as can be seen from the table or the graph, the amplitude of the components will gradually diminish, until at some point they can be ignored. The process is slower for large values of m, so the number of sets of sidebands that has to be considered is greater for larger modulation indices. A practical rule of thumb is to ignore sidebands with a Bessel coefficient whose absolute value is less than 0.01. The bandwidth, for practical purposes, is equal to twice the number of the highest significant Bessel coefficient, multiplied by the modulating frequency.

EXAMPLE 8.7

An FM signal has a deviation of 3 kHz and a modulating frequency of 1 kHz. Its total power P_T is 5 W, developed across a 50 Ω resistive load. The carrier frequency is 160 MHz.

(a) Calculate the RMS signal voltage V_T.

(b) Calculate the RMS voltage at the carrier frequency and each of the first three sets of sidebands.

(c) For the first three sideband pairs, calculate the frequency of each sideband.

(d) Calculate the power at the carrier frequency and at each of the sideband frequencies found in Part (c).

(e) Determine what percentage of the total signal power is unaccounted for by the components described above.

(f) Sketch the signal in the frequency domain, as it would appear on a spectrum analyzer. The vertical scale should be power in dBm and the horizontal scale should be frequency.

Solution

(a) The signal power does not change with modulation, and neither does the voltage, which can easily be found from the power equation.

$$P_T = \frac{V_T^2}{R_L}$$

$$\begin{aligned} V_T &= \sqrt{P_T R_L} \\ &= \sqrt{5 \text{ W} \times 50 \text{ } \Omega} \\ &= 15.8 \text{ V (RMS)} \end{aligned}$$

(b) The modulation index must be found in order to use Bessel functions to find the carrier and sideband voltages.

$$\begin{aligned} m_f &= \frac{\delta}{f_m} \\ &= \frac{3 \text{ kHz}}{1 \text{ kHz}} \\ &= 3 \end{aligned}$$

From the Bessel function table, the coefficients for the carrier and the first three sideband pairs are:

$$J_0 = -0.26 \qquad J_1 = 0.34 \qquad J_2 = 0.49 \qquad J_3 = 0.31$$

These are normalized voltages, so they will have to be multiplied by the total RMS signal voltage to get the RMS sideband and carrier-frequency voltages.

For the carrier,

$$V_c = J_0 V_T$$

Now, J_0 has a negative sign. This simply indicates a phase relationship between the components of the signal. It would be required if we wanted to add together all the components to get the resultant signal. For our present purpose, however, it can simply be ignored, and we can use

$$\begin{aligned} V_c &= |J_0| V_T \\ &= 0.26 \times 15.8 \text{ V} \\ &= 4.11 \text{ V} \end{aligned}$$

Similarly, we can find the voltage for each of the three sideband pairs. Note that these are voltages for individual components. There will be a lower and an upper sideband with each of these calculated voltages.

$$V_1 = J_1 V_T$$
$$= 0.34 \times 15.8 \text{ V}$$
$$= 5.37 \text{ V}$$

$$V_2 = J_2 V_T$$
$$= 0.49 \times 15.8 \text{ V}$$
$$= 7.74 \text{ V}$$

$$V_3 = J_3 V_T$$
$$= 0.31 \times 15.8 \text{ V}$$
$$= 4.9 \text{ V}$$

(c) The sidebands are separated from the carrier frequency by multiples of the modulating frequency. Here, $f_c = 160$ MHz and $f_m = 1$ kHz, so there are sidebands at each of the following frequencies:

$$f_{USB_1} = 160 \text{ MHz} + 1 \text{ kHz}$$
$$= 160.001 \text{ MHz}$$

$$f_{USB_2} = 160 \text{ MHz} + 2 \text{ kHz}$$
$$= 160.002 \text{ MHz}$$

$$f_{USB_3} = 160 \text{ MHz} + 3 \text{ kHz}$$
$$= 160.003 \text{ MHz}$$

$$f_{LSB_1} = 160 \text{ MHz} - 1 \text{ kHz}$$
$$= 159.999 \text{ MHz}$$

$$f_{LSB_2} = 160 \text{ MHz} - 2 \text{ kHz}$$
$$= 159.998 \text{ MHz}$$

$$f_{LSB_3} = 160 \text{ MHz} - 3 \text{ kHz}$$
$$= 159.997 \text{ MHz}$$

(d) Since each of the components of the signal is a sinusoid, the usual equation can be used to calculate power. All the components appear across the same 50 Ω load.

$$P_c = \frac{V_c^2}{R_L}$$
$$= \frac{4.11^2}{50}$$
$$= 0.338 \text{ W}$$

$$P_1 = \frac{V_1^2}{R_L} \qquad\qquad P_2 = \frac{V_2^2}{R_L} \qquad\qquad P_3 = \frac{V_3^2}{R_L}$$

$$= \frac{5.37^2}{50} \qquad\qquad = \frac{7.74^2}{50} \qquad\qquad = \frac{4.9^2}{50}$$

$$= 0.576 \text{ W} \qquad\qquad = 1.2 \text{ W} \qquad\qquad = 0.48 \text{ W}$$

(e) To find the total power P_T in the carrier and the first three sets of sidebands, it is only necessary to add the powers calculated above, counting each of the sideband powers twice, because each of the calculated powers represents one of a pair of sidebands. We only count the carrier once, of course.

$$P_T = P_c + 2(P_1 + P_2 + P_3)$$
$$= 0.338 + 2(0.576 + 1.2 + 0.48) \text{ W}$$
$$= 4.85 \text{ W}$$

This is not quite the total signal power, which was given as 5 W. The remainder is in the additional sidebands. To find how much is unaccounted for by the carrier and the first three sets of sidebands, we can subtract. Call the difference P_x.

$$P_x = 5 - 4.85$$
$$= 0.15 \text{ W}$$

As a percentage of the total power this is

$$P_x \, (\%) = \left(\frac{0.15}{5}\right)100$$

$$= 3\%$$

(f) All the information we need for the sketch is on hand, except that the power values have to be converted to dBm using the equation

$$P \text{ (dBm)} = 10 \log \frac{P}{1 \text{ mW}}$$

This gives

$$P_c \text{ (dBm)} = 10 \log 338$$
$$= 25.3 \text{ dBm}$$
$$P_1 \text{ (dBm)} = 10 \log 576$$
$$= 27.6 \text{ dBm}$$
$$P_2 \text{ (dBm)} = 10 \log 1200$$
$$= 30.8 \text{ dBm}$$
$$P_3 \text{ (dBm)} = 10 \log 480$$
$$= 26.8 \text{ dBm}$$

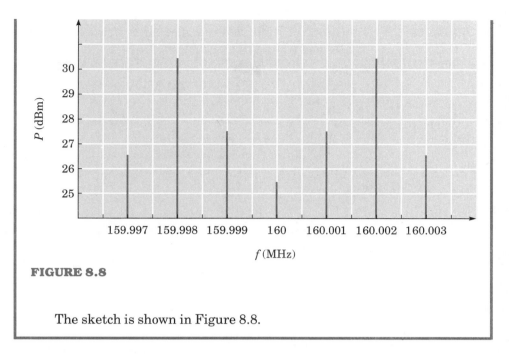

FIGURE 8.8

The sketch is shown in Figure 8.8.

8.5.2 Bandwidth

For PM, the bandwidth varies directly with the modulating frequency, since doubling the frequency doubles the distance between sidebands. It is also roughly proportional to the maximum phase deviation, since increasing m_p increases the number of sidebands. For FM, however, the situation is complicated by the fact that

$$m_f = \frac{\delta}{f_m}$$

For a given amount of frequency deviation, the modulation index is inversely proportional to the modulating frequency. Recall that the frequency deviation is proportional to the amplitude of the modulating signal. Then, if the amplitude of the modulating signal remains constant, increasing the frequency reduces the modulating index. Reducing m_f reduces the number of sidebands with significant amplitude. On the other hand, the increase in f_m means that the sidebands will be further apart in frequency, since they are separated from each other by f_m. These two effects are in opposite directions. The result is that the bandwidth does increase somewhat with increasing modulating-signal frequency, but the bandwidth is not directly proportional to the frequency. Sometimes FM is called a *constant-bandwidth* communications mode for this reason, though the bandwidth is not really constant. Figure 8.9 provides a few examples that show the relationship between modulating frequency and bandwidth. For this example, the deviation remains constant at 10 kHz as the modulating frequency varies from 2 kHz to 10 kHz.

One other point must be made about the sidebands: with AM, restricting the bandwidth of the receiver has a very simple effect on the signal. Since the

FIGURE 8.9   Variation of Effective FM Signal Bandwidth with Modulating Frequency
                                      (Deviation is 10 kHz in all cases)

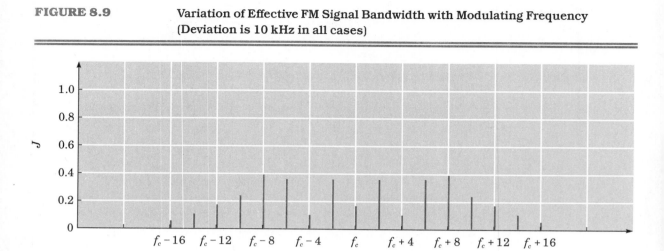

(a) $f_m = 2$ kHz $m_f = 5.0$

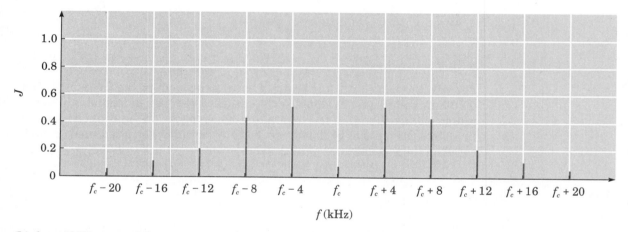

(b) $f_m = 4$ kHz $m_f = 2.5$

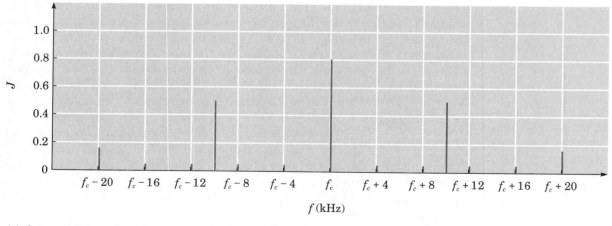

(c) $f_m = 10$ kHz $m_f = 1.0$

side frequencies furthest from the carrier contain the high-frequency base-band information, restricting the receiver bandwidth reduces its response to high-frequency baseband signals, leaving all else unchanged. When reception conditions are poor, bandwidth can be restricted to the minimum necessary for intelligibility. For FM, the situation is more complicated, since even low-frequency modulating signals can generate sidebands that are far removed from the carrier frequency. FM receivers must be designed to include all the significant sidebands that are transmitted; otherwise, severe distortion, not just limited frequency response, will result.

8.5.3 Carson's Rule

The calculation of the bandwidth of an FM signal from Bessel functions is easy enough, since the functions are available in a table, but it can be a bit tedious. There is an approximation, known as *Carson's rule*, that can be used to find the bandwidth of an FM signal. It is not as accurate as using Bessel functions, but can be applied almost instantly, without using tables or even a calculator.

Here is Carson's rule.

$$B = 2(\delta_{max} + f_{m\ (max)})$$ (8.16)

Equation (8.16) assumes that the bandwidth is proportional to the sum of the deviation and the modulating frequency. This is not strictly true. Carson's rule also makes the assumption that maximum deviation occurs with the maximum modulating frequency. Sometimes this leads to errors in practical situations, where often the highest baseband frequencies have much less amplitude than lower frequencies, and therefore do not produce as much deviation.

EXAMPLE 8.8

Use Carson's rule to calculate the bandwidth of the signal used in Example 8.7.

Solution

Here there is only one modulating frequency, so

$$\begin{aligned} B &= 2(\delta + f_m) \\ &= 2(3\ \text{kHz} + 1\ \text{kHz}) \\ &= 8\ \text{kHz} \end{aligned}$$

In the previous example, we found that 97% of the power was contained in a bandwidth of 6 kHz. An 8 kHz bandwidth would contain more of the signal power. Carson's rule gives quite reasonable results in this case, with very little work.

8.5.4 Narrowband and Wideband FM

It was mentioned earlier that there are no theoretical limits to the modulation index, or the frequency deviation, of an FM signal. The limits are practical, and result from a compromise between signal-to-noise ratio and bandwidth. In general, larger values of deviation result in an increased signal-to-noise ratio, while also resulting in greater bandwidth. The former is desirable, but the latter is not, especially in regions of the spectrum where frequency space is in short supply. It is also necessary to have some agreement about deviation, since receivers must be designed for a particular signal bandwidth.

For these reasons, the bandwidth of FM transmissions is generally limited by government regulations that specify the maximum frequency deviation and the maximum modulating frequency, since both of these affect bandwidth. In general, relatively narrow bandwidth (on the order of 15 kHz) is used for voice communication, with wider bandwidths for such services as FM broadcasting (about 200 kHz) and satellite television (36 MHz for one system).

It seems logical to distinguish between **narrowband FM** (NBFM) signals, used for voice transmission, and **wideband FM** (WBFM), used for most other transmissions. This is, in fact, the terminology that is in daily use. The reader should be aware that there is a more restrictive use of the term *narrowband FM*, found mostly in textbooks, to refer to a signal that has m_f less than about 0.5. Such a signal has a bandwidth about the same as that of an AM signal, and only one pair of sidebands with significant power. Even communications transmitters that are normally called NBFM do not qualify under this definition. For example, a typical communications transmitter has a maximum deviation of 5 kHz, with a voice-frequency response from about 300 Hz to 3 kHz. The maximum value of m_f occurs for maximum deviation with minimum modulating frequency. In this case,

$$\text{Maximum } m_f = \frac{5000}{300}$$
$$= 16.7$$

This hardly qualifies as NBFM in the strict, textbook sense, though it is universally called NBFM in practice.

8.6 FM AND NOISE

The original reason for developing FM was to give improved performance in the presence of noise, and that is still one of its main advantages over AM. This improved noise performance can actually result in a better signal-to-noise ratio at the output of a receiver than is found at its input.

One way to approach the problem of FM and noise is to think of the noise voltage as a phasor having random amplitude and phase angle. The noise will add to the signal, causing random variations in both the amplitude and phase angle of the signal as seen by the receiver. Figure 8.10 shows this vector addition.

The amplitude component of noise is easily dealt with in a well-designed

FIGURE 8.10

Effect of Noise on an FM Signal

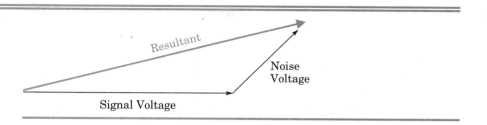

FM system. Since FM signals do not depend on an envelope for detection, the receiver can employ limiting to remove any amplitude variations from the signal. That is, it can use amplifiers whose output amplitude is the same for a wide variety of input signal levels. The effect of this on the amplitude of a noisy signal is shown in Figure 8.11. As long as the signal amplitude is considerably larger than the noise to begin with, the amplitude component of the noise will not be a problem.

It is not possible for the receiver to ignore phase shifts, however. A PM receiver obviously must respond to phase changes, but so will an FM receiver, because, as we have seen, phase shifts and frequency shifts always occur together. Therefore, phase shifts due to noise will be associated with frequency shifts that will be interpreted by the receiver as part of the modulation.

Figure 8.12 shows the situation at the input to the receiver. The circle represents the fact that the noise phasor will have a constantly changing angle with respect to the signal. Its greatest effect, and thus the peak phase shift to the signal, will occur when the noise phasor is perpendicular to the resultant, E_R. At that time, the phase shift due to noise will be

$$\phi_N = \sin^{-1} (E_N/E_S) \tag{8.17}$$

where E_N/E_S is the reciprocal of the voltage signal-to-noise ratio at the input. A little care is needed here, as S/N is usually given as a power ratio, in decibels, and will have to be converted to a voltage ratio before being used in Equation (8.17).

Equation (8.17) can be simplified as long as the signal is much larger than the noise. This will cause the phase deviation to be small, and for small

FIGURE 8.11   **Removal of Amplitude Noise Component by Limiting**

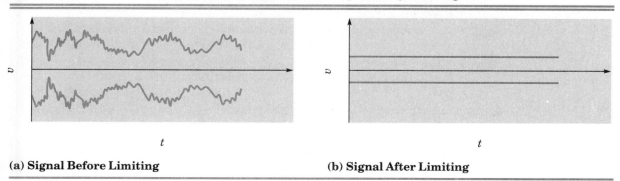

(a) Signal Before Limiting   **(b) Signal After Limiting**

FIGURE 8.12

Phase Shift due to Noise

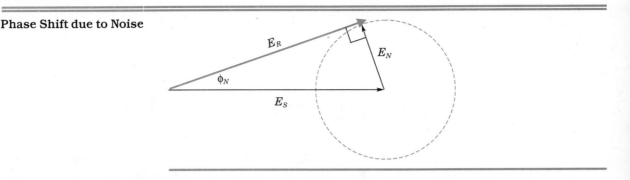

angles the sine of the angle is approximately equal to the angle itself, in radians. Thus in a practical situation we can use

$$\phi_N \approx E_N / E_S \tag{8.18}$$

The effect of noise can be reduced by making the signal voltage, relative to the noise voltage, as large as possible. Figure 8.12 demonstrates this. This, of course, requires increased transmission power, a better receiver noise figure, or both. Perhaps less obvious is the fact that the relative importance of phase shifts due to noise can be reduced by having the phase shifts in the signal as large as possible. This is accomplished by keeping the value of m_f high, since m_f represents the peak phase shift in radians. It would seem that the ratio of signal voltage to noise voltage at the output would be proportional to m_f, and this is approximately true for wideband FM under strong-signal conditions.

EXAMPLE 8.9

An FM signal has a frequency deviation of 5 kHz and a modulating frequency of 1 kHz. The signal-to-noise ratio at the input to the receiver detector is 20 dB. Calculate the approximate signal-to-noise ratio at the detector output.

Solution

First, notice the word "approximate" above. Our analysis is obviously a little simplistic, since noise exists at more than one frequency. We are also going to assume that the detector is completely unresponsive to amplitude variations, and that it adds no noise of its own. Our results will not be precise, but they will show the process that is involved.

Let us first convert 20 dB to a voltage ratio:

$$E_S / E_N = \log^{-1} \frac{(S/N) \ (\text{dB})}{20}$$

$$= \log^{-1} \frac{20}{20}$$

$$= 10$$

$$E_N/E_S = 1/10$$
$$= 0.1$$

Since $E_S \gg E_N$, we can use Equation (8.18).

$$\phi \approx E_N/E_S$$
$$= 0.1 \text{ rad}$$

Remembering that the receiver will interpret the noise as an FM signal with a modulation index equal to ϕ_N, we find

$$m_{fN} = 0.1$$

This can be converted into an equivalent frequency deviation δ_N due to the noise.

$$\delta_N = m_f f_m$$
$$= 0.1 \times 1 \text{ kHz}$$
$$= 100 \text{ Hz}$$

The frequency deviation due to the signal is given as 5 kHz, and the receiver output voltage will be proportional to the deviation. Therefore the output S/N as a voltage ratio will be equal to the ratio between the deviation due to the signal and that due to the noise.

$$(E_S/E_N)_o = \delta_S/\delta_N$$
$$= 5 \text{ kHz}/100 \text{ Hz}$$
$$= 50$$

Since S/N is nearly always expressed in decibels, it would be well to change this to decibels.

$$(S/N)_o \text{ (dB)} = 20 \log 50$$
$$= 34 \text{ dB}$$

This is an improvement of 14 dB over the S/N at the input.

8.6.1 The Threshold Effect and the Capture Effect

An FM signal can produce a better signal-to-noise ratio at the output of a receiver than an AM signal with a similar input S/N, but this is not always the case. The superior noise performance of FM depends on there being a sufficient input S/N ratio. There exists a threshold S/N below which the performance is no better than AM. In fact, it will actually be worse, because the greater bandwidth of the FM signal requires a wider receiver noise bandwidth.

The FM **threshold effect** can easily be observed by driving away from an FM broadcast transmitter while listening to it on the car radio. The recep-

FIGURE 8.13

S/N Improvement with
Wideband FM

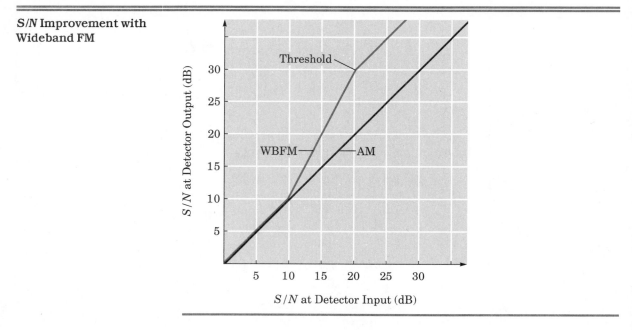

tion remains essentially noise-free for a considerable distance, then degrades
very quickly as the limit of the station's coverage is reached. Shortly before
reception of the station is lost completely, there will be some very loud clicks
in the audio. This effect is caused by noise which causes a rapid 360° phase
shift in the signal. AM reception, by contrast, will degrade gradually and continuously as the signal strength is reduced.

When the signal strength is above the threshold, the noise performance
of FM can be more than 20 dB better than for AM, as shown in Figure 8.13.
The numbers in this figure are approximate: the exact values of the threshold
and *S/N* improvement depend on the modulation index. Because wideband
FM allows more noise to enter the receiver than narrowband FM, the threshold is higher for WBFM. For signal strengths above the threshold, the performance of WBFM is superior to that of NBFM; in fact, the *S/N* improves with
the square of the modulation index.

The analysis of noise that was conducted in the previous section would
apply equally well to an interfering signal. As long as the desired signal is
considerably stronger than the interference, the ratio of desired to interfering
signal strength will be greater at the output of the detector than at the input.
We could say that the stronger signal *captures* the receiver, and in fact this
property of FM is usually called the **capture effect**. It is very easy to demonstrate with any FM system. For example, it is the reason that there is less
interference between cordless telephones, which all share a few channels in
the 46–49 MHz bands, than one might expect.

8.6.2 Pre-emphasis and De-emphasis

The phase component of noise will create phase modulation of the signal.
Assuming that the noise power is evenly distributed across the channel at

the receiver input (*white noise*), the phase modulation will be evenly distributed over the receiver bandwidth. With a phase modulation system, the resulting noise after demodulation would also be evenly distributed over the baseband spectrum.

With an FM receiver, the situation is different. The phase modulation due to noise is interpreted as frequency modulation by the receiver. As we have already discovered, phase and frequency deviation are related by Equation (8.7):

$$m_f = \frac{\delta}{f_m}$$

which can be restated as

$$\delta = m_f f_m$$

Remember that the modulation index m_f is simply the peak phase deviation in radians. The frequency deviation is proportional to the modulating frequency. This tells us that, if the phase deviation due to thermal noise is randomly distributed over the baseband spectrum, then the amplitude of the demodulated noise will be proportional to frequency. This relationship between noise voltage and frequency is shown in Figure 8.14. Since power is proportional to the square of voltage, the noise power will have the parabolic spectrum that is also shown in Figure 8.14.

In order to improve the noise performance of an FM system, it is useful to increase the deviation. There are limits to how much this can be done, because of considerations of channel width and threshold carrier-to-noise ratio. However, the foregoing shows that large deviation is more important for high modulating frequencies. Unfortunately, for most common analog signals (including voice, music, and video signals), the high-frequency components generally have lower amplitude than low and medium frequencies. An

FIGURE 8.14

Spectrum of Demodulated Noise in an FM System (without De-emphasis)

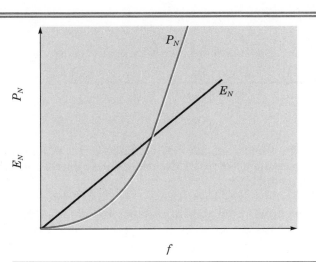

FM Stereo Broadcasting

FM is now the preeminent mode of local radio broadcasting because of its better performance compared with AM. In part, this enhanced performance is due to the S/N improvement inherent in FM, but there are other advantages that are unrelated to the modulation scheme. The higher carrier frequencies used with FM broadcasting allow it to avoid nighttime interference from distant stations, which often ruins AM reception. The FM broadcast system, as implemented, allows a wider audio frequency response as well.

Another factor that established FM as the preferred broadcast medium for music was the availability of FM stereo. AM stereo is a relative newcomer that so far shows little sign of challenging the predominance of FM for music programming.

FM stereo broadcasting, using the present multiplexing system, began in 1961. Before that time, and even after, there were a number of experimental stereo broadcasts using two stations, one for each channel. Most used one FM and one AM station, since few households had more than one FM receiver. This was an interesting novelty but, besides the inherent waste of spectrum space, the differences in the two transmission systems and the two receivers usually resulted in stereo sound that would be rather disconcerting to listeners today. Some 1960s vintage AM/FM tuners can still be found that attempt to make the best of a bad idea by incorporating separate tuning dials and separate outputs so that AM and FM can be heard on separate channels of a stereo system. At least in this case, the audio amplifiers and speakers would be the same for both channels.

The current system allows high fidelity by the standards of the early 1960s. Present-day audiophiles would prefer a wider frequency range and a better S/N. The only way now to get a wider frequency range would be to move the pilot carrier and the L-R signal to a higher frequency. Since that would make all current FM receivers obsolete, such a move is unlikely. An attempt to improve the S/N by using DOLBY™ noise reduction was made in the 1970s, but since the result was only partially compatible with ordinary receivers, the system never became popular.[1] Interestingly, the new television stereo sound system does incorporate noise reduction, using the dBx™ system, for the L-R signal only.[2] More about that in Chapter 14.

[1] "DOLBY" is a trademark of Dolby Laboratories Licensing Corporation.
[2] "dBx" is a trademark of dBx Technology Licensing.

improvement in S/N can be made by boosting (*pre-emphasizing*) these high frequencies before modulation, with a corresponding cut in the receiver after demodulation.

Obviously, it is necessary to use similar filter characteristics for **pre-emphasis** and **de-emphasis**. In FM broadcasting, for instance, a 75 μs standard is used. The number refers to the time constants of the high-pass filter at the transmitter and the low-pass filter at the receiver. This gives a

FIGURE 8.15   **Typical Pre-emphasis and De-emphasis Circuits**

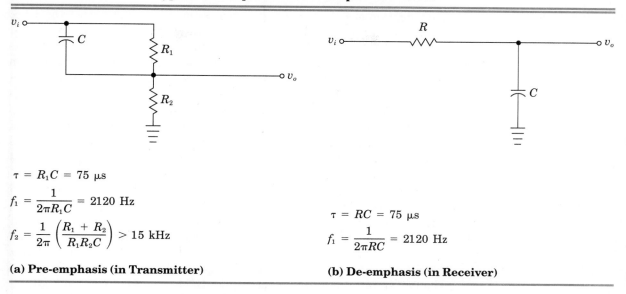

$\tau = R_1C = 75\ \mu\text{s}$

$f_1 = \dfrac{1}{2\pi R_1 C} = 2120\ \text{Hz}$

$f_2 = \dfrac{1}{2\pi}\left(\dfrac{R_1 + R_2}{R_1 R_2 C}\right) > 15\ \text{kHz}$

(a) Pre-emphasis (in Transmitter)

$\tau = RC = 75\ \mu\text{s}$

$f_1 = \dfrac{1}{2\pi RC} = 2120\ \text{Hz}$

(b) De-emphasis (in Receiver)

turnover frequency of 2.12 kHz. Depending on the program material, pre-emphasis can make a considerable difference. The improvement in *S/N* is about 12 dB for FM broadcasting.

Figure 8.15 gives examples of simple circuits that will provide pre-emphasis and de-emphasis, and Figure 8.16 shows the frequency response of these circuits.

FIGURE 8.16   **Frequency Response of Pre-emphasis and De-emphasis Circuits**

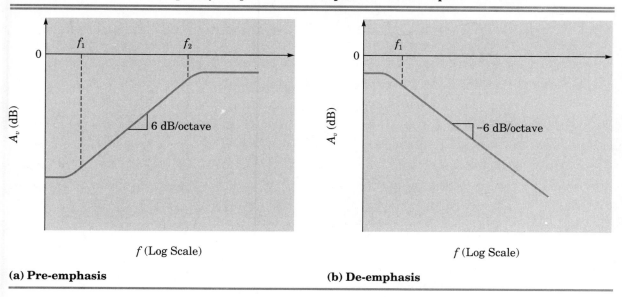

(a) Pre-emphasis

(b) De-emphasis

8.7 FM STEREO

The introduction of FM stereo in 1961 was accomplished in such a way as to maintain compatibility with the monaural system already in use. It was essential that the mono receiver should reproduce, not the left channel or the right, but a combination of the two, formed by adding the two signals to get a left-plus-right (L+R) signal. It was also required that the existing 200 kHz channel spacing be preserved. This was accomplished by using a form of frequency-division multiplexing, modified to maintain compatibility.

Figure 8.17(a) shows the baseband spectrum of an FM stereo signal. This is the signal that is applied to the transmitter. It is also the signal that appears at the output of the receiver detector. For comparison, Figure 8.17(b) shows the baseband of a monaural FM broadcast transmission.

The mono L+R signal occupies the frequencies from about 50 Hz to 15 kHz in the stereo signal. This maintains the compatibility spoken of above, since the monaural receiver will ignore the rest of the stereo signal. Even if frequencies above 15 kHz make their way through the audio amplifier, they are not likely to be reproduced by the loudspeaker; even if they are, 19 kHz is at or outside the limits of hearing for most people.

The two stereo channels can be reproduced at the receiver if, in addition to the sum L+R signal, a difference signal L−R is transmitted. This signal is sent using DSBSC AM, at a (suppressed) **subcarrier** frequency of 38 kHz. The DSBSC AM requires a pilot carrier for proper detection; this is sent at half the subcarrier frequency, at 19 kHz. The pilot carrier causes a 10% (7.5 kHz) deviation of the main carrier frequency.

FIGURE 8.17

FM Broadcasting: Baseband Spectra

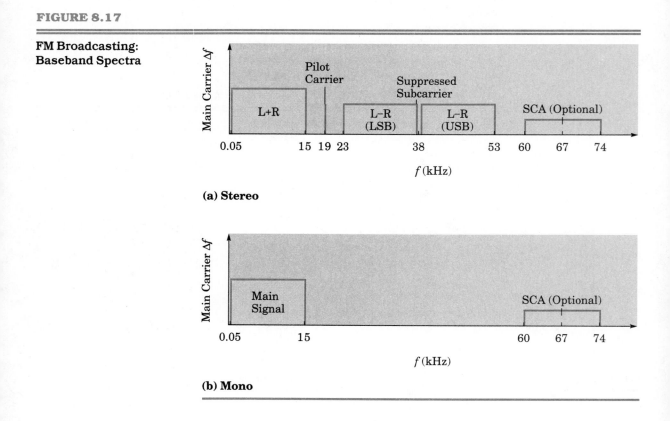

(a) Stereo

(b) Mono

It might be expected that a stereo FM signal would have a much wider bandwidth than would the monaural version. This is what would be predicted by Carson's rule. The maximum deviation in both cases is 75 kHz, but the maximum baseband frequency is 15 kHz for monaural, and 53 kHz for stereo FM. Carson's rule would predict, for monaural FM, a bandwidth of

$$B = 2(\delta_{max} + f_{m\ (max)})$$
$$= 2(75\ \text{kHz} + 15\ \text{kHz})$$
$$= 180\ \text{kHz}$$

For stereo, the prediction from Carson's rule would be

$$B = 2(\delta_{max} + f_{m\ (max)})$$
$$= 2(75\ \text{kHz} + 53\ \text{kHz})$$
$$= 256\ \text{kHz}$$

In practice, the increase is not as drastic. It is unlikely that maximum frequency deviation will result from $L-R$ signals with maximum frequency. The $L-R$ signal will usually have lower amplitude than the $L+R$ signal, and, in any case, frequencies at the upper end of the audio range are likely to have low amplitude, even after pre-emphasis has been applied.

There is a noise penalty with stereo FM. Pre-emphasis is applied to the left and right signals. This is effective with the $L+R$ signal, but the $L-R$ signal does not get the full advantage of the pre-emphasis, as is made clear in Figure 8.18. The upper sideband receives some benefit from pre-emphasis, but the pre-emphasis actually works in reverse on the lower sideband. Matters are made even worse by the fact that the $L-R$ signal has been shifted upward in frequency before modulation onto the main FM carrier. The noise voltage in the demodulated signal is proportional to frequency, as we have seen. This means that the $L-R$ signal will be noisier than the $L+R$ signal. When the DSBSC $L-R$ signal is demodulated, this noise will be shifted down in frequency into the audible range. The result of all this is an S/N degradation of about 22 dB for stereo FM compared to the monaural signal. The dif-

FIGURE 8.18   **Noise Performance of FM Stereo**

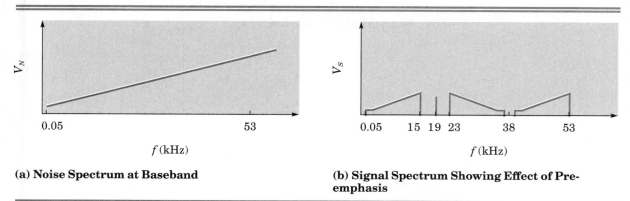

(a) Noise Spectrum at Baseband

(b) Signal Spectrum Showing Effect of Pre-emphasis

ference can easily be heard by tuning in a relatively weak station, then switching a stereo receiver between stereo and monaural reception. In fact, many receivers switch automatically to monaural mode in the presence of weak signals.

There is one part of the baseband signal shown in Figure 8.17(a) that has not yet been explained. The Subsidiary Carrier Authorization (SCA) signal, centered on a subcarrier at 67 kHz, is used for services such as background music for stores and offices. The SCA signal is monaural, with a maximum audio frequency of 7 kHz. The subcarrier can be modulated using either AM or (usually) NBFM, but the total deviation of the main carrier due to the SCA subcarrier is limited to 10% (7.5 kHz). Not all stations use SCA.

8.8 TEST EQUIPMENT AND PROCEDURES: MEASUREMENT OF FM INDEX AND DEVIATION

The maximum frequency deviation of FM transmitters is restricted by law, not by any physical constraint (as is the AM index). Setting a particular value of deviation is not quite as simple as determining 100% modulation from an oscilloscope display of an AM signal. In fact, oscilloscope displays of FM signals are not very enlightening at all. There is no envelope variation, and the frequency variations are usually such a small percentage of the carrier frequency that they are difficult to see—let alone measure—with an oscilloscope.

The spectrum analyzer is a more useful instrument for the observation of FM signals. It allows the power in the carrier and each sideband to be measured, and it allows the modulating frequency to be found by measuring the separation between sidebands. At least this is true for single-tone modulation; with several modulating frequencies, the pattern of sidebands becomes very confusing, as each tone produces multiple sets of sidebands.

It would, in theory, be possible to measure the value of m_f for single-tone modulation by measuring the amplitude of each sideband pair and comparing them with a table of Bessel functions. This is tedious and seldom attempted. There are several values of m_f, however, which are very easily recognized. These are the values where the term J_0 is equal to zero. As m_f increases from zero, this happens for m_f equal to 2.4, 5.5, 8.65, 11.8, and 14.9, to list the first few points. At each of these values of m_f, the carrier-frequency component of the spectrum disappears, an occurrence that is easy to observe on a spectrum analyzer. This provides several reference points for measuring m_f.

Usually, in adjusting a transmitter, for instance, it is deviation rather than modulation index that must be measured. This can be done using the relation between deviation, modulation index, and modulating frequency given in Equation (8.7):

$$m_f = \frac{\delta}{f_m}$$

An example will show how this is done. The method is often called the *Bessel-zero method*, for obvious reasons.

EXAMPLE 8.10

Show how the deviation of a voice-communications FM transmitter can be set to 5 kHz using the Bessel-zero method with a spectrum analyzer.

Solution

Set up the equipment as in Figure 8.19. The frequency counter is not an absolute requirement, but is recommended because the accuracy of the deviation measurement depends on the accuracy with which the modulating frequency can be set. The dial calibrations on many function generators are not very accurate.

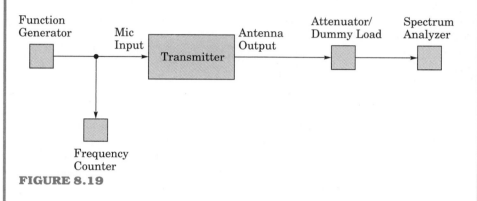

FIGURE 8.19

Since the transmitter is designed for voice frequencies, it is necessary to find a frequency within the range of 300 Hz to 3 kHz, approximately, that will produce a carrier null for a deviation of 5 kHz. This can be done by trial and error, starting with the first null for $m_f = 2.4$. It is easy to rearrange Equation (8.7) to find f_m.

$$m_f = \frac{\delta}{f_m}$$

$$f_m = \frac{\delta}{m_f}$$

$$= \frac{5 \text{ kHz}}{2.4}$$

$$= 2.08 \text{ kHz}$$

This frequency is within the acceptable range. If it had not been, the next carrier null, for $m_f = 5.5$, could have been tried.

The function generator is set to 2.08 kHz, and the deviation is increased gradually from zero until the carrier disappears for the first time. The deviation is now 5 kHz.

This method of measuring deviation is elegant and simple, but it does require an expensive spectrum analyzer and complete control over the modulating frequency. Thus, it is unsuitable for continuous monitoring of trans-

FIGURE 8.20   **FM Deviation Meter**

Courtesy Belar Electronics Laboratory, Inc.

mitter operation. For that purpose, the fact that the output of an FM detector is proportional to deviation can be used. A detector can have a meter attached and then be calibrated (using a spectrum analyzer, for instance). The result is an instrument that can provide instant readings, with reasonable accuracy, for any type of modulating signal. Figure 8.20 shows an example of this type of instrument, which is used for continuous monitoring of FM broadcast transmitters.

SUMMARY OF CHAPTER 8

Here are the main points to remember from this chapter.

1. Angle modulation includes frequency and phase modulation, which are closely related.
2. FM is widely used for analog communications, while PM sees greatest application in data communications.
3. The power of an angle modulation signal does not change with modulation, but the bandwidth increases due to the generation of multiple sets of sidebands.
4. The voltage and power of each sideband can be calculated using Bessel functions. An approximate bandwidth is given by Carson's rule.
5. FM has a significant advantage compared with AM in the presence of noise or interference, provided the deviation is relatively large and the signal is reasonably strong.
6. The S/N for FM can be improved considerably by using pre-emphasis and de-emphasis. This involves greater gain for the higher baseband frequencies before modulation, with a corresponding reduction after demodulation.
7. Stereo FM transmission is accomplished using a multiplexing scheme that preserves compatibility with monaural receivers and requires only a slight increase in bandwidth. A noise penalty is paid with stereo FM, however.

> **8.** A spectrum analyzer is much more useful than an oscilloscope for the analysis of FM signals. It can also be used to calibrate other FM test equipment, such as deviation meters.

IMPORTANT EQUATIONS

$$k_f = \frac{\Delta f}{e_m} \tag{8.1}$$

$$f_{sig}(t) = f_c + k_f e_m(t) \tag{8.2}$$

$$f_{sig}(t) = f_c + k_f E_m \sin \omega_m t \tag{8.4}$$

$$\delta = k_f E_m \tag{8.5}$$

$$f_{sig}(t) = f_c + \delta \sin \omega_m t \tag{8.6}$$

$$m_f = \frac{\delta}{f_m} \tag{8.7}$$

$$f_{sig}(t) = f_c + m_f f_m \sin \omega_m t \tag{8.8}$$

$$k_p = \frac{\phi}{e_m} \tag{8.9}$$

$$\theta(t) = \theta_c + k_p e_m(t) \tag{8.10}$$

$$\theta(t) = \theta_c + m_p \sin \omega_m t \tag{8.12}$$

$$\begin{aligned}
v(t) &= A \sin (\omega_c t + m \sin \omega_m t) \\
&= A\{J_0(m) \sin \omega_c t \\
&\quad - J_1(m)[\sin (\omega_c - \omega_m)t - \sin (\omega_c + \omega_m)t] \\
&\quad + J_2(m)[\sin (\omega_c - 2\omega_m)t + \sin (\omega_c + 2\omega_m)t] \\
&\quad - J_3(m)[\sin (\omega_c - 3\omega_m)t + \sin (\omega_c + 3\omega_m)t] \\
&\quad + \cdots\}
\end{aligned} \tag{8.14}$$

$$P_T = J_0^2 + 2J_1^2 + 2J_2^2 + \cdots \tag{8.15}$$

$$B = 2(\delta_{max} + f_{m\ (max)}) \tag{8.16}$$

$$\phi = \sin^{-1} (E_N/E_S) \tag{8.17}$$

$$\phi_n \approx E_N/E_S \tag{8.18}$$

GLOSSARY

angle modulation   general term that includes frequency and phase modulation

capture effect   ability of an FM receiver to receive the stronger of two signals, ignoring the weaker

de-emphasis   low-pass filter in a receiver to remove the effect of pre-emphasis on the frequency response

frequency deviation   amount by which the frequency of an FM signal shifts to each side of the carrier frequency

modulation index   in FM and PM, the peak amount in radians by which the phase of a signal deviates from its resting value

narrowband FM (NBFM)   FM with a relatively low modulation index

pre-emphasis   high-pass filter used in an FM transmitter to improve the signal-to-noise ratio. Always used with de-emphasis at the receiver

subcarrier   secondary carrier that can carry an additional modulating signal and is itself modulated onto the main carrier

threshold effect   noise-reduction effect which occurs with strong FM signals

wideband FM (WBFM)   FM with a relatively large modulation index

QUESTIONS

1. What two types of modulation are included in the term *angle modulation*?
2. What is the relationship between phase and frequency for a sine wave?
3. Define and compare the modulation index for FM and PM.
4. Compare, in general terms, the bandwidth and signal-to-noise ratio of FM and AM.
5. Describe and compare two ways to determine the practical bandwidth of an FM signal.
6. What is pre-emphasis, and how is it used to improve the *S/N* of FM transmissions?
7. For FM, what characteristic of the modulating signal determines the instantaneous frequency deviation?
8. What is the capture effect?
9. Draw a diagram showing the baseband spectrum of an FM stereo signal, and explain why this system is used.
10. What is the purpose of the pilot carrier in an FM stereo signal?
11. Where is PM used?
12. Explain why the signal-to-noise ratio of FM can increase with the bandwidth. Is this always true for FM? Compare with the situation for AM.
13. Compare the effects of modulation on the carrier power and the total signal power in FM and AM.
14. What is the threshold effect?
15. Why is the *S/N* for FM stereo transmissions worse than for mono?
16. Explain how limiting reduces the effect of noise on FM signals.
17. Explain how noise affects FM signals even after limiting.
18. Explain the fact that there is no simple relationship between modulating frequency and bandwidth for an FM signal.
19. Define NBFM in two different ways, and explain the use of each.
20. Why does limiting the receiver bandwidth to less than the signal bandwidth cause more problems with FM than with AM?

PROBLEMS

Section 8.2

21. An FM signal has a deviation of 10 kHz and a modulating frequency of 2 kHz. Calculate the modulation index.

22. Calculate the frequency deviation for an FM signal with a modulating frequency at 5 kHz and a modulation index of 2.

23. A sine-wave carrier at 100 MHz is modulated by a 1 kHz sine wave. The deviation is 100 kHz. Draw a graph showing the variation of instantaneously modulated signal frequency with time.

24. An FM modulator has k_f = 50 kHz/V. Calculate the deviation and modulation index for a 3 kHz modulating signal of 2 V (RMS).

Section 8.3

25. A phase modulator with k_p = 3 rad/V is modulated by a sine wave with an RMS voltage of 4 V at a frequency of 5 kHz. Calculate the phase modulation index.

26. A PM signal has a modulation index of 2, with a modulating signal that has an amplitude of 100 mV and a frequency of 4 kHz. What would be the effect on the modulation index of:

    (a) changing the frequency to 5 kHz?
    (b) changing the voltage to 200 mV?

Section 8.4

27. A sine wave of frequency 1 kHz phase-modulates a carrier at 123 MHz. The peak phase deviation is 0.5 rad. Calculate the maximum frequency deviation.

28. What is the maximum phase deviation that can be present in an FM radio broadcast signal, assuming it transmits a baseband frequency range of 50 Hz to 15 kHz, with a maximum deviation of 75 kHz?

Section 8.5

29. An FM signal has a deviation of 10 kHz, and is modulated by a sine wave with a frequency of 5 kHz. The carrier frequency is 150 MHz, and the signal has a total power of 12.5 W, operating into an impedance of 50 Ω.

    (a) What is the modulation index?
    (b) How much power is present at the carrier frequency?
    (c) What is the voltage level of the second sideband below the carrier frequency?
    (d) What is the bandwidth of the signal, ignoring all components that have less than 1% of the total signal voltage?

30. An FM transmitter operates with a total power of 10 W, a deviation of 5 kHz, and a modulation index of 2.

    (a) What is the modulating frequency?
    (b) How much power is transmitted at the carrier frequency?
    (c) If a receiver has a bandwidth sufficient to include the carrier and the first two sets of sidebands, what percentage of the total signal power will it receive?

31. An FM transmitter has a carrier frequency of 220 MHz. Its modulation index is 3 with a modulating frequency of 5 kHz. The total power output is 100 W into a 50 Ω load.

    (a) What is the deviation?

(b) Sketch the spectrum of this signal, including all sidebands with more than 1% of the signal voltage.

(c) What is the bandwidth of this signal, according to the criterion used in Part (b)?

(d) Use Carson's rule to calculate the bandwidth of this signal, and compare with the result found in Part (c).

32. An FM transmitter has a carrier frequency of 160 MHz. The deviation is 10 kHz and the modulation frequency is 2 kHz. A spectrum analyzer shows that the carrier-frequency component of the signal has a power of 5 W. What is the total signal power?

33. Use Carson's rule to compare the bandwidth that would be required to transmit a baseband signal with a frequency range from 300 Hz to 3 kHz using:

(a) NBFM with maximum deviation of 5 kHz

(b) WBFM with maximum deviation of 75 kHz

Section 8.6

34. An FM receiver operates with an S/N of 30 dB at its detector input, and is operating with $m_f = 10$.

(a) If the received signal has a voltage of 10 mV, what is the amplitude of the noise voltage?

(b) Find the maximum phase shift that could be given to the signal by the noise voltage.

(c) Calculate the S/N at the detector output, assuming the detector is completely insensitive to amplitude variations.

35. The circuits shown in Figure 8.21 are 75 μs pre-emphasis and de-emphasis networks. Identify which is which, and calculate values for the missing components.

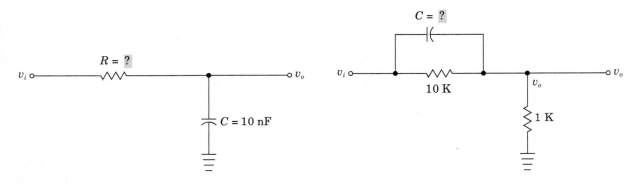

FIGURE 8.21

Section 8.7

36. Sketch the baseband spectrum for each of the following stereo FM signals. A frequency scale is required, but an amplitude scale is not.

(a) A 1 kHz sine wave, of equal amplitude and phase, on both channels.

(b) As in Part (a), except that the left and right channels are 180° out of phase.

(c) A 1 kHz sine wave on the left channel, no signal on the right channel.

Section 8.8

37. Using a function generator and spectrum analyzer, how could the deviation of an FM generator be set to 75 kHz, to simulate a broadcast FM signal?

Comprehensive

38. A certain full-carrier DSB AM signal has a bandwidth of 20 kHz. What would be the approximate bandwidth required if the same information signal were to be transmitted using:

    (a) DSBSC AM
    (b) SSB
    (c) FM with 10 kHz deviation
    (d) FM with 50 kHz deviation

39. Demonstrate, using the table of Bessel functions, that the total power in an FM signal is equal to the power in the unmodulated carrier, for $m = 2$. Compare with the situation for full-carrier AM and for SSB AM.

40. Suppose you were called upon to recommend a modulation technique for a new communication system for voice frequencies. State which of the techniques studied so far you would recommend, and why, in each of the following situations:

    (a) simple, cheap receiver design is of greatest importance
    (b) narrow signal bandwidth is of greatest importance
    (c) immunity to noise and interference is of greatest importance

9

FM Equipment

Objectives

After studying the material in this chapter, you should be able to:

1. Describe the ways in which FM transmitters and receivers differ from those for AM, and explain why they differ.
2. Explain the difference between direct- and indirect-FM generation, and draw block diagrams for both systems.
3. Analyze the operation of both direct and indirect frequency modulators.
4. Explain the use of frequency multipliers and mixers in FM transmitters.
5. Calculate the carrier frequency and deviation for FM transmitters.
6. Analyze and compare the operation of each of the following types of FM detector: discriminator, ratio detector, phase-locked loop, and quadrature.
7. Explain the reasons for the use of limiters in FM receivers, and describe the operation of a limiter.
8. Explain the operation of FM transceivers, including those with split-frequency operation.
9. Perform sensitivity and channel-separation measurements on FM receivers.
10. Describe the operation of FM stereo transmitters and receivers.

9.1 INTRODUCTION

Now that you are familiar with the basics of FM signals, it is time to consider how such signals are created and employed. The fact that both transmitters and receivers are covered in a single chapter is certainly not intended to imply that these circuits are of less importance than those for AM. If anything, the contrary is true. However, there are many similarities between transmitters for different modulation types, and the same is true for receivers. In this chapter, it will be necessary to describe only the differences between FM and AM equipment.

The processes of modulation and demodulation are very different for FM than for AM. In addition, there is more variety in the circuits used for the modulation and demodulation of FM than there is for AM. Consequently, most of this chapter will actually be devoted to modulators and demodulators, and their associated circuitry.

9.2 FM TRANSMITTERS

The differences between FM and AM transmitters are easier to understand if we start by considering the differences between FM and AM signals. To begin, FM requires the transmitter frequency to be varied. That in turn implies that the modulation must be applied early, probably at the carrier-oscillator stage. (The "probably" is intended to leave room for *indirect* FM, to be described shortly.)

Another difference is that FM signals have no amplitude variations. This means that Class C amplifiers can be used with FM signals. Since Class C cannot be used to amplify AM signals, most AM transmitters use high-level modulation, which takes place in the final stage of power amplification. High-level modulation allows Class C to be used for all RF amplifier stages. FM transmitters can employ Class C throughout, even after modulation.

Frequency multipliers are used in AM transmitters, before the modulator stage, when very high carrier frequencies are required. Multiplier stages can be used for the same reason in FM transmitters, but they have another function as well: frequency multiplication of an FM signal also multiplies the deviation. Since the amount of frequency deviation that can be achieved with some types of FM modulators is quite small, frequency multiplication is often used as a way of increasing deviation.

Figure 9.1 is a simplified block diagram of a typical FM transmitter. There are many variations on this, of course, and some of them will be described in the following sections. For instance, frequency multiplication is not always necessary, depending on the modulator design and the amount of deviation required.

You will notice that the transmitter in Figure 9.1 is divided into two sections, the *exciter* and the *power amplifier*. This is typical of large transmitters such as those used in broadcasting, where the exciter is a separate unit containing the audio and low-power RF stages. The power amplifier stages—involving high voltages, large amounts of power, forced-air or water cooling, and sometimes vacuum-tube technology—are in a separate cabinet. Small transmitters like those used in mobile communications are, of course, self-contained. Those in current production are completely solid-state.

Typical FM Transmitter

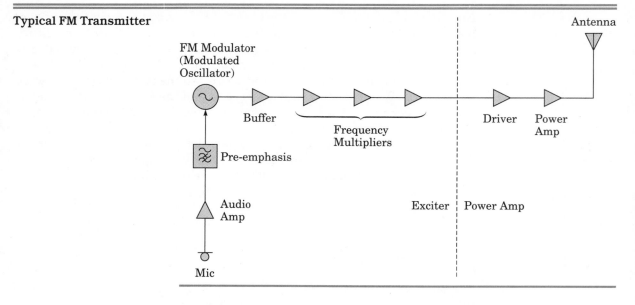

9.2.1 Direct-FM Modulators

FM signals can be generated either *directly*, by varying the frequency of the carrier oscillator, or *indirectly*, by converting phase modulation to frequency modulation. Both techniques are in common use. In this section, we discuss techniques for **direct FM**; **indirect FM** will be covered later in the chapter.

Direct FM requires that the frequency of the carrier oscillator be varied in accordance with the instantaneous amplitude of the modulating signal (after any required pre-emphasis has been added). The *reactance modulator* is the simplest way to do this. It works by using the modulating signal to vary a reactance in the frequency-determining circuit. One common way to build a reactance modulator is to put a varactor into the frequency-determining circuit of the carrier oscillator. Figure 9.2 shows two simple circuits: in Figure 9.2(a), an LC oscillator is used, while that in Figure 9.2(b) is crystal-controlled. In both circuits, the varactor is reverse-biased, and the bias is varied by the modulating signal.

The circuit in Figure 9.2(a) should be familiar from Chapter 2 as a Clapp oscillator. The frequency is determined mainly by L_1 and the varactor. Capacitor C_3 has a large value, and exists to isolate the varactor from V_{CC}; C_1 and C_2 determine the feedback fraction, and have some small effect on the frequency. The tuning potentiometer adjusts the dc bias on the varactor. To this is added the ac modulating signal, causing the bias voltage, and therefore the frequency, to vary around its nominal value.

Quite large values of frequency deviation are possible with varactor modulation of an LC oscillator. In fact, there are broadcast FM transmitter designs that generate the FM signal at the carrier frequency, achieving the full 75 kHz of deviation without any frequency multiplication at all. On the other hand, an LC oscillator at VHF frequencies will be quite unstable, and will require some form of **automatic frequency control (AFC)**.

As for the crystal-controlled version in Figure 9.2(b), its frequency can be changed only very slightly by modulation: usually only a few tens of hertz. A

FIGURE 9.2

Direct FM Varactor
Modulators

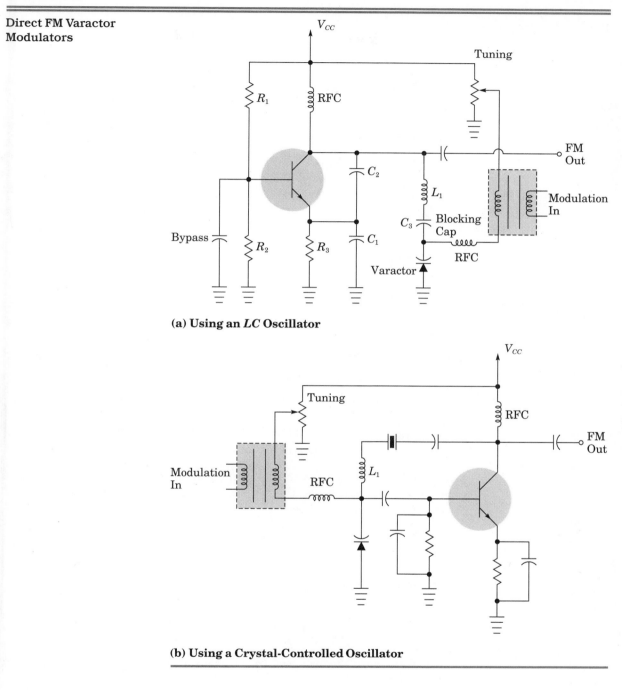

(a) Using an *LC* Oscillator

(b) Using a Crystal-Controlled Oscillator

great deal of frequency multiplication will always be needed with crystal-controlled direct-FM modulators.

An older type of reactance modulator that is similar in principle (though not in circuit) to the varactor modulator is shown in Figure 9.3. A transistor is connected across the frequency-determining tuned circuit, and is connected so that it acts as a capacitive reactance across the circuit. A JFET is shown, but bipolar transistors (and tubes, for that matter) have also been

FIGURE 9.3

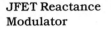

**JFET Reactance
Modulator**

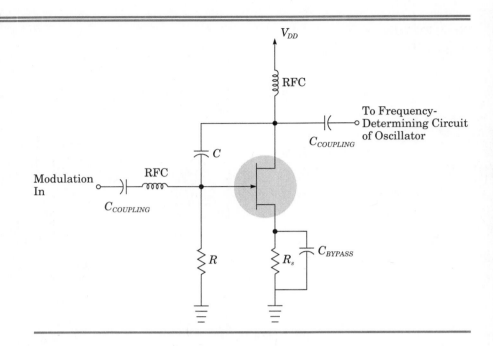

used. It can be shown that the transistor, connected in this way, appears as a capacitance with a value equal to

$$C_{eq} = g_m RC \tag{9.1}$$

where C_{eq} = equivalent capacitance between drain and source
       g_m = transconductance of the JFET

and R and C are as shown in Figure 9.3. The modulating signal varies the gate voltage, which changes the transconductance of the transistor, and thus the capacitive reactance. Recall that for a JFET,

$$g_m = g_{m_0}\left(1 - \frac{V_{gs}}{V_P}\right) \tag{9.2}$$

where g_m = transconductance as a function of V_{gs}
      V_{gs} = gate-source voltage
      g_{m_0} = transconductance for $V_{gs} = 0$
      V_P = pinch-off voltage

Another, more modern way to generate direct FM is to use either an integrated-circuit VCO or an oscillator chip with an external varactor. For low-power applications such as cordless phones, it is even possible to get an entire transmitter on a chip. The MC2833, shown in Figure 9.4, is an example.

9.2.2 Frequency Multipliers

Frequency multipliers were discussed in Chapter 2. You will recall that they are essentially Class C amplifiers with the output circuit tuned to a harmonic

FIGURE 9.4   An IC Transmitter: The MC2833

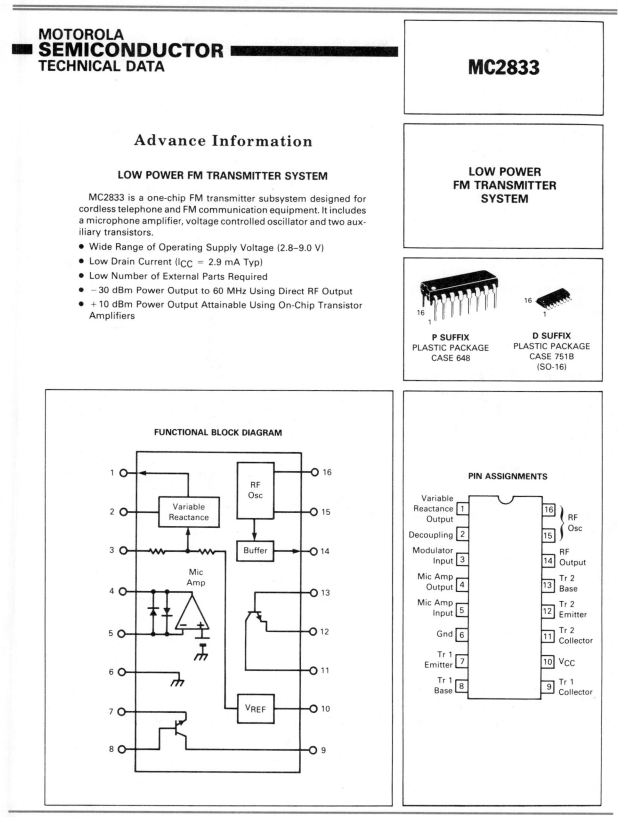

MOTOROLA
SEMICONDUCTOR
TECHNICAL DATA

MC2833

Advance Information

LOW POWER FM TRANSMITTER SYSTEM

MC2833 is a one-chip FM transmitter subsystem designed for cordless telephone and FM communication equipment. It includes a microphone amplifier, voltage controlled oscillator and two auxiliary transistors.

- Wide Range of Operating Supply Voltage (2.8–9.0 V)
- Low Drain Current (I_{CC} = 2.9 mA Typ)
- Low Number of External Parts Required
- − 30 dBm Power Output to 60 MHz Using Direct RF Output
- + 10 dBm Power Output Attainable Using On-Chip Transistor Amplifiers

LOW POWER
FM TRANSMITTER
SYSTEM

P SUFFIX
PLASTIC PACKAGE
CASE 648

D SUFFIX
PLASTIC PACKAGE
CASE 751B
(SO-16)

FUNCTIONAL BLOCK DIAGRAM

PIN ASSIGNMENTS

1 — Variable Reactance Output
2 — Decoupling
3 — Modulator Input
4 — Mic Amp Output
5 — Mic Amp Input
6 — Gnd
7 — Tr 1 Emitter
8 — Tr 1 Base

16 — RF Osc
15 — RF Osc
14 — RF Output
13 — Tr 2 Base
12 — Tr 2 Emitter
11 — Tr 2 Collector
10 — V_{CC}
9 — Tr 1 Collector

Courtesy Motorola, Inc.

(continued)

FIGURE 9.4   (*Continued*)

MAXIMUM RATINGS

Ratings	Symbol	Value	Unit
Power Supply Voltage	V_{CC}	10 (max)	V
Operating Supply Voltage Range	V_{CC}	2.8–9.0	V
Junction Temperature	T_J	+150	°C
Operating Ambient Temperature	T_A	−30 to +75	°C
Storage Temperature Range	T_{stg}	−65 to +150	°C

ELECTRICAL CHARACTERISTICS (V_{CC} = 4.0 V, T_A = 25°C, unless otherwise noted)

Characteristics	Symbol	Pin	Min	Typ	Max	Unit
Drain Current (No input signal)	I_{CC}	10	1.7	2.9	4.3	mA
FM MODULATOR						
Output RF Voltage (f_o = 16.6 MHz)	V_{out} RF	14	60	90	130	mVrms
Output DC Voltage (No input signal)	Vdc	14	2.2	2.5	2.8	V
Modulation Sensitivity (f_o = 16.6 MHz) (V_{in} = 0.8 V to 1.2 V)	SEN	3.0 14	7.0 —	10 —	15 —	Hz/mVdc
Maximum Deviation (f_o = 16.6 MHz) (V_{in} = 0 V to 2.0 V)	Fdev	3.0 14	3.0 —	5.0 —	10 —	kHz
MIC AMPLIFIER						
Closed Loop Voltage Gain (V_{in} = 3.0 mVrms) (f_{in} = 1.0 kHz)	A_v	4.0 5.0	27 —	30 —	33 —	dB
Output DC Voltage (No input signal)	V_{out} dc	4.0	1.1	1.4	1.7	V
Output Swing Voltage (V_{in} = 30 mVrms) (f_{in} = 1.0 kHz)	V_{out} p-p	4.0	0.8	1.2	1.6	Vp-p
Total Harmonic Distortion (V_{in} = 3.0 mVrms) (f_{in} = 1.0 kHz)	THD	4.0	—	0.15	2.0	%

AUXILIARY TRANSISTOR STATIC CHARACTERISTICS

Characteristics	Symbol	Min	Typ	Max	Unit
Collector Base Breakdown Voltage (I_C = 5.0 μA)	$V_{(BR)CBO}$	15	45	—	V
Collector Emitter Breakdown Voltage (I_C = 200 μA)	$V_{(BR)CEO}$	10	15	—	V
Collector Substrate Breakdown Voltage (I_C = 50 μA)	$V_{(BR)CSO}$	—	70	—	V
Emitter Base Breakdown Voltage (I_E = 50 μA)	$V_{(BR)EBO}$	—	6.2	—	V
Collector Base Cut Off Current (V_{CB} = 10 V) (I_E = 0)	I_{CBO}	—	—	200	nA
DC Current Gain (I_C = 3.0 mA) (V_{CE} = 3.0 V)	h_{FE}	40	150	—	

AUXILIARY TRANSISTOR DYNAMIC CHARACTERISTICS

Current Gain Bandwidth Product (V_{CE} = 3.0 V) (I_C = 3.0 mA)	f_T	—	500	—	MHz
Collector Base Capacitance (V_{CE} = 3.0 V) (I_C = 0)	C_{CB}	—	2.0	—	pF
Collector Substrate Capacitance (V_{CS} = 3.0 V) (I_C = 0)	C_{CS}	—	3.3	—	pF

FIGURE 1 — TEST CIRCUIT

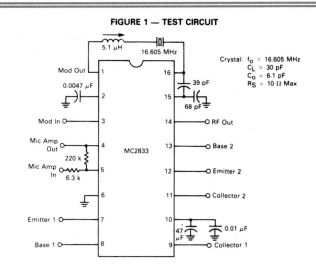

Crystal: f_O = 16.605 MHz
C_L = 30 pF
C_O = 6.1 pF
R_S = 10 Ω Max

FIGURE 2 — SINGLE CHIP VHF NARROWBAND FM TRANSMITTER

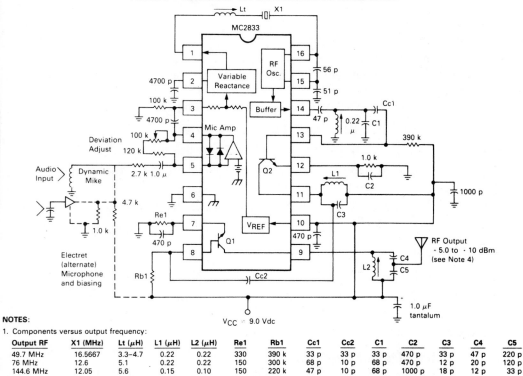

NOTES:

1. Components versus output frequency:

Output RF	X1 (MHz)	Lt (µH)	L1 (µH)	L2 (µH)	Re1	Rb1	Cc1	Cc2	C1	C2	C3	C4	C5
49.7 MHz	16.5667	3.3–4.7	0.22	0.22	330	390 k	33 p	33 p	33 p	470 p	33 p	47 p	220 p
76 MHz	12.6	5.1	0.22	0.22	150	300 k	68 p	10 p	68 p	470 p	12 p	20 p	120 p
144.6 MHz	12.05	5.6	0.15	0.10	150	220 k	47 p	10 p	68 p	1000 p	18 p	12 p	33 p

2. Crystal X1 is fundamental mode, calibrated for parallel resonance with a 32 pF load. The final output frequency is generated by frequency multiplication within the MC2833 IC. The RF output buffer (Pin 14) and Q2 transistor are used as a frequency tripler and doubler, respectively, in all three transmitters. The Q1 output transistor is a linear amplifier in the 49.7 MHz and 76 MHz transmitters, and a frequency doubler in the 144 MHz transmitter.

3. All coils used are 7 mm tunable shielded inductors, Toko B199SN-T10XXZ, B199KN-T10XXZ or equivalent.

4. Power output is ≈ +10 dBm for 49.7 MHz and 76 MHz transmitters, and ≈ +5.0 dBm for the 144 MHz transmitter at V_{CC} = 8.0 V. Power output drops with lower V_{CC}.

5. All capacitors in microfarads, inductors in Henries and resistors in Ohms unless otherwise specified.

of the input frequency. Doublers and triplers are the most common types of frequency multiplier, though greater multiplication is possible at the expense of reduced efficiency.

Frequency multipliers are particularly useful in FM transmitters because they multiply the deviation of an FM signal by the same factor as the carrier frequency. This is easy to show. First, suppose that an unmodulated carrier with frequency f_c is applied to the input of a circuit that multiplies the frequency by N. The frequency of the output signal will of course be Nf_c. Now, suppose that due to modulation the frequency of the input changes by an amount δ to $(f_c + \delta)$. The output signal will have frequency

$$
\begin{aligned}
f_o &= N(f_c + \delta) \\
    &= Nf_c + N\delta
\end{aligned}
\qquad (9.3)
$$

which corresponds to a carrier at N times the original carrier frequency, with a frequency deviation N times the original deviation. Thus a frequency multiplier can increase the deviation obtained at the modulator by any required amount.

EXAMPLE 9.1

A direct-FM transmitter has a varactor modulator with $k_f = 2$ kHz/V and a maximum deviation of 300 Hz. This modulator is followed by a buffer and three stages of frequency multiplication: a tripler, a doubler, and another tripler, followed by a driver and power amplifier.

(a) Draw a block diagram of this transmitter.

(b) Will this transmitter be capable of 5 kHz deviation at the output?

(c) What should the oscillator frequency be if the transmitter is to operate at a carrier frequency of 150 MHz?

(d) What audio voltage will be required at the modulator input to obtain full deviation?

Solution

(a) See Figure 9.5 for a block diagram of this transmitter.

(b) The frequency multipliers multiply the deviation by a factor of

$$
3 \times 2 \times 3 = 18
$$

The maximum deviation at the oscillator must then be

$$
\begin{aligned}
\delta_{osc} &= \frac{\delta_o}{18} \\
             &= \frac{5 \text{ kHz}}{18} \\
             &= 278 \text{ Hz}
\end{aligned}
$$

Since the modulator is capable of 300 Hz deviation, it will enable the output deviation to reach 5 kHz.

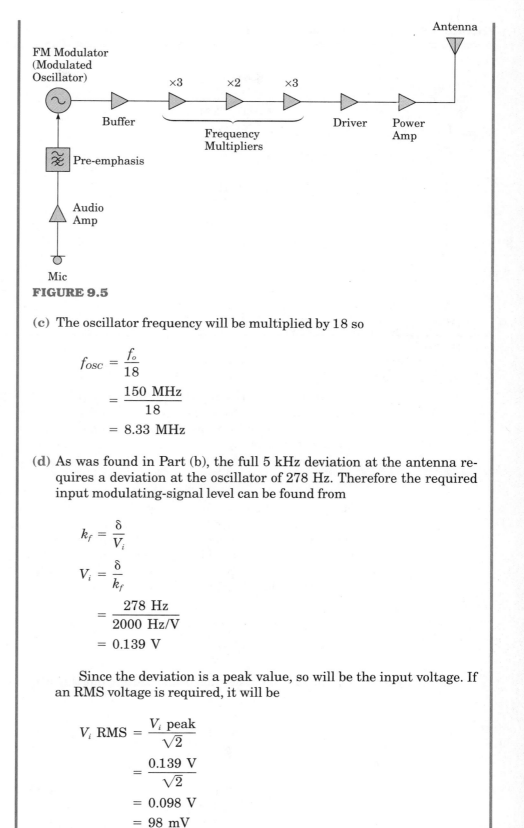

FIGURE 9.5

(c) The oscillator frequency will be multiplied by 18 so

$$f_{osc} = \frac{f_o}{18}$$

$$= \frac{150 \text{ MHz}}{18}$$

$$= 8.33 \text{ MHz}$$

(d) As was found in Part (b), the full 5 kHz deviation at the antenna requires a deviation at the oscillator of 278 Hz. Therefore the required input modulating-signal level can be found from

$$k_f = \frac{\delta}{V_i}$$

$$V_i = \frac{\delta}{k_f}$$

$$= \frac{278 \text{ Hz}}{2000 \text{ Hz/V}}$$

$$= 0.139 \text{ V}$$

Since the deviation is a peak value, so will be the input voltage. If an RMS voltage is required, it will be

$$V_i \text{ RMS} = \frac{V_i \text{ peak}}{\sqrt{2}}$$

$$= \frac{0.139 \text{ V}}{\sqrt{2}}$$

$$= 0.098 \text{ V}$$

$$= 98 \text{ mV}$$

FIGURE 9.6   **FM Transmitter with Multiplication and Mixing**

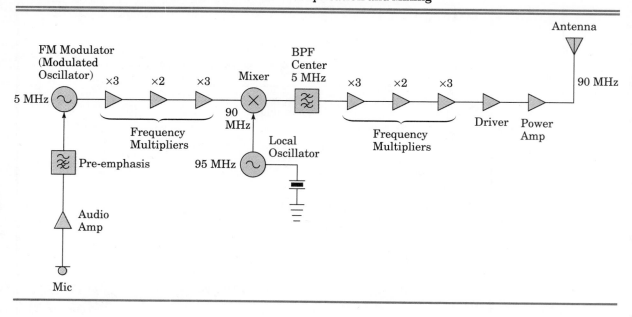

Of course, increasing the deviation by frequency multiplication also increases the carrier frequency. What about the case where sufficient multiplication to achieve the required amount of deviation results in a carrier frequency that is too high? The solution to that problem is actually quite simple: the carrier frequency can be lowered by mixing, as shown in Figure 9.6.

The mixing process can change the carrier frequency to any required value, but will have no effect on the deviation. Again, this is easy to show. Let one input to the mixer have frequency $(f_c + \delta)$, and let the other input frequency be f_{LO}, provided by a local oscillator. The sum and difference frequencies will be produced by the mixer, but, in this case, it is the difference that will be used, since the object is to lower the carrier frequency. Here there are two possibilities, depending on which frequency is higher. If the carrier frequency is higher than the local oscillator frequency, the output will be

$$f_o = (f_c + \delta) - f_{LO}$$
$$= (f_c - f_{LO}) + \delta$$

The carrier frequency has been reduced, but the deviation has not changed. Alternatively, if $f_{LO} > f_c$,

$$f_o = f_{LO} - (f_c + \delta)$$
$$= (f_{LO} - f_c) - \delta$$

This time, the carrier frequency has been decreased, and the deviation has changed in sign but not in magnitude. The change in sign is equivalent to a 180° phase shift in the modulating signal, and makes no practical difference.

EXAMPLE 9.2

For Figure 9.6, calculate the required frequency deviation of the oscillator if the output frequency is to deviate by 75 kHz.

Solution

In this transmitter, a modulated oscillator operating at 5 MHz has its frequency multiplied by 18, just as in the previous example. Then the signal is mixed back down to 5 MHz and multiplied another 18 times. The net multiplication is 18 for the carrier frequency, but the deviation has been multiplied by $18 \times 18 = 324$. A deviation of 75 kHz at the antenna would require the oscillator frequency to deviate by only $(75 \text{ kHz})/324 = 232 \text{ Hz}$.

9.2.3 Automatic Frequency Control Systems

As an alternative to the use of crystal-controlled oscillators and a great deal of frequency multiplication, it is possible to achieve a reasonable amount of deviation at the oscillator frequency by using an LC oscillator operated as a VCO. As mentioned before, however, stability is a serious problem with such oscillators, and some form of AFC is always required.

The classical way to do this is known as the **Crosby system**, and it is illustrated in Figure 9.7. Essentially, the transmitter carrier frequency (or some submultiple of it) is mixed with a crystal-controlled reference signal. The difference frequency is sent to a frequency discriminator. The operation of a discriminator circuit will be discussed in connection with receivers; for now all we need to know is that it produces an output voltage proportional to the difference between the frequency at its input and the center frequency to which the circuit has been tuned. The center frequency of the discriminator is set so that the transmitter will have the correct carrier frequency when the mixer output is at that frequency.

FIGURE 9.7

The Crosby AFC System

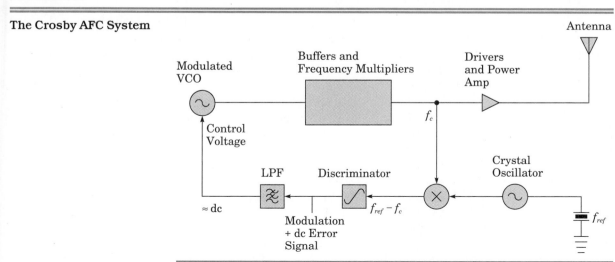

The output of the discriminator is sent to a low-pass filter so that rapid changes of voltage due to modulation will be removed, leaving only dc levels due to slow drift of the carrier oscillator away from its correct frequency. When the carrier is at exactly the correct frequency, the output voltage from the discriminator will be zero. The dc level is applied to the carrier oscillator, where it causes the oscillator frequency to move toward the correct value. The whole network thus exhibits negative feedback. The AFC circuit will not give perfect stability, because the control voltage is only present when there is an error between the oscillator frequency and the correct value. In addition, the discriminator itself contains an LC tuned circuit whose components are subject to some variation.

EXAMPLE 9.3

A Crosby-type FM transmitter has the block diagram of Figure 9.7. The transmitter is to have a carrier frequency of 99.9 MHz. The crystal oscillator has a frequency of 105 MHz. What should be the center frequency of the discriminator?

Solution

The difference frequency output from the mixer will be

$$f_{ref} - f_c = 105.0 \text{ MHz} - 99.9 \text{ MHz}$$
$$= 5.1 \text{ MHz}$$

This is the frequency to which the discriminator should be tuned. By the way, 105 MHz is not a very practical frequency for a crystal oscillator. Frequency multipliers would actually have to be used between the oscillator and the mixer. Alternatively, the transmitter signal could be brought out to the mixer at a lower frequency, partway along the chain. (See Problem 26 at the end of this chapter for a situation like this.)

9.2.4 Phase-Locked Loop FM Generators

Another, more modern way to solve the stability problem is to make the modulator VCO part of a phase-locked loop (PLL). See Figure 9.8 for an example of this method of stabilization. As in a frequency synthesizer, the VCO is locked to some multiple N of a crystal-controlled reference frequency. In fact, the VCO can be part of a frequency synthesizer for the carrier frequency. The loop filter will allow the system to ignore the rapid variations of frequency associated with modulation, while preventing drift of the center frequency of the VCO away from its nominal value. Unlike the Crosby AFC system, the PLL FM transmitter is capable of locking the transmitting frequency exactly to the crystal-controlled reference frequency. With a well-designed VCO, the PLL system is also capable of producing wideband FM without frequency multiplication. These advantages make the use of PLLs very popular in new designs for both communications and broadcast transmitters.

The modulator sensitivity with this system is determined by the VCO. Recall from Chapter 7 that the VCO in a PLL has a constant k_f that was defined as

FIGURE 9.8

PLL FM Transmitter

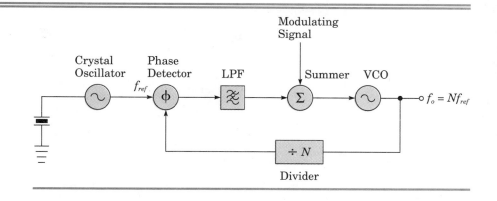

$$k_f = \frac{\Delta f}{\Delta v} \tag{9.4}$$

where k_f = VCO sensitivity in hertz per volt
  Δf = change in frequency
  Δv = change in control voltage required for the change in frequency

It is no coincidence at all that k_f was also used, in Chapter 8, for the deviation sensitivity of a frequency modulator, and defined in a similar way. In fact, k_f for the VCO becomes k_f for the modulator when a VCO is used as a frequency modulator.

EXAMPLE 9.4

A PLL FM generator has the block diagram of Figure 9.8, with f_{ref} = 100 kHz, N = 200, and k_f = 50 kHz/V.

(a) Calculate the carrier frequency of the output signal.

(b) What RMS modulating voltage will be required for a deviation of 10 kHz at the carrier frequency?

Solution

(a) $f_c = N f_{ref}$
    $= 200 \times 100 \text{ kHz}$
    $= 20 \text{ MHz}$

(b) $k_f = \dfrac{\delta}{V_p}$

    $V_p = \dfrac{\delta}{k_f}$

    $= \dfrac{10 \text{ kHz}}{50 \text{ kHz/V}}$

    $= 0.2 \text{ V}$

This is peak voltage, since deviation is given as a peak value. The RMS voltage will be

$$V_{RMS} = \frac{V_p}{\sqrt{2}}$$

$$= \frac{0.2 \text{ V}}{\sqrt{2}}$$

$$= 0.141 \text{ V}$$

$$= 141 \text{ mV}$$

Many, perhaps most, FM transmitters now use this technique, which makes frequency multiplication unnecessary. The modulator is simply followed by enough stages of amplification to achieve the desired power output.

9.2.5 Indirect-FM Modulators

We devoted considerable time in Chapter 8 to showing that FM and PM are closely related. To review, the modulation index in frequency modulation, with a sinusoidal modulating signal, is

$$m_f = \frac{\delta}{f_m}$$

where δ = maximum frequency deviation
   f_m = frequency of the modulating signal
   m_f = modulation index, and also the maximum phase deviation in radians

Therefore, there is more phase deviation in an FM signal for lower modulating frequencies. Frequency modulation can be produced using a phase modulator if the modulating signal is passed through a suitable low-pass filter before it reaches the modulator, so that lower modulating frequencies will produce greater phase deviation.

For readers familiar with the terminology of calculus, another way to look at the situation is to note that, since phase is essentially the integral of frequency, it is possible to integrate the baseband signal and then apply it to a phase modulator. The resulting signal will be exactly the same as if the original baseband signal had been applied to a frequency modulator. An integrator circuit is in fact a low-pass filter, so the two explanations lead to the same result. Figure 9.9 shows the general idea, which is called *indirect FM*.

One reason for using indirect FM is that it is easier to change the phase than the frequency of a crystal oscillator. Figure 9.10 shows one circuit that can do this. The changing voltage across the varactor, due to the modulating signal, changes the resonant frequency of the tuned circuit. This will cause the constant-frequency output from the crystal oscillator to be alternately below and above the changing resonant frequency of the tuned circuit. The phase shift due to the tuned circuit will be zero at resonance, and will vary from that value for a nonresonant condition. Thus, phase modulation is achieved, without any interference with the stability of the crystal oscillator.

FIGURE 9.9

Indirect FM Transmitter

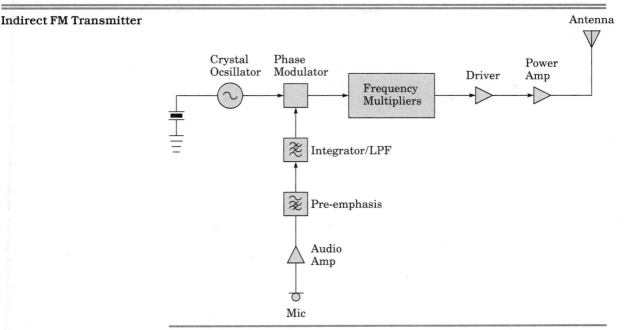

Unfortunately, the amount of phase shift that can be achieved in this way is relatively small. With more than a few degrees of phase shift, the variation of phase with frequency becomes nonlinear, and the variation of varactor capacitance with voltage is not linear, either. To avoid serious distortion of the modulating signal, the phase shift must be small. This means that only a small modulation index is possible, and frequency multipliers will be needed with this system.

Another, rather roundabout, way of producing phase modulation uses a phase-shift network and a balanced modulator, as illustrated in Figure 9.11. It is often referred to as the **Armstrong method**, after Edwin Armstrong, with whom we are already familiar. It works because of a mathematical similarity between narrowband phase modulation and AM for low modulation indices. Essentially, at these low indices, phase modulation is like AM except for a 90° phase shift of the carrier component. Consequently, PM can be

FIGURE 9.10

Phase Modulator

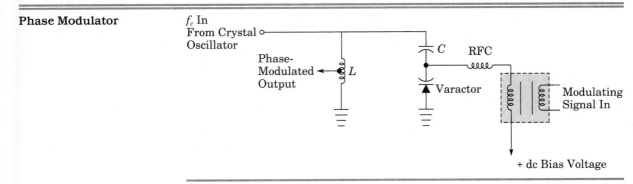

FIGURE 9.11

Armstrong Method of Indirect FM

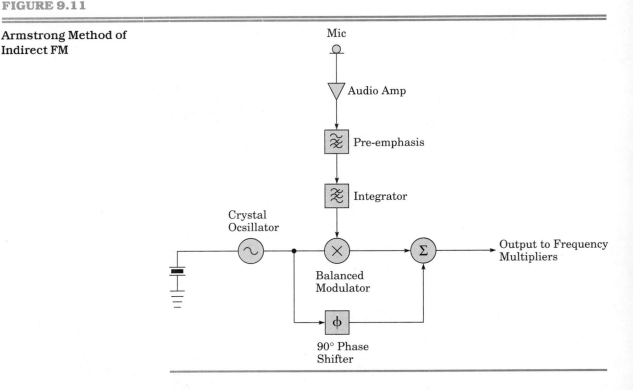

"faked" by using a balanced modulator to generate the sidebands, with no carrier, then adding in a carrier that has been shifted 90° beforehand. Once again, the amount of phase shift (and, therefore, frequency deviation) that can be obtained is very limited, and frequency multiplication will be required. As with the previous system, however, the carrier oscillator can be crystal-controlled, and its stability is not impaired by the modulator.

9.3 FM RECEIVERS

Although FM receivers are similar to AM receivers in basic design, this discussion will emphasize the differences. These are mainly in the IF and detector stages. Of course, the demodulator must work on different principles from that in an AM receiver, since it must respond to changes in frequency rather than amplitude. In addition, the IF amplifiers need not respond linearly to amplitude changes; in fact, it would be better if they did not, for the purpose of reducing the effect of noise. Those in an AM receiver, on the other hand, must be as linear as possible, since any amplitude distortion in the RF or IF stages appears directly as distortion of the demodulated baseband signal.

9.3.1 Demodulators

The FM demodulators must convert frequency variations of the input signal into amplitude variations at the output. To demodulate FM properly, the amplitude of the output must be proportional to the frequency deviation of the

FIGURE 9.12

**S-Curve Characteristic
of FM Detectors**

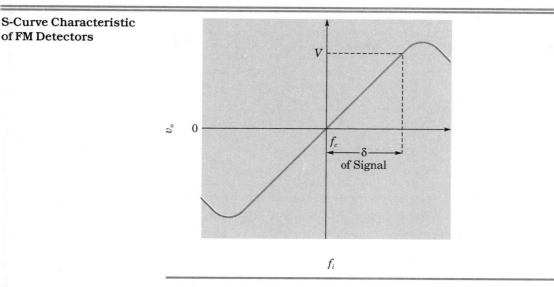

input. This results in a characteristic *S-curve* for many FM detectors (see
Figure 9.12). The output voltage is proportional to frequency deviation over
a range at least equal to 2δ, since the deviation is the distance the signal
frequency moves above and below the carrier frequency. Beyond that, the
shape of the curve is not important. This type of curve explains why the tuning is relatively critical with FM receivers: once the detector begins to operate
in the nonlinear portion of the S-curve, severe distortion results.

   The sensitivity of an FM detector can be given as

$$k_d = \frac{V_o}{\delta} \qquad (9.5)$$

where k_d = detector sensitivity in volts per hertz
   V_o = output voltage
   δ   = frequency deviation required for the output voltage

Note the similarity to the sensitivity of a frequency modulator, which is given
in hertz per volt. The sensitivity of a detector is the slope of the straight-line
portion of the S-curve of Figure 9.12.

EXAMPLE 9.5

An FM detector produces a peak-to-peak output voltage of 1.2 V from an
FM signal that is modulated to 10 kHz deviation by a sine wave. What is
the detector sensitivity?

Solution

First, the peak-to-peak voltage must be changed to peak, since that is the
way deviation is specified.

$$V_o \text{ peak} = \frac{V_o \text{ peak-to-peak}}{2}$$

$$= \frac{1.2 \text{ V}}{2}$$

$$= 0.6 \text{ V}$$

Now the detector sensitivity is

$$k_d = \frac{0.6 \text{ V}}{10 \text{ kHz}}$$

$$= 60 \text{ }\mu\text{V/Hz}$$

There are three major types of FM detector. The Foster-Seeley **discriminator**, along with its variation the **ratio detector**, are obsolescent but still very commonly found in older receivers and new receivers built to older designs. These detectors are effective but require a good number of discrete components, including a specially designed transformer, and must be ad-

Using an AM Receiver for FM: Slope Detection

Although hardly a practical system, the slope detector is perhaps the easiest type of FM detector to understand. Its operation can be demonstrated by tuning an AM receiver to an FM signal, or rather, mistuning it slightly. As shown in Figure 9.13, this mistuning can put the FM carrier frequency about halfway down the skirt of the IF amplifier frequency-response curve. In that case, as the frequency deviates in one direction, the amplitude of the output signal from the IF amplifier increases; in the other direction it decreases. Frequency variations have thus produced amplitude variations, which can be demodulated by an envelope detector. Of course, the output is likely to be distorted, since the side of the IF amplifier frequency-response curve is not likely to be a straight line. This type of reception can be demonstrated by using a VHF receiver (such as a scanner) set to AM, to receive FM signals (mobile-radio signals, for example).

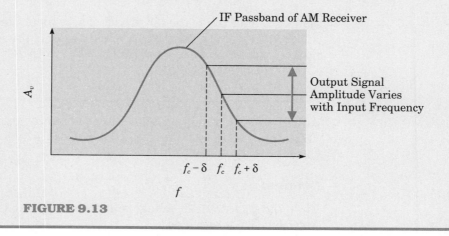

FIGURE 9.13

justed by hand. More modern types of demodulator include the **quadrature detector** and the PLL, both of which are well adapted to IC construction.

Foster-Seeley Discriminator and Ratio Detector. Both of these circuits convert frequency changes first to phase shifts and then to amplitude variations. The resulting AM signal is then demodulated by a combination of two diode detectors. The conversion from frequency to phase modulation is achieved by using the fact that the phase angle between voltage and current in a tuned circuit will change as the applied frequency goes through resonance. Recall that a series-tuned circuit is capacitive for frequencies below resonance, resistive at resonance, and inductive above resonance. At resonance, the current through the circuit will be in phase with the voltage across it. Below resonance, the current leads the voltage, and above resonance, the current lags the voltage.

In the Foster-Seeley discriminator, shown in Figure 9.14(a), the transformer is double-tuned, with both the primary (L_1–C_1) and secondary (L_2–C_2) circuits resonant at the carrier frequency. The secondary voltage at resonance is 90° out of phase with the primary voltage. The primary voltage is also applied, in phase, to the center of the secondary winding through capacitor C_3, which is chosen to have low reactance at the carrier frequency. Capacitors C_4 and C_5 are chosen in the same way as the filters for AM detec-

FIGURE 9.14　　　**Foster-Seeley Discriminator**

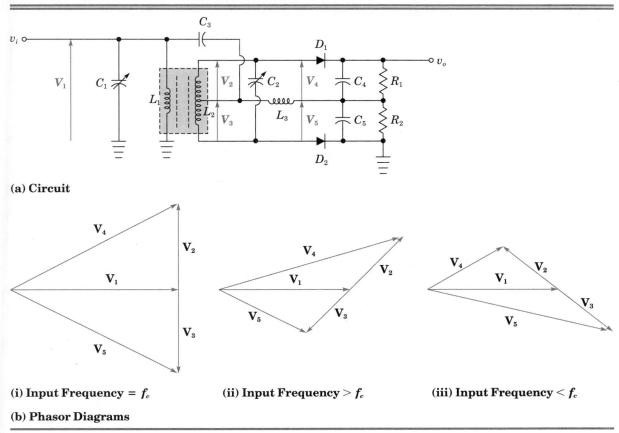

(a) Circuit

(i) Input Frequency = f_c　　　**(ii) Input Frequency > f_c**　　　**(iii) Input Frequency < f_c**

(b) Phasor Diagrams

tors: they should have low reactance at the carrier frequency, but high reactance at the modulating frequency, so that D_1–C_4 and D_2–C_5 act as peak detectors. Inductor L_3 is an RF choke with high reactance at the carrier frequency, and low reactance at the modulating frequency. It can be seen that the low-reactance connection made by C_3 causes virtually the entire primary voltage to appear across L_3.

Figure 9.14(b) shows how the primary and secondary voltages add. The vector sum of the primary voltage and one-half the secondary voltage is applied to one diode detector (consisting of D_1 and C_4), and the vector difference between the primary voltage and one-half the secondary voltage goes to the other detector (D_2 and C_5). The output will be the difference between the rectified and filtered voltages at the output of the two detectors, so it will be the difference between the amplitudes of the two phasors representing the voltages applied to the two detectors.

When the incoming signal frequency is equal to the resonant frequency of the tuned circuits, the voltages applied to the two detectors are equal in magnitude. This can be seen by observing that the vectors $\mathbf{V}_4$ and $\mathbf{V}_5$ in Figure 9.14(b)(i) are the same length. Therefore, the net output voltage is zero. If the frequency increases, the secondary voltage has a leading phase angle, and the relative length of these vectors changes. Vector $\mathbf{V}_4$ is now greater than $\mathbf{V}_5$, as shown in Figure 9.14(b)(ii). The output voltage becomes positive. Reducing the frequency below resonance causes a similar but opposite phase change, causing $\mathbf{V}_5$ to be less than $\mathbf{V}_4$ and producing a negative output voltage. The result is an output voltage that follows the modulation.

A change in primary voltage amplitude (due to noise, for example) will also cause a change in output voltage. This undesirable effect can be reduced by using limiters before the detector. However, a detector that is less sensitive to amplitude variations would increase the effectiveness of the limiters.

The *ratio detector* is a variation on the discriminator that has greatly reduced sensitivity to amplitude variations, at the cost of a 50% reduction in output voltage. This 6 dB loss can easily be made up elsewhere in the receiver. See the circuit in Figure 9.15. It is similar to that of the discriminator, but can be recognized at a glance by the fact that one of the diodes has been reversed, so that the two outputs across C_4 and C_5 add rather than subtract. Rather than going to the output, the sum of the two voltages is applied to an additional capacitor, shown as C_6 in the diagram. This capacitor has a long discharge time constant in combination with R_1 and R_2, and provides a reference level. Its voltage will only change with long-term variations in signal

FIGURE 9.15

Ratio Detector

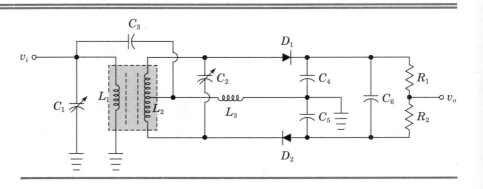

strength, and not with brief amplitude variations, such as those due to noise. As the vectors change in the way described above, the ratio between the two voltages produced by the diode detectors across C_4 and C_5 changes, but their sum is fixed by the capacitor voltage.

The output of this circuit cannot be taken between the two diodes as for the discriminator, since this voltage does not vary with modulation. Instead, it is taken between the center point of the load resistor and the junction of C_4 and C_5. The output voltage will be one-half of the difference between the voltages across C_4 and C_5.

PLL Method. The use of a PLL to demodulate FM signals is very straightforward. See Figure 9.16 for a typical PLL detector. The incoming FM signal is used to control the frequency of the VCO. As the incoming frequency varies, the PLL will generate a control voltage to change the VCO frequency, which will follow that of the incoming signal. This control voltage varies at the same rate as the frequency of the incoming signal, and so it can be used directly as the output of the circuit. Unlike the PLLs used in transmitter modulator circuits, this PLL must have a short time constant so that it can follow the modulation. The capture range of the PLL is not important, since the free-running frequency of the VCO will be set equal to the signal's carrier frequency at the detector (that is, to the center of the IF passband). The lock range must be at least twice the maximum deviation of the signal. If it is deliberately made wider, the detector will be able to function in spite of a small amount of receiver mistuning or local oscillator drift. Amplitude variations of the input signal will not affect the operation of this detector, unless they are so great that it stops working altogether.

It is easy to calculate the output voltage for a PLL detector, provided k_f for the VCO is known. Since the loop stays locked as it follows the modulation, the VCO frequency follows the signal frequency, varying over the range from $(f_c - \delta)$ to $(f_c + \delta)$. Therefore, the peak output voltage is the voltage necessary to move the local oscillator by the amount of the deviation δ. That is,

$$V_o \text{ peak} = \frac{\delta}{k_f} \tag{9.6}$$

where V_o = output voltage from the detector
   δ = deviation of the signal in hertz
   k_f = VCO proportionality constant in hertz per volt

FIGURE 9.16

PLL FM Detector

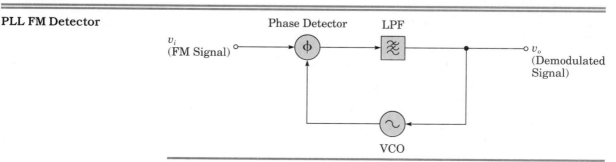

VCO

EXAMPLE 9.6

A PLL FM detector uses a VCO with $k_f = 100$ kHz/V. If it is receiving an FM signal with a deviation of 75 kHz and sine-wave modulation, what is the RMS output voltage from the detector?

Solution

From Equation (9.6),

$$V_o \text{ peak} = \frac{\delta}{k_f}$$

$$= \frac{75 \text{ kHz}}{100 \text{ kHz/V}}$$

$$= 0.75 \text{ V}$$

For sine-wave modulation,

$$V_o \text{ RMS} = \frac{V_o \text{ peak}}{\sqrt{2}}$$

$$= \frac{0.75 \text{ V}}{\sqrt{2}}$$

$$= 0.53 \text{ V}$$

Quadrature Detector. Like the PLL detector, the *quadrature detector* is adapted to integrated circuitry. Like the discriminator, it uses the phase shift in a resonant circuit. In the quadrature detector (see Figure 9.17), the incoming signal is applied to one input of a phase detector. The signal is also applied to a phase-shift network. This consists of a capacitor (C_1 in the figure) with high reactance at the carrier frequency, which will cause a 90° phase shift (this is where the term *quadrature* arises). The tuned circuit consisting of L_1 and C_2 is resonant at the carrier frequency. Therefore, it will cause no phase shift at the carrier frequency, but will cause a phase shift at other frequencies that will add to or subtract from the basic 90° shift caused by C_1.

FIGURE 9.17

Quadrature FM Detector

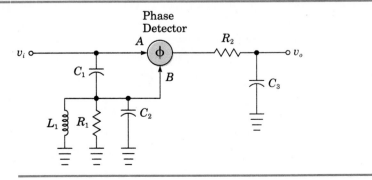

The output of the phase-shift network is applied to the second input of the phase detector. When the input frequency changes, the angle of phase shift in the quadrature circuit will vary, as the resonant circuit becomes inductive or capacitive. The output from the phase detector will vary at the signal frequency but will have an average value proportional to the amount the phase angle differs from 90°. Low-pass filtering the output will recover the modulation. Figure 9.17 shows this function accomplished by a simple first-order filter consisting of R_2 and C_3. As usual with detector low-pass filters, the cutoff frequency should be well above the highest modulating frequency, and well below the receiver intermediate frequency.

The phase detector is the same as is used for PLLs, and can take any of the forms discussed in Chapter 7: that is, it can be an analog multiplier (product detector), or a digital gate (either an AND or an exclusive-OR gate).

9.3.2 Limiters

FM obtains some immunity to noise because amplitude variations of the signal do not carry information. This advantage will be lost if the receiver responds to variations in signal amplitude. Some detectors, like the ratio detector and PLL, are inherently insensitive to amplitude variations. The Foster-Seeley discriminator and quadrature detector, on the other hand, respond to amplitude as well as to frequency changes.

Since only the frequency of the signal carries information, distortion of the waveform is of no consequence. In our study of Class C amplifiers, we noticed that the output signal level from such an amplifier is almost independent of the input signal amplitude, provided it is above the level needed to drive the transistor into saturation on peaks. From this, it might seem that, just as Class C amplifiers are satisfactory in FM transmitters, they might be useful as limiters in the receiver IF section.

In fact, a receiver limiter is usually designed to operate Class A with small signals, switching to Class C as the signal becomes larger. That is, it is biased in the active region, unlike a Class C amplifier, which is biased beyond cutoff.

The operating range is deliberately made narrow, however, so that large signals will drive the transistor into saturation on one signal peak and cutoff on the other. This allows the receiver amplifier to amplify weak signals, whereas the straight Class C amplifier would not respond at all to these signals, causing the sensitivity of the receiver to be reduced.

Figure 9.18 shows the difference between a straightforward Class C amplifier and a typical limiter stage. A Class C amplifier, as described in Chapter 2, is shown in Figure 9.18(a). With no signal, there is no transistor bias. The presence of a signal with sufficient peak voltage to cause base current to flow causes C_B to charge and put a negative bias on the base, which ensures that the transistor continues to conduct only on peaks. With a weak signal, the transistor is never driven into conduction, and there is no output.

The limiter in Figure 9.18(b) is very similar. The designation of R_B has been changed to R_{B_2} and resistors R_{B_1} and R_E (bypassed by C_E) have been added to form a more conventional bias circuit that causes the transistor to conduct with no signal. Capacitor C_B is still present, so that large signals cause a negative bias as before, causing the transistor to be cut off except on signal peaks. In addition, there is a resistor R_c in the collector circuit, which

FIGURE 9.18   **Class C Amplifier and Limiter**

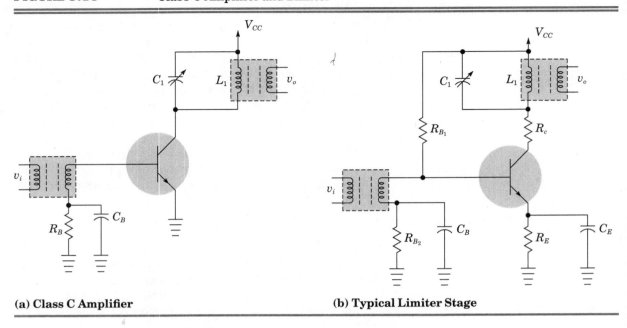

(a) Class C Amplifier   **(b) Typical Limiter Stage**

limits the current flow in the circuit, making it easier for the signal to saturate the stage. Of course, this resistor also reduces the efficiency of the amplifier, but that is not important in a receiver.

Figure 9.19(a) shows the transfer curve for a limiter. For an input voltage less than the threshold value, the output is proportional to input. Above the threshold, the output rises hardly at all. Another way of saying this is that the stage gain reduces sharply above the threshold. Since the output remains constant for input levels above the threshold, the gain, defined as the ratio of output to input voltage, reduces as the input increases. Figure 9.19(b) illustrates this idea. The threshold point is often defined as the input signal level where the gain is 3 dB down from its small-signal value.

For very large input signals, the output will actually reduce, because the

FIGURE 9.19   **Limiters**

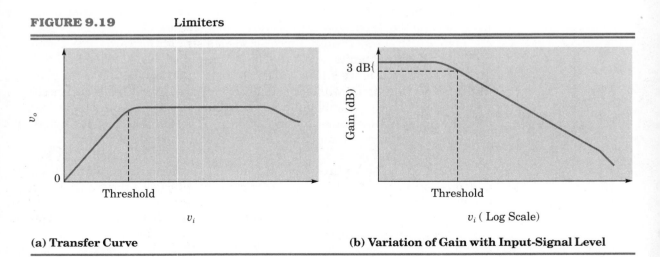

(a) Transfer Curve   **(b) Variation of Gain with Input-Signal Level**

transistor will conduct for only a very short time during each cycle. This can be prevented by applying automatic gain control (AGC) to an earlier stage in the receiver. If a sharper transition is required, two or more limiting amplifiers can be used in cascade.

In many modern receiver designs, the limiting function is built into a specially designed FM IF integrated circuit. In fact, ICs are commonly available that include both the IF amplifier stages and a quadrature FM detector. The CA3089 is one such circuit. A block diagram of its internal structure is shown in Figure 9.20, along with some of its specifications.

It is even possible to obtain an entire FM receiver, except for a few passive components, in a single IC. The MC3362 (see Figure 9.21), for instance, is a complete double-conversion receiver on a chip.

FIGURE 9.20

The CA3089

Features

- For FM IF Amplifier Applications in High–Fidelity, Automotive, and Communications Receivers
- Includes: IF Amplifier, Quadrature Detector, AF Preamplifier, and Specific Circuits for AGC, AFC, Muting (Squelch), and Tuning Meter
- Exceptional Limiting Sensitivity .... 12μV (Typ.) @ –3dB Point
- Low Distortion: (with Double–Tuned Coil) .... 0.1% (Typ.)
- Single–Coil Tuning Capability
- High Recovered Audio .... 400mV (Typ.)
- Provides Specific Signal for Control of Interchannel Muting (Squelch)
- Provides Specific Signal for Direct Drive of a Tuning Meter
- Provides Delayed AGC Voltage for RF Amplifier
- Provides a Specific Circuit for Flexible AFC
- Internal Supply-Voltage Regulators

Description

Harris CA3089E is a monolithic integrated circuit that provides all the functions of a comprehensive FM-IF system. Figure 1 is a block diagram showing the CA3089E features, which include a three-stage FM-IF amplifier/ limiter configuration with level detectors for each stage, a doubly-balanced quadrature FM detector and an audio amplifier that features the optional use of a muting (squelch) circuit.

The advanced circuit design of the IF system includes desirable deluxe features such as delayed AGC for the RF tuner, and AFC drive circuit, and an output signal to drive a tuning meter and/or provide stereo switching logic. In addition, internal power supply regulators maintain a nearly constant current drain over the voltage supply range of +8.5 to +16 volts.

The CA3089E is ideal for high-fidelity operation. Distortion in a CA3089E FM-IF System is primarily a function of the phase linearity characteristic of the outboard detector coil.

The CA3089E utilizes the 16-lead dual-in-line plastic package and can operate over the ambient temperature range of –40°C to +85°C.

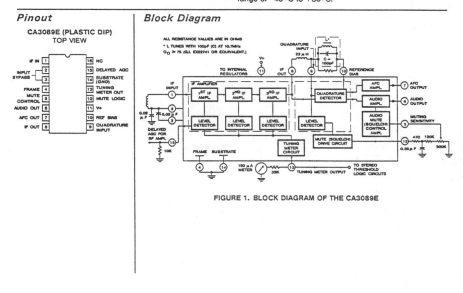

Pinout

CA3089E (PLASTIC DIP)
TOP VIEW

IF IN	1	16	NC
INPUT BYPASS	2	15	DELAYED AGC
	3	14	SUBSTRATE (GND)
FRAME	4	13	TUNING METER OUT
MUTE CONTROL	5	12	MUTE LOGIC
AUDIO OUT	6	11	V+
AFC OUT	7	10	REF BIAS
IF OUT	8	9	QUADRATURE INPUT

Block Diagram

FIGURE 1. BLOCK DIAGRAM OF THE CA3089E

CAUTION: These devices are sensitive to electrostatic discharge. Proper I.C. handling procedures should be followed.

Courtesy Harris Corporation.

FIGURE 9.21   The MC3362

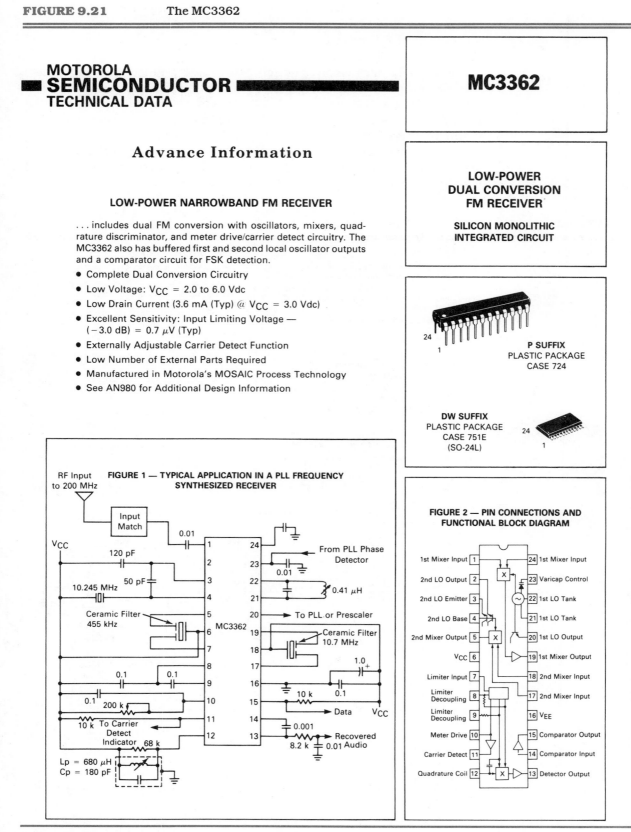

MOTOROLA
■■ **SEMICONDUCTOR** ■■■■■
TECHNICAL DATA

MC3362

Advance Information

LOW-POWER
DUAL CONVERSION
FM RECEIVER

SILICON MONOLITHIC
INTEGRATED CIRCUIT

LOW-POWER NARROWBAND FM RECEIVER

. . . includes dual FM conversion with oscillators, mixers, quad-rature discriminator, and meter drive/carrier detect circuitry. The MC3362 also has buffered first and second local oscillator outputs and a comparator circuit for FSK detection.

- Complete Dual Conversion Circuitry
- Low Voltage: V_{CC} = 2.0 to 6.0 Vdc
- Low Drain Current (3.6 mA (Typ) @ V_{CC} = 3.0 Vdc)
- Excellent Sensitivity: Input Limiting Voltage —
  (– 3.0 dB) = 0.7 μV (Typ)
- Externally Adjustable Carrier Detect Function
- Low Number of External Parts Required
- Manufactured in Motorola's MOSAIC Process Technology
- See AN980 for Additional Design Information

P SUFFIX
PLASTIC PACKAGE
CASE 724

DW SUFFIX
PLASTIC PACKAGE
CASE 751E
(SO-24L)

FIGURE 1 — TYPICAL APPLICATION IN A PLL FREQUENCY SYNTHESIZED RECEIVER

RF Input to 200 MHz

Input Match

V_{CC}
120 pF
10.245 MHz
50 pF
Ceramic Filter 455 kHz
MC3362
0.1 0.1
0.1 200 k
10 k To Carrier Detect Indicator 68 k
L_p = 680 μH
C_p = 180 pF

From PLL Phase Detector
0.01
0.41 μH
To PLL or Prescaler
Ceramic Filter 10.7 MHz
1.0
10 k 0.1
Data V_{CC}
0.001
8.2 k 0.01 Recovered Audio

FIGURE 2 — PIN CONNECTIONS AND FUNCTIONAL BLOCK DIAGRAM

Pin		Pin
1st Mixer Input [1]		[24] 1st Mixer Input
2nd LO Output [2]		[23] Varicap Control
2nd LO Emitter [3]		[22] 1st LO Tank
2nd LO Base [4]		[21] 1st LO Tank
2nd Mixer Output [5]		[20] 1st LO Output
V_{CC} [6]		[19] 1st Mixer Output
Limiter Input [7]		[18] 2nd Mixer Input
Limiter Decoupling [8]		[17] 2nd Mixer Input
Limiter Decoupling [9]		[16] V_{EE}
Meter Drive [10]		[15] Comparator Output
Carrier Detect [11]		[14] Comparator Input
Quadrature Coil [12]		[13] Detector Output

Courtesy Motorola, Inc.

MAXIMUM RATINGS (T_A = 25°C, unless otherwise noted)

Rating	Pin	Symbol	Value	Unit
Power Supply Voltage (See Diagram)	6	$V_{CC(max)}$	7.0	Vdc
Operating Supply Voltage Range (Recommended)	6	V_{CC}	2.0 to 6.0	Vdc
Input Voltage ($V_{CC} \geqslant 5.0$ Vdc)	1, 24	V_{1-24}	1.0	Vrms
Junction Temperature	—	T_J	150	°C
Operating Ambient Temperature Range	—	T_A	−40 to +85	°C
Storage Temperature Range	—	T_{stg}	−65 to +150	°C

ELECTRICAL CHARACTERISTICS (V_{CC} = 5.0 Vdc, f_o = 49.7 MHz, Deviation = 3.0 kHz, T_A = 25°C, Test Circuit of Figure 3 unless otherwise noted)

Characteristic	Pin	Min	Typ	Max	Units
Drain Current (Carrier Detect Low — See Figure 5)	6	—	4.5	7.0	mA
Input for −3.0 dB Limiting	—	—	0.7	2.0	μVrms
Recovered Audio (RF signal level = 10 mV)	13	—	350	—	mVrms
Noise Output (RF signal level = 0 mV)	13	—	250	—	mVrms
Carrier Detect Threshold (below V_{CC})	10	—	0.64	—	Vdc
Meter Drive Slope	10	—	100	—	nA/dB
Input for 20 dB (S + N)/N (See Figure 7)	—	—	0.7	—	μVrms
First Mixer 3rd Order Intercept (Input)	—	—	−22	—	dBm
First Mixer Input Resistance (Rp)	—	—	690	—	Ω
First Mixer Input Capacitance (Cp)	—	—	7.2	—	pF
First Mixer Conversion Voltage Gain	—	—	18	—	dB
Second Mixer Conversion Voltage Gain	—	—	21	—	dB
Detector Output Resistance	13	—	1.4	—	kΩ

FIGURE 3 — TEST CIRCUIT

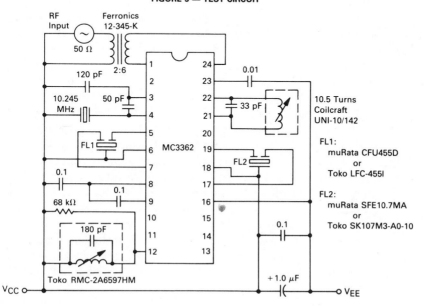

One characteristic of limiting amplifiers is that they have large gain for small signals. This can be quite annoying, and for that reason *squelch* (called *muting* in broadcast receivers) is almost universal in FM receivers. Turning off the muting on an FM broadcast receiver, where this is possible, will demonstrate the effect of limiting. The loud hiss that is heard between stations is amplified noise. The hiss disappears when a strong station is received and the gain of the limiters is reduced. Squelch systems were described in connection with AM receivers. The same techniques are used with FM.

9.3.3 Automatic Frequency Control Systems

As in any receiver, it is important that the local oscillator frequency be stable so that the receiver will not drift away from the station to which it is tuned. A small amount of drift can cause the detector to operate in the nonlinear portion of its S-curve, with resulting distortion; a greater amount of mistuning will cause reception to be lost altogether. The need for accurate tuning with FM is generally not quite as critical as for SSB, but more so than for AM. The problem of stability is made more difficult by the fact that FM is generally transmitted at relatively high frequencies (VHF and up). The LC oscillators at VHF and UHF are not likely to have sufficient stability.

There are two basic solutions to the problem. The first is to use some form of crystal control. Direct crystal control of the local oscillator is possible when the receiver has from one to a few channels. For receivers that must tune to many different frequencies, a frequency synthesizer can be employed. The other method is to use automatic frequency control (AFC) to keep the frequency of an LC local oscillator close to the correct value in the presence of a received signal. Most modern receiver designs use frequency synthesis, rendering AFC unnecessary, but there are still many receivers that incorporate AFC.

The AFC circuits in receivers operate in a similar manner to the Crosby transmitter system already discussed. One major difference is that there is no need to add a discriminator, since the receiver's FM detector circuit will serve that purpose.

A block diagram for a receiver AFC is shown in Figure 9.22(a). The S-curve representing the discriminator output is shown in Figure 9.22(b). When the tuning is correct, the discriminator output varies symmetrically about the zero point, and there is no dc output. When the receiver is tuned too high, most of the signal-frequency variation will be in the lower-frequency portion of the curve, and there will be a negative dc component to the output voltage. The opposite effect will occur for a receiver that is tuned to too low a frequency. This dc offset can be applied to a varactor diode in the frequency-determining circuit of the local oscillator, in such a way that a negative voltage will cause the frequency to which the receiver is tuned to decrease, and vice versa. This negative feedback will result in the local oscillator moving to a frequency that is almost correct (though not exactly correct, since the needed control voltage for the oscillator can only be generated when there is a frequency error).

AFC is not needed with synthesized receivers, because their local oscillators are stable enough without it. It is found in virtually all receivers of reasonable quality that use VFO, rather than synthesizer, tuning. Since the

FIGURE 9.22   Receiver AFC System

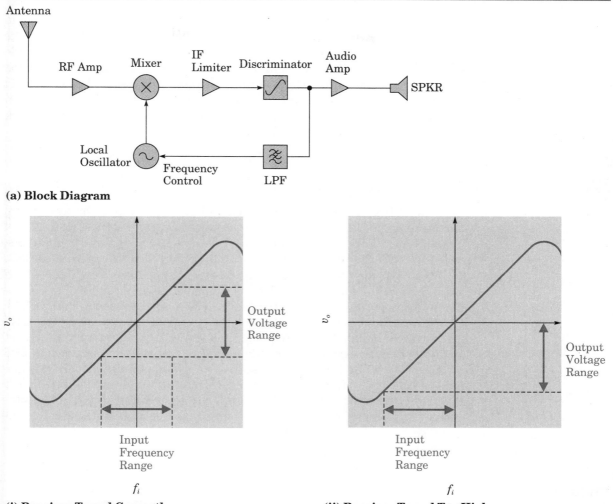

(a) Block Diagram

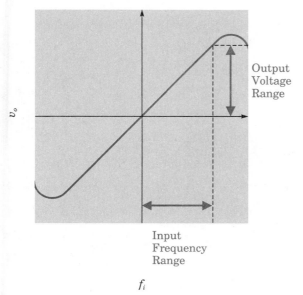

(i) Receiver Tuned Correctly

(ii) Receiver Tuned Too High

(iii) Receiver Tuned Too Low
(b) Effect of Mistuning on Discriminator Output

effect of AFC is to *pull* the receiver to almost the correct frequency even when it is tuned a relatively long way from the station's frequency, AFC can make a receiver difficult to tune accurately. Therefore, AFC is sometimes made switchable, so that it can be disabled to allow precise tuning, then switched on to counteract drift in the local oscillator circuit. Another reason for making the AFC switchable is to allow reception of weak signals that are close in frequency to strong ones. Sometimes there is a tendency for the AFC to lock on to the nearest strong station, ignoring the weaker signals.

9.4 FM TRANSCEIVERS

As with SSB systems, a major use for FM is in two-way communications, particularly mobile radio. In such an application, compactness and ease of installation and operation are important. Transmitter-receiver combinations, called *transceivers*, are therefore very common for FM voice communications.

Much FM mobile communication is done by means of *repeaters*, which are receiver-transmitter combinations connected to an antenna or set of antennas located well above local terrain. The idea is illustrated in Figure 9.23. Communication between mobile units is by way of the repeater, which receives transmissions on one frequency and simultaneously retransmits them on another. For instance, the mobile units can transmit on frequency f_1 and receive on f_2, while the repeater receives on f_1 and simultaneously retransmits on f_2. The two frequencies are generally a few hundred kilohertz apart. It would be difficult to design the repeater to receive and transmit simultaneously on frequencies that were very close together without the receiver experiencing interference from the transmitter.

FIGURE 9.23

FM Mobile Radio System with Repeater

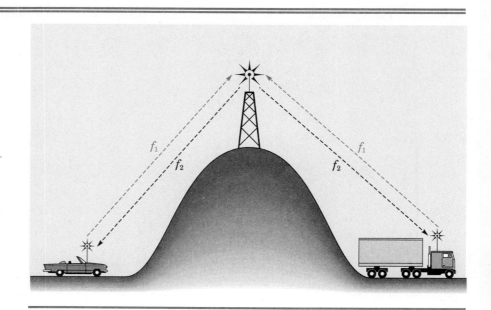

Consequently, while there is no need for the mobile transceivers, unlike the repeaters themselves, to be able both to transmit and to receive at the same time, it is often essential that they be able to transmit and receive on different, accurately determined frequencies. Often a single frequency synthesizer will be used for the transmitter carrier oscillator and the receiver local oscillator; in that case, it must be able to move quickly from one frequency to another and to settle rapidly at the new frequency without instability.

Other than the main frequency synthesizer and some of the audio circuitry, there is not too much that can be shared between transmitter and receiver. The receiver may incorporate a crystal or ceramic filter, but there is no need to share it with the transmitter, as is done with SSB transceivers. The combination does offer convenience of operation, however.

9.4.1 Example: A VHF Marine Radio Transceiver

Figure 9.24 shows a photograph and a block diagram of a typical FM communications transceiver. This one is used for VHF marine radio.

As in most modern designs, the central feature of the transceiver is a microprocessor-controlled frequency synthesizer. The same PLL controls two VCOs, one each for transmitting and receiving. The transmitting VCO is modulated directly at the carrier frequency; there are a total of four stages of transmitter amplification but no frequency multipliers. The transmitter output power is switchable between 1 W and 25 W, the low-power setting being used for short-range communication, such as between boats in harbor.

The receiver is double conversion. There is one stage of RF amplification before the first mixer, which works with the receive VCO as its local oscillator. There is a crystal filter at the first IF of 16.9 MHz. A crystal-controlled local oscillator and second mixer lower the signal frequency to that of the second IF, which is the familiar 455 kHz. The second IF amplifier chain and the detector are contained in an IC. The receiver is also equipped with an adjustable squelch circuit.

9.5 FM STEREO TRANSMITTERS AND RECEIVERS

In the previous chapter, the signals associated with FM stereo were examined, but no attempt was made to discuss hardware for producing or decoding these signals. To review, the baseband spectrum of an FM stereo signal is shown in Figure 9.25. The optional SCA signal has been removed for simplicity. Recall that this entire signal, spanning the frequency range from about 30 Hz to 53 kHz, is used to modulate the FM transmitter. No modifications to the transmitter are required, assuming that it is capable of handling this frequency range, as, by now, all FM broadcast transmitters are. All the circuitry for generating the FM stereo signal is contained in the baseband (audio) portion of the system.

Figure 9.26 is a block diagram of a system that can be used to encode an FM stereo signal. Both channels are given pre-emphasis in the usual way. Then a simple matrix arrangement adds and subtracts the two signals. This process can easily be accomplished by a couple of operational amplifiers (op-

FIGURE 9.24

VHF FM Transceiver

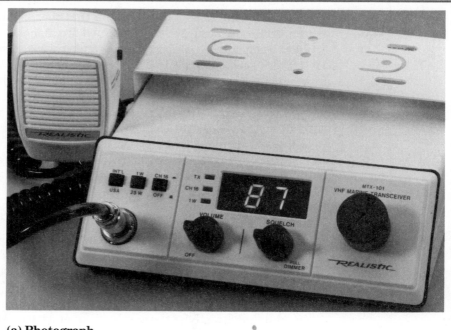

(a) Photograph

Diagram and photo courtesy of Radio Shack, a division of Tandy Corporation.

amps). The $L-R$ signal is applied to a balanced modulator, along with the 38 kHz subcarrier, to create a DSBSC signal. The $L+R$ signal is passed through a delay network to compensate for the time taken for the $L-R$ signal to pass through the balanced modulator, so that they arrive at the summing amplifier (summer) at the same time. In the summer, the $L+R$ signal is added to the DSBSC $L-R$ signal, and a 19 kHz pilot carrier is added as well. This last signal must be phase-coherent with the suppressed 38 kHz subcarrier, a requirement that is easily satisfied by deriving both from the same source. Either the 38 kHz signal can be divided down to produce the 19 kHz pilot, as shown, or the 19 kHz signal can be sent through a frequency doubler to produce the 38 kHz subcarrier.

The entire baseband stereo signal is applied to the input of an FM broadcast transmitter. Figure 9.27 is a partial data sheet for a typical FM broadcast transmitter. Notice that the frequency response for the modulating signal is specified as flat to 53 kHz. Note also that this transmitter uses direct FM at the operating frequency, followed by enough amplification to achieve its specified power output of 10 kW. All the amplification stages are solid-state, except for the power amplifier, which uses a tube. As with AM, all-solid-state FM broadcast transmitters are also available.

Figure 9.28 shows one popular way to decode the stereo signal at the receiver. The stereo decoder receives a signal at a point after the FM detector in the receiver, but before de-emphasis. The $L+R$ and modulated $L-R$ signals are easily separated by filtering, using a low-pass filter with a cutoff frequency of about 15 kHz for $L+R$, and a bandpass filter with a passband extending from 23 to 53 kHz for the $L-R$ signal. A small time delay will once

BLOCK DIAGRAM

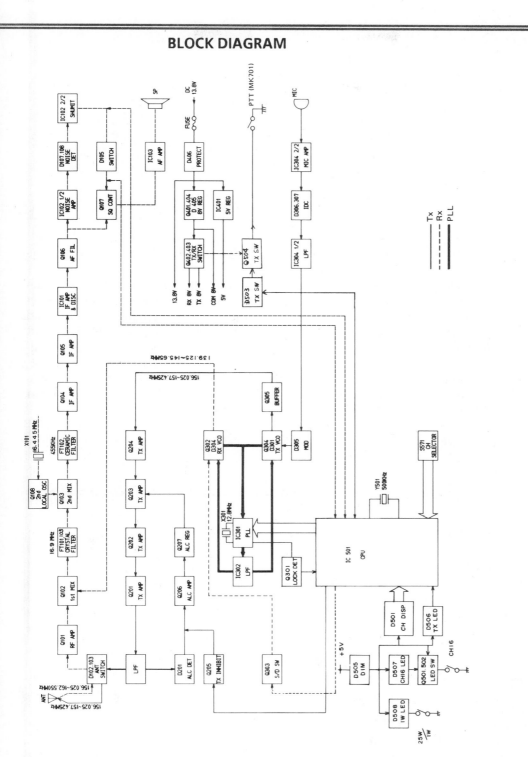

(b) Block Diagram

FIGURE 9.25

Baseband Spectrum for FM Stereo (without SCA)

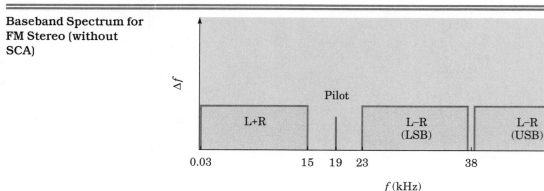

again be added to the L+R signal to compensate for the time taken for the L−R signal to go through a demodulator.

The DSBSC L−R signal must be demodulated in a product detector before it can be used. For this, the original 38 kHz subcarrier must be recovered. The 19 kHz pilot carrier can be recovered using a bandpass filter, and then either sent through a frequency doubler or used to stabilize a PLL that provides the required subcarrier. The latter method is the one shown in Figure 9.28.

Once the L+R and demodulated L−R signals are available, they can simply be added and subtracted to produce left and right signals, which must then go through individual de-emphasis networks.

A stereo decoder will generally have an output for a light or LED to indicate when a stereo signal is being received. A stereo signal is identified by

FIGURE 9.26

Generation of FM Stereo Signal

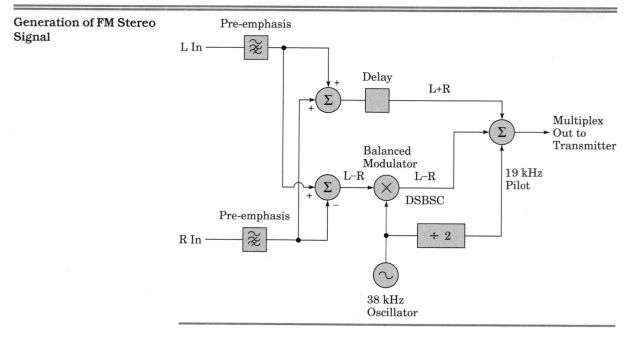

FIGURE 9.27 Data Sheet for the Harris FM Broadcast Transmitter

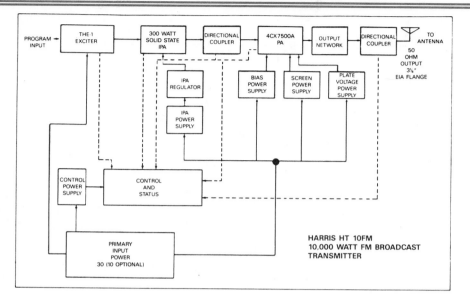

HARRIS HT 10FM SPECIFICATIONS

GENERAL
POWER OUTPUT: 5,000 to 10,000 watts (FCC type-notified range)
FREQUENCY RANGE: 87.5 to 108 MHz in 50 kHz steps. Tuned to single frequency.
EXCITATION: Harris THE-1 High Power FM Exciter
TYPE OF MODULATION: Direct carrier frequency modulation (DCFM)
MODULATION CAPABILITY: $\pm$200 kHz
RF LOAD IMPEDANCE: 50 ohms
RF OUTPUT TERMINATION: 3⅛″ EIA flange, female
PA MATCHING RANGE: 1.7:1 VSWR, maximum for full output power; automatic power reduction into high VSWR's
RF HARMONIC/SPURIOUS OUTPUT: Suppression meets or exceeds FCC/DOC/CCIR specifications
AC INPUT POWER: 197-250 VAC, 50/60 Hz, three phase, 3-wire closed delta or 360-415 VAC 4-wire WYE. OPTIONAL: 197-250 VAC, 50/60 Hz, single phase, 2-wire.
POWER CONSUMPTION: 15.7 kW typical at 10 kW RF output
POWER FACTOR: 0.95 (3-phase), 0.85 (1-phase)
AMBIENT TEMPERATURE RANGE: −20 to +50°C at sea level; derated 2°C/1000 ft altitude
MAXIMUM ALTITUDE: 10,000 ft (60 Hz), 7,500 ft (50 Hz)
MAXIMUM HUMIDITY: To 95% non-condensing
AIR REQUIREMENTS: 400 CFM (60 Hz), 350 CFM (50 Hz); no back pressure
CABINET SIZE: 33″W (84 cm) × 34″D (99 cm) × 72″H (183 cm)
WEIGHT/VOLUME: 1,125 lbs/46.7 cu. ft., domestic packed

WIDEBAND COMPOSITE OPERATION (STANDARD)
INPUTS: Two; one balanced, floating and one unbalanced.
INPUT IMPEDANCE: 2000 ohms resistive.
INPUT CONNECTORS: Female BNC (rear panel).
INPUT LEVEL: 1.0 volt RMS nominal for $\pm$75 kHz deviation.
AMPLITUDE RESPONSE: $\pm$0.1 dB, 30 Hz to 53 kHz; −0.2 dB at 100 kHz.
FM SIGNAL TO NOISE: 80 dB below 100% modulation (reference 400 Hz at $\pm$75 kHz deviation with 75 microsecond de-emphasis, 20 Hz to 200 kHz bandwidth).
HARMONIC DISTORTION: 0.08%
INTERMODULATION DISTORTION: 0.02% (60 Hz/7 kHz 1:1 tone pair).
CCIF INTERMODULATION DISTORTION: All distortion products below 80 dB (reference 14 kHz/15 kHz test tone pair).

ASYNCHRONOUS AM SIGNAL TO NOISE: 55 dB below equivalent 100% amplitude modulation
SYNCHRONOUS AM SIGNAL TO NOISE: 50 dB below equivalent 100% amplitude modulation of output carrier with 75 microsecond de-emphasis (FM modulation $\pm$75 kHz @ 400 Hz).
PHASE RESPONSE: +0.5/−1.0 degrees from linear phase, 20 Hz to 53 kHz.
TRANSIENT INTERMODULATION DISTORTION: 0.05%, 2.96 kHz square wave/14 kHz sine wave modulation.
TEST INPUT (FRONT PANEL): Nominal 1.0 volt for $\pm$75 kHz deviation at 400 Hz (10,000 ohm input impedance, BNC female connector)
TEST OUTPUT (FRONT PANEL): Nominal 1.0 volt for $\pm$75 kHz deviation at 400 Hz (200 ohm source impedance, BNC female connector)

MONAURAL OPERATION (STANDARD)
AUDIO INPUT IMPEDANCE: 600 ohms, balanced, resistive, transformerless.
AUDIO INPUT LEVEL: +10 dBm, $\pm$1 dB for $\pm$75 kHz deviation at 400 Hz.
AUDIO FREQUENCY RESPONSE: Standard 75 microsecond FCC pre-emphasis curve $\pm$0.5 dB, 30 Hz-15 kHz. Selectable: flat, 25, 50 or 75 microsecond pre-emphasis.
HARMONIC DISTORTION: 0.08%, 30 Hz to 15 kHz, de-emphasized.
INTERMODULATION DISTORTION: 0.04%, 60 Hz/7 kHz test tone pair, 4:1 ratio.
CCIF INTERMODULATION DISTORTION: All distortion products down 70 dB (reference 14 kHz/15 kHz test tone pair).
TRANSIENT INTERMODULATION DISTORTION: 0.05%, 2.96 kHz square wave/14 kHz sine wave modulation
FM SIGNAL TO NOISE RATIO: At least 80 dB below 100% modulation (reference 400 Hz @ $\pm$75 kHz deviation, measured 20 Hz to 200 kHz bandwidth, 75 microsecond de-emphasis).

SCA INPUTS (STANDARD)
EXTERNAL SCA GENERATOR INPUTS: Two
INPUT CONNECTORS: BNC female (rear panel).
INPUT IMPEDANCE: 10,000 ohms, unbalanced.
INPUT LEVEL: 0.1V (nominal) for 10% injection.
RANGE OF SUBCARRIER FREQUENCIES: 57 kHz to 92 kHz (25 kHz to 92 kHz in monaural operation).
AMPLITUDE RESPONSE: +0.1 dB, −0.2 dB; 20 kHz to 100 kHz.

SPECIFICATIONS ARE REFERENCED TO 10 kW OPERATION, EXCEPT WHEN NOTED.

**HARRIS MAINTAINS A POLICY OF CONTINUOUS IMPROVEMENTS ON ITS EQUIPMENT
AND THEREFORE RESERVES THE RIGHT TO CHANGE SPECIFICATIONS WITHOUT NOTICE.**

Courtesy Harris Corporation.

FIGURE 9.28   FM Stereo Decoder

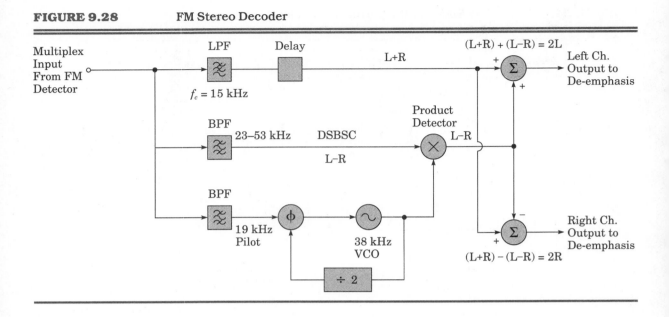

the presence of the 19 kHz pilot carrier. There will also be provision for manual and/or automatic switching between mono and stereo reception. As previously mentioned, there is a noise penalty for stereo, so its use is inappropriate with weak signals. Recall that many stereo receivers will automatically switch to mono if the signal strength is insufficient for good stereo reception. In addition, when the *S/N* is marginal for stereo reception, some receivers are capable of operating in stereo at low and midrange audio frequencies, while blending the high-frequency audio components of the two channels. The theory here is that noise will be more troublesome at high frequencies, and that blending the two channels can reduce this noise, since noise signals that are different on the two channels will tend to cancel each other. At the same time, the listener will experience some stereo effect from the lower frequencies.

Modern stereo decoders are generally contained, except for some filter and frequency-determining components, on a single IC. One example is the CA3090, which works in the way just described.

9.6 TEST EQUIPMENT AND PROCEDURES: FM RECEIVER MEASUREMENTS

Selectivity can be measured for FM receivers in a manner similar to that for AM. Measurement techniques for sensitivity are different, however. There are two popular methods for measuring the sensitivity of FM receivers. One of them, known as **usable sensitivity** or *SINAD sensitivity*, can also be used with AM receivers, and occasionally is. The other technique, called **quieting sensitivity**, is suitable only for FM receivers. In addition, this section will describe the measurement of **channel separation** in stereo systems. This description would apply equally well to FM and AM stereo receivers; in fact, it would also apply to other stereo audio components, such as phonograph turntables and tape recorders.

9.6.1 Usable Sensitivity

Usable sensitivity is the signal level required for a given SINAD. SINAD, you will probably remember, stands for the ratio of signal-plus-noise-and-distortion to noise-and-distortion. A value of 12 dB is common for communications systems, while FM broadcast receivers are rated for a SINAD of 30 dB. The reason for the difference is that, while a 12 dB SINAD is sufficient for the understanding of speech, it is not enough for enjoyable music listening. The test signal is modulated at 60% of full deviation. The sensitivity may be expressed in terms of voltage (in microvolts) or power (in dBm or dBf, where the term dBf means decibels referenced to 1 femtowatt, which is 1×10^{-15} W). The receiver must be capable of producing at least one-half its rated audio output power with this signal; if not, the (higher) signal level required to produce one-half its rated power is the usable sensitivity.

Figure 9.29 shows a setup for measuring usable sensitivity. It is important that all cables be shielded, and that all covers be on the receiver. Remember that any noise that gets into the receiver through faulty connections or unshielded leads will detract from the sensitivity reading. If the generator output impedance does not match the receiver input impedance, a suitable matching network must be used, and its loss must be taken into account. It will not do just to connect, for instance, a 50 Ω unbalanced generator output to the 300 Ω balanced input of an FM broadcast receiver. Any results obtained will be meaningless if this is done. Similarly, the receiver output must be connected to a suitable load, otherwise the output power and distortion readings may be incorrect. For example, a receiver designed for use with an 8 Ω loudspeaker can be connected to an 8 Ω resistor with a power rating at least equal to the receiver's rated output power.

The generator and receiver must be tuned to the same frequency. Use a 1 kHz modulating signal that produces 60% of the rated deviation for the system in use. For example, with a communications system that uses 5 kHz maximum deviation, the deviation at the generator would be set to 3 kHz.

The distortion analyzer connected to the receiver audio output is used to measure SINAD. The procedure is identical to that used to measure harmonic distortion. The meter is first adjusted for a full-scale reading, then its notch filter is used to tune out as much of the fundamental 1 kHz signal as possible. What remains consists of noise and distortion, and the ratio between the two readings, expressed in decibels, is the SINAD. In most cases, the analyzer will have a decibel scale that allows SINAD to be measured directly in decibels.

FIGURE 9.29

**Test Setup for
Measurement of
Receiver Sensitivity**

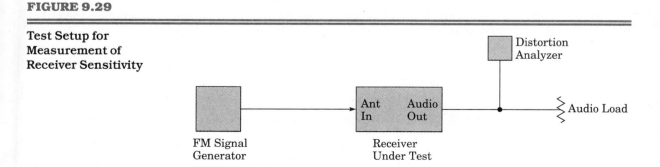

Normally, the nominal sensitivity of the receiver will be known, so that is a logical RF signal level at which to begin. Measure the SINAD with this signal. If it is greater than 12 dB (for a communications receiver), reduce the RF level; if the SINAD is less than 12 dB, increase the level. Continue until a SINAD of 12 dB is obtained. Then measure the audio power output. This is easily done by using the voltmeter function of the distortion analyzer to measure the voltage across the load resistor, then calculating the power from the equation

$$P = \frac{V^2}{R} \qquad\qquad\qquad\qquad\qquad (9.7)$$

Assuming that this power is at least one-half the rated power output specification, the generator RF output, after any matching networks are taken into account, is the usable sensitivity of the receiver. If not, simply increase the generator output until the audio output power increases to one-half the specified value, and use the generator output level required to achieve this as the usable sensitivity.

The procedure outlined above uses general-purpose test equipment. In applications where a great deal of receiver testing is done, a dedicated SINAD meter is used. This is like a distortion analyzer, except that it operates only at 1 kHz and is self-nulling. It can read out the SINAD in decibels without any adjustment, and therefore speeds up the process, though it does not improve the accuracy of the measurement.

9.6.2 Quieting Sensitivity

Quieting sensitivity is a measure of the effectiveness of FM limiting. The receiver is adjusted to produce noise power of 25% of rated output with no input signal (squelch or muting off). An unmodulated carrier is applied to the receiver and its strength is increased until the noise output decreases by a specified amount (20 dB for communications receivers, 50 dB for broadcast receivers). The level of carrier required for this is the *quieting sensitivity*. As for usable sensitivity, the quieting sensitivity can be specified in microvolts, dBm, or dBf.

The test setup for measuring quieting sensitivity can be the same as that described above for usable sensitivity. A distortion analyzer is not required, however; an audio voltmeter will do as well. For the best accuracy, a meter that reads true-RMS voltage for any waveshape is preferred, since the noise waveform will not be a sine wave. Since only a carrier is required, it is not necessary that the generator be capable of FM. Some older generators can produce only AM and CW signals; they will be quite satisfactory provided that the output level can be set accurately, and that the amount of signal leakage through the cabinet of the generator is low. Most low-cost generators do not meet these requirements, and cannot be used for sensitivity measurements, though they can be of some use in troubleshooting.

9.6.3 Stereo Channel Separation

A stereo audio system should ideally have no right-channel output when a signal is applied to the left channel only (with the right-channel input shorted to prevent hum and noise pickup). Of course, the converse is also true: any signal applied to the right-channel input should appear only at the right-channel output. This applies to any stereo system, whether phonograph, tape, compact disc, or AM or FM stereo radio is the signal source. This ideal situation is never attained in practice, of course. There is always some leakage from the driven channel to the unused channel. This is called *crosstalk*. A small amount of crosstalk is not a problem, but large amounts will limit the maximum possible stereo separation with actual program material.

Channel separation is given in decibels, and is easy to measure. With only one channel driven, the output voltage from both channels is measured, and the separation is found as

$$\text{Separation (dB)} = 20 \, \log \frac{V_1}{V_2} \qquad (9.8)$$

where V_1 = output voltage from the driven channel
        V_2 = output voltage from the other channel

The measurement of channel separation in an FM (or AM) stereo broadcast receiver requires a specialized generator that is capable of producing a stereo signal that is switchable between left and right channels. Such a generator will also have a mono setting and/or a "stereo" or "L − R" setting, where the two channels have signals of equal amplitude but opposite phase. The generator will be capable of providing a stereo baseband signal, suitable for the input of the stereo decoder, and a modulated RF signal that can be applied to the antenna input of a stereo receiver or tuner. In high-fidelity audio systems, a tuner is simply a receiver that contains no audio power amplifiers. Its output circuitry is designed to supply a signal on the order of 1 V to a high-impedance load, such as the input to a separate power amplifier.

Figure 9.30 shows a test setup for measuring the channel separation of an FM receiver or tuner. With the generator producing a signal on both channels, set the gain and/or balance controls on the receiver so that the two channels have equal output levels. Then, switch the generator to one channel, measure the output voltage of each channel, and apply Equation (9.8). The measurement should be repeated with the other channel active. Ideally, it should also be repeated for several audio frequencies. Lastly, the results should be compared with the receiver specifications. A high-quality stereo receiver should be capable of a separation of at least 40 dB, while an inexpensive portable or clock radio may have less than 20 dB of separation.

FIGURE 9.30

**Test Setup to Measure
Stereo Separation**

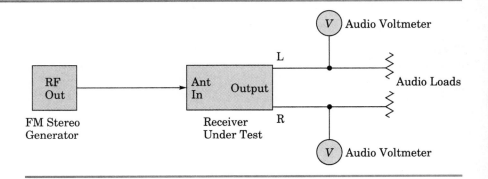

SUMMARY OF CHAPTER 9

Here are the main points to remember from this chapter.

1. FM transmitters and receivers have many similarities with their AM counterparts, but there are some significant differences. The differences are mainly due to the fact that FM signals have constant amplitude.

2. FM can be generated in two basic ways. Direct FM requires that the carrier oscillator be frequency modulated. In indirect FM, the modulating signal is integrated and then applied to a phase modulator.

3. Frequency multipliers are used in some FM transmitters to increase the frequency deviation. Mixers can be used to move the signal to a different frequency without affecting the modulation.

4. PLLs can be used to achieve wideband FM directly at the operating frequency, and, when incorporated as part of a frequency synthesizer, they can have excellent stability.

5. Class C amplifiers can be used to amplify FM signals, as amplitude distortion is not important with FM.

6. FM demodulators typically convert frequency variations to phase shifts, and then to amplitude variations. The exception to this is the PLL detector, in which the control voltage of a loop locked to the incoming signal becomes the output signal.

7. Limiters are used in FM receivers to remove amplitude variations due to noise, improving the output signal-to-noise ratio.

8. Both transmitters and receivers commonly use AFC circuits to achieve stability approximating that of crystal control with *LC* oscillator circuits. This is not necessary when PLL techniques are used in transmitter carrier oscillators and receiver local oscillators.

9. Transceivers are common for two-way mobile operation. Transceivers can use a common frequency synthesizer and can share audio components.

10. Stereo FM broadcasting is accomplished by synthesizing the composite baseband signal before modulation onto the main carrier. Similarly, stereo receivers first demodulate the FM signal, then demodulate the subcarrier, and finally combine the main and subcarrier signals.

IMPORTANT EQUATIONS

$$k_f = \frac{\Delta f}{\Delta v} \tag{9.4}$$

$$k_d = \frac{V_o}{\delta} \tag{9.5}$$

$$V_o \text{ peak} = \frac{\delta}{k_f} \tag{9.6}$$

$$\text{Separation (dB)} = 20 \log \frac{V_1}{V_2} \tag{9.8}$$

GLOSSARY

Armstrong method   a technique for generating FM indirectly using a balanced modulator and a phase shifter

automatic frequency control (AFC)   a scheme for keeping a transmitter or a receiver tuned to the correct frequency by applying a correction voltage to a VCO when the operating frequency is incorrect

channel separation   in a stereophonic audio system, the difference, in decibels, in the output level of the two channels with only one channel driven

Crosby system   a method of generating FM signals directly, which uses a discriminator for automatic frequency control

direct FM   any system which generates FM without going through a preliminary stage of phase modulation

discriminator   a circuit whose output is proportional to the difference between its input frequency and a predetermined center frequency

indirect FM   any method which generates FM using a phase modulator and an integrator applied to the modulating signal

quadrature detector   an FM detector circuit that makes use of the phase shift in a resonant circuit, as a signal moves through resonance

quieting sensitivity   the strength of an unmodulated carrier that reduces the noise output of an FM receiver by a specified amount

ratio detector   a type of discriminator that has the advantage of being insensitive to amplitude variations of the signal

SINAD   ratio of signal-plus-noise-and-distortion to noise-plus-distortion (Closely related to signal-to-noise ratio.)

usable sensitivity   strength of an FM signal, with defined deviation, required to produce a specified SINAD in a receiver

QUESTIONS

1. Why can an FM transmitter use low-level modulation followed by Class C amplification, while this is impossible with AM?
2. What limits the amount of frequency deviation that can be obtained with a varactor modulator for direct FM?
3. How can the deviation of an FM signal be increased?
4. Draw a block diagram of an FM transmitter using the Crosby system, and explain how this system stabilizes the carrier frequency.
5. What is meant by indirect FM?
6. Draw a block diagram showing an FM transmitter using the Armstrong method.

7. What method of direct FM allows wideband FM to be generated directly at the carrier frequency? Draw a block diagram for this type of modulator.

8. What is a limiter? Why and where are limiters used in FM receivers?

9. Why is AGC a requirement in an AM receiver, but not always used with FM?

10. What is the function of an AFC circuit?

11. Why is AFC more common in FM than in AM broadcast receivers?

12. Why is AFC unnecessary with receivers using a frequency-synthesized local oscillator?

13. Compare the Foster-Seeley discriminator and the ratio detector, giving one advantage of each.

14. Draw and explain the S-curve for a typical FM detector.

15. Describe two modern types of FM detectors, and explain how each works. Use diagrams to help with the explanations.

16. How does an FM stereo decoder detect that a stereo signal is present?

17. What is the reason for high-frequency blending in FM stereo receivers?

18. Why are FM transceivers often designed to transmit and receive on different frequencies?

19. Define two types of sensitivity measurement for FM receivers, and describe the procedure for each.

20. Why are the sensitivity specifications defined in different ways for broadcast than for communications receivers?

21. How would you measure the channel separation of a stereo FM receiver? About what amount of separation would you expect to find for a high-fidelity FM tuner? A clock radio?

PROBLEMS

Section 9.2

22. A direct-FM transmitter has a block diagram as shown in Figure 9.31.

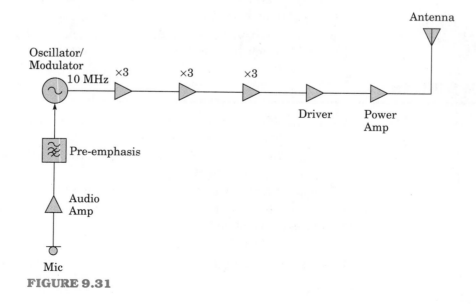

FIGURE 9.31

(a) What is the carrier frequency of the output signal?

(b) If the modulator has a sensitivity of 5 kHz/V, what modulating voltage would be required for a deviation, at the output, of 50 kHz?

23. Calculate the carrier frequency and deviation at the output of the direct-FM transmitter shown in Figure 9.32.

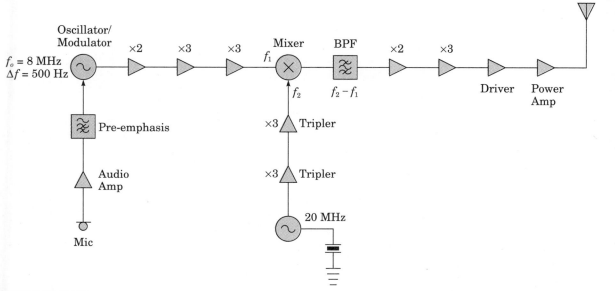

FIGURE 9.32

24. An indirect-FM transmitter has the block diagram of Figure 9.33. The total output power is 1 W into 50 Ω.

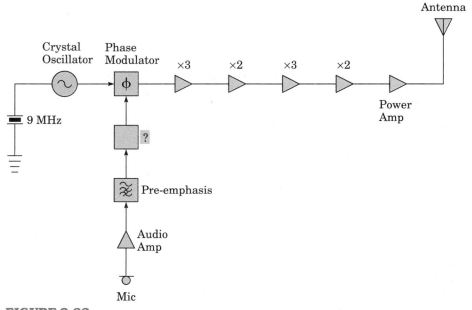

FIGURE 9.33

(a) Label the block that has been left blank.

(b) What phase deviation will be necessary at the modulator in order that the output frequency deviate by 5 kHz with a 1 kHz modulating signal?

(c) Use Bessel functions to sketch the spectrum of the output signal. Be sure to include suitable scales. It is only necessary to include the carrier and the first three sidebands on each side.

25. A block diagram for an FM transmitter using indirect FM is shown in Figure 9.34.

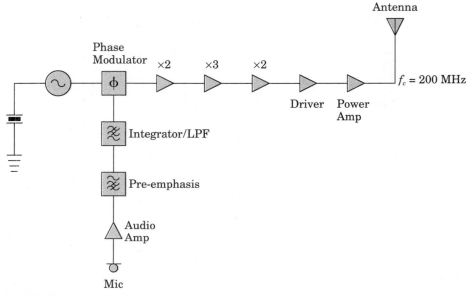

FIGURE 9.34

(a) What is the frequency of the carrier oscillator?

(b) What is the maximum phase deviation the phase modulator must be capable of supplying, if the transmitter is to produce FM with a maximum deviation of 25 kHz with a modulating-signal frequency of 10 kHz?

(c) Why is the integrator necessary?

26. The block diagram of a Crosby-type transmitter is shown in Figure 9.35. Calculate the output frequency and deviation.

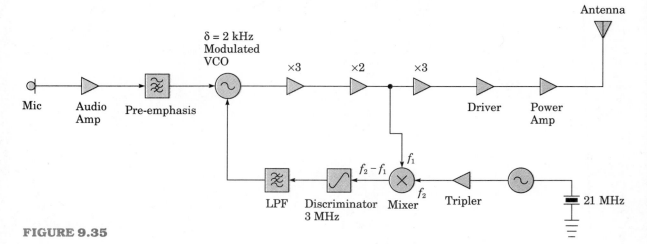

FIGURE 9.35

27. Figure 9.36 is the block diagram of an FM transmitter using a PLL.

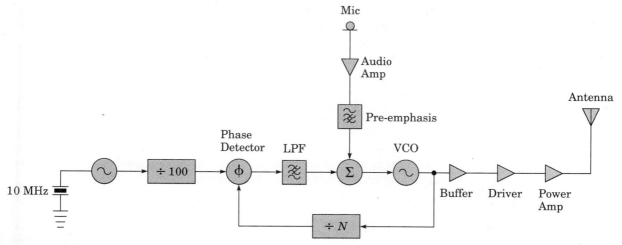

FIGURE 9.36

(a) Calculate the required value of N for a carrier frequency of 91.1 MHz.

(b) Calculate the minimum acceptable value of k_f for the VCO if the transmitter is to produce a 75 kHz deviation with an input signal of 2 V RMS.

Section 9.3

28. (a) Draw the block diagram for an FM broadcast receiver. It is to have one RF stage and an IF of 10.7 MHz. The local oscillator will operate above the signal frequency. Indicate on the diagram the frequency or frequencies at which each stage operates when the receiver is receiving a station at 94.5 MHz.

(b) What is the image frequency of the receiver described above?

(c) Give two ways in which the image rejection of the receiver could be improved.

29. A PLL FM detector uses a VCO with $k_f = 20$ kHz/V. If it is receiving an FM signal with a deviation of 30 kHz and sine-wave modulation, what is

(a) the output voltage from the detector?

(b) the minimum value for the lock range of the loop, for satisfactory operation of the detector?

30. An FM detector has a sensitivity of 100 mV/kHz. How much deviation, assuming a sinusoidal modulating signal, is required for an output of 2 V RMS?

31. A limiter has a limiting threshold of 225 mV at the input.

(a) How much voltage gain is needed before the limiter if the receiver is to start limiting with a signal of 0.4 μV at the antenna?

(b) Would it be possible to receive weaker signals with this receiver? Explain.

32. An FM receiver is double conversion with a first IF of 10.7 MHz and a second IF of 455 kHz. High-side injection is used in both mixers. The first local oscillator is a VFO, while the second is crystal-controlled. There is one RF stage, one stage of IF amplification at the first IF, and three stages of combined IF amplification and limiting at the second IF. A Foster-Seeley discriminator is used as the detector. The receiver is tuned to a signal with a carrier frequency of 160 MHz.

(a) Draw a block diagram for this receiver.

(b) Calculate the Q that will be necessary in the input circuits to achieve a rejection of 50 dB at the first image frequency, assuming that there are two tuned circuits, with no coupling between them, before the first mixer. [*Hint*: you will find the necessary equations in Chapter 5.]

(c) If AGC were used with this receiver, to which stages could it be applied?

(d) If AFC were necessary, from which stage would it be derived, and to which stage would it be applied?

Section 9.4

33. Refer to the block diagram for an FM transceiver shown in Figure 9.24.

    (a) What frequencies would the synthesizer produce while transmitting and receiving, respectively, on marine channel 14 (156.7 MHz)?

    (b) Explain the absence of frequency multipliers in the transmitter section.

    (c) The transmitter section uses a maximum deviation of 5 kHz and operates with modulating frequencies in the range of approximately 300 Hz to 3 kHz. Suggest an appropriate receiver bandwidth.

Section 9.5

34. Show mathematically how the left and right signals can be combined into L+R and L−R signals, and then separated back into left and right, using only addition and subtraction.

35. Suppose that a receiver has a usable sensitivity of 50 μV in stereo. What is likely to be the approximate sensitivity when the receiver is switched to mono?

Section 9.6

36. A stereo FM receiver is tested for channel separation. With a stereo generator producing a signal on the left channel only, the receiver has an output of 2 V on the left channel and 30 mV on the right. What is the stereo separation in decibels?

37. You are making a usable sensitivity test of an FM communications receiver and have just measured a SINAD of 10 dB with an input signal level of 0.5 μV. What do you do next?

38. In making a quieting sensitivity test on an FM broadcast receiver, you have measured the noise output voltage, with no signal, as 2 V. You turn on the carrier and, as you increase the carrier level, the output voltage decreases.

    (a) You continue to increase the carrier level until the output voltage declines to what value?

    (b) Once the output voltage level found in Part (a) is reached, what do you do next?

Comprehensive

39. Commercial FM radio stations often try to sound "louder" than the competition. Within the primary coverage area (that is, the area where the signal is quite strong at the receiver antenna), which of the following approaches would be more effective in increasing loudness? Explain your answer.

    (a) increasing the transmitter carrier power

    (b) using audio compression to increase the average deviation

40. Draw a block diagram of a communications receiver that could be used to receive AM, SSB, and FM transmissions.

10

Transmission Lines

Objectives

After studying this chapter, you should be able to:

1. Give several examples of transmission lines, and explain what parameters of a transmission line must be considered as the frequency increases.
2. Define characteristic impedance, and calculate the impedance of a coaxial or open-wire transmission line.
3. Describe the responses of a matched and a mismatched transmission line to a step or pulse input.
4. Define reflection coefficient and standing-wave ratio, and calculate them in practical situations.
5. Explain the importance of impedance matching with respect to transmission lines, and describe several methods of matching lines.
6. Perform the necessary calculations to achieve an impedance match using a quarter-wave transformer and a single stub.
7. Describe the function and use of several types of transmission-line test equipment.

▰▰▰▰ 10.1 INTRODUCTION

So far in our study of communications systems, we have essentially looked at both ends: the transmitter and the receiver. In the next few chapters, we will study the middle, referred to in Chapter 1 as the *channel*. The signal can proceed from transmitter to receiver by a variety of means, including metallic cable, optical fiber, and radio transmission. We begin with the metallic cable, which is referred to as a **transmission line**.

Almost any configuration of two or more conductors can operate as a transmission line, but there are several types that have the virtues of being relatively easy to study as well as being very commonly used in practice.

Figure 10.1 shows some **coaxial lines**, in which the two conductors are concentric, separated by an insulating dielectric. Figure 10.1(a) shows a solid-dielectric cable; the cable in Figure 10.1(b) uses air (and occasional plastic spacers) for the dielectric. Sometimes the conductors are separated by a single helical (spiral) spacer, as shown in Figure 10.1(c). When high power is used, it is important to keep the interior of the line dry, so these lines are sometimes pressurized with nitrogen to keep out moisture. Coaxial cables are referred to as *unbalanced lines* because of their lack of symmetry with respect to ground (usually the outer conductor is grounded).

A number of parallel-line cables are pictured in Figure 10.2. Figure 10.2(a) shows television twin-lead. The two conductors are separated by a thin ribbon of plastic, but actually air forms a good part of the dielectric, because the electric field fills the space around the wires, not just the region directly between them. Figure 10.2(b) is a depiction of **open-wire line**, in which the amount of solid dielectric is greatly reduced, only a few spacers being required to separate the conductors.

Parallel lines are usually operated as *balanced lines*; that is, the impedance to ground from each of the two wires is equal. This ensures that the currents in the two wires in the line will be equal in magnitude and opposite in sign, with the result that radiation from the cable, and also its susceptibility to outside interference, will be reduced. If required, a grounded shield can be placed around the cable, as shown in Figure 10.2(c). The shield plays no part in transferring the signal from source to destination; its only function is to reduce noise and interference.

At first glance, it may seem that there is very little to say about transmission lines. The wire connections familiar from everyday life, or from basic electricity and electronics for that matter, are often considered ideal. Like the lines representing wires on a schematic diagram, they are assumed to have no effect on the signals that pass through them.

When lines used at low frequencies are not considered ideal, they are usually represented as having only resistance. It is the resistance of a household extension cord that causes the voltage, under load, to be lower at the load than at the source. Resistance is easy to calculate and to measure, and its effects are very predictable.

Transmission lines become more complex in their behavior as the frequency increases. Besides resistance in the conductors, inductive and capacitive effects become important. As indicated in Chapter 2, these factors are more important the higher the frequency and the longer the line.

Transmission lines are seldom the most interesting part of a communications system. In fact, the best transmission line is often the least interesting one: the one that approaches most closely the ideal of a line with no loss

FIGURE 10.1

Coaxial Cables

(a) Solid Dielectric

(b) Air Dielectric

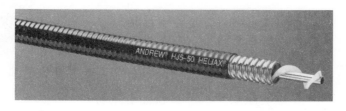

(c) Heliax®™

(a) Courtesy Cooper Industries/Belden Division. (b) and (c) Courtesy Andrew Corporation.

and no reactance. However, it is necessary to understand transmission lines in order to achieve this pleasant, if unexciting result, which allows signals to move from source to load with minimal attenuation or distortion.

Transmission lines can have other uses as well. They can be employed to provide phase shifts and time delays, and to match impedances. A length of transmission line can even be used as a filter, or as a tuned circuit in a receiver or transmitter stage. All of this makes the study of transmission lines important and necessary.

FIGURE 10.2

Parallel-Line Cables

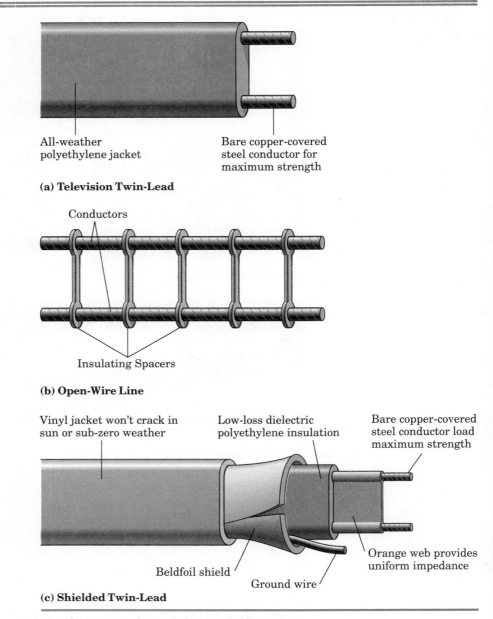

All-weather
polyethylene jacket

Bare copper-covered
steel conductor for
maximum strength

(a) Television Twin-Lead

Conductors

Insulating Spacers

(b) Open-Wire Line

Vinyl jacket won't crack in
sun or sub-zero weather

Low-loss dielectric
polyethylene insulation

Bare copper-covered
steel conductor load
maximum strength

Beldfoil shield

Ground wire

Orange web provides
uniform impedance

(c) Shielded Twin-Lead

(a) and (c) courtesy Cooper Industries, Belden Division.

10.2 ELECTRICAL MODEL OF A TRANSMISSION LINE

In the analysis of transmission lines, it is necessary to use distributed rather
than lumped constants. The factors that must be considered are as follows.

First, there is the resistance of the line. This is familiar from low-
frequency applications, but there is one difference: resistance increases with
frequency. Any current flow in a conductor is associated with a magnetic
field, both within the conductor and in the space surrounding it. At high fre-

Other Transmission Lines

Coaxial and parallel-wire transmission lines are the most common for high-frequency communications, but they are not the only possibilities. Twisted-pair line, shown in Figure 10.3(a), is popular for audio-frequency circuits, such as telephone and intercom cables. When hum pickup from power wiring and crosstalk from nearby circuits are problems, a shield may be provided around the pair. The twisting results in quite good cancellation of interfering fields, especially if the line is operated in a balanced configuration with respect to ground. Twisted pairs are quite lossy at radio frequencies, but are sometimes used at frequencies of several MHz (in local-area networks for computers, for example). Twisted-pair lines in these applications have the advantage of low cost; in fact, many office buildings have extra telephone lines already installed that can be used for this application.

A single wire can operate as a transmission line, provided that ground is used as a second, *return* connection. The early telegraph systems worked this way, with each circuit on a single wire. In fact, iron rather than copper was used, because the lower cost and greater strength of iron wire offset its lower conductivity.

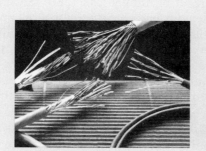

(a) Twisted-Pair Line　　　　　**(b) Electric Power Line**
FIGURE 10.3
(a) Courtesy Cooper Industries/Belden Division.

Long-distance power cables, as in Figure 10.3(c), are also transmission lines.

quencies, the magnetic field within the conductor causes most of the current to flow near its surface. As the frequency increases, the region of high current density becomes thinner, reducing the effective cross-sectional area and increasing the resistance of the conductor. Because most of the current flows in a thin region resembling a "skin" near the surface of the wire, this phenomenon is called the *skin effect*. Skin effect is the reason that hollow tubing works just as well as a solid conductor for such applications as television antennas and the coils in radio transmitters for VHF and higher frequencies. We came across the skin effect once before, in Chapter 5, where we studied its effect on the variation of the Q-factor of inductors with frequency.

FIGURE 10.4   **Model of a Short Transmission-Line Section**

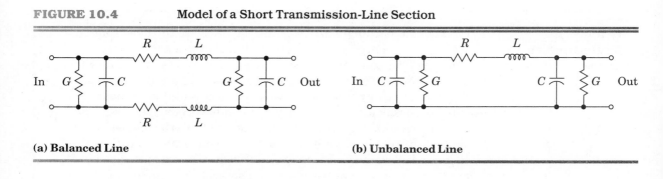

(a) Balanced Line   **(b) Unbalanced Line**

In addition to the resistance of the wires, there is the conductance of the dielectric. At low frequencies, this is very small and can be neglected. Dielectrics, however, tend to become more lossy as the frequency increases.

As well as resistance, any conductor or combination of conductors has inductance. There is also capacitance between any two conductors separated by a dielectric. Any transmission-line model will therefore have to include both inductance and capacitance.

It is difficult to visualize and to work with distributed constants. One way to approach the problem is to consider a short section of line, and assign to this section a number of lumped constants. Such a scheme is the basis for Figure 10.4. Models for both balanced (such as open-wire) and unbalanced (such as coaxial) line are illustrated. The figure allows for R, the resistance of the wire; G, the conductance of the dielectric; L, the series inductance; and C, the shunt capacitance. All of these will be given per unit length; for example, the resistance can be expressed in ohms per meter. However, the length of line we are considering is certainly much smaller than a meter. In fact, the idea is to allow the section to shrink until its length is infinitesimal.

At dc and low frequencies, the inductance has no effect because its reactance is very small compared with the resistance of the line. Similarly, the reactance of the shunt capacitance is very large, so the effect of the capacitance is negligible, as well. The line is characterized by its resistance, and possibly by the conductance of the dielectric, though this can usually be neglected.

As the frequency increases, the inductance and capacitance will have an effect. The higher the frequency, the larger the series inductive reactance and the lower the parallel capacitive reactance. In fact, it is often possible at high frequencies to simplify calculations by neglecting the resistive elements and considering only the inductance and capacitance of the line. Such a line is called *lossless*, since the inductive and capacitive reactances store, but do not dissipate, energy.

10.3 STEP AND PULSE RESPONSE OF LINES

Although transmission lines are used for a wide variety of signals, it is useful to begin looking at the response of transmission lines by considering some very simple signals. It is also useful, as a starting assumption, to suppose that the line is lossless and infinite in length. That way, we will not have to

FIGURE 10.5

**Step Input Applied to an
Infinitely Long Line**

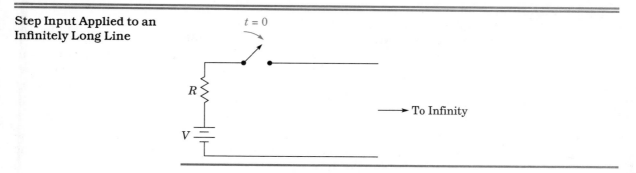

worry about what happens to the signal when it reaches the end of the line. Let us take such a line, shown schematically in Figure 10.5, and apply a step input to one end; that is, let the voltage applied to the line be zero until some time that we can arbitrarily call $t = 0$. At that time, the switch closes, connecting a source with voltage V to the line, through a resistance R.

Current begins to flow into the line, charging the capacitances. A surge of energy moves along the line at some finite speed. It takes time for the current to build up in the inductances, and for the capacitances to charge. If the line is infinitely long, the surge will continue along the line forever.

10.3.1   Characteristic Impedance

Returning to the source for a moment, we might ask what happens to the voltage and current at that point. Since the line has capacitance to be charged, the initial current will not be zero, as we might have expected. Even though the line is open-circuited, it does not look like an open circuit to the source. Nor does it look like a short circuit at any time, as a completely discharged capacitor would, because the inductance of the line limits the initial current. Rather, it appears that the current will have some definite, finite value that will not change as long as the surge continues to move down the line. Though all we have attempted here is to show that this is a plausible idea, it turns out to be true. There is a definite ratio between the voltage and current for any transmission line under the conditions just described. Since the ratio of voltage to current is generally called impedance, and since the ratio we are talking about is a characteristic of the type of line used, we call it the **characteristic impedance** of the line. Sometimes it is also called the **surge impedance**. The impedance is a real number (that is, resistive) for a line with no losses. Note carefully that, even though the characteristic impedance of a line has the units of ohms, it does not represent an actual resistance. That is, calling a transmission line a "50 Ω line" does not say anything about the resistance of the wire in the line. The hypothetical line that we are considering does not have any resistance, since it is lossless. However, since it has infinite length, energy put into the line continues to move along it forever, and, looking into the line from the source, the line looks like a resistance. Instead of being dissipated as heat, the energy from the source continues to move along the line, away from the source, forever. The effect is the same, as seen from the source: energy put into the line disappears.

If this is the case, then rather than continuing to deal with an infinite line which must remain a figment of the imagination, it should be possible to replace that line with one of finite length, terminated at the destination end with a resistance equal to the characteristic impedance of the line. Looking down the line from the source, it would be impossible to distinguish this finite, terminated line from the infinite line discussed earlier. Instead of moving down the line forever, the electrical energy would be converted into heat in the resistor, but no difference would be apparent from the source. A transmission line that is terminated in its characteristic impedance is called a *matched line*. Figure 10.6 shows a matched line of finite length.

The characteristic impedance depends on the electrical properties of the line. In particular, it can be shown that, for any transmission line,

$$Z_0 = \sqrt{\frac{R + j\omega L}{G + j\omega C}} \tag{10.1}$$

where Z_0 = characteristic impedance of the line in ohms

R = conductor resistance in ohms per unit length

j = $\sqrt{-1}$

L = inductance in henrys per unit length

G = dielectric conductance in siemens per unit length

C = capacitance in farads per unit length

ω = operating frequency in radians per second

This equation has several interesting mathematical properties. In general, the impedance is complex and is a function of frequency, as well as the physical characteristics of the line. For a lossless line, however, R and G would be zero and Equation (10.1) would simplify to

$$Z_0 = \sqrt{\frac{j\omega L}{j\omega C}}$$

$$= \sqrt{\frac{L}{C}} \tag{10.2}$$

FIGURE 10.6

Step Input Applied to a Matched Line of Finite Length

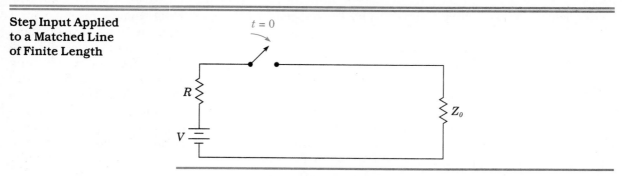

Of course, there is no such thing as a completely lossless line, but many practical lines approach the ideal closely enough that the characteristic impedance can be approximated by Equation (10.2). This is especially true at high frequencies: as ω gets larger, the values R and G become less significant in comparison with L and C. For this reason, Equation (10.2) is often referred to as the *high-frequency model of a transmission line*. Equation (10.2) gives a characteristic impedance that is a real number and does not depend on frequency, but only on such characteristics as the geometry of the line and the permittivity of the dielectric. Remember that L and C are the inductance and capacitance per unit length. The unit of length does not matter, so long as it is the same for both L and C.

EXAMPLE 10.1

A coaxial cable has a capacitance of 90 pF/m and a characteristic impedance of 50 Ω. Find the inductance of a 1 m length.

Solution

We can do a little basic algebra with Equation (10.2):

$$Z_0 = \sqrt{\frac{L}{C}}$$

$$Z_0^2 = \frac{L}{C}$$

$$L = Z_0^2 C$$

$$= 50^2 \times 90 \times 10^{-12} \text{ H/m}$$

$$= 225 \text{ nH/m}$$

The characteristic impedance can be calculated for any type of transmission line by calculating the capacitance and inductance per unit length, but it can be a tedious business. Two equations are presented here, which will allow the majority of practical cases to be dealt with. First, consider a parallel line with air dielectric, as shown in cross-section in Figure 10.7. The impedance of such a line can be found from

FIGURE 10.7

Air-Dielectric Parallel Line: Cross Section

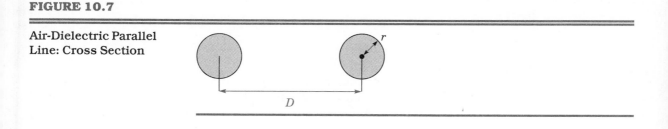

FIGURE 10.8

Coaxial Cable:
Cross Section

$$Z_0 \approx 276 \ \log \frac{D}{r} \tag{10.3}$$

where Z_0 = characteristic impedance of the line
        D = spacing between the centers of the conductors
        r = conductor radius

Equation (10.3) is valid only for $D \gg r$. It also neglects the effect of spacers on the impedance. For open-wire ladder line, this is reasonable, but for open-wire line with a solid ribbon joining the conductors, such as TV twin-lead, the dielectric will cause the characteristic impedance to be less than that given by Equation (10.3).

For a coaxial cable like the one shown in cross-section in Figure 10.8, the impedance can be found from

$$Z_0 \approx \frac{138}{\sqrt{\epsilon_r}} \ \log \frac{D}{d} \tag{10.4}$$

where Z_0 = characteristic impedance of the line
        D = inside diameter of the outer conductor
        d = diameter of the inner conductor
        ϵ_r = relative permittivity of the dielectric, compared with that of
             free space (ϵ_r is often called the *dielectric constant*.)

Here it is quite obvious that increasing the dielectric constant reduces the impedance of the cable.

EXAMPLE 10.2

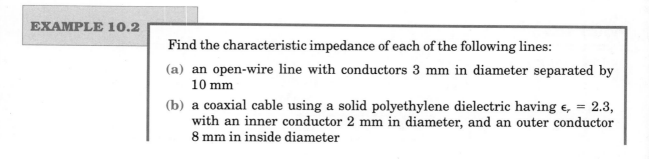

Find the characteristic impedance of each of the following lines:

(a) an open-wire line with conductors 3 mm in diameter separated by 10 mm

(b) a coaxial cable using a solid polyethylene dielectric having $\epsilon_r = 2.3$, with an inner conductor 2 mm in diameter, and an outer conductor 8 mm in inside diameter

Solution

(a) From Equation (10.3),

$$Z_0 \approx 276 \log \frac{D}{r}$$

$$= 276 \log \frac{10 \text{ mm}}{1.5 \text{ mm}}$$

$$= 227 \ \Omega$$

(b) From Equation (10.4),

$$Z_0 = \frac{138}{\sqrt{\epsilon_r}} \log \frac{D}{d}$$

$$= \frac{138}{\sqrt{2.3}} \log \frac{8}{2}$$

$$= 54.8 \ \Omega$$

In practice, it is usually not necessary to use Equation (10.4) to find the impedance of a coaxial cable, since this is part of the cable specifications. In fact, the impedance is often marked right on the insulating jacket. For coaxial cable, there are a few standard impedances that fulfill most requirements. Some examples are shown in Table 10.1. The type numbers in Table 10.1 began as U.S. military specifications, but have since come into general use and are employed by a number of manufacturers.

Practical open-wire lines usually have relatively high impedance. Television twin-lead is rated at 300 Ω, and open-wire lines are often 600 or 900 Ω.

10.3.2 Velocity Factor

It was mentioned above that a signal moves down the line at some finite rate. This is obvious; even in free space, nothing can move at a velocity greater than the speed of light, which is given approximately by

$$c = 3 \times 10^8 \text{ m/s} \tag{10.5}$$

TABLE 10.1

Coaxial Cable Applications	Impedance (ohms)	Application	Typical Type Numbers
	50	radio transmitters	RG-8/U
		communications receivers	RG-58/U
	75	cable television	RG-59/U
		TV antenna feedlines	
	93	computer networks	RG-62/U

The speed at which energy is propagated along a transmission line is always less than the speed of light. It varies from about 66% of this velocity on coaxial cable with solid polyethylene dielectric, through 78% for polyethylene foam dielectric, to about 95% for air-dielectric cable. Rather than specify the actual velocity of propagation, it is normal for manufacturers to specify the **velocity factor**, which is given simply by

$$v_f = \frac{v_p}{c} \qquad (10.6)$$

where v_f = velocity factor, as a decimal fraction
       v_p = propagation velocity on the line
       c = speed of light in free space

The velocity factor is also commonly expressed as a percentage, found by simply multiplying the value found from Equation (10.6) by 100.

The velocity factor for a transmission line depends almost entirely on the dielectric used in the line. It is given by

$$v_f = \frac{1}{\sqrt{\epsilon_r}} \qquad (10.7)$$

where ϵ_r = line's dielectric constant

EXAMPLE 10.3

Find the velocity factor and propagation velocity for a cable with a Teflon dielectric ($\epsilon_r = 2.1$).

Solution

From Equation (10.7),

$$v_f = \frac{1}{\sqrt{\epsilon_r}}$$

$$= \frac{1}{\sqrt{2.1}}$$

$$= 0.69$$

From Equation (10.6),

$$v_f = \frac{v_p}{c}$$

$$v_p = v_f c$$

$$= 0.69 \times 3 \times 10^8 \text{ m/s}$$

$$= 2.07 \times 10^8 \text{ m/s}$$

FIGURE 10.9

Step Input to an
Open-Circuited Line

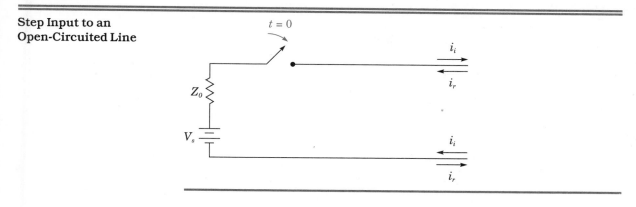

10.3.3 Reflections

Suppose that we have a line of finite length, but terminated in an open circuit (as shown in Figure 10.9) rather than in its characteristic impedance. This mismatched line will behave in exactly the same way as the matched line, when the switch is first closed. A surge of energy will move down the line, and the ratio between voltage and current at the source will be equal to the characteristic impedance Z_0. Since the source impedance is equal to the characteristic impedance of the line, one-half the source voltage will appear across the input end of the line. The other one-half appears across the source resistance.

Now, let us follow the surge down the line. The voltage and current will continue to be related by Z_0 until the surge reaches the end of the line. At that point, the total current must be zero since there is an open circuit. On the other hand, the incoming surge of energy cannot simply disappear, because there is nothing capable of dissipating energy at this point. What happens is that the energy reflects from the open end of the line. The reflected voltage is the same as the incident voltage, but the current is equal in magnitude and opposite in direction to the incident current. Thus, the total voltage at the destination end of the line is twice the incident voltage, and the total current is zero. The incident voltage is one-half the source voltage, so the total voltage is simply equal to the source voltage.

As the reflected surge moves back toward the input, the current gradually becomes zero all along the line. Once the input is reached, the final conditions at the far end apply all along the line: the voltage is equal to the source voltage, and the current is zero. It is interesting that this final condition is exactly what anyone would predict: when a voltage is connected across an open-circuited line, the voltage across the line becomes equal to the source voltage, and the current becomes zero. The only difference is that, with a long transmission line, the process takes time. Once the transient has died down, there are no surprises.

Figure 10.10 shows what happens. Figure 10.10(a) shows how the voltage at the source end of the line varies with time. The voltage rises from zero to one-half the supply voltage at $t = 0$, then rises again, to the full supply voltage, once the surge has had time to move down the line to the open end and

FIGURE 10.10   Voltage on an Open-Circuited Line with Step Input

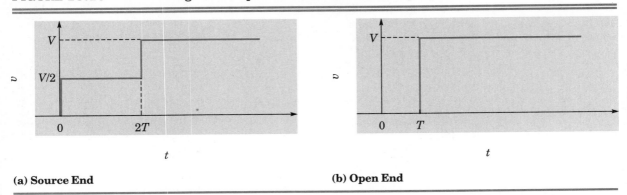

(a) Source End   **(b) Open End**

back again. As shown in Figure 10.10(b), there is no voltage at the open end until the surge has reached it. At that instant, the voltage rises from zero to the full supply voltage. The time taken for the signal to move down the line is given simply by

$$T = \frac{L}{v_p} \tag{10.8}$$

where T = time for a signal to travel the length of the line
L = length of the line
v_p = propagation velocity on the line

Next, let us consider what happens if a step input is applied to a transmission line that is shorted at the far end, as shown in Figure 10.11(a). At first, there is no perceptible difference. While the incident surge is still moving along the line toward the shorted end, the voltage and current at the input end will be related by the characteristic impedance of the line, just as

FIGURE 10.11   Voltage Step Applied to a Shorted Line

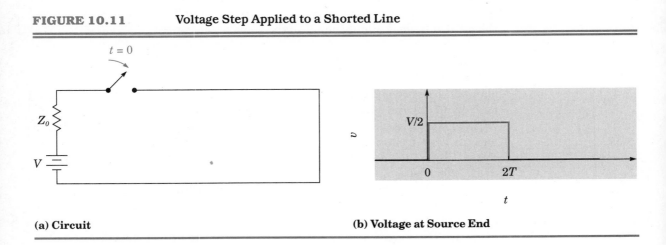

(a) Circuit   **(b) Voltage at Source End**

before. However, once the surge reaches the shorted end, a different boundary condition applies. With an open end, the total current was compelled to be zero, but there was no restriction on the voltage. A short circuit, on the other hand, must have a voltage of zero across it. The current can have any value, as determined by other constraints on the system. Once again, the signal will be reflected, but this time the current will be in the same direction as the incident current, and the voltage will have the opposite polarity. When the reflected surge reaches the input of the line, the voltage all along the line will be zero, and the current will be limited only by the source resistance. This is exactly what we would have expected previously, for a short-circuited line connected to a dc source. Figure 10.11(b) shows the sequence of events.

Instead of a step input, a pulse could be applied to the line, as in Figure 10.12. Once again, if the line were terminated in its characteristic impedance, the pulse would simply proceed down the line and be dissipated. With an open circuit at the far end, the pulse would be reflected with the same voltage polarity; with a shorted line, the reflected pulse would have opposite polarity.

The situations that have been described so far are simple, limiting cases, where there is either complete reflection or no reflection at all. Now suppose that the line is terminated in an impedance, but one that is not equal to the characteristic impedance of the line. For instance, consider a 50 Ω line terminated by a 25 Ω resistor. One might guess that there would be partial reflection of the incident surge or pulse, and that is in fact what happens. At

FIGURE 10.12   **Pulse Input to Transmission Lines: Voltages at the Source End**

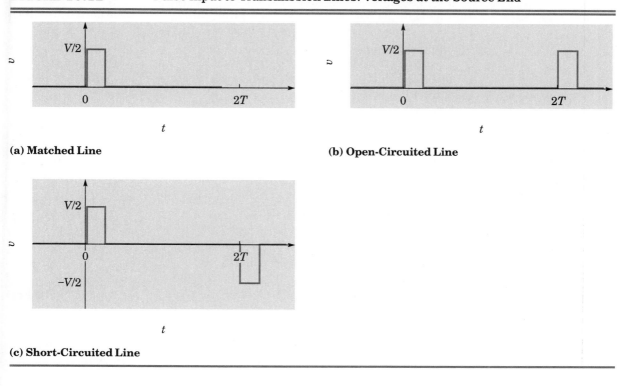

(a) Matched Line

(b) Open-Circuited Line

(c) Short-Circuited Line

the termination, the ratio of voltage to current is given by the load resistance Z_L, which in this case is 25 Ω. That is, Ohm's Law must be satisfied at the termination. On the other hand, the voltage and current in the incident surge are related by the Z_0 of the line, in this case 50 Ω.

A line that is terminated by an impedance other than Z_0 is said to be *mismatched*. Mismatches, while usually undesirable, are very common with transmission lines. It is often difficult to design a load, such as an antenna, to present precisely the right impedance to the line, especially when the frequency must vary. When performing calculations with mismatched lines, it is convenient to define a reflection coefficient

$$\Gamma = \frac{V_r}{V_i} \tag{10.9}$$

where Γ = voltage reflection coefficient, a dimensionless number (In
             general, it is a complex number.)
       V_r = reflected voltage
       V_i = incident voltage

Note that Γ is formally defined as the *voltage reflection coefficient*. This is to distinguish it from the reflection coefficient for power, which turns out to be equal to Γ^2. However, in common use, Γ is usually referred to simply as the **reflection coefficient**.

From the work done so far, it can be seen that a matched line has a reflection coefficient of zero, since there is no reflected voltage. For an open-circuited line, $\Gamma = 1$, since the reflected voltage has the same magnitude and sign as the incident voltage. On the other hand, a short-circuited line has $\Gamma = -1$, because the incident and reflected voltages are equal in magnitude but opposite in sign.

The reflection coefficient is determined by Z_L and Z_0. It is quite possible to derive a relation among these three parameters. First, note that at the load, the total voltage will be

$$V_T = V_r + V_i \tag{10.10}$$

and the total current, that is, the current into the load, will be given by Ohm's Law:

$$I_T = \frac{V_T}{Z_L} \tag{10.11}$$

Also, the total current at the load will be

$$I_T = I_i + I_r \tag{10.12}$$

where I_i and I_r are the incident and reflected currents, respectively.

On the line, the incident voltage and current will be related by the characteristic impedance Z_0, and so will the reflected voltage and current.

$$I_i = \frac{V_i}{Z_0} \tag{10.13}$$

$$I_r = -\frac{V_r}{Z_0} \tag{10.14}$$

The negative sign in Equation (10.14) is there because the reference direction of current is initially defined as toward the load on one of the conductors. A reflected current flowing away from the load will then be negative.

Now, at the termination, combining Equations (10.11) and (10.12), we have

$$\frac{V_T}{Z_L} = I_i + I_r$$

Substituting in the expression for V_T found in Equation (10.10) gives

$$\frac{V_r + V_i}{Z_L} = I_i + I_r$$

Next, substitute the expressions for I_i and I_r found in Equations (10.13) and (10.14), respectively:

$$\frac{V_r + V_i}{Z_L} = \frac{V_i}{Z_0} - \frac{V_r}{Z_0}$$

$$\frac{V_r + V_i}{Z_L} = \frac{V_i - V_r}{Z_0} \tag{10.15}$$

Remembering that $\Gamma = V_r/V_i$, Equation (10.15) can be simplified by cross-multiplication.

$$V_i Z_L - V_r Z_L = V_i Z_0 + V_r Z_0$$

$$V_r(Z_0 + Z_L) = V_i(Z_L - Z_0)$$

$$\frac{V_r}{V_i} = \frac{Z_L - Z_0}{Z_L + Z_0}$$

$$\Gamma = \frac{Z_L - Z_0}{Z_L + Z_0} \tag{10.16}$$

Once again, it should be pointed out that any or all of the quantities in Equation (10.16) can be complex.

EXAMPLE 10.4

The switch shown in Figure 10.13(a) closes at time $t = 0$, applying a 1 V source through a 50 Ω resistor to a 50 Ω line that is terminated by a 25 Ω resistor. The line is 10 m in length, with a velocity factor of 0.7. Draw graphs showing the variation of voltage with time at each end of the line.

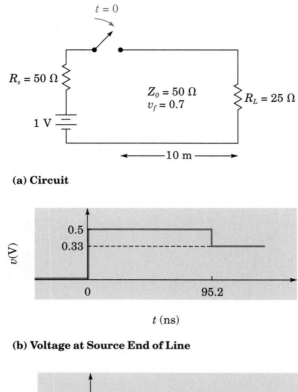

(a) Circuit

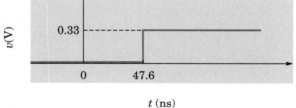

t (ns)

(b) Voltage at Source End of Line

0.33

0   47.6

t (ns)

(c) Voltage at Load End of Line
FIGURE 10.13

Solution

At $t = 0$, the source voltage will divide between R_s and the Z_0 of the line. Since the two resistances are equal, the voltage will divide equally. Therefore, at $t = 0$, the voltage at the source end of the line will rise from zero to 0.5 V.

The voltage at the load will remain zero until the surge reaches it. The time for this is

$$T = \frac{L}{v_p}$$

$$= \frac{L}{v_f c}$$

$$= \frac{10 \text{ m}}{0.7 \times 3 \times 10^8 \text{ m/s}}$$

$$= 47.6 \text{ ns}$$

After 47.6 ns, the voltage at the load will rise. The new value can be found by calculating the reflection coefficient. From Equation (10.16),

$$\Gamma = \frac{Z_L - Z_0}{Z_L + Z_0}$$

$$= \frac{25\ \Omega - 50\ \Omega}{25\ \Omega + 50\ \Omega}$$

$$= \frac{-25\ \Omega}{75\ \Omega}$$

$$= -0.333$$

The reflected voltage can be found from Equation (10.9).

$$\Gamma = \frac{V_r}{V_i}$$

$$V_r = \Gamma V_i$$

$$= -0.333 \times 0.5\ \text{V}$$

$$= -0.1665\ \text{V}$$

The total voltage at the load is given by Equation (10.10).

$$V_T = V_r + V_i$$

$$= -0.1665\ \text{V} + 0.5\ \text{V}$$

$$= 0.3335\ \text{V}$$

The reflected voltage will propagate back along the line, reaching the source at time $2T = 95.2$ ns. From then on, the voltage will be 0.3335 V all along the line.

The astute reader may have noticed that there is an easier way to calculate the final voltage. The final condition, after all transients have died down, must be the same as the dc situation, ignoring transmission-line characteristics. This situation is shown in Figure 10.13(a). It is simply a voltage divider. The voltage across the line, and the load, will be

$$V_L = 1\ \text{V} \times \frac{25\ \Omega}{50\ \Omega + 25\ \Omega}$$

$$= 0.3333\ \text{V}$$

Allowing for round-off error, this is the same answer as before. The reflection coefficient thus seems to provide an explanation of the mechanism by which the final condition is reached.

Figures 10.13(b) and (c) show the variation over time of the voltages at the source and load ends of the line, respectively.

10.4 WAVE PROPAGATION ON LINES

We began the discussion of transmission-line characteristics with step and pulse inputs mainly to simplify matters, though both types of signals are in fact used with transmission lines. Pulses in particular are common in some digital applications. It is time now to consider the effect of transmission lines on sine waves.

Let us begin, as before, with a matched line. A sinusoidal wave is applied to one end, through a source resistance that is equal to Z_0. From our experience with step and pulse inputs, we would expect that the signal would simply move down the line and disappear into the load, and that is what does happen. Such a signal is called a *travelling wave*. We would also expect that the process would take time. Once the sine wave has been operating long enough so that the first part of the signal has already reached the far end, a steady-state situation will exist. The signal at any point along the line will be the same as that at the source, except for a time delay.

With a sine wave, of course, a time delay is equivalent to a phase shift. A time delay of one period will cause a phase shift of 360°, or one complete cycle; a wave that has been delayed that much will be indistinguishable from one that has not been delayed at all. The length of line L that causes a delay of one period is known as a *wavelength*, for which the usual symbol is λ. If we could look at the voltage along the line at one instant of time, the resulting "snapshot" would look like the input sine wave, except that the horizontal axis would be distance rather than time, and one complete cycle of the wave would occupy one wavelength instead of one period. See Figure 10.14 for an example.

To arrive at an expression for wavelength on a line, we start with the definition of velocity

$$v = \frac{d}{t} \tag{10.17}$$

where v = velocity in meters per second
       d = distance in meters
       t = time in seconds

Since we are trying to find wavelength, we substitute the period T of the signal for t, the wavelength λ for d, and the propagation velocity along the line v_p for v. This gives us

$$v_p = \frac{\lambda}{T} \tag{10.18}$$

More commonly, frequency is used rather than period. Since

$$f = \frac{1}{T}$$

Equation (10.18) becomes

$$v_p = f\lambda \tag{10.19}$$

By the way, Equation (10.19) is valid for all types of waves, in any medium.

FIGURE 10.14 **Travelling Waves on a Matched Line**

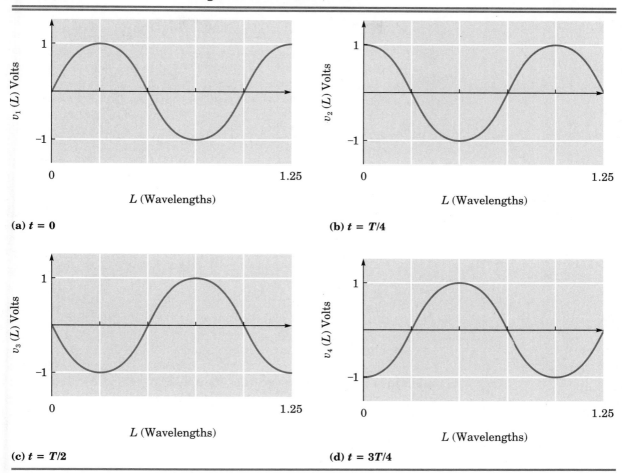

(a) $t = 0$

(b) $t = T/4$

(c) $t = T/2$

(d) $t = 3T/4$

Lengths of line that are not equal to a wavelength will provide a phase delay proportional to their length. Since a length λ produces a phase shift of 360°, the phase delay produced by a given line is simply

$$\phi = (360)\frac{L}{\lambda} \qquad (10.20)$$

where ϕ = phase shift in degrees
       L = length of the line
       λ = wavelength on the line

In many applications, the time delay and phase shift due to a length of transmission line are of no concern. There are times, however, as when two signals must arrive at a given point in phase with each other, when it is important to consider phase shift. Transmission lines can also be used deliberately to introduce phase shifts and time delays where required.

EXAMPLE 10.5

What length of standard RG-8/U coaxial cable would be required to obtain a 45° phase shift at 200 MHz?

Solution

The velocity factor for this line is 0.66, so from Equation (10.6),

$$v_p = v_f c$$
$$= 0.66 \times 3 \times 10^8 \text{ m/s}$$
$$= 1.98 \times 10^8 \text{ m/s}$$

The wavelength on the line of a 200 MHz signal is found from Equation (10.19):

$$v_p = f\lambda$$

$$\lambda = \frac{v_p}{f}$$

$$= \frac{1.98 \times 10^8 \text{ m/s}}{200 \times 10^6 \text{ Hz}}$$

$$= 0.99 \text{ m}$$

The required length for a phase shift of 45° would be

$$L = 0.99 \text{ m} \times \frac{45°}{360°}$$

$$= 0.124 \text{ m}$$

10.4.1 Standing Waves

It has been shown that a sine wave applied to a matched line results in an identical sine wave, except for phase, appearing at every point on the line, as the incident wave travels down it. If the line is unmatched, we would probably expect a reflected wave from the load to add to the incident wave from the source. This is exactly what happens.

As we did for the step input, let us begin with a rather simple situation: a transmission line terminated in an open circuit. Assume that the line is reasonably long, say one wavelength. The situation is shown in Figure 10.15. The reflection of a sine wave is harder to visualize than that of a pulse or a step of voltage, but this figure shows the situation at several points in the cycle. As before, the reflected voltage will have the same amplitude and polarity as the incident voltage at the load, since the open circuit cannot dissipate any power. The reflected voltage will propagate down the line until it is dissipated in the source impedance, which has been chosen to match the line at the source. (If there is a mismatch at both ends, the situation will be complicated by multiple reflections.) At every point on the line, the instantaneous values of incident and reflected voltage will add algebraically to give the total

FIGURE 10.15

Incident and Reflected Waves on an Open-Circuited Line

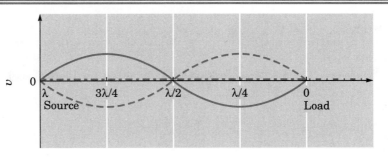

(a) 0°

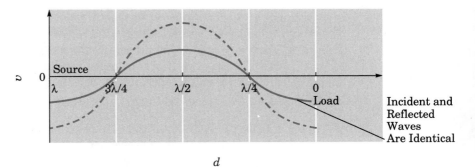

Incident and Reflected Waves Are Identical

(b) 90°

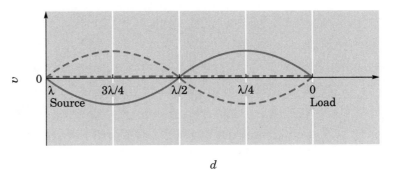

(c) 180°

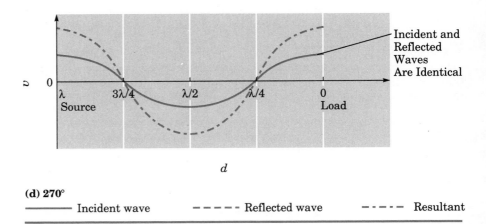

Incident and Reflected Waves Are Identical

(d) 270°

——— Incident wave   - - - - Reflected wave   -·-·- Resultant

FIGURE 10.16

Standing Waves on an
Open-Circuited Line

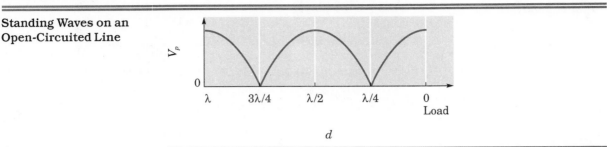

voltage. From the figure, it can be inferred that the voltage at every point on the line will vary sinusoidally, but that, due to constructive and destructive interference between the incident and reflected waves, the amplitude of the voltage will vary greatly. At the open-circuited end of the line, the peak voltage will be maximum. One-quarter wavelength away from that end, the incident and reflected voltages will exactly cancel, because the two signals will have equal amplitude and opposite phase. At a distance of one-half wavelength from the open-circuited end, there will be another voltage maximum, and the process continues for every half-wavelength segment of line.

It is possible to make a sketch showing the variation of peak (or RMS) voltage along the line, as shown in Figure 10.16. It is important to realize that this figure does not represent either instantaneous or dc voltages. There is no dc on this line at all. Rather, it shows the way in which the amplitude of a sinusoidal voltage varies along the line.

The interaction between the incident and reflected waves, which are both travelling waves, causes what appears to be a stationary pattern of waves on the line. It is customary to call these *standing waves* because of this appearance.

For comparison, Figure 10.17 shows the standing waves of voltage on a line with a shorted end. Naturally, there is no voltage at the shorted end. A voltage maximum occurs one-quarter wavelength from the end, another null at one-half wavelength, and so on.

The current responds in just the opposite way to the voltage. For the open line, the current must be zero at the open end. It will be a maximum one-quarter wavelength away, zero again at a distance of one-half wavelength, and so on. In other words, the graph of current as a function of position for an

FIGURE 10.17

Standing Waves on a
Short-Circuited Line

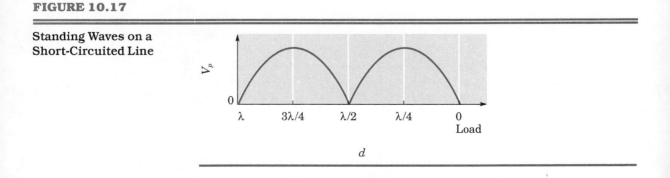

open-circuited line looks just like the graph of voltage versus position for a short-circuited line (see Figure 10.17). Similarly, the current on a short-circuited line has a maximum at the termination, a minimum one-quarter wavelength away, and another maximum at a distance of one-half wavelength, just like the voltage curve for the open-circuited line (Figure 10.16).

What happens when the line is mismatched, but not so drastically as discussed above? There will be a reflected wave, but it will not have as large an amplitude as the incident wave. Its phase angle will depend on the load impedance compared to that of the line. Recall from Equation (10.9) that the incident and reflected voltages are related by the coefficient of reflection:

$$\Gamma = \frac{V_r}{V_i}$$

and that, from Equation (10.16),

$$\Gamma = \frac{Z_L - Z_0}{Z_L + Z_0}$$

In general, Γ is complex, but for a lossless line it will be a real number if the load is resistive. This number will be positive for $Z_L > Z_0$ and negative for $Z_L < Z_0$. For $Z_L = Z_0$, the reflection coefficient will of course be zero. A positive coefficient means that the incident and reflected voltages are in phase at the load.

When a reflected signal is present but of lower amplitude than the incident wave, there will be standing waves of voltage and current, but there will be no point on the line where the voltage or current remains zero over the whole cycle. See Figure 10.18 for an example. It is possible to define the voltage **standing-wave ratio** (VSWR, or just SWR) as follows:

$$\text{SWR} = \frac{V_{max}}{V_{min}} \tag{10.21}$$

The SWR concerns magnitudes only, and is thus a real number. It must be positive and greater than or equal to 1. For a matched line, the SWR will be 1 (sometimes expressed as 1:1 to emphasize that it is a ratio), and the closer the line is to being matched, the lower will be the SWR. The SWR has the advantage of being easier to measure than the reflection coefficient, but the

FIGURE 10.18

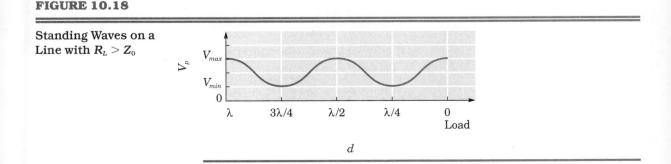

Standing Waves on a
Line with $R_L > Z_0$

latter is more useful in many calculations. Since both are essentially measures of the amount of reflection on a line, it is possible to find a relationship between them.

The maximum voltage on the line will occur where the incident and reflected signals are in phase, and the minimum voltage will be found where they are out of phase. Therefore, using absolute value signs to emphasize the lack of a need for phase information, we can write

$$V_{max} = |V_i| + |V_r| \tag{10.22}$$

and

$$V_{min} = |V_i| - |V_r| \tag{10.23}$$

Combining Equations (10.21), (10.22), and (10.23), we get

$$
\begin{aligned}
\text{SWR} &= \frac{V_{max}}{V_{min}} \\[2mm]
&= \frac{|V_i| + |V_r|}{|V_i| - |V_r|} \\[2mm]
&= \frac{1 + \dfrac{|V_r|}{|V_i|}}{1 - \dfrac{|V_r|}{|V_i|}} \\[2mm]
&= \frac{1 + |\Gamma|}{1 - |\Gamma|}
\end{aligned}
\tag{10.24}
$$

A little algebra will show that $|\Gamma|$ can also be expressed in terms of SWR:

$$|\Gamma| = \frac{\text{SWR} - 1}{\text{SWR} + 1} \tag{10.25}$$

For the special, but important, case of a lossless line terminated in a resistive impedance, it is possible to find a simple relationship between SWR and the load and line impedances. First, suppose that $Z_L > Z_0$. Then, from Equation (10.24),

$$
\begin{aligned}
\text{SWR} &= \frac{1 + |\Gamma|}{1 - |\Gamma|} \\[2mm]
&= \frac{1 + \dfrac{Z_L - Z_0}{Z_L + Z_0}}{1 - \dfrac{Z_L - Z_0}{Z_L + Z_0}} \\[2mm]
&= \frac{Z_L + Z_0 + Z_L - Z_0}{Z_L + Z_0 - Z_L + Z_0} \\[2mm]
&= \frac{2Z_L}{2Z_0} \\[2mm]
&= \frac{Z_L}{Z_0}
\end{aligned}
\tag{10.26}
$$

It is easy to show, in a similar way, that if $Z_0 > Z_L$, then

$$\text{SWR} = \frac{Z_0}{Z_L} \tag{10.27}$$

The use of the appropriate equation will always give a positive SWR that is greater than or equal to one.

EXAMPLE 10.6

A 50 Ω line is terminated in a 25 Ω resistance. Find the SWR.

Solution

In this case, $Z_0 > Z_L$ so the solution is given by Equation (10.27).

$$\text{SWR} = \frac{Z_0}{Z_L}$$

$$= \frac{50}{25}$$

$$= 2$$

The presence of standing waves causes the voltage at some points on the line to be higher than it would be with a matched line, while at other points the voltage is low but the current is higher than with a matched line. This situation will result in increased losses. In a transmitting application, standing waves put additional stress on the line and can result in failure of the line, or of equipment connected to it. For instance, if the transmitter happens to be connected at or near a voltage maximum, the output circuit of the transmitter may be subjected to a dangerous overvoltage condition. This is especially likely to damage solid-state transmitters, which for this reason are often equipped with circuits to reduce the output power in the presence of an SWR greater than about 2:1.

Reflections cause the power dissipated in the load to be less than it would be with a matched line, for the same source. This is because some of the power is reflected back to the source. Since power is proportional to the square of voltage, the fraction of the power that is reflected is Γ^2; that is,

$$P_r = \Gamma^2 P_i \tag{10.28}$$

where P_r = power reflected from the load
P_i = incident power at the load
Γ = voltage reflection coefficient

Sometimes Γ^2 is referred to as the *power reflection coefficient*.

The amount of power absorbed by the load is the difference between the incident power and the reflected power; that is,

$$P_L = P_i - \Gamma^2 P_i$$

$$= P_i(1 - \Gamma^2) \tag{10.29}$$

EXAMPLE 10.7

A generator sends 50 mW down a 50 Ω line. The generator is matched to the line, but the load is not. If the coefficient of reflection is 0.5, how much power is reflected and how much is dissipated in the load?

Solution

The amount of power that is reflected is, from Equation (10.28),

$$P_r = \Gamma^2 P_i$$
$$= 0.5^2 \times 50 \text{ mW}$$
$$= 12.5 \text{ mW}$$

The remainder of the power reaches the load. This amount is

$$P_L = P_i - P_r$$
$$= 50 \text{ mW} - 12.5 \text{ mW}$$
$$= 37.5 \text{ mW}$$

Alternatively, the load power can be calculated directly from Equation (10.29):

$$P_L = P_i(1 - \Gamma^2)$$
$$= 50 \text{ mW} \times (1 - 0.5^2)$$
$$= 37.5 \text{ mW}$$

Since SWR is easier to measure than the reflection coefficient, an expression for the power absorbed by the load in terms of the SWR would be useful. It is easy to derive such an expression by using the relationship between Γ and SWR given in Equation (10.25). The derivation is left as an exercise; the result is as follows.

$$P_L = \frac{4\text{SWR}}{(1 + \text{SWR})^2} P_i \qquad (10.30)$$

EXAMPLE 10.8

A transmitter supplies 50 W to a load through a line with an SWR of 2:1. Find the power absorbed by the load.

Solution

From Equation (10.30),

$$P_L = \frac{4\text{SWR}}{(1 + \text{SWR})^2} P_i$$
$$= \frac{4 \times 2}{(1 + 2)^2} \times 50 \text{ W}$$
$$= 44.4 \text{ W}$$

Reflections on transmission lines can cause problems in receiving applications as well. For instance, reflections on a television antenna feedline can cause a double image or "ghost" to appear. In data transmission, reflections can distort pulses, causing errors.

10.4.2 Variation of Impedance Along a Line

A matched line presents its characteristic impedance to a source located any distance from the load. If the line is not matched, however, the impedance seen by the source can vary greatly with its distance from the load. At those points where the voltage is high and the current low, the impedance will be higher than at points on the line with the opposite current and voltage characteristics. In addition, the phase angle of the impedance can vary. At some points, a mismatched line may look inductive, at others capacitive, and, at a few points, resistive. Very near the load, the impedance looking into the line will be very close to Z_L. This is one reason why transmission-line techniques need to be used only with relatively high frequencies and/or long lines: a line shorter than about one-sixteenth of a wavelength can usually be ignored.

The impedance that a lossless transmission line presents to a source will vary in a periodic way. We have already noticed that the standing-wave pattern repeats itself every one-half wavelength along the line; the impedance varies in the same fashion. At the load, and at distances from the load that are multiples of one-half wavelength, the impedance looking into the line will be that of the load.

The impedance at any point on a lossless transmission line is given by the equation

$$Z = Z_0 \frac{Z_L \cos \theta + jZ_0 \sin \theta}{Z_0 \cos \theta + jZ_L \sin \theta} \tag{10.31}$$

where Z = impedance looking toward the load
$\quad\quad\quad Z_L$ = load impedance
$\quad\quad\quad Z_0$ = characteristic impedance of the line
$\quad\quad\quad \theta$ = distance to the load in degrees (For example, a quarter-wavelength would be 90°.)

Provided that $\cos \theta$ is not equal to zero, this simplifies to

$$Z = Z_0 \frac{Z_L + jZ_0 \tan \theta}{Z_0 + jZ_L \tan \theta} \tag{10.32}$$

EXAMPLE 10.9

Calculate the impedance looking into a 50 Ω line 1 m long, terminated in a load impedance of 100 Ω, if the line has a velocity factor of 0.8 and operates at a frequency of 30 MHz.

Solution

First, we need the length of the line in degrees. At 30 MHz with a velocity factor of 0.8, the wavelength is

$$\lambda = \frac{v}{f}$$

$$= \frac{v_f c}{f}$$

$$= \frac{0.8 \times 3 \times 10^8 \text{ m/s}}{30 \times 10^6 \text{ Hz}}$$

$$= 8 \text{ m}$$

The length of the line in degrees is

$$\theta = \frac{1 \text{ m}}{8 \text{ m}} \times 360°$$

$$= 45°$$

Substituting this value and the given values into Equation (10.32),

$$Z = Z_0 \frac{Z_L + jZ_0 \tan \theta}{Z_0 + jZ_L \tan \theta}$$

$$= 50 \ \Omega \ \frac{100 \ \Omega + j(50 \ \Omega) \tan 45°}{50 \ \Omega + j(100 \ \Omega) \tan 45°}$$

$$= 50 \ \Omega \ \frac{100 \ \Omega + j(50 \ \Omega)}{50 \ \Omega + j(100 \ \Omega)}$$

$$= \frac{100 \ \Omega + j(50 \ \Omega)}{1 + j2}$$

$$= \frac{[100 \ \Omega + j(50 \ \Omega)](1 - j2)}{(1 + j2)(1 - j2)}$$

$$= \frac{100 \ \Omega + j(50 \ \Omega) - j(200 \ \Omega) + 100 \ \Omega}{1 + 4}$$

$$= \frac{200 \ \Omega - j(150 \ \Omega)}{5}$$

$$= 40 \ \Omega - j(30 \ \Omega)$$

The calculation of transmission-line impedances is usually done in one of two ways: graphically, with the aid of a **Smith chart**™ (see Section 10.6), or by computer.

10.4.3 Characteristics of Open and Shorted Lines

Though a section of transmission line that is terminated in an open or short circuit is useless for transmitting power, it can serve other purposes. Such a line can be used as an inductive or capacitive reactance, or even as a resonant circuit. In practice, short-circuited sections are more common, because open-circuited lines tend to radiate energy from the open end.

The impedance of a short-circuited line can be found from Equation (10.32), by setting Z_L equal to zero. The impedance looking toward the short circuit is

$$Z = jZ_0 \tan \theta \qquad\qquad (10.33)$$

Note that the impedance has no resistive component. This is logical, since there is no resistive element that can dissipate any power. For short lengths, less than one-quarter wavelength or 90°, the impedance is inductive. At one-quarter wavelength, the line looks like an open circuit, and for lengths between one-quarter and one-half wavelength, the line is capacitive, since the tangent of the corresponding angle is negative. For a length of one-half wavelength, or 180°, tan θ is zero, and the line once again behaves like a short circuit. For longer lines, the cycle repeats. Figure 10.19(a) shows graphically how the impedance varies with length. It is easy to remember that a short length of shorted line is inductive by visualizing the shorted end as a loop, or coil, hence an inductance. At exactly one-quarter wavelength, the line behaves as a parallel-resonant circuit; that is, it has a very high impedance. As the line is made still longer, it becomes capacitive, finally becoming series-resonant at a length of one-half wavelength. For still longer lines, the process repeats.

The open-circuited line, on the other hand, is capacitive in short lengths. You can remember this by thinking of the two parallel lines at the end of the transmission line as a capacitor. It is series-resonant (that is, it behaves like a short circuit) at a length of one-quarter wavelength, inductive between one-quarter and one-half wavelength, and parallel-resonant at a length of one-half wavelength. For longer lines, the cycle repeats. Figure 10.19(b) shows graphically how the impedance varies with length for this line.

Short transmission-line sections, called *stubs*, can be substituted for capacitors, inductors, or tuned circuits in applications where lumped-constant components (conventional inductors or capacitors) would be inconvenient or impractical. At VHF and UHF frequencies, the required values of inductance

FIGURE 10.19   **Variation of Impedance with Length on Shorted and Open Lines**

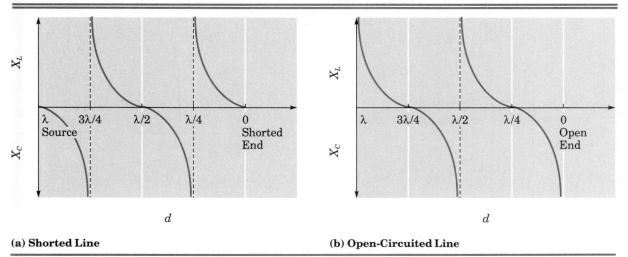

(a) Shorted Line   **(b) Open-Circuited Line**

and capacitance are often very small. It would be difficult to build a 1 pF capacitor, for instance; the capacitance between its leads might well be more than that. In addition, physically small components are difficult to use where large amounts of power are involved, as in transmitters, because of the large voltages and currents that must be handled. A short section of transmission line can avoid all these problems. In addition, an air-dielectric transmission line can have much higher Q than a typical lumped-constant resonant circuit.

Though either short-circuited or open-circuited stubs could be used, in practice the short-circuited version is much more useful. Open-circuited lines tend to radiate energy from the end of the line, and in addition, it is easier to adjust a short-circuited stub by means of a moving short circuit. Any extra line that extends past the short circuit will have no effect on the operation of the stub.

EXAMPLE 10.10

A series tuned circuit operating at a frequency of 1 GHz is to be constructed from a shorted section of air-dielectric coaxial cable. What length should be used?

Solution

The velocity factor of an air-dielectric line is about 0.95, so the propagation velocity is, from Equation (10.6),

$$v_p = v_f c$$
$$= 0.95 \times 3 \times 10^8 \text{ m/s}$$
$$= 2.85 \times 10^8 \text{ m/s}$$

The wavelength on the line is given by Equation (10.19).

$$v_p = f\lambda$$
$$\lambda = \frac{v_p}{f}$$
$$= \frac{2.85 \times 10^8 \text{ m/s}}{1000 \times 10^6 \text{ Hz}}$$
$$= 0.285 \text{ m}$$

Since this is a shorted stub, a half-wavelength section will be series-resonant. Therefore the length will be

$$L = \frac{\lambda}{2}$$
$$= \frac{0.285 \text{ m}}{2}$$
$$= 0.143 \text{ m}$$

10.5 TRANSMISSION-LINE LOSSES

Up to this point, we have concentrated on lossless lines. No real transmission line is completely lossless, of course, but the approximation is often valid. This is particularly true when the section of line is short. In that case, a matched line can be ignored, and a mismatched line will be noteworthy mainly for the reactance that it introduces, the reflections (which can cause distortion), and the standing waves (which increase losses and can damage components).

In order to simplify our discussion of lossy lines, we will begin with matched lines. Then we will look briefly at the general case, where a line is both mismatched and lossy.

10.5.1 Loss Mechanisms

The most obvious form of loss in a transmission line is that due to the resistance of the conductors. Sometimes this is called *I²R loss* because it is proportional to the square of the current, or *copper loss* because copper is the most common material for the conductors. All other things being equal, this loss will be less with higher characteristic impedances, because the same power can be delivered with less current when the impedance is increased. The *I²R* loss will increase with frequency due to the skin effect, which causes the resistance of the conductors to rise as the frequency increases.

The dielectric of a transmission line will also have some loss. Usually, the conductance of the dielectric increases with frequency. Dry air has low loss, so air-dielectric lines, like open-wire line and air-dielectric coaxial cable, should have better performance than solid-dielectric coaxial cable in this regard. Because open-wire line is exposed to the weather, it does not always have the lower loss that might be expected. Foam-dielectric coaxial lines typically have lower loss than those that use solid polyethylene.

Open-wire lines can radiate energy. This type of loss becomes more significant as the frequency increases, and is worse for greater spacing between the conductors. The tendency of these cables to radiate can be reduced by carefully balancing the line to ground.

Radiation is not a problem for coaxial cables that are properly terminated and grounded. The signal currents flow near the outside surface of the inner conductor, and the inside surface of the outer conductor. The outside of the outer conductor acts as a shield. This assumes, of course, that the shield completely covers the cable. Some low-cost cables use a twisted or braided shield with less than 100% coverage, with the result that there can be some radiation, especially at VHF and UHF frequencies.

Radiation of energy not only contributes to cable losses, but can also be a source of interference and crosstalk. Any cable that radiates can also absorb radiation.

10.5.2 Loss in Decibels

Transmission-line losses are usually given in decibels per 100 feet or per 100 meters. When choosing a transmission line for a given application, attention must be paid to losses. Remember, for instance, that a 3 dB loss in a feedline between a transmitter and its antenna means that only one-half the transmitter power actually reaches the antenna. The rest of the power goes to heat up the transmission line. Losses are just as important in receivers, where a low noise figure depends on minimizing the losses before the first stage of amplification. In some receiving applications, notably satellite receivers, it is common to install a low-noise amplifier right at the antenna to reduce the adverse effect of the transmission line on noise performance. In fact, with satellite receivers, it is very common to down-convert the signal to a lower frequency at the antenna in order to take advantage of the fact that cable losses are less at lower frequencies.

EXAMPLE 10.11

A transmitter is required to deliver 100 W to an antenna through 45 m of coaxial cable with a loss of 4 dB/100 m. What must be the output power of the transmitter, assuming the line is matched?

Solution

The loss in decibels is

$$\text{Loss (dB)} = 45 \text{ m} \times \frac{4 \text{ dB}}{100 \text{ m}}$$
$$= 1.8 \text{ dB}$$

The relation between input and output power is

$$\frac{P_{in}}{P_{out}} = \text{antilog} \frac{1.8}{10}$$
$$= 1.51$$

The transmitter power must be

$$P_{in} = 1.51 \times 100 \text{ W}$$
$$= 151 \text{ W}$$

10.5.3 Mismatched Lossy Lines

The technologist is often spared the problem of dealing with losses and mismatches simultaneously. Most transmission lines that are long enough to have significant losses are reasonably well matched. One thing that is worth noting, though, is the effect that losses have on SWR. When the line is lossy, the SWR at the source will be lower than that at the load. Figure 10.20 shows why. Incident and reflected signals are each subject to the same amount of loss in decibels, so the ratio between them is greater at the source than at the load. Therefore, the reflection coefficient and standing-wave ratio both have larger magnitudes at the load. Consequently, when using SWR measurements to aid in matching a load to a line, it is better to take the measurements at the load, rather than at the source, particularly if the line is long.

Any of the computer programs that are available for transmission-line

FIGURE 10.20

SWR on a Lossy Line

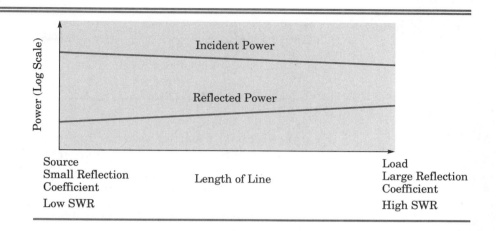

analysis are capable of handling losses and mismatches simultaneously. Recourse to one of these programs is suggested for such calculations.

10.5.4 Power Ratings

The maximum power that can be applied to a transmission line is limited by one of two things: power dissipation in the line, which causes heating and can raise the temperature of the line to the point at which damage occurs; and voltage, which, when it exceeds a maximum value, can cause breakdown of the dielectric. Both power dissipation and maximum voltage increase with the standing-wave ratio, so a transmission line will be capable of handling more power when it is properly matched than when it is mismatched. Published power ratings for commercial cables generally specify a maximum SWR at which they are valid. Since losses increase with frequency, a derating curve may apply that reduces the maximum power-handling capability as the operating frequency increases.

This is a good point at which to consider the effect of changing the characteristic impedance of a transmission line. Increasing the impedance increases the voltage required for a given power level, and makes dielectric breakdown more likely. On the other hand, increasing the impedance reduces the current for a given power level, and consequently the I^2R losses will be less with high-impedance lines. The two most common impedances for coaxial cables in RF circuits are 50 and 75 Ω. An impedance of 50 Ω is more common in transmitting applications, while an impedance of 75 Ω is used with television antennas and cable-TV systems, where power levels are low. This makes sense, since—other things being equal—the 75 Ω cable will have lower loss, while the 50 Ω line will have greater power-handling capability.

10.6   IMPEDANCE MATCHING

In most transmission-line applications, it is highly desirable that the load match the impedance of the line, in order to avoid the problems associated with reflections and a high SWR. Transmission lines are readily available with only a few standard impedance values, and most equipment is designed

FIGURE 10.21

Balun Transformer

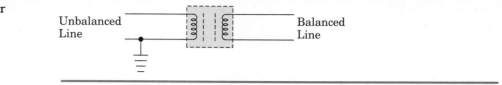

to work with only a narrow range of impedances. For example, a typical transmitter will require a 50 Ω load impedance with an SWR of 2:1 or less. If the antenna connected to such a transmitter has an impedance close to 50 Ω, resistive, there is no problem. Often, however, the impedance of the load will differ, being neither resistive nor equal in magnitude to that of the line. In that case, some means must be provided for impedance matching.

Sometimes this is done by adding inductance or capacitance at the load as required to remove the reactance, and using a conventional or toroidal transformer to change the magnitude of the impedance. Lumped-constant matching networks such as the *T*, *L*, and pi networks, are also common at relatively low frequencies. It is also possible, however, to use transmission-line sections as matching networks. The transmission-line matching sections themselves will not usually be matched, and may have quite a high SWR, which must be taken into account whenever large power levels are present.

Another requirement for proper transmission-line operation is that balanced and unbalanced lines should be terminated by balanced and unbalanced loads, respectively. The sources should also have the same configuration as the line. If, for instance, a balanced load is connected to a coaxial cable, signal current is likely to flow on the outer surface of the shield, leading to radiation from the cable or the pickup of interference by the cable. On the other hand, connecting an unbalanced source or load to a balanced line will unbalance the entire line, causing similar problems. Balanced and unbalanced lines, loads, and sources can be connected together by using a transformer called a **balun** (see Figure 10.21). The balun can have a 1:1 turns ratio, or it can also be used to match impedances. An example of the latter is the common television-receiver balun. It is used to match a 75 Ω unbalanced coaxial cable to an antenna input designed for 300 Ω balanced twin-lead. Baluns can also be constructed from sections of transmission line, as described later in this chapter.

10.6.1 Quarter-Wave Transformers

A section of transmission line that is one-quarter wavelength long can act as an impedance transformer. This can be useful when the load impedance is resistive, but has the wrong magnitude. This can be seen by looking at Equation (10.31):

$$Z = Z_0 \frac{Z_L \cos \theta + jZ_0 \sin \theta}{Z_0 \cos \theta + jZ_L \sin \theta}$$

For a quarter-wave line, $\theta = 90°$, so $\cos \theta = 0$ and $\sin \theta = 1$. Equation (10.31) simplifies to

$$Z = Z_0 \frac{0 + jZ_0}{0 + jZ_L} \tag{10.34}$$

$$= Z_0 \frac{jZ_0}{jZ_L}$$

$$= \frac{Z_0^2}{Z_L} \tag{10.35}$$

The impedance looking into the quarter-wave section is Z, and Z_0 is the characteristic impedance of the **quarter-wave transformer**. To avoid confusion when the transformer is inserted into the main line, let us replace Z with the term Z_0, since the idea is to have the input impedance of the transformer equal to the characteristic impedance of the main line. The term Z_0 in the above equation can be replaced by Z_0' to emphasize that this is the impedance of the transformer section. Equation (10.35) then becomes

$$Z_0 = \frac{Z_0'^2}{Z_L}$$

This means that, in order to match a load impedance Z_L to a line impedance Z_0, the quarter-wave section must have an impedance given by

$$Z_0' = \sqrt{Z_0 Z_L} \tag{10.36}$$

This technique usually requires a custom-made line section. It can be used with either open-wire or coaxial lines, though it is easier to construct with the former. A match can only be obtained over a small range of frequencies, as the length of the matching section will no longer be a quarter-wavelength if the frequency changes.

The quarter-wave transformer is only useful, by itself, in matching impedances that are resistive. If the load is reactive, however, it may be possible to add the quarter-wave section at a point along the line, at such a distance from the load that the impedance, looking down the line to the load, is real. Such a point can be found using a Smith chart, by a method to be shown later.

EXAMPLE 10.12

Find the impedance and length of a quarter-wave transformer to match a 75 Ω load to a 50 Ω line at a frequency of 300 MHz.

Solution

The impedance of the matching section can be found from Equation (10.36).

$$Z_0' = \sqrt{Z_0 Z_L}$$

$$= \sqrt{50 \ \Omega \times 75 \ \Omega}$$

$$= 61.2 \ \Omega$$

The length of the line section will depend on the dielectric. Assuming an air dielectric with a velocity factor of 0.95, the propagation velocity will be, from Equation (10.6),

$$v_p = v_f c$$
$$= 0.95 \times 3 \times 10^8 \text{ m/s}$$
$$= 2.85 \times 10^8 \text{ m/s}$$

The wavelength is given by Equation (10.19).

$$v_p = f\lambda$$
$$\lambda = \frac{v_p}{f}$$
$$= \frac{2.85 \times 10^8 \text{ m/s}}{300 \times 10^6 \text{ Hz}}$$
$$= 0.95 \text{ m}$$

The line must be one-quarter wavelength long, so its length is given by

$$L = 0.25 \times 0.95 \text{ m}$$
$$= 0.238 \text{ m}$$

10.6.2 Other Series Matching Sections

Though it is the simplest to analyze, the quarter-wave matching section is not the only possibility. It is possible to match complex impedances by using a section of a different length, or two sections with different impedances. In order to avoid tedious calculations, it is logical to use a computer to design such sections. In fact, there is a free program, called APPCAD™ and available from Hewlett-Packard, that will handle these problems as well as matching using conventional *LC* networks. In addition, it will perform calculations in several areas as diverse as noise figure and thermal analysis of semiconductors, among others. The author recommends it highly, especially considering the price!

10.6.3 Transmission-Line Baluns

An alternative to a balun transformer, for use within a relatively narrow frequency range, is shown in Figure 10.22. The half-wavelength section has a phase shift of 180° along its length, causing the currents in the two wires of the balanced line to be equal in magnitude and opposite in phase, as required. By Kirchhoff's current law, at Node *A*, the sum of the magnitudes of the two currents on the balanced line must be equal to the current on the center conductor of the unbalanced line. That is, the current on the balanced line is one-half that on the unbalanced line. Assuming a lossless line section is used to make the balun, the power on both lines must be the same. Since

$$P = I^2 R$$

FIGURE 10.22

**Transmission-Line
Balun**

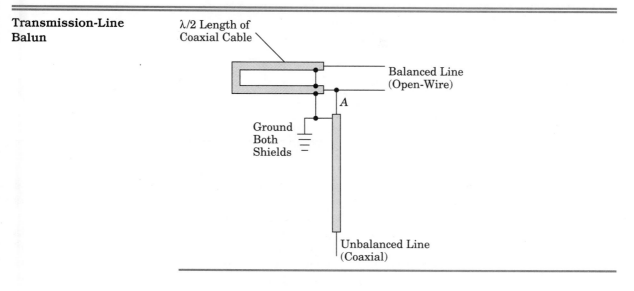

λ/2 Length of
Coaxial Cable

Balanced Line
(Open-Wire)

A

Ground
Both
Shields

Unbalanced Line
(Coaxial)

then

$$R = \frac{P}{I^2}$$

That is, for constant power, the impedance is inversely proportional to the square of current. Dividing the current by two will therefore multiply the impedance by four. This balun, then, effects a 4:1 impedance transformation as well as conversion from unbalanced to balanced lines. It would be quite suitable for matching 75 Ω coaxial cable to 300 Ω twin-lead, for example. It would not be a satisfactory replacement for the television-receiver matching transformer described earlier, because it would not operate over the wide frequency range required. It would, however, be a simple and economical balun for narrowband use at VHF and higher frequencies.

10.6.4 Stub Matching

Stub matching makes use of two properties of transmission lines that have already been noted. The impedance varies with position along a mismatched line, and a shorted section of line, called a **stub**, can simulate a capacitance or an inductance, depending on its electrical length. Figure 10.23(a) shows the idea. Theoretically, an open stub could also be used, but this is not practical because of radiation from the open end.

Single-stub matching consists of two steps. First, a point is found where the resistive component of the impedance looking into the line, toward the load, is equal to the characteristic impedance of the line. In general, the impedance at this point will be complex, of course, but the reactive component can be cancelled by adding an appropriate reactance to the line. Actually, admittance and susceptance are generally used in these calculations instead of impedance and reactance because it is easier to connect the stub in parallel with the line than in series. The second step, then, is to calculate the length

FIGURE 10.23   Stub Matching

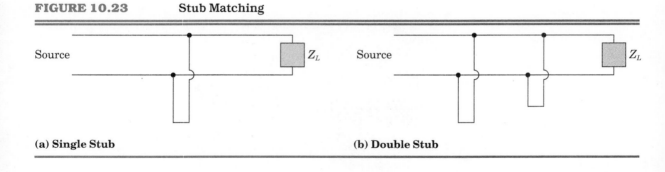

(a) Single Stub   (b) Double Stub

of a stub with the correct susceptance. These calculations can be performed using a *Smith chart*, a calculator, or a computer. Smith chart calculations will be described in the next section.

Single-stub matching requires that the position of the stub, as well as its length, be changed if any adjustments are required. This is relatively simple with open-wire line, but difficult for coaxial cable. Another technique uses two stubs in fixed positions, and allows all adjustments to be made by changing the lengths of the stubs. This allows easier adjustment, but it is somewhat less flexible in the range of impedances that can be matched.

Figure 10.23(b) shows a double-stub tuner. The spacing of the stubs is usually an odd number of eighth-wavelengths. The stubs work together to achieve the impedance match.

10.6.5 The Smith Chart

All of the techniques for impedance matching described above require some way of calculating impedances and/or admittances at various points on a transmission line. This is not particularly difficult; it is only necessary to solve Equation (10.31), or its simplified form in Equation (10.32).

In general, of course, Z_L is complex, so the equation becomes quite messy. If, in addition, the line is very lossy, Equation (10.32) will no longer be sufficiently accurate, and it will be necessary to use a more complicated equation involving hyperbolic functions. However, even the more complex equations are no problem for a computer, or even a programmable calculator. These tools were unavailable in 1938, though, when P. H. Smith developed a graphical transmission-line calculator. The improved form shown in Figure 10.24 dates from 1944, and is still in use. It is not as accurate as a good computer program, but it does provide a great deal of insight into transmission-line operation. In fact, many computer programs, as well as test instruments for RF components and systems, are capable of providing output in Smith chart format, because many people find this graphical presentation easier to interpret than a list of numbers. Many technical papers also use Smith charts to illustrate the properties of transmission lines and other RF components. Therefore, while graphical computations are passing from general use, a brief study of the Smith chart will prove useful to anyone working with transmission lines.

The most obvious thing about the Smith chart is that it is circular. We noticed earlier that the impedance of a lossless transmission line repeats every one-half wavelength; that is, the impedance one-half wavelength from

FIGURE 10.24   Smith Chart

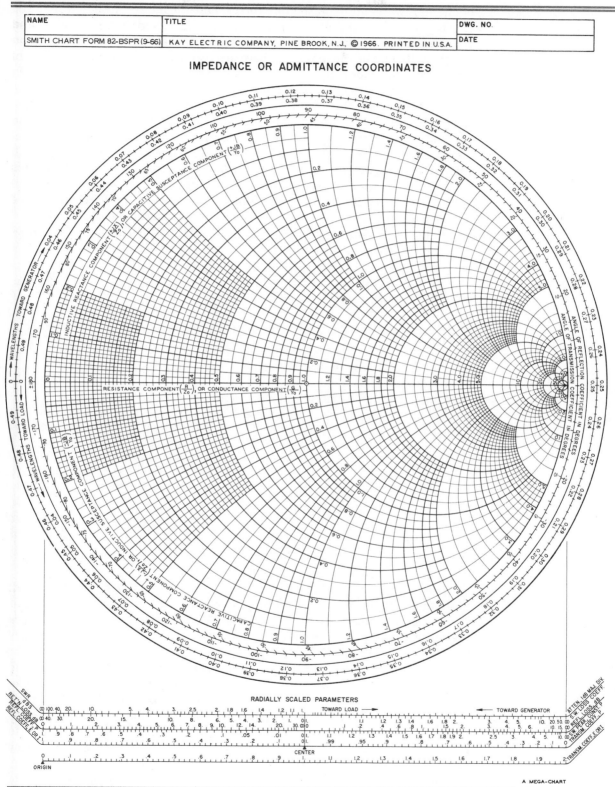

IMPEDANCE OR ADMITTANCE COORDINATES

NAME	TITLE	DWG. NO.	
SMITH CHART FORM 82-BSPR (9-66)	KAY ELECTRIC COMPANY, PINE BROOK, N.J., © 1966. PRINTED IN U.S.A.	DATE	

RADIALLY SCALED PARAMETERS

A MEGA-CHART

"Smith" is a trademark, and the Smith chart is copyright by Analog Instruments Co., Box 808, New Providence, NJ, 07974. Used by permission.

the load is the same as that of the load itself. On the Smith chart, a distance of one-half wavelength on the line corresponds to one revolution around the chart. Clockwise rotation represents movement toward the generator, and counterclockwise rotation represents progress toward the load, as shown on the chart itself. For convenience, there are two scales, in decimal fractions of a wavelength, around the outside of the chart, one in each direction. Each scale runs from zero to 0.5 wavelength.

The body of the chart is made up of families of orthogonal circles; that is, they intersect at right angles. The impedance or admittance at any point on the line can be plotted by finding the intersection of the real component (resistance or conductance), which is indicated along the horizontal axis, with the imaginary component (reactance or susceptance), shown above the axis for positive values and below for negative. Because of the wide variation of transmission-line impedances, the Smith chart uses *normalized* impedance and admittance to reduce the range of values that have to be shown.

To normalize an impedance, simply divide it by the characteristic impedance of the line.

$$z = \frac{Z}{Z_0} \tag{10.37}$$

where z = normalized impedance at a point on the line
$\quad\quad\quad Z$ = actual impedance at the same point
$\quad\quad\quad Z_0$ = characteristic impedance of the line

Since z is actually the ratio of two impedances, it is dimensionless.

For example, a load impedance of $100 + j25\ \Omega$, on a 50 Ω line, would be normalized to

$$z = \frac{Z_L}{Z_0}$$

$$= \frac{100 + j25\ \Omega}{50\ \Omega}$$

$$= 2 + j0.5$$

Figure 10.25 shows how this would be plotted on the chart.

Once the normalized impedance at one point on the line has been plotted, that at any other point can be found very easily. Draw a circle with its center in the center of the chart, which is at the point on the horizontal axis where the resistive component is equal to 1. Set the radius so that the circle passes through the point just plotted. Then draw a radius through that point, right out to the outside of the chart, as shown in Figure 10.25. Move around the outside in the appropriate direction, using the wavelength scale as a guide. Just follow the arrows. If the first point plotted is the load impedance, then move in the direction of the generator. On the other hand, if the impedance looking into the line from the generator has been plotted, it is necessary to move toward the load. Once the new location on the line has been found, draw another radius. The normalized impedance at the new position is the intersection of the radius with the circle. In other words, the circle is the locus of

FIGURE 10.25   **Smith Chart Operations**

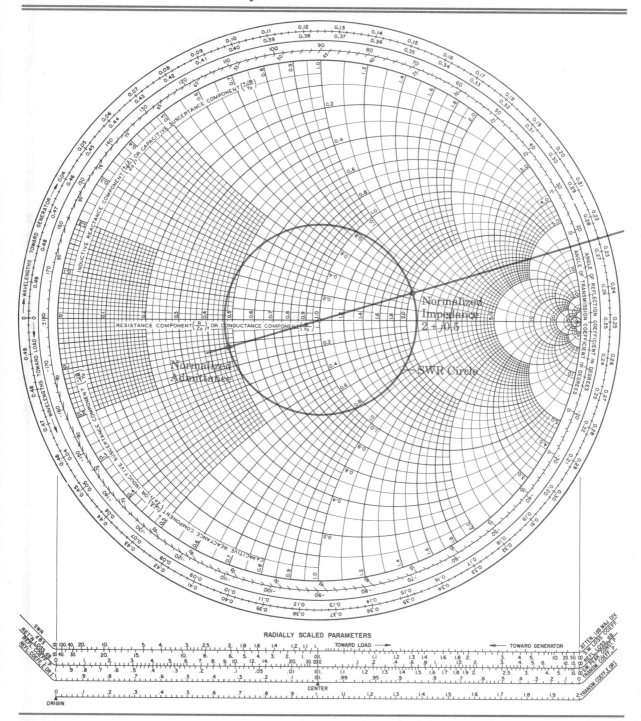

"Smith" is a trademark, and the Smith chart is copyright by Analog Instruments Co., Box 808, New Providence, NJ, 07974. Used by permission.

the impedance of the line. Every point on the circle represents the impedance at some point on the line.

The radius of the circle represents the SWR on the line. In fact, it is usually referred to as the *SWR circle*. The SWR can easily be found by reading the normalized resistance value where the circle crosses the horizontal axis to the right of the center of the chart. Another way is to mark the radius of the circle on the standing-wave voltage ratio scale at the bottom of the chart. If required, the SWR can also be found in decibels by using the adjoining scale. For this example, the SWR is read as approximately 2.15. (Calculation gives a value of 2.163.)

We noted before that the SWR depends on the magnitude of the reflection coefficient. There is another scale on the chart that allows this magnitude to be found directly. See the reflection coefficient voltage scale at the lower left of the chart. Marking off the radius of the SWR circle on this scale will give the magnitude of Γ, which can be read as 0.37. The power reflection coefficient, which is simply Γ^2, is also provided, as are the reflection and transmission losses in decibels.

Many transmission-line calculations are easier to perform using admittance. Converting from impedance to admittance is easy enough using a calculator, especially one that does complex arithmetic, but it is even easier on the Smith chart. All that is necessary is to draw a diameter that intersects the SWR circle at the impedance that must be converted to admittance. Follow the diameter across the circle to the other point at which it intersects the SWR circle. This represents the normalized admittance at the same point on the line. From Figure 10.25, the normalized admittance is found to be approximately $0.48 - j0.13$. (Calculation gives $0.44 - j0.18$.)

Normalized admittance, by the way, has the same relationship to the characteristic admittance of the line as normalized impedance has to characteristic impedance. That is,

$$y = \frac{Y}{Y_0} \tag{10.38}$$

where y = normalized admittance at a point on the line

$\quad\quad\quad Y$ = admittance at the same point

$\quad\quad\quad Y_0$ = characteristic admittance

Normalized admittance, like normalized impedance, is dimensionless. Since admittance is the reciprocal of impedance,

$$Y_0 = \frac{1}{Z_0}$$

and Equation (10.38) can also be written

$$y = YZ_0 \tag{10.39}$$

The Smith chart can be used to solve many types of transmission-line problems. Several examples will be given to illustrate the way in which such problems can be approached. The first shows the calculation of SWR and input impedance.

EXAMPLE 10.13

A 50 Ω line operating at 100 MHz has a velocity factor of 0.7. It is 6 m long, and is terminated with a load impedance of $50 + j50\ \Omega$. Find:

(a) the load admittance Y_L

(b) the SWR

(c) the input impedance of the line Z_i

Solution

(a) Refer to Figure 10.26. First, normalize the load impedance, using Equation (10.37):

$$z_L = \frac{Z_L}{Z_0}$$

$$= \frac{50 + j50\ \Omega}{50\ \Omega}$$

$$= 1 + j1$$

Enter this on the chart as shown, and draw a circle with this radius. This is the SWR circle.

To find the normalized admittance, draw a diameter at the z_L point. It intersects the circle, on the opposite side, at the point $0.5 - j0.5$. This is the normalized admittance. To find the actual admittance, use Equation (10.39):

$$y = YZ_0$$

$$Y = \frac{y}{Z_0}$$

$$= \frac{0.5 - j0.5}{50\ \Omega}$$

$$= 0.01 - j0.01\ \text{S}$$

(b) To find the SWR, simply note where the SWR circle crosses the horizontal axis to the right of the center of the chart, or scale the radius on the VSWR scale at the bottom of the chart. Read the SWR as 2.6.

(c) In order to find the input impedance, it is necessary to know the length of the line in wavelengths. To do that, first find the wavelength on the line.

$$\lambda = \frac{v_p}{f}$$

$$= \frac{v_f c}{f}$$

$$= \frac{0.7 \times 3 \times 10^8\ \text{m/s}}{100 \times 10^6\ \text{Hz}}$$

$$= 2.1\ \text{m}$$

FIGURE 10.26   Smith Chart Calculations for Example 10.13

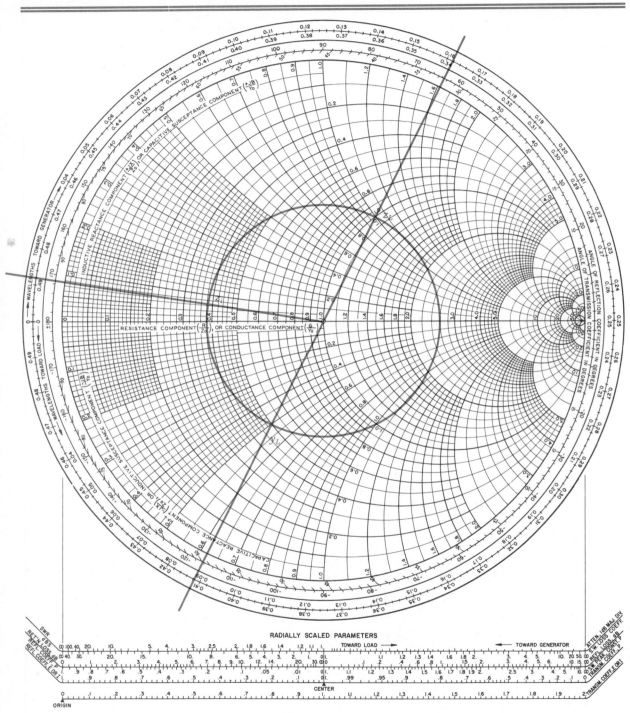

"Smith" is a trademark, and the Smith chart is copyright by Analog Instruments Co., Box 808, New Providence, NJ, 07974. Used by permission.

The length of the line in wavelengths is

$$L = \frac{6 \text{ m}}{2.1 \text{ m}}$$
$$= 2.85$$

Mentally go five times around the Smith chart toward the generator, to account for 2.5 wavelengths. That is, any whole number of half-wavelengths can be subtracted from the length of the line before starting. Now go the remaining distance of 0.35 wavelength toward the generator. There are two scales around the outside of the chart. The one closer to the center increases as we move toward the load, and the other one, at the very outside of the chart, increases as we go toward the generator. Choose the outside scale, since we are going toward the generator. At the load, this scale reads 0.162. Add 0.35, the number of wavelengths remaining before we get to the generator, to get 0.512. The maximum value for the scale is 0.5, so go to that point and then go on a further distance equal to the distance,

$$0.512 - 0.5 = 0.012$$

Draw a radius at this point. The place where it intersects the SWR circle represents the input impedance. Read the normalized impedance from the chart as

$$z_i = 0.38 + j0.064$$

To de-normalize this, multiply by Z_0.

$$Z_i = z_i Z_0$$
$$= (0.38 + j0.064)50 \text{ } \Omega$$
$$= 19.2 + j3.2 \text{ } \Omega$$

The next example concerns the placement of a quarter-wave transformer. Recall that this must be added to the line at a point where the input impedance is resistive. It is easy to find such a point on a Smith chart by noting where the SWR circle crosses the horizontal axis. At such points, the reactive component of the impedance is zero, so the impedance is resistive. The value of the impedance can be calculated from the chart and used to calculate the impedance of the quarter-wave transformer section. This example also demonstrates the fact that it is possible to move along the Smith chart in either direction. For example, the load impedance can be found when the input impedance of the line is known.

EXAMPLE 10.14

The measured input impedance of a 10 m length of 300 Ω line is $120 - j650 \text{ } \Omega$. The velocity factor is 0.9, and the frequency is 50 MHz. Find:

(a) the load impedance

(b) a suitable place to install a quarter-wave transformer

(c) the characteristic impedance that the transformer section should have

Solution

(a) The first step is to normalize the input impedance.

$$z_i = \frac{Z_i}{Z_0}$$

$$= \frac{120 - j650 \ \Omega}{300 \ \Omega}$$

$$= 0.4 - j2.17$$

Plot this on the chart and draw the SWR circle, as shown in Figure 10.27.

Next, find the wavelength on the line. This can be found as before, or a slight shortcut can be taken:

$$\lambda = \frac{v_p}{f}$$

$$= \frac{v_f c}{f}$$

$$= \frac{0.9 \times 3 \times 10^8 \text{ m/s}}{50 \times 10^6 \text{ Hz}}$$

$$= 5.4 \text{ m}$$

The length of the line in wavelengths is $10/5.4 = 1.85$ wavelengths.

To find the load impedance, start at the point already plotted, which corresponds to the input, and go round the circle three times, then a further 0.35 wavelength. Be sure to go counterclockwise: follow the arrow marked "wavelengths toward load." Following the scale that increases toward the load, the source is at 0.182. Adding 0.35 to this gives 0.532, so go to 0.5 and then keep going in the same direction for the difference of 0.032.

Now the normalized impedance can be read from the intersection of a line drawn from the center to 0.032 on the wavelength scale, with the SWR circle. The normalized impedance is $z_L = 0.075 - j0.21$. This can easily be converted to the actual load impedance.

$$Z_L = z_L Z_0$$

$$= (0.075 - j0.21)300 \ \Omega$$

$$= 22.5 - j63 \ \Omega$$

(b) The quarter-wave transformer must be installed in a place where the impedance is real. It is easy to find such a place on a Smith chart: the impedance is real wherever the SWR circle crosses the horizontal axis, which it does once every quarter-wavelength. Normally, the closest such place to the load would be used, to minimize the length of mismatched line in the system. Remember that the transformer itself,

FIGURE 10.27   Smith Chart Calculations for Example 10.14

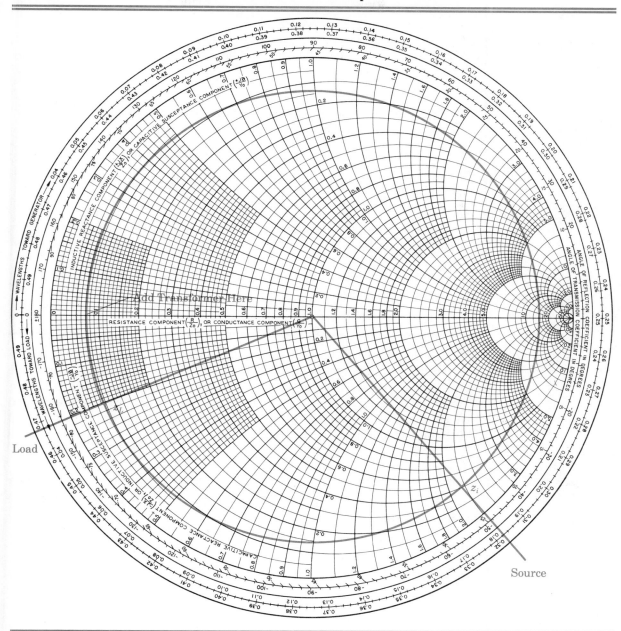

"Smith" is a trademark, and the Smith chart is copyright by Analog Instruments Co., Box 808, New Providence, NJ, 07974. Used by permission.

and the line between the transformer and the load, will not be matched.

Proceeding from the load in the direction of the generator, it is necessary to go a distance of 0.032 wavelength before the circle crosses the x axis. In meters, the distance will be

$$0.032 \times 5.4 = 0.173 \text{ m}$$

(c) The impedance of the line at this point is real, and equal to 0.07 in normalized form. The actual impedance is

$$Z_L = z_L Z_0$$
$$= 0.07 \times 300 \ \Omega$$
$$= 21 \ \Omega$$

The required impedance of the quarter-wave transformer section can be found from Equation (10.36):

$$Z_0' = \sqrt{Z_0 Z_L}$$
$$= \sqrt{300 \ \Omega \times 21 \ \Omega}$$
$$= 79.4 \ \Omega$$

For our final Smith chart example, let us look at single-stub matching. Since the stub cannot change the resistive part of the impedance, we look for a spot on the line where this is the correct value (that is, equal to Z_0), which, when normalized, appears as a value of 1 on the chart. Then, we use the stub to cancel the reactive component. For practical reasons, we will use a shorted stub connected in parallel with the main line.

EXAMPLE 10.15

Match a load with impedance $Z_L = 120 - j100 \ \Omega$ to a line with $Z_0 = 72 \ \Omega$, using a shorted stub.

Solution

As usual, the first step is to normalize Z_L.

$$z_L = \frac{Z_L}{Z_0}$$
$$= \frac{120 - j100 \ \Omega}{72 \ \Omega}$$
$$= 1.67 - j1.39$$

Draw the SWR circle (see Figure 10.28).

Since the stub will be in parallel with the line, it is easier to use admittances than impedances, since admittances in parallel add. Therefore, the next step is to change the load impedance to an admittance. Do this by drawing a diameter on the SWR circle, from the load impedance, through the center. The point where this line intersects the circle on the opposite side represents the load admittance. Call it y_L. From the chart, $y_L = 0.35 + j0.3$.

FIGURE 10.28   Smith Chart Calculations for Example 10.15

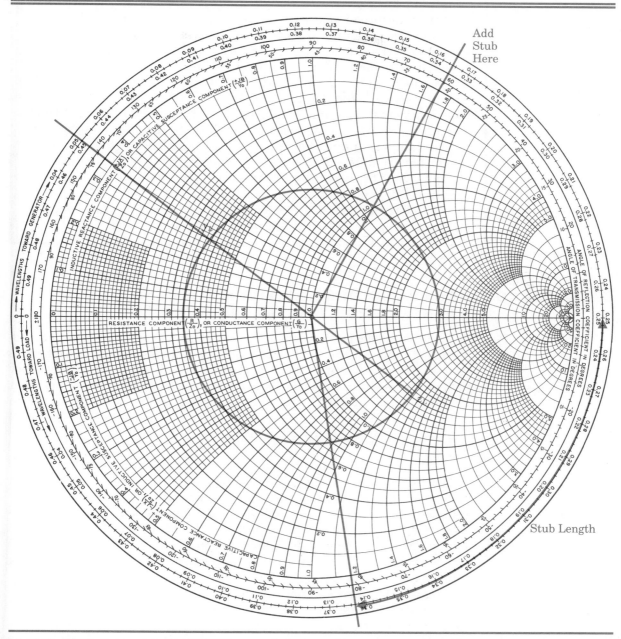

"Smith" is a trademark, and the Smith chart is copyright by Analog Instruments Co., Box 808, New Providence, NJ, 07974. Used by permission.

The next step is to move toward the generator until the real part of the admittance has the correct normalized value of 1. Move from y_L toward the generator until the SWR circle intersects the $g = 1$ circle. At this point, the normalized admittance is $1 + j1.18$. We have moved a distance of 0.116 wavelength. The stub will have to have a normalized admittance of $-j1.18$ so that it will cancel the susceptance of the line.

The stub will be entirely reactive, and will have infinite SWR. Its SWR circle, therefore, is the outside of the chart. To find the correct stub length, we will have to start at the "load" end of the stub (the short-circuited end) and proceed around the outside of the chart in the direction of the generator, until the susceptance has the required value. The shorted end will, of course, have infinite admittance, so we start at the right end of the chart. Moving toward the generator a distance of 0.112 wavelength will bring us to a point with a normalized admittance of $-j1.18$.

Therefore the load can be matched by inserting a stub with a length of 0.112 wavelength at a point 0.116 wavelength from the load.

10.7 TEST EQUIPMENT AND PROCEDURES: TRANSMISSION-LINE MEASUREMENTS

There is a good deal of specialized test equipment for transmission-line measurements in both the time and frequency domains. The time domain is represented in this section by *time-domain reflectometry*. In the frequency domain, we will examine the *slotted line*, and look briefly at two simpler methods for measuring the SWR on a transmission line.

10.7.1 Time-Domain Reflectometry

The techniques of time-domain reflectometry (TDR) are essentially practical applications of the methods used at the beginning of the chapter. A step input or a pulse is applied to the input of a transmission line. By examining the reflected signal, a good deal of information can be gained about such things as the length of the line, the way in which it is terminated, and the type and location of any impedance discontinuities on the line. These discontinuities can be caused by such things as faulty splices or connectors, water leaking into the cable, kinks, and so forth.

Figure 10.29 shows a typical TDR setup. Figure 10.29(a) is a photograph of the equipment, and Figure 10.29(b) is a block diagram showing its operation. In reality, the step input will be a low-frequency pulse, so that it can be repetitive to allow easier display. However, this will not affect measurements made near the leading edge of the step. If a dedicated TDR unit is not available, a pulse generator and a high-speed oscilloscope can be connected as shown in Figure 10.29(b).

The most important use of TDR is to determine the position and type of defects on a line. Often, these are not easy to locate by other means, especially in an underground or underwater cable. With TDR, the distance to the defect is easy to determine from the time taken for the reflection to return to the source. The type of defect can be gauged to some extent from the nature of the reflection.

FIGURE 10.29

Time-Domain
Reflectometry

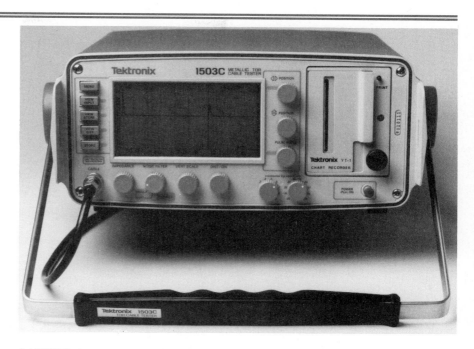

(a) TDR Setup

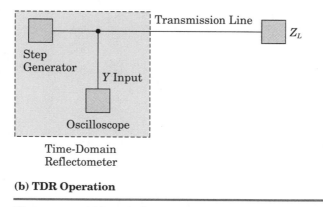

(b) TDR Operation

Photo courtesy Tektronix, Inc.

Figure 10.30 shows several types of reflections. The line in Figure
10.30(a) is open-circuited, and the voltage rises to double the initial value.
In Figure 10.30(b), the termination is a short circuit, causing the final voltage
to fall to zero. Figure 10.30(c) shows $Z_L > Z_0$.

Reactive terminations can also be distinguished using TDR. An inductance will initially appear like an open circuit, then gradually draw more
current until after a time it appears as a short. Figure 10.30(d) shows how an
inductive termination would appear using TDR. On the other hand, a capacitor would initially appear as a short circuit, becoming more like an open circuit as it charges, as can be seen in Figure 10.30(e).

Figure 10.31 shows two reflections. The first is from an impedance discontinuity in the line, the second from a mismatched load.

FIGURE 10.30   Typical Waveform Displays Using TDR

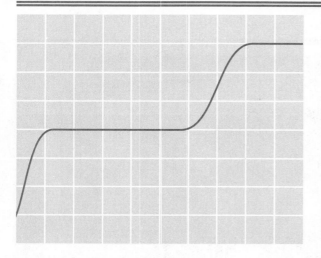

(a) Open-Circuited Line

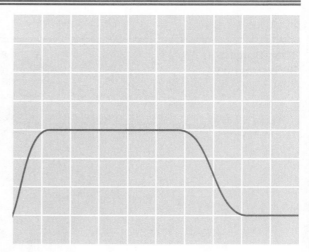

(b) Short-Circuited Line

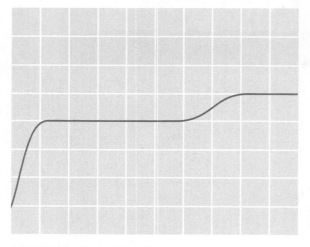

(c) Z_L Resistive and $Z_L > Z_0$

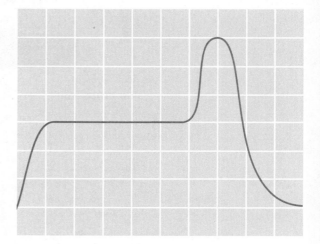

(d) Z_L Inductive

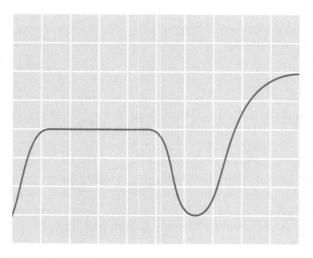

(e) Z_L Capacitive

FIGURE 10.31

TDR Display Showing
an Impedance
Discontinuity

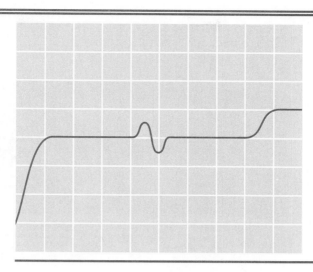

EXAMPLE 10.16

A TDR display shows a discontinuity at a time of 1.4 μs from the start. If the line has a velocity factor of 0.8, how far is the fault from the reflectometer?

Solution

The time shown on the screen is the time taken for the signal to travel to the defect and back. Therefore, one-half that time should be used to calculate the distance.

$$
\begin{aligned}
d &= \frac{v_p t}{2} \\
&= \frac{v_f c t}{2} \\
&= \frac{0.8(3 \times 10^8 \text{ m/s})(1.4 \times 10^{-6} \text{ s})}{2} \\
&= 168 \text{ m}
\end{aligned}
$$

10.7.2 The Slotted Line

The slotted line is a very straightforward way of conducting transmission-line measurements. See Figure 10.32 for a photograph. Essentially, it is a short section of air-dielectric coaxial line, with a slot in the outer conductor through which a probe is inserted. The probe does not actually touch the inner conductor; it is capacitively coupled, so as to disturb the signal on the line as little as possible. The probe is connected to a diode detector that rectifies and filters the RF voltage. The resulting signal is sent to a sensitive voltmeter. Usually, an amplitude-modulated signal is used with the slotted line, so the output from the diode will be an audio-frequency signal that is easy to

FIGURE 10.32

A Slotted Line

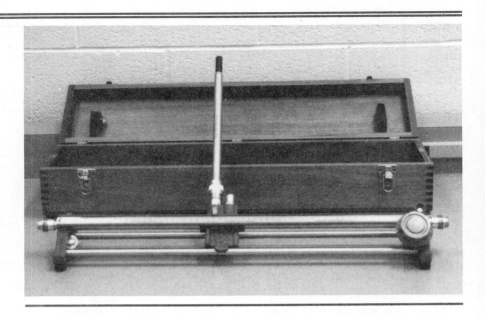

amplify and measure. A means is provided of accurately positioning the probe at a given point on the line.

The length of a slotted line must be at least one-half wavelength. Due to the limited physical length of practical slotted lines, they are most commonly used at UHF and higher frequencies. The line must have the same impedance as the transmission line with which it is used; generally this will be 50 Ω.

In operation, the slotted line is inserted into the transmission line. It allows the wavelength on the line and the SWR to be measured very easily. Since the SWR is, by definition, V_{max}/V_{min}, it is easily measured by sliding the probe along the line and recording the maximum and minimum voltages found. The meters used with slotted lines allow the maximum to be adjusted to a set point. Then, the SWR can be read directly from the meter, when the line is adjusted for a minimum reading. The measurement of wavelength is equally simple; since voltage minima occur at half-wavelength intervals on a line, all that is required is to note the position of two adjacent minimum voltage readings. Voltage maxima could be used instead, but in practice the minima can be located more precisely.

EXAMPLE 10.17

Two adjacent minima on a slotted line are 23 cm apart. Find the wavelength, and the frequency, assuming a velocity factor of 95%.

Solution

The minima are separated by one-half wavelength so

$$\lambda = 2 \times 23 \text{ cm}$$
$$= 46 \text{ cm}$$

The frequency is given by

$$v_p = f\lambda$$
$$f = \frac{v_p}{\lambda}$$
$$= \frac{v_f c}{\lambda}$$
$$= \frac{0.95 \times 3 \times 10^8 \text{ m/s}}{0.46 \text{ m}}$$
$$= 620 \times 10^6 \text{ Hz}$$
$$= 620 \text{ MHz}$$

10.7.3 Standing-Wave Ratio Meters and Directional Wattmeters

Though the slotted line allows the SWR to be measured in a very straight-forward way, it is a cumbersome piece of equipment for field use, and is unsuitable for frequencies below UHF. There are other methods that are more convenient.

The direct measurement of SWR requires a *directional coupler*. This device allows the measurement of power moving along the line in each direction; that is, it is possible to measure incident and reflected power separately. Once this has been done, it is easy to calculate the reflection coefficient and, from that, the SWR.

There are several ways to build a directional coupler. Figure 10.33 shows one method. The idea is to insert a short section of air-dielectric coaxial line with the same characteristic impedance into the transmission line on which measurements are to be made. An additional conductor is placed between the center conductor and the shield, as a sensing element. The current on the transmission line is sensed by using the mutual inductance between the center conductor of the line and the added conductor, and the voltage is sensed by capacitive coupling between the added conductor and the center conductor. The phase of the current relative to the voltage will depend on the direction in which the wave is travelling along the line. It is possible to sum the current-derived and voltage-derived signals across a resistor in such a way that they cancel for a signal moving in one direction and add for a signal moving the other way. The resulting signal is rectified and filtered and can then be observed with an ordinary dc microammeter or milliammeter. With careful design and calibration, the power moving along the line can be measured fairly accurately.

Probably the best-known example of an in-line directional wattmeter is the Bird Thruline™ meter, which was pictured in Chapter 4. This device allows power to be measured in either direction by simply turning a removable element. The power travelling in the direction of the arrow is measured. A number of interchangeable elements calibrated for different power and frequency ranges can be used with this meter.

FIGURE 10.33

Directional Coupler

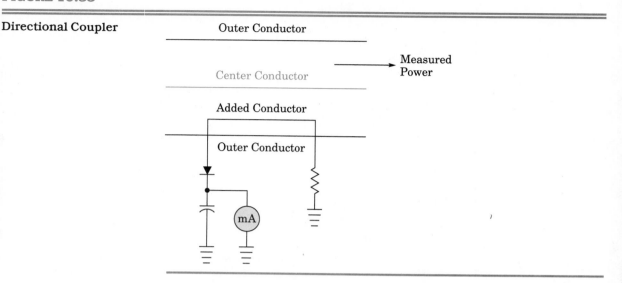

In order to measure SWR with a meter of this type, the power moving in each direction is measured, and the SWR is calculated from the power ratio. The SWR is given by Equation (10.24):

$$SWR = \frac{1 + |\Gamma|}{1 - |\Gamma|}$$

The ratio of reflected power to incident power is the square of the voltage ratio, or Γ^2. Therefore,

$$SWR = \frac{1 + \sqrt{P_r/P_i}}{1 - \sqrt{P_r/P_i}} \qquad\qquad (10.40)$$

EXAMPLE 10.18

The forward power in a transmission line is 150 W, and the reverse power is 20 W. Calculate the SWR on the line.

Solution

The easiest way is to first calculate the quantity $\sqrt{P_r/P_i}$.

$$\sqrt{\frac{P_r}{P_i}} = \sqrt{\frac{20 \text{ W}}{150 \text{ W}}}$$

$$= 0.365$$

Then, substituting into Equation (10.40),

$$SWR = \frac{1 + 0.365}{1 - 0.365}$$

$$= 2.15$$

An easier way to calculate the SWR using a directional wattmeter is to use a nomograph, such as the one in Figure 10.34. Redoing the previous example with the nomograph will yield the same result. Try it.

An even more convenient method for measuring SWR uses two directional couplers, one each for forward and reflected power. This allows the use of a meter calibrated directly in SWR. Figure 10.35 is a photograph of a typi-

FIGURE 10.34

SWR Nomograph

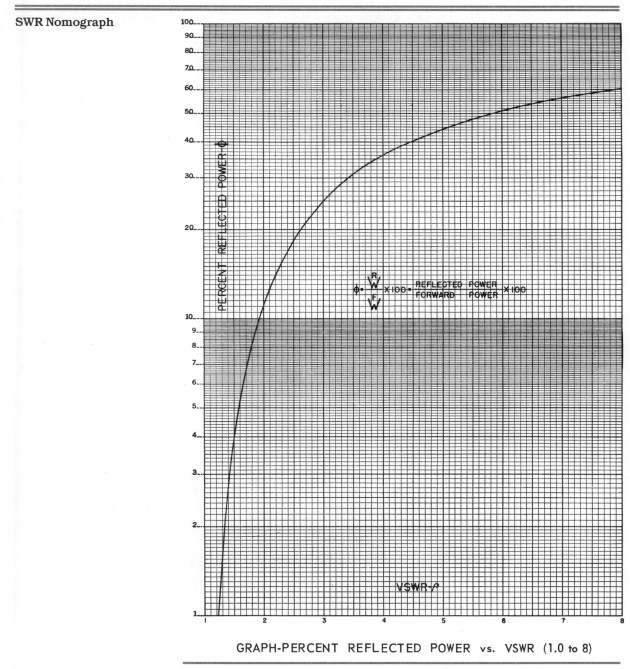

GRAPH-PERCENT REFLECTED POWER vs. VSWR (1.0 to 8)

Courtesy Bird Electronic Corp.

FIGURE 10.35

SWR meter

MFJ Enterprises, Inc., P.O. Box 494, Mississippi State, MS 39762.

cal instrument of this type. The meter is set for full-scale deflection in the "forward" position of the switch; then the switch is moved to the "reverse" position, and the SWR read directly from the meter. When any adjustment is made to the line, the forward reading must be readjusted to the set point. Also, meters of this type are suitable only for SWR measurements; they are not usable for the measurement of actual power levels.

The ultimate in convenience is the cross-needle SWR meter shown in Figure 10.36. This device responds to forward and reverse power simultaneously, with a separate meter for each. The SWR is found from the point at which the two meter needles intersect. No switching is required, and no readjustment is needed to compensate for changes in forward power as adjustments are made to the line or the load.

FIGURE 10.36

**Cross-needle
SWR meter**

MFJ Enterprises, Inc., P.O. Box 494, Mississippi State, MS 39762.

**SUMMARY
OF
CHAPTER 10**

Here are the main points to remember from this chapter.

1. Any pair of conductors can act as a transmission line. The methods of this chapter need to be used whenever the line is longer than approximately one-sixteenth wavelength at the highest frequency in use.

2. Any transmission line has a characteristic impedance determined by its geometry and its dielectric. If a transmission line is terminated by an impedance that is different from its characteristic impedance, part of a signal travelling down the line will be reflected. Usually, this is undesirable.

3. Reflections on lines are characterized by the reflection coefficient and the standing-wave ratio. The latter is easier to measure. A matched line has a reflection coefficient of 0 and an SWR of 1.

4. Lines that are terminated by either a short or an open circuit can be used as reactances or as either series- or parallel-resonant circuits, depending on their length.

5. Lines can be matched using lumped constants, transmission-line transformers, or shorted stubs.

6. The Smith chart allows graphical transmission-line calculations, and shows transmission-line parameters in an intuitive way.

7. Most transmission-line calculations are now done using special computer programs.

8. Specialized test equipment used with transmission lines includes slotted lines, time-domain reflectometers, and SWR meters.

IMPORTANT EQUATIONS

$$Z_0 = \sqrt{\frac{L}{C}} \tag{10.2}$$

$$Z_0 \approx 276 \log \frac{D}{r} \tag{10.3}$$

$$Z_0 \approx \frac{138}{\sqrt{\epsilon_r}} \log \frac{D}{d} \tag{10.4}$$

$$v_f = \frac{v_p}{c} \tag{10.6}$$

$$v_f = \frac{1}{\sqrt{\epsilon_r}} \tag{10.7}$$

$$T = \frac{L}{v_p} \tag{10.8}$$

$$\Gamma = \frac{V_r}{V_i} \tag{10.9}$$

$$\Gamma = \frac{Z_L - Z_0}{Z_L + Z_0} \tag{10.16}$$

$$v_p = f\lambda \tag{10.19}$$

$$\phi = (360)\frac{L}{\lambda} \tag{10.20}$$

$$\text{SWR} = \frac{V_{max}}{V_{min}} \tag{10.21}$$

$$\text{SWR} = \frac{1 + |\Gamma|}{1 - |\Gamma|} \tag{10.24}$$

$$|\Gamma| = \frac{\text{SWR} - 1}{\text{SWR} + 1} \tag{10.25}$$

$$\text{SWR} = \frac{Z_L}{Z_0}, \text{ if } Z_L > Z_0 \tag{10.26}$$

$$\text{SWR} = \frac{Z_0}{Z_L}, \text{ if } Z_0 > Z_L \tag{10.27}$$

$$P_r = \Gamma^2 P_i \tag{10.28}$$

$$P_L = P_i(1 - \Gamma^2) \tag{10.29}$$

$$P_L = \frac{4\text{SWR}}{(1 + \text{SWR})^2} P_i \tag{10.30}$$

$$Z = Z_0 \frac{Z_L \cos\theta + jZ_0 \sin\theta}{Z_0 \cos\theta + jZ_L \sin\theta} \tag{10.31}$$

$$Z = Z_0 \frac{Z_L + jZ_0 \tan\theta}{Z_0 + jZ_L \tan\theta} \tag{10.32}$$

$$Z = jZ_0 \tan\theta \text{ for shorted line} \tag{10.33}$$

$$Z_0' = \sqrt{Z_0 Z_L} \text{ for quarter-wave shorted stub} \tag{10.36}$$

$$z = \frac{Z}{Z_0} \tag{10.37}$$

$$y = \frac{Y}{Y_0} \tag{10.38}$$

$$\text{SWR} = \frac{1 + \sqrt{P_r/P_i}}{1 - \sqrt{P_r/P_i}} \tag{10.40}$$

GLOSSARY

balun   a device for coupling balanced and unbalanced lines

characteristic impedance   the ratio between voltage and current on an infinitely long transmission line

coaxial line   a transmission line with concentric conductors

open-wire line   a transmission line with parallel conductors, separated by spacers

propagation velocity   the speed at which signals travel down a transmission line

quarter-wave transformer   a section of transmission line, electrically a quarter-wavelength in length, that is used to change impedances on a transmission line

reflection coefficient   the ratio of reflected to incident voltage on a transmission line

Smith chart   a graphical transmission-line calculator

standing-wave ratio (SWR)   the ratio of maximum to minimum voltage on a transmission line

stub   a short section of line, usually short-circuited at one end, used for impedance matching

surge impedance   see characteristic impedance

transmission line   any pair of conductors used to conduct electrical energy

velocity factor   ratio of the speed of propagation on a line to that of light in free space

QUESTIONS

1. Why is an ordinary extension cord not usually considered as a transmission line, while a television antenna cable of the same length would be?

2. Explain the difference between balanced and unbalanced lines, and give an example of each.

3. Draw the equivalent circuit for a short section of transmission line, and explain the physical meaning of each circuit element.

4. What is meant by the characteristic impedance of a transmission line?

5. Define the velocity factor for a transmission line, and explain why it can never be greater than one.

6. Explain what is meant by the SWR on a line, and state its value when a line is perfectly matched.

7. Why is a high SWR generally undesirable?

8. Why are shorted stubs preferred to open ones for impedance matching?

9. Draw a sketch showing how the impedance varies with distance along a lossless shorted line.

10. What would be the effect of placing a shorted, half-wave stub across a matched line?

11. What are the major contributors to transmission-line loss?

12. How does transmission-line loss vary with frequency, and why?

13. How does the SWR at the source differ from that at the load with a lossy line, and why?

14. Draw circuits for a balun using a transformer, and using a transmission-line segment. Which of these has greater bandwidth? Which is likely to have a larger ratio of power-handling capacity to cost?

15. Draw a sketch showing how a quarter-wave transformer can be used for impedance matching. Is the transformer itself matched?

16. Compare single- and double-stub matching, giving one advantage of each.

17. Why is a Smith chart circular?

18. Why are admittances rather than impedances used for stub-matching calculations?

19. Explain how TDR can be used to find a defect in an underground cable.

20. Describe two ways in which the standing-wave ratio on a transmission line can be measured.

PROBLEMS

Section 10.3

21. A 12 V dc source is connected to a 93 Ω lossless line through a 93 Ω source resistance at time $t = 0$. The line is 85 m long and is terminated in a resistance.

    (a) What is the voltage across the input of the line immediately after $t = 0$?
    (b) At time $t = 1.0$ μs, the voltage at the input end of the line changes to 7.5 V, with the same polarity as before. What is the resistance that terminates the line?
    (c) What is the velocity factor of the line?

22. A 10 V positive-going pulse is sent down 50 m of lossless 50 Ω cable with a velocity factor of 0.8. The cable is terminated with a 150 Ω resistor. Calculate the length of time it will take the reflected pulse to return to the start, and the amplitude of the reflected pulse.

23. Calculate the characteristic impedance of an open-wire transmission line consisting of two wires with diameter 1 mm and separation 1 cm.

24. Calculate the characteristic impedance of a coaxial line with a polyethylene dielectric, if the diameter of the inner conductor is 3 mm and the inside diameter of the outer conductor is 10 mm.

Section 10.4

25. A transmitter delivers 50 W into a 600 Ω lossless line that is terminated with an antenna that has an impedance of 275 Ω, resistive.

    (a) What is the coefficient of reflection?
    (b) How much of the power actually reaches the antenna?

26. A generator is connected to a short-circuited line 1.25 wavelengths long.

    (a) Sketch the waveforms for the incident, reflected, and resultant voltages at the instant the generator is at its maximum positive voltage.
    (b) Sketch the pattern of voltage standing waves on the line.
    (c) Sketch the variation of impedance along the line. Be sure to note whether the impedance is capacitive, inductive, or resistive at each point.

27. A transmission line is to be used to shift the phase of a 10 MHz signal by 90°. If RG-8/U with solid polyethylene dielectric is used, how long should the line be?

28. An open-circuited line is 0.75 wavelength long.

    (a) Sketch the incident, reflected, and resultant voltage waveforms at the instant the generator is at its peak negative voltage.
    (b) Draw a sketch showing how RMS voltage varies with position along the line.

29. A 75 Ω source is connected to a 50 Ω load (a spectrum analyzer) with a length of 75 Ω line. The source produces 10 mW. All impedances are resistive.

    (a) Calculate the SWR.

(b) Calculate the voltage reflection coefficient.

(c) How much power will be reflected from the load?

Section 10.5

30. A properly matched transmission line has a loss of 1.5 dB/100 m. If 10 W are supplied to one end of the line, how many watts reach the load, 27 m away?

31. A receiver requires 0.5 μV of signal for satisfactory reception. How strong (in microvolts) must the signal be at the antenna if the receiver is connected to the antenna by 25 m of matched line having an attenuation of 6 dB per 100 m?

32. A transmitter with an output power of 50 W is connected to a matched load by means of 32 m of matched coaxial cable. It is found that only 35 W of power is dissipated in the load. Calculate the loss in the cable in decibels per 100 meters.

Section 10.6

33. A 75 Ω transmission line is terminated with a load having an impedance of 45 − j30 Ω. Use the Smith chart to find:

   (a) the distance (in wavelengths) from the load to the closest place at which a quarter-wave transformer could be used to match the line

   (b) the characteristic impedance that should be used for the quarter-wave transformer

34. Draw a dimensioned sketch of a transmission-line balun that could be used to match RG-59/U type coaxial cable, with foam dielectric, to television twin-lead. Design it to work in the center of the FM broadcast band (98 MHz).

35. Find the length and position of the transformer in Problem 33, given that the operating frequency is 20 MHz and the velocity factor for both lines is 0.66.

36. A transmitter supplies 100 W to a 50 Ω lossless line that is 5.65 wavelengths long. The other end of the line is connected to an antenna with a characteristic impedance of 150 + j25 Ω.

   (a) Use the Smith chart to find the SWR and the magnitude of the reflection coefficient.

   (b) How much of the transmitter power reaches the antenna?

   (c) Use the Smith chart to find the best place at which to insert a shorted matching stub on the line. (Give the answer in wavelengths from the load.)

   (d) Use the Smith chart to find the proper length for the stub (in wavelengths).

Section 10.7

37. A slotted line having an impedance of 50 Ω and a velocity factor of 0.95 is used to conduct measurements with a generator and a resistive load, which is known to be more than 50 Ω. It is found that the maximum voltage on the line is 10 V, and the minimum is 3 V. The distance between two minima is measured as 75 cm.

   (a) What is the wavelength on the line?

   (b) What is the frequency of the generator?

   (c) What is the standing wave ratio?

   (d) What is the load resistance?

38. A time-domain reflectometer produces a 1 V step input. Sketch the signal that would be viewed, if the line is lossless, 50 m in length, with a velocity factor of 0.8, and terminated in an open circuit.

Comprehensive

39. A transmission line of unknown impedance is terminated with two different resistances, and the SWR is measured each time. With a 75 Ω termination, the SWR

measures 1.5. With a 300 Ω termination, it measures 2.67. What is the impedance of the line?

40. Lumped constants can be used instead of transmission-line stubs to match lines. Solve Problem 36 in this way, by calculating the correct value of inductance or capacitance that could be placed across the line, instead of using the shorted stub. The frequency is 20 MHz.

11 Radio-Wave Propagation

Objectives

After studying this chapter, you should be able to:

1. Describe the nature and behavior of radio waves, and compare them to other forms of electromagnetic radiation.
2. Calculate power density and electric and magnetic field intensity for waves propagating in free space.
3. Explain the meaning of wave polarization, and differentiate between vertical, horizontal, circular, and elliptical polarization.
4. Calculate free-space attenuation.
5. Describe reflection, refraction, and diffraction, and calculate angles of reflection and refraction.
6. Describe the most common methods of terrestrial propagation, decide on the most suitable method for a given frequency and distance, and perform the necessary calculations to determine the communication range.
7. Calculate propagation delay and path loss for communication links involving geostationary communications satellites, and describe the types of applications for which satellites are suitable.

11.1 INTRODUCTION

Radio waves are one form of electromagnetic radiation. Other forms include infrared, visible light, ultraviolet, X rays, and gamma rays. Figure 11.1 shows the place of radio waves at the low end of the frequency spectrum.

Scientists believe that electromagnetic radiation has a dual nature. Under some circumstances, it acts like a set of waves, while at other times its behavior is more easily explained by considering it to be a stream of particles called **photons**. Which description is better depends on frequency; at radio frequencies, the wave model is generally more appropriate, while light is sometimes better modelled by photons.

Electromagnetic radiation, as the name implies, involves the creation of electric and magnetic fields in free space or in some physical medium. The waves that propagate are known as *transverse electromagnetic* (TEM) waves. This means that the electric field, the magnetic field, and the direction of travel of the wave are all mutually perpendicular. The sketch in Figure 11.2 is an attempt to represent this three-dimensional process in two dimensions.

Electromagnetic radiation can be generated by many different means, but all of them involve the movement of electrical charges. In the case of radio waves, the charges are electrons moving in a conductor, or set of conductors, called an *antenna*. Antennas are the subject of the next chapter.

Once launched, electromagnetic waves can travel through **free space** (that is, through a vacuum), and through many materials. Any good dielectric will pass radio waves; the material does not have to be transparent to light. The waves do not travel well through lossy conductors, such as sea water, because the electric fields cause currents to flow that dissipate the energy of the wave very quickly. Radio waves will reflect from good conductors, such as copper or aluminum, and will be refracted as they pass from one medium to another, just as light is.

The speed of **propagation** of the wave in free space is the same as that of light, approximately 3×10^8 m/s. This should be no surprise, since the two forms of energy are very similar. In other media, the velocity is lower. The propagation velocity is given by

$$v = \frac{c}{\sqrt{\epsilon_r}} \tag{11.1}$$

where v = propagation velocity in the medium
c = 3×10^8 m/s, the propagation velocity in free space
ϵ_r = relative permittivity of the medium

FIGURE 11.1   **Electromagnetic Spectrum**

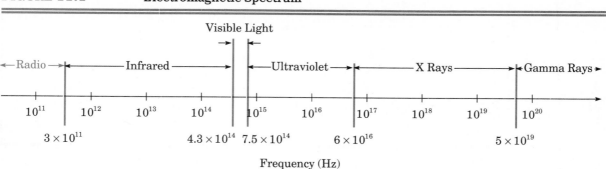

FIGURE 11.2

**Transverse
Electromagnetic Waves**

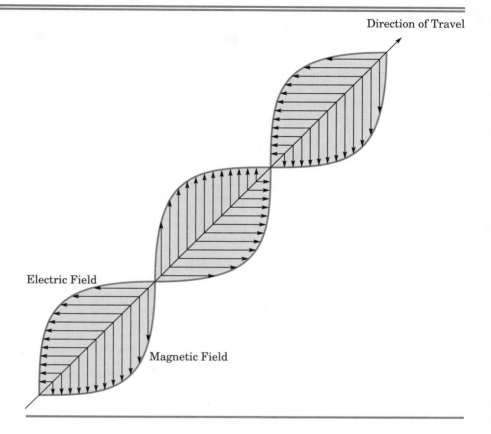

This is the same equation given in Chapter 10 for the velocity of signals along a coaxial line. One way of thinking of such a line is to imagine an electromagnetic wave propagating through the dielectric, guided by the conductors.

11.2   ELECTROMAGNETIC WAVES

In this chapter, we will look at transverse electromagnetic waves propagating through space and other media. We should remember that these waves are characterized by frequency and wavelength.

Wavelength was discussed in Chapter 10. It is related to frequency and propagation velocity by a simple equation, repeated here for easy reference.

$$v = f\lambda \tag{11.2}$$

where v = propagation velocity in meters per second
f = frequency in hertz
λ = wavelength in meters

Radio waves differ from other waves found in nature, such as water waves and sound waves, in that there is no physical motion of the medium, and, in fact, no need of any physical medium. The waves consist only of time-varying electric and magnetic fields. This makes them somewhat harder to visualize, but it is necessary to make the attempt in order to understand the way they propagate, and the ways in which they are generated and intercepted by antennas.

11.2.1 Electric and Magnetic Fields

An electromagnetic wave propagating through space consists of electric and magnetic fields, perpendicular both to each other and to the direction of travel of the wave (see Figure 11.2). The fields vary together, both in time and space, and there is a definite ratio between the intensities of the electric and magnetic fields. This ratio is called the *characteristic impedance* of the medium, and is expressed in ohms.

The relationship between the electric and magnetic field intensities is analogous to the relation between voltage and current in circuits using lumped constants. For circuits, we have the familiar Ohm's Law:

$$Z = \frac{V}{I}$$

where Z = impedance in ohms
V = electromotive force in volts
I = current in amperes

Ohm's Law for electromagnetic waves is very similar:

$$\mathscr{Z} = \frac{\mathscr{E}}{\mathscr{H}} \tag{11.3}$$

where $\mathscr{Z}$ = impedance of the medium in ohms
$\mathscr{E}$ = electric field strength in volts per meter
$\mathscr{H}$ = magnetic field strength in amperes per meter

For a lossless medium, this is equivalent to

$$\mathscr{Z} = \sqrt{\frac{\mu}{\epsilon}} \tag{11.4}$$

where μ = permeability of the medium in henrys per meter
ϵ = permittivity of the medium in farads per meter

For free space,

$$\mu_0 = 4\pi \times 10^{-7} \text{ H/m} \quad \text{and} \quad \epsilon_0 = 8.854 \times 10^{-12} \text{ F/m}$$

so the impedance of free space is

$$\mathscr{Z}_0 = \sqrt{\frac{\mu_0}{\epsilon_0}}$$

$$= 377 \ \Omega \tag{11.5}$$

For most media in which electromagnetic waves can propagate, the permeability is the same as that of free space. The permittivity is likely to be given as a dielectric constant, which is simply the permittivity of the medium relative to that of free space; that is,

$$\epsilon_r = \frac{\epsilon}{\epsilon_0} \tag{11.6}$$

where ϵ_r = dielectric constant (relative permittivity)
ϵ = permittivity of the medium
ϵ_0 = permittivity of free space

It follows from Equations (11.4), (11.5), and (11.6) that the impedance of a nonmagnetic medium is

$$\begin{aligned}
Z &= \sqrt{\frac{\mu}{\epsilon}} \\
&= \sqrt{\frac{\mu_0}{\epsilon_r \epsilon_0}} \\
&= \sqrt{\frac{\mu_0}{\epsilon_0}} \times \frac{1}{\sqrt{\epsilon_r}} \\
&= \frac{377}{\sqrt{\epsilon_r}}
\end{aligned} \tag{11.7}$$

EXAMPLE 11.1

Find the characteristic impedance of polyethylene, which has a dielectric constant of 2.3.

Solution

From Equation (11.7),

$$\begin{aligned}
Z &= \frac{377}{\sqrt{\epsilon_r}} \\
&= \frac{377}{\sqrt{2.3}} \\
&= 249 \ \Omega
\end{aligned}$$

11.2.2 Power Density

In lumped-constant circuits, power is given by the equation

$$P = \frac{V^2}{R}$$

where P = power in watts
V = voltage in volts
R = resistance in ohms

The analogous equation for electromagnetic waves is

$$P_D = \frac{\mathscr{E}^2}{\mathscr{Z}} \qquad (11.8)$$

where P_D = power density in watts per square meter
$\mathscr{E}$ = electric field strength in volts per meter
$\mathscr{Z}$ = impedance of the medium in ohms

The **power density** can also be expressed in two other forms that are analogous to equations from circuit theory:

$$P_D = \mathscr{H}^2 \mathscr{Z} \qquad (11.9)$$

$$P_D = \mathscr{E}\mathscr{H} \qquad (11.10)$$

In physical terms, power density in space is the amount of power that flows through each square meter of a surface perpendicular to the direction of travel. To find the power in the whole wave, it would be necessary to integrate (that is, sum) the power density over the surface area.

EXAMPLE 11.2

The **dielectric strength** of air is about 3 MV/m. Arcing is likely to take place at field strengths greater than that. What is the maximum power density of an electromagnetic wave in air?

Solution

From Equation (11.8),

$$P_D = \frac{\mathscr{E}^2}{\mathscr{Z}}$$

$$= \frac{(3 \times 10^6)^2}{377}$$

$$= 23.9 \text{ GW/m}^2$$

Power densities of this order of magnitude will be found very close to antennas that are physically small and operated at very high power. Some radar antennas fit this description. Such densities can also be found in resonant cavities and waveguides, which will be discussed in Chapter 13.

11.2.3 Plane and Spherical Waves

Conceptually, the simplest source of electromagnetic waves would be a point in space. Waves would radiate equally from this source in all directions. A

FIGURE 11.3

Isotropic Radiator

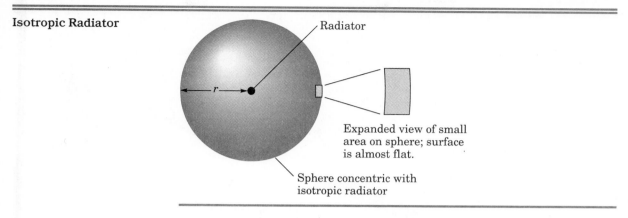

Radiator

r

Expanded view of small area on sphere; surface is almost flat.

Sphere concentric with isotropic radiator

wavefront, that is, a surface on which all the waves have the same phase, would be the surface of a sphere. Such a source is called an **isotropic radiator**, and is shown in Figure 11.3.

Of course, an actual point source is not a practical possibility, but at distances from a real source that are large compared with the dimensions of the source, the approximation will be good. This is usually the case with radio propagation at reasonable distances from the antenna.

If only a small area on the sphere shown in Figure 11.3 is examined, and if the distance from its center is large, the area in question will resemble a plane. In the same way, we experience the earth as flat, though we know it is roughly spherical. Consequently, many practical cases of wave propagation can be studied in terms of plane waves, which are often simpler to deal with than spherical waves. Reflection and refraction are examples of problems that can be simplified by assuming plane waves.

11.2.4 Polarization

The **polarization** of a plane wave is simply the direction of its electric field vector (see Figure 11.2). If this is unvarying, the polarization is described as linear. It is also customary to refer the polarization axis to the horizon. Polarization is important because the polarization of the receiving antenna must be the same as that of the wave for best reception.

Sometimes, the polarization axis rotates as the wave moves through space, rotating 360° for each wavelength of travel. (See Figure 11.4.) In that case, the polarization is circular if the field strength is equal at all angles of polarization, and elliptical if the field strength varies as the polarization changes. The wave can rotate in either direction, and is called *right-handed* if it rotates in a clockwise direction as it recedes. The wave shown in Figure 11.4 has right-handed circular polarization. Circularly polarized waves can be received reasonably well by antennas using either horizontal or vertical polarization, as well as by circularly polarized antennas. For example, FM radio broadcasting commonly uses circular polarization, producing a signal which can be received by either horizontal or vertical antennas.

FIGURE 11.4   **Circular Polarization**

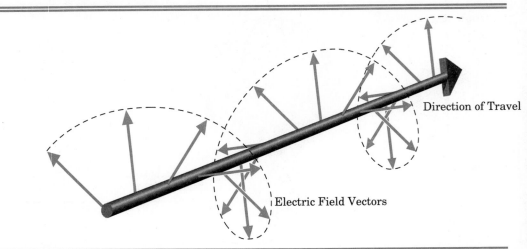

Direction of Travel

Electric Field Vectors

11.3 FREE-SPACE PROPAGATION

Radio waves, like light, propagate through free space in a straight line with a velocity of 3×10^8 m/s. There is no loss of energy in free space, but there is attenuation due to the spreading of the waves. Though the earth's atmosphere is definitely not free space, it can sometimes be treated as if it were. In addition, free-space propagation is of interest in satellite communications.

11.3.1 Attenuation of Free Space

Consider an isotropic radiator as introduced in Section 11.2.3, that is, an antenna that radiates equally well in all directions, and is perfectly efficient. While not exactly a practical construction project, the isotropic radiator has a simple and predictable radiation pattern that will be used in this and the next chapter as an aid to understanding. A real situation can often be approximated either by an isotropic radiator or by some simple modification to it.

An isotropic radiator would produce spherical waves like those shown in Figure 11.5. If a sphere were drawn at any distance from the source and concentric with it, all the energy from the source would pass through the surface of the sphere. Since no energy would be absorbed by free space, this would be true for any distance, no matter how large. The energy would be spread over a larger surface as the distance from the source increased, however.

Since the isotropic radiator radiates equally in all directions, the power density, in watts per square meter, would be simply the total power divided by the surface area of the sphere. Put mathematically,

$$P_D = \frac{P_t}{4\pi r^2} \tag{11.11}$$

where P_D = power density in watts per square meter
 P_t = total power in watts
 r = distance from the antenna in meters

FIGURE 11.5

Spherical Waves

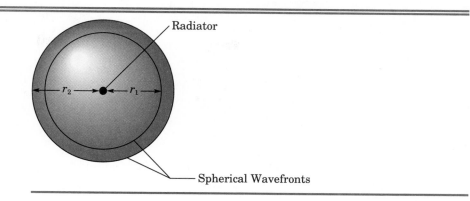

Not surprisingly, this is the same *square-law* attenuation that applies to light, sound, and in fact to any form of radiation. It is important to realize that this attenuation is not due to any loss of energy in the medium, but only to the spreading out of the energy as it moves further from the source. Any actual losses will be in addition to this.

EXAMPLE 11.3

A power of 100 W is supplied to an isotropic radiator. What is the power density at a point 10 km away?

Solution

From Equation (11.11),

$$P_D = \frac{P_t}{4\pi r^2}$$

$$= \frac{100 \text{ W}}{4\pi \,(10 \times 10^3 \text{ m})^2}$$

$$= 79.6 \text{ nW/m}^2$$

By the way, this is actually quite a strong signal, in radio terms. Radio, while very useful for transmitting information, is obviously not an efficient method of power transmission.

The strength of a signal is more often given in terms of its **electric field intensity** rather than power density, perhaps because the former is easier to measure. It is quite easy to derive an equation relating electric field strength at a distance to radiated power. From Equation (11.8),

$$P_D = \frac{\mathscr{E}^2}{\mathscr{Z}}$$

and from Equation (11.11),

$$P_D = \frac{P_t}{4\pi r^2}$$

Equating these two expressions for power density gives

$$\frac{\mathscr{E}^2}{\mathscr{Z}} = \frac{P_t}{4\pi r^2}$$

$$\mathscr{E}^2 = \frac{P_t \mathscr{Z}}{4\pi r^2}$$

$$\mathscr{E} = \sqrt{\frac{P_t \mathscr{Z}}{4\pi r^2}} \tag{11.12}$$

For free space, $\mathscr{Z} = 377 \ \Omega$ and the equation simplifies to

$$\mathscr{E} = \sqrt{\frac{377 P_t}{4\pi r^2}}$$

$$\mathscr{E} = \frac{\sqrt{30 P_t}}{r} \tag{11.13}$$

where $\mathscr{E}$ = electric field strength in volts per meter
   P_t = total power in watts
   r = distance from the source in meters

EXAMPLE 11.4

Find the electric field strength for the signal in the previous example.

Solution

From Equation (11.13),

$$\mathscr{E} = \frac{\sqrt{30 P_t}}{r}$$

$$= \frac{\sqrt{30 \times 100}}{10 \times 10^3}$$

$$= 5.48 \ \text{mV/m}$$

11.3.2 Transmitting Antenna Gain

In a practical communication system, it is very important to know the signal strength at the receiver input. This depends on the transmitter power and the distance from transmitter to receiver, of course, but there are two other very important determinants: the transmitting and receiving antennas. The details of antenna design will be dealt with in the next chapter, but it is useful to look at two important antenna characteristics at this point, because they have an effect on propagation calculations. These are *gain* for a transmitting antenna, and **effective area** for a receiving antenna.

Until now, we have been assuming an isotropic antenna, that is, one that radiates equally in all directions. Many practical antennas are designed to radiate more power in some directions than others. They are said to have *gain* in those directions in which the most power is radiated. This is not gain in the sense in which amplifiers have gain. The antenna is a passive device, so the total output power cannot be more than the power input. It will, in fact, be somewhat less, because of losses. The gain in some directions is more than compensated for by a loss in others. The antenna can nonetheless be thought of as having gain in its direction(s) of maximum radiation, when compared with an isotropic source.

Antenna gain can be visualized by an analogy with light. Consider a flashlight bulb suspended by its connecting wires. The amount of light radiated is approximately equal in most directions, ignoring the light that is blocked by the base of the bulb. Now put a reflector behind the bulb, as in a flashlight. The bulb emits the same amount of light as before, but its intensity is greater in the beam of the flashlight. Of course, this is compensated for by the fact that no light at all is radiated to the back of the flashlight.

If the transmitting antenna has gain in a given direction, then the power density in that direction is increased by the amount of the gain, and the equation for power density becomes

$$P_D = \frac{P_T G_T}{4\pi r^2} .$$
(11.14)

where P_D = power density in watts per square meter
P_T = total transmitter power in watts
G_T = gain of the transmitter antenna
r = distance from transmitter to receiver, in meters

Another way of looking at the gain of a transmitting antenna is to note that, in a given direction, the power density is the same as it would be if the transmitting antenna were replaced with an isotropic radiator, and the transmitter power multiplied by the antenna gain. From a distant point, it makes no difference whether a signal comes from a powerful transmitter and an isotropic radiator, or a less powerful transmitter used with an antenna with gain. A common practice is to speak of the *effective isotropic radiated power* (EIRP), which is very easily found:

$$\text{EIRP} = P_T G_T$$
(11.15)

11.3.3 Receiving Antenna Gain

A receiving antenna absorbs some of the energy from radio waves that pass it. Since the power in the wave is proportional to the area through which it passes, it seems reasonable that a large antenna will intercept more energy than a smaller one, because it intercepts a larger area. It also seems logical that some antennas, at least, will be more efficient at absorbing power from some directions than from others. For instance, a satellite dish would not be very efficient if it were pointed at the ground instead of the satellite. In other words, receiving antennas have gain, just as transmitting antennas do. In fact, the gain will be the same whether the antenna is used for receiving or transmitting.

The power extracted from the wave by a receiving antenna, then, ought to depend both on its physical size and on its gain. The effective area of an antenna can be defined as

$$A_{eff} = \frac{P_R}{P_D} \tag{11.16}$$

where A_{eff} = effective area of the antenna in square meters

       P_R = power delivered to the receiver in watts

       P_D = power density of the wave in watts per square meter

Equation (11.16) simply tells us that the effective area of an antenna is the area from which all the power in the wave is extracted and delivered to the receiver. Combining Equation (11.16) with Equation (11.14) gives

$$\begin{aligned} P_R P_R &= A_{eff} P_D \\ &= \frac{A_{eff} P_T G_T}{4\pi r^2} \end{aligned} \tag{11.17}$$

It can be shown that the effective area of a receiving antenna is

$$A_{eff} = \frac{\lambda^2 G_R}{4\pi} \tag{11.18}$$

where G_R = antenna gain, as a power ratio

      λ = wavelength of the signal

11.3.4 Path Loss

Combining Equation (11.18) with Equation (11.17) gives an expression for receiver power in terms of the gains of the two antennas and the wavelength.

$$\begin{aligned} P_R &= \frac{A_{eff} P_T G_T}{4\pi r^2} \\ &= \frac{\lambda^2 G_R P_T G_T}{(4\pi)(4\pi r^2)} \\ &= \frac{\lambda^2 P_T G_T G_R}{16\pi^2 r^2} \end{aligned} \tag{11.19}$$

It is more common to express this equation in terms of the **attenuation of free space**, that is, the ratio of received power to transmitter power:

$$\frac{P_R}{P_T} = \frac{\lambda^2 G_T G_R}{16\pi^2 r^2} \tag{11.20}$$

While accurate, this equation is not very convenient. Gain and attenuation are usually expressed in decibels rather than directly as power ratios; the distance between transmitter and receiver is more likely to be given in kilometers than meters; and the frequency of the signal, in megahertz, is more commonly used than its wavelength. It is quite easy to perform the necessary conversions to arrive at a more useful equation. The work involved is left as a problem for the reader; the solution follows.

$$\frac{P_R}{P_T} \text{ (dB)} = [G_T \text{ (dBi)}] + [G_R \text{ (dBi)}]$$
$$- (32.44 + 20 \log d + 20 \log f) \qquad (11.21)$$

where d = distance between transmitter and receiver in kilometers
f = frequency in megahertz

The term dBi indicates that the antenna gains are given with respect to an isotropic radiator; for instance,

$$G_T \text{ (dBi)} = 10 \log \text{ (power density in the direction of the receiver from}$$
the transmitting antenna divided by power density in
the same direction from an isotropic radiator with the
same power input)

All the other quantities are as in the previous equations.

Equation (11.21) is expressed as a decibel gain between transmitting and receiving antennas. Of course, the received signal will be weaker than the transmitted signal, so this gain will always be negative. Negative gains are more commonly called *losses*, and Equation (11.21) can be written as a loss by simply changing the signs. The loss thus found is called *free-space loss* or **path loss**:

$$L_{fs} = 32.44 + [20 \log d \text{ (km)}] + [20 \log f \text{ (MHz)}]$$
$$- [G_T \text{ (dBi)}] - [G_R \text{ (dBi)}] \qquad (11.22)$$

where $L_{fs} = 10 \log \dfrac{P_T}{P_R}$

and everything else is as in Equation (11.21)

Note that P_T and P_R are the power levels at the transmitting and receiving antennas, respectively. Attenuation due to transmission-line losses or mismatch is not included, but these losses can be found separately and added to the result given above (assuming all the losses are found in decibels).

For those working in miles rather than kilometers, the equivalent equation is:

$$L_{fs} = 36.58 + [20 \log d \text{ (mi)}] + [20 \log f \text{ (MHz)}]$$
$$- [G_T \text{ (dBi)}] - [G_R \text{ (dBi)}] \qquad (11.23)$$

The following examples show the use of Equation (11.22) in calculating received signal strength. First, let us consider a very straightforward application, then one that is a little more complex, involving transmission-line loss and mismatch.

EXAMPLE 11.5

A transmitter has a power output of 150 W at a carrier frequency of 325 MHz. It is connected to an antenna with a gain of 12 dBi. The receiving antenna is 10 km away and has a gain of 5 dBi. Calculate the power delivered to the receiver, assuming free-space propagation. Assume also that there are no losses or mismatches in the system.

Solution

From Equation (11.22),

$$L_{fs} = 32.44 + [20 \log d \text{ (km)}] + [20 \log f \text{ (MHz)}]$$
$$- [G_T \text{ (dBi)}] - [G_R \text{ (dBi)}]$$
$$= 32.44 + 20 \log 10 + 20 \log 325 - 12 - 5$$
$$= 85.7 \text{ dB}$$

$$10 \log \frac{P_T}{P_R} = 85.7$$

$$\log \frac{P_T}{P_R} = \frac{85.7}{10}$$

$$\frac{P_T}{P_R} = \text{antilog} \frac{85.7}{10}$$

$$P_R = \frac{P_T}{\text{antilog} (85.7/10)}$$

$$= \frac{150 \text{ W}}{372 \times 10^6}$$

$$= 404 \times 10^{-9} \text{ W}$$

$$= 404 \text{ nW}$$

EXAMPLE 11.6

A transmitter has a power output of 10 W at a carrier frequency of 250 MHz. It is connected by 10 m of a transmission line having a loss of 3 dB/100 m to an antenna with a gain of 6 dBi. The receiving antenna is 20 km away and has a gain of 4 dBi. There is negligible loss in the receiver feedline, but the receiver is mismatched: the antenna and line are designed for a 50 Ω impedance, but the receiver input is 75 Ω. Calculate the power delivered to the receiver, assuming free-space propagation.

Solution

When dealing with a relatively complex situation like this, it is a good idea to sketch the system. Then each loss can be added to the sketch as it is found, and a quick check will determine whether anything has been left out. We begin with the sketch of Figure 11.6.

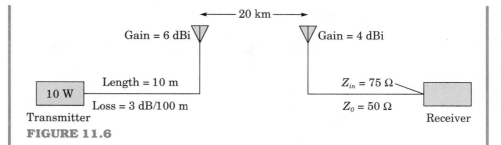

FIGURE 11.6

Next, it is necessary to find the loss at each stage of the system. If we find them all in decibels, it will only be necessary to add them to get the total loss.

First, we use Equation (11.22) to find the path loss.

$$L_{fs} = 32.44 + [20 \log d \text{ (km)}] + [20 \log f \text{ (MHz)}] - [G_T \text{ (dBi)}]$$
$$- [G_R \text{ (dBi)}]$$
$$= 32.44 + 20 \log 20 + 20 \log 250 - 6 - 4$$
$$= 96.42 \text{ dB}$$

For the transmitter feedline, the loss is

$$L_{TX} = 10 \text{ m} \times 3 \text{ dB/100 m}$$
$$= 0.3 \text{ dB}$$

The receiver feedline is lossless, but some of the power will reflect from the receiver back into the antenna due to the mismatch. This power will be re-radiated by the antenna, and will never reach the receiver. Therefore, for our purposes it is a loss. Remember from Chapter 10 that the proportion of power reflected is the square of the reflection coefficient, and the reflection coefficient is given by

$$\Gamma = \frac{Z_L - Z_0}{Z_L + Z_0}$$
$$= \frac{75 - 50}{75 + 50}$$
$$= 0.2$$
$$\Gamma^2 = 0.2^2$$
$$= 0.04$$

The proportion of the incident power that reaches the load is

$$1 - \Gamma^2 = 0.96$$

In decibels, the loss due to mismatch is

$$L_{RX} = -10 \log 0.96$$
$$= 0.177 \text{ dB}$$

The total loss is

$$L_t = L_{fs} + L_{TX} + L_{RX}$$
$$= 96.42 \text{ dB} + 0.3 \text{ dB} + 0.177 \text{ dB}$$
$$= 96.9 \text{ dB}$$

This represents a power ratio of

$$\frac{P_T}{P_R} = \text{antilog } \frac{96.9}{10}$$
$$= 4.90 \times 10^9$$

So

$$P_R = \frac{10 \text{ W}}{4.90 \times 10^9}$$
$$= 2.04 \text{ nW}$$

11.4   REFLECTION, REFRACTION, AND DIFFRACTION

The three properties listed in the title of this section should be familiar from the behavior of light. Radio waves are identical to light waves except for frequency, and we should expect them to behave in similar ways. The lower frequency is associated with longer wavelength, of course, and this will have some effect in practical situations. For both reflection and refraction, it is assumed that the surfaces involved are much larger than the wavelength. If this is not the case, diffraction will occur.

11.4.1   Reflection

Figure 11.7 shows the reflection of plane waves from a smooth surface (*specular reflection*). The angle of incidence is equal to the angle of reflection, with both angles measured from a line **normal** (that is, perpendicular) to the reflective surface. In the case of radio waves, the reflection will be complete if the reflector is an ideal conductor. This is never quite the case, of course, and some of the wave will propagate for a short distance into a lossy material before being completely absorbed.

Reflection can also take place from dielectrics. Usually, some of the energy is reflected and some refracted, but sometimes all the energy is reflected. This is really a special case of refraction, and it will be considered in that context.

The reflecting surface does not have to be a single plane. Figure 11.8 shows a corner reflector, for instance. This scheme is often used for antennas.

Also very useful in antenna design, of course, is the parabolic reflector, shown in Figure 11.9. Any plane wave entering along the axis of the antenna will be reflected in such a way that all of its energy passes through a single

FIGURE 11.7

Specular Reflection

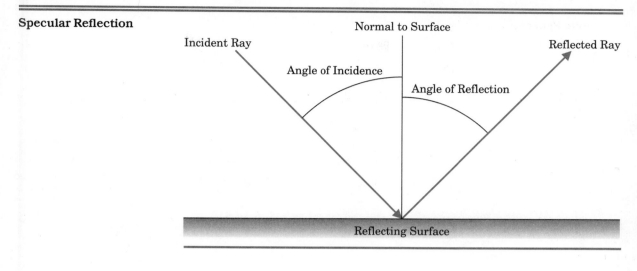

point called the *focus* of the reflector. Both the corner reflector and parabolic reflector will be considered in more detail later in this book, when we look at antennas.

Not all reflections are from smooth surfaces. Radio waves often reflect from the earth, for example. Figure 11.10 shows what happens with these diffuse reflections. The angles of incidence and reflection for each part of the surface are equal as before, but since each part has a different orientation, the reflected wave is scattered.

FIGURE 11.8

Corner Reflector

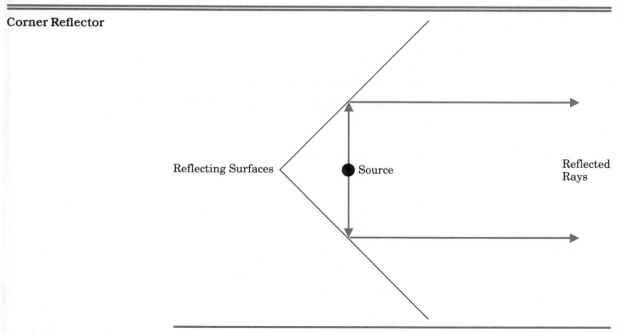

FIGURE 11.9

Parabolic Reflector

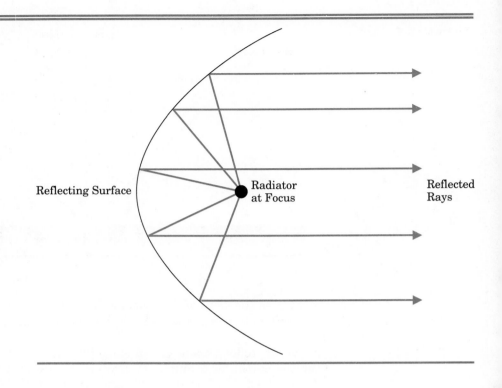

Reflecting Surface   Radiator at Focus   Reflected Rays

11.4.2 Refraction

A transition from one medium to another often results in the bending, or refraction, of radio waves, just as it does with light. In optics, the angles involved are given by Snell's Law:

$$n_1 \sin \theta_1 = n_2 \sin \theta_2 \qquad (11.24)$$

where θ_1 = angle of incidence
        θ_2 = angle of refraction
        n_1 = index of refraction in the first medium
        n_2 = index of refraction in the second medium

FIGURE 11.10

Diffuse Reflection

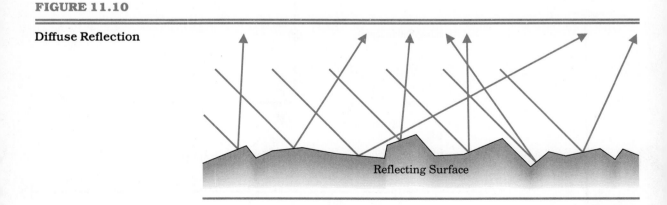

Reflecting Surface

FIGURE 11.11

Refraction

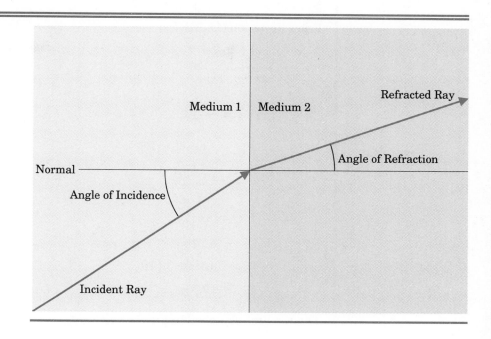

Figure 11.11 shows how the angles are measured. Once again, they are measured from a line normal to the surface and not from the surface itself.

Now if Snell's Law works for radio waves, and it does, how do we find the **index of refraction**? This is actually quite simple: for a given medium,

$$n = \sqrt{\mu_r \epsilon_r} \qquad (11.25)$$

where μ_r = relative permeability of the medium
    ϵ_r = relative permittivity of the medium

and, since μ_r is almost always 1 for the medium of interest, in practical terms

$$n = \sqrt{\epsilon_r} \qquad (11.26)$$

Substituting Equation (11.26) into Equation (11.24) gives

$$\frac{\sin \theta_1}{\sin \theta_2} = \frac{\sqrt{\epsilon_{r2}}}{\sqrt{\epsilon_{r1}}} \qquad (11.27)$$

$$= \sqrt{\frac{\epsilon_{r2}}{\epsilon_{r1}}}$$

EXAMPLE 11.7

A radio wave moves from air ($\epsilon_r = 1$) to glass ($\epsilon_r = 7.8$). Its angle of incidence is 30°. What is the angle of refraction?

Solution

From Equation (11.27),

$$\frac{\sin \theta_1}{\sin \theta_2} = \sqrt{\frac{\epsilon_{r2}}{\epsilon_{r1}}}$$

$$\sin \theta_2 = \frac{\sin \theta_1}{\sqrt{\dfrac{\epsilon_{r2}}{\epsilon_{r1}}}}$$

$$= \frac{\sin 30°}{\sqrt{\dfrac{7.8}{1}}}$$

$$= 0.179$$

$$\theta_2 = \arcsin 0.179$$

$$= 10.3°$$

The result is sketched in Figure 11.12.

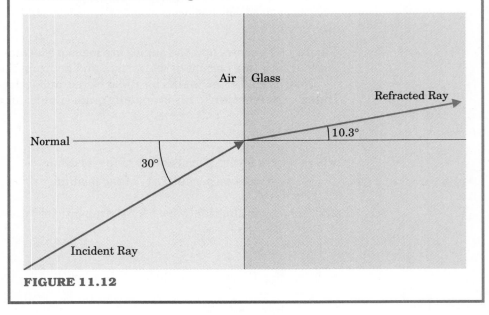

FIGURE 11.12

It can be seen from this example that, when a wave enters a region of higher dielectric constant, and thus lower propagation velocity, it bends toward the normal. This can be remembered by thinking of a wavefront perpendicular to the direction of travel, as shown in Figure 11.13. When this wavefront enters a slower medium, the part that enters the medium slows down, causing the whole front to swing toward the normal, as shown in Figure 11.13(a). On the other hand, if the second medium has a lower dielectric constant (and a greater propagation velocity), the wave will be deflected away from the normal, as shown in Figure 11.13(b).

FIGURE 11.13 **Refraction as a Function of Dielectric Constant**

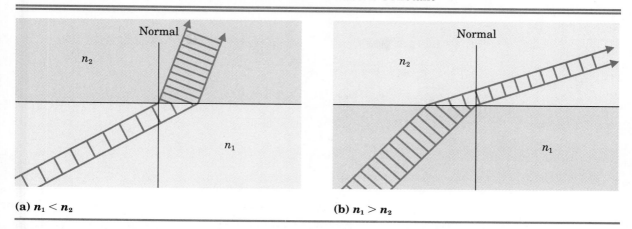

(a) $n_1 < n_2$ **(b) $n_1 > n_2$**

In extreme cases, where the angle of incidence is large and the wave travels into a region of considerably lower dielectric constant, the angle of refraction can be greater than 90°, so that the wave comes out of the second medium and back into the first. Refraction becomes a form of reflection, called *total internal reflection*, under these circumstances. Figure 11.14 shows this. The angle of incidence that results in an angle of refraction of exactly 90° (so that the wave propagates along the boundary between the two media) is known as the *critical angle*, and is given by

$$\theta_c = \arcsin \frac{n_2}{n_1}$$ (11.28)

FIGURE 11.14

Total Internal Reflection

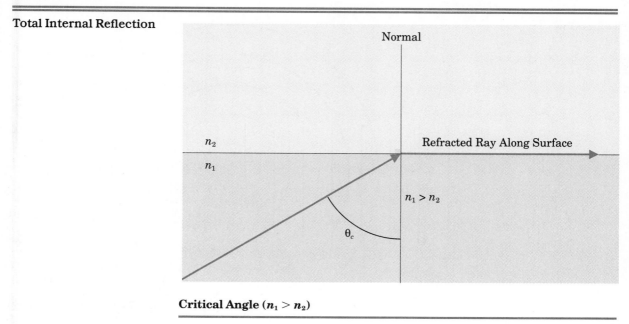

Critical Angle ($n_1 > n_2$)

FIGURE 11.15

Optical Fiber ($n_1 > n_2$)

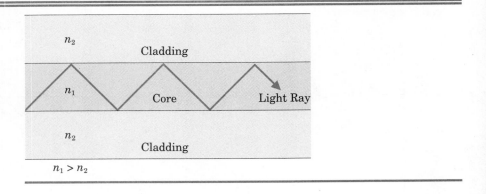

Total internal reflection is found in optical fibers. The light reflects from the boundary between the core of the fiber and a cladding material with lower refractive index. Figure 11.15 shows the basic idea, which will be developed in more detail in Chapter 16.

11.4.3 Diffraction

It is a common saying that light travels in straight lines, but it does appear to go around corners occasionally. This is caused by diffraction, and there are numerous examples of its use in optics (such as diffraction gratings). Diffraction also occurs for radio waves, and can, for example, allow reception from a transmitting antenna on the far side of a mountain.

Figure 11.16 illustrates diffraction. It can be described by assuming that each point on a wavefront acts as an isotropic source of radio waves. This

FIGURE 11.16

Diffraction

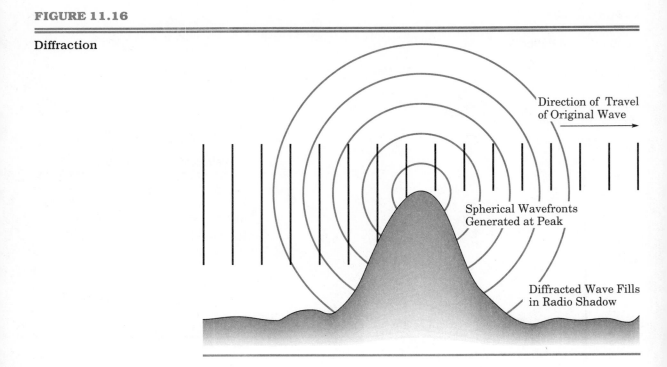

Early Experiences with Propagation

Radio propagation was predicted mathematically by James C. Maxwell in 1865, but first demonstrated experimentally by Heinrich R. Hertz in 1887. Hertz demonstrated transmission from one spark gap to another, and also showed the presence of standing waves. The frequency he used is not known, but the size of his laboratory is. In order to set up standing waves, he must have been operating in the VHF part of the spectrum.

When radio communication became practical, around the turn of the century, the frequencies used were much lower. In fact, the terms "medium," "low," and "high" frequency date from these early days. Beginning in the medium-frequency range, experimenters went lower and lower in frequency in search of more reliable long-distance propagation, using ground waves. Frequencies as low as 60 kHz were popular. These low frequencies had the additional advantage that they could be generated by mechanical alternators, which were more efficient than spark gaps. (Vacuum tubes were not available until later.)

In the early days of radio, frequencies in the HF range and up were considered useless for long-distance communication.

would allow some of the wavefronts that pass beside an obstruction to radiate into the area behind it.

Diffraction is more pronounced when the object that causes it has a sharp edge, that is, when its dimensions are small in comparison with the wavelength.

11.5 TERRESTRIAL PROPAGATION

Most of the time, the radio waves we use are not quite in free space. Their behavior near the earth is influenced by factors that include the properties of the earth itself, and of the various layers of the atmosphere. Most of these factors are frequency-dependent. For convenience, we will proceed roughly in order of frequency, from lower to higher, and look briefly at some of the more important **terrestrial propagation** modes. In addition to direct, straight-line radiation from transmitter to receiver (called *line-of-sight* or, sometimes, **space-wave** propagation), we will discover **ground waves**, which follow the surface of the earth, and **sky waves**, which refract from ionized layers in the atmosphere. We will also look at some other less common but still important modes of propagation.

11.5.1 Ground-Wave Propagation

At frequencies up to approximately 2 MHz, the most important method of propagation is by *ground waves*. These are vertically polarized waves that follow the ground, and can therefore follow the curvature of the earth to propagate far beyond the horizon. They must have vertical polarization to minimize currents induced in the ground itself, which result in losses. Nonetheless, there is a tendency for the waves to "tilt" toward the horizontal, increasing losses, as the distance from the transmitter increases. See Figure 11.17 for an illustration of ground-wave propagation.

FIGURE 11.17

**Ground-Wave
Propagation**

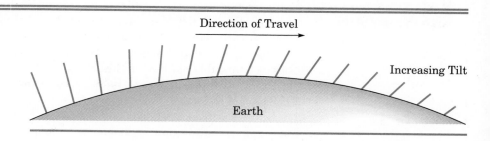

At close range, some energy is also transmitted to the receiver directly
from the transmitter through the air. However, this signal is often nearly cancelled by another signal reflected from the ground, as shown in Figure 11.18. At low frequencies, the earth is highly reflective, and also provides
a 180° phase shift. The transmitting and receiving antennas are likely to be
close to the ground, so the direct path from transmitter to receiver will be
almost equal in length to the path involving a reflection from the earth. Since
the direct and reflected waves are out of phase, there will be partial cancellation. This leaves the ground wave as the main mode of propagation at low
frequencies.

Ground waves provide very reliable communication that is almost independent of weather and solar activity, both of which can greatly affect some
other propagation modes. With sufficient power at a low enough frequency,
round-the-world communication is possible, and some military transmitters
operate at frequencies as low as 15 kHz. There is a United States government
time-and-frequency station, WWVL, at 60 kHz, and an international navigation system called LORAN-C operates at 100 kHz. In addition, of course,
the standard AM broadcast band relies mainly on ground-wave propagation.

Ground waves are attenuated very quickly above about 2 MHz, so that
the whole spectrum that is usable for this propagation mode has only one-
third the width of a single television channel. This lack of spectrum space is
one of the main disadvantages to ground-wave propagation. Ground waves
require relatively high power, and the low frequencies are associated with
long wavelengths that require physically large antennas for good efficiency.
Ground-wave propagation will certainly remain in use, but it has little or no
room for expansion.

FIGURE 11.18

**Cancellation Due to
Ground Reflections**

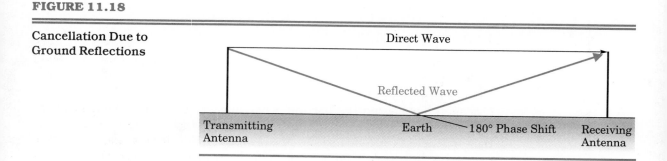

11.5.2 Ionospheric Propagation

Many people, including the author of this book, first became interested in radio communications when they listened to a high-frequency (shortwave) radio receiver, and heard stations from all over the world. Long-range communication in the high-frequency band is possible because of refraction in a region of the upper atmosphere called the **ionosphere**, where some of the air molecules are ionized by solar radiation. Figure 11.19 illustrates the idea. Note that the drawing is not to scale: the ionospheric layers vary in height from about 60 to 400 km above the earth's surface, while the radius of the earth is approximately 6400 km.

The ionosphere can be divided into three regions known as the D, E, and F layers. The F layer itself is divided into two parts, called F_1 and F_2. The level of ionization increases with height above the earth, and is greater in the daytime. At night, when solar radiation is not received, the D and E regions disappear, and the F_1 and F_2 layers combine into a single F layer. The F layer remains during the night because the atmosphere is so rarified at this height that ions take longer to recombine.

Ionization levels change with the amount of solar activity. This varies greatly over an eleven-year cycle known as the *sunspot cycle*. Peaks in this cycle are associated with greater activity in the ionosphere, and generally more favorable conditions for ionospheric propagation, especially toward the high end of the HF region of the spectrum. The most recent peak occurred in 1990.

The signal is returned from the ionosphere by a form of refraction. The ionized air contains free electrons, which can move in the presence of electromagnetic waves. The interaction is complex, but the net effect is to cause an effective decrease in the dielectric constant, which causes the waves to bend toward the earth. The higher the frequency, the more ionization is necessary for refraction. If the signal is not refracted enough to reach the earth, it may be absorbed, or may pass right through the atmosphere into space.

In the daytime, the D and E layers absorb frequencies below around 8 or 10 MHz. Frequencies above this, up to about 30 MHz, are refracted by the F_1

FIGURE 11.19

Ionospheric Layers

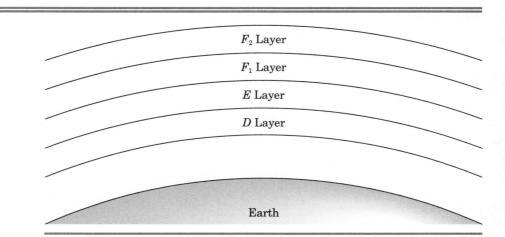

F_2 Layer

F_1 Layer

E Layer

D Layer

Earth

and F_2 layers, and may return to earth. At night, the D and E layers virtually disappear. This allows lower frequencies to reach the F layer without being absorbed. They are refracted by the F layer, causing propagation at the lower frequencies to be better at night than during the day. The higher frequencies pass right through all the layers at night, so propagation at frequencies above about 10 MHz tends to be better during the daylight hours.

It is possible for the signal to reflect from the ground and make two or more "hops" before reaching its final destination, as shown in Figure 11.20, but each reflection from the ground or refraction in the ionosphere greatly reduces its strength.

If all this sounds rather vague and approximate, it is because the ionosphere is constantly changing, from hour to hour, month to month, and year to year. It is generally possible to find some frequency at which the ionosphere will return signals in the desired direction and at the correct distance, but any predictions must be approximate.

Large users of high-frequency communications, such as shortwave broadcasters and the military, employ *frequency diversity;* that is, they transmit on a number of different frequencies that span the HF spectrum, in the hope that at least one of these will work at a given time. Of course, they also use the kind of general observations that have just been made to reduce the amount of wasted energy expended. Usually, there is no point in trying to use 3 MHz in the daytime, or 30 MHz at night.

It is possible to make some measurements on which to base propagation predictions. This can be done by *ionospheric sounding,* shown in Figure 11.21. A signal is sent straight up, and the frequency is gradually increased. At low frequencies, the signal will be absorbed. As the frequency is increased, the signal will be returned to earth, and can be picked up by the receiver. As the frequency is increased still further, a frequency will be reached at which the signal ceases to return to earth, but instead passes right through the ionosphere. The highest frequency that is returned to earth in the vertical direction is called the *critical frequency.*

FIGURE 11.20

Double-Hop Propagation

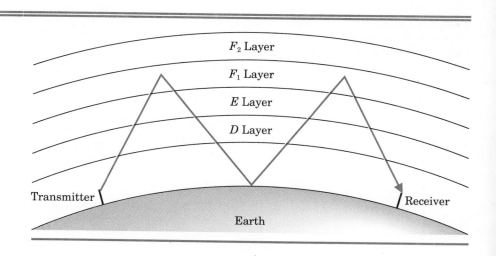

FIGURE 11.21

Ionospheric Sounding

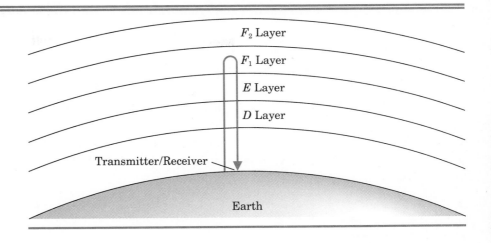

Practical communication takes place over a path more like that shown in Figure 11.22. Since the angle at which the signal meets the ionosphere is smaller, the signal will not have to refract as much to return to earth, and we would expect that a higher frequency would be usable. We would also expect that longer paths would be operable at higher frequencies, since they require smaller angles. The highest frequency that will return to earth over a given path is called the **maximum usable frequency** (MUF). Since absorption decreases with increasing frequency, it would be expected that a frequency at or just below the MUF would give best results. This is true, but because of the general instability of the ionosphere, it is usually better to operate at a lower frequency, perhaps 85% of the MUF. This is sometimes called the *optimum working frequency* (OWF). Of course, a particular station will not have access to every frequency, and will have to choose one of those for which it is authorized.

FIGURE 11.22

Ionospheric Propagation

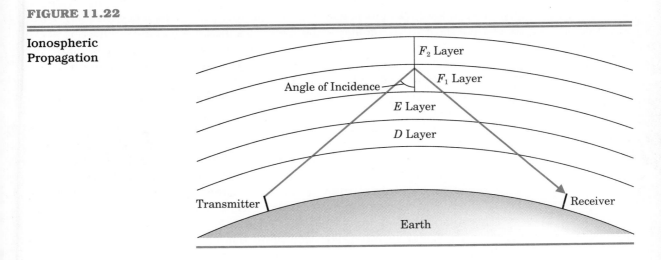

In theory, at least, it is possible to predict the MUF from the critical frequency. The equation is sometimes called the *secant law*:

$$f_m = f_c \sec \theta_1 \tag{11.29}$$

where f_m = MUF

f_c = critical frequency

θ_1 = angle of incidence, as shown in Figure 11.22

A more convenient way to use this law, since the secant is not directly available on most calculators, is to rewrite it as

$$f_m = \frac{f_c}{\cos \theta_1} \tag{11.30}$$

EXAMPLE 11.8

The critical frequency at a particular time is 11.6 MHz. What is the MUF for a transmitting station if the required angle of incidence for propagation to a desired destination is 70°?

Solution

The MUF will be, from Equation (11.30),

$$
\begin{aligned}
f_m &= \frac{f_c}{\cos \theta_1} \\
&= \frac{11.6 \text{ MHz}}{\cos 70°} \\
&= 33.9 \text{ MHz}
\end{aligned}
$$

Figure 11.22 hints at an interesting phenomenon. As the angle of elevation becomes larger, the distance covered becomes smaller, and the MUF becomes lower. To put it another way, for any given frequency that is greater than the critical frequency, there will be a maximum angle of elevation above the horizon for which the signal will be reflected. For frequencies above the critical frequency, then, there may be a region relatively close to the transmitter where the signal cannot be received, even though it can easily be picked up at much greater distances. This region is called the *skip zone*, and is pointed out in Figure 11.23.

Ionospheric propagation allows communication over great distances with relatively simple equipment and reasonable power levels. Most commercial transmitters range in power from about 100 W to a few kilowatts. Broadcasters are the exception, often using very large power levels (into the megawatt range) in an attempt to overwhelm the competition. Radio amateurs have achieved worldwide communications with very low power levels, sometimes less than 1 W.

On the other hand, HF communication via the ionosphere is noisy and uncertain. It is also prone to phase shifting and frequency-selective fading.

FIGURE 11.23

Skip Zone

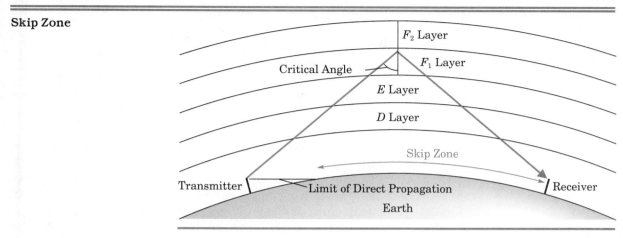

For instance, the phase shift and signal attenuation may be different for the upper and lower sidebands of the same signal. This makes high-fidelity music transmission impossible, and restricts data transmission to very low rates, typically under 100 bits per second. There is little doubt that ionospheric propagation will remain in use, but it is being superseded for many applications by other technologies—for example, communications satellites and optical-fiber cables. It is still in use to a limited extent for telephony, and quite extensively for ship and aircraft communications, international newswire services, military communication links, communication with outlying settlements (in the far north, for instance), and amateur radio. In addition, of course, there is shortwave broadcasting, which is of interest mainly to hobbyists in North America, but is an important source of information programming in much of the world.

11.5.3 Line-of-Sight Propagation

Signals in the VHF range and higher are not usually returned to earth by the ionosphere, though during peaks of solar activity frequencies in the low VHF range, even as high as television Channel Two (54–60 MHz) have been propagated that way. Most terrestrial communication at these frequencies uses the direct radiation from the transmitting antenna to the receiving antenna. There may be reflection from the ground, but that is more likely to cause problems than to increase the signal strength. This type of propagation is variously referred to as *space-wave, line-of-sight,* or *tropospheric propagation* (because the lowest layer of the atmosphere is known as the **troposphere**).

The practical communication distance for line-of-sight propagation is limited by the curvature of the earth. In spite of the title of this section, the maximum distance is actually greater than the eye can see. This is because refraction in the atmosphere tends to bend radio waves slightly toward the earth. The dielectric constant of air usually decreases with increasing height, because of the reduction in pressure, temperature, and humidity with increasing distance from the earth. The effect varies with weather conditions,

FIGURE 11.24

Line-of-Sight Propagation (Antenna heights are greatly exaggerated)

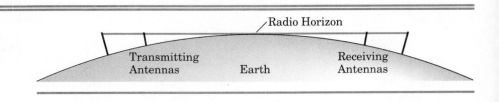

but it usually results in radio communications being possible over a distance approximately one-third greater than the visual line of sight.

Just as one can see further from a high place, the height above average terrain of both the transmitting and receiving antennas is very important in calculating the maximum distance for space-wave radio communications. Note the difference compared with the other types of propagation studied so far. Antennas for ground-wave propagation are usually vertically polarized arrays with one end on the ground. For sky-wave (ionospheric) propagation, the antenna height is important only insofar as reflections from the ground may change the radiation pattern of the antenna. With space-wave propagation, however, height above ground is important, and the higher the better. Figure 11.24 shows the advantage of increased antenna height.

An approximate value for the maximum distance between transmitter and receiver, over reasonably level terrain, is given by the following equation:

$$d = \sqrt{17h_T} + \sqrt{17h_R} \qquad\qquad (11.31)$$

where d = maximum distance in kilometers

h_T= height of the transmitting antenna in meters

h_R= height of the receiving antenna in meters

EXAMPLE 11.9

A taxi company uses a central dispatcher, with an antenna at the top of a 15 m tower, to communicate with taxicabs. The taxi antennas are on the roofs of the cars, approximately 1.5 m above the ground. Calculate the maximum communication distance:

(a) between the dispatcher and a taxi

(b) between two taxis

Solution

(a) $\quad d = \sqrt{17h_T} + \sqrt{17h_R}$

$\qquad = \sqrt{17 \times 15} + \sqrt{17 \times 1.5}$

$\qquad = 21.0 \text{ km}$

(b) $\quad d = \sqrt{17h_T} + \sqrt{17h_R}$

$\qquad = \sqrt{17 \times 1.5} + \sqrt{17 \times 1.5}$

$\qquad = 10.1 \text{ km}$

The line-of-sight range can sometimes be extended by diffraction, particularly if there is a relatively sharp obstacle, such as a mountain peak, in the way. Diffraction, however, will greatly reduce the signal strength, requiring more powerful transmitters and more sensitive receivers than are required for line-of-sight communications.

The attenuation of free space, as expressed by Equation (11.22), is usually the most important factor in determining the signal power at the receiver. There are other factors, however, arising from the fact that propagation takes place near the ground, and from the fact that the actual medium is air.

Although line-of-sight propagation uses a direct path from transmitter to receiver, the receiver can also pick up signals that have been reflected or diffracted. For instance, the signal can be reflected from the ground, as shown in Figure 11.25. If the ground is rough, the reflected signal will be scattered, and its intensity will be low in any given direction. If, on the other hand, the reflecting surface is relatively smooth—a body of water, for instance—the reflected signal at the receiver can have a strength comparable to that of the incident wave, and the two signals will interfere. Whether the interference will be constructive or destructive depends on the phase relationship between the signals: if they are in phase, the resulting signal strength will be increased, but if they are 180° out of phase, there will be partial cancellation. When the surface is highly reflective, the reduction in signal strength can be 20 dB or more. This effect is called *fading*. The exact phase relationship depends on the difference, expressed in wavelengths, between the lengths of the transmission paths for the direct and reflected signals. In addition, there is usually a phase shift of 180° at the point of reflection.

In a practical situation where the transmitter and receiver locations are fixed, the effect of reflections can often be reduced by carefully surveying the proposed route, and adjusting the transmitter and receiver antenna heights so that any reflection takes place in wooded areas or rough terrain; the reflection will be diffuse and, therefore, weak. Where most of the path is over a reflective surface such as desert or water, fading can be reduced by using either *frequency diversity* or *spatial diversity*. In the former method, more than one frequency is available for use; the difference, in wavelengths, between the direct and reflected path lengths will be different for the two frequencies. In spatial diversity, there are two antennas, usually mounted one above the other on the same tower. The difference in direct and reflected path length will be different for the two antennas.

FIGURE 11.25

**Ground Reflections
in Line-of-Sight
Propagation**

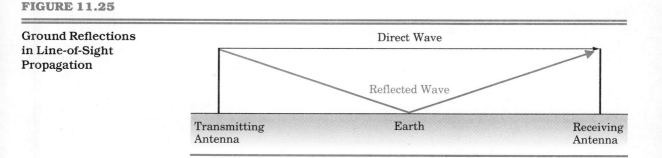

FIGURE 11.26

Fading Due to Diffraction

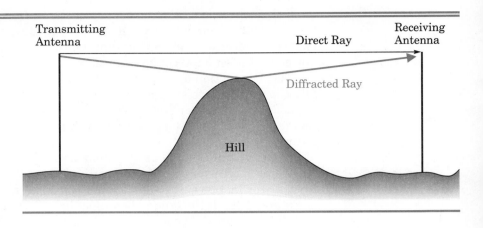

Diffraction from obstacles in the path can also be a problem for line-of-sight radio links. Earlier, the beneficial effects of diffraction in allowing reception on the far side of an obstruction were mentioned. Diffraction can also cause problems when the direct wave and the diffracted wave have opposite phase and tend to cancel. Figure 11.26 shows this effect of diffraction.

The solution to the problem of interference due to diffraction is to arrange for the direct and refracted signal to be in phase. Again, this requires a careful survey of the proposed route, and adjustment of the transmitting and receiving antenna heights to achieve this result.

Problems can also occur when the signal reflects from large objects, like hills or buildings, as shown in Figure 11.27. There may be not only phase cancellation, but also significant time differences between the direct and reflected waves. These can cause a type of distortion called (not surprisingly) **multipath distortion** in FM radio reception. The "ghosts" that appear in television reception have the same cause. Directional receiving antennas, aimed in the direction of the direct signal, can reduce the problem of reflections for fixed receivers, but they are not very practical on vehicles. In the case of mobile radio, the path length changes constantly, and the transmitted power must be sufficient to achieve usable signal strength at the receiver, even with cancellation due to reflections. Spacial diversity, with two or more

FIGURE 11.27

Multipath Reception

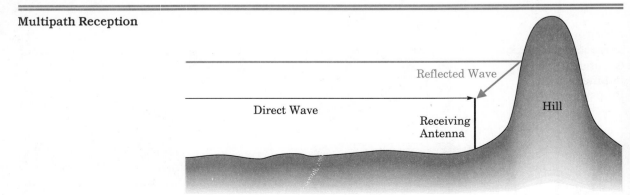

antennas mounted on one vehicle, is occasionally used to reduce multipath distortion in mobile radio reception.

There will also be some absorption of the signal by the atmosphere. At frequencies under 20 GHz, this absorption is quite small compared with the square-law attenuation of free space.

11.6   PROPAGATION VIA SATELLITE

There are many different types of satellites, at varying heights above the earth, with orbits that range from nearly circular to highly elliptical. Some orbits are above the equator; others pass over both poles. Satellites operate at frequencies ranging from the lower end of the VHF band to well above 10 GHz. Each of the types of satellites has its own characteristics and requirements. New types of satellites are constantly being developed. For instance, satellites have recently been deployed for navigation, marine radio communication, and even mobile communication between dispatchers and truck fleets.

11.6.1 Geostationary Satellites

There is one class of satellite that has proven extremely useful for communications, and has become the most important for this function. That is the **geostationary** (also called **geosynchronous**) **satellite**. Such an orbit is shown in Figure 11.28.

The period of a satellite's rotation depends on its distance from the earth. The closer it is, the more rapidly it must rotate, in order that the centrifugal force due to its rotation will exactly balance the force of gravity. It happens that a satellite must have a distance above the earth's surface of about 36,000 km in order to rotate once per day, that is, to be *geosynchronous*. If it is directly above the equator and rotates in the same direction as the earth, it will appear to remain fixed above one spot on the earth, and can also be called *geostationary*. The distance above the earth is farther than one might like, in terms of signal loss and time delay, but for many applications the advantages of a satellite that is always visible and always in the same spot far outweigh these drawbacks. Lower satellites require antennas that can move quickly and automatically to track them, and have the annoying habit of disappearing over the horizon from time to time.

Although there have been experiments with satellites as passive reflectors, present-day communications satellites act as *repeaters;* that is, they receive signals in one frequency band, and retransmit them in another. Typically, the transmitters in satellites have relatively low power, on the order of 10 W, in order to conserve energy, but the use of high-gain antennas and low-noise receivers allows very reliable communications.

The main loss factor in satellite communications is usually free-space attenuation, as given by Equation (11.22). This loss is very great due to the enormous distances involved.

An isotropic antenna would obviously radiate nearly all its power uselessly into space. The antennas used on satellites are highly directional, varying from those that illuminate half the earth's surface, to focused "spot beams" that are used for more specific areas.

Figure 11.29 shows the antenna patterns for one satellite. The dBW in-

FIGURE 11.28

Geostationary Satellite
Orbit

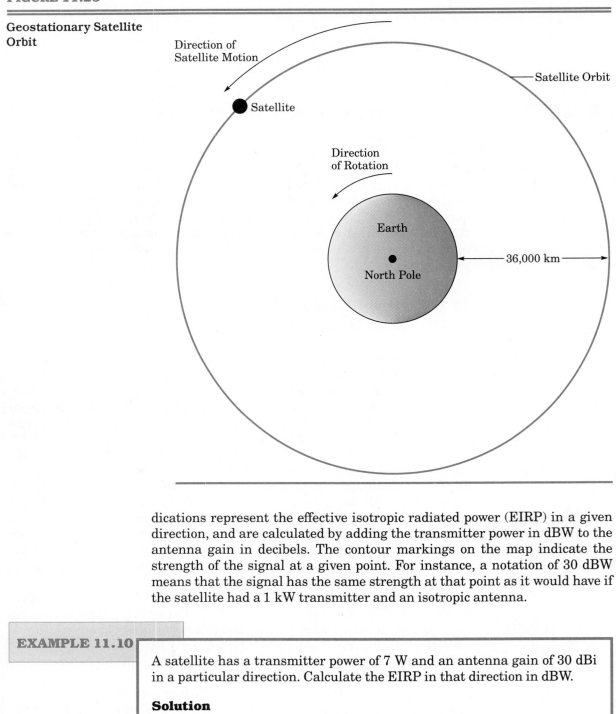

dications represent the effective isotropic radiated power (EIRP) in a given direction, and are calculated by adding the transmitter power in dBW to the antenna gain in decibels. The contour markings on the map indicate the strength of the signal at a given point. For instance, a notation of 30 dBW means that the signal has the same strength at that point as it would have if the satellite had a 1 kW transmitter and an isotropic antenna.

EXAMPLE 11.10

A satellite has a transmitter power of 7 W and an antenna gain of 30 dBi in a particular direction. Calculate the EIRP in that direction in dBW.

Solution

$$\text{Transmitter power in dBW} = 10 \log 7$$
$$= 8.45 \text{ dBW}$$

$$\text{EIRP} = 8.45 \text{ dBW} + 30 \text{ dBi}$$
$$= 38.45 \text{ dBW}$$

FIGURE 11.29   Satellite Footprint

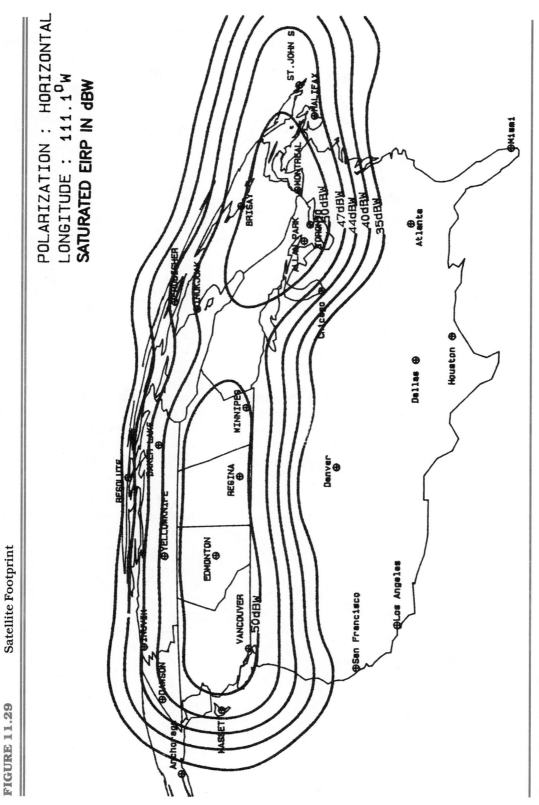

POLARIZATION : HORIZONTAL
LONGITUDE : 111.1°W
SATURATED EIRP IN dBW

Courtesy Telesat Canada.

FIGURE 11.30

Variation in Satellite-Antenna Elevation Angle with Latitude

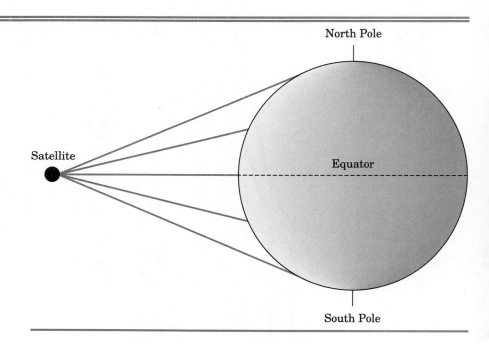

There are some disadvantages to an **equatorial orbit**, especially if the satellite is to be accessed from a point near one of the poles. See Figure 11.30 for an illustration. As the location of a transmitter or receiver moves further toward one of the poles, the required angle of elevation of the antenna becomes lower. This makes the system more susceptible to blocking of the an-

FIGURE 11.31   **Spreading of Satellite Beam at High Latitudes**

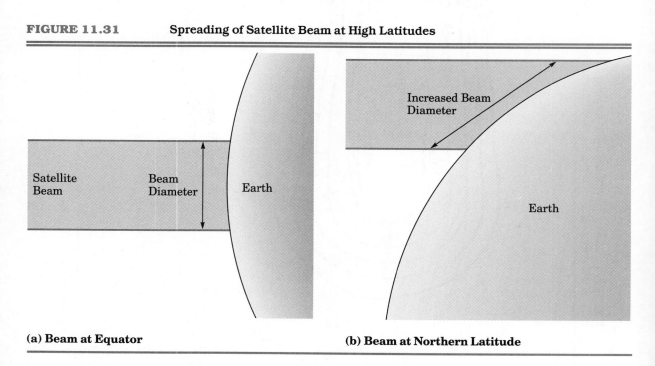

(a) Beam at Equator   **(b) Beam at Northern Latitude**

tenna beam by trees, buildings, hills, and so forth. The signal must also travel a greater distance through the atmosphere, where it is subject to absorption and scattering. In addition, the beam pattern of a satellite antenna will tend to spread out at higher latitudes, as shown in Figure 11.31. This makes the design of the transmitting antenna more complex, and may result in a *footprint* (the outline of the antenna pattern on the earth) that is larger than desired.

Since there is only one possible orbit, the number of geostationary communications satellites is limited. The orbit is large, of course, with a circumference that can easily be calculated as follows.

Height of satellite above earth:	36.0×10^3 km
Radius of earth:	$+\ \ 6.4 \times 10^3$ km
Radius of orbit:	42.4×10^3 km
Circumference of orbit:	$2\pi(42.4 \times 10^3$ km$) = 266 \times 10^3$ km

There does not appear to be much danger of satellites bumping into each other. The real problem is that even directional antennas take in a substantial angle. With a beamwidth of 2°, for instance, there would be room for only 180 satellites, operating in the same frequency band. With more satellites, there would be interference between adjacent satellites. Of course, some locations are more desirable than others, leading to competition for the most desirable spots.

There are two ways to increase the number of satellites allowed. One is to improve the directional characteristics of ground-station antennas, and the other is to increase the portion of the radio-frequency spectrum used for satellite communication. Both of these methods are currently being explored. In addition, it is common practice to make use of two transmissions with their waves polarized at 90° to each other. This orthogonal polarization allows receivers to ignore one of the two signals and virtually doubles usable spectrum space.

11.6.2 Time Delay

A round trip from earth to satellite and back is at least 72,000 km in length for a geosynchronous satellite. The time taken for the trip, neglecting any delay in the satellite itself, is

$$t = \frac{d}{c}$$

$$= \frac{72 \times 10^6 \text{ m}}{3 \times 10^8 \text{ m/s}}$$

$$= 0.24 \text{ s}$$

In other words, it takes about one-quarter of a second to communicate one way between two points on earth via satellite. This **propagation delay** is insignificant for applications like broadcasting, but consider a two-way telephone conversation. When the first speaker stops talking, it is one-quarter second before the second speaker is aware of the fact. If the second person then starts talking immediately, it is another one-quarter second before the first hears the reply. Thus there is a delay of one-half second between asking a question and hearing the answer, even if the other person responds immediately. This delay, while acceptable, is annoying, and satellites are not the method of first choice for long-distance telephony. Cables and terrestrial microwave links offer much faster response. A similar situation exists for full-duplex data transmission.

11.6.3　Faraday Rotation

Most satellites operate at heights well above the ionosphere, which means that signals must pass through it on their way to and from the satellite. At the microwave frequencies generally used for geostationary satellites, the ionosphere has virtually no effect on signals. On the other hand, it is obvious that signals in the MF and HF range, which are absorbed or refracted by the ionosphere, would not be very suitable for satellite communication. In the VHF and UHF ranges, however, the ionosphere can change the polarization of signals without refracting them. This process is called **Faraday rotation**, and results from the interaction of the earth's magnetic field with the ionosphere.

Faraday rotation has two major effects: It prevents the use of dual polarization at the affected frequencies as a means of gaining increased communication bandwidth, and it requires the use of receiving antennas at both ends that can make use of waves with random polarization. In practice, this requires the use of circular polarization for both transmitting and receiving antennas.

11.7　OTHER PROPAGATION MODES

The types of propagation discussed so far represent the majority of communication systems. There are, however, some other methods that, while not found as commonly as the others, still represent useful ways of getting radio waves—and the information they carry—from one place to another.

11.7.1　Tropospheric Scatter

The *troposphere* is the lowest layer of the atmosphere. We have already seen that it refracts radio waves slightly, providing communication at distances somewhat beyond the visual line of sight. The troposphere also absorbs radiation at some frequencies. What is not quite so well known is that irregularities in the troposphere can also scatter radio waves. Exactly what causes these irregularities is not known, but one theory is that they are caused by the effect of variations in temperature on the water vapor content in the atmosphere.

Figure 11.32 shows the basic idea behind tropospheric scatter (often called *troposcatter*). The transmitting antenna is aimed in the direction of the receiver, but the receiver is over the horizon. Most of the transmitted energy simply continues on into space, but a small portion of it is scattered. A small fraction of the scattered energy reaches the receiver.

Troposcatter can give reliable communication over distances of about 80 to 800 km, at frequencies from about 250 MHz to around 5 GHz. It is an inefficient system, requiring larger transmitter power, antennas with higher gain, and more sensitive receivers than line-of-sight systems. It is also sub-

FIGURE 11.32

Tropospheric Scatter

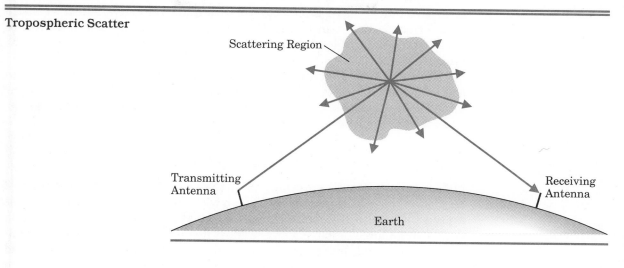

ject to fading, which can be reduced by using spatial diversity at both ends; that is, each station has at least two antennas separated by 100 wavelengths or more.

On the other hand, troposcatter can operate at much greater range than line-of-sight communication, reducing the requirement for repeater stations. This is a great benefit when the communications path is over water, over difficult terrain such as mountains, or when a foreign, possibly unfriendly government controls the territory between ends of the link. Government and commercial point-to-point radio links are the main users of tropospheric scatter.

11.7.2 Ducting

Under certain conditions, especially over water, a *superrefractive* layer (one with a lower refractive index) can form in the troposphere that returns signals to earth. The signals can then propagate over long distances by alternately reflecting from the earth (or water) and refracting from the superrefractive layer. Figure 11.33(a) shows this phenomenon.

FIGURE 11.33   **Tropospheric Ducting**

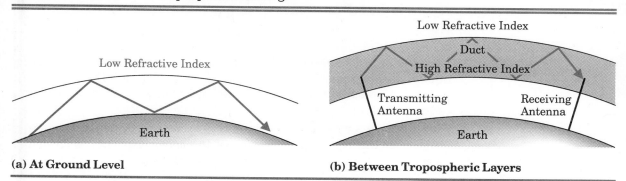

(a) At Ground Level   **(b) Between Tropospheric Layers**

A related condition involves a thin layer in the troposphere with a high refractive index, and a layer above it with a low refractive index, so that a *duct* forms, as shown in Figure 11.33(b). This too will propagate waves for long distances, provided that both the transmitting and receiving antennas are in the duct.

Unfortunately **ducting** is not reliable enough for commercial use. In fact, it is more likely to cause problems. By carrying signals far from their intended destinations, this phenomenon can cause fading of desired signals, and interference from signals at great distances.

11.7.3 Meteor-Trail Propagation

Meteors are constantly entering the earth's atmosphere and being destroyed because of the heat generated by friction with the air. A few of the larger ones are visible, and occasionally enough material survives to strike the earth with enough force to make a crater. Most meteors are the size of dust particles, however, and their paths are not visible. They do leave a trail of ionized air behind, and it is possible to use it for communication, though only for a few minutes. Although it may seem far-fetched, meteor-trail propagation is in use for data communications, especially in the North. It is not suitable for voice because of its intermittent nature.

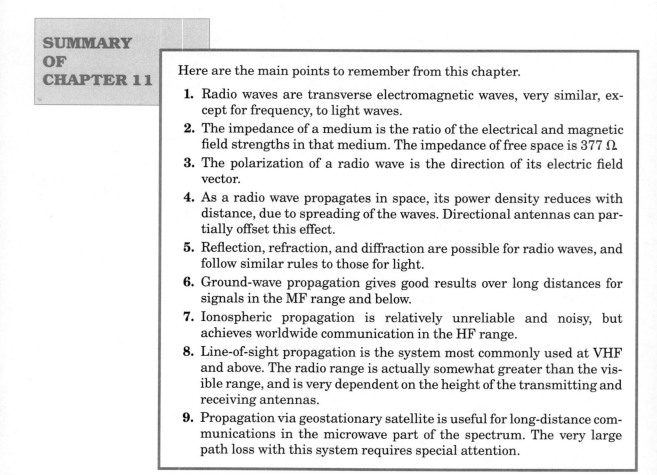

SUMMARY OF CHAPTER 11

Here are the main points to remember from this chapter.

1. Radio waves are transverse electromagnetic waves, very similar, except for frequency, to light waves.

2. The impedance of a medium is the ratio of the electrical and magnetic field strengths in that medium. The impedance of free space is 377 Ω.

3. The polarization of a radio wave is the direction of its electric field vector.

4. As a radio wave propagates in space, its power density reduces with distance, due to spreading of the waves. Directional antennas can partially offset this effect.

5. Reflection, refraction, and diffraction are possible for radio waves, and follow similar rules to those for light.

6. Ground-wave propagation gives good results over long distances for signals in the MF range and below.

7. Ionospheric propagation is relatively unreliable and noisy, but achieves worldwide communication in the HF range.

8. Line-of-sight propagation is the system most commonly used at VHF and above. The radio range is actually somewhat greater than the visible range, and is very dependent on the height of the transmitting and receiving antennas.

9. Propagation via geostationary satellite is useful for long-distance communications in the microwave part of the spectrum. The very large path loss with this system requires special attention.

IMPORTANT EQUATIONS

$$v = \frac{c}{\sqrt{\epsilon_r}} \tag{11.1}$$

$$\mathscr{Z} = \frac{\mathscr{E}}{\mathscr{H}} \tag{11.3}$$

$$\mathscr{Z}_0 = 377\ \Omega \tag{11.5}$$

$$\epsilon_r = \frac{\epsilon}{\epsilon_0} \tag{11.6}$$

$$\mathscr{Z} = \frac{377}{\sqrt{\epsilon_r}} \tag{11.7}$$

$$P_D = \frac{\mathscr{E}^2}{\mathscr{Z}} \tag{11.8}$$

$$P_D = \mathscr{H}^2 \mathscr{Z} \tag{11.9}$$

$$P_D = \mathscr{E}\mathscr{H} \tag{11.10}$$

$$P_D = \frac{P_t}{4\pi r^2} \tag{11.11}$$

$$\mathscr{E} = \frac{\sqrt{30 P_t}}{r} \tag{11.13}$$

$$P_D = \frac{P_T G_T}{4\pi r^2} \tag{11.14}$$

$$\text{EIRP} = P_T G_T \tag{11.15}$$

$$A_{eff} = \frac{P_R}{P_D} \tag{11.16}$$

$$P_R = \frac{A_{eff} P_T G_T}{4\pi r^2} \tag{11.17}$$

$$A_{eff} = \frac{\lambda^2 G_R}{4\pi} \tag{11.18}$$

$$L_{fs} = 32.44 + [20 \log d\ (\text{km})] + [20 \log f\ (\text{MHz})] \\ - [G_T\ (\text{dBi})] - [G_R\ (\text{dBi})] \tag{11.22}$$

$$n_1 \sin \theta_1 = n_2 \sin \theta_2 \tag{11.24}$$

$$n = \sqrt{\epsilon_r} \tag{11.26}$$

$$\frac{\sin \theta_1}{\sin \theta_2} = \sqrt{\frac{\epsilon_{r2}}{\epsilon_{r1}}} \tag{11.27}$$

$$\theta_c = \arcsin \frac{n_2}{n_1} \tag{11.28}$$

$$f_m = \frac{f_c}{\cos \theta_1} \tag{11.30}$$

$$d = \sqrt{17 h_T} + \sqrt{17 h_R} \tag{11.31}$$

GLOSSARY

attenuation of free space   reduction in signal strength due to spreading of the waves at a distance from the transmitter

dielectric strength   magnitude of the electric field required to cause breakdown and arcing in a dielectric

ducting   a means of propagation where the waves are confined within a refractive region of the troposphere, or between such a region and the ground

effective area   the area from which a receiving antenna can be considered to extract all the energy in an electromagnetic wave

electric field intensity (or strength)   ratio of the electric force on a charge to the charge, at a given point (Units are volts per meter.)

equatorial orbit   a satellite path that passes over the equator of the earth at all points on its rotation

Faraday rotation   a change in the polarization of an electromagnetic wave as it passes through a magnetic field (The ionosphere causes Faraday rotation of VHF and UHF signals.)

free space   a vacuum that allows radio waves to propagate without any obstruction

geostationary (geosynchronous) satellite   an artificial satellite whose orbital period is equal to that of the earth's rotation, so that to an observer on earth the satellite appears to remain stationary (A geostationary satellite must have an equatorial orbit.)

ground wave   a vertically polarized electromagnetic wave that propagates along the surface of the earth

index of refraction   ratio of the phase velocity of a wave in free space to that in the medium under consideration

ionosphere   ionized region of the earth's atmosphere

isotropic radiator   a hypothetical antenna having zero physical size, no loss, and radiating equally in all directions

magnetic field intensity (or strength)   magnitude of the magnetic field vector (Units are amperes per meter.)

maximum usable frequency (MUF)   the highest frequency that will be returned by the ionosphere at a given point

multipath reception   situation where a signal arrives at a receiving antenna via two or more paths (Usually one of these paths is direct from the transmitting antenna and the other(s) involve reflections.)

normal   a line drawn perpendicular to the interface between two media

path loss   the ratio between the signal appearing at the receiving antenna terminals, and that at the transmitting antenna terminals

photon   a quantum of electromagnetic radiation

polarization   the direction of the electric field vector of an electromagnetic wave

power density   power flowing through a unit cross-sectional area normal to the direction of travel of an electromagnetic wave

propagation   process by which waves travel through a medium

propagation delay   time taken for a signal to propagate from transmitting to receiving antenna

sky wave   electromagnetic wave returned to earth by the ionosphere

space wave   electromagnetic wave propagating directly from transmitting to receiving antenna

terrestrial propagation   propagation along or near the surface of the earth

troposphere   closest region of the atmosphere to the earth

QUESTIONS

1. What are the similarities between radio waves and light waves?
2. What is meant by the characteristic impedance of a medium? What is the characteristic impedance of free space?
3. State the difference between power and power density, and explain why power density decreases with the square of the distance from a source.
4. A radio wave propagates in such a way that its magnetic field is parallel with the horizon. What is its polarization?
5. What is an isotropic radiator? Could such a radiator be built? Explain.
6. State three factors that determine the amount of power extracted from a wave by a receiving antenna.
7. Distinguish between specular and diffuse reflections. For wavelengths on the order of 1 m, state which type is more likely from:
   (a) a calm lake
   (b) a field strewn with large boulders
8. For waves passing from one medium to another, what is meant by the critical angle of incidence, and what happens when the angle of incidence exceeds the critical value?
9. What phenomenon accounts for the fact that radio waves from a transmitter on one side of a mountain can sometimes be received on the other side?
10. Why do stations in the AM standard broadcast band always use vertically polarized antennas?
11. Why is the ionosphere more highly ionized during the daylight hours than it is at night?
12. When the critical frequency is 12 MHz, what will happen to a 16 MHz signal that is radiated straight up? Repeat for a 10 MHz signal.
13. Sometimes an HF radio transmission can be heard at a distance of 1000 km from the transmitter, but not 100 km away. Explain why.
14. Why is antenna height much more important for an FM broadcast-band antenna than for one designed for the AM broadcast band?
15. State two undesirable effects that can be caused by reflections in line-of-sight communications, and explain how they arise.
16. Why are all geostationary communications satellites at the same distance from the earth?

17. Why are other means such as optical fibers preferred over satellites for telephony?

18. What is Faraday rotation, and under what circumstances does it cause problems?

19. Why are high-gain antennas needed with geostationary satellites?

20. Which mode of propagation is normally used for each of the following services? Explain your answer.

    (a) FM radio broadcasting
    (b) shortwave radio broadcasting
    (c) cellular telephones (frequency of about 800 MHz)
    (d) LORAN-C navigation beacons (frequency 100 kHz)

PROBLEMS

Section 11.1

21. Find the propagation velocity of radio waves in glass, with a relative permittivity of 7.8.

Section 11.2

22. Find the wavelength, in free space, for radio waves at each of the following frequencies:

    (a) 50 kHz
    (b) 1 MHz
    (c) 23 MHz
    (d) 300 MHz
    (e) 450 MHz
    (f) 12 GHz

23. Find the characteristic impedance of glass, with a relative permittivity of 7.8.

24. An isotropic source radiates 100 W of power in free space. At a distance from the source of 15 km, calculate the power density and the electric field intensity.

Section 11.3

25. A certain antenna has a gain of 7 dB with respect to an isotropic radiator.

    (a) What is its effective area if it operates at 200 MHz?
    (b) How much power would it absorb from a signal with a field strength of 50 μV/m?

26. A transmitter has an output power of 50 W. It is connected to its antenna by a feedline that is 25 m long and properly matched. The loss in the feedline is 5 dB/100 m. The antenna has a gain of 8.5 dBi.

    (a) How much power reaches the antenna?
    (b) What is the EIRP in the direction of maximum antenna gain?
    (c) What is the power density 1 km from the antenna, in the direction of maximum gain, assuming free-space propagation?
    (d) What is the electric field strength at the same place as in Part (c)?

27. A satellite transmitter operates at 4 GHz with an antenna gain of 40 dBi. The receiver, 40,000 km away, has an antenna gain of 50 dBi. If the transmitter has a power of 8 W, find, ignoring feedline losses and mismatch:

    (a) the EIRP in dBW
    (b) the power delivered to the receiver

Section 11.4

28. Sketch the path of the reflected waves in each of the diagrams in Figure 11.34.

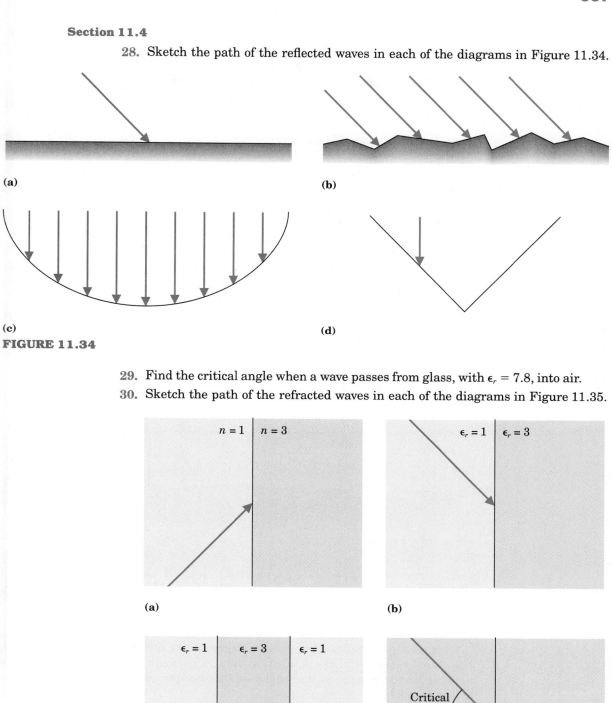

(a)

(b)

(c)

(d)

FIGURE 11.34

29. Find the critical angle when a wave passes from glass, with $\epsilon_r = 7.8$, into air.

30. Sketch the path of the refracted waves in each of the diagrams in Figure 11.35.

$n = 1$ | $n = 3$

(a)

$\epsilon_r = 1$ | $\epsilon_r = 3$

(b)

$\epsilon_r = 1$ | $\epsilon_r = 3$ | $\epsilon_r = 1$

(c)

Critical Angle

(d)

FIGURE 11.35

Section 11.5

31. At a certain time, the MUF for transmissions at an angle of incidence of 75° is 17 MHz. What is the critical frequency?

32. If the critical frequency is 12 MHz, what is the critical angle at 15 MHz?

33. An FM broadcast station has a transmitting antenna located 50 m above average terrain. How far away could the signal be received:

    (a) by a car radio with an antenna 1.5 m above the ground?
    (b) by a rooftop antenna 12 m above the ground?

34. A boat is equipped with a VHF marine radio, which it uses to communicate with other nearby boats and shore stations.

    (a) Name the mode of propagation.
    (b) If the antenna on the boat is 2.3 m above the water, calculate the maximum distance for communication with:
        (i) another similar boat
        (ii) a shore station with an antenna on a tower 22 m above the water level
        (iii) another boat, but using the shore station as a repeater

35. An FM broadcast signal arrives at an antenna via two paths, as shown in Figure 11.36. Calculate the difference in arrival time for the two paths.

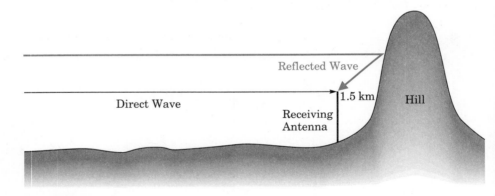

FIGURE 11.36

36. A transmitter and receiver are separated by water. If the difference in the length of the direct and reflected paths is 2 m, calculate the phase difference at 400 MHz.

Section 11.6

37. Two points 1000 km apart on earth can communicate using satellite, coaxial cable, or terrestrial microwave link. Calculate the one-way time delay for each mode, assuming that the satellite is 38,000 km from each station, and that the speed of transmission on the coaxial cable is two-thirds that in free space. Ignore any time delay in the electronics.

38. A map indicates that a certain spot on the earth is in the 40 dBW contour of a satellite beam. What is the actual signal strength at the earth's surface, in W/m², if the satellite is 37,000 km away?

Comprehensive

39. Explain each of the following:

    (a) AM radio broadcast stations must often reduce power at night to avoid interference.

(b) CB radio at 27 MHz is intended for local communication, but can often communicate for hundreds of kilometers.

(c) HF communications are often severely disrupted when the aurora borealis (northern lights) is visible.

(d) Sometimes VHF signals appear, and cause interference, hundreds of kilometers away from their intended route.

(e) It is possible to communicate, using UHF signals, somewhat further than the visible horizon.

40. Construct a table showing all the propagation methods discussed in this chapter, with the frequency ranges and distances for which they are useful. Include comments concerning economy and reliability of operation.

12

Antennas

Objectives

After studying this chapter, you should be able to:

1. Explain the basic principles of operation of antenna systems.
2. Define antenna gain and beamwidth, and find the gain and beamwidth of an antenna, given a plot of its radiation pattern.
3. Calculate effective isotropic radiated power and effective radiated power for an antenna-transmitter combination, and explain the difference between the two terms.
4. Calculate the effective area for a receiving antenna, and use it to calculate the power delivered to a receiver.
5. Calculate the dimensions of simple practical antennas for a given frequency.
6. Identify, explain the operation of, and sketch the approximate radiation pattern for common types of antennas and antenna arrays.
7. Calculate the gain and beamwidth for parabolic antennas.

12.1 INTRODUCTION

So far, we have considered the ways in which electromagnetic waves can propagate along transmission lines and through space. The **antenna** is the interface between these two media, and is a very important part of the communications path. In this chapter, we will study the basic operating principles of antennas, and look at some of the parameters that describe their performance. Some representative examples of practical antennas will be analyzed.

Before we begin, it would be well to have two ideas firmly in mind. First, antennas are passive devices. Therefore, the power radiated by a transmitting antenna cannot be greater than the power entering from the transmitter. In fact, it will be less because of losses. We will speak of antenna gain, but we must remember that gain in one direction results from a concentration of power, and will be accompanied by a loss in other directions. Antennas achieve gain the same way a flashlight reflector increases the brightness of the bulb: by concentrating energy.

By the way, you may hear someone talk about an **active antenna** for a high-frequency communications receiver or an FM or television broadcast receiver. This term simply describes the combination of a receiving antenna with a low-noise preamplifier. The antenna part of the combination is still a passive device.

The second concept to keep in mind is that antennas are reciprocal; that is, the same design works equally well as a transmitting or a receiving antenna, and in fact will have the same gain. That does not mean that transmitting and receiving antennas are necessarily identical. For instance, the conductors in a transmitting antenna must be sized to handle larger currents. However, the designs are quite similar, and many of the calculations are identical.

Essentially, the task of a transmitting antenna is to convert the electrical energy travelling along a transmission line into electromagnetic waves in space. This process, while difficult to analyze in mathematical detail, should not be hard to visualize. The energy in the transmission line is contained in the electric field between the conductors and in the magnetic field surrounding them. All that is needed is to "launch" these fields, and the energy they contain, into space.

At the receiving antenna, the electric and magnetic fields in space will cause current to flow in the conductors that make up the antenna. Some of the energy will thereby be transferred from these fields to the transmission line connected to the receiving antenna.

12.2 SIMPLE ANTENNAS

In order to understand the operation of antennas, two simple antennas will be described first. The **isotropic radiator** was introduced in the previous chapter. Though merely a theoretical construct, it serves as a way to describe the functions of an antenna, and as a reference for other antennas. The *half-wave dipole antenna*, on the other hand, is very practical and in common use. An understanding of the half-wave dipole is important both in its own right and as a basis for the study of more complex antennas.

FIGURE 12.1

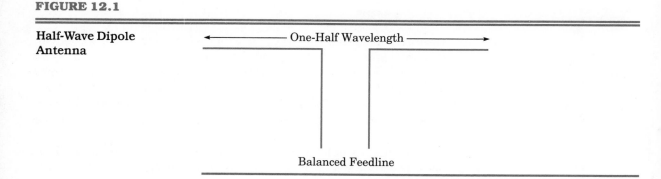

Half-Wave Dipole
Antenna

12.2.1 The Isotropic Radiator

An ideal isotropic radiator would radiate all the electrical power supplied to it, and would do so equally in all directions. It would also be a point source; that is, it would have zero size.

It would not be a good idea to try to build such an antenna. Of course, the zero size is not physically realizable, and neither is the complete losslessness or complete nondirectionality of the isotropic antenna. Nonetheless, we will use the isotropic antenna as a standard of comparison. That is because, even though this antenna cannot be built and tested, its characteristics are simple and easy to derive.

12.2.2 The Half-Wave Dipole

A more practical construction project is the **dipole**, shown in Figure 12.1. The word *dipole* simply means it has two parts, as shown. A dipole antenna does not have to be one-half wavelength in length like the one shown in the figure, but this length is handy for impedance matching, as we shall see. Actually, in practice its length will be slightly less than one-half the free-space wavelength, to allow for capacitive effects. A half-wave dipole is sometimes called a Hertz antenna, though strictly speaking the term *Hertzian dipole* refers to a dipole of infinitesimal length. This, like the isotropic radiator, is a theoretical construct, and is used in the calculation of antenna radiation patterns.

Typically, the length of a half-wave dipole, assuming that the conductor diameter is much less than the length of the antenna, is 95% of one-half wavelength measured in free space. The free-space wavelength is given by

$$\lambda = \frac{c}{f} \tag{12.1}$$

where λ = free-space wavelength in meters
$c = 3 \times 10^8$ m/s
f = operating frequency in hertz

Therefore, the length L of a half-wave dipole, in meters, is

$$L = 0.95 \times 0.5\left(\frac{c}{f}\right)$$

$$= .475\left(\frac{c}{f}\right)$$

$$= \frac{.475(3 \times 10^8)}{f}$$

$$= \frac{142.5 \times 10^6}{f} \tag{12.2}$$

In the above equation, L is the length in meters and f is the frequency in hertz. Very often, megahertz are a more convenient unit for frequency, in which case Equation (12.2) becomes

$$L = \frac{142.5}{f} \tag{12.3}$$

where L = length of a half-wave dipole in meters
f = operating frequency in megahertz

For length measurements in feet, the equivalent equation is

$$L = \frac{468}{f} \tag{12.4}$$

where L = length of a half-wave dipole in feet
f = operating frequency in megahertz

EXAMPLE 12.1

Calculate the length of a half-wave dipole for an operating frequency of 20 MHz.

Solution

From Equation (12.3),

$$L = \frac{142.5}{f}$$

$$= \frac{142.5}{20}$$

$$= 7.13 \text{ m}$$

One way to think about the half-wave dipole is to consider an open-circuited length of parallel-wire transmission line, as shown in Figure 12.2(a). The line will have a voltage maximum at the open end, a current maximum one-quarter wavelength from the end, and a very high standing-

FIGURE 12.2   **Development of the Half-Wave Dipole**

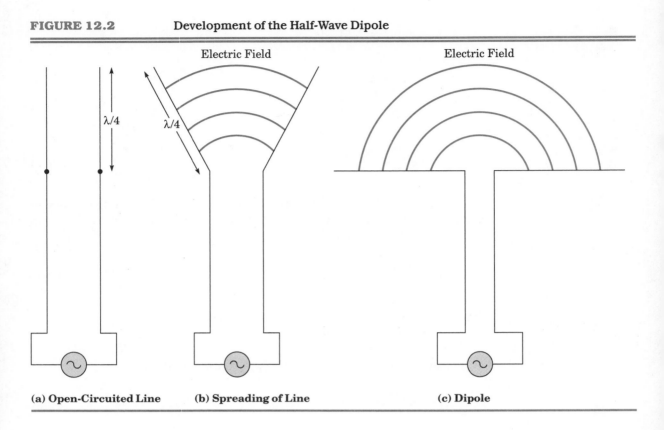

(a) Open-Circuited Line   **(b) Spreading of Line**   **(c) Dipole**

wave ratio (SWR). In fact, we noted in our discussion of transmission lines in Chapter 10 that the SWR for this open-circuited stub would be infinite, except for the fact that there would be some radiation from the open end. That is, the stub will act as an antenna, though a very inefficient one.

Now, suppose that the two conductors are separated at a point one-quarter wavelength from the end, as in Figure 12.2(b). The drawing shows how the electric field seems to stretch away from the wires. If the process continues, as in Figure 12.2(c), some of the field detaches itself from the antenna, and helps to form electromagnetic waves that propagate through space.

12.2.3 Radiation Resistance

The radiation of energy from a dipole is quite apparent if we measure the impedance at the feedpoint in the center of the antenna. An actual open-circuited lossless line, as described in Chapter 10, would look like a short circuit at a distance of one-quarter wavelength from the open end. At distances slightly greater than or less than one-quarter wavelength, the line would appear reactive. There would never be a nonzero resistive component to the feedpoint impedance, since an open-circuited line has no way of dissipating power.

The half-wave dipole will not dissipate power either, assuming the material of which it is made is lossless, but it will radiate power into space. The effect on the feedpoint impedance is the same as if a loss had taken place. Whether power is dissipated or radiated, it disappears from the antenna, and therefore causes the input impedance to have a resistive component. The half-wave dipole, for instance, will look like a resistance of about 70 Ω at its feedpoint.

The portion of an antenna's input impedance that is due to power radiated into space is known, appropriately, as the **radiation resistance**. It is important to understand that this does not represent losses in the conductors that make up the antennas.

The idealized antenna just described will radiate all the power supplied to it into space. A real antenna will, of course, have ohmic losses in the conductor. It will therefore have an efficiency less than 1. This efficiency can be defined as

$$eff = \frac{P_r}{P_T}$$

where P_r = radiated power
        P_T = total power supplied to the antenna

Recalling that $P = I^2R$, we have

$$eff = \frac{I^2R_r}{I^2R_T}$$

$$= \frac{R_r}{R_T} \tag{12.5}$$

where R_r = radiation resistance, as seen from the feedpoint
        R_T = total resistance, as seen from the feedpoint

EXAMPLE 12.2

A dipole antenna has a radiation resistance of 67 Ω and a loss resistance of 5 Ω, measured at the feedpoint. Calculate the efficiency.

Solution

From Equation (12.5),

$$eff = \frac{R_r}{R_T}$$

$$= \frac{67}{67 + 5}$$

$$= 0.93 \quad \text{or} \quad 93\%$$

The half-wave dipole does not radiate uniformly in all directions. The field strength is at its maximum along a line at a right angle to the antenna and is zero off the ends of the antenna.

12.3 ANTENNA DIRECTIONAL CHARACTERISTICS

Now that two simple antennas have been described, it is already apparent that antennas differ in the amount of radiation they emit in various directions. This would be a good time to introduce some terms to describe and quantify the directional characteristics of antennas, and to demonstrate methods of graphing some of them. These will be applied to the isotropic and half-wave dipole antennas at once, and will also be applied to other antenna types as each is introduced.

12.3.1 Radiation Pattern

The diagrams used in this book follow the three-dimensional coordinate system shown in Figure 12.3. As shown in the figure, the x–y plane is horizontal, and the angle **phi** (ϕ) is measured from the x axis in the direction of the y axis. The z axis is vertical, and the angle **theta** (θ) is measured from the horizontal plane toward the **zenith**. This is not quite the same as the standard polar coordinate system used in geometry texts. In the "standard" system, θ is measured down from the z axis. Antenna manufacturers generally prefer to measure the vertical angle upward from the ground, rather than downward from the zenith, so that is the method that will be used here. The vertical angle, measured upward from the ground, is called the *angle of elevation*.

Figure 12.4 shows two ways in which the radiation pattern of a dipole can be represented. The three-dimensional picture in Figure 12.4(a) is useful in showing the general idea, and in getting a feel for the characteristics of the antenna. The two views in Figures 12.4(b) and (c), on the other hand, are less intuitive, but can be used to provide quantitative information about the performance of the antenna. They are also much easier to draw. It is the latter

FIGURE 12.3

Three-Dimensional Coordinate System

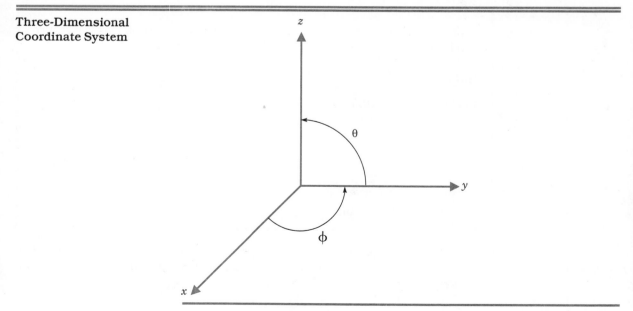

FIGURE 12.4   Radiation Pattern of Horizontal Half-Wave Dipole

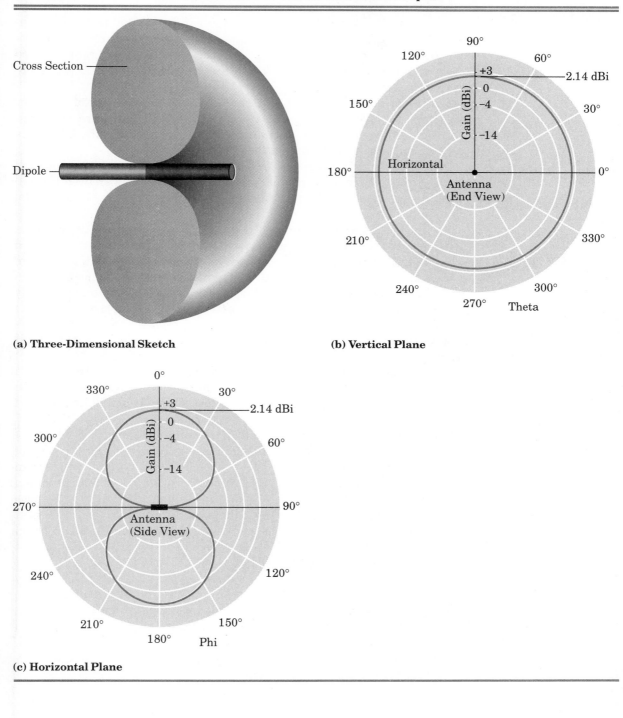

(a) Three-Dimensional Sketch

(b) Vertical Plane

(c) Horizontal Plane

type of antenna pattern that is usually found in manufacturers' literature, for instance, and it is this type of depiction that will be used in this book.

In looking at Figures 12.4(b) and (c), you will notice that polar graph paper is used. The angle is measured in one plane from a reference axis. Because the three-dimensional space around the antenna is being represented in two dimensions, at least two views are required to give the complete picture.

It is important to choose the axes carefully to take advantage of whatever symmetry exists. Very often, the horizontal and vertical planes are used; usually the antenna will be mounted either vertically or horizontally, so the axis of the antenna itself can be used as one of the reference axes.

The two graphs in Figures 12.4(b) and (c) can be thought of as slices of the three-dimensional pattern of Figure 12.4(a). If the axes are well chosen, two such slices will usually be sufficient to describe the three-dimensional radiation pattern.

In Figure 12.4, the dipole itself is drawn to help visualize the antenna orientation. Please note, however, that these radiation patterns are only valid in the **far-field region;** that is, an observer must be far enough away from the antenna that any local capacitive or inductive coupling is negligible. In practice, this means a distance of at least several wavelengths, and generally an actual receiver will be at a much greater distance than that. From this distance, the antenna would be more accurately represented as a dot in the center of the graph. The area close to the antenna is called the **near-field region** and does not have the same directional characteristics.

The directions follow the coordinate system of Figure 12.3. The antenna is considered to be in free space, so its radiation is plotted in all directions. In many practical situations, radiation below the horizon is not plotted. In that case, the bottom half of Figure 12.4(b) would be omitted.

The distance out from the center of the graph represents the strength of the radiation in a given direction. The scale is usually in decibels with respect to some reference. Here, there are a lot of choices. For instance, the scale can be arbitrary, with the outside circle representing the maximum radiation from the antenna. Often, the reference is an isotropic radiator, as is the case in Figure 12.4. Note that the furthest point on the graph from the center is at 2.14 dB; that is, we can say that the gain of a lossless dipole, in its direction of maximum radiation, is 2.14 dB with respect to an isotropic radiator. This gain is usually expressed as 2.14 **dBi**.

The half-wave dipole itself is sometimes used as a reference, especially for high-frequency antennas. In that case, the gain of an antenna may be expressed in decibels with respect to a half-wave dipole, or **dBd** for short. Since the gain of the dipole is known to be 2.14 dBi, the gain of any antenna in dBd is 2.14 dB less than the gain of the same antenna expressed in dBi. Obviously, when comparing antennas, it is important to know which reference antenna was used in gain calculations.

EXAMPLE 12.3

Two antennas have gains of 5.3 dBi and 4.5 dBd respectively. Which has greater gain?

Solution

Convert both gains to the same standard. In this case, let us use dBi. Then, for the second antenna,

$$Gain = 4.5 \text{ dBd}$$
$$= 4.5 + 2.14 \text{ dBi}$$
$$= 6.64 \text{ dBi}$$

Therefore, the second antenna has higher gain.

12.3.2 Gain and Directivity

The sense in which a half-wave dipole antenna can be said to have gain can be seen from Figure 12.5. This sketch shows the pattern of a dipole, from Figure 12.4(c), superimposed on that of an isotropic radiator. It can be seen that, while the dipole has a gain of 2.14 dBi in certain directions, in others its gain is negative. If the antenna were to be enclosed by a sphere that would absorb all the radiated power, the total radiated power would be found to be the same for both antennas. Remember that, for antennas, power gain in one direction is at the expense of losses in others.

Sometimes the term **directivity** is used. This is not quite the same as gain. Directivity is the gain calculated assuming a lossless antenna. Real antennas have losses, and gain is simply the directivity multiplied by the efficiency of the antenna.

FIGURE 12.5

Isotropic and Dipole Antennas

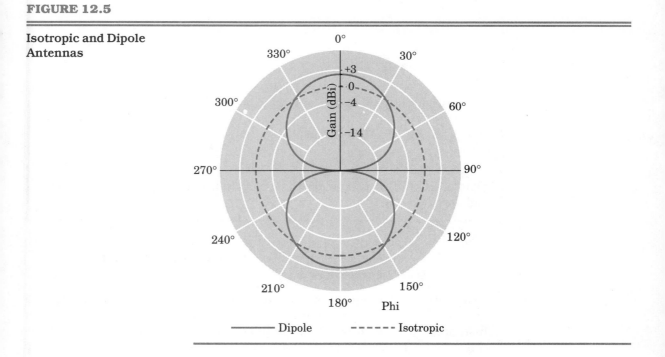

When an antenna is used for transmitting, the total power emitted by the antenna will be somewhat less than that delivered to it by the feedline. In fact,

$$P_x = P_T(eff) \tag{12.6}$$

where P_x = total radiated power
 P_T = power supplied to the antenna
 eff = antenna efficiency

The figure of 2.14 dBi we have been using for the gain of a lossless dipole is also the directivity for *any* dipole. To find the gain of a real (lossy) dipole, it is necessary first to convert the decibel directivity to a power ratio, then to multiply by the efficiency.

EXAMPLE 12.4

A dipole antenna has an efficiency of 85%. Calculate its gain in decibels.

Solution

The directivity of 2.14 dBi can be converted to a power ratio:

$$D = \text{antilog } \frac{2.14}{10}$$
$$= 1.638$$

Now, find the gain:

$$G = D(eff)$$
$$= 1.637 \times 0.85$$
$$= 1.39$$

If required, this gain can be converted to decibels.

$$G \text{ (dB)} = 10 \log 1.39$$
$$= 1.43 \text{ dBi}$$

(The "i" in "dBi" is a reminder that the gain is with respect to an isotropic radiator.)

12.3.3 Effective Isotropic Radiated Power and Effective Radiated Power

In a practical situation, we are usually more interested in the power emitted in a particular direction than in the total radiated power. Looking from a distance, it is impossible to tell the difference between a high-powered transmitter using an isotropic antenna, and a transmitter of lower power working into an antenna with gain. In Chapter 11, we defined an **effective isotropic radiated power** (EIRP), which is simply the actual power going into the antenna multiplied by its gain with respect to an isotropic radiator:

$$\text{EIRP} = P_T G_T \tag{12.7}$$

Another similar term that is in common use is **effective radiated power** (ERP), which represents the power input multiplied by the antenna gain measured with respect to a half-wave dipole. Since an ideal half-wave dipole has a gain of 2.14 dBi, the EIRP is 2.14 dB greater than the ERP for the same antenna-transmitter combination.

12.3.4 Effective Area

As mentioned in Chapter 11, the signal power delivered by a receiving antenna depends not only on its gain but also on its effective physical size. The signal power available is simply the power density multiplied by the **effective area**, and the effective area is given by

$$A_{eff} = \frac{\lambda^2 G_R}{4\pi} \tag{12.8}$$

where A_{eff} = effective area

G_R = antenna gain, as a power ratio

λ = wavelength of the signal

12.3.5 Beamwidth

Just as a flashlight emits a beam of light, a directional antenna can be said to emit a beam of radiation in one or more directions. The width of this beam is defined as the angle between its half-power points. These are also the points at which the power density is 3 dB less than it is at its maximum point. An inspection of Figure 12.4(b) and (c) will show that the half-wave dipole has a **beamwidth** of about 78° in one plane and 360° in the other. Many antennas are much more directional than this, with a narrow beamwidth in both planes.

12.3.6 Impedance

The radiation resistance of a half-wave dipole, situated in free space and fed at the center, is approximately 70 Ω. The impedance will be completely resistive at resonance, which takes place for a physical length of about 95% of the calculated free-space half-wavelength value. The exact length depends on the diameter of the antenna conductor relative to the wavelength. If the frequency is above resonance, the feedpoint impedance will have an inductive component; if the frequency is lower than resonance, the antenna impedance will be capacitive. Another way of saying the same thing is that an antenna that is too short will appear capacitive, while one that is too long will be inductive. Figure 12.6 shows graphically how reactance varies with frequency.

FIGURE 12.6

**Variation of Dipole
Reactance with
Frequency**

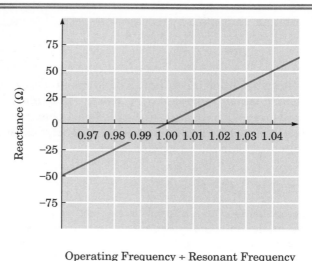

Operating Frequency ÷ Resonant Frequency

A center-fed dipole is a balanced device, and should be used with a balanced feedline. If coaxial cable is used, a *balun* (balanced-to-unbalanced) transformer should be connected between the cable and the antenna. (Balun transformers were described in Chapter 10.)

A half-wave dipole does not have to be fed at its midpoint. It is possible, for instance, to connect the feedline at one end, as shown in Figure 12.7. As you might expect from our earlier analogy with an open-circuited transmission line, this will result in a higher impedance at the feedpoint. It is also possible to feed the antenna at some distance from the center in both directions, as shown in Figure 12.8. This system is called a *delta match*, and allows the impedance to be adjusted to match a transmission line. Just as for a center-fed antenna, these two systems will result in a resistive impedance only if the antenna length is accurate.

FIGURE 12.7

**End-Fed Half-Wave
Dipole**

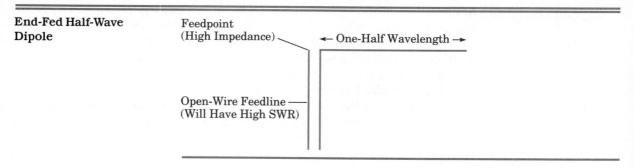

Feedpoint
(High Impedance)   ← One-Half Wavelength →

Open-Wire Feedline
(Will Have High SWR)

FIGURE 12.8

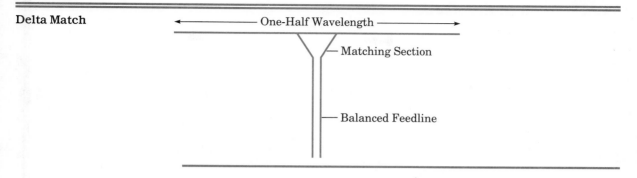

Delta Match

12.3.7 Polarization

The **polarization** of a half-wave dipole is easy to determine: It is the same as the axis of the wire. That is, a horizontal antenna produces horizontally polarized waves, and a vertical antenna gives vertical polarization.

The choice of polarization is sometimes dictated by the method of propagation, as explained in Chapter 11. For instance, ground-wave propagation requires vertical polarization. For HF communication, polarization is not very important because the ionosphere will randomize it; horizontal polarization is more common because a horizontal wire is usually easier to install. At VHF and above, it is important only that the polarization be the same at both ends of a communications path.

12.3.8 Ground Effects

When an antenna is installed within a few wavelengths of the ground, the earth acts as a reflector and has a considerable influence on the radiation pattern of the antenna. Ground effects are important up through the HF range. At VHF and above, the antenna is usually far enough above the earth that reflections from the ground near the antenna are not significant. Reflections at a considerable distance from the antenna can still be very important, since they can cause *fading*. (This problem was discussed in Chapter 11.)

Ground effects are complex because the characteristics of the ground are so variable. In particular, the conductivity varies over a wide range. Any detailed analysis of ground effects requires a computer, and there are several programs that will do the job. An intuitive understanding of the process, without calculations, can be gained, however, by considering the earth to be a perfectly conductive sheet under the antenna. This will impart a 180° phase shift to the reflected wave, and the reflected wave will have the same amplitude as the incident wave. Just as with an ordinary mirror, it is possible to imagine an "image" of the real antenna, and to use it to understand the changes that ground reflections make in the antenna pattern.

Figure 12.9 shows the effects of average ground on the vertical radiation pattern of a horizontal dipole. Figure 12.9(a) shows the dipole in free space. This figure is the same as Figure 12.4(b) except that the radiation below the horizon has not been shown. Figure 12.9(b) shows the dipole one-quarter wavelength above the ground. Radiation in the upward direction is increased,

FIGURE 12.9   Effect of Ground on Radiation Pattern

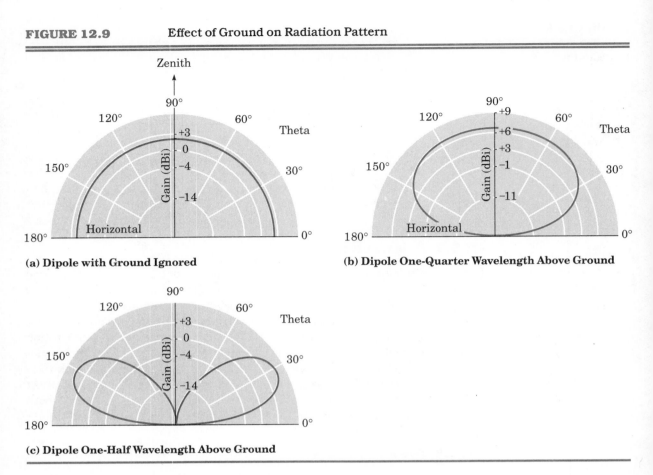

(a) Dipole with Ground Ignored

(b) Dipole One-Quarter Wavelength Above Ground

(c) Dipole One-Half Wavelength Above Ground

because the reflected wave reinforces the incident wave. On the other hand, the situation in Figure 12.9(c), where the antenna is one-half wavelength above the ground, results in cancellation of radiation toward the zenith, with a corresponding increase in low-angle radiation. The latter is generally more useful than high-angle radiation for HF communications.

12.4 OTHER SIMPLE ANTENNAS

The half-wave dipole is simple, useful, and very common, but it is by no means the only type of antenna in use. In this section, some other simple antennas will be introduced. Later, ways of combining antenna elements into arrays with specific characteristics will be examined.

12.4.1 The Folded Dipole

Figure 12.10 shows a folded dipole. It is the same length as a standard half-wave dipole, but it is made with two parallel conductors, joined at both ends and separated by a distance that is short compared with the length of the antenna. One of the conductors is broken in the center and connected to a balanced feedline.

FIGURE 12.10

Folded Dipole

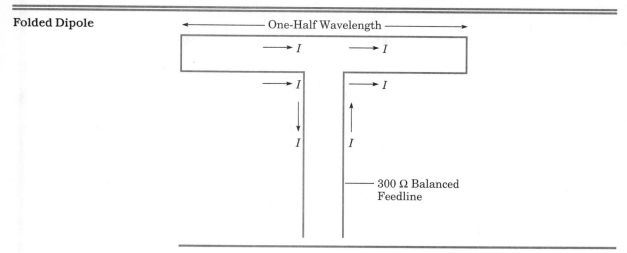

The folded dipole differs in two ways from the ordinary half-wave dipole described above. It has a wider bandwidth; that is, the range of frequencies within which its impedance remains approximately resistive is larger than for the single-conductor dipole. For this reason, it is often used—alone or with other elements—for television and FM broadcast receiving antennas. It also has approximately four times the feedpoint impedance of an ordinary dipole. This accounts for the extensive use of 300 Ω balanced line (known as *twin-lead*) in TV and FM receiving installations.

It is easy to see why the impedance of a folded dipole is higher than that of a standard dipole. First, suppose that a voltage V and a current I are applied to an ordinary, nonfolded dipole. Then, at resonance when the feedpoint impedance is resistive, the power supplied is

$$P = VI \tag{12.9}$$

where P = average power
V = RMS voltage
I = RMS current

Now reconsider the folded dipole of Figure 12.10. Looking at the center of the dipole, the length of the path from the center of the lower conductor to the center of the upper conductor is one-half wavelength. Therefore, the currents in the two conductors will be equal in magnitude. If the points were one-half wavelength apart on a straight transmission line, we would say that they were equal in magnitude but out of phase. Here, however, because the wire has been folded, the two currents, flowing in opposite directions with respect to the wire, actually flow in the same direction in space, and contribute equally to the radiation from the antenna.

If a folded dipole and a regular dipole radiate the same amount of power, the total current must be the same in both. However, the current at the feedpoint of a folded dipole is only one-half the total current. If the feedpoint current is reduced by one-half, yet the power remains the same, the feedpoint voltage must be doubled. That is,

$$P = VI$$
$$= 2V\left(\frac{I}{2}\right)$$

Assuming equal power is provided to both antennas, the feedpoint voltage must be twice as great for the folded dipole.

The resistance at the feedpoint of the ordinary dipole is, by Ohm's Law,

$$R = \frac{V}{I}$$

For the folded dipole, it will be

$$R' = \frac{2V}{I/2}$$
$$= \frac{4V}{I}$$
$$= 4R$$

Since the current has been divided by two, and the voltage multiplied by two, the folded dipole has four times the feedpoint impedance as the regular version.

It is also possible to build folded dipoles with different-size conductors, and with more than two conductors. In this way, a wide variety of feedpoint impedances can be produced.

12.4.2 The Monopole Antenna

For low- and medium-frequency transmission, it is necessary to use vertical polarization to take advantage of ground-wave propagation. A vertical half-wave dipole would be possible, of course, but rather long. For instance, at 1 MHz, the wavelength is 300 m. Similar results can be obtained by using a quarter-wave **monopole** antenna, fed at one end with an unbalanced feed-line, with the ground conductor of the feedline connected to a good earth ground. In practice, this usually means a fairly extensive array of **radials** (conductors buried in the ground and extending outward from the antenna). Such an antenna is often called a *Marconi antenna* (though there is some doubt as to whether Marconi was actually the first to use it), and usually takes the form of a guyed tower. The tower can be insulated from the ground, as shown in Figure 12.11; alternatively, it can be grounded and fed at a point above ground using a *gamma match*, as in Figure 12.12. Moving the feedline connection higher up on the tower increases the feedpoint impedance.

When connected to an ideal ground (or a good radial system) the radiation pattern of a quarter-wave monopole in the vertical plane has the same shape as that of a vertical half-wave dipole in free space, except that only one-half the pattern is present, since there is no underground radiation. The ver-

FIGURE 12.11

**Monopole Antenna
Using Insulated Tower**

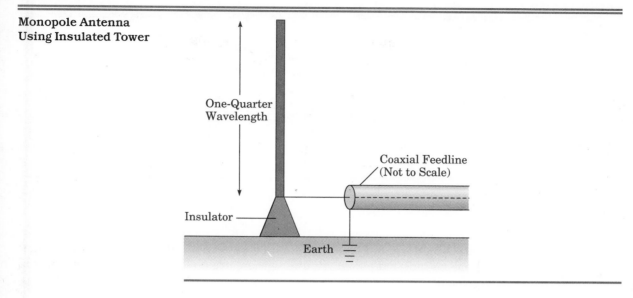

FIGURE 12.12

tical pattern is shown in Figure 12.13. In the horizontal plane, of course, a
vertical monopole will be omnidirectional. Since, assuming no losses, all of
the power is radiated into one-half the pattern of a dipole, this antenna has a
power gain of two (or 3 dB) over a dipole in free space.

    The impedance at the base of a quarter-wave monopole is one-half that of
a dipole. This can be explained as follows. With the same current, the an-

**Monopole Antenna
Using Grounded Tower**

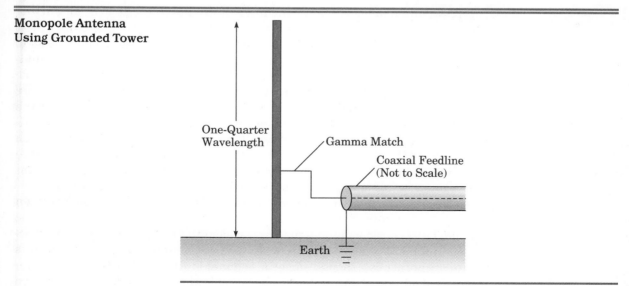

FIGURE 12.13

Vertical Monopole

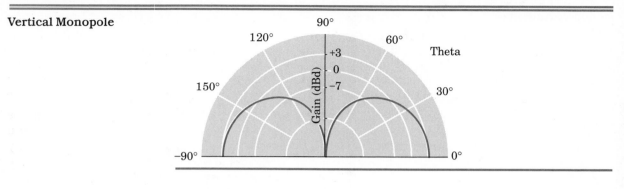

tenna produces one-half the radiation pattern of a dipole, and therefore one-half the radiated power. Assuming there are no losses, the radiated power is given by

$$P_r = I^2 R_r$$

where P_r = radiated power
   I = antenna current at the feedpoint
   R_r = radiation resistance measured at the feedpoint

If the radiated power decreases by a factor of two for a given current, then so must the feedpoint radiation resistance.

12.4.3 Ground-Plane Antennas

Vertical antennas need not be mounted at ground level. At VHF and above, where line-of-sight propagation is the most common method, antenna height is very important. An antenna mounted at ground level would be ineffective. It is possible to preserve the simplicity and low radiation angle of the ground-mounted monopole by, in effect, constructing an artificial ground at the base of the antenna. This **ground plane** can be a conductive sheet, but is more likely to be constructed of four or more metal rods radiating outward from the base of the antenna, and made at least as long as the antenna itself. Ground-plane antennas are often seen with CB base stations. Since the wavelength in the 27 MHz CB band is about 11 m, the quarter-wave antenna will need to be almost 3 m long. Such an antenna is obviously difficult enough to mount on a tower; a full-length dipole at twice the length would be much clumsier.

Mobile antennas are usually ground-plane antennas, with the car itself acting as the ground plane. Thus, a simple whip antenna on an automobile would be expected to be one-quarter wavelength in length. This is quite practical in the FM broadcast band, where the wavelength is about 3 m, but it is rather impractical (though sometimes done) in the 27 MHz CB band (wavelength about 11 m), and out of the question in the AM broadcast band, where the wavelength is on the order of 300 m.

12.4.4 Loop Antennas

The lengths of the antennas studied so far are all an appreciable fraction of a wavelength. Sometimes, particularly for receiving, a much smaller antenna is required. These will not be very efficient, but can perform adequately for such tasks as the reception of local AM broadcasts. They are also used for marine radio direction finders.

Figure 12.14 shows two versions of the popular loop antenna. The one in Figure 12.14(a) is an older design, using an air-wound coil. This antenna is bidirectional, with its greatest sensitivity in the plane of the loop, as shown by the arrow. This would be a multi-turn coil for the AM broadcast band; only one turn is shown for clarity. Figure 12.14(b) is a representation of a ferrite "loopstick" antenna, such as is found in practically every AM broadcast receiver (except those for automobiles). Here, the directionality is still in the plane of the individual coil turns, but this is broadside to the axis of the ferrite core. As we saw in Chapter 5, the loopstick antenna usually doubles as the coil in the input tuned circuit of a receiver.

12.4.5 The Five-Eighths Wavelength Antenna

This antenna is often used vertically as either a mobile or base antenna in VHF and UHF systems. Like the quarter-wave vertical antenna, it has om-

FIGURE 12.14   Loop Antennas

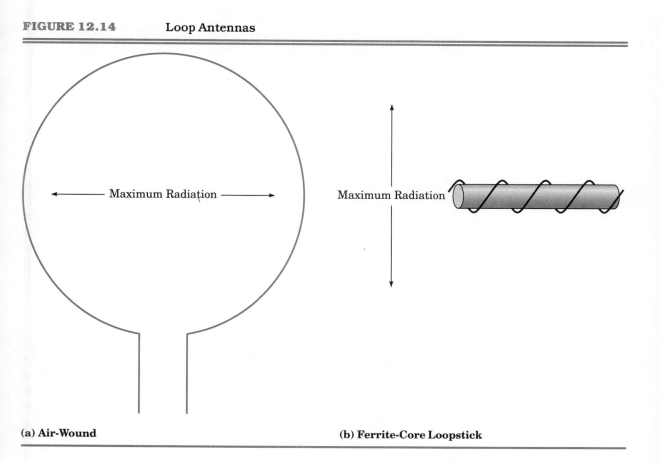

(a) Air-Wound   **(b) Ferrite-Core Loopstick**

FIGURE 12.15

**Commercial Five-
Eighths Wavelength
Antenna**

Courtesy Cushcraft Corporation.

nidirectional response in the horizontal plane. However, the radiation is concentrated at a lower angle. This results in gain in the horizontal direction, which is often more useful. In addition, it has a higher feedpoint impedance and so does not require as good a ground, because the current at the feedpoint is less. The impedance is typically lowered to match that of a 50 Ω feedline by the use of an impedance-matching section. The circular section at the base of the antenna in the photograph in Figure 12.15 is an impedance-matching device.

12.4.6 The Discone Antenna

The rather unusual-looking antenna shown in Figure 12.16 is known appropriately as the *discone*. It is characterized by very wide bandwidth, covering approximately a 10:1 frequency range, and an omnidirectional pattern in the horizontal plane. The signal is vertically polarized, and the gain is comparable to that of a dipole. The feedpoint impedance is approximately 50 Ω; the

FIGURE 12.16

**Commercial Discone
Antenna**

Courtesy of Radio Shack, a division of Tandy Corporation.

feedpoint is located at the intersection of the disk and the cone. The disk-cone combination acts as a transformer to match the feedline impedance to the impedance of free space, which is 377 Ω. Typically, the length measured along the surface of the cone is about one-quarter wavelength at the lowest operating frequency.

The wide bandwidth of the discone makes it a very popular antenna for general reception in the VHF and UHF ranges. It is a favorite for use with *scanners*. These receivers can tune automatically to a large number of channels in succession, and are often used for monitoring emergency services. The discone can be used for transmitting, but seldom is. Most transmitting stations operate at one frequency or over a narrow band of frequencies. Simpler

FIGURE 12.17

Helical Antenna

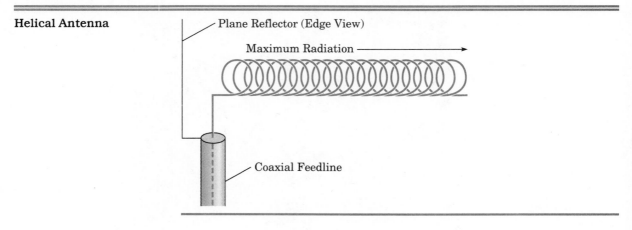

antennas with equivalent performance, or equally elaborate antennas with better performance, are available when wide bandwidth is not required.

12.4.7 The Helical Antenna

There are actually several different types of antennas that are described as *helical*. A **helix**, of course, is simply a spiral. A quarter-wave monopole antenna can be shortened and wound into a helix. This is the common *rubber ducky* antenna used with many handheld transceivers. Sometimes it is called a *helical antenna*, and it certainly is helical in shape.

The antenna that is usually referred to as helical, however, is much longer, usually several wavelengths long. Such an antenna is shown in Figure 12.17. It is used with a plane reflector, as shown, to improve its directional characteristics. Typically, the circumference of each turn is about one wavelength, and the turns are about one-quarter wavelength apart.

Helical antennas of the type shown in Figure 12.17 produce circularly polarized waves whose sense is the same as that of the helix. A helical antenna can be used to receive circularly polarized waves with the same sense, and can also receive plane-polarized waves with the polarization in any direction. The gain is proportional to the number of turns, and can be several decibels greater than a dipole.

Helical antennas are often used with VHF satellite transmissions. Since they respond to any polarization angle, they avoid the problem of Faraday rotation, which makes the polarization of waves received from a satellite impossible to predict.

There is yet another type of antenna that can be described as helical. The one illustrated in Figure 12.18 resembles a half-wave dipole, except that the ends are bent. This antenna is used to produce both horizontally and vertically polarized waves simultaneously, with one-half the input power going into each polarization. Such antennas are very common in FM broadcasting, and enable the signal to be received with both vertical antennas (such as those on cars) and horizontal antennas (like the television antennas installed on the roofs of houses). Mounted as shown, the antenna is approximately omnidirectional in the horizontal plane.

FIGURE 12.18

Commercial FM
Broadcasting
Transmitting Antenna

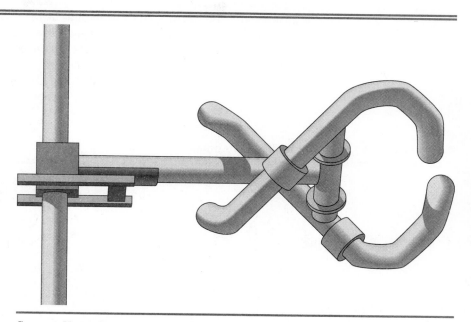

Courtesy Harris Corp.

12.5 ANTENNA MATCHING

Sometimes a resonant antenna is too large to be convenient. The question then arises as to what can be done to allow reasonably efficient transmission and/or reception with a shorter antenna. On other occasions, the same antenna may be required to work at several widely different frequencies, and cannot be expected to be of resonant length at all of them. At some frequencies it may be too short, at others too long.

To begin, it must be noted that there is no necessity for an antenna to be resonant for it to radiate or receive signals, which requires only that there be current flowing in the antenna. A nonresonant antenna will not be matched to its feedline, however, and the mismatch will result in a high SWR and a good deal of reflected power. The antenna will still radiate the power that is supplied to it, but it will be difficult to supply that power. In addition, for short antennas the radiation resistance will be very low, resulting in reduced efficiency.

The problem of mismatch can be rectified by matching the antenna to the feedline. Perhaps the most obvious solution is an *LC* matching network. It should be installed as close to the antenna as possible, because any feedline between the network and the antenna will have a high SWR. Transmission-line matching techniques using shorted stubs can also be used. (These were described in Chapter 10.) Whether lumped- or distributed-constant techniques are used is largely a function of frequency. At VHF and higher, the lumped-constant approach may involve impractically small component values, while at low frequencies transmission-line techniques may require inconveniently long stubs.

The antenna-matching network is separate from the transmitter output, or receiver input, circuit. The latter matches the transmitter or receiver to the transmission line, and the former matches the line to the antenna.

With the proper matching network, it is possible to use any random length of wire as an antenna. The directional properties of such an antenna will not be the same as those of a half-wave dipole, however. As a rough rule of thumb, the longer the wire, the more **lobes**, and the closer the radiation

FIGURE 12.19　　　　**Long-Wire Antennas**

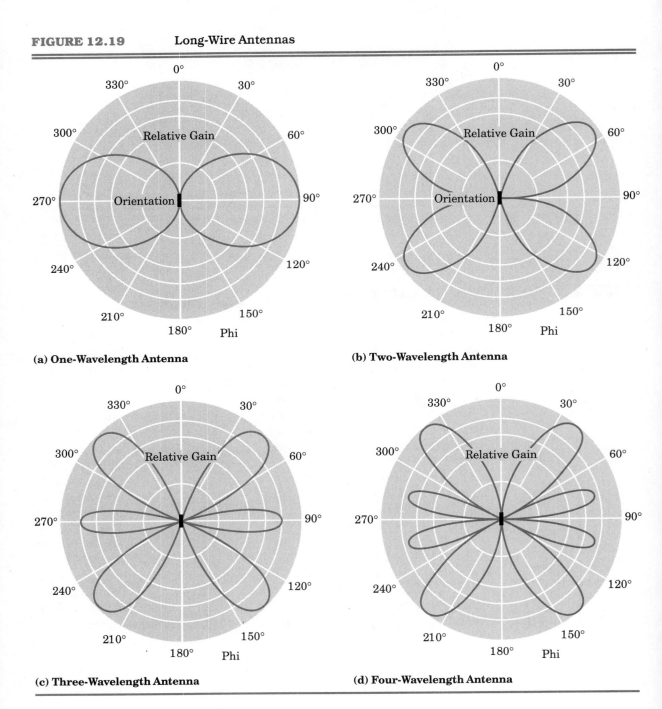

(a) One-Wavelength Antenna

(b) Two-Wavelength Antenna

(c) Three-Wavelength Antenna

(d) Four-Wavelength Antenna

maximum is to the axis of the antenna. Figure 12.19 shows the patterns for a few selected *long-wire* antennas, in the plane of the antenna.

12.5.1 Inductive and Capacitive Loading

One simple but effective technique to match a short antenna to a feedline is to increase its electrical length. For instance, a mobile whip antenna that is less than one-quarter wavelength can have an inductance added at its base, as shown in Figure 12.20. This inductance, called a **loading** coil, cancels the capacitive effect of the too-short antenna, and, when carefully adjusted, can result in an antenna that looks electrically like a quarter-wave monopole. There are two drawbacks to this solution: the coil will raise the Q of the antenna and narrow its bandwidth; and the coil resistance will increase the losses.

The problem of coil losses is made more serious by the fact that the coil is installed at the point in the antenna where the current is maximum. In addition, short antennas have very low radiation resistance, so they require high antenna currents which lead to significant losses.

The I^2R losses in the coil could be reduced by moving it away from the feedpoint, as shown in Figure 12.21. Unfortunately, this also reduces the effectiveness of the coil, requiring a greater inductance. Nonetheless this center-loading technique is often seen with mobile antennas.

Another way to employ inductive loading is to construct the whole antenna in the form of a coil, or helix. The rubber-ducky antennas often found on handheld transceivers are generally of this type. Figure 12.22 shows an example.

It is also possible to increase the antenna's electrical length by adding capacitance to the end away from the feedpoint. This takes the form of a "hat" consisting of a metallic disk or a collection of rods or wires, as shown in Figure 12.23. Though not suitable for mobile antennas because it would cause increased wind resistance, this technique can be seen on fixed antennas (for instance, those used in AM broadcasting).

FIGURE 12.20

Loaded Whip Antenna

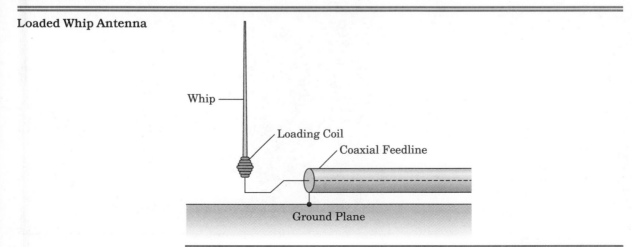

FIGURE 12.21

Center-Loaded Whip Antenna

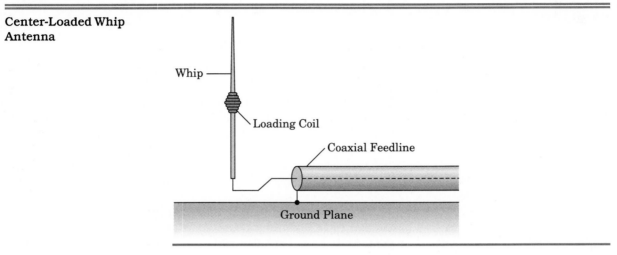

FIGURE 12.22

Commercial Rubber-Duckie Antenna for Portable Transceiver

Icom America Inc.

FIGURE 12.23

Capacitive Loading

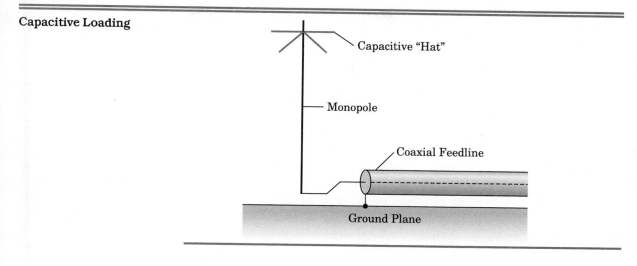

12.6 ANTENNA ARRAYS

It is possible to combine the simple elements described above to build a more elaborate antenna. The radiation from the individual **elements** will combine, resulting in reinforcement in some directions and cancellation in others. This can give greater gain and better directional characteristics. For instance, it is often desirable to have high gain in only one direction, something that is not possible with the simple antennas previously described.

Arrays can be classified as *broadside* or *end-fire*, according to their direction of maximum radiation. If the maximum radiation is along the main axis of the antenna (which may or may not coincide with the axis of its individual elements), the antenna is an end-fire array. If the maximum radiation is at right angles to this axis, the array has a broadside configuration. Figure 12.24 shows the axes of each.

Another way to classify antenna arrays is according to the way in which the elements are connected. A *phased array* has all its elements connected to the feedline. There may be phase-shifting, power-splitting, and impedance-matching arrangements for individual elements, but all receive power from the feedline (assuming a transmitting antenna). Since the transmitter can be

FIGURE 12.24

Broadside and End-Fire Arrays. Arrows Represent Direction(s) of Maximum Radiation.

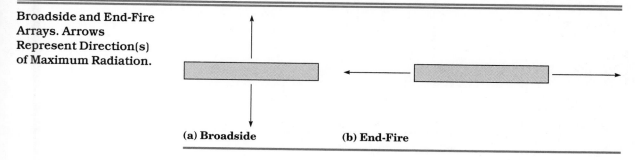

(a) Broadside   (b) End-Fire

said to *drive* each element by supplying power, these are also called *driven arrays*. On the other hand, in some arrays, only one element is connected to the feedline. The others work by absorbing and re-radiating power radiated from the driven element. These are called *parasitic elements*, and the antennas are known as *parasitic arrays*.

The following sections will describe several common types of arrays, and give some general characteristics of each. Detailed analysis of antenna arrays is usually done with the aid of a computer.

12.6.1 The Yagi Array

The Yagi array shown in Figure 12.25 is a parasitic end-fire array. It has one driven element, one *reflector* behind the driven element, and one or more *directors* in front of the driven element. The driven element is a half-wave dipole or folded dipole. The reflector is slightly longer than one-half wavelength, and the directors are slightly shorter. The spacing between elements varies, but is typically about 0.2 wavelength. The Yagi antenna is more formally referred to as the *Yagi-Uda array*.

The Yagi antenna is unidirectional, with a single **main lobe** in the direction shown in Figure 12.25, as well as several **minor lobes**. The antenna pattern for a typical Yagi with eight elements—one driven, one reflector, and six directors—is shown in Figure 12.26. Yagis are often constructed with five or six directors for a gain of about 10 dBi, but higher gains, up to about 16 dBi, can be achieved by using more directors.

The antenna pattern shown in Figure 12.26 has a beamwidth for the main lobe, at the 3 dB down points, of approximately 40°. In addition, it has four lobes out to the sides of the pattern, called, appropriately enough, **sidelobes**, and a lobe to the back of the pattern as well. In addition to their gain and beamwidth, antennas such as this are characterized by their **front-to-back ratio**, which is the ratio (in decibels) between the power density in the direction of maximum radiation, and that radiated in a direction 180° away from it. In the figure, this ratio is approximately 11 dB.

FIGURE 12.25

Yagi Array

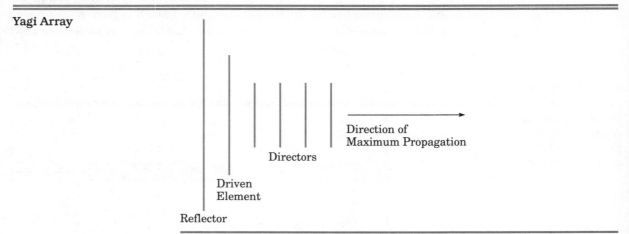

Direction of
Maximum Propagation

Directors

Driven
Element

Reflector

FIGURE 12.26

Radiation Pattern for Eight-Element Yagi

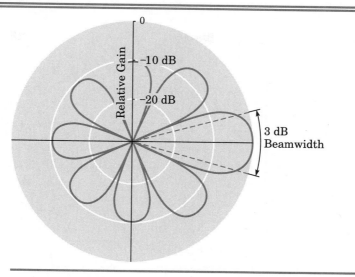

The Yagi is a relatively narrow-band antenna. When optimized for gain, its usable bandwidth is only about two percent of the operating frequency. Wider bandwidth can be obtained by varying the length of the directors, making them shorter as the distance from the driven element increases. This is necessary, for instance, when the Yagi is used for television reception. In fact, it is common to build two or three Yagis on one support (called a *boom*), with the elements interspersed, for the low-VHF, high-VHF, and/or UHF television bands. A folded dipole is generally used for the driven element in a TV antenna, because its bandwidth is wider than that of an ordinary dipole.

12.6.2 The Log-Periodic Dipole Array

The log-periodic antenna derives its name from the fact that the feedpoint impedance is a periodic function of the operating frequency. Although log-periodic antennas take many forms, perhaps the simplest and most common is the dipole array, illustrated in Figure 12.27. The log-periodic dipole array (LPDA) is probably the most common antenna for television reception.

The elements are dipoles, with the longest at least one-half wavelength in length at the lowest operating frequency, and the shortest less than one-half wavelength at the highest. The ratio between the highest and lowest frequencies can be 10:1 or more. A balanced feedline is connected to the narrow end, and power is fed to the other dipoles via a network of crossed connections as shown. The operation is quite complex, with the dipoles that are closest to resonance at the operating frequency doing most of the radiation. The gain is reasonable, typically about 8 dBi, but not as good as that of a well-designed Yagi with the same number of elements.

The design of a log-periodic antenna is based on several equations. A parameter τ is chosen, with a value that must be less than 1 and is typically

FIGURE 12.27

Log-Periodic Dipole Array

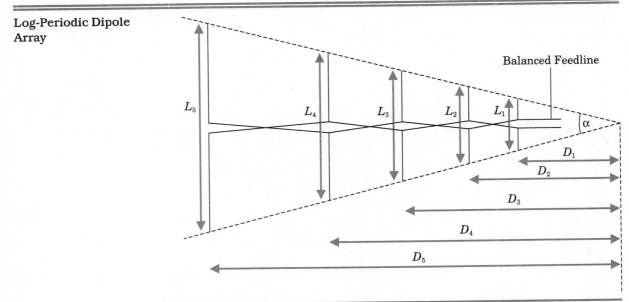

between 0.7 and 0.9. A value toward the larger end of the range gives an antenna with better performance but more elements. This is the ratio between the lengths and spacing of adjacent elements. That is,

$$\tau = \frac{L_1}{L_2} = \frac{L_2}{L_3} = \frac{L_3}{L_4} = \cdots \qquad (12.10)$$

where $L_1, L_2, \ldots$ = lengths of the elements, in order from shortest to longest

and

$$\tau = \frac{D_1}{D_2} = \frac{D_2}{D_3} = \frac{D_3}{D_4} = \cdots \qquad (12.11)$$

where $D_1, D_2, \ldots$ = spacings between the elements and the apex of the angle between them, in order from shortest to longest

This angle is usually designated α, and is typically about 30°. From simple trigonometry it can be shown that

$$\frac{L_1}{2D_1} = \tan \frac{\alpha}{2} \qquad (12.12)$$

The following example shows how an LPDA can be designed. Computer programs are available to automate the procedure.

EXAMPLE 12.5

Design a log-periodic antenna to cover the frequency range from 100 to 300 MHz. Use $\tau = 0.7$ and $\alpha = 30°$.

Solution

In order to get good performance across the frequency range of interest, it is advisable to design the antenna for a slightly wider bandwidth. For the longest element, we can use a half-wave dipole cut for 90 MHz, and for the shortest, one designed for 320 MHz.

From Equation (12.3),

$$L = \frac{142.5}{f}$$

For a frequency of 90 MHz,

$$L = \frac{142.5}{90}$$
$$= 1.58 \text{ m}$$

For 320 MHz,

$$L = \frac{142.5}{320}$$
$$= 0.445 \text{ m}$$

Because of the way the antenna is designed, it is unlikely that elements of both these two exact lengths will be present, but we can start with the shorter one and simply make sure that the longest element has at least the length calculated above.

Starting with the first element and using Equation (12.12),

$$\frac{L_1}{2D_1} = \tan \frac{\alpha}{2}$$

We have specified $L_1 = 0.445$ m and $\alpha = 30°$, so

$$D_1 = \frac{L_1}{2 \tan \dfrac{\alpha}{2}}$$

$$= \frac{0.445}{2 \tan 15°}$$

$$= 0.830 \text{ m}$$

From Equation (12.10),

$$\tau = \frac{L_1}{L_2} = \frac{L_2}{L_3} = \frac{L_3}{L_4} = \cdots$$

τ and L_1 are known, so L_2 can be calculated.

$$L_2 = \frac{L_1}{\tau}$$

$$= \frac{0.445}{0.7}$$

$$= 0.636 \text{ m}$$

Continue this process until an element length is obtained that is greater than 1.58 m.

$$L_3 = \frac{L_2}{\tau}$$

$$= \frac{0.636}{0.7}$$

$$= 0.909 \text{ m}$$

Similarly, $L_4 = 1.30$ m and $L_5 = 1.85$ m. This is longer than necessary, so the antenna will need five elements.

The spacing between the elements can be found from Equation (12.11):

$$\tau = \frac{D_1}{D_2} = \frac{D_2}{D_3} = \frac{D_3}{D_4} = \cdots$$

Since τ and D_1 are known, it is easy to find D_2, and then the rest of the spacings, in the same way as the lengths were found above. We get $D_2 = 1.19$ m, $D_3 = 1.69$ m, $D_4 = 2.42$ m, and $D_5 = 3.46$ m.

12.6.3 The Turnstile Array

Antenna arrays are not always designed to give directionality and gain. The turnstile array illustrated in Figure 12.28(a) is a simple combination of two dipoles designed to give omnidirectional performance in the horizontal plane, with horizontal polarization. The dipoles are fed 90° out of phase.

The gain of a turnstile antenna is actually about 3 dB less than that of a single dipole in its direction of maximum radiation, because each of the elements of the turnstile receives only one-half the transmitter power. Figure 12.28(b) shows the radiation pattern for a typical turnstile antenna.

Turnstile antennas are often used for FM broadcast reception, where they give reasonable performance in all directions without the need for a rotor. Variations of the turnstile are also used for television and FM broadcasting.

12.6.4 The Monopole Phased Array

A single vertical quarter-wave monopole antenna has an omnidirectional radiation pattern in the horizontal plane. Often, however, the AM broadcast

FIGURE 12.28   Turnstile Antenna

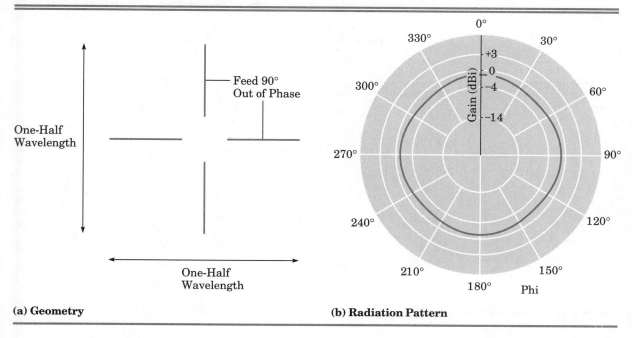

(a) Geometry   **(b) Radiation Pattern**

stations that use this type of antenna must have a directional pattern, to avoid interference with other stations. In fact, it is common for a station to be required to use two different radiation patterns, one for day and one for night. (Interference is more likely at night because of ionospheric propagation.)

Two or more monopole antennas can be arranged in an array. One possibility, using only two antennas, is shown in Figure 12.29. Both elements

FIGURE 12.29

Monopole Array

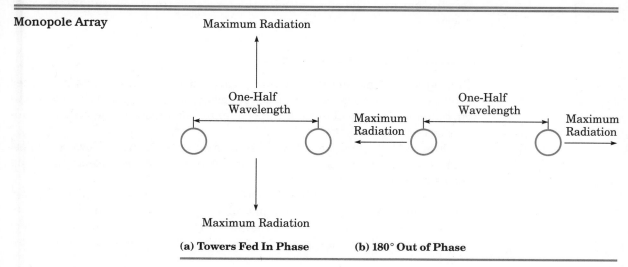

(a) Towers Fed In Phase   **(b) 180° Out of Phase**

FIGURE 12.30   **Radiation Patterns for Monopole Arrays**

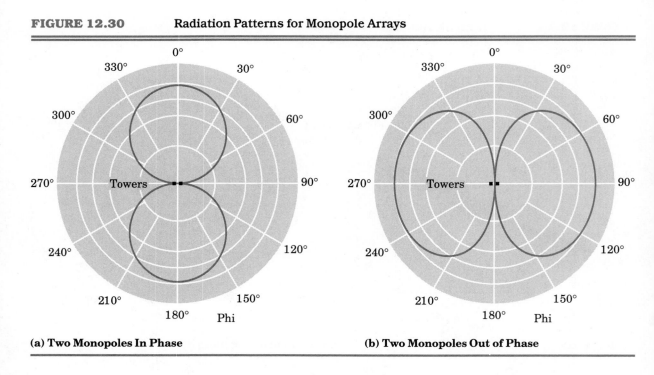

(a) Two Monopoles In Phase   **(b) Two Monopoles Out of Phase**

are driven, and the radiation pattern depends on the phase angle between the elements. Two possibilities are illustrated in the figure. When the antennas are fed in phase, and are one-half wavelength apart as shown, the radiation will be at a maximum broadside to a line drawn between the two towers. On the other hand, if the antennas are fed 180° out of phase, the radiation will cancel broadside to the array, and will be at a maximum off the ends. Figure 12.30 shows the patterns. A larger number of possible radiation patterns can be created by varying the number, spacing, and phase angle of the elements. In addition, the current distribution among the elements will affect the pattern. The element with the largest current will radiate the most power.

12.6.5 Other Phased Arrays

Phased arrays can be made by connecting any of the simple antenna types already discussed. Depending on the geometry of the array and the phase and current relationships between the elements, the array can be either broadside or end-fire.

Figure 12.31 shows one type of broadside array using half-wave dipoles. This is called a *collinear array*, because the axes of the elements are all along the same line.

Suppose that the collinear antenna in the figure is used for transmitting, and imagine a receiving antenna placed along the main axis of the antenna. None of the individual elements radiates any energy in this direction, so of course there will be no signal from the array in this direction, either. Now

FIGURE 12.31

Collinear Array (All
Elements One-Half
Wavelength, Fed In
Phase)

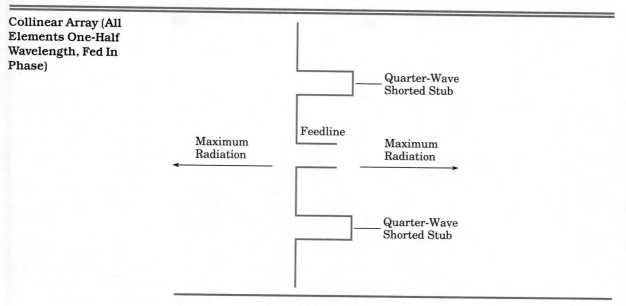

move the hypothetical receiving antenna to a point straight out to one side of
the antenna. All of the individual elements will radiate a signal in this direction, but it is necessary to determine whether these signals add construc-
tively or destructively. If the elements are in phase, the signals will add. A
quick check shows that this is indeed the case. The half-wave dipoles are
linked by quarter-wave transmission-line sections that provide a phase re-
versal between adjacent ends. Therefore, all the dipoles will be in phase.

Collinear antennas are often mounted with the main axis vertical. They
will then be omnidirectional in the horizontal plane, but will have a narrow
angle of radiation in the vertical plane. Thus, they make good base-station
antennas for mobile radio systems.

Another broadside array using dipoles is shown in Figure 12.32. This
time, the elements are not collinear, but they are still in phase. That may not

FIGURE 12.32

Broadside Array (Half-
Wave Dipoles Separated
by One-Half Wavelength;
Maximum Radiation is
into and Out of the
Paper)

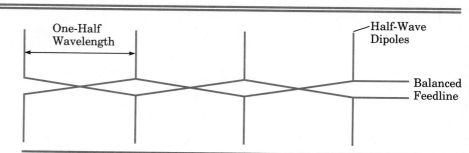

be obvious because of the crossed lines connecting them, but notice that the separation between the elements is one-half wavelength. This would cause a 180° phase shift as the signal travels along the feedline from one element to the next. This phase shift is cancelled by the crossing of the transmission-line section that joins the dipoles.

Although this antenna, like the previous one, is a broadside array, its pattern is not identical. There is no radiation off the end of any of the elements, so of course there is no radiation in the equivalent direction from the array. If this antenna were erected with its main axis vertical, it would *not* be omnidirectional in the horizontal plane. In addition to having a narrow vertical pattern, it would have a bidirectional pattern in the horizontal plane. As for radiation off the end of the antenna, each dipole provides a signal in this direction, but the signals from adjacent elements cancel due to the one-half wavelength difference in path lengths between adjacent elements. Figure 12.33 is a comparison of the radiation patterns for the two types of antennas in the horizontal plane.

The helical dipoles described above for FM broadcast use are often combined in a broadside array. Figure 12.34 is an example of such an antenna. Gains of several decibels in the horizontal plane can be achieved in this way.

Figure 12.35 shows an end-fire array using dipoles. It is identical to the broadside array in Figure 12.32 except for the fact that the feedline between elements is not crossed this time. This causes alternate elements to be 180° out of phase. Therefore, the radiation from one element will cancel that from the next in the broadside direction. Off the end of the antenna, however, the radiation from all the elements will be additive, since the 180° phase shift

FIGURE 12.33

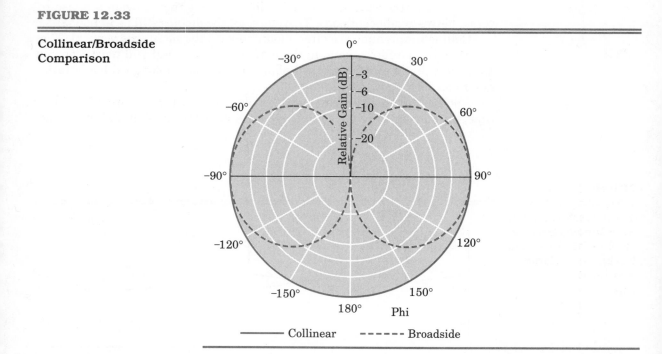

Collinear/Broadside
Comparison

FIGURE 12.34

Commercial FM
Broadcast Array

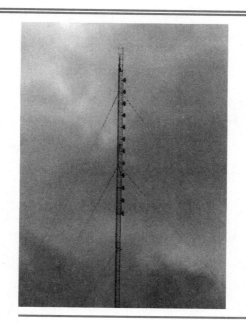

between adjacent elements is neatly cancelled by the one-half wavelength physical separation between them.

Phased arrays are not restricted to combinations of simple dipoles. For example, Figure 12.36 shows a pair of "stacked" Yagis; that is, the Yagis are mounted one above the other and fed in phase. This gives better directivity than a single Yagi. Ideally, the gain will be 3 dB greater than for a single Yagi of the same type. Of course, it is also possible to combine larger numbers of Yagis, and similar arrays can be constructed from other types of antennas, such as log-periodic arrays or helical antennas.

FIGURE 12.35

End-Fire Array (Half-
Wave Dipoles Separated
by One-Half Wavelength)

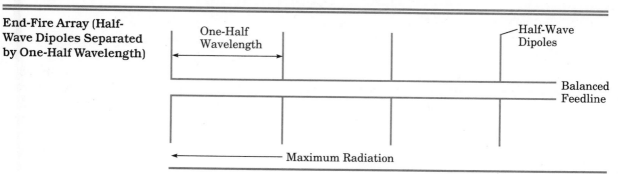

FIGURE 12.36

Stacked Yagis

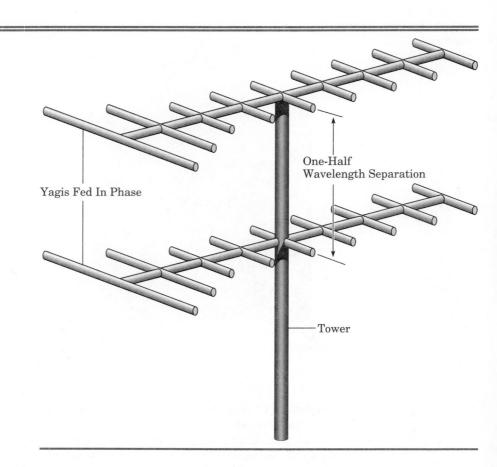

One-Half
Wavelength Separation

Yagis Fed In Phase

Tower

12.7 REFLECTING SURFACES

In addition to the reflecting element found in a Yagi antenna, it is possible to construct a conductive surface that will reflect antenna power in the desired direction. The surface may consist of one or more planes, or may be parabolic in shape.

12.7.1 Plane and Corner Reflectors

A plane reflector acts in a similar way to an ordinary mirror. Like a mirror, its effects can be predicted by supposing that there is an "image" of the antenna on the opposite side of the reflecting surface at the same distance from it as the source. Reflection changes the phase angle of a signal by 180°. Whether the image antenna's signal aids or opposes the signal from the real antenna depends on the spacing between the antenna and the reflector, and the location of the receiver. In Figure 12.37, the antenna is one-quarter wavelength from the reflector, and the signals aid in the direction shown. The reflected signal experiences a 180° phase shift on reflection, and another 180° shift because it must travel an additional one-half wavelength to reach the

FIGURE 12.37

Plane Reflector and
Image

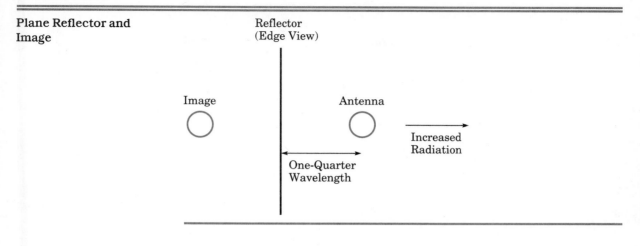

receiver. The magnitude of the electric field in the direction shown is thus
increased by a factor of two. The power density in this direction is increased
by a factor of four, or 6 dB, because power is proportional to the square of
voltage.

It is possible to use a plane reflector with almost any antenna. For example, a reflector can be placed behind a collinear antenna, as shown in Figure 12.38. The antenna becomes directional in both the horizontal and vertical planes. Base antennas for cellular radio systems are often of this type.

FIGURE 12.38

Collinear Array with
Plane Reflector (All
Elements One-Half
Wavelength, Fed In
Phase)

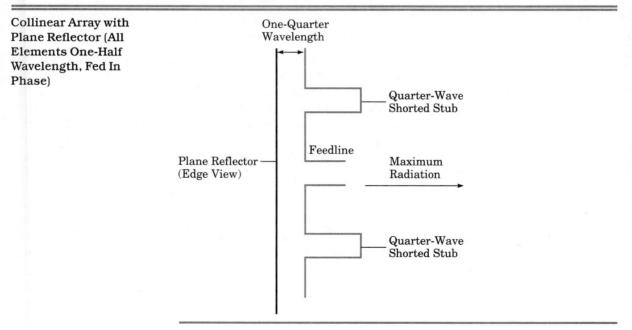

FIGURE 12.39

Corner Reflector and Images

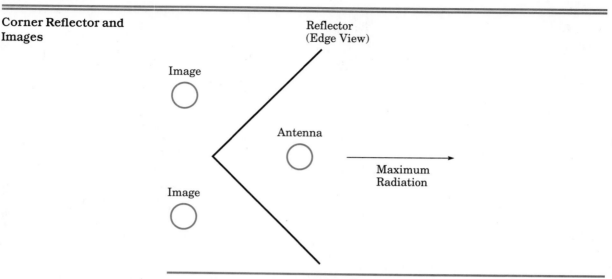

The plane does not have to be solid. In fact, it is often made of wire mesh to reduce wind resistance. It can also be made as a series of metal rods or tubes, oriented in the same way as the antenna. Reflectors of this type will work well provided the separation between the components of the reflector is much less than a wavelength at the operating frequency.

The corner reflector creates two images, as shown in Figure 12.39, for a somewhat sharper pattern. Corner reflectors are often combined with Yagi arrays in UHF television antennas.

12.7.2 The Parabolic Reflector

Parabolic reflectors have the useful property that any ray that originates at a point called the *focus* and strikes the reflecting surface will be reflected parallel to the axis of the parabola; that is, a *collimated* beam of radiation will be produced. The parabolic "dish" antenna, familiar from backyard satellite-receiver installations, consists of a small antenna at the focus of a large parabolic reflector, which focuses the signal in the same way as the reflector of a searchlight focuses a light beam. Figure 12.40 shows a typical example. Of course, the antenna is reciprocal: radiation entering the dish along its axis will be focused by the reflector.

Ideally, the antenna at the feedpoint should illuminate the entire surface of the dish with the same intensity of radiation, and should not spill any radiation off the edges of the dish or in other directions. If that were the case, the gain and beamwidth of the antenna could easily be calculated. The equation for beamwidth is

$$\theta = \frac{70\lambda}{D}$$

(12.13)

where   θ = beamwidth in degrees at the 3 dB points
          λ = free-space wavelength
          D = diameter of the dish

The radiation pattern for a typical parabolic antenna is shown in Figure 12.41. The width of the beam measured between the first nulls is approximately twice the 3 dB beamwidth.

For gain, the equation is

$$G = \frac{\pi^2 D^2}{\lambda^2} \qquad\qquad (12.14)$$

where G = gain as a power ratio (not in decibels)
          D = diameter of the dish
          λ = free-space wavelength

The effect of uneven illumination of the antenna, of losses, or of any radiation spilling off at the edges, is to reduce the gain. To include these effects in gain calculations, it is necessary to include a constant *eff*, which is known as the efficiency of the antenna. This constant can theoretically have a value

FIGURE 12.40

Parabolic Antenna

Courtesy Andrew Corporation.

FIGURE 12.41

Polar Pattern for Typical
Parabolic Antenna

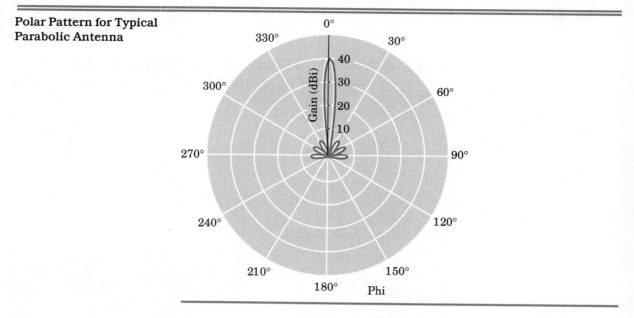

between zero and one, but is between 0.5 and 0.7 for a typical antenna. When this constant is included, Equation (12.14) becomes

$$G = \frac{eff\pi^2 D^2}{\lambda^2}$$

(12.15)

where G = gain as a power ratio (not in decibels) with respect to an
           isotropic radiator
       D = diameter of the dish
       λ = free-space wavelength
       eff = efficiency

EXAMPLE 12.6

A parabolic antenna has a diameter of 3 m, an efficiency of 60%, and operates at a frequency of 4 GHz. Calculate its gain and beamwidth.

Solution

The free-space wavelength is

$$\lambda = \frac{c}{f}$$

$$= \frac{3 \times 10^8}{4 \times 10^9}$$

$$= 0.075 \text{ m}$$

Substituting this into Equation (12.13), the beamwidth is

$$\theta = \frac{70\lambda}{D}$$
$$= \frac{70 \times 0.075}{3}$$
$$= 1.75°$$

The gain is given by Equation (12.15):

$$G = \frac{eff\pi^2 D^2}{\lambda^2}$$
$$= \frac{0.6\pi^2 \times 3^2}{0.075^2}$$
$$= 9475$$

In decibels (relative to an isotropic radiator), the gain is

$$G = 10 \log 9475$$
$$= 39.8 \text{ dBi}$$

Any type of antenna can be used with a parabolic reflector. In the microwave portion of the spectrum, where parabolic reflectors are most useful because they can have a practical size, a *horn antenna* provides a simple and efficient method to feed power to the antenna. Because horn antennas are essentially an extension of a waveguide, their operation will be described in the next chapter.

Besides the simple horn feed shown in Figure 12.40, there are several ways to get power to the parabolic reflector. For example, the Gregorian feed shown in Figure 12.42(a) uses a feedhorn in the center of the dish itself, which radiates to a reflector at the focus of the antenna. This reflects the signal to the main parabolic reflector. By removing the feedhorn from the focus, this system allows any waveguide or electronics associated with the feedpoint to be placed in a more convenient location. The strange-looking antenna shown in Figure 12.42(b) is a combination of horn and parabolic antennas called a *hog-horn;* it is often used for terrestrial microwave links.

12.8 TEST EQUIPMENT: THE ANECHOIC CHAMBER

Measurement of antenna gain and directivity requires the ability to set up an antenna in a location that is free from reflecting surfaces and large enough that a receiving antenna can be located in the far field of the antenna under test. This requires a spacing of at least several wavelengths. It is also

FIGURE 12.42

Parabolic Antenna
Variations

(a) Gregorian Feed

(b) Hog-Horn Antenna

Courtesy Andrew Corporation.

desirable to electrically shield the whole test setup from the outside world to avoid interference from nearby radio transmitters.

These requirements can be difficult or impossible to meet when tests of large, low-frequency antennas are required. Sometimes an open field called an *antenna range* is used. For higher frequencies, however, it is possible to build an enclosed space called an **anechoic chamber** that will meet all the requirements listed above.

The required size of an anechoic chamber is dictated by the physical size of the antennas to be tested and the frequency of operation. The size of the chamber can be reduced by scaling down the antenna and scaling up the frequency, as long as the effects of ground do not have to be considered. Two representative samples of anechoic chambers are shown in Figure 12.43. The one in Figure 12.43(a) is a large commercial installation, while that in Figure 12.43(b) was built as a student project at Niagara College in Welland, Ontario. Though small, the student-built chamber gives quite satisfactory results, at least for demonstration purposes, at frequencies of 1 GHz and above.

An anechoic chamber must be lined with a material that absorbs radio waves. Figure 12.44 shows the material used: a foam plastic that is impregnated with carbon. This will absorb any radio waves originating within the chamber. If these waves were reflected from the walls, they could cause false readings. The absorbent material has the additional advantage of preventing signals originating outside the chamber from reaching the receiving antenna inside.

Inside the chamber, there must be provision for mounting and rotating the antenna under test and connecting it to a signal source, which can be a laboratory RF generator. There must also be a receiving antenna, which can be a simple arrangement such as a dipole, connected to a receiver and a readout device such as a chart recorder.

It is easy to obtain a plot of relative gain by simply rotating the antenna under test while observing the output from the receiving antenna. Actual measurements of gain with respect to an isotropic antenna or a dipole are a little more difficult. Of course, an isotropic antenna is impossible to build, but a half-wave dipole is quite simple. The signal strength from the dipole can be measured and compared with that from the test antenna. The only difficulty is to make sure that both dipole and test antenna have the same impedance (using matching devices, if necessary) so that each will absorb the same amount of power from the generator. If a matching network is used, the loss in the network must also be taken into account.

Figure 12.45 shows a Yagi antenna used for tests, and Figure 12.46 shows a plot of its output made in the anechoic chamber of Figure 12.43(b). Notice that this plot is not on polar graph paper, because it was made with a conventional x–y plotter. The x axis represents antenna position in degrees, and the y axis shows output power on a logarithmic scale. This plot can easily be transferred to polar graph paper if desired.

A method similar to the one just described can be used to measure the performance of an antenna system in real-world conditions. The antenna under test is used as a transmitting antenna, and driven with a source of known power output. A device called a field-strength meter is then used to measure the field strength in a number of directions at a distance from the antenna that is great enough to ensure that the far-field performance is being mea-

FIGURE 12.43   Anechoic Chambers

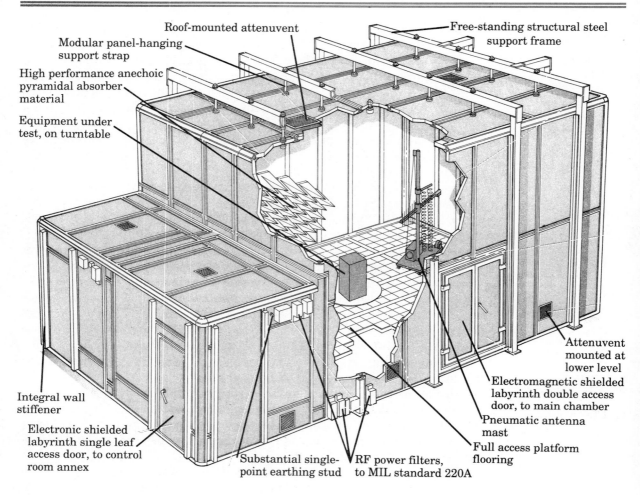

Roof-mounted attenuvent

Free-standing structural steel support frame

Modular panel-hanging support strap

High performance anechoic pyramidal absorber material

Equipment under test, on turntable

Integral wall stiffener

Electronic shielded labyrinth single leaf access door, to control room annex

Substantial single-point earthing stud

RF power filters, to MIL standard 220A

Attenuvent mounted at lower level

Electromagnetic shielded labyrinth double access door, to main chamber

Pneumatic antenna mast

Full access platform flooring

(a) Commercial

(b) Student-Built

(a) Courtesy Rainford Corporation.

FIGURE 12.44

Radiation-Absorbing
Material

FIGURE 12.45

Yagi Test Antenna

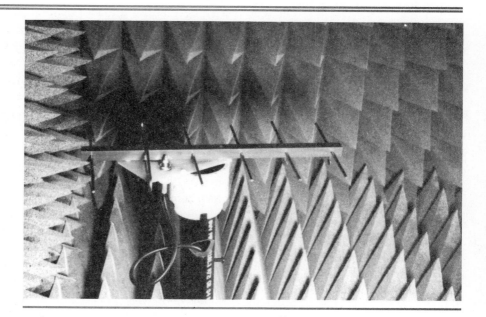

sured. A field-strength meter is basically a calibrated antenna-receiver combination. This method is especially useful at lower frequencies where the
wavelength is large and the effect of ground is important. It is also the only
practical way to measure the performance of an actual installation, where the
effect of obstacles such as buildings on field strength may be important.

FIGURE 12.46

Plot from Antenna in
Figure 12.45

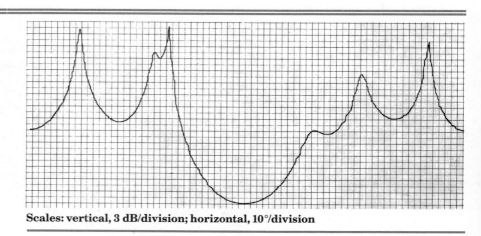

Scales: vertical, 3 dB/division; horizontal, 10°/division

Here are the main points to remember from this chapter.

1. Antennas are reciprocal passive devices that couple electrical energy between transmission lines and free space.

2. The isotropic radiator is convenient for calculations because it emits all the energy supplied to it equally in all directions. Antenna gain is usually calculated with reference to an isotropic radiator.

3. Gain is obtained in an antenna by emitting more power in certain directions than in others.

4. The beamwidth of a directional antenna is the angle between the half-power points of the main lobe.

5. The half-wave dipole is a simple practical antenna with a bidirectional pattern, a small gain over an isotropic radiator, and a feedpoint impedance of about 70 Ω. The dipole is sometimes used instead of an isotropic antenna as a reference for antenna gain.

6. The radiation resistance of an antenna is the equivalent resistance that appears at its feedpoint due to the radiation of energy into space.

7. The polarization of most simple antennas is the same as the axis of the antenna.

8. Antennas must often be matched to the feedline in order to achieve efficient transfer of power. Means for doing this include lumped-constant matching networks, transmission-line sections, and loading coils.

9. Simple antennas can be used as elements of arrays, in order to obtain specified values of gain and directivity.

10. Plane and parabolic reflectors can be used to increase the gain and directivity of antennas.

IMPORTANT EQUATIONS

$$\lambda = \frac{c}{f} \tag{12.1}$$

$$L = \frac{142.5}{f} \tag{12.3}$$

$$eff = \frac{R_r}{R_T} \tag{12.5}$$

$$P_x = P_T(eff) \tag{12.6}$$

$$\text{EIRP} = P_T G_T \tag{12.7}$$

$$A_{eff} = \frac{\lambda^2 G_R}{4\pi} \tag{12.8}$$

$$\tau = \frac{L_1}{L_2} = \frac{L_2}{L_3} = \frac{L_3}{L_4} = \cdots \tag{12.10}$$

$$\tau = \frac{D_1}{D_2} = \frac{D_2}{D_3} = \frac{D_3}{D_4} = \cdots \tag{12.11}$$

$$\frac{L_1}{2D_1} = \tan\frac{\alpha}{2} \tag{12.12}$$

$$\theta = \frac{70\lambda}{D} \tag{12.13}$$

$$G = \frac{eff\pi^2 D^2}{\lambda^2} \tag{12.15}$$

GLOSSARY

active antenna   a receiving antenna with a built-in preamplifier

anechoic chamber   an enclosure lined with material which absorbs electromagnetic radiation

antenna   a device to radiate or receive electromagnetic radiation at radio frequencies

array   an antenna system composed of two or more simpler antenna elements

beamwidth   the angle between the points on the major lobe of an antenna at which the radiated power density is one-half its maximum value

dBd   a measure of antenna gain: decibels with respect to a lossless half-wave dipole

dBi   a measure of antenna gain: decibels with respect to an ideal isotropic radiator

dipole   any antenna consisting of a single conductor with zero current only at its two ends

directivity   the ratio of the maximum to the average radiation intensity for an antenna

effective area   for a receiving antenna, the ratio of the available output power to the power density of the received wave

effective isotropic radiated power (EIRP)   product of the power supplied to a transmitting antenna and the gain of the antenna with respect to an isotropic radiator

effective radiated power (ERP)   product of the power supplied to a transmitting antenna and the gain of the antenna with respect to a lossless half-wave dipole

element   in an antenna array, an individual conductor or group of conductors

far-field region   a distance far enough from an antenna that local inductive and capacitive effects are insignificant

front-to-back ratio   the ratio between the radiation intensity in an antenna's direction of maximum radiation, and the intensity at an angle of 180° to this direction

ground plane   an artificial ground consisting of a conducting surface or an equivalent (such as wire mesh or a group of wires) at the base of a vertical antenna

helix   a spiral

isotropic radiator   a hypothetical antenna that would radiate all the energy supplied to it, with equal intensity in all directions

loading   the process of increasing the electrical length of an antenna by the addition of inductance or capacitance

lobe   the portion of an antenna pattern between two nulls

main lobe   the lobe in the direction of maximum radiation

minor lobe   a lobe with less intensity than the main lobe

monopole   an antenna with a current null at one end and a maximum at the other, with no other nulls in between

near-field region   the region close to an antenna, where local inductive and capacitive effects predominate

null   a place where some quantity (such as radiation or current) is zero

phi (ϕ)   in an antenna pattern, the Greek letter phi denotes the angle in the horizontal plane, from the x axis toward the y axis

polarization   the direction of the electric field vector of an electromagnetic wave

radial   In a monopole antenna, a wire extending along the surface of the ground or just below it, away from the antenna (A set of radials is used to improve the effective conductivity of the ground.)

radiation resistance   equivalent resistance at the feedpoint corresponding to the radiation of energy by an antenna

sidelobe   a minor lobe at an angle of approximately 90° to the main lobe

theta (θ)   in an antenna pattern, the Greek letter theta refers to the angle from the horizontal (x–y) plane toward the zenith, represented by the z axis

zenith the direction straight up from the horizontal plane

QUESTIONS

1. Explain how an antenna can have a property called *gain*, even though it is a passive device.

2. Sketch the radiation patterns of an isotropic antenna and of a half-wave dipole in free space, and explain why a dipole has gain over an isotropic radiator.

3. What is the significance of radiation resistance, and what would be the effect on the antenna's efficiency of reducing the radiation resistance, all other things being equal?

4. Suggest two reasons for using a directional antenna.

5. Distinguish between the near and far field of an antenna. Why is it necessary to use the far field for all antenna measurements?

6. Draw three types of impedance-matching circuit that can be used to match an antenna to a transmission line.

7. Why is the effect of the ground significant for MF and HF antennas, but not usually for UHF and microwave antennas?

8. Monopole antennas have better performance if they are installed on damp ground such as swampland. Sometimes they are even installed in shallow water just off the shore of a lake or other body of water. Explain.

9. How can the effect of damp ground be simulated if it is not available, for a monopole antenna?

10. How does a ground plane differ from an actual ground connection, and why is it used?

11. How can an antenna's electrical length be increased without increasing its physical length?

12. Why are small loop antennas seldom used for transmitting?

13. What advantages does a five-eighths wavelength antenna have over a quarter-wave antenna for mobile use? Does it have any disadvantages?

14. Name two types of antennas that are noted for having particularly wide bandwidth. Which of these is directional?

15. Name one type of antenna that produces circularly polarized waves. Under what circumstances is circular polarization desirable?

16. Distinguish between end-fire and broadside arrays. Name and sketch one example of each.

17. Distinguish between parasitic and phased arrays. Name and sketch one example of each. What is another term for a phased array?

18. What is the purpose of using a plane reflector with an antenna array?

19. What is meant by a collinear antenna? Sketch such an antenna and its radiation pattern, and classify it as broadside or end-fire, phased or parasitic.

20. Describe how an anechoic chamber could be used to measure the front-to-back ratio for an antenna.

PROBLEMS

Section 12.2

21. Calculate the length of a practical half-wave dipole for a frequency of 15 MHz.

22. Calculate the efficiency of a dipole with a radiation resistance of 68 Ω and a total feedpoint resistance of 75 Ω.

Section 12.3

23. Given that a half-wave dipole has a gain of 2.14 dBi, calculate the electric field strength at a distance of 10 km in free space in the direction of maximum radiation, from a half-wave dipole that is fed, by means of lossless, matched line, by a 15 W transmitter.

24. Sketch the radiation pattern, in the vertical plane, for a half-wave dipole at each of the following heights above perfectly conducting ground:

    (a) one-quarter wavelength
    (b) one-half wavelength
    (c) three-quarter wavelength
    (d) one wavelength

25. Refer to the plot in Figure 12.47, and find the gain and beamwidth for the antenna shown.

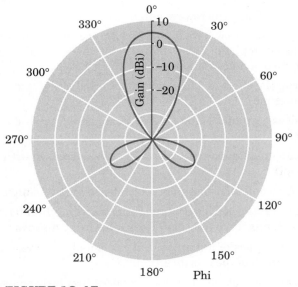

FIGURE 12.47

26. Calculate the EIRP in dBW for a 25 W transmitter operating into a dipole with 90% efficiency.

Section 12.4

27. Calculate the length of a quarter-wave monopole antenna for a frequency of 1000 kHz.

28. Calculate the optimum length of an automobile FM broadcast antenna, for operation at 100 MHz.

29. The loop antenna shown in Figure 12.48 produces its maximum output when

oriented in the direction shown, when in each of the two locations shown. Locate the transmitting station.

Ferrite Loopstick

Ferrite Loopstick

Ferrite Loopstick

FIGURE 12.48

30. Draw a dimensioned sketch of a discone antenna that will cover the VHF range from 30 to 300 MHz.

Section 12.5

31. **(a)** How could the antenna of Problem 28 be made to work on the 27 MHz CB band?
   (b) Would the modified antenna of Part (a) be as efficient as a full-size quarter-wave CB antenna? Explain.

Section 12.6

32. Characterize each of the arrays sketched in Figure 12.49 as phased or parasitic, and as broadside or end-fire. Indicate the direction(s) of maximum output for each antenna.

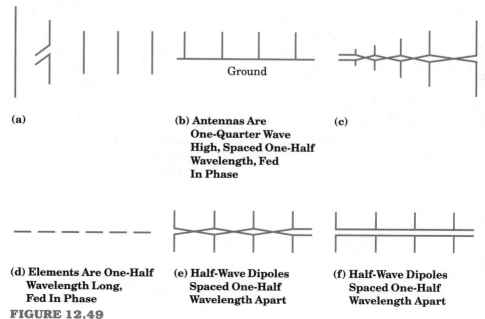

(a)

(b) Antennas Are One-Quarter Wave High, Spaced One-Half Wavelength, Fed In Phase

(c)

(d) Elements Are One-Half Wavelength Long, Fed In Phase

(e) Half-Wave Dipoles Spaced One-Half Wavelength Apart

(f) Half-Wave Dipoles Spaced One-Half Wavelength Apart

FIGURE 12.49

33. Design a log-periodic antenna to cover the low-VHF television band, from 54 to 88 MHz. Use $\alpha = 30°$, $\tau = 0.7$, $L_1 = 1.5$ m.

34. A monopole array consists of four towers spaced at one-half-wavelength intervals along a north–south line. Sketch the radiation pattern when the towers are fed:

    (a) in phase
    (b) with adjacent towers 180° out of phase

35. Two horizontal Yagis are stacked with one of them one-half wavelength above the other and fed in phase. What will be the effect of the stacking on the radiation

    (a) in the horizontal plane?
    (b) in the vertical plane?

Section 12.7

36. Calculate the minimum diameter for a parabolic antenna with a beamwidth of 2° at a frequency of:

    (a) 4 GHz   **(b)** 12 GHz

37. Calculate the gain of each of the antennas in the previous problem, assuming an efficiency of 65%.

38. Calculate the effective area of a 3 m dish, with an efficiency of 0.7, at 3 GHz and at 12 GHz. Explain your result.

Section 12.8

39. An anechoic chamber is 5 m in length. How far is this in wavelengths at 1 GHz?

40. Redraw the x–y plot in Figure 12.46 on polar graph paper. Then calculate:

    (a) the front-to-back ratio
    (b) the beamwidth
    (c) the strength of the minor lobes compared to the main lobe

Comprehensive

41. A Yagi antenna has a gain of 10 dBi and a front-to-back ratio of 15 dB. It is located 15 km from a transmitter with an ERP of 100 kW at a frequency of 10 MHz. The antenna is connected to a receiver via a matched feedline with a loss of 2 dB. Calculate the signal power supplied to the receiver if the antenna is:

    (a) pointed directly toward the transmitting antenna
    (b) pointed directly away from the transmitting antenna

42. A lossless half-wave dipole is located in a region with a field strength of 150 μV/m. Calculate the power that this antenna can deliver to a receiver if the frequency is:

    (a) 10 MHz   **(b)** 500 MHz

43. Calculate the gain, with respect to an isotropic antenna, of a half-wave dipole, with efficiency of 90%, that is placed one-quarter wavelength from a plane reflector that reflects 100% of the signal striking it.

13 Microwave Devices

13.1 INTRODUCTION

There is no sharp distinction between **microwaves** and other radio-frequency signals. Conventionally, the lower boundary for microwave frequencies is set at 1 GHz. We will examine microwaves separately because many conventional techniques for generating, amplifying, and transmitting signals become less effective as the frequency increases, while other techniques, impractical at lower frequencies, become more useful.

As the frequency increases, many of the simplifying assumptions that work at lower frequencies become less accurate. Here are a few examples.

At low frequencies, the inductance and capacitance of component leads can be ignored. At microwave frequencies, even short connecting leads have significant capacitive and inductive reactance, and the physical design of components must change.

At frequencies up to about the UHF range, the time taken for charge carriers to move through such devices as diodes, transistors, and vacuum tubes can usually be ignored. As the period of signals becomes shorter, however, this *transit time* becomes a significant fraction of a complete cycle. Some conventional components have been redesigned to minimize transit time, and other active devices have been specially designed to incorporate transit-time effects in their operation.

Because of the short wavelengths of microwave signals, antennas of reasonable physical size can have very high gain, and parabolic reflectors become practical.

At microwave frequencies, the losses in conventional transmission lines are quite large. Waveguides, which will be introduced in the next section, have much lower losses, but are impractical at lower frequencies due to their large size.

13.2 WAVEGUIDES

As discussed in Chapter 10, both dielectric and conductor losses in conventional transmission lines increase with frequency. **Waveguides** provide an alternative at microwave frequencies. A waveguide is essentially a pipe through which an electromagnetic wave travels. As it travels along the guide, it reflects from the walls. Figure 13.1 shows the general idea of waveguides and waveguide propagation. Rectangular waveguides of brass or aluminum, sometimes silver-plated on the inside, are most common, but elliptical and circular cross sections are also used.

It is possible to build a waveguide for any frequency, but waveguides operate essentially as high-pass filters; that is, for a given waveguide cross section, there is a cutoff frequency below which waves will not propagate. At frequencies below the gigahertz range, waveguides are too large to be practical for most applications.

As the electric and magnetic fields are completely contained within the guide, waveguides have no radiation loss. Dielectric losses are very small, since the dielectric is usually air. There are some losses in the conductive walls of the waveguide, but due to the large surface area of the walls, they are much smaller than the losses in coaxial or open-wire line.

FIGURE 13.1

Waveguides

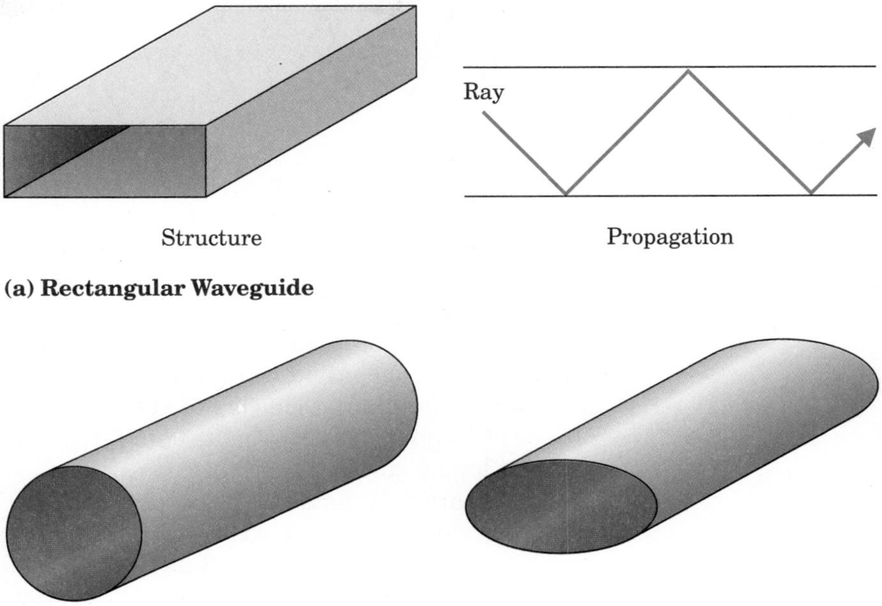

Structure   Propagation

(a) Rectangular Waveguide

(b) Circular Waveguide   **(c) Elliptical Waveguide**

13.2.1 Modes and Cutoff Frequency

There are a number of ways (called **modes**) in which electrical energy can propagate along a waveguide. All of these modes must satisfy certain boundary conditions. For instance, assuming an ideal conductor for the guide, there cannot be any electric field along the wall of the waveguide. If there were such a field, there would have to be a voltage gradient along the wall. That would be impossible since there cannot be any voltage across a short circuit.

It may help to understand modes by thinking of a wave moving through the guide as if it were a ray of light. Figure 13.2 shows the idea. For each different mode, the ray will strike the walls of the waveguide at a different angle. As the angle a ray makes with the wall of the guide becomes larger, the ray must travel a greater distance to reach the far end of the guide. Though the propagation in the guide is at the speed of light, the greater distance travelled causes the effective velocity down the guide to be reduced.

It is desirable to have only one mode propagating in a waveguide. To see the effect of more than one mode propagating at a time (called *multimode propagation*), consider a brief pulse of microwave energy applied to one end of a waveguide. The pulse will arrive at the far end at several different times, one for each mode. Thus a brief pulse will be spread out over time, becoming longer. If another pulse follows close behind, there may be interference between the two.

The effect just described is called **dispersion**. Dispersion limits the usefulness of waveguides with pulsed signals and other types of modulation. For this reason, it is undesirable to have more than one mode propagating.

FIGURE 13.2

Multimode Propagation

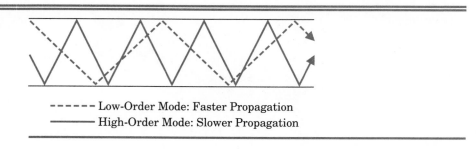

------- Low-Order Mode: Faster Propagation
———— High-Order Mode: Slower Propagation

Each mode has a cutoff frequency below which it will not propagate. Single-mode propagation can be achieved by using only the mode with the lowest cutoff frequency. This is called the *dominant mode*, and the waveguide is used at frequencies between its cutoff frequency and that of the mode with the next lowest cutoff frequency.

Modes are designated as transverse electric (TE) or transverse magnetic (TM) according to the pattern of electric and magnetic fields within the waveguide. Recall from Chapter 11 that electromagnetic waves in free space are known as transverse electromagnetic (TEM) waves, because both the electric and magnetic fields are perpendicular to the direction of travel. When these waves travel diagonally along a waveguide, reflecting from wall to wall, only one component—either the electric or magnetic field—can remain transverse to the direction of travel. The term "TE" means that there is no component of the electric field along the length of the guide.

Figure 13.3 shows several examples of TE modes in a rectangular waveguide. The electric-field strength is represented by the arrows, with the length of the arrows proportional to the field strength. Note that in all cases

FIGURE 13.3

TE Modes in
Rectangular Waveguide
(Arrows Represent
Electric-Field Strength)

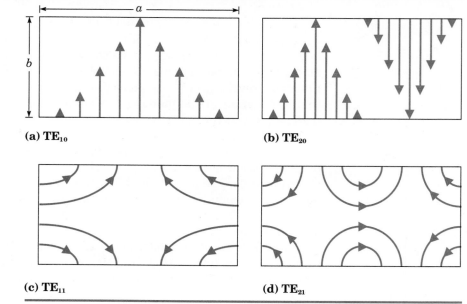

(a) TE$_{10}$ **(b) TE$_{20}$**

(c) TE$_{11}$ **(d) TE$_{21}$**

the field strength is zero along the walls of the guide. It is possible to have an electric field at the guide surface, but only if the field is perpendicular to the surface.

The field strength varies sinusoidally across the guide cross section. The numbers that follow the TE designation represent first, the number of half-cycles of the wave along the long dimension (a) of the rectangular guide, and second, the number of variations along the short dimension (b). Of course, a and b refer to inside dimensions.

In a rectangular waveguide, TE_{10} is the dominant mode (that is, the mode with the lowest cutoff frequency). In a typical rectangular waveguide with $a = 2b$, the TE_{01} and TE_{20} modes each have a cutoff frequency twice that of the TE_{10} mode, giving an approximate 2:1 frequency range for the waveguide in its dominant mode. In what follows, a rectangular waveguide operating in the TE_{10} mode will be assumed unless otherwise stated.

The cutoff frequency for the TE_{10} mode can easily be found. For this mode to propagate, there must be at least one-half wavelength along the wall. Therefore, at the cutoff frequency,

$$a = \frac{\lambda_c}{2} \tag{13.1}$$

where a = longer dimension of the waveguide cross section
λ_c = cutoff wavelength in the dielectric material that fills the waveguide (usually air)

Assuming that the waveguide has an air dielectric, the cutoff frequency can be found as follows. From Equation (13.1),

$$a = \frac{\lambda_c}{2}$$
$$\lambda_c = 2a$$

From earlier work we know that, for propagation in free space (and as a close approximation for an air dielectric),

$$\lambda = \frac{c}{f}$$

where λ = free-space wavelength in meters
$c = 3 \times 10^8$ meters per second
f = frequency in hertz

Therefore, at the cutoff frequency f_c,

$$\lambda_c = \frac{c}{f_c}$$
$$2a = \frac{c}{f_c}$$
$$f_c = \frac{c}{2a} \tag{13.2}$$

EXAMPLE 13.1

Find the cutoff frequency for the TE_{10} mode in an air-dielectric waveguide with an inside cross section of 2 cm by 4 cm. Over what frequency range is the dominant mode the only one that will propagate?

Solution

The larger dimension, 4 cm, is the one to use in calculating the cutoff frequency. From Equation (13.2), the cutoff frequency is

$$f_c = \frac{c}{2a}$$

$$= \frac{3 \times 10^8 \text{ m/s}}{2 \times 4 \times 10^{-2} \text{ m}}$$

$$= 3.75 \times 10^9 \text{ Hz}$$

$$= 3.75 \text{ GHz}$$

The dominant mode is the only mode of propagation over a 2:1 frequency range, so the waveguide will be usable to a maximum frequency of $3.75 \times 2 = 7.5$ GHz.

The dominant mode depends on the shape of the waveguide. Figure 13.4 shows some of the modes in a circular guide. Here, the same boundary conditions result in different modes because of the changed geometry. For a circular guide, the dominant mode is the TE_{11} mode, but the TM_{01} mode is also used, because it has circular symmetry which allows its use in rotating joints. These are necessary with rotating radar antennas. See Figure 13.5 for an example.

FIGURE 13.4

Modes in Circular
Waveguide (Arrows
Represent Electric-Field
Strength)

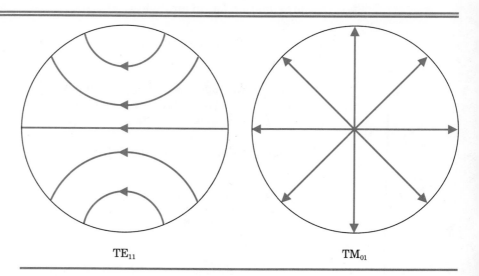

FIGURE 13.5

Rotating Waveguide
Joint

13.2.2 Group and Phase Velocity

Assuming an air dielectric, the wave travels inside the waveguide at the speed of light. However, it does not travel straight down the guide, but reflects back and forth from the walls. The actual speed of travel of a signal down the guide is called the **group velocity**, and is considerably less than the speed of light. The group velocity in a rectangular waveguide is given by the equation

$$v_g = c\sqrt{1 - \left(\frac{\lambda}{2a}\right)^2} \tag{13.3}$$

where v_g = group velocity
  λ = free-space wavelength
  a = larger dimension of the interior cross section

There is an equivalent form of this equation which is more useful in specific situations. Since, from Equation (13.2),

$$f_c = \frac{c}{2a}$$

the wavelength and dimension in Equation (13.3) can be replaced by two frequencies, the operating frequency and the cutoff frequency. This results in the equation

$$v_g = c\sqrt{1 - \left(\frac{f_c}{f}\right)^2} \tag{13.4}$$

FIGURE 13.6

Variation of Group
Velocity with Frequency

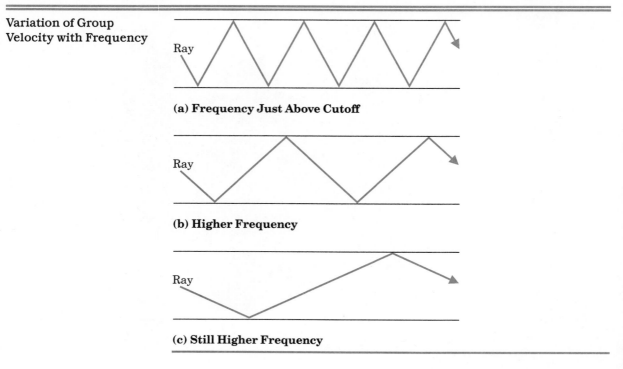

(a) Frequency Just Above Cutoff

(b) Higher Frequency

(c) Still Higher Frequency

where f_c = cutoff frequency
       f = operating frequency

Generally, manufacturers provide the cutoff frequency for waveguides, so this is a useful form of the equation. The two equivalent radicals in Equations (13.3) and (13.4) appear often in waveguide calculations, and are worth remembering.

From the above equation, it can be seen that the group velocity is a function of frequency, becoming zero at the cutoff frequency. At frequencies below cutoff, of course, there is no propagation, so the equation does not apply. The physical explanation of the variation of group velocity is that the angle the wave makes with the wall of the guide varies with frequency. At frequencies near the cutoff value, the wave moves back and forth across the guide more often while travelling a given distance down the guide than it does at higher frequencies. Figure 13.6 gives a qualitative idea of the effect.

EXAMPLE 13.2

Find the group velocity for the waveguide of Example 13.1, at a frequency of 5 GHz.

Solution

From Example 13.1, the cutoff frequency is 3.75 GHz. Therefore, from Equation (13.4), the group velocity is

$$v_g = c \sqrt{1 - \left(\frac{f_c}{f}\right)^2}$$

$$= (3 \times 10^8 \text{ m/s}) \sqrt{1 - \left(\frac{3.75}{5}\right)^2}$$

$$= 1.98 \times 10^8 \text{ m/s}$$

When we first looked at waveguide propagation, we noted that waveguides are generally used with only one mode, in order to reduce dispersion (that is, the tendency of signals to spread in time). Since the group velocity varies with frequency, it now appears that some dispersion exists even for single-mode propagation. If two signals of different frequencies start to travel down a guide at the same instant, the higher-frequency signal will arrive at the other end first. This can be a real problem for pulsed and other wideband signals. For instance, the upper sideband of an FM or AM signal travels faster than the lower sideband.

EXAMPLE 13.3

A waveguide has a cutoff frequency for the dominant mode of 10 GHz. Two signals with frequencies of 12 and 17 GHz respectively propagate down a 50 m length of the guide. Calculate the group velocity for each and the difference in arrival time for the two.

Solution

The group velocities can be calculated from Equation (13.4).

$$v_g = c \sqrt{1 - \left(\frac{f_c}{f}\right)^2}$$

For the 12 GHz signal,

$$v_g = (3 \times 10^8 \text{ m/s}) \sqrt{1 - \left(\frac{10}{12}\right)^2}$$

$$= 165.8 \times 10^6 \text{ m/s}$$

Similarly, the 17 GHz signal has $v_g = 242.6 \times 10^6$ m/s. The difference in speeds is 76.8×10^6 m/s. Over a distance of 50 m this will cause a time difference of

$$t = \frac{d}{v}$$

$$= \frac{50 \text{ m}}{76.8 \times 10^6 \text{ m/s}}$$

$$= 651 \times 10^{-9} \text{ s}$$

$$= 651 \text{ ns}$$

It is often necessary to calculate the wavelength of a signal in a wave-guide. It may be required for impedance matching, for instance. It might seem that the wavelength along the guide could be found using the group velocity, in much the same way that the velocity factor of a transmission line is used. However, this common-sense approach does not work. The reason is that what is really important when doing impedance-matching calculations is the change in phase angle along the line. Figure 13.7 shows how the angle varies along the guide. The guide wavelength shown represents 360° of phase variation. The guide wavelength is always larger than the free-space wave-length. Interestingly, the more slowly the waves in the guide propagate along it, the more quickly the phase angle varies along the guide.

It is possible to define a quantity called **phase velocity** to describe the variation of phase along the wall of the guide. Phase velocity is the rate at which the wave *appears* to move along the wall of the guide, based on the way the phase angle varies along the walls. Surprisingly, the phase velocity in a waveguide is always greater than the speed of light. Of course, the laws of physics prevent anything from actually moving that fast (except in science-fiction stories). A phase velocity greater than the speed of light is pos-sible because phase velocity is not really the velocity of anything. Phase ve-locity is used when calculating the wavelength in a guide, but the group velocity must be used to determine the length of time it takes for a signal to move from one end to the other.

A similar effect can be seen with water waves at a beach. If the waves approach the shore at an angle, as in Figure 13.8, the crest of the wave will appear to run along the shore at a faster rate than that at which the waves approach the beach. Once again, there is nothing physically moving at the faster velocity.

The relationship between phase velocity and group velocity is very simple. The speed of light is the geometric mean of the two, that is,

$$v_g v_p = c^2 \qquad\qquad (13.5)$$

where v_p = phase velocity

FIGURE 13.7

Variation of Phase Angle Along a Waveguide

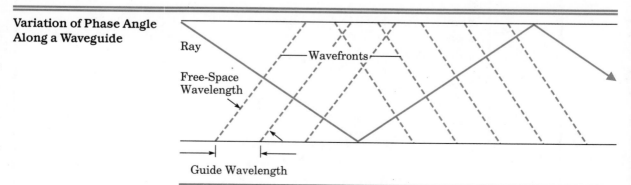

FIGURE 13.8

Water Waves at a
Shoreline

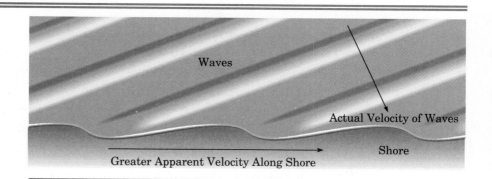

It is easy to modify the equations given earlier for group velocity, so that
the phase velocity can be calculated without first finding the group velocity.
The derivation is left as an exercise; the results are given below.

$$v_p = \frac{c}{\sqrt{1 - \left(\dfrac{\lambda}{2a}\right)^2}} \tag{13.6}$$

$$= \frac{c}{\sqrt{1 - \left(\dfrac{f_c}{f}\right)^2}} \tag{13.7}$$

EXAMPLE 13.4

Find the phase velocity for the waveguide used in Examples 13.1 and 13.2,
at a frequency of 5 GHz.

Solution

The answer can be found directly from Equation (13.7).

$$v_p = \frac{c}{\sqrt{1 - \left(\dfrac{f_c}{f}\right)^2}}$$

$$= \frac{3 \times 10^8 \text{ m/s}}{\sqrt{1 - \left(\dfrac{3.75}{5}\right)^2}}$$

$$= 4.54 \times 10^8 \text{ m/s}$$

Alternatively, it can be found from Equation (13.5), since the group velocity is already known.

$$v_g v_p = c^2$$

$$v_p = \frac{c^2}{v_g}$$

$$= \frac{(3 \times 10^8 \text{ m/s})^2}{1.98 \times 10^8 \text{ m/s}}$$

$$= 4.54 \times 10^8 \text{ m/s}$$

13.2.3 Impedance

Like any transmission line, a waveguide has a characteristic impedance. Unlike wire lines, however, the waveguide impedance is a function of frequency. As has been pointed out, the impedance of free space is 377 Ω. It might be expected that the impedance of a waveguide with an air dielectric would have some relationship to this value, and that is true. The actual impedance Z_0 of a waveguide is given by

$$Z_0 = \frac{377}{\sqrt{1 - \left(\dfrac{\lambda}{2a}\right)^2}} \ \Omega \tag{13.8}$$

or

$$Z_0 = \frac{377}{\sqrt{1 - \left(\dfrac{f_c}{f}\right)^2}} \ \Omega \tag{13.9}$$

EXAMPLE 13.5

Find the characteristic impedance of the waveguide used in the previous examples, at a frequency of 5 GHz.

Solution

From Equation (13.9),

$$Z_0 = \frac{377}{\sqrt{1 - \left(\dfrac{f_c}{f}\right)^2}} \ \Omega$$

$$= \frac{377}{\sqrt{1 - \left(\dfrac{3.75}{5}\right)^2}} \ \Omega$$

$$= 570 \ \Omega$$

13.2.4 Impedance Matching with Waveguides

Smith® charts, and the computer programs that emulate them, can be used with waveguides in much the same way as they can with conventional transmission lines. There are only a few differences. There are two different velocities in a waveguide, and both change with frequency. For calculating wavelength in the guide, the phase velocity must be used. To calculate the wavelength in the guide, we can start with the basic wave equation

$$v = f\lambda$$

Here, the velocity in question is the phase velocity v_p, and the wavelength is the guide wavelength λ_g. Rearranging the equation allows λ_g to be found:

$$\lambda_g = \frac{v_p}{f} \tag{13.10}$$

Another way to find the guide wavelength, given the free-space wavelength, is to use the equivalent equations

$$\lambda_g = \frac{\lambda}{\sqrt{1 - \left(\dfrac{\lambda}{2a}\right)^2}} \tag{13.11}$$

and

$$\lambda_g = \frac{\lambda}{\sqrt{1 - \left(\dfrac{f_c}{f}\right)^2}} \tag{13.12}$$

where λ = free-space wavelength
  λ_g = guide wavelength

EXAMPLE 13.6

Find the guide wavelength for the waveguide used in the previous examples.

Solution

Since the phase velocity is already known, the easiest way to solve this problem is to use Equation (13.10).

$$\lambda_g = \frac{v_p}{f}$$

$$= \frac{4.54 \times 10^8 \text{ m/s}}{5 \times 10^9 \text{ Hz}}$$

$$= 0.0908 \text{ m}$$

$$= 9.08 \text{ cm}$$

FIGURE 13.9

Tuning Screw

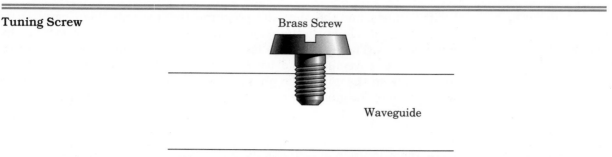

Techniques for matching impedances using waveguides also differ from those used with conventional transmission lines. Shorted stubs of adjustable length can be used, but a simpler method is to add capacitance or inductance by inserting materials into the guide. In practice, a **tuning screw** is used, as shown in Figure 13.9. As the screw is inserted further into the guide, the effect is first capacitive, then series-resonant, and finally inductive.

13.2.5 Coupling Power into and Out of Waveguides

Now that the propagation of energy down a waveguide has been described, it is time to consider how power can be put into and taken out of the guide. This is not as simple as it is with conventional transmission lines. Any attempt to connect wires to the guide will merely result in a short circuit.

There are three basic ways to launch a wave down a guide. Figure 13.10 shows all three. In Figure 13.10(a), there is a probe, resembling a quarter-wave monopole antenna. The probe will couple to the electric field in the guide, and therefore should be located at an electric-field maximum. For the TE_{10} mode, it should be in the center of the a (wide) dimension. The probe will launch a wave along the guide in both directions. Assuming that propagation in only one direction is desired, it is only necessary to place the probe a quarter-wavelength from the closed end of the guide. This closed end represents a short circuit, and following the same logic as for transmission lines, there will be an electric-field maximum at a distance of one-quarter wavelength from the end of the guide. This is the waveguide equivalent of the voltage maximum one-quarter wavelength from the shorted end of a transmission line.

The wave emitted by the probe will reflect from the shorted end of the guide. As usual, there will be a 180° phase shift at the reflecting surface, and this, combined with the 180° shift due to the total path length of one-half wavelength, will result in the reflected wave being in phase with, and adding to, the direct wave in the direction along the length of the guide. The one-quarter wavelength to the end of the guide will have to be calculated using the phase velocity described earlier, of course.

Figure 13.10(b) shows another way to couple power to a guide. A loop is used to couple with the magnetic field in the guide. It is placed in a location of maximum magnetic field, which for the TE_{10} mode occurs close to the end wall of the guide. It may help in understanding this to think of the magnetic field in a waveguide as equivalent to current in a conventional trans-

FIGURE 13.10   Coupling Power to a Waveguide

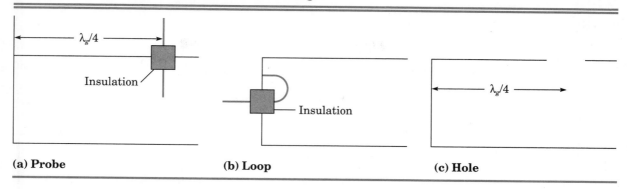

(a) Probe   **(b) Loop**   **(c) Hole**

mission line, and to consider the electric field as the equivalent of voltage. There is, of course, a current maximum at the short-circuited end of a transmission line.

A third way of coupling energy, shown in Figure 13.10(c), is simply to put a hole in the waveguide, so that electromagnetic energy can propagate into or out of the guide from the region exterior to it.

Waveguides are reciprocal devices, just like transmission lines and antennas; that is, the same means can be used to couple power into and out of a waveguide.

Variations on the above methods are also possible. For instance, it is possible to use two probes spaced one-quarter wavelength apart to launch a wave in only one direction. Figure 13.11 shows how this works. The probes are fed 90° out of phase. The waves from the two probes add in phase to the right, because the phase shift between Probe 1 and Probe 2 along the transmission line is equal to the phase shift along the waveguide, and in the same direction. The signal from Probe 1 arrives at Probe 2 in phase with the wave being generated at Probe 2. The signals cancel to the left because the phase shifts add to 180°. Similarly, if this arrangement is used to transfer power from a waveguide to a transmission line, the line will only accept energy travelling along the guide from right to left. Devices of this type are called **directional couplers**.

FIGURE 13.11

Directional Coupler Using Probes

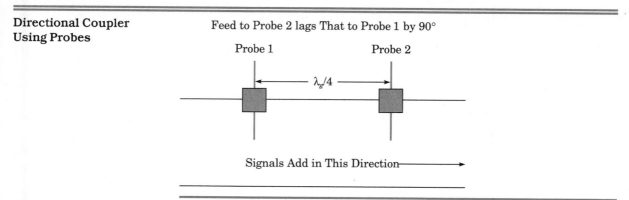

FIGURE 13.12

Two-Hole Directional
Coupler

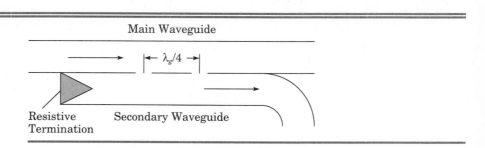

Often, holes are used to make directional couplers. In the example shown in Figure 13.12, a signal moving down the main guide in the direction of the arrow will be coupled to the secondary guide, where it will again move in the direction indicated by the arrow. A wave propagating down the main guide in the opposite direction will also be coupled to the secondary guide, but it will propagate in the opposite direction and will be absorbed by the resistive material in the end of the secondary guide. Though only two holes are shown, in practice there is often a greater number to provide more coupling between the guides.

Directional couplers are characterized by their insertion loss, coupling, and directivity. All three are normally specified in decibels. The *insertion loss* is the amount by which a signal in the main guide will be attenuated. The *coupling* specification gives the amount by which the signal in the main guide is greater than that coupled to the secondary waveguide. The *directivity* refers to the ratio between the power coupled to the secondary guide, for signals travelling in the two possible directions along the main guide.

EXAMPLE 13.7

A signal with a level of 20 dBm enters the main waveguide of a directional coupler in the direction of the arrow. The coupler has an insertion loss of 1 dB, coupling of 20 dB, and directivity of 40 dB. Find the strength of the signal emerging from each guide. Also, find the strength of the signal that would emerge from the secondary guide, if the signal in the main guide were propagating in the other direction.

Solution

The signal level in the main guide is

$$20 \text{ dBm } - 1 \text{ dB } = 19 \text{ dBm}$$

The signal level in the secondary guide is

$$20 \text{ dBm } - 20 \text{ dB } = 0 \text{ dBm}$$

If the signal direction in the main guide were reversed, the signal level in the secondary guide would be reduced by 40 dB to

$$0 \text{ dBm } - 40 \text{ dB } = -40 \text{ dBm}$$

FIGURE 13.13   **Stripline and Microstrip Lines: Cross Sections**

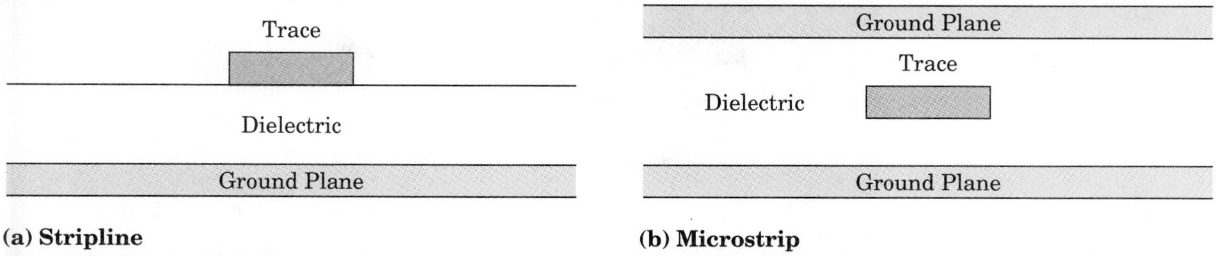

Trace		Ground Plane
Dielectric	Dielectric	Trace
Ground Plane		Ground Plane

(a) Stripline   **(b) Microstrip**

13.2.6 Striplines and Microstrips

Striplines and **microstrips** are transmission lines that can be constructed on a printed-circuit board. Figure 13.13 shows the construction of these two types of line in cross-section. They are very useful for making interconnections between components. They are included here because it is only in the UHF and microwave regions of the spectrum that circuit-board traces must be considered as transmission lines.

Striplines and microstrips, like waveguides, have a critical frequency. Below the critical frequency, they resemble conventional transmission lines. The impedance depends only on the geometry and the dielectric constant of the line. Above the critical frequency, their characteristic impedance becomes a function of frequency, as it is for a waveguide.

These lines are usually used below the critical frequency. Since the fields are not entirely enclosed within the line, radiation is a problem as the frequency increases.

13.3 PASSIVE COMPONENTS

The use of waveguides requires redesign of some of the ordinary components that are used with feedlines. The lowly tee connector is an example. In addition, there are components, such as resonant *cavities*, that are too large to be practical at lower frequencies. Several other varieties of microwave passive components will be described, as well.

13.3.1 Bends and Tees

Anything that changes the shape or size of a waveguide will have an effect on the electric and magnetic fields inside. If the disturbance is sufficiently large, there will be a change in the characteristic impedance of the guide. However, as long as any bend or twist is gradual, the effect will be minimal. Figure 13.14 shows examples of bends. The bends are designated as E-plane bends, as in Figure 13.14(a), or H-plane bends, shown in Figure 13.14(b). Since the rectangular guide shown normally operates in the TE_{10} mode, the electric-field lines are perpendicular to the long direction. Therefore, the E-plane bend is the one that changes the direction of the electric-field lines. Similar logic holds for the H-plane bend.

Rigid waveguide, with its carefully designed gradual bends, resembles

FIGURE 13.14

Waveguide Bends

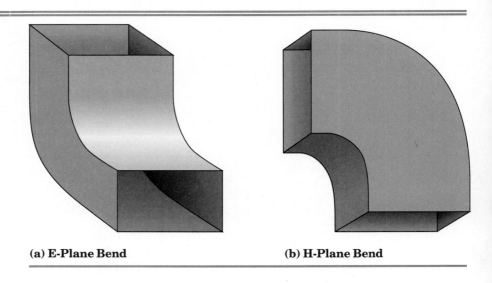

(a) E-Plane Bend   **(b) H-Plane Bend**

plumbing and is just as tricky to install. Flexible waveguide exists, however, as shown in Figure 13.15, and is used for awkward installations.

One of the more common components used with ordinary transmission line is the tee, which allows one line to branch into two. A typical tee for coaxial cable is shown in Figure 13.16. Of course, when using tees, special attention must be paid to impedance matching. Connecting a 50 Ω load to each of the branches of the tee in Figure 13.16, for example, results in a 25 Ω load being applied to the cable.

Tees can also be built for waveguides. Figure 13.17 shows E-plane and H-plane tees, named in the same way as the bends described above. A signal applied to Port 1 appears at each of the others. For the H-plane tee, the signal will be in phase at the two outputs, while the E-plane tee produces two out-of-phase signals. Sometimes the H-plane tee is referred to as a *shunt tee*, and the E-plane tee is called a *series tee*.

The **hybrid** or **magic tee** shown in Figure 13.18, is a combination of E-plane and H-plane tees. It has some interesting features. In particular, it

FIGURE 13.15

Flexible Waveguide

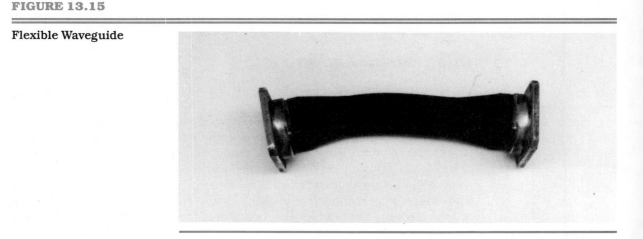

FIGURE 13.16

Coaxial Tee

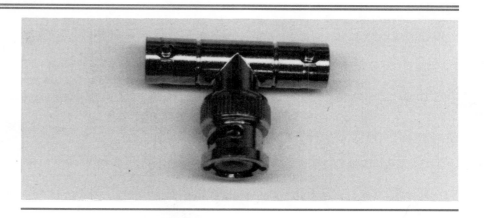

FIGURE 13.17

Waveguide Tees

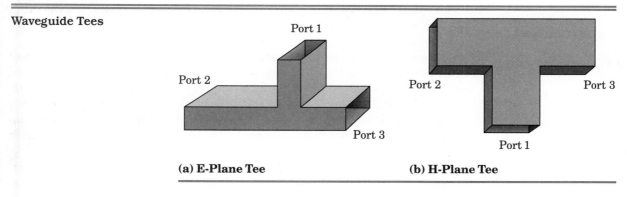

(a) E-Plane Tee   **(b) H-Plane Tee**

FIGURE 13.18

Hybrid Tee

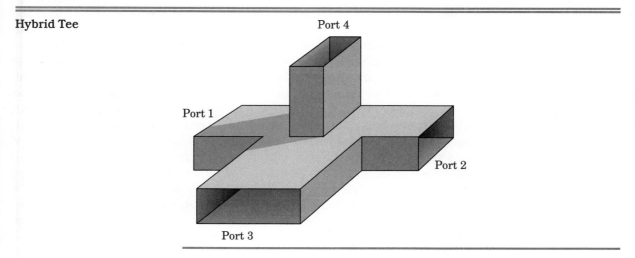

FIGURE 13.19

Receiver Front-End Using Hybrid Tee

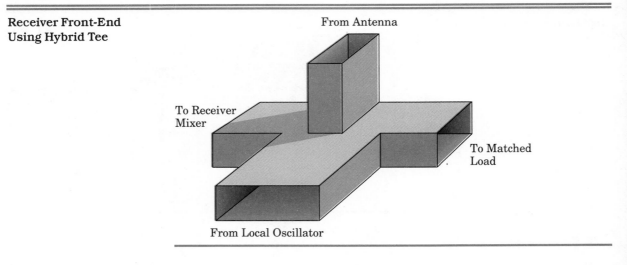

From Antenna

To Receiver Mixer

To Matched Load

From Local Oscillator

can provide isolation between signals. An input at Port 3 will result in equal and in-phase outputs at Ports 1 and 2, but no output at Port 4. On the other hand, a signal entering via Port 4 will produce equal and out-of-phase outputs at Ports 1 and 2, but no output at Port 3.

One example of the applicability of the hybrid tee is the receiver front end shown in Figure 13.19. In any receiver, it is desirable to mix the received signal and the local oscillator signal without allowing the local oscillator signal to reach the antenna. If it does, it may be radiated, causing interference to other communication services. The system of Figure 13.19 allows the incoming signal at Port 4 to be combined with the local oscillator signal that is supplied to Port 3. The combined output appears at Ports 1 and 2. One of these ports will be connected to the mixer input, and the other must be terminated with a matched load to prevent reflections. This means that the input signal suffers a 3 dB loss in the tee.

13.3.2 Cavity Resonators

In our discussion of waveguides, it was noted that the waves reflect from the walls as they proceed down the guide. Suppose that instead of using a continuous waveguide, we were to launch a series of waves in a short section of guide called a **cavity**, as illustrated in Figure 13.20. The waves would, of course, reflect back and forth from one end to the other. A cavity of random size and shape would have random reflections with a variety of phase angles and a good deal of cancellation. However, suppose that the cavity had a length of exactly one-half wavelength. Waves would reflect from one end to the other, and would be in phase with the incident signal. There would be a buildup of field strength within the cavity. In a perfectly lossless cavity, this could continue forever. Of course, losses in the walls of a real cavity would result in the signal eventually dying out unless sustained by new energy input.

This description should sound familiar: it is a description of resonance. Like any other resonant device, a waveguide cavity has a Q; the Q for resonant cavities is very high, on the order of several thousand. Cavities can be

FIGURE 13.20

Rectangular Cavity

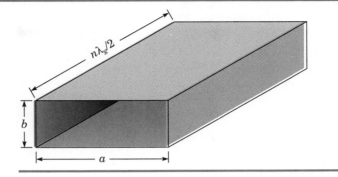

tuned by changing their size, for instance by moving a short-circuiting plate at one end.

The rectangular cavity described above is not the only possible type. Figure 13.21 shows several other types. Resonant cavities of various sorts are found in many types of microwave devices (for example, the **magnetron** and **klystron** to be described later).

One useful application of the resonant cavity is the wavemeter shown in Figure 13.22. The wavemeter is a kind of frequency meter consisting of a cavity with an adjustable plunger. The adjusting screw is attached to a fre-

FIGURE 13.21

Resonant Cavities

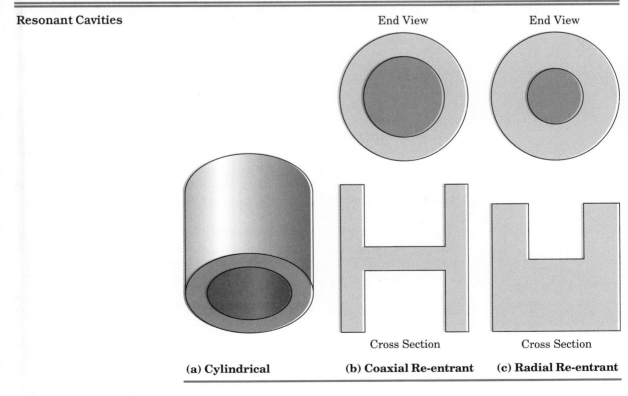

(a) Cylindrical   (b) Coaxial Re-entrant   (c) Radial Re-entrant

FIGURE 13.22

Wavemeter

quency scale. This cavity is attached to a waveguide through which the signal of unknown frequency passes from its source to a power meter. When the cavity is adjusted to resonance, it will absorb some of the power in the guide, causing a reduction in the indication of the power meter.

13.3.3 Attenuators and Loads

At lower frequencies, a load is simply a resistor, and an attenuator is a combination of resistors designed to preserve the characteristic impedance of a system while reducing the amount of power applied to a load. Once again, this lumped-constant approach becomes less practical as the frequency increases. At microwave frequencies, an ordinary carbon or metal-film resistor would have a complex equivalent circuit involving a good deal of distributed inductance and capacitance. Nonetheless, the idea of using resistive material to absorb energy is still valid.

Figure 13.23 shows waveguide versions of attenuators. The flap attenuator in Figure 13.23(a) uses a carbon flap that can be inserted to a greater or lesser extent into the waveguide. The fields inside the guide are, of course, present in the carbon as well. A current flows in the flap, causing power loss.

The attenuator shown in Figure 13.23(b) uses a rotating vane. When the vane is rotated so that the electric field is perpendicular to its surface, little loss occurs, but when the field runs along the surface of the vane, a much larger current is induced, causing greater loss. Both fixed and variable vane attenuators are available with a variety of attenuation values.

Figure 13.24 illustrates a terminating load for a waveguide. The carbon insert is designed to dissipate the energy in the guide without reflecting it.

FIGURE 13.23

Waveguide Attenuators

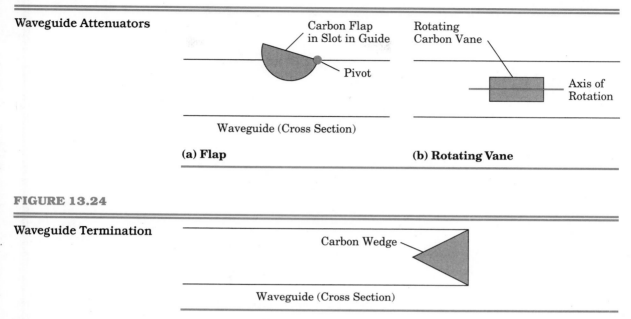

Carbon Flap
in Slot in Guide

Rotating
Carbon Vane

Pivot

Axis of
Rotation

Waveguide (Cross Section)

(a) Flap   **(b) Rotating Vane**

FIGURE 13.24

Waveguide Termination

Carbon Wedge

Waveguide (Cross Section)

13.3.4 Circulators and Isolators

Isolators and **circulators** are useful microwave components that generally make use of ferrites in their operation. The theory of their operation will be described shortly; first, a brief description of their operating characteristics is in order.

An *isolator* is a device that allows a signal to pass in only one direction. In the other direction, it is greatly attenuated. An isolator can be used to shield a source from a mismatched load. Energy will still be reflected from the load, but instead of reaching the source, the reflected power is dissipated in the isolator. Figure 13.25 illustrates this application of an isolator.

The *circulator*, shown schematically in Figure 13.26, is a very useful de-

FIGURE 13.25

Operation of Isolator

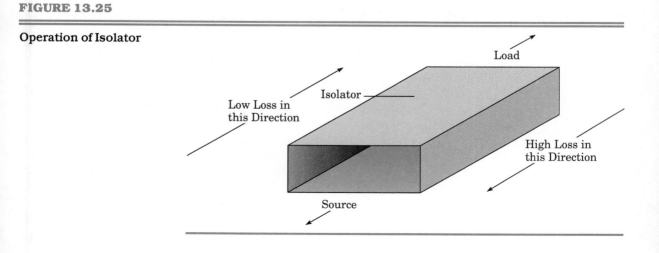

Load

Low Loss in
this Direction

Isolator

High Loss in
this Direction

Source

FIGURE 13.26

Circulator (Top View)

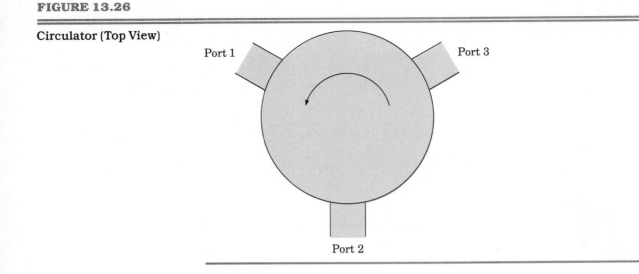

vice that allows the separation of signals. The three-port circulator of Figure 13.26 allows a signal introduced at any port to appear at, and only at, the next port in counterclockwise rotation. For instance, a signal entering at Port 1 appears at Port 2 but not at Port 3. Circulators can have any number of ports from three up, but three- and four-port versions are the most common.

One simple example of an application for a circulator is as a transmit-receive switch. Figure 13.27 shows the idea. The transmitter output is connected to Port 1, the antenna to Port 2, and the receiver input to Port 3. The transmitter output signal is applied to the antenna, and a received signal from the antenna reaches the receiver. The transmitter signal does not reach the receiver, however; if it did, it would probably burn out the receiver front end. Another feature of this setup is that any local oscillator signal that leaks out the receiver input will appear only at the transmitter port, where it will

FIGURE 13.27

Circulator Transmit-
Receive Switch

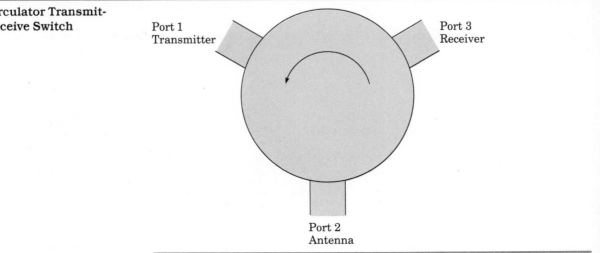

do no harm. It will not reach the antenna, where it could cause undesired radiation and possible interference.

The operation of both the isolator and the circulator is based on the magnetic properties of ferrites. These ceramic compounds of iron oxide with other metals are ferromagnetic but nonconductive. They are very common as cores for radio-frequency inductors and transformers, where they provide a permeability greater than that of air. Microwave applications of ferrites also make use of the ferromagnetic properties of the material, but in a different way.

In all materials, the electrons spin about their axes while orbiting around the nucleus of an atom. The magnetic properties of ferrites result from the fact that most of their electrons spin in the same direction.

If a ferrite material is subjected to a magnetic field from a permanent magnet, the electron spins take on a characteristic called **precession**: the axis about which an electron spins, itself begins to spin. This precession can be demonstrated with a child's top, as shown in Figure 13.28. With the top, the precession is caused by gravity rather than by a magnetic field, however. Adjusting the magnetic field can cause the precession in a ferrite to occur at a frequency in the microwave range. If an electromagnetic wave is allowed to propagate through the ferrite, and if the frequency of the wave is equal to the precession frequency, the precession can be either increased or decreased by the magnetic field associated with the signal. Which occurs depends on the relative directions of the signal and the fixed magnetic field. If the interaction is such as to increase the precession, energy will be extracted from the signal; if the interaction goes in the other direction, little or no loss will occur. This meets the requirements for an *isolator*, which can consist of a slab of ferrite, located in a waveguide and subject to a constant magnetic field from a permanent magnet, as shown in Figure 13.29.

The theory behind the ferrite *circulator* is a little more complex and will not be gone into in detail. The interaction between an electromagnetic wave

FIGURE 13.28

Precession

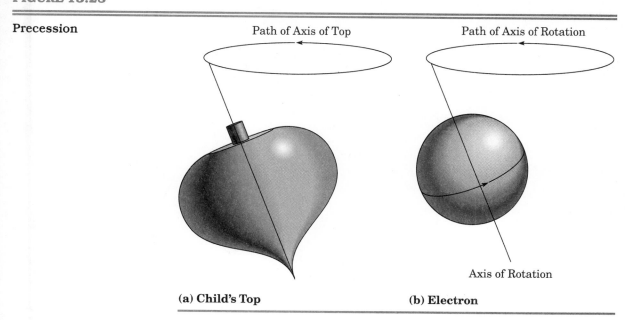

Path of Axis of Top   Path of Axis of Rotation

Axis of Rotation

(a) **Child's Top**   (b) **Electron**

FIGURE 13.29

Ferrite Isolator

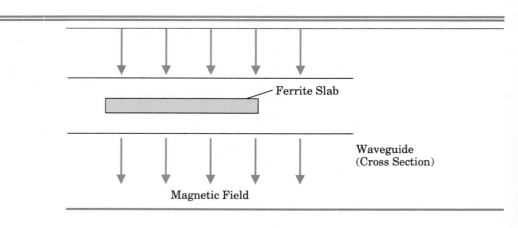

FIGURE 13.30

and the ferrite results in a phase shift as the wave propagates through the material. This shift is called *Faraday rotation*. It is the same phenomenon that causes the polarization of waves of certain frequencies to change as they move through the ionosphere, as described in Chapter 11.

The amount of phase shift depends on the length of the ferrite and the strength of the dc magnetic field to which it is subjected. The shift is nonreciprocal; that is, it depends on the direction of propagation of the signal. By careful choice of phase shifts, a circulator can be devised that will achieve addition of the waves along two paths at the desired port, and cancellation at the other port or ports.

It is possible to build a circulator without using ferrites. Figure 13.30 shows how one can be constructed using two hybrid tees. A 180° phase-shift

Circulator Using Two
Hybrid Tees

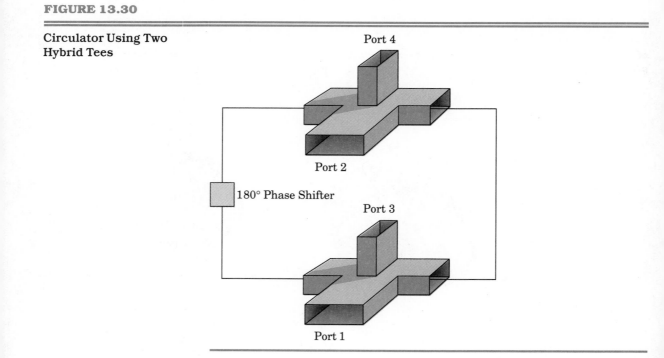

section of guide is also needed, but since the phase shift can be the same in both directions, ferrites are not required.

13.4 MICROWAVE SOLID-STATE DEVICES

When transistors were first invented, they were restricted to low-frequency applications. The author can remember hybrid AM broadcast-band car radios, which used tubes in the RF section and transistors for the audio portion. Gradually, the usefulness of transistors has been extended through the radio-frequency spectrum, until at present there are transistors that are usable in the microwave range. At the same time, other solid-state devices have been designed that take advantage of some of the factors that reduce transistor efficiency as the frequency increases. In this section, we will first take a brief look at the ways in which conventional transistors can be adapted for microwave use, and then examine some of these ingenious alternatives.

13.4.1 Microwave Transistors

Conventional silicon transistors suffer from two major problems as the frequency increases. First, all the components of the transistor, including the leads and the silicon elements themselves, exhibit stray capacitance and inductance. As the frequency increases, the inductive reactances get larger and the capacitive reactances get smaller, so that the transistor begins to look like the circuit of Figure 13.31. Eventually, there is so much feedback from collector to base that the transistor is useless.

It is impossible to eliminate stray capacitance and inductance, but they can be reduced. Figure 13.32 shows a typical microwave transistor. Note from the photograph of the package in Figure 13.32(a) that the wire leads have been replaced with metal strips that can be soldered directly to striplines. Figure 13.32(b) is a sketch of the internal structure. The "interdigital" design, with emitter and base formed from a series of "fingers," is designed to reduce capacitance to a minimum. The collector is formed from the substrate.

A more fundamental problem with any conventional transistor, and with many other types of devices as well, is transit time. In order for a transistor, diode, or other similar device to operate, charge carriers must move from one

FIGURE 13.31

Equivalent Circuit of Transistor at High Frequency

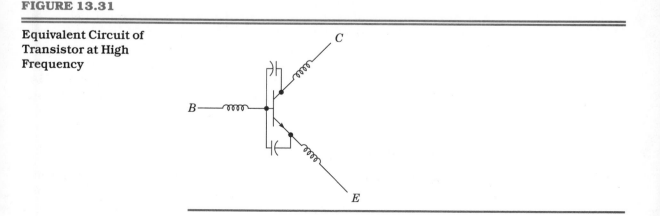

FIGURE 13.32

Microwave Transistor

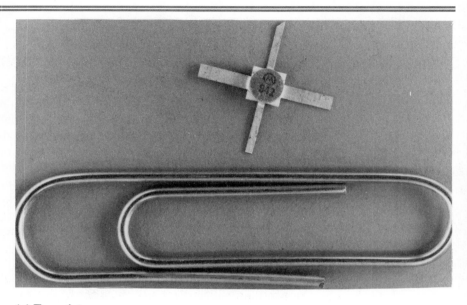

(a) Transistor

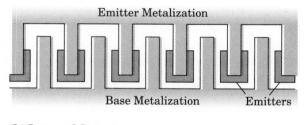

(b) Internal Structure

Photograph courtesy Motorola Inc.

region to another. This takes a small but finite amount of time, depending on the type of carrier, the **mobility** of the carriers in the material, and the distance they have to move. If the transit time is of the same order as the signal period, it will result in phase shifts that can seriously degrade performance. In general, free electrons move more quickly than holes, which accounts for the preponderance of NPN bipolar transistors and N-channel field-effect transistors in high-frequency applications. Gallium arsenide is "faster" than silicon, and is preferred in microwave applications.

Transistors modified for microwave use are quite common in the lower part of the microwave spectrum. For instance, the low-noise amplifier (LNA) that is the first stage of a satellite-reception system is practically always a gallium arsenide field-effect transistor (GaAsFET).

13.4.2 Gunn Devices

One solution to the problem of transit time that appears in a good many microwave devices could be loosely described as "if you can't beat it, join it." That

FIGURE 13.33

Gunn Device (Cross Section)

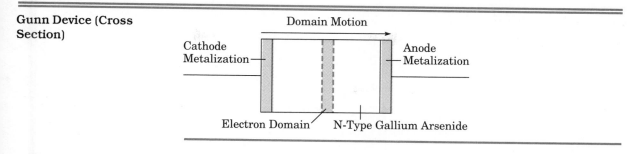

is, rather than attempting to keep the transit time as small as possible to minimize its undesired effects, the designer includes a phase shift due to transit time as one of the device characteristics. The **Gunn device**, also known as the **transferred-electron device** (TED), is one of the simpler transit-time devices.

The Gunn device is sometimes called a *Gunn diode* because it has two terminals. It has no junction, however, but is just a slab of N-type gallium arsenide, as shown in Figure 13.33. Gallium arsenide has the interesting and unusual property that the mobility of the electrons actually decreases as the electric-field strength increases, over a certain range. Normally, we would expect that the larger the electric-field strength, the faster electrons would move through the conductor, and that is in fact the case with most conductors and semiconductors. With gallium arsenide and a few other semiconductors, such as indium arsenide and gallium phosphide, there is a region where the mobility actually decreases as the field increases. The graph in Figure 13.34 shows this phenomenon.

A curve like the one in Figure 13.34 is said to have a negative-resistance region. This does not actually mean that the material has less than zero resistance. What it does mean is that, in some part of the curve, the cur-

FIGURE 13.34

Drift Velocity in Gallium Arsenide

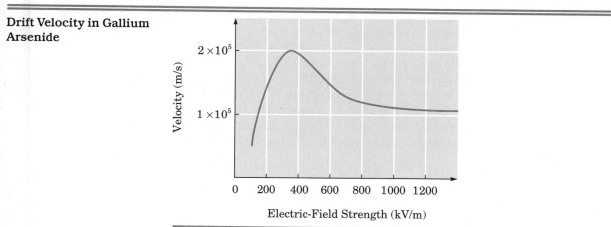

FIGURE 13.35

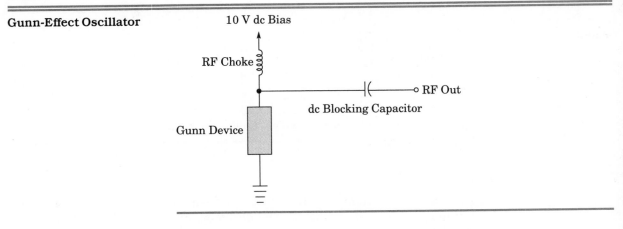

Gunn-Effect Oscillator

rent decreases as the voltage decreases. A differential resistance could be defined as

$$r = \frac{\Delta v}{\Delta i} \tag{13.13}$$

and it is this differential resistance that is negative over a certain range of field strength in gallium arsenide. A device with negative resistance can be made to work as an oscillator or, with a bit more work, as an amplifier.

This negative-resistance effect was discovered by J. B. Gunn (b. 1928) in 1963. He found that when a voltage was placed across a small slab of gallium arsenide, as shown in Figure 13.35, oscillations occurred. The frequency depended on the thickness of the slab, and seemed to have a period equal to the time taken for an electron to pass through the slab, at a drift velocity of about 1×10^5 m/s. Gunn eventually discovered that the reason for the oscillations was that a **domain** with a large electric field formed in the material, and moved toward the positive terminal. The domain forms when, possibly due to some local impurity in the material, a small region exists with an electric field greater than in the rest of the material. When this field reaches the threshold value, the electrons in the region move more slowly than elsewhere. A concentration of charge builds up, forming a *domain*, which moves through the device, collecting most of the charge as it does so. When the domain reaches the positive terminal, all of its electrons leave at once, causing a pulse of current. Simultaneously, another domain forms and the process continues. For obvious reasons, this mode of operation is called the *transit-time mode*. It can also be referred to as the *domain mode*, and it is sometimes called the *Gunn mode*, as it was first discovered by Mr. Gunn.

EXAMPLE 13.8

A Gunn device has a thickness of 7 μm. At what frequency will it oscillate in the transit-time mode?

Solution

The transit time of an electron can be found from the basic velocity relationship

$$v = \frac{d}{t}$$

where v = velocity in meters per second
d = distance in meters
t = time in seconds

Rearranging gives

$$t = \frac{d}{v}$$
$$= \frac{7 \times 10^{-6} \text{ m}}{1 \times 10^{5} \text{ m/s}}$$
$$= 7 \times 10^{-11} \text{ s}$$

This will be the period T of the oscillation, so the frequency is

$$f = \frac{1}{T}$$
$$= \frac{1}{7 \times 10^{-11} \text{ s}}$$
$$= 14.3 \times 10^{9} \text{ Hz}$$
$$= 14.3 \text{ GHz}$$

The Gunn mode provides oscillations at a frequency that depends on the device geometry, and not on the external circuit. There are several other modes of operation that allow the device to be tuned by placing it in a resonant cavity. The effect of the cavity is to remove, or *quench*, the domain before it reaches the anode terminal, causing the device to operate at a higher frequency than it would otherwise. The limited-space-charge-accumulation (LSA) mode is the most common. In this mode, the device is biased in the negative-resistance region, but the voltage swing is such that the device moves out of this region once per cycle, so that the domain is quenched.

The Gunn device can also be used as an amplifier, but, like all two-terminal negative-resistance devices, it requires a circulator to separate the input from the output. This mode of operation is illustrated in Figure 13.36.

13.4.3 IMPATT Diodes

The acronym **IMPATT** stands for IMPact Avalanche and Transit Time. Unlike the Gunn device, the IMPATT has a P–N junction; in fact, it is often a four-layer device, as shown in Figure 13.37. As suggested by its name, the device operates in the reverse-breakdown (avalanche) region.

The width of the intrinsic region in the IMPATT diode is made such that

FIGURE 13.36

Use of Gunn Device as an Amplifier

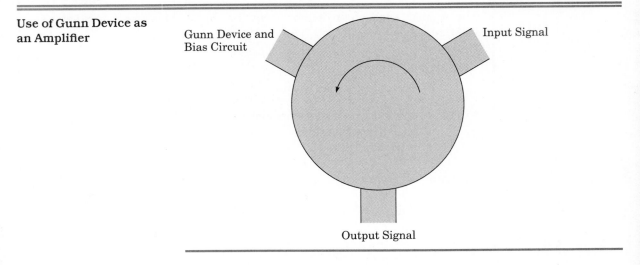

Gunn Device and Bias Circuit

Input Signal

Output Signal

the transit time of an electron across this region is equal to one-half of the period at the operating frequency. The device is reverse-biased at just below its breakdown voltage. The ac waveform superimposed upon this bias causes breakdown to occur once per cycle. When the junction breaks down, a surge of electrons enters the intrinsic region, emerging one-half cycle later. This has an effect similar to that of the domains in the Gunn device, and results in oscillation at a frequency that depends on the dimensions of the IMPATT diode and the resonant frequency of the cavity in which it is installed.

IMPATT diodes can achieve greater efficiency and higher power levels than Gunn devices (on the order of 10 W compared to 1 W for a Gunn device), but their use of avalanche breakdown causes them to be considerably noisier. A variation of the IMPATT called the **TRAPATT** (for TRApped Plasma Avalanche Triggered Transit) can operate with still higher power (about 100 W in pulsed operation) and efficiency levels, but it is expensive, requires complex circuitry, and is noisier than the IMPATT.

13.4.4 PIN Diodes

The structure of a P-intrinsic-N (**PIN**) diode, shown in Figure 13.38, is similar to that of an IMPATT diode, but its operation is quite different. This device is used as an electronic switch and an attenuator. When reverse biased, it represents a large resistance in parallel with a small capacitance, but for-

FIGURE 13.37

IMPATT Diode

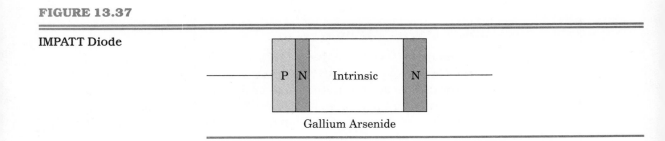

Gallium Arsenide

FIGURE 13.38

PIN Diode

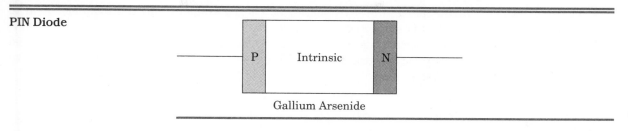

ward biasing the diode results in a variable resistance that can approach zero when the diode is completely turned on. Ordinary silicon diodes are used as switches in just this way at lower frequencies. The PIN diode is preferred in the microwave region of the spectrum because of its small capacitance when reverse biased. At microwave frequencies, there will be enough stored charge in the intrinsic region, when the diode is forward biased, for good conduction.

13.4.5 Varactor Diodes

The varactor diode is familiar from lower-frequency work as a means of providing a capacitance that can be changed by varying the voltage that reverse biases the diode. Varactors are found in such devices as frequency synthesizers and frequency modulators. They can serve the same function at microwave frequencies.

There is, however, another use to which varactors can be put in the microwave part of the spectrum. They are often used in frequency multipliers. These, too, should be familiar from earlier chapters, but they are even more useful at microwave frequencies. They can be used to generate microwave signals with crystal-controlled stability, something that is otherwise difficult to accomplish at such high frequencies. The frequency multipliers described previously used transistor Class C amplifiers, but of course this is not necessary. All that is required is that there be a nonlinear circuit that will generate harmonics of the input signal, and a tuned circuit to emphasize the desired harmonic and reject the fundamental and any undesired harmonics.

All diodes are nonlinear in the "knee" region of their characteristic curve, but varactors, due to their variable capacitance, can be used with reverse bias as nonlinear devices, achieving greater efficiency. A variation of the varactor, known as the *step-recovery* or *snap diode*, can also be used as a multiplier. It is operated with forward bias for most of the cycle of the fundamental. During this period, a charge is stored in the diode. When the diode becomes reverse biased at the input cycle peak, the charge flows back across the junction as a brief pulse with a high harmonic content.

Since varactor multipliers are passive devices, they obviously have a gain of less than unity. They can be quite efficient, though: an efficiency of 50% is not unusual. This compares quite favorably with generating a microwave signal directly.

13.4.6 Yttrium-Iron-Garnet Devices

Yttrium-iron-garnet (**YIG**) is a type of ferrite. A YIG sphere can be used in place of a resonant cavity as a microwave resonant circuit. Such a sphere is often used as the frequency-determining circuit in a microwave oscillator. Since the resonant frequency of a ferrite depends on the magnetic field to which it is subjected, the frequency of the oscillator can be changed by varying the dc current to an electromagnet that generates the field. This method of frequency setting is much more convenient than varying the physical size of a resonant cavity.

13.4.7 Dielectric Resonators

Another innovative frequency-determining technique that is often used in solid-state microwave sources is the dielectric resonator. This is essentially a resonant cavity that is made of a solid slab of a dielectric material such as alumina. Microwaves propagate within the medium, and are reflected at the interface between the dielectric and air. The result is a stable, low-cost resonant device. Compared to the YIG sphere, it has the disadvantage of being fixed in frequency, but it is inexpensive and needs no magnet.

13.5 MICROWAVE TUBES

As at lower frequencies, vacuum tubes are still the preferred technology when large amounts of microwave power are required. Many of the problems that reduce the efficiency of transistors, such as transit time and stray reactance, also affect tubes. Lead inductance can be reduced by making connections to rings on the tube, rather than pins at one end. Transit time can be reduced by making tubes smaller, but this increases stray capacitance and reduces the tube's ability to dissipate power. Such "fixes" work only up to a point. This section will look at some vacuum-tube designs that avoid the shortcomings of conventional tubes by such techniques as incorporating transit time into the design. There are many other types not represented here, but these examples illustrate the logic behind microwave tube designs.

13.5.1 Magnetrons

The *magnetron*, invented in 1921, is the oldest microwave tube design. The magnetron and its variations are high-power, fixed-frequency oscillators, not noted for stability or ease of modulation, but simple, rugged, and relatively efficient (about 40% to 70%). Magnetrons are commonly used in radar transmitters, where they can generate peak power levels in the megawatt range. The smaller ones used in microwave ovens produce several hundred watts continuously.

Figure 13.39 shows a cross-sectional view of a typical cavity magnetron. As might be expected from its name, the magnetron needs a magnet: a powerful permanent magnet. The tube itself is situated between the poles of the magnet, and only the magnetic-field lines are shown.

A hot cathode is surrounded by an anode in which there are a number of resonant cavities. Were it not for the magnetic field, electrons emitted from the cathode would simply move radially across to the anode. However, the interaction between the fixed magnetic field and the magnetic field generated by a moving electron produces a force that causes the electron to travel in a curved path. The size of the loop made by the electrons depends on the relation between the magnetic-field strength and that of the electric field be-

FIGURE 13.39   **Cavity Magnetron**

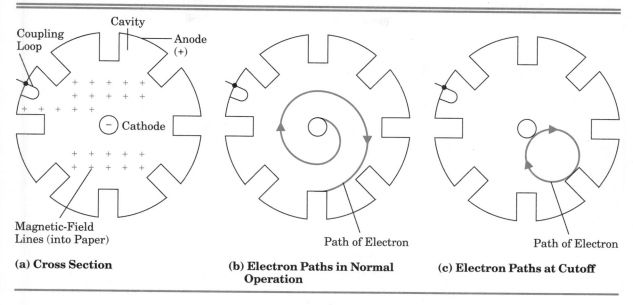

(a) Cross Section   **(b) Electron Paths in Normal Operation**   **(c) Electron Paths at Cutoff**

tween anode and cathode. For a large anode-cathode voltage, as in Figure 13.39(b), the electrons will reach the anode, and anode current will flow. If the voltage is reduced, as in Figure 13.39(c), electrons will circle around and return to the cathode. In this case, there will be no anode current and the tube will be cut off. The rate at which electrons move around the cathode is called the *cyclotron frequency*. Since the magnetic and electric fields applied to the tube are at right angles, magnetrons are also known as **crossed-field tubes**.

Now it is necessary to consider the function of the cavities around the outside of the magnetron. The motion of electrons past the opening of a cavity starts oscillations that will be sustained if the conditions are right for the continual transfer of energy from the electron stream to the fields in the cavities. In most magnetrons, this involves the presence of a travelling wave that moves around the tube, with one-half wavelength representing the distance between adjacent cavities. This requires a movement of the wave around the tube at a rate much slower than the speed of light, and for this reason the circular arrangement of resonant cavities is called a **slow-wave structure**.

A slow-wave structure is required because the electrons must move around the structure at almost the same rate as the electromagnetic wave, in order for the electrons to give up energy to the wave. Of course, the electrons in the tube move with a velocity much less than that of light, so the propagation velocity of the wave must be reduced.

In order to get power from the magnetron, one of the cavities must be coupled electrically to the outside. The tube shown uses a loop for this. The energy provided by the tube comes from the wave travelling around the tube, and that energy in turn comes from the electron stream. The electrons, of course, get their energy from the applied anode-to-cathode potential. For the tube to work, this voltage must be adjusted so that most of the electrons cycle around the tube. As they spiral from cathode to anode, they gain energy from the dc electric field and give up most of it to the travelling wave. Finally, they

Microwave Ovens   There is very likely a magnetron in your kitchen, since every microwave oven contains one. The magnetron produces several hundred watts at 2.45 GHz, a frequency that is chosen because water molecules are resonant at that frequency. If subjected to a microwave field at that frequency, they absorb power and become hot, causing any food containing water to cook. The output of the magnetron is coupled to a resonant cavity, the oven itself. The several kilovolts required by the tube are supplied by a step-up transformer and a voltage-doubler rectifier circuit. In variable-power ovens, a thyristor circuit pulses the power so that the magnetron switches rapidly on and off.

Now that you know something about microwaves, some of the well-known cautions about microwave ovens should be easy to understand. Operating the oven without food is undesirable, since this is equivalent to operating a transmitter without a load. Most of the power will be reflected back to the tube. Similarly, the use of metal containers is frowned on because currents will be induced in the metal, heating it and also distorting the field in the oven. Some special microwave cookware contains resistive material, designed to absorb power from the microwaves and become hot, as an aid to browning certain foods. Some microwaves rotate the food or use a "stirrer" to disturb the fields in the cavity. This is because the oven, like any resonant cavity, has standing waves, so that the electric-field strength, and therefore the cooking ability, is not evenly distributed.

Lastly, a knowledge of microwaves can be applied to safety concerns. It is important that the door seal properly, and that the interlocks (there are usually two, for backup) disconnect the power when the door is opened. The microwaves that can cook your food can cook you too, if allowed to escape. (The eyes are especially vulnerable, because they have poor circulation which does not allow heat to be removed quickly by the blood.) But no, the food itself is not "radioactive" in any way. We are dealing with radio waves, at frequencies far below those of ordinary light. Giving radioactivity to substances requires photons with much more energy. "Nuking" food in the microwave is just a figure of speech!

land on the anode. Some electrons will, on the other hand, cycle around and land on the cathode. This *back-bombardment* of the cathode reduces the life of the tube and lowers efficiency.

The magnetron described above is essentially a fixed-tuned oscillator. Changing the frequency requires changing the physical size of the cavities, which can only be done mechanically. Voltage-controlled versions are available, but with much lower efficiency. There is also an amplifier tube that works on similar principles and is known as the *crossed-field amplifier* (CFA). The operating bandwidth is very limited because of the high-Q resonant cavities.

Many magnetrons are used to generate pulses of radio-frequency energy, in radar transmitters for instance. Typically these are low-duty-cycle applications. Recall that the duty cycle for any device is simply

$$D = \frac{T_{on}}{T_T} \tag{13.14}$$

where D = duty cycle as a decimal fraction
T_{on} = on time per operating cycle
T_T = total time per operating cycle

Duty cycle can also be expressed as a percentage, by multiplying D from Equation (13.14) by 100. The power averaged over time for a magnetron, or any other device that generates pulses, is the power in the pulses, which is generally called (somewhat misleadingly) *peak power*, multiplied by the duty cycle. That is,

$$P_{avg} = P_P D \tag{13.15}$$

where P_{avg} = average power
P_P = pulse power
D = duty cycle

Equation (13.15) is equally applicable to input power, output power, or power dissipation.

EXAMPLE 13.9

A pulsed magnetron operates with an average power of 1.2 kW and a peak power of 18.5 kW. One pulse is generated every 10 ms. Find the duty cycle and the length of a pulse.

Solution

From Equation (13.15),

$$P_{avg} = P_P D$$

$$D = \frac{P_{avg}}{P_P}$$

$$= \frac{1.2}{18.5}$$

$$= 0.065 \quad \text{or} \quad 6.5\%$$

From Equation (13.14),

$$D = \frac{T_{on}}{T_T}$$

$$T_{on} = D T_T$$

$$= 0.065 \times 10 \text{ ms}$$

$$= 0.65 \text{ ms}$$

13.5.2 Klystrons

The *klystron* is the preferred tube for high-power, high-stability amplification of signals at frequencies from UHF to about 30 GHz. In fact, it is commonly found in UHF television transmitters. There are actually two quite different types of klystrons: the reflex klystron, a small tube used as an oscillator; and the multicavity klystron, employed as a power amplifier. The reflex klystron is nearly obsolete, having been replaced in most applications by the solid-state techniques described above, so this section will deal with the multi-cavity klystron, some examples of which are pictured in Figure 13.40(a).

FIGURE 13.40

Multicavity Klystron Tubes

(a) Klystron Tubes

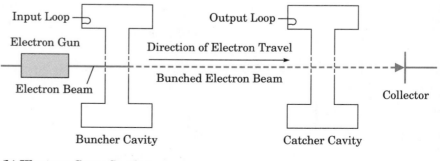

(b) Klystron Cross Section

Photograph courtesy of Varian Associates.

Like the magnetron, the klystron works by transferring energy from a stream of electrons to an electromagnetic wave moving in a slow-wave structure. In the klystron, the electrons are formed into a beam that moves in a straight line past a series of resonant cavities. For this reason, the klystron, like the travelling-wave tube to be described next, is known as a **linear-beam tube**.

The electron beam is produced in the conventional manner by a hot cathode. An electron gun, aided by one or more external magnets, focuses the electrons into a beam. As shown in Figure 13.40(b), the beam passes two or more resonant cavities before reaching the anode, which in this tube is called the **collector**.

The input signal is applied to the cavity closest to the cathode. The beam is velocity-modulated, or **bunched**, by the microwave field in the input cavity, which is also called the **buncher**. The output signal is taken from the cavity closest to the collector, which is sometimes called the **catcher**. In between, there may be one or more intermediate cavities.

The velocity modulation, or bunching, takes place due to the interaction between the electric field at the input cavity and the electrons in the beam. Figure 13.41 shows the process. When the electric field due to the input signal is in the same direction as the accelerating field from the main power supply, it causes the velocity of the electrons to increase. On the other hand, when the two fields are in opposition, the beam is slowed. The result is that, at a certain distance from the cavity, the electrons bunch together, as the figure shows. A simple analogy would be with cars at a traffic light. Some cars are stopped at the light, some are let through, and the result is that the cars tend to travel in groups.

The bunched electrons produce an alternating microwave field at the output cavity that has the same frequency as the input wave but a greater amplitude, thus fulfilling the requirements for an amplifier. The electrons move on to the collector, but not before much of their energy has been given up. Klystrons can produce very high powers, up to the megawatt range.

13.5.3 Travelling-Wave Tube

The **travelling-wave tube** (TWT) can be used as a moderate-power amplifier or, with modifications, as an oscillator. It is distinguished by its wide

FIGURE 13.41

Velocity Modulation

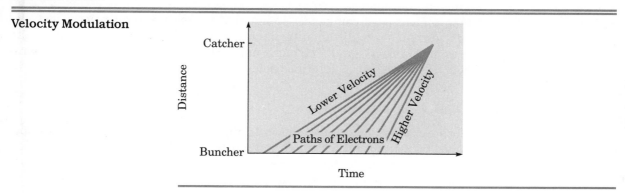

FIGURE 13.42

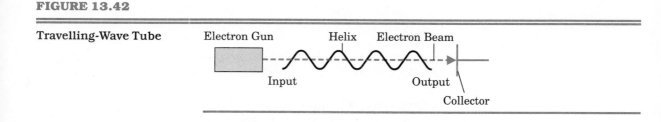

Travelling-Wave Tube

bandwidth. The transmitters in communications satellites and their ground stations, for example, often use TWT amplifiers. Like the klystron and unlike the magnetron, the TWT is a linear-beam tube. Like both of these tubes, the TWT has a slow-wave structure. It may consist of a series of coupled cavities, reminiscent of the klystron, but it often consists of a helix, as shown in the diagram of Figure 13.42. The helix can be thought of as a sort of waveguide, along which a wave travels by going around and around the turns of the spiral. The progress of the wave along the length of the helix is much slower than its speed as it moves around the turns. An electron beam travels from the cathode down the center of the helix to the collector, giving up some of its energy to the wave as it does so. Since the helix, unlike a series of cavities, is nonresonant, the bandwidth of the TWT can be much greater than that of the klystron. On the other hand, it is more difficult to remove heat from the helix than from the cavities of the klystron, so the helix TWT is a low- to medium-power device, for power levels up to about a kilowatt.

Like the klystron, the TWT needs a magnetic structure for the purpose of focusing the electron beam so that it passes within the helical slow-wave structure.

13.6 MICROWAVE ANTENNAS

There is no theoretical difference between microwave antennas and those for lower frequencies. The differences are practical: at microwave frequencies it is possible to build elaborate, high-gain antennas of reasonable physical size. Certainly dipoles, Yagis, and log-periodic antennas are also possible, but this section will look at antennas whose construction would be impractical at lower frequencies.

The parabolic reflector has already been described in Chapter 12. Though occasionally used at lower frequencies (for radio telescopes, for instance), it is most commonly employed at microwave frequencies. The parabolic dish is not really an antenna: it is a reflector, and needs an antenna to provide it with a signal. The most common feed antenna for use with a parabolic dish is the horn, described below.

13.6.1 Horn Antennas

Horn antennas, like those shown in Figure 13.43, can be viewed as impedance transformers that match waveguide impedances to that of free space. The examples in the figure represent the most common types. The E- and H-plane sectoral horns are named for the plane in which the horn flares; the

FIGURE 13.43

Horn Antennas

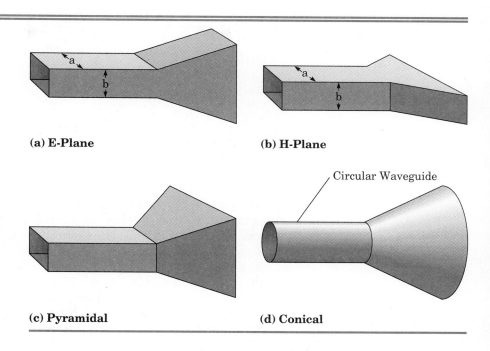

(a) E-Plane

(b) H-Plane

(c) Pyramidal

(d) Conical

pyramidal horn flares in both planes. With circular waveguide the conical horn is most appropriate.

The gain and directivity of horn antennas depends on the type of horn and its dimensions. Gain is often in the vicinity of 20 dBi for practical horns. The bandwidth is about the same as that of the associated waveguide; that is, it works over a frequency range of approximately 2:1.

As well as being used with a parabolic reflector, the horn antenna can be—and often is—used alone as a simple, rugged antenna with moderate gain.

13.6.2 Other Microwave Antennas

There are many other types of microwave antennas in common use. This section will give a brief introduction to some of them.

Figure 13.44 shows a slot antenna, which is actually just a hole in a waveguide. The length of the slot is generally one-half wavelength. Its radiation pattern is similar to that of a dipole with a plane reflector behind it. It therefore has much less gain than, for instance, a horn antenna. It is seldom used alone, but is usually combined with many other slots to make a phased array. Phased arrays with several elements were discussed in Chapter 12; at microwave frequencies, a great many slot antennas can be used to form a narrow, high-gain beam whose direction can be changed electronically, by changing the phase of the signals to the individual elements. Such antennas are especially useful for airborne radar. In this case, the slots would be filled with a dielectric material to present a smooth surface to the air. Large aircraft actually have several slot arrays, for communications as well as radar.

FIGURE 13.44

Slot Antenna

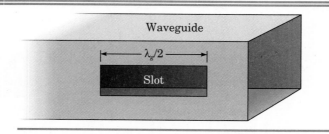

Of course, phased arrays can also be constructed from more conventional antenna elements, such as dipoles. The dipoles are sometimes built in a slightly unconventional way, by laying them out on a printed-circuit board, in the form of a stripline antenna.

Just as the optical property of reflection can be put to good use in microwave antennas, so can refraction. At microwave frequencies, it is practical to build a lens for radio waves. The lens does not have to be optically transparent, but must be made from a good dielectric: teflon is popular. To reduce the physical size of the lens, the **Fresnel lens** is usually chosen, as shown in Figure 13.45. The lens, like the parabolic dish, is not really an antenna; it must be fed by a radiating element such as a horn antenna. One common application of the dielectric lens is in police radar systems.

FIGURE 13.45

Dielectric Lens

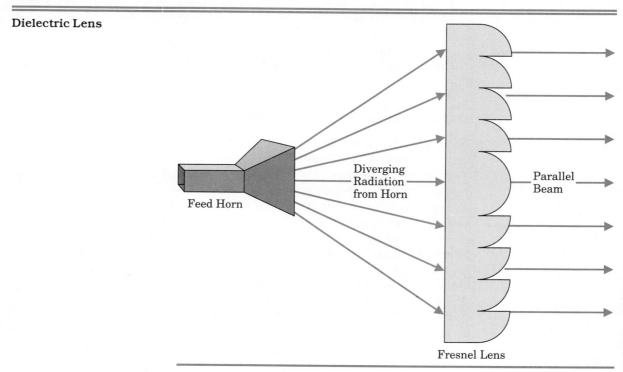

13.7 RADAR

Radar is an acronym for RAdio Detection And Ranging, and had its origins in World War II. Essentially, radar requires that a transmitter emit a signal using a directional antenna toward some object called the **target**. The signal reflects from the target back to the source, where it is received and interpreted. Information about the target can be obtained by analyzing the reflection, or echo.

Early radars often used signals in the HF or VHF regions of the spectrum. Microwaves are more common now, however. They can be focused into narrower beams, and they can detect smaller targets, since the target must be relatively large compared with the wavelength to provide a good reflection. An exception is long-range over-the-horizon radar, which uses frequencies in the HF range to take advantage of ionospheric propagation.

Radar equipment can be divided into two main categories. Pulse radar works by transmitting a short burst of microwaves called a *pulse*, which reflects from the target and is received at a later time. The time between transmitted and received pulses gives distance information. Continuous wave (CW) radar transmits continuously, and compares the frequency of the received echo with that of the transmitted signal. Relative motion between the radar and the target can cause a change of frequency, from which velocity information can be obtained. This is called the **Doppler effect**. There are also radars that combine pulse and Doppler techniques.

Because a radar target does not transmit a signal, but merely reflects it, the returning signal may be quite weak. Its strength is subject to square-law attenuation both as it travels to the target and on the return trip to the receiver. Thus, for radar, the received signal power is inversely proportional to the fourth power of the distance, rather than the second power as for a normal communications system.

The power of a return signal is also affected by the size, shape, and composition of the target. Targets are assigned a **radar cross section** that is defined as the area of a perfectly conducting flat plate, facing the source, that would reflect the same amount of power toward the receiver. Since real targets are neither perfect conductors nor flat planes with the correct orientation, the radar cross section will be smaller than the actual cross-sectional area of the target, as seen from the radar installation.

The information given above can be quantified and expressed in a propagation equation called the *radar equation*.

$$P_R = \frac{\lambda^2 P_T G^2 \sigma}{(4\pi)^3 r^4} \tag{13.16}$$

where P_R = received power in watts
  λ = free-space wavelength
  P_T = transmitted power in watts
  G = antenna gain as a power ratio
  σ = radar cross section of the target in square meters
  r = range (distance to the target) in meters

Note the fourth-power variation of received signal strength with distance. The transmitting and receiving antennas will normally be the same for a radar system, so their gains have been combined in the equation.

EXAMPLE 13.10

A radar transmitter has a power of 10 kW and operates at a frequency of 9.5 GHz. Its signal reflects from a target 15 km away with a radar cross section of 10.2 m². The gain of the antenna is 20 dBi. Calculate the received signal power.

Solution

First, since wavelength, rather than frequency, is required, and antenna gain must be expressed as a power ratio, some preliminary calculations are necessary.

$$\lambda = \frac{c}{f}$$

$$= \frac{3 \times 10^8}{9.5 \times 10^9}$$

$$= 0.0316 \text{ m}$$

$$G = \text{antilog} \frac{\text{dB}}{10}$$

$$= \text{antilog} \frac{20}{10}$$

$$= 100$$

Next, substitute into Equation (13.16):

$$P_R = \frac{\lambda^2 \, P_T G^2 \sigma}{(4\pi)^3 r^4}$$

$$= \frac{0.0316^2 (10 \times 10^3)(100^2)(10.2)}{(4\pi)^3 (15 \times 10^3)^4}$$

$$= 10.1 \times 10^{-15} \text{ W}$$

$$= 10.1 \text{ fW}$$

13.7.1 Determination of Direction and Range with Pulse Radar

Determination of the direction from the radar antenna to a target can be very simple. Either the antenna is moved physically, or, in the case of a phased-array antenna, its radiation pattern is moved electronically. When the received signal is strongest, the main lobe of the antenna is pointed at the target. The type of antenna and the way in which it is moved depend on the target characteristics. For instance, radar on a ship intended to locate shoreline features and other ships need only scan in azimuth (that is, horizontally), while radar on an airplane designed to track other airplanes will have to scan in both azimuth and elevation. The angular resolution depends on the beamwidth of the antenna.

The distance, or range, to the target can be found by a method analogous to that used with time-domain reflectometry. (See Chapter 10.) A brief pulse

FIGURE 13.46

Radar Range Determination

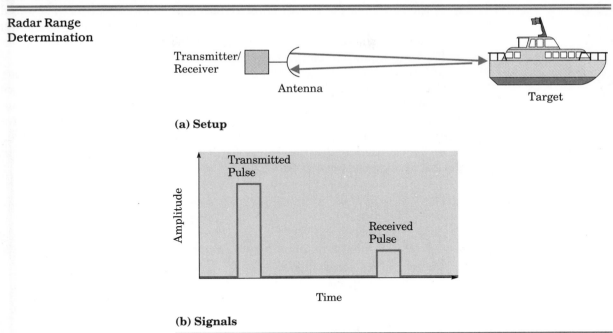

(a) Setup

(b) Signals

of radio-frequency energy is emitted, and the time that elapses before its return is measured. From this elapsed time, the range can easily be calculated. (See Figure 13.46.) Of course, the measured time is that for a round trip to the target and back, so the one-way distance is found by dividing by two. Assuming free-space propagation,

$$R = \frac{ct}{2} \tag{13.17}$$

where R = distance to the target
c = velocity of light
t = time taken for the echo to return

EXAMPLE 13.11

A pulse sent to a target returns after 15 μs. How far away is the target?

Solution

From Equation (13.17),

$$R = \frac{ct}{2}$$
$$= \frac{(3 \times 10^8)(15 \times 10^{-6})}{2}$$
$$= 2250 \text{ m}$$
$$= 2.25 \text{ km}$$

Of course, some of the time taken may be time spent by the signal getting from the electronics to the antenna and back. If significant, this time would have to be subtracted from the total time to get the time taken for the signal to travel from the antenna to the target and back. It is this net time that is called for in Equation (13.17).

A problem can arise with pulse radar if the period between pulses is less than the time taken for a pulse to return from the target. Figure 13.47 shows the problem. There is an ambiguity: the radar cannot distinguish between this target and a much closer one. This means that any pulse radar will have a maximum unambiguous range that is limited by the pulse repetition rate. It is easy to calculate this range: it is the distance a signal can travel between pulses, divided by two, of course.

$$R_{max} = \frac{cT}{2}$$

(13.18)

where R_{max} = maximum unambiguous range
c = velocity of light
T = pulse period

Often, the pulse repetition rate (frequency) is used instead of the period. Equation (13.18) becomes

$$R_{max} = \frac{c}{2f}$$

(13.19)

where f = pulse repetition rate

There is another possible difficulty. The transmitted pulse will have a finite pulse duration, also called *pulse width*. If the echo returns while the pulse is still being transmitted, it will not be detected by the receiver. The minimum usable range for the radar, then, is one-half the distance the signal can travel during the time it takes to transmit the pulse:

$$R_{min} = \frac{cT_P}{2}$$

(13.20)

FIGURE 13.47

Radar Range Ambiguity

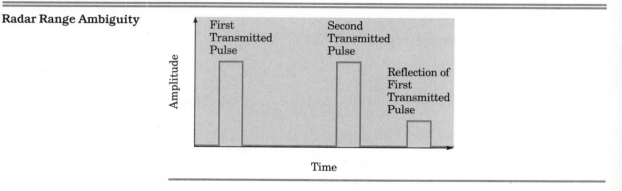

where R_{min} = minimum usable range
      c = velocity of light
      T_P = pulse duration

EXAMPLE 13.12

A pulse radar emits pulses with a duration of 1 μs and a repetition rate of 1 kHz. Find the maximum and minimum range for this radar.

Solution

From Equation (13.19), the maximum unambiguous range is

$$R_{max} = \frac{c}{2f}$$
$$= \frac{3 \times 10^8}{2 \times 1000}$$
$$= 150 \times 10^3 \text{ m}$$
$$= 150 \text{ km}$$

Of course, this range is subject to the signal being able to propagate that far. If this radar were installed on a ship for the purpose of observing objects on the water, the range would actually be limited to the line-of-sight communication distance.

The minimum range is found from Equation (13.20):

$$R_{min} = \frac{cT_P}{2}$$
$$= \frac{(3 \times 10^8)(1 \times 10^{-6})}{2}$$
$$= 150 \text{ m}$$

Obviously, the use of short pulses improves the performance of the radar at short range. It also improves the ability of the system to separate targets that are in the same direction but at different ranges, as illustrated by Figure 13.48. Short pulses also result in a lower duty cycle for the transmitter, which reduces the average power requirement for a given pulse power. The disadvantage of short pulses is that they increase the bandwidth of the signal. A wider signal bandwidth requires a wider receiver bandwidth, reducing the signal-to-noise ratio, and also increases the congestion of the spectrum.

13.7.2 Doppler Radar

Pulse radar can measure velocity only in an indirect way: by finding the position of a target at two different times and calculating how far it has moved in a given time interval. It is possible to measure velocity directly using radar, with certain limitations. Radars that do this use the Doppler effect. This causes the frequency of an echo to differ from that of the transmitted signal

FIGURE 13.48

Discrimination of
Targets in Same
Direction

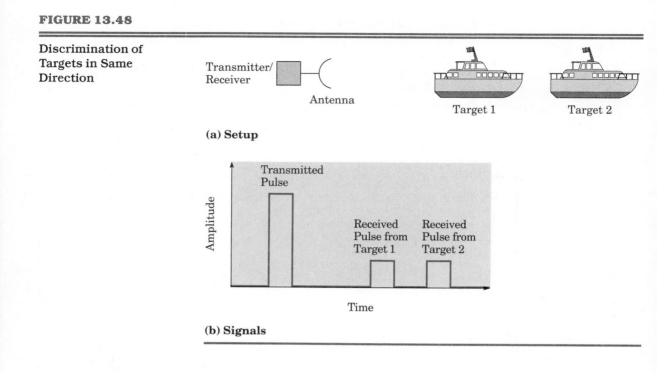

(a) Setup

(b) Signals

when there is relative motion between the radar and the target, along a line joining the two. Motion that closes the gap between source and target raises the frequency of the reflection, and motion in the other direction reduces it.

To understand the Doppler effect, consider a plane wave, emitted from a stationary source, impinging on a reflecting plane that is moving toward the source, as sketched in Figure 13.49. The wavefronts are shown as dashed lines. As soon as front 1 reaches the surface, it reflects. Since the surface is moving in the opposite direction to that of the wave, peak 2 will reach the surface after a shorter interval than it would have had the target been stationary. As soon as peak 2 reaches the target, it reflects. Thus, the period of the reflected wave has been reduced. Since the speed of propagation does not

FIGURE 13.49

Doppler Effect

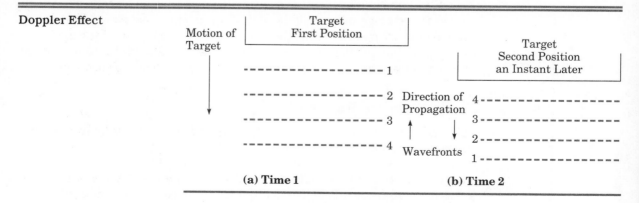

change, this means that the frequency of the reflected wave has correspond-
ingly increased. A similar logic shows that the frequency of the reflection will
decrease if the target is moving away from the source.

The equation governing the Doppler effect is quite simple:

$$f_D = \frac{2v_r f_i}{c}$$

(13.21)

where f_D = Doppler shift in hertz

v_r = relative velocity of the source and target in meters per second
along a line between them, and is positive if the two are closing
(getting closer)

f_i = incident frequency in hertz

c = velocity of light in meters per second

EXAMPLE 13.13

Find the Doppler shift caused by a vehicle moving toward a radar at 60
mph, if the radar operates at 10 GHz.

Solution

It is possible to derive equations relating speed in miles per hour and kilometers per hour to Doppler shift (see Problem 31 at the end of this chapter), but let us do this example from first principles.

A logical first step is to convert miles per hour to kilometers per hour:

$$60 \text{ mph} = 60 \times 1.6 \text{ km/h}$$
$$= 96 \text{ km/h}$$

Now, convert kilometers per hour to meters per second. It is necessary to
multiply by 1000 to convert kilometers to meters, and to divide by 3600 to
convert hours to seconds:

$$96 \text{ km/h} = \frac{96 \times 1000}{3600} \text{ m/s}$$
$$= 26.7 \text{ m/s}$$

Finally, the Doppler shift can be found from Equation (13.21),

$$f_D = \frac{2v_r f_i}{c}$$
$$= \frac{2(26.7)(10 \times 10^9)}{3 \times 10^8}$$
$$= 1.778 \text{ kHz}$$

Doppler radar can only measure the component of velocity along a line
that joins the source and the target. This is, of course, not always the true

FIGURE 13.50

**Doppler Radar Used in a
Speed Trap**

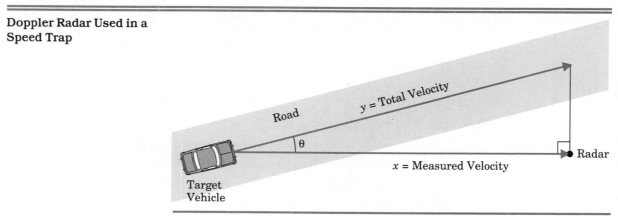

velocity. For instance, Figure 13.50 shows a police radar setup at the side of
a road. It is apparent that the speed measured by the radar is actually less
than the true speed of the vehicle. In fact, a little trigonometry will show that

$$x = y \cos \theta \qquad\qquad (13.22)$$

where x = component of velocity measured by the radar
y = total velocity
θ = angle between the direction of travel and the direction of a line
from the target to the radar

Since the transmitted and received frequencies are different, Doppler radars can use either CW or pulse techniques. Police speed radars, for instance,
are CW. The transmitted and received signals are separated by a circulator,
as shown in Figure 13.51. A small portion of the transmitted signal is allowed
to mix with the received signal to produce a difference signal at the Doppler
frequency. A frequency counter measures this frequency, and the result is
converted into a speed readout in kilometers or miles per hour. Circuitry can
also be provided to take the speed of the vehicle in which the radar is
mounted into account, since of course it is the relative speed between police
and target vehicle that is measured by the radar.

When applied to pulse radar, Doppler techniques allow the velocity of a
target, as well as its position and direction, to be estimated. They also allow
the elimination of stationary objects called *clutter* from the display.

13.7.3 Transponders

Radar echoes are generally simple reflections of the original pulse, not by any
means images of the target. By using narrow beams, frequency sweeps during the pulse, and a large amount of computing power, it is sometimes possible to make some inferences about the target from its radar *signature*. In
the case of friendly targets, however, the whole process can be made much

FIGURE 13.51

CW Doppler Radar

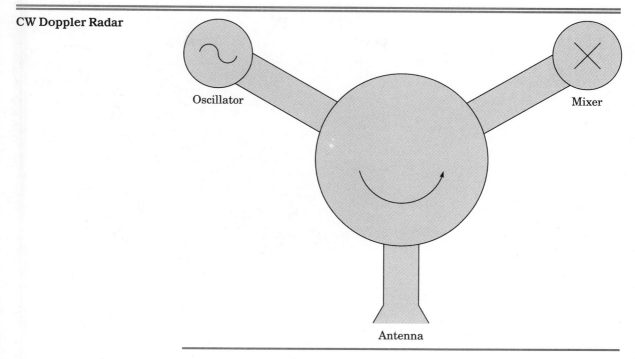

Oscillator

Mixer

Antenna

more efficient by installing a transmitter-receiver, called a **transponder**, on the target, usually an airplane. The transponder will respond to a radar pulse by transmitting a signal that can identify the aircraft, even down to its flight number and destination. In the case of military aircraft, transponders are part of an "identification friend or foe" (IFF) system designed to prevent friendly aircraft from being attacked.

13.7.4 Stealth

There are times when, for example, it would be better for an airplane not to appear on radar: if the airplane is military and the radar belongs to an enemy anti-aircraft battery, for instance. The techniques for avoiding detection by radar are collectively known as *stealth*, and the details are, of course, kept secret. Still, the general ideas behind stealth are well known and easily understood.

The two basic ways of escaping radar detection are, first, to absorb radar waves rather than reflect them, and, second, to scatter any reflected signals as widely as possible to avoid returning a strong signal to the hostile radar. To achieve the first result involves the use of resistive materials in the body of the aircraft; sharp angles will help achieve the second. Figure 13.52 shows a stealth fighter developed by the United States military. Note the rather ungainly angular appearance.

In practice, stealth will also involve reducing the *heat signature* of the aircraft, since some missiles use the hot exhaust gas from a jet engine to track their target.

FIGURE 13.52

Stealth Fighter

AP/Wide World Photos

SUMMARY OF CHAPTER 13

Here are the main points to remember from this chapter.

1. As frequency increases into the gigahertz range, distributed capacitance and inductance become very important everywhere in a circuit.

2. Waveguides are a very practical means of transmitting electrical energy at microwave frequencies, as they have much lower losses than coaxial cable. They are not very useful at lower frequencies because they must be too large in cross-section.

3. Waveguides are generally useful over only a 2:1 frequency range. They have a lower cutoff frequency that depends on their dimensions, and they exhibit dispersion due to multimode propagation at high frequencies.

4. Two velocities must be calculated for waveguides. The group velocity, lower than the speed of light, is the speed at which signals travel down the guide. The phase velocity, which is greater than the speed of light, is used for calculating the wavelength in the guide.

5. Power can be coupled into and out of waveguides using probes, loops, or holes in the guide.

6. Ferrites have special properties that make them just as useful at microwave frequencies as at lower frequencies. Applications of ferrites include circulators, isolators, attenuators, and resonators.

7. At microwave frequencies, conventional active components such as tubes and transistors suffer from excessive carrier transit time as well as stray capacitance and inductance. Solutions to these problems in-

clude redesigning standard component types, as well as innovative solutions designed to include transit time as part of the device operation.

8. The short wavelengths in the microwave region allow antennas to be built with high gain and narrow beamwidth. The combination of a horn antenna with a parabolic reflector is very popular.

9. Radar systems are of two basic types. Pulse radar works by reflecting pulses from a target, gauging its distance from the time taken for pulses to return. Doppler radar estimates the velocity of a target by measuring the difference in frequency between transmitted and received signals.

IMPORTANT EQUATIONS

$$a = \frac{\lambda_c}{2} \tag{13.1}$$

$$f_c = \frac{c}{2a} \tag{13.2}$$

$$v_g = c\sqrt{1 - \left(\frac{\lambda}{2a}\right)^2} \tag{13.3}$$

$$v_g = c\sqrt{1 - \left(\frac{f_c}{f}\right)^2} \tag{13.4}$$

$$v_g v_p = c^2 \tag{13.5}$$

$$v_p = \frac{c}{\sqrt{1 - \left(\frac{f_c}{f}\right)^2}} \tag{13.7}$$

$$Z_0 = \frac{377}{\sqrt{1 - \left(\frac{f_c}{f}\right)^2}}\ \Omega \tag{13.9}$$

$$\lambda_g = \frac{v_p}{f} \tag{13.10}$$

$$\lambda_g = \frac{\lambda}{\sqrt{1 - \left(\frac{f_c}{f}\right)^2}} \tag{13.12}$$

$$D = \frac{T_{on}}{T_T} \tag{13.14}$$

$$P_{avg} = P_P D \tag{13.15}$$

$$P_R = \frac{\lambda^2 P_T G^2 \sigma}{(4\pi)^3 r^4} \tag{13.16}$$

$$R = \frac{ct}{2} \tag{13.17}$$

$$R_{max} = \frac{cT}{2} \tag{13.18}$$

$$R_{max} = \frac{c}{2f} \tag{13.19}$$

$$R_{min} = \frac{cT_P}{2} \tag{13.20}$$

$$f_D = \frac{2v_r f_i}{c} \tag{13.21}$$

$$x = y \cos \theta \tag{13.22}$$

GLOSSARY

buncher   in a klystron, a cavity that velocity-modulates the electron beam

bunching   velocity modulation of an electron beam

catcher   in a klystron, a cavity that removes some of the energy from the electron beam and transfers it in the form of microwave energy to the output

cavity   a space in which microwaves can resonate by means of in-phase reflections from the walls

circulator   a device with three or more ports that allows an input to one port to emerge only at the next port in order

collector   in a klystron or travelling-wave tube, the element that receives the electron beam; in a conventional tube, this element is called the anode

crossed-field tube   a microwave tube in which the electric and magnetic fields are at right angles

directional coupler   a device which launches, or receives, a wave in a transmission line or waveguide, in one direction only

dispersion   variation of velocity as a function of frequency in a waveguide or medium

domain   in a semiconductor, a concentration of charge

Doppler effect   change in frequency when a wave reflects from a moving object

Fresnel lens   a lens that is stepped in order to reduce its size

group velocity   the speed of transmission of a signal along a waveguide

Gunn device   a slab of N-type gallium arsenide that can operate as an oscillator or amplifier by means of domain formation

hybrid tee   a combination of E-plane and H-plane tees

IMPATT diode   a junction device that can operate as an oscillator or amplifier, by means of avalanche breakdown

isolator   a waveguide device that has low loss in one direction and high loss in the other

klystron   a type of linear-beam microwave tube that uses velocity modulation of the electron beam

linear-beam tube   a microwave tube in which electrons travel in a straight line down the length of the tube

magic tee   see hybrid tee

magnetron   a crossed-field microwave-tube oscillator in which electrons circle around the cathode under the influence of a magnetic field

microstrip   a microwave transmission line constructed on a printed-circuit board, consisting of a single conductor on one side of the board and a ground plane on the other side

microwave   conventionally, electromagnetic radiation in the range above approximately 1 GHz

mobility   the speed of electron drift in a conductor or semiconductor

mode   in a waveguide, a specific configuration of electric and magnetic fields that allows a wave to propagate

phase velocity   the apparent speed of propagation along a waveguide based on the distance between wavefronts along the walls of the guide

PIN diode   a three-layer diode (P-intrinsic-N) that can be used as a switch and an attenuator at microwave frequencies

precession   in a ferrite, rotation of the axis of rotation of the electrons

radar cross section   the equivalent size of a radar target, compared with a perfectly conducting flat plate oriented toward the receiver

slow-wave structure   in a microwave tube, any device that causes a wave to propagate at less than the speed of light, so that the electron beam and the wave will be moving at approximately the same speed

stripline   a microwave transmission line that consists of a conductor inside a circuit board, working against two ground planes, one on top and one on the bottom of the board

target   in radar, the object whose range, direction, and/or velocity is to be measured

transferred-electron device (TED)   see Gunn device

transponder   a transmitter-receiver combination

TRAPATT diode   a variation of the IMPATT designed for high-power operation

travelling-wave tube (TWT)   a linear-beam microwave tube in which an electron beam gives up energy to a slow-wave structure

tuning screw   a metal object threaded into a waveguide to add capacitance or inductance

waveguide   a hollow structure with no center conductor which allows waves to propagate down its length

YIG   yttrium-iron-garnet, a type of ferrite

QUESTIONS

1. Give the approximate limiting frequencies for the microwave portion of the spectrum.

2. Explain what is meant by the dominant mode for a waveguide, and why the dominant mode is usually the one used.

3. Name the dominant mode for a common rectangular waveguide, and explain what is meant by the numbers in the designation.

4. Explain the difference between phase velocity and group velocity in a waveguide. State which one of these is greater than the speed of light, and explain how this is possible.

5. State what is meant by dispersion, and show how it can arise in two different ways in a waveguide.

6. Give one difference between the use of characteristic impedance in a coaxial line and a waveguide.

7. Draw diagrams showing the correct positions in which to install a probe and a loop to launch the dominant mode in a rectangular waveguide.

8. Explain the function of a directional coupler, and draw a sketch of a directional coupler for a waveguide.

9. Compare stripline and microstrip construction techniques, and draw a sketch of each type.

10. Explain the operation of a hybrid tee.

11. The wavemeter shown in Figure 13.53 is based on a resonant cavity. Which way should the plunger be moved to tune it to a higher frequency?

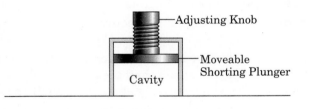

FIGURE 13.53

12. Sketch a four-port circulator and show what happens to a signal entering at each port.

13. Microwave transistors are always NPN if bipolar and N-channel if FETs. Why?

14. Explain what is meant by negative resistance, and give two examples of solid-state microwave devices that have this characteristic.

15. Give two applications each for PIN diodes and varactor diodes.

16. Characterize the microwave tubes studied with respect to the following factors: linear-beam or crossed-field; two- or three-terminal device; amplifier or oscillator; narrow- or wideband.

17. A popular indoor antenna designed for VHF and UHF television reception incorporates a dish-shaped element about 25 cm in diameter. Will this really operate as a parabolic reflector? Explain.

18. Explain how ambiguities can arise when measuring distance with a pulse radar installation.

19. Assuming that a Doppler radar has been correctly calibrated for targets moving directly toward the antenna, will its readings be high or low for targets moving toward it at an angle? Explain.

PROBLEMS

Section 13.2

20. RG-52/U waveguide is rectangular with an inner cross section of 22.86 by 10.16 mm. Calculate the cutoff frequency for the dominant mode and the range of frequencies for which single-mode propagation is possible.

21. For a 10 GHz signal in RG-52/U waveguide (described in Problem 20), calculate:
    (a) the phase velocity
    (b) the group velocity
    (c) the guide wavelength
    (d) the characteristic impedance

Section 13.3

22. (a) Draw a sketch showing how a hybrid tee could be used to connect a transmitter and a receiver to the same antenna in such a way that the transmitter power would not reach the receiver.
    (b) Draw a sketch showing how a circulator could be arranged to do the same thing.
    (c) What advantage does the circulator version of this device have over the hybrid-tee version?

Section 13.4

23. What would be the approximate thickness of a Gunn device intended to operate in the Gunn mode at a frequency of 20 GHz?

24. A crystal oscillator operates at a frequency of 10 MHz with a stability of $\pm 0.002\%$. A series of varactor multipliers is used to multiply this frequency to a nominal frequency of 6 GHz. By how much could the output frequency vary?

Section 13.5

25. A radar transmitter uses a magnetron to generate pulses with a power level of 1 MW. The pulses have a duration of 2 μs and the pulse repetition rate is 500 Hz. The magnetron has an efficiency of 60%. Calculate:
    (a) the duty cycle of the tube
    (b) the peak power input to the magnetron
    (c) the average power output from the magnetron
    (d) the average power input to the magnetron
    (e) the average power dissipated in the magnetron

Section 13.7

26. Calculate the time between the emission of a pulse and the detection of its reflection for a target at a distance of:
    (a) 100 m   (b) 3 km   (c) 75 km

27. Find the maximum and minimum useful ranges for a radar with a pulse duration of 50 μs and a pulse repetition rate of 200 Hz.

28. Calculate the received signal strength if a radar transmits a pulse with a power of 50 kW, which is reflected from a target 20 km away, having a radar cross section of 3 m². The signal frequency is 20 GHz, and the antenna has a gain of 25 dB.

29. A Doppler radar at 15 GHz has a return signal with a frequency 50 kHz higher than the transmitted signal. Assume that the radar installation is stationary.

   (a) What is the component of the target velocity along a line joining the radar and the target?
   (b) Is the target moving toward the radar or away from it?

30. Find the required transmission frequency for a Doppler radar if the frequency difference in hertz is to be equal to the target velocity in meters per second.

31. Rewrite Equation (13.21) so that the frequency difference in hertz is expressed in terms of the target velocity in kilometers per hour and the radar frequency in gigahertz.

32. A Doppler radar is set up beside a highway as shown in Figure 13.54, and measures the speed of a vehicle as 110 km/hr. What is its actual speed?

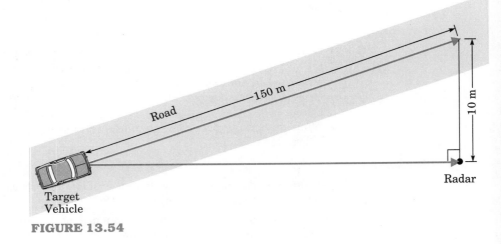

FIGURE 13.54

Comprehensive

33. A geosynchronous communications satellite transmits at 4 GHz with a TWT amplifier having an output power of 7 W. The transmitting antenna is a parabolic dish with an efficiency of 0.8 and a gain of 50 dB. The receiving antenna, located at the equator, is a 3 m dish with an efficiency of 0.65. Calculate:

   (a) the diameter of the transmitting antenna
   (b) the EIRP of the satellite in dBW
   (c) the gain of the receiving antenna in dBi
   (d) the signal strength at the receiving antenna terminals, in dBf

34. Convert the radar equation (Equation 13.16) to a more practical form: antenna gain should be expressed in dBi, distance in kilometers, and frequency in gigahertz.

35. An isolator has a forward loss of 0.7 dB and a return loss of 26 dB. A source provides 1 W to the isolator, and the load is resistive with an SWR of 3.

   (a) How much power is dissipated in the load?
   (b) How much power returns to the source?

36. Which of the active devices studied in this chapter would be most appropriate for each of the following applications, and why?
    (a) a radar transmitter with pulse power of 100 kW
    (b) a low-noise preamplifier operating at 4 GHz
    (c) a low-cost oscillator for low-power applications
    (d) an amplifier with a frequency range of 10 to 20 GHz and a power output of 100 W.

14

Television

Objectives

After studying this chapter, you should be able to:

1. Describe the characteristics of an NTSC television signal (monochrome and color).
2. Calculate horizontal and vertical resolution for video signals.
3. Explain the composite color video system, and compare it with RGB and component systems.
4. Describe the system used for terrestrial television broadcasting.
5. Explain the system used for stereo sound in television, and compare it with the system used in FM broadcasting.
6. Draw block diagrams for monochrome and color television receivers, and explain their operation.
7. Diagnose faults in monochrome and color television receivers.
8. Sketch a typical satellite-television receiving installation and explain its operation.
9. Calculate signal levels at various points in a cable-television system.
10. Explain the need for and operation of cable-television components, such as amplifiers, directional couplers, and converters.
11. Describe some of the problems with the current television standard, and suggest ways in which it could be improved.

Historical Development of Television

It is perhaps surprising that the development of television followed closely after that of radio. Regular radio broadcasting began in 1920; by 1928 there were experimental television systems in existence. Regular broadcasting began in the 1930s in both North America and Europe. It was suspended in most countries during World War II, and became popular immediately after the war. Color television was introduced in North America in 1953.

Early television systems were about equally divided between electromechanical and all-electronic systems. The former used a single photodetector at the camera which scanned the image to be transmitted by means of a complex mechanical system of disks containing holes and slots. At the receiver, another mechanical system, synchronized with the first, directed light from a single source to reproduce an image on a screen.

Electronic television, which with refinements became the system in use today, worked by scanning electron beams by means of magnetic fields. At the camera, the beam scanned a photosensitive surface, creating a current proportional to the intensity of light at a given time at a particular point on the image. At the receiver, the electron beam scanned a phosphorescent material, producing a moving beam of light that drew the image on the screen of a **cathode-ray tube** (CRT). The **luminance** (brightness) of the original image was reproduced by intensity-modulating the electron beam in the receiver CRT. Once again, the transmitter and receiver had to be synchronized.

Figure 14.1 shows an early electronic system. Though they worked, the mechanical systems showed less detail, and they were also subject to mechanical failures. The electronic systems soon prevailed.

FIGURE 14.1
Photo courtesy Zenith Electronics Corporation.

14.1 INTRODUCTION

Television and video systems form a very important part of the communications environment. This chapter will introduce the video system used in conventional television broadcasting, and will also provide insights into the kinds of changes that are made for such applications as high-definition video and computer video displays. In addition, an introduction will be provided to the hardware involved in television receivers and video monitors.

Video systems form pictures by a scanning process. The image is divided into a number of horizontal lines, which are traced out in synchronism at both camera and receiver. The number of lines is arbitrary, but increasing it gives better resolution in the vertical direction. The North American standard uses 525 lines, compared to 625 in Europe. Not all the lines are visible on the television screen: in North America, the number of scan lines actually used to form the image is about 483. The remaining lines are transmitted during the time interval between images.

In order that the image should be visible all at once, rather than as a series of consecutively drawn lines, it is necessary to complete the process quickly. Also, in order to provide the illusion of motion, many images must be drawn in quick succession. The more quickly the images follow one another, the less flicker is present. On the other hand, this requires more information to be transmitted, increasing the required bandwidth. In North America, the images, or **frames**, are sent at the rate of approximately 30 per second; the equivalent number for European systems is 25.

Frame rates of 25 or 30 Hz will cause noticeable flicker. To reduce this, television systems use a technique called *interlaced scan*. This involves transmitting alternate lines of the picture, then returning and filling in the missing lines. Figure 14.2 shows the idea. Each half of the picture thus sent is called a **field**, and of course the field rate is twice the frame rate, or 60 Hz in the North American system. The result is reduced flicker without increased bandwidth.

From Figure 14.2, it can be seen that the picture is scanned from left to right and from top to bottom. The electron beam that traces the picture is

FIGURE 14.2

Interlaced Video

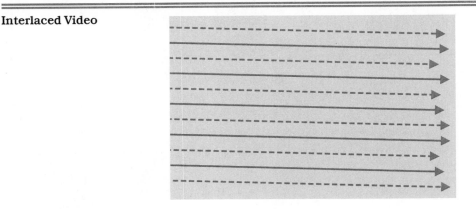

Scan Lines
Field 1 ------- Field 2 ─────

blanked during the time intervals in which the beam **retraces** its path, from right to left and from bottom to top. These are called the *horizontal* and *vertical blanking intervals* respectively.

The ratio of width to height, or **aspect ratio**, is 4:3. If the frame rate is 30 Hz, then the rate at which lines are sent will be 525 multiplied by 30, or 15,750 Hz. These frequency specifications are somewhat arbitrary, but obviously the same standards must be in use at both ends of the communication path.

Good color reproduction can be achieved by mixing three primary colors: red, green, and blue. This requires considerably more information to be transmitted for color television than for monochrome, and it also requires a rather elaborate picture tube containing three electron guns.

14.2 BASEBAND VIDEO SIGNAL

In this section, we shall look at the video signal as it might emerge from a camera and enter a video monitor. Later, we shall see how this signal can be modulated on a carrier and transmitted along with a sound signal by a television broadcasting station.

The signal we will examine most closely is called a **composite signal**, because it combines all the picture information, along with synchronizing pulses. This allows the whole signal to be transmitted on the same cable, or the same radio channel. It is also possible to separate the various components of a video signal, and this is often done when the transmission distance is not great—with computer monitors, for instance.

14.2.1 Luminance Signal

Figure 14.3 shows one line of a monochrome video signal that conforms to the North American standard. Note that the duration of the line is 63.5 μs. This is simply the period corresponding to the horizontal line frequency of 15.75 kHz. About 10 μs of this is used by the horizontal synchronizing (sync) pulse, which will be discussed later. The rest of the time is occupied by an analog signal that represents the variation of the *luminance* (brightness) level along the line.

The video signal shown in Figure 14.3 has negative sync; that is, the sync pulses are in the negative direction. An inverted version of this signal, called *positive sync*, is also possible. The polarity of the luminance portion depends on that of the sync pulses, which are always in the direction of black. Both polarities and a variety of amplitudes are in use within video equipment, but for the interconnection of equipment, the standard is negative sync with an amplitude of 1 V peak-to-peak, into a 75 Ω terminating resistance.

Because of the variety of possible levels and polarities for the video signal, there exists a scale of relative amplitudes that can be used for setting the correct proportions between the level of the synchronizing pulses and the luminance signal. It is called the *IRE scale*, for the Institute of Radio Engineers, a precursor to the Institute of Electrical and Electronic Engineers (IEEE), with which you may be familiar. On this scale, which is also shown in Figure 14.3, zero represents **blanking** level, and -40 represents the level of the sync pulses. The maximum luminance level, called **peak white**, is 100,

FIGURE 14.3   **Monochrome Video Signal**

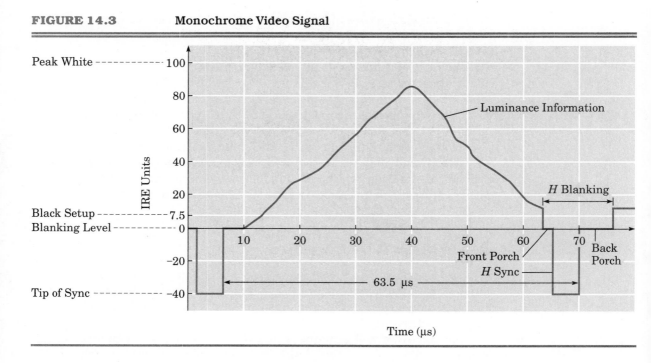

and 7.5 is a level that should represent black at the receiver. From this, it can be seen that the blanking and synchronizing pulses are "blacker than black" and will ensure that the CRT electron beam is completely turned off while it retraces from right to left.

EXAMPLE 14.1

A video signal has 50% of the maximum luminance level. Find its level in IRE units.

Solution

The **black setup** level of 7.5 IRE represents zero luminance, and 100 IRE is maximum brightness. Therefore, the range from minimum to maximum luminance has $100 - 7.5 = 92.5$ units. We must add 50% of this to the setup level of 7.5. Therefore, the level is

$$IRE = 7.5 + 0.5 \times 92.5$$
$$= 53.75 \text{ IRE units}$$

14.2.2 Synchronizing Pulses

In order to synchronize the scanning of a scene at the camera and the receiver, two types of synchronizing pulses are used: horizontal sync, at the end of each line; and vertical sync, at the end of each field. Interlaced scan is accomplished by using two slightly different types of vertical sync, one for odd-numbered and one for even-numbered fields.

While the electron beam in the receiver CRT is returning from right to left and from bottom to top, it must be turned off, or *blanked*. To this end, the sync pulses extend into the blacker-than-black region of the video signal, and are part of the blanking interval, which is also in the blacker-than-black region. Modern receivers also blank the electron beam automatically during retrace.

Figure 14.3 shows the horizontal blanking interval, at the end of each line, for a monochrome signal. The duration of the blanking pulse is approximately 10 µs, of which about one-half is used for the sync pulse itself. The period of blanking before the sync pulse is called the **front porch**, and, naturally, the **back porch** follows the sync pulse. The back porch is longer to allow the electron beam time to move from the right to the left side of the screen.

The vertical blanking interval is shown in Figure 14.4. There are two diagrams, corresponding to the two video fields in each frame. The vertical blanking interval occupies the time required for approximately 21 horizontal lines: about 1.3 ms, compared with 10 µs for horizontal blanking. The time difference allows the receiver to distinguish between horizontal and vertical sync on the basis of duration. It also allows more time for the electron beam to move from bottom to top of the CRT, which simplifies the receiver deflection circuitry.

The structure of the vertical blanking interval is relatively complex. It begins with six **equalizing pulses**, spaced $H/2$ apart, where H is the length of one horizontal line, or 63.5 µs. This takes a total time of $3H$, or 190.5 µs. **Interlace** is accomplished by putting the first equalizing pulse in the middle of a line for one field and at the end of a line for the next.

FIGURE 14.4   **Vertical Blanking Interval**

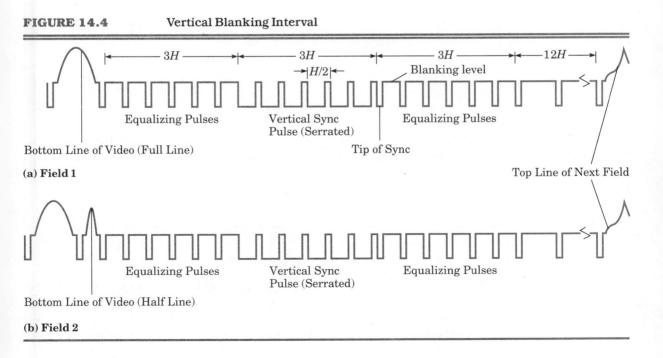

(a) Field 1

(b) Field 2

Since the vertical blanking interval is several lines long, it is necessary to continue to send horizontal sync information during the vertical blanking interval. Otherwise, the horizontal scanning in the receiver might lose synchronization during vertical retrace. For this reason, the vertical sync pulse is *serrated*; that is, the pulse is interrupted at intervals of $H/2$, throughout its duration of $3H$. Following the vertical sync pulse, there are six more equalizing pulses, spaced $H/2$ apart, for a total duration so far of nine horizontal lines. This time may not be sufficient for the electron beam in the receiver CRT to return to the top of the screen, so a further period of about $12H$ consists of blanking level with normal horizontal sync pulses. Often, some of this extra blanking time is used for such purposes as the sending of test signals and closed-caption information.

EXAMPLE 14.2

Calculate the total percentage of the signal time that is occupied by:

(a) horizontal blanking

(b) vertical blanking

(c) active video

Solution

(a) Horizontal blanking occupies approximately 10 μs of the 63.5 μs duration of each line. Therefore,

$$\text{Horizontal Blanking (\%)} = \frac{10}{63.5} \times 100$$
$$= 15.7\%$$

(b) Vertical blanking normally occupies 21 lines per field, or 42 per frame. A frame has 525 lines altogether, so

$$\text{Vertical Blanking (\%)} = \frac{42}{525} \times 100$$
$$= 8.0\%$$

(c) Since 8% of the time is lost in vertical blanking, 92% remains as active lines. Each line loses 15.7% to horizontal blanking, leaving

$$\text{Active Video (\%)} = (100 - 15.7) \times 0.92$$
$$= 77.6\%$$

14.2.3 Resolution and Bandwidth

The resolution of a video system is a function of the number of details that can be seen both horizontally and vertically. Horizontal and vertical resolution are determined quite differently for a television system. The vertical resolution is perhaps more straightforward, since it depends directly on the number of scanning lines. The maximum possible number of details in the vertical direction is the number of visible scan lines, about 483 for the North American system. However, in an ordinary picture, only about 70% of the vertical resolution can be used, because details in the scene that is photographed do not line up exactly with the scan lines in the camera and receiver. This factor of about 0.7 is called the **utilization factor**. The number of details that can be seen vertically is thus about

$$N_V = 483 \times 0.7$$
$$= 338 \tag{14.1}$$

where N_V = number of details in the vertical direction

Horizontal resolution for an analog video system depends on the bandwidth that is allowed. This can be seen by referring to Figure 14.5. Each cycle of the highest-frequency component of the video signal represents two picture details, assuming that a detail is any change in the luminance level.

For a baseband signal, there is no theoretical limit on the bandwidth. A limit is imposed on signals to be broadcast, however, because of the finite width of the broadcast channel. The upper limit for a broadcast video signal is 4.2 MHz. It is easy to find how many details this allows horizontally. First, multiply by the scan time for the visible portion of one line to find the number of cycles per line, then multiply by two. This gives the following result:

$$N_H = (4.2 \times 10^6)(53.5 \times 10^{-6}) \times 2$$
$$= 449 \text{ details} \tag{14.2}$$

where N_H = number of details in the horizontal direction

FIGURE 14.5

Horizontal Resolution

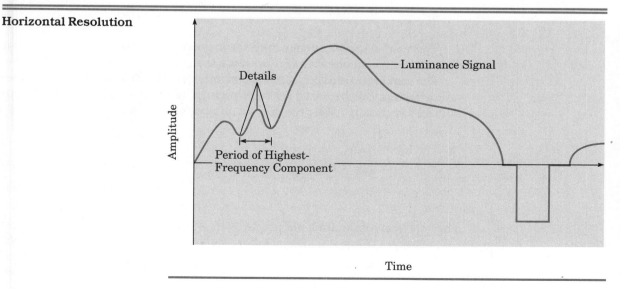

In practice, horizontal resolution for a television system is usually specified in lines rather than details. The aspect ratio of the picture is $4:3$; that is, the picture is one-third wider than it is high. Picture elements will be the same size horizontally as they are vertically if the number of details horizontally is four-thirds of the vertical number. The horizontal resolution in lines is defined as the number of details across three-quarters of the screen. This will give numbers for horizontal and vertical resolution that can easily be compared with each other. For a 4.2 MHz signal, the horizontal resolution in lines is

$$
\begin{aligned}
L_H &= 449 \times 0.75 \\
&= 337 \text{ lines}
\end{aligned}
\tag{14.3}
$$

where L_H = horizontal resolution in lines

This is almost exactly the same as the vertical resolution found above.

Horizontal resolution depends only on bandwidth once the other parameters of a video system, such as line rate and blanking time, are fixed. Therefore it is possible to derive a simple relation between **resolution** and bandwidth. Resolution is proportional to bandwidth; therefore, if 4.2 MHz of bandwidth gives 337 lines of resolution,

$$
\begin{aligned}
\frac{L_H}{B} &= \frac{337 \text{ lines}}{4.2 \text{ MHz}} \\
&= 80 \text{ lines/MHz}
\end{aligned}
$$

where B = bandwidth in megahertz

This relation is usually expressed:

$$
L_H = B \text{ (MHz)} \times 80
\tag{14.4}
$$

There is no low-frequency limit for a video signal. In fact, there will be a dc or near-dc component that represents the average brightness of the scene.

The maximum number of picture elements (**pixels**) in the broadcast television system can be found by multiplying the numbers of details horizontally and vertically. This gives a total number of details of

$$
\begin{aligned}
N_P &= N_H N_V \\
&= 449 \times 338 \\
&= 151{,}761
\end{aligned}
\tag{14.5}
$$

where N_P = total number of pixels
 N_H = number of details horizontally
 N_V = number of details vertically

EXAMPLE 14.3

A typical low-cost monochrome receiver has a video bandwidth of 3 MHz. What is its horizontal resolution in lines?

Solution

From Equation (14.4),

$$L_H = B \text{ (MHz)} \times 80$$
$$= 3 \times 80$$
$$= 240 \text{ lines}$$

The frequency spectrum of a monochrome video signal is not uniform. It should be obvious that there will be strong components at the horizontal and vertical scanning rates, and at harmonics of these frequencies, due to the synchronizing pulses. What is not quite so obvious is that most of the energy due to the picture information is also at multiples of these frequencies. This can be seen by referring to Figure 14.6, which shows a section of a picture and several horizontal scanning lines. For a typical image, one scan line is much like the next. Therefore, the whole signal repeats at the horizontal rate, and the smaller details along the lines cause frequency components at multiples of that rate. Diagonal lines, and areas with motion, do create components at one-half the horizontal line rate, as well as at odd multiples of that frequency.

A similar argument can be made for repetition at the vertical scanning rate. Usually, the two fields that make up one frame are very similar since they are two views of the same image shifted by one horizontal line. Therefore, we expect a spectrum like that shown in Figure 14.7, with peaks at multiples of the horizontal scan rate, and with a finer line structure showing peaks at 60 Hz intervals. The pattern is not perfect, of course: horizontal lines in the picture, abrupt changes of scene, and motion in the picture all

FIGURE 14.6

Components at
Multiples of Horizontal
Frequency

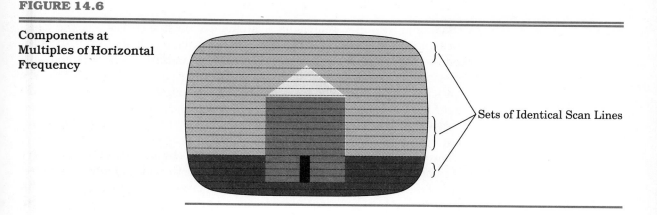

Sets of Identical Scan Lines

FIGURE 14.7

Baseband Spectrum of a
Video Signal

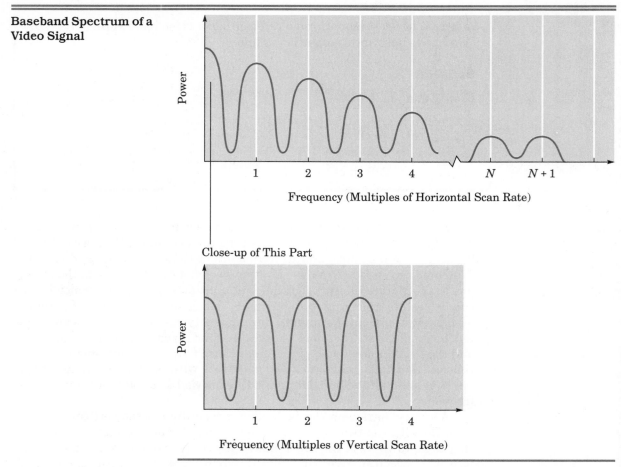

disturb it. Nonetheless, the frequency distribution of the signal follows the suggested pattern quite closely. This fact can be used to advantage in the transmission of color, to be considered in the next section.

14.2.4 The NTSC Color System

In 1953, The National Television Systems Committee (**NTSC**) of the Electronics Industries Association (**EIA**), an American industry group, established the first color television standard. This standard is still used today in North and Central America, most of South America, and Japan. There are two other standards, known as *PAL* (Phase Alternation by Line) and *SECAM* (Sequential Color and Memory) that were developed in Germany and France, respectively. SECAM is used in France and eastern Europe, and PAL is employed in most of the rest of the world.

As was mentioned in the introduction to this chapter, it is not necessary for a color video system to actually reproduce all the colors found in nature. In fact, due to the way in which color is perceived by the human eye-brain combination, it is only necessary to transmit information for three colors: red, green, and blue. Even this is a rather daunting requirement, since at first glance it would appear to triple the bandwidth required for a color, as compared to a monochrome, video signal. Indeed, there are video systems that transmit red, green, and blue signals separately on three separate conductors. These are called **RGB** systems. Most color computer monitors use three separate lines included in one cable for the three colors, with additional lines for synchronizing information.

RGB video is not suitable for conventional television broadcasting because of bandwidth limitations. There was also a requirement, when the NTSC color system was developed, to maintain compatibility with monochrome; that is, it was required that an existing monochrome receiver should be able to receive a color television signal.

In order to maintain compatibility, the composite color system includes a luminance signal. Two other signals are multiplexed onto it to provide color. Only two are needed because the original luminance signal is a combination of red, green, and blue. Two other signals, each a linear, independent combination of the three colors, provide enough information to reconstruct the original red, green, and blue signals. If the red, green, and blue signals are normalized (that is, they have amplitude values between zero and one), then the equations for the three signals that make up the composite color signal are as follows.

$$Y = 0.30R + 0.59G + 0.11B$$
$$I = 0.60R - 0.28G - 0.32B \qquad (14.6)$$
$$Q = 0.21R - 0.52G + 0.31B$$

where Y = luminance signal
I = in-phase component of the color signal
Q = quadrature component of the color signal

The terms *in-phase* and *quadrature* refer to the process by which the **chrominance** (color) signal is combined with the luminance signal, and will be made clear very shortly.

There are several noteworthy things about these equations. First, the luminance signal is not composed of equal amounts of red, green, and blue, as you might suppose. This is because the frequency response of the human eye is not flat; we are much more sensitive to green light than to red or blue. By assigning the three primary colors the weights given above, the system designers ensured that monochrome pictures would replace the various colors with shades of grey whose luminance on the television screen corresponds to the brightness they would have when viewed "live." The two chrominance signals will have zero amplitude when a grey or white object is viewed; that is, when the values of R, G, and B are all equal.

EXAMPLE 14.4

An RGB video signal has normalized values of $R = 0.2$, $G = 0.4$, $B = 0.8$. Find the values of Y, I, and Q.

Solution

From Equation (14.6),

$$Y = 0.30R + 0.59G + 0.11B$$
$$= (0.30 \times 0.2) + (0.59 \times 0.4) + (0.11 \times 0.8)$$
$$= 0.384$$

$$I = 0.60R - 0.28G - 0.32B$$
$$= (0.60 \times 0.2) - (0.28 \times 0.4) - (0.32 \times 0.8)$$
$$= -0.248$$

$$Q = 0.21R - 0.52G + 0.31B$$
$$= (0.21 \times 0.2) - (0.52 \times 0.4) + (0.31 \times 0.8)$$
$$= 0.082$$

The I and Q signals are modulated onto a subcarrier at approximately 3.58 MHz, using suppressed-carrier quadrature AM. The subcarrier frequency is chosen to be an odd multiple of one-half the horizontal frequency ($227.5f_H$), so that the chrominance sidebands will tend to fit between luminance sidebands, and so that interference between color and luminance will be less visible on the screen. The diagram in Figure 14.8 shows how this works. Recall from the previous section that the spectrum of a video signal has most of its energy at multiples of the horizontal line rate. Setting the color subcarrier at an odd multiple of one-half the horizontal frequency ensures that the maxima of the luminance and chrominance signals (often called **luma** and **chroma**) will not occur at the same frequencies. This allows the chroma and luma frequency ranges to overlap, and reduces the total bandwidth required for the signal.

If the chroma signal were allowed to extend right down to zero frequency, there would be excessive interference between luma and chroma. On the

FIGURE 14.8

Spectrum of a Color
Video Signal

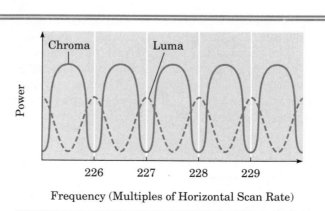

Frequency (Multiples of Horizontal Scan Rate)

other hand, if the signal is to be broadcast, its upper frequency limit must be limited to 4.2 MHz. Therefore, the chroma signal must be band-limited. The Q signal is cut off at 0.5 MHz and transmitted using both sidebands, while the I signal has a cutoff at about 1.3 MHz for the lower sideband and 0.5 MHz for the upper; that is, the I signal uses vestigial sideband (VSB) modulation. This allows the receiver to take advantage of the extra I bandwidth to improve the color resolution. The combination of colors used for the I signal places it in a region where the acuity of the eye for color, never as good as for monochrome, is maximum.

Even if the receiver takes advantage of the entire I bandwidth (and most do not), the horizontal resolution for color is much less than for the luminance signal. With a bandwidth of 1.3 MHz, the color resolution is

$$L_H \text{ (color)} = 80 \times 1.3$$
$$= 104 \text{ lines} \tag{14.7}$$

and, for the more common situation where the color bandwidth is restricted to 0.5 MHz, the resolution is

$$L_H \text{ (color)} = 80 \times 0.5$$
$$= 40 \text{ lines} \tag{14.8}$$

This is much lower than the luminance resolution; the difference can readily be seen when observing stationary alphanumeric characters on a television screen. The colors will appear to "bleed" between the characters and the background, assuming that the background color is different from that of the lettering. The lack of color resolution is usually less apparent with ordinary program material. The vertical resolution depends on the number of scanning lines, so of course it is the same for color as for luminance.

As with any suppressed-carrier system, the receiver is required to regenerate the color subcarrier. This requires information about the subcarrier frequency and phase to be transmitted. In color television, this is accomplished by adding a color burst consisting of eight to eleven cycles of the subcarrier to the back porch of each horizontal sync pulse, as shown in Figure 14.9. The receiver also uses the color burst to sense the presence of a color signal.

FIGURE 14.9

Color Burst

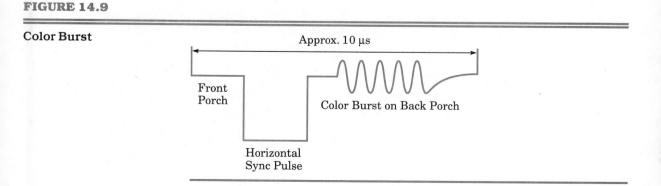

▄▄▄ 14.3 TELEVISION BROADCASTING

In ordinary VHF and UHF television broadcasting, the picture and sound are transmitted using separate modulation schemes, on separate carriers (sometimes with separate transmitters). The two carriers are 4.5 MHz apart, and the whole channel has a width of 6 MHz. See Figure 14.10 for the spectrum of channel 2, the lowest in frequency. (See Appendix H for a complete list of television channel frequencies.)

The picture signal uses vestigial-sideband AM, with a full upper sideband and a band-limited lower sideband. This reduces the bandwidth compared with double-sideband AM, while avoiding the problems that single-sideband systems have in transmitting low-frequency information. The picture carrier is not suppressed so that critical tuning is avoided and a simple envelope detector can be used for the video signal. The color modulation appears on the upper sideband only of the main picture carrier.

The sound signal uses wideband FM and employs a stereo system similar, but not identical, to that used for stereo FM radio broadcasting.

14.3.1 Picture Signal

Figure 14.11 shows the AM picture signal in the time domain. As can be seen from the figure, sync pulses are transmitted with the maximum signal amplitude, and hence the maximum transmitter power, and black represents a higher amplitude than white. This is called *negative picture transmission*. Table 14.1 shows the relative levels of the various parts of the signal, in terms of percentage of the maximum amplitude.

Negative picture transmission uses less transmitter power than would the reverse polarity, since the duration of the sync pulses is relatively short. It also ensures that the sync pulses will be received in the presence of noise. This is important since, without sync, the picture is completely unwatchable. Impulse noise will tend to cause black spots in the picture, which are thought to be less disturbing than white spots. Note that the signal level never

FIGURE 14.10

Spectrum of Channel 2

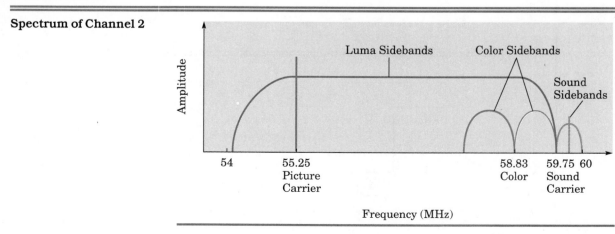

FIGURE 14.11   Television Picture Signal in the Time Domain

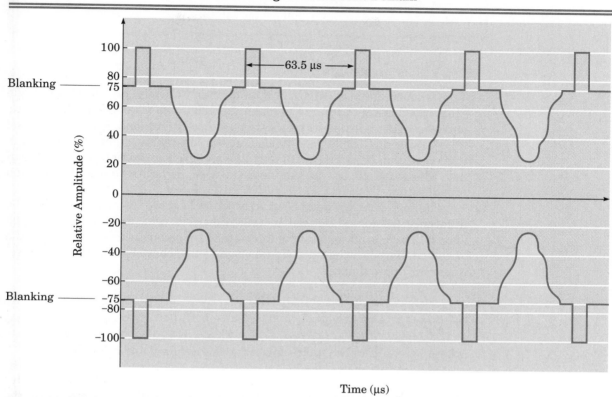

Time (μs)

TABLE 14.1

Broadcast Video Levels

Signal	IRE Level	Carrier Level (Percent)
Tip of sync	−40	100
Blanking	0	75
Black setup	7.5	67.5
Peak white	100	12.5

reaches zero; this would disturb the operation of the intercarrier sound system in modern receivers, which will be described shortly.

EXAMPLE 14.5

What proportion of the maximum transmitter power is used to transmit a black setup level?

Solution

The proportions in Table 14.1 are voltage levels and will have to be squared for power. Therefore, the power level, as a fraction of the maximum, is

$$P \text{ (black setup)} = 0.675^2$$
$$= 0.456 \quad \text{or} \quad 45.6\%$$

14.3.2 Sound Signals

Television sound is wideband FM, similar to FM radio broadcasting. While radio broadcasting uses a total maximum deviation of 75 kHz for both mono and stereo, the television specification is 25 kHz for mono. For multichannel television sound (MTS), the television stereo sound system, the maximum total deviation is 75 kHz, with only 25 kHz in the mono (left-plus-right) signal. The same 75 μs pre-emphasis is used in both radio and television broadcasting.

The sound carrier must be exactly 4.5 MHz above the picture carrier: the difference frequency is used in television receivers, as will be explained shortly. There is a possibility of interference in receivers due to mixing of the sound carrier with the color signal on the same channel. The difference frequency is approximately

$$f_d = 4.5 \text{ MHz} - 3.58 \text{ MHz}$$
$$= 0.92 \text{ MHz} \tag{14.9}$$

where f_d = difference frequency between the sound and color signals

This 920 kHz signal will cause visible interference if it gets into the video signal path in the receiver. To reduce the visibility of this interference, the designers of the NTSC color system changed the horizontal scanning rate slightly so that the difference between the color subcarrier frequency and the sound carrier frequency would be an odd multiple of one-half the horizontal scanning frequency. The color subcarrier is set at 3.579545 MHz. Now the difference frequency is

$$f_d = 4.5 \text{ MHz} - 3.579545 \text{ MHz}$$
$$= 0.920455 \text{ MHz}$$

The horizontal scanning frequency f_H is changed so that the color subcarrier is still at $227.5f_H$:

$$f_H = \frac{3.579545 \times 10^6}{227.5}$$
$$= 15.734 \times 10^3 \text{ Hz}$$
$$= 15.734 \text{ kHz}$$

The ratio between f_d and the horizontal scanning frequency is

$$\frac{f_d}{f_H} = \frac{920.455 \times 10^3}{15.734 \times 10^3}$$
$$= 58.5$$

That is, the interference frequency is an odd multiple of $f_H/2$.

14.3.3 Multichannel Television Sound

The television stereo sound system is called multichannel television sound. It is similar to the system used for FM radio broadcasting with the following differences. The spectra illustrated in Figure 14.12 will help in understanding the differences. Figure 14.12(a) shows the spectrum of a stereo FM radio signal, before modulation, and Figure 14.12(b) shows a typical multichannel television sound signal.

Perhaps the most obvious difference between radio and television stereo sound is that the pilot frequency for television is equal to the horizontal scanning frequency of 15.734 kHz, instead of the 19 kHz frequency used for FM radio. As before, the left-minus-right (L − R) signal is modulated on a subcarrier at twice the pilot-carrier frequency, using double-sideband suppressed-carrier (DSBSC) modulation.

Another change for television sound is that noise reduction is used for the L − R channel. Recall from Chapter 8 that stereo FM radio broadcasting has a severe noise penalty compared with mono. This results from shifting the L − R signal upward in frequency during the modulation process. With any FM system, the demodulated noise voltage is directly proportional to frequency. The additional noise is moved downward in frequency, into the audible range, during demodulation. A noise reduction scheme is used to reduce this problem in the MTS system.

FIGURE 14.12

Stereo FM Radio and Television Sound

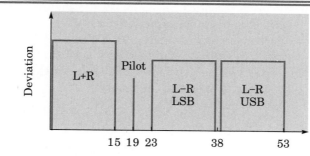

(a) FM Radio Stereo Multiplex Spectrum

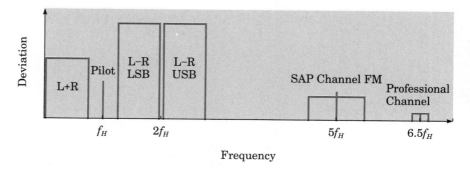

(b) Typical MTS Television Sound Spectrum

The noise reduction scheme involves gain compression at the transmitter, followed by expansion at the receiver. A *compressor* is an amplifier that has more gain for low-level signals than for those with greater amplitude. The result of applying compression is to produce a greater modulation index for the low-level signals, and therefore to produce a greater signal-to-noise ratio for these signals. In order to restore the original relationship between loud and quiet sounds, the receiver must reverse the compression process. The amplifier that does this is called an *expander*. It has more gain for high-level than for low-level audio signals. The combination of compressor and expander should have the same gain at all frequencies in the audio range.

Noise reduction is not used for the mono L+R channel in television sound, in order to maintain compatibility with receivers that are not equipped for MTS.

The multichannel television sound system also provides for a separate audio program (SAP) and one or more professional channels. The SAP can be received by the public and is intended to be used for services such as second-language translation. It can also be used for a completely separate audio program such as background music or weather information. The professional channels are for use by the television station for its own purposes, such as sending messages to employees in the field.

14.4 TELEVISION RECEIVERS

As we have seen, the color television signal evolved from the monochrome signal. In a similar way, color television receivers evolved from their monochrome predecessors. Consequently, the easiest way to understand television receivers is to begin with monochrome, then add the color circuitry once the monochrome receiver is understood. The procedure with the monochrome receiver will be first to examine the stages that process the incoming signal, then to look at the circuits that scan and synchronize the CRT electron beam and that provide power to the receiver.

14.4.1 Signal Processing in a Monochrome Receiver

Figure 14.13 is the block diagram of a monochrome receiver, with the signal-processing circuitry highlighted. Note the similarity to the superheterodyne radio receivers studied earlier.

The incoming signal is applied to the *tuner*, which is a separate box that includes the RF amplifier, mixer, and local oscillator circuitry. In fact, the traditional way is to use two boxes, one each for VHF and UHF. In this system, which can be recognized by the set of two channel-switching knobs on the front panel, the UHF tuner has no RF stage, and the VHF tuner is configured as an additional stage of intermediate frequency (IF) amplification when the receiver is receiving a UHF channel. Figure 14.14 shows how this works. The IF is the same for VHF and UHF. The signal from the UHF tuner is fed to the VHF tuner for additional IF amplification, but not frequency conversion. The local oscillator in the VHF tuner is off when UHF is selected.

The standard IF is 45.75 MHz for the picture carrier. Since high-side injection of the local oscillator is used, the sound carrier will be converted to an IF of 41.25 MHz, 4.5 MHz below the picture carrier. Figure 14.15 shows the spectrum of a television signal in the IF circuitry.

FIGURE 14.13   Monochrome Television Receiver

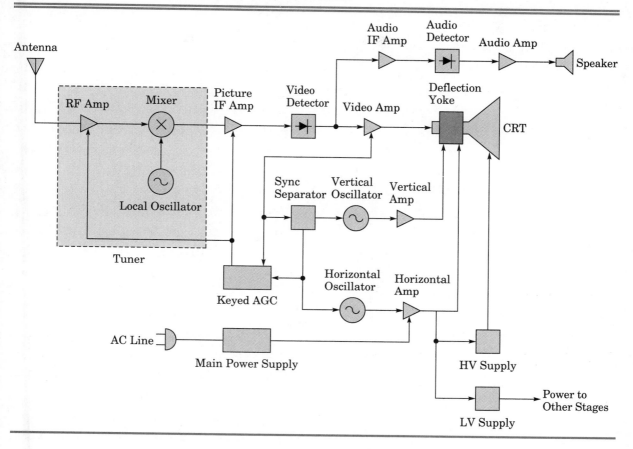

Most modern receivers use some form of varactor tuning, rather than a mechanical switch, to change channels. Varactor tuning allows easy remote control and programmability, as all tuning can be accomplished using variable dc voltages. Varactor tuners also use diode switching to change among three frequency bands: VHF-Low (channels 2–6), VHF-High (channels 7–13), and UHF. The most advanced tuners use frequency synthesis to generate the local oscillator signal.

Following the tuner, the signal passes to the picture IF amplifier. The term is actually a misnomer, for this section of the receiver also handles the sound signal. The IF amplifier looks much like those studied earlier, except that its bandwidth of about 6 MHz is much wider. The shape of the IF passband is not flat: the spectral response has the form of a "haystack curve" as shown in Figure 14.16. It has more gain for those parts of the video signal that are transmitted on only one sideband, and less gain for the sound signal, to reduce interference between the sound and picture signals. Notch filters, or *traps*, are often present, tuned to the adjacent-channel picture and sound carrier frequencies. These filters reduce adjacent-channel interference.

Older designs have several stages using discrete transistors, employing several tuned circuits to shape the IF amplifier response, but newer versions often use a single integrated circuit with a surface-acoustic-wave (SAW) filter. A SAW filter consists of two sets of electrodes mounted on a piezoelectric sub-

FIGURE 14.14

Television Tuner

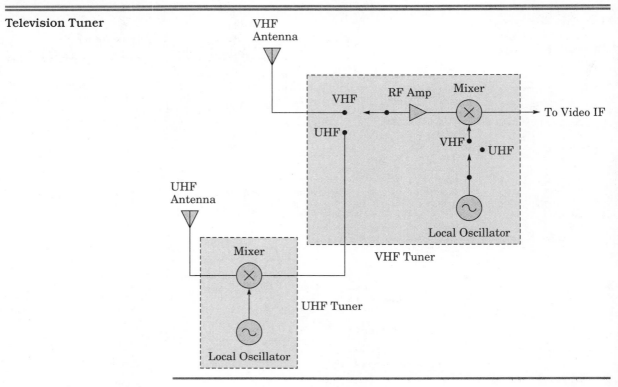

strate. Voltage applied to the input causes vibrations in the substrate, which are transmitted acoustically through the crystal and converted back to electrical energy at the output. The frequency response of the filter depends both on the substrate and the electrode configurations. Unlike multiple tuned circuits, SAW filters do not need alignment.

FIGURE 14.15

Television Signal
Spectrum at Tuner
Output

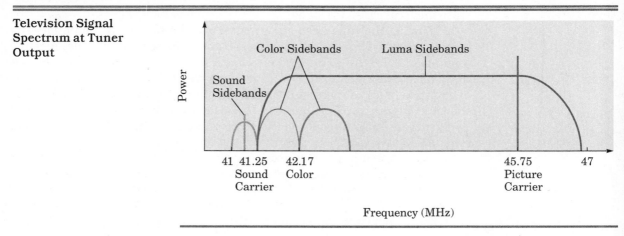

FIGURE 14.16

Television Receiver
Picture IF Amplifier
Frequency Response

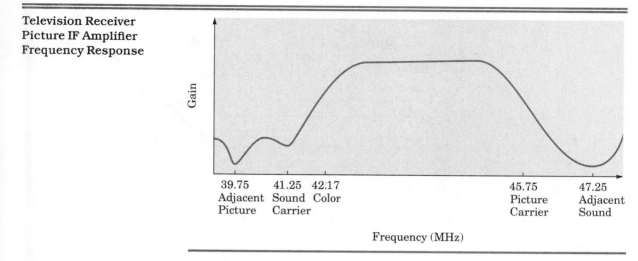

Frequency (MHz)

Just as in the receivers studied earlier, there will be an automatic gain control (AGC) circuit to allow the receiver to cope with the dynamic range of available signals. In television receivers, however, the AGC is described as *gated* or *keyed*, which simply means that the AGC looks at the signal level during the blanking interval, monitoring either the blanking or sync-pulse level, and adjusting the IF gain accordingly. The advantage of this is that the relative levels of blanking and sync are fixed, and vary only with the strength of the signal. Other video levels are affected by picture content.

The stage following the video IF amplifier is the *video detector*. This is usually a diode envelope detector followed by a low-pass filter. As well as demodulating the video signal, the diode provides a 4.5 MHz sound IF signal, because of the mixing between sound and picture carriers that occurs in the diode. This technique is called **intercarrier sound**. Since no separate local oscillator is required for the sound signal, the sound will automatically be correctly tuned whenever the receiver is tuned for the best picture.

Following the video detector, the sound and picture signals go their separate ways. The sound is still modulated and must go through a sound IF amplifier, operating at 4.5 MHz, followed by an FM detector. The video, on the other hand, is now a baseband signal and can proceed to the video amplifier and then to the CRT.

The video amplifier provides a positive-sync video signal at the CRT cathode. It has a 4.5 MHz trap at its input to reduce interference from the sound signal. The contrast control is a video gain control.

14.4.2   The Monochrome Cathode-Ray Tube

The basic structure of a monochrome CRT is shown in Figure 14.17. The electron gun emits a beam of electrons the intensity of which is controlled by the video signal. Inside the electron gun, a hot cathode emits electrons by thermionic emission. The control grid (G_1) is actually a cylinder, maintained at a negative potential with respect to the cathode. In modern circuits, the grid is

FIGURE 14.17   Monochrome CRT

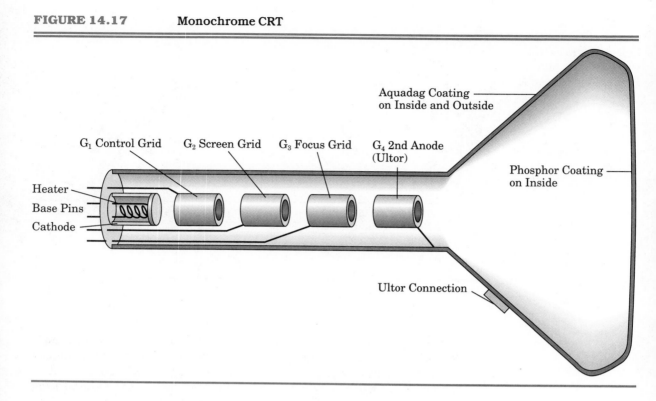

kept at ground potential, and the cathode is about 50 V positive. The brightness control adjusts the bias, and the positive-sync video signal is also applied to the cathode. As the video signal becomes more positive, the intensity of the electron beam is reduced.

The screen grid (G_2), sometimes called the *first anode*, is maintained at a positive voltage of about 400 V and serves to accelerate the electron beam. The screen grid and the focus grid (G_3) form an electrostatic lens to focus the electron beam on the screen. The second anode (G_4) is connected to the high-voltage supply, which in a monochrome receiver is about 10 to 12 kV, and accelerates the beam to its final velocity. The beam current is low, generally less than 1 mA.

The second anode is often called the **ultor**. Its connection is on the flared part of the tube, called the *bell*, to avoid arcing, which could occur if the high voltage were applied to the pins at the end of the tube. Both the inside and outside of the tube have a conductive coating called the *aquadag*. The inside coating is at the second anode potential, and the outside is grounded. The coatings provide shielding, and the capacitance between the two coatings is used as a filter in the high-voltage power supply.

The electron beam strikes a phosphor coating on the inside of the CRT faceplate. For a monochrome tube, the phosphor is white with a persistence of about 5 ms. There is a thin aluminum coating on the inside of the faceplate, to provide a return path for electrons.

In order to trace out the lines that form an image on the screen of the tube, the beam is deflected by electromagnets that surround the tube, and are collectively called the **yoke**. They are provided with sawtooth current waveforms by special circuits which will be described shortly. Permanent magnets are also used for beam centering and to correct distortion around the edges of the screen.

14.4.3 Scanning and Synchronization

A television receiver needs some means of producing the required sawtooth current waveforms in the vertical and horizontal deflection coils. In addition, these waveforms must be synchronized with the incoming signal. The means of doing this differ from one receiver to another, with modern designs using one integrated circuit that separates the sync pulses from the rest of the signal, further separates horizontal from vertical sync, and generates horizontal and vertical scanning signals that are properly synchronized. All that remains are the horizontal and vertical output stages, which use too much power to be included in the integrated circuit. Older designs use a number of discrete transistors (or tubes, in still older circuits) to perform the same functions.

Whether part of an integrated circuit or made up of discrete components, the sync separator uses the difference in amplitude between the sync pulses and the rest of the video signal to detect sync pulses. Then, it uses the difference in duration to separate horizontal from vertical sync. That can be accomplished by a simple integrator like the one shown in Figure 14.18. The capacitor does not have time to charge very far during the brief horizontal sync pulses, and it discharges completely between pulses. The vertical sync pulses are much longer, however, and allow the capacitor time to charge completely.

The separated horizontal and vertical sync pulses are used to control the frequency of the horizontal and vertical oscillators, which must also be free-running in the absence of sync pulses so that there will be a **raster** on the screen in the absence of a signal. (*Raster* is the term used to describe the pattern of scanning lines on the screen.)

The output from each oscillator must be amplified and shaped to produce the required sawtooth current waveform in the deflection coils. In the vertical coils, the current and voltage waveforms will have roughly the same shape, and the amplifier circuitry will resemble an ordinary push-pull Class B amplifier.

The horizontal output circuitry is less conventional. The deflection coils will have considerable inductive reactance, causing the voltage waveform across them to resemble a train of pulses more than it does a sawtooth. There is also a good deal of power required, since the coils store energy in their magnetic fields. The circuitry used for horizontal deflection circuits is de-

FIGURE 14.18

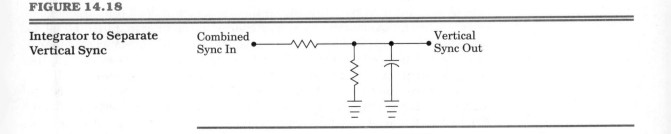

Integrator to Separate Vertical Sync

Combined Sync In — Vertical Sync Out

FIGURE 14.19   **Simplified Horizontal Output Circuit**

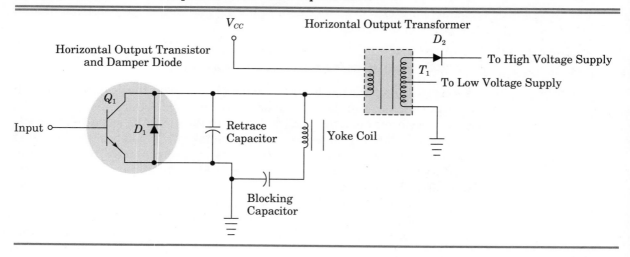

signed to increase efficiency by recycling some of the stored energy, using a resonant circuit to do so.

Figure 14.19 is a simplified schematic diagram of a typical horizontal output circuit. The transistor Q_1 is a specialized type with very high breakdown voltage. The damper diode D_1 is often combined with the transistor in a single package, as shown. T_1 is the horizontal output transformer, often called the *flyback transformer*. Often D_2, the high-voltage rectifier, is really a voltage-tripler circuit that is enclosed within the package containing T_1.

The input to this circuit is in the form of pulses that occur once per horizontal line. The output consists of a sawtooth current waveform in the horizontal deflection coils. The voltage pulses necessary to provide this are also used, suitably changed in amplitude, to provide both the high voltage for the CRT and low voltages for many of the other circuits in the receiver.

Figure 14.20 illustrates some of the important waveforms that occur in this circuit. The simplest way to analyze the circuit is to begin in the center of the trace. When the beam is in the center of the screen, there is no current flowing in the deflection coils. A current pulse flows into the base of Q_1 (Figure 14.19) at this time, saturating it. The current through the transistor and the deflection coils increases gradually, because of the inductance of the coils. The graph of current with respect to time is exponential, of course, but it appears to be almost a straight line because we are looking at only the beginning of the curve.

As the deflection-coil current increases the electron beam moves at a constant rate toward the right of the screen. When it gets there, the base current into Q_1 is reduced to zero, turning off the transistor. In fact, the current goes negative briefly to remove any stored charge in the base and turn Q_1 off as quickly as possible. This is necessary because the transistor must dissipate a good deal of heat when it is in the active region. When it is saturated, the power dissipation is small, and when cut off, it dissipates no power at all.

Of course, the current through the deflection coils cannot be reduced to zero instantaneously, because of the energy stored in their magnetic fields. Current continues to flow, charging the retrace capacitor C_1. After several

FIGURE 14.20   **Horizontal Output Circuit Waveforms**

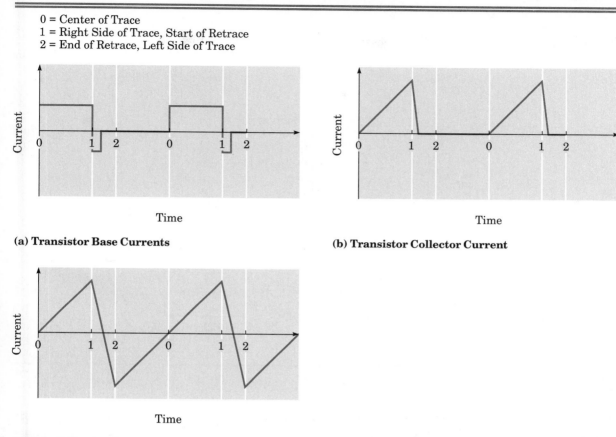

0 = Center of Trace
1 = Right Side of Trace, Start of Retrace
2 = End of Retrace, Left Side of Trace

(a) Transistor Base Currents

(b) Transistor Collector Current

(c) Yoke Inductor Current

microseconds, the current has been reduced to zero, at which point the capacitor begins to discharge into the coil. In other words, there is a resonant circuit. Damped oscillations would continue for some time except for the fact that, just as the capacitor finishes discharging, and the electron beam reaches the left of the screen, the damper diode begins to conduct, and the coils discharge the remainder of their energy through the diode. By the time this has occurred, the electron beam is once more in the center of the screen and the process is ready to repeat.

The presence of a large amount of power at a frequency of about 16 kHz is used by receiver manufacturers to simplify power-supply design. Figure 14.21 is a representative example. Typically, there will be either a full-wave bridge or a half-wave rectifier operating directly from the ac line voltage, without a power transformer. After filtering, this will produce a voltage on the order of 160 V, which can be used to power some stages, notably the horizontal output. The rest of the voltages are derived from the horizontal output transformer. The whole operation somewhat resembles a switching power supply, in that advantage is taken of the reduced requirements for transformer iron and filter capacitance at the higher frequency. The use of a transformerless supply does represent a safety hazard when working on television

FIGURE 14.21

**Television Receiver
Power Supply**

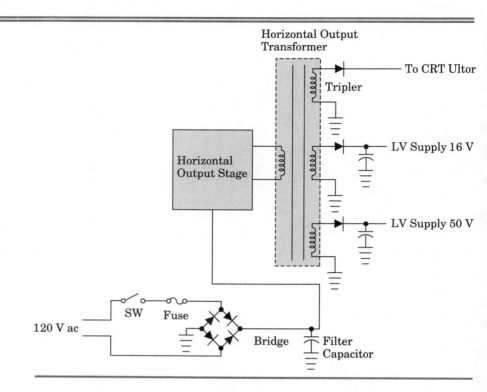

receivers, since one side of the power line is often connected directly to the
chassis. An isolation transformer, which is simply a transformer with a 1:1
turns ratio and sufficient power capacity to supply the power required by the
receiver, should always be used when working on television receivers.

Since the horizontal output circuit supplies much of the power used in
the receiver, a failure in the horizontal circuitry is likely to result in a television that shows little if any sign of life. This is a useful thing to remember
when troubleshooting a television receiver.

14.4.4 Color-Signal Processing

Thus far, we have been looking at monochrome receivers. They are still made,
of course, but mainly for use as second receivers, for kitchens, as portables,
and so on. However, we will see that what we have learned in our study of
monochrome receivers is directly applicable to color television.

Figure 14.22 is the block diagram of a typical color receiver. Some similarities should be immediately apparent. The tuner and video IF sections are
unchanged from before, on the block diagram. Actually there are likely to be
a few differences in circuitry. Color receivers are far more likely to use more
advanced technology such as varactor tuners, frequency synthesis, and remote control. Likewise, the audio section is similar, but color receivers are
much more likely to incorporate stereo sound. Raster and synchronizing circuitry is the same for both types, except that color receivers use a higher
accelerating voltage for the CRT.

FIGURE 14.22   Color Television Receiver

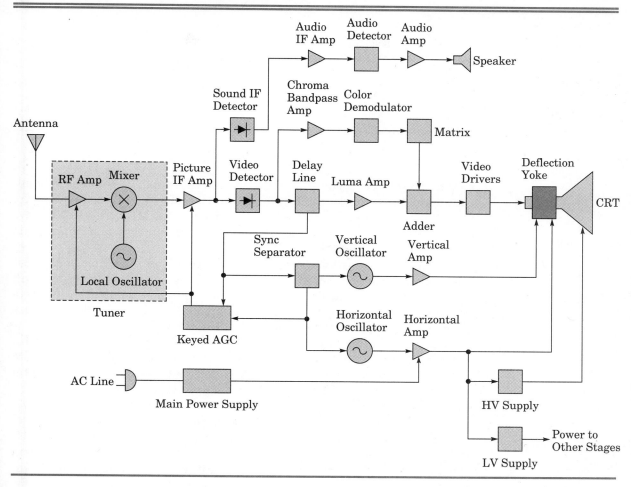

What, then, are the essential differences between monochrome and color receivers? Mainly they have to do with the way in which the chroma signal is demodulated and the result combined with the luma. The composite color signal must be converted back into an RGB signal before it is applied to the color CRT.

Proceeding along the signal chain from the antenna, the first notable difference is that there is a separate sound IF detector, which separates out the 4.5 MHz sound IF signal. In the monochrome receiver, this function was quite adequately served by the video detector, but this is inadvisable in a color receiver. If the sound signal were allowed to enter the video detector, undesirable interference could result from the mixing of the sound carrier and the color signal. This is avoided by taking off the sound at an earlier point and installing a 41.25 MHz trap to remove the sound signal before the video detector.

Following the video detector, the chroma signal must be separated from the luma and sent to the color demodulators. There are two common ways to

FIGURE 14.23

Comb Filter

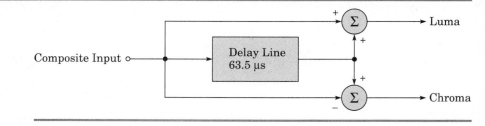

do this. The simplest, shown in Figure 14.22, is to apply the signal to a band-pass amplifier with its center frequency at the color subcarrier frequency of approximately 3.58 MHz, and a bandwidth of about 1 MHz. This will remove most of the luma, though of course the highest-frequency luma components will pass through the filter.

Similarly, there will be some chroma information in the luma signal. Because of the frequency interleaving described earlier, any interference will not be very serious. The delay line shown in the luma signal path is there to compensate for the delay that the chroma signal experiences in the bandpass filter, so that chroma and luma arrive at the CRT at the same time.

There is a better way to separate luma from chroma, by using a **comb filter**. Figure 14.23 shows how this works. The composite signal is applied to the input of a $1H$ delay line, that is, one with a delay of exactly one horizontal line period, or 63.5 μs. The original and delayed signals are added to give the luminance output, and subtracted to produce the chroma.

The operation of a comb filter can be understood by considering a signal with a period of $1H$ (63.5 μs). The delay will result in a phase shift of 360°, and the delayed signal will be indistinguishable from the original. Assuming that the amplitudes are made equal, the output of the summer will be enhanced, while the subtractor will have zero output. The same logic holds for any harmonic of the horizontal scanning frequency. If we recall that luminance signals have most of their power at or near the harmonics of the horizontal line rate, it is apparent that the luma signal will emerge from the filter almost unchanged.

Now consider a signal with a frequency one-half that of the horizontal scanning frequency; that is, its period will be $2H$, and the delay line will shift its phase by 180°. Summing the delayed and original signals will result in cancellation, but subtracting will result in reinforcement. It can easily be shown that the same is true for odd multiples of one-half the horizontal line rate. Recalling that the chroma signal is modulated on a subcarrier at 227.5 f_H, it is apparent that the subtractor output will have most of the chroma, and very little of the luma. Comb filters are used in most new designs.

Once the chroma signal has been separated from the luma, it must still be demodulated. The obvious solution would be to demodulate the QUAM color signal into the original I and Q signals using two balanced modulators, each supplied with a carrier of the correct phase. Of course, those carriers have to be synchronized with the original subcarrier, a relatively easy job since the **color burst** provides a sample once per line.

FIGURE 14.24   Color Demodulation Using *I* and *Q* Axes

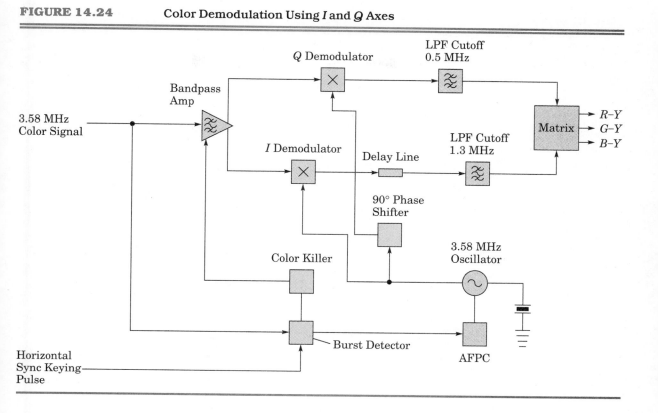

Unfortunately the reality of a demodulator using the I and Q axes is fairly complex. Figure 14.24 shows how it can be done. The extra filters take into account the different bandwidths of the I and Q signals, and the extra delay line compensates for the different delays in filters of different bandwidth. The initials AFPC, by the way, stand for *automatic frequency and phase control*, a system that phase-locks a crystal oscillator to the exact frequency and phase of the color burst. This in turn requires that the burst be separated from the rest of the signal. Its position right after horizontal sync makes this fairly easy. The block marked *color killer* does what it says: it turns off the color circuitry when there is no color burst.

It is possible to simplify the color circuitry somewhat by using two different axes to demodulate the color signal. This is possible as long as the axes chosen are at a phase angle of 90° to one another. Of course, the signals that result are combinations of I and Q, but by judicious choice of axes, it is possible to end up with the two signals $R-Y$ and $B-Y$.

Figure 14.25 shows a color demodulator using the $R-Y$ and $B-Y$ axes. The number of filters is reduced, and there is no need for a delay line. This is because, since both of the demodulated signals are combinations of I and Q, there is no point in using a bandwidth larger than that of the Q signal, which is 500 kHz. The relative simplicity of the $R-Y/B-Y$ demodulator is paid for in reduced horizontal resolution for color.

Whichever color demodulator system is chosen, the output will consist of three signals, $R-Y$, $B-Y$, and $G-Y$. These can simply be added to the lu-

FIGURE 14.25

Color Demodulation
Using R − Y and B − Y
Axes

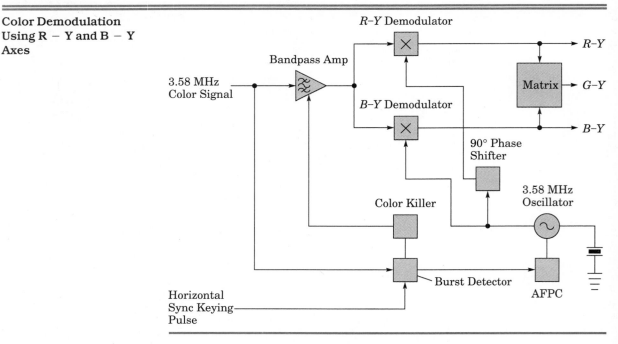

minance signal to produce the three primary signals (red, green, and blue) that are required by the CRT.

The color and tint controls adjust the color circuitry. The tint color alters the phase of the injected carriers to the demodulators, and the color control varies the gain of the color bandpass amplifier.

14.4.5 Color Cathode-Ray Tubes

In order to reproduce the three primary colors of the color television system, a color CRT requires three electron beams. Most tubes use three electron guns, arranged either in a triangle (delta guns) or in a straight line (in-line guns). Most modern tubes use the in-line system, which makes adjustment easier at the expense of requiring a large neck diameter. An exception is the Sony Trinitron™, which has a single large gun with three cathodes. Figure 14.26 shows a color CRT using in-line guns.

The three video signals are applied at the cathodes, one to each electron gun. The faceplate is covered with a dot pattern using three types of phosphor that glow red, green, and blue, respectively, when bombarded with electrons. Each gun illuminates a different type of phosphor. Because the electron beams are large enough to illuminate more than one phosphor dot, a *shadow mask*, consisting of a sheet of steel containing thousands of tiny holes, is placed near the faceplate. Excess electrons that would have hit the wrong phosphor dot strike the shadow mask instead. Of course, the efficiency of the tube is much reduced since about 70% of the electrons strike the mask rather than the tube. Consequently the accelerating voltage must be higher than for a monochrome tube: on the order of 20 to 30 kV for a typical color tube.

FIGURE 14.26   Color CRT

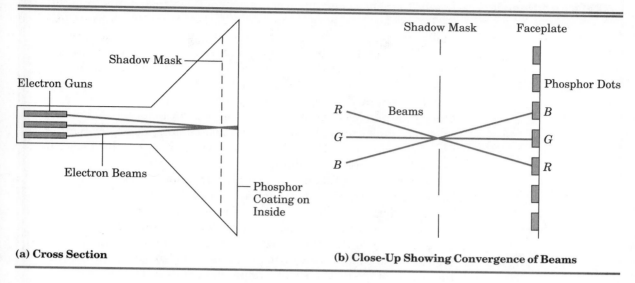

(a) Cross Section   (b) Close-Up Showing Convergence of Beams

Color CRTs require careful adjustment. First, the beam from each gun must land on the proper color dots; this adjustment is called **purity**. If the purity is not adjusted correctly, patches of incorrect color can be seen. Second, all three beams must land on the same triad, or group of three dots. This is **convergence**, and misadjusted convergence results in color fringes. The means of adjustment of purity and convergence varies according to whether the tube employs delta or in-line guns. For in-line guns, purity is adjusted using permanent magnets mounted around the neck of the tube. So is convergence near the center of the screen, using a different set of magnets. Convergence near the center is called *static convergence*. The term *dynamic convergence* refers to convergence around the edges and in the corners; this is adjusted by tilting the entire **yoke** that contains the deflection coils.

Because of the steel shadow mask, color CRTs can be affected by external magnetic fields. They can magnetize the shadow mask so that it deflects the electrons a small amount, causing them to land on phosphor dots of the wrong color. The problem is avoided by installing a coil around the faceplate of the tube, which is energized by 60 Hz ac for a few seconds each time the power to the receiver is turned on. The function of this *degaussing coil* is to remove any magnetism that has built up on the shadow mask since the last time the set was used. Even the magnetic field of the earth can cause problems if the receiver is moved.

14.5 SATELLITE TELEVISION SYSTEMS

The ordinary terrestrial broadcasting system described in Section 14.3 is, of course, not the only way to transmit television signals. Communication satellites have become the major way in which networks send programs to their affiliated stations, and in which specialized services such as news and movie channels send material to cable-television systems. They are also used to a limited extent for direct-to-home broadcasting.

CRTs and Health   Lately, there has been much concern about health hazards from the CRTs in video display terminals (VDTs) and television receivers. Actually, most of the talk is about VDTs, but electrically there is very little difference between a television receiver and a computer monitor. People do tend to sit closer to computers, increasing their exposure to whatever radiation is emitted.

It is certainly true that CRTs emit radiation. The most obviously dangerous and least controversial form is X-ray emission. X-rays are known as ionizing radiation because each photon has sufficient energy to ionize molecules. X-rays are known to cause cancer. They can be generated when high-energy electrons strike the screen or shadow mask in a color CRT. Only color receivers and monitors emit X-rays; the voltages used with monochrome CRTs are too low to generate them. Modern color CRTs emit very low levels of X-rays, provided the high voltage level is correct, and all modern monitors and receivers have circuitry to shut them down if the voltage gets too high due to a malfunction. Older receivers using tubes can also emit X-rays from the high-voltage rectifier and damper diode tubes. The levels of X-rays from modern equipment are generally considered so low as to represent negligible danger, but with ionizing radiation there really is no "safe" limit. The background radiation due to cosmic rays and uranium deposits is already "dangerous" in the sense that it causes some cancers and mutations.

CRTs also produce a strong dc electric field due to the accelerating voltage, and fairly strong electric and magnetic fields at the vertical and horizontal sweep frequencies. Whether there is any danger from these fields is highly controversial. Some people have claimed that they contribute to everything from headaches to miscarriages to cancer. Studies have been and are being done, but nothing has really been proven as of this writing. Still, reputable scientists, who used to scoff at claims of any danger from non-ionizing radiation, have begun to take these fields seriously. It is certainly true that reducing exposure to them would do no harm. Some monitor designers have already begun to take steps to improve electric and magnetic shielding.

One obvious solution to these problems is to move to solid-state liquid-crystal displays. These emit no X-rays at all, and the electric fields are very weak due to the low voltages involved. Unfortunately, so far the cost is higher, and the resolution and contrast are not as good as for CRTs.

A more mundane problem with CRTs is eyestrain. This can result from too low a resolution (hence is worse with LCD displays), and from glare from the screen. Whether or not electromagnetic fields cause headaches, eyestrain certainly does, and the VDT user is wise to eliminate sources of glare. Screens with antireflective coating also help.

TABLE 14.2

C-Band Downlink Transponder Frequencies	Channel	Center Frequency (Megahertz)	Channel	Center Frequency (Megahertz)	Channel	Center Frequency (Megahertz)
	1	3720	9	3880	17	4040
	2	3740	10	3900	18	4060
	3	3760	11	3920	19	4080
	4	3780	12	3940	20	4100
	5	3800	13	3960	21	4120
	6	3820	14	3980	22	4140
	7	3840	15	4000	23	4160
	8	3860	16	4020	24	4180

14.5.1 C-Band and Ku-Band Satellites

All the satellites used for television are of the geostationary type described in Chapter 11. Thus, it is necessary to aim a receiving antenna at the satellite only once, after which it can be left in place as long as required with no tracking necessary. There are two frequency ranges in use for these satellites. The C-band satellites have an uplink in the 6 GHz range, and a downlink at about 4 GHz. The other frequency allocation is in the Ku band, with uplink at about 14 GHz and downlink at 12 GHz.

Table 14.2 shows the downlink transponder frequencies in the C band. Each has a bandwidth of 36 MHz with a 4 MHz guard band between transponders. Because orthogonal polarization is used, with odd and even channel numbers having their polarization 90° apart, the frequency ranges can overlap.

14.5.2 Modulation Schemes

The fact that satellite transponders have much greater bandwidth than ordinary television channels allows frequency modulation to be used for the video as well as the audio. In fact, the usual way is for the audio to occupy one or more subcarriers on the video signal, with the subcarriers themselves using frequency modulation. The entire signal is then modulated onto one carrier using FM. The sound can be either mono or stereo, and there are several ways to construct a stereo signal. One channel can be put on each of two subcarriers, or a matrix system similar to that used in FM radio can be used. In addition, there is room for additional audio channels (that in some cases may have nothing to do with the video programming) on the same transponder.

14.5.3 Receiving Installations

Satellite receivers universally use a parabolic dish antenna. The geostationary satellites are all in a single arc along the equator, so a type of mounting called a *polar mount* can be used. This allows the dish to rotate about only one axis, so it is much simpler than a mounting that would allow the dish to be set to any azimuth and elevation. Once the arc of rotation is set up and the position of each satellite along the arc has been found, it is possible to return the antenna to a given satellite at any time. In fact, the correct position of the dish for each satellite is often programmed into the receiver. After that, the satellite designation can merely be supplied to the receiver by keying it in, and the dish will automatically be aimed at the satellite.

The size of the dish depends on the frequency band, the location of the receiver, and the quality of reception required. Since antenna gain is higher at 12 GHz, dishes for Ku-band reception do not have to be as large as those for C-band satellites. Of course there is no reason why a reflector sized for the C band cannot be used on the Ku band as well, provided the appropriate feedhorn is used. Receiving installations located well toward the poles need higher gain because of the additional atmospheric attenuation caused by the low angle of the satellite. In addition, of course, a stronger, more noise-free signal is required for rebroadcast by a television station or cable system than for a private in-home receiving system.

Figure 14.27 shows several typical television receive-only (TVRO) satellite installations. Next in the signal chain after the antenna, and mounted right at the feedhorn, is a low-noise amplifier (LNA). In fact, the LNA is often combined with a downconverter which reduces the signal frequency from the microwave to the VHF or UHF range. Tunable converters, called LNCs, which are controlled by the receiver, are sometimes used, but block downconverters are more common. These convert the entire range of signal frequencies on the satellite to a range of equal width at a lower frequency. The combination of LNA and block downconverter is called an LNB. The reason for putting all this hardware at the antenna is to increase the signal amplitude and (usually) lower its frequency before it encounters the lossy coaxial cable that connects the antenna to the receiver.

The receiver itself, then, is really a VHF or UHF receiver rather than a microwave receiver, and it can use rather conventional technology. A simplified block diagram is shown in Figure 14.28. First, the signal is converted to an intermediate frequency, which is often about 70 MHz. The entire signal is demodulated and then the subcarriers are removed, demodulated, and combined as necessary to get a stereo signal, if required. The output can be provided as baseband audio and video signals, for use with a monitor, or remodulated onto a carrier in the VHF TV band, using conventional AM for the video and FM for the audio, so that an ordinary television receiver can be used. Cable-television companies use this method to place the satellite-derived signal on a convenient television channel for distribution.

14.6 CABLE TELEVISION

Community-antenna television (CATV) systems, usually known as cable television, have several advantages over individual rooftop antennas. Besides removing what some people (though not usually communications specialists) think of as unsightly antennas from rooftops, the system often allows better reception, since cable companies can choose a good location and erect a relatively elaborate antenna system. CATV systems can also provide services unavailable on regular broadcast television, both locally generated channels and specialized services delivered by satellite, such as news, weather, sports, and movie channels.

FIGURE 14.27

Typical TVRO Satellite Installations

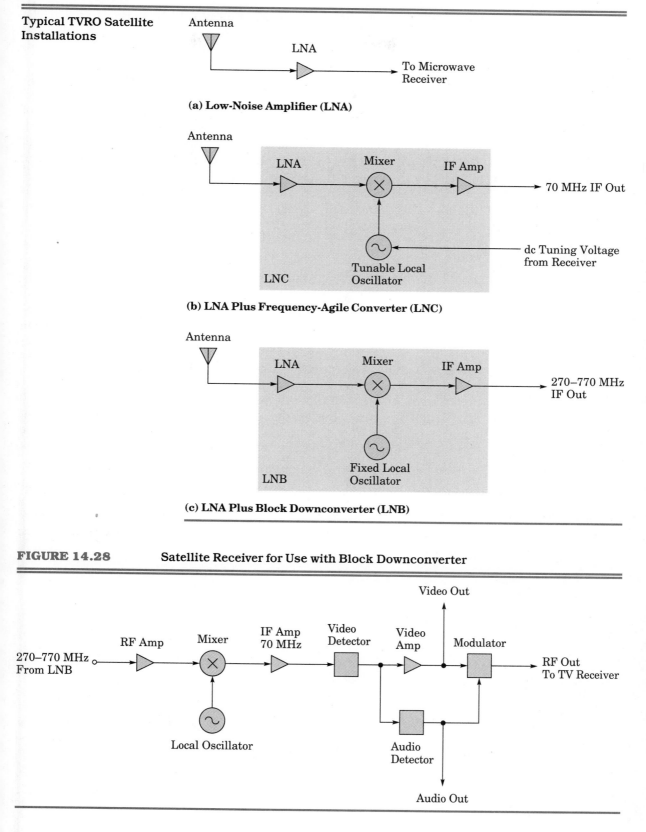

(a) Low-Noise Amplifier (LNA)

(b) LNA Plus Frequency-Agile Converter (LNC)

(c) LNA Plus Block Downconverter (LNB)

FIGURE 14.28   Satellite Receiver for Use with Block Downconverter

14.6.1 Structure

Figure 14.29 shows the structure of a typical cable-television system. The origination point for the signals, called the *head-end*, is located at a suitable antenna sight, usually just out of town and preferably on a hill. The structure is that of a tree, with trunk lines using large-diameter low-loss coaxial cable, and smaller-diameter branches leading to *drop lines* which provide the signal to individual buildings.

Because of the losses in the cable, amplifiers are necessary at intervals of about 0.65 km. The actual distance is governed by the loss on the line, which varies with the maximum frequency used by the system and with the type of line. A typical cable system has amplifiers with approximately 20 dB gain, followed by a section of line with about 20 dB of loss. Since the loss on the cable increases with frequency, so does the gain of the amplifiers. Temperature also affects the cable losses, so cable system amplifiers use a form of automatic gain control that monitors the levels of at least two pilot carriers sent down the cable from the head-end, keeping them at the appropriate value.

Signal levels on cable-television systems are usually given in dBmV (that is, decibels with respect to 1 mV). Mathematically, this is

$$\text{dBmV} = 20 \log \frac{V}{1 \text{ mV}} \tag{14.10}$$

where V = voltage on the cable

Signal levels at the subscriber drop vary from 0 dBmV (1 mV) to about 10 dBmV (3 mV) or more; that is, the levels at the receiver are similar to what would be obtained from an antenna with a reasonably strong signal. On the trunk and branch cables there is somewhat more variation, in the range of

FIGURE 14.29

Cable-Television System

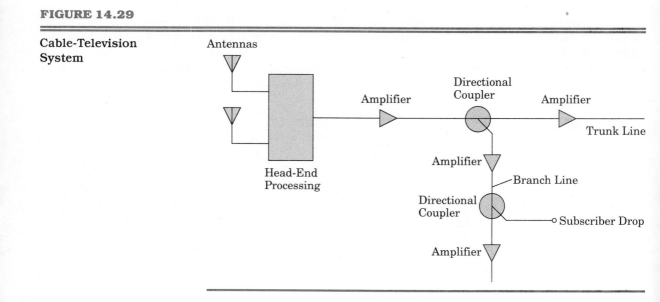

about 10 to 40 dBmV. The standard impedance for cable used in CATV systems is 75 Ω.

Directional couplers are used where the cable branches, to avoid problems due to reflection. Mismatches that cause reflection of signals can be caused by defects in the cable or by faulty connections. The subscriber drops sometimes use directional couplers, and sometimes simple resistive taps. In either case, there must be at least 18 dB of isolation between subscribers.

EXAMPLE 14.6

Refer to the portion of a CATV system shown in Figure 14.30 and calculate the signal level, in dBmV, at the input and output of each amplifier and at the subscriber drop. Also calculate the signal level at the subscriber drop in microvolts and in dBm.

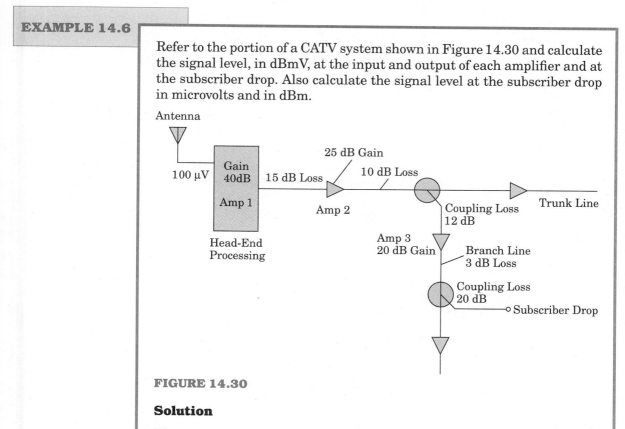

FIGURE 14.30

Solution

The first step is to convert the 100 μV level from the antenna to dBmV. From Equation (14.10),

$$In_1 \text{ (dBmV)} = 20 \log \frac{V}{1 \text{ mV}}$$

$$= 20 \log 0.1$$

$$= -20 \text{ dBmV}$$

This signal is applied to the input of the first amplifier, actually the head-end signal processing, which has a gain of 40 dB. One of the pleasant things about the decibel system is the fact that, once inputs are converted to a decibel form, addition of gains and subtraction of losses are all that is required to find output levels. Thus, the output level of Amp 1 is

$$Out_1 \text{ (dBmV)} = -20 + 40$$

$$= 20 \text{ dBmV}$$

The input level of Amp 2 is this less the loss in the intervening cable:

$$In_2 \text{ (dBmV)} = 20 - 15$$
$$= 5 \text{ dBmV}$$

At the output of this amplifier, we have

$$Out_2 \text{ (dBmV)} = 5 + 25$$
$$= 30 \text{ dBmV}$$

This is followed by 10 dB of loss in the cable, and another 12 in the directional coupler, for a total of 22 dB loss before the next amplifier.

$$In_3 \text{ (dBmV)} = 30 - 22 \qquad Out_3 \text{ (dBmV)} = 8 + 20$$
$$= 8 \text{ dBmV} \qquad\qquad\qquad = 28 \text{ dBmV}$$

There is a further 3 dB cable loss and 20 dB loss in the subscriber tap, leaving a signal strength at the subscriber drop of

$$V_{drop} \text{ (dBmV)} = 28 - 23$$
$$= 5 \text{ dBmV}$$

As a voltage, this is

$$V_{drop} = \text{antilog} \frac{5}{20} \text{ mV}$$
$$= 1.78 \text{ mV}$$
$$= 1780 \text{ } \mu V$$

The power into 75 Ω is

$$P_{drop} = \frac{V_{drop}^2}{75}$$
$$= \frac{(1.78 \times 10^{-3})^2}{75}$$
$$= 42.16 \times 10^{-9} \text{ W}$$
$$= 42.16 \text{ nW}$$

In dBm, this is

$$P_{drop} \text{ (dBm)} = 10 \log \frac{P_{drop}}{1 \text{ mW}}$$
$$= 10 \log (42.16 \times 10^{-6})$$
$$= -43.8 \text{ dBm}$$

14.6.2 Midband and Superband Channels

The loss in any type of coaxial cable increases with frequency. It is certainly possible to transmit UHF television frequencies along a cable, but more amplifiers would be required. Since every amplifier adds noise, the signal-to-noise ratio would be impaired. A less expensive and more practical alternative is to take advantage of the gap in the spectrum between channels 6 and 7, and to use the frequencies above channel 13. Of course, these regions are occupied by other services, but there is no need to carry most of them on the cable. The one exception is the FM broadcast band, located immediately above channel 6, which most cable systems do carry. (Appendix I shows the most common way of numbering the cable-television channels.) The numbers 2 through 13 are used for the channels that coincide with the normal VHF broadcast frequencies. This allows an ordinary receiver to be used with the cable without modification. The midband channels 14 through 22 occupy the spectrum between the top of the FM broadcast band and the bottom of channel 13. The superband channels start with number 23 just above the normal broadcast channel 13, and continue as far as necessary to handle all the requirements of the system.

Obviously, UHF television stations must be converted to lower frequencies in order to be carried on cable. Often, VHF stations are moved from their broadcast channels as well. There are many reasons for this. Sometimes it is to avoid interference between the channel on cable and signals which may be picked up on the internal wiring of the television receiver. Some channels are also subject to local oscillator interference from other receivers connected to the cable. Another factor that influences channel allocation is the fact that cable loss increases with frequency. The loss is made up by amplifiers along the cable, but the signal-to-noise ratio will become worse as the frequency gets higher.

For all these reasons and others, cable companies frequently move stations from one channel to another. This can be done either by mixing or by demodulating the signal and remodulating it on another carrier.

EXAMPLE 14.7

Suppose a television receiver tuned to channel 6 radiates a local oscillator signal into the cable. What channel will be interfered with?

Solution

All modern receivers use a picture IF of 45.75 MHz with high-side injection of the local oscillator signal. The picture carrier of channel 6 is at a frequency of 83.25 MHz, so

$$f_{LO} = 83.25 + 45.75$$
$$= 129 \text{ MHz}$$

This could interfere with midband cable channel 15 (126–132 MHz).

FIGURE 14.31

Cable-Television
Converter

Photo courtesy of Jerrold Communications.

14.6.3 Converters and Cable-Ready Receivers

Now that we have seen how the midband and superband channels are configured, it is possible to make sense of the term *cable-ready* as it applies to receivers. Such a receiver simply has a tuner that is capable of tuning to the midband and at least some of the superband channels. Exactly how high the superband channels go varies both with the cable system and the receiver design.

When an ordinary, non-cable-ready receiver is used with a CATV system, it can tune in only channels 2 to 13. Converters can be obtained that allow the reception of midband and superband channels with such a receiver. Usually, these are tunable converters that can move any channel on the cable to a single VHF channel. Figure 14.31 shows a photograph of a typical converter, and Figure 14.32 is its block diagram. The signal is first converted up

FIGURE 14.32   Cable-Television Converter Block Diagram

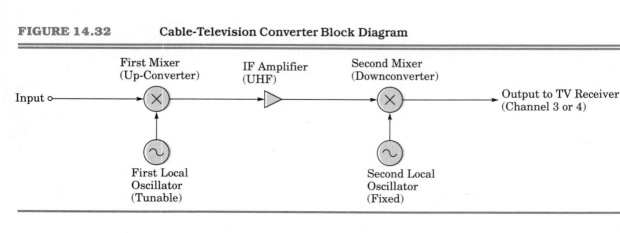

in frequency, then back down to the VHF output channel. This is to ensure that the first local oscillator frequency is higher than any frequency used by the cable. That way, any signal leakage into the cable from the local oscillator will cause no interference. Virtually all converters use varactor tuning, and some have quite elaborate remote control and programming features.

One feature of some cable converters that adds considerably to the complexity (and the price) is a volume control. The reason is that the only way to control the volume is to demodulate the sound, adjust the volume, then re-modulate it onto a new carrier.

14.7   TEST EQUIPMENT AND SIGNALS

Television and video systems require a wide variety of specialized test equipment, only some of which will be described in this section. In the next section, we will briefly investigate troubleshooting techniques, some of which will use some of these signals and generators.

14.7.1   Color Bar Pattern

See Figure 14.33 for a sketch of a color test signal. Figure 14.33(a) shows what the signal looks like on the television screen, and Figure 14.33(b) shows the appearance of the signal on an oscilloscope. There are seven color bars, from left to right: white, yellow, cyan, green, magenta, red, and blue. All are

FIGURE 14.33   Color Bars

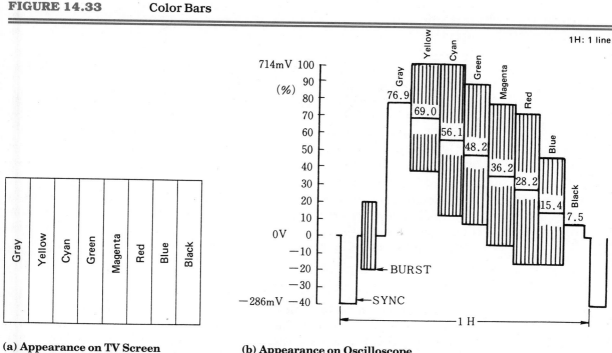

(a) Appearance on TV Screen

(b) Appearance on Oscilloscope

Courtesy Leader Instruments Corporation.

FIGURE 14.34

Color Bar Generator

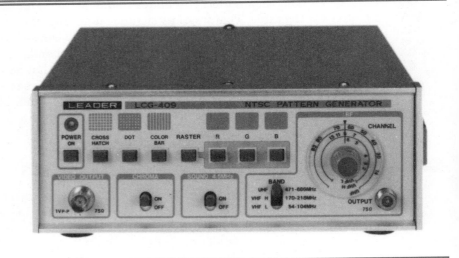

Courtesy Leader Instruments Corporation.

modulated at 75% (that is, the color component is 75% of its maximum amplitude) except for white, which is just 75% of maximum luminance. When sent in this order, the luminance levels form a signal whose amplitude decreases from left to right. Such a signal is called a *staircase signal*, because of its appearance when viewed on an oscilloscope at the horizontal rate.

When the color modulation is turned on, the color modulation appears on an oscilloscope as a high-frequency (approximately 3.58 MHz) ac signal superimposed on the luminance level. Because of the frequency interleaving used for color, the color modulation will not be synchronized on the scope, and will appear as a blur, like the envelope of an AM signal.

Figure 14.34 shows a generator that generates this pattern. It also generates red, green, and blue rasters, useful for checking CRT purity, and a crosshatch pattern that can be used for convergence adjustment.

There is also a more elaborate version of the color bar pattern, used in broadcasting studios, that has extra bars across the bottom of the picture, including 100% white, $-I$ and Q color signals, and black level.

Color bars can be used to set up the color response of video systems by eye, and, in fact, the eye is the most suitable instrument for overall evaluation of a video system. However, some quantitative results can be obtained by using a **vectorscope**, as described in the next section.

14.7.2 Vectorscope Display

For the chroma signal, magnitude represents color intensity (saturation) and phase represents the hue. If the color signal is displayed on an oscilloscope in such a way that two signals with a phase difference of 90° are applied, one to the vertical and one to the horizontal plates of the oscilloscope, then a color-

FIGURE 14.35

Vectorscope/Waveform
Monitor

Courtesy Leader Instruments Corporation.

bar signal with predetermined levels and phases will appear the same each time. The screen can be marked with the correct values for use in adjustment. This is called a *vectorscope display*. Any axes 90° apart can be used, but $B - Y$ and $R - Y$ are common. Note that looking at the bars on an ordinary scope gives a check on amplitude but not phase. Figure 14.35 is a photograph of a combination **vectorscope** and waveform monitor. The latter is simply an oscilloscope calibrated with IRE levels and equipped with a triggering arrangement that can allow it to scan any line as required.

14.8 TELEVISION RECEIVER TROUBLESHOOTING

A detailed discussion of television repair would merit a book in itself, and in fact many books have been written on the subject. What this brief section will do is to point out how a knowledge of the basic operation of a color television receiver can be used to localize problems. Following that, signal injection or tracing can be used to narrow the search further. At this point, of course, reference to a complete circuit diagram, and preferably a service manual, is required.

The first step in the process should be to examine carefully the symptoms of a fault. A color television is a complex device, but it also has a complex display that can reveal a good deal of information about troubles. Attention should be paid to the presence or absence, normality or abnormality, of raster, picture, sound, and color. Next, with reference to the block diagram of a typical color television receiver, an attempt can be made to localize the problem. The following sections will show how the symptoms can lead to at least an approximate diagnosis of the fault.

14.8.1 No Raster

Lack of any light on the screen points to a problem with either the main power supply or the horizontal deflection circuit. The former is obvious; the latter conclusion depends on a knowledge that the high-voltage supply for the CRT, as well as various low voltages for signal-processing stages, are derived from the horizontal output transformer.

If there is only one horizontal line on the screen, it is a safe guess that the problem is in the vertical oscillator or amplifier.

14.8.2 No Video and/or Audio

Assuming that there is light on the screen but no picture, a good deal can be learned from noticing whether there is audio. Recall that the video and audio are combined through the tuner and the picture IF amplifier. Therefore, if both video and audio disappear, the problem is probably in one of these circuits. On the other hand, if the audio is present but the video is not, the problem is after the point at which the sound is separated from the video. Check the video detector and the circuitry past that point. In a color receiver, the problem is likely to be between the video detector and the point at which the chroma and luma are separated. A failure in the luminance amplifier after that point will allow a picture to appear, but it will consist only of saturated colors, while a fault in the color circuitry may result in a monochrome picture or one with incorrect color. On the other hand, a fault in a monochrome receiver anywhere between the video detector and the CRT can cause the picture to be absent.

Another possibility is that the picture may be normal while the sound is absent. This points to the circuitry between the first sound detector (or the video detector in a monochrome receiver) and the loudspeaker.

14.8.3 Synchronization Problems

When the receiver displays a picture but synchronization is lacking, it is important to notice whether the problem is with vertical or horizontal synchronization or both. Lack of vertical synchronization is shown by a picture that moves up or down. With no horizontal sync, the picture will not move sideways; rather it will break up into diagonal bars. Lack of both types of synchronization in the presence of an otherwise strong image points to the sync separator. On the other hand, presence of one and absence of the other could result from a fault in either the sync separator or one of the oscillators: horizontal or vertical, depending on which is out of sync.

Color receivers have a third type of synchronization: color sync. This is what keeps all the colors stable, and its failure will probably result in a rainbow effect in the picture. The trouble could be in the color oscillator itself or in the AFPC circuit (in modern receivers these will be included in the same integrated circuit anyway).

14.8.4 Signal Tracing and Injection

Once the trouble has been localized to a relatively small area of the receiver, and in the absence of any obvious faults (such as overheating components or loose wires), it is necessary to make further checks. The simplest test is to look for the proper signals and voltages at various test points, following manufacturers' instructions. However, even when the fault has been localized to some extent, there may be a great number of such points to check.

There are two basic methods for following a signal through a series of stages. One is signal tracing, where a normal television signal is applied to the antenna terminals and followed through the receiver, or more likely through that part of the receiver where the fault is suspected to lie. The instrument most commonly employed for signal tracing is the oscilloscope, though for audio circuits a sensitive amplifier with a small loudspeaker is sometimes used.

In television receivers, the oscilloscope can be used in all the baseband video circuitry, the audio IF and baseband circuitry, and the horizontal and vertical sync, oscillator, and amplifier circuits. Be careful in the horizontal output stage, however: because of the high voltages and brief pulses in this stage, there are often points on the circuit marked "do not measure." For your own safety and that of your equipment, please obey that injunction.

Depending on the high-frequency response of the oscilloscope, it may detect the presence of a signal in the video IF circuits. As long as the signal level is high enough, a demodulator probe can eliminate the necessity for a high-frequency scope. In the tuner, however, the signals are likely to be too weak and at too high a frequency for measurement. For these areas, signal substitution is useful. For instance, applying a simulated video IF signal to the input of a video IF amplifier can determine whether a tuner is at fault. Tuners, by the way, are seldom repaired except by returning them to the manufacturer in exchange for a new or rebuilt tuner. The main exception is that conventional switch-type tuners sometimes become intermittent due to dirt and oxidation on the contacts. A liberal application of contact cleaner will often improve them considerably.

Other circuits where signal injection is useful include the horizontal and vertical sweep circuits. In fact, sometimes a horizontal drive signal is applied to the horizontal output transistor in order to get it working and allow the set to generate high and low power-supply voltages. Sometimes, it is only in this way that a problem can be found that shuts down the horizontal output signal, since without power it is very difficult to determine which stages are operable. Figure 14.36 shows a specialized test instrument that is designed to supply these types of signals, as well as RF, IF, and baseband video signals. It can also provide signals at specific frequencies for aligning the notch filters commonly found in television receivers.

FIGURE 14.36

Video Analyzer

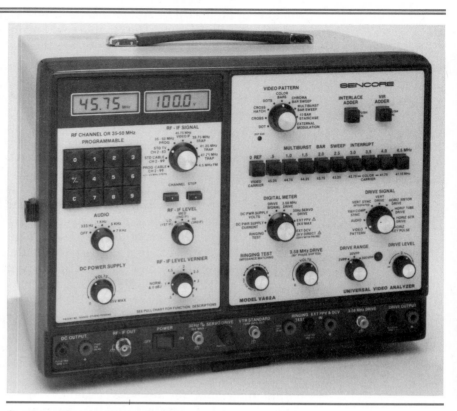

Courtesy of Sencore Electronics Inc.

14.9 HIGH-DEFINITION TELEVISION

The NTSC television system has served well for many years. However, it is far from ideal. The following are some of its more important deficiencies:

1. Luma resolution is inadequate for large-screen use.
2. Chroma resolution, while adequate in the vertical direction, is insufficient in the horizontal direction. This causes color smearing.
3. A wider aspect ratio, similar to commercial motion pictures, would be more suitable for many types of programming. The current 4:3 ratio was set in the early days of television, and was compatible with the motion pictures of the day.
4. The interleaving of luma and chroma signals conserves bandwidth but causes interference between the two signals. Cross luminance results from the luminance component interfering with the chroma signal. It occurs whenever there is diagonal detail in the picture, and gets worse with motion. It is visible as spurious color in the picture. Cross color is caused by color information being interpreted as luminance, causing a structure of fine dots.
5. Interlaced scanning conserves bandwidth for a given flicker rate with most pictures. However, certain types of image, alphanumerics for in-

stance, still appear to flicker at a 30 Hz rate. Interlacing also appears to lower the resolution for motion, and makes the scanning-line structure more obvious.

In recent years, there have been many attempts to solve these problems. They can be divided roughly into three categories: systems designed to get the best performance possible from the existing NTSC system; systems that change the signal to achieve higher definition, but in such a way as to maintain compatibility with existing receivers; and completely new schemes, requiring new equipment from beginning to end of the signal chain. Obviously, there are possible compromises; for instance, an incompatible system might be made compatible by adding a specially designed converter to an ordinary NTSC receiver.

Compatibility between any new system and present systems is very desirable for economic reasons, but impractical without special efforts that either increase the costs of the improved system, or lower its quality, or both. One possibility that is being explored by several groups is to use two channels, transmitting an ordinary NTSC signal on one, and whatever additional information is required for HDTV on the other. An ordinary receiver would work as before, and an HDTV receiver would simply receive both channels simultaneously. Obviously, this reduces the number of available channels, though some of the most recent systems claim to be able to use the "taboo" channels (currently unassigned channels adjacent to occupied channels) for the extra information, without creating undue interference.

The entire field of improved television systems is in a state of flux at this writing, and tests are being performed in an attempt to find a system, or combination of systems, for use in North America. Rather than look at all of the proposals here, an attempt will be made to outline some of the ways in which shortcomings of the current system can be overcome, and to outline a few of the specific proposals.

14.9.1 Improved Performance with the Current System

The cheapest and least disruptive approach to improving the television image is to work with the existing NTSC system to extract maximum performance. There are a few things that can be done at the transmitter without sacrificing compatibility, and many more that can be done at the receiver. In general, techniques that work with the NTSC signal are known as *improved-definition television* (IDTV) rather than *high-definition television* (HDTV).

Comb filtering can be used to reduce interference between the luma and chroma signals. The use of a comb filter in the receiver has already been described. The signal can also be *pre-combed* by using a comb filter at the transmitter in the creation of the color signal.

The impression of greater vertical resolution can be achieved by deriving additional scan lines from those that are transmitted. For instance, with enough memory in the receiver to store two lines, it is possible to insert new lines which are the average of the two adjacent lines.

A way to create the illusion of greater horizontal resolution is to emphasize edges. Since vertical edges involve high-frequency components in the video signal, a simple peaking circuit that emphasizes the high frequencies in the luminance signal will have this effect. This has been done in most receivers for many years.

More elaborate signal processing can be accomplished if the video signal is first digitized in the receiver. Once the signal has been digitized, it is relatively easy to use digital signal processing to sharpen edges (giving the impression of greater resolution) and to reduce noise by correlating consecutive lines. Since adjacent lines tend to be quite similar, any sharp difference between one line and the next is likely to be caused by noise. Of course, it can also be caused by a horizontal edge in the picture, so some care is needed and it is better to compare more than two lines. Similarly, ghosts can be eliminated by looking for repeating patterns along a line.

Picture-in-picture is another possibility; for this, more than one source is digitized, and the sources can be combined on the screen.

Non-interlaced scanning, called **progressive scan**, can be implemented at the receiver. This can be accomplished by including a field store, that is, a memory that can store one complete video field, in the receiver. You will recall that the complete television frame includes two fields, each sent in 1/60 s. Normally, the fields are shown as they are received, and no memory is required at the receiver. However, if one of the fields were to be stored, lines from that field could be interspersed with lines from the field currently being received, allowing the whole frame to be displayed at once. As lines are clocked out of memory and displayed, the vacated memory locations can be used to store the current field, for use during the next field. This requires that the receiver be able to convert the signal to digital form for storage and back to analog for display. The receiver also has to have at least one megabit of memory, assuming that data compression is used.

The only way to achieve a wider aspect ratio with the current system is to use fewer active scan lines. This is called *letterboxing*. It is done occasionally by stations broadcasting widescreen movies, and is also seen with some music videos. Of course, this leaves a blank band at the top and bottom of the screen with current receivers, and also results in decreased vertical resolution since some of the available scan lines are unused. It is therefore not a very satisfactory arrangement.

14.9.2 High-Definition Television Techniques

If compatibility with the NTSC system is not required, there are many things that can be done to improve the system. The aspect ratio can be changed to any desired value; 16:9 is a popular choice. Horizontal and vertical resolution can be improved by increasing the number of scan lines and also increasing the number of details per line. Both of these changes result in increased bandwidth, unless some form of compression is used.

Improved color resolution requires the use of a **component color** system, in which luma and chroma are transmitted separately. This also avoids the interference between luma and chroma that was mentioned above. Several groups have proposed a multiplexed analog component (MAC) system, wherein luma and chroma for each line would be sent sequentially, one at a time. This is a form of analog time-division multiplexing. An example of how this could be done is shown in Figure 14.37. Note that there are still two color signals, and that more time is reserved for luma than chroma. The resolution will still be less for chroma than luma, but the difference will be less than

FIGURE 14.37

Multiplexed Analog Component Video Signal

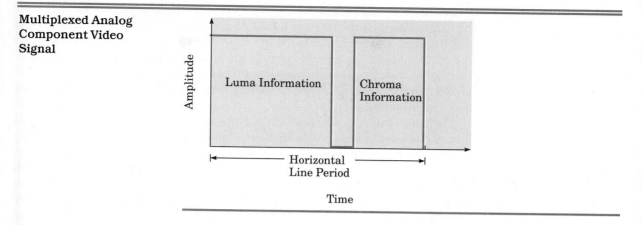

before. Each signal has to be time-compressed, which results in a corresponding increase in bandwidth. Essentially, the signal is clocked into a memory at the normal rate, then clocked out at a faster rate. Even though the storage will most likely be digital, the transmission is generally analog to reduce bandwidth.

MAC systems can be used with normal or high-definition signals. Besides removing interference between chroma and luma signals, the MAC system avoids the increase in color noise that results when a normal video signal is transmitted using FM, as with satellite transmission.

MAC systems require memory in the receiver, since it has to store a whole line, then recombine the luma and chroma for simultaneous display.

Until recently, all HDTV broadcast systems used analog transmission in order to conserve bandwidth, even though much of the signal processing at both ends might be digital. Recent advances in data-compression techniques, however, have made digital transmission a practical possibility.

14.9.3 High-Definition Television Proposals

This section will look at a few specific HDTV systems. Tests by the FCC in the United States are currently underway involving these and others. Unfortunately, the results are not available as of this writing, but should be available by the time this book is published. See Table 14.3 for a brief summary of the specifications for these systems.

TABLE 14.3

Some Proposed HDTV Systems	Name	Scan Lines	Interlace?	Analog or Digital?
	MUSE	1125	Yes	Analog
	ACTV	525	No	Analog
	SC-HDTV	787.5	No	Analog
	DigiCypher	1050	Yes	Digital

MUSE System. The MUSE system was developed by NHK Corporation in Japan. MUSE stands for *multiple sub-Nyquist encoding*, which implies that the information transmitted is less than that theoretically necessary for the resolution obtained. The resulting picture has high definition, with 1125 lines and 60 fields per second, using 2:1 interlace and a 16:9 aspect ratio. Each active line has 1920 picture elements for luminance and half that number for chroma. Luminance and color are sent separately using a multiplexed analog component system.

Only one-quarter of the picture information is transmitted during each field and stored in memory in the receiver. After four fields have been transmitted, the complete frame is available. From then on, one-quarter of the picture elements are updated each field. The result is that the full resolution is obtained for still pictures, but motion is slightly blurred. Motion of the whole frame (due, for instance, to camera panning), is compensated for in the receiver, and does not result in blurring. The receiver needs enough memory to store a whole frame.

The normal MUSE signal requires a baseband bandwidth of 8.1 MHz and is incompatible with NTSC receivers unless an elaborate adapter is used. It has been demonstrated with FM microwave transmission and also with VSB transmission using two adjacent UHF channels. Japan has already begun direct satellite broadcasting (one hour per day) with the MUSE system. Acceptance has been slow, even in Japan, because of the high cost of MUSE receivers, which is on the order of $30,000 U.S. A reduced-bandwidth version of MUSE, which fits in a 6 MHz TV channel, is under consideration for North American use. It would be simulcast with a normal NTSC signal on a separate channel.

Advanced Compatible TV (ACTV). This system is a joint effort by three companies in the United States: RCA, NBC, and GE. It has two forms: one, called ACTV-1, uses one ordinary channel; the other, ACTV-2, uses two channels. ACTV divides the picture into three sections, as shown in Figure 14.38: a center section, with the same 4:3 aspect ratio as for the NTSC system, and two side panels, which give the increased aspect ratio necessary for HDTV. The system involves time-compressing the signal for the side panels and

FIGURE 14.38

Advanced Compatible TV Picture

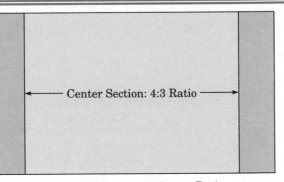

Center Section: 4:3 Ratio

Complete Image: 16:9 Aspect Ratio

sending it at reduced definition. In addition, extra information is sent in the same frequency range as the main signal using QUAM. The total system would have a wide aspect ratio of 16:9 but maintain the standard 30 Hz interlaced frame rate.

Spectrum Compatible TV (SC-HDTV). This system from Zenith Corporation uses a standard NTSC signal on one channel. Another, low-power transmitter on a separate channel sends the extra data needed for HDTV, using a partly digital, partly analog system, with the low-frequency information sent digitally in the vertical blanking interval. Since most of the energy in a conventional video signal is in the low-frequency portion, it is claimed that the second channel can be transmitted at low enough power that normally unusable adjacent channels can be used; for example, channel 3 in an area that already uses channel 4.

DigiCypher. At present, this system from General Instruments is the only digital system scheduled for FCC tests, though others are under development. This is a simulcast system, where a standard NTSC signal would be transmitted on one channel, with a complete digitized HDTV signal on the other. Like the Zenith signal above, it is claimed that this system could make use of the "taboo" channels that cannot presently be used because of adjacent-channel interference. This system is different from all those described above, however, in that an elaborate data-compression scheme allows the whole HDTV video signal, along with four sound channels, to be digitized and then transmitted, using 16-level QAM (quadrature AM with 4 phases and 4 amplitude levels). (See the next chapter for details on digital QAM schemes.) The system fits into a 6 MHz bandwidth.

SUMMARY OF CHAPTER 14

Here are the main points to remember from this chapter.

1. The North American NTSC television system transmits 30 frames per second, each consisting of 525 lines. A 2:1 interlace is used so that 60 fields are transmitted per second.

2. Vertical resolution depends on the number of scan lines, while horizontal resolution depends on the maximum frequency that can be transmitted.

3. Composite color video uses two color signals modulated on a subcarrier at 3.58 MHz using DSBSC QUAM. This results in no increase in bandwidth for color, but also means that the horizontal resolution for color is much less than for monochrome.

4. Terrestrial television broadcasting uses a channel 6 MHz wide. The picture is transmitted using VSB AM with carrier, and the sound uses FM on a separate carrier.

5. Picture and sound travel together through the tuner and picture IF sections of a television receiver. The sound is then converted to a separate 4.5 MHz IF, in a system known as intercarrier sound.

6. Color receivers use synchronous demodulation to retrieve the color sig-

nal. The frequency and phase reference for the color demodulators are provided by a color burst on the back porch of the horizontal synchronizing signal.

7. The CRTs used in television receivers generate an electron beam that is intensity-modulated by the video signal, and deflected horizontally and vertically by coils that surround the tube. The beam produces an image by striking a phosphor coating on the tube face. Color CRTs use three electron beams striking three different types of phosphors which glow red, green, and blue, respectively.

8. Satellite television systems use geostationary satellites with uplink/ downlink frequencies of approximately 6/4 and 14/12 Ghz. Transmission is FM for both picture and sound.

9. Satellite-television receiving installations use parabolic antennas. A low-noise amplifier at the antenna is usually followed by a downconverter (also installed at the antenna).

10. Cable-television systems reduce losses by moving UHF signals to the VHF range. They make use of frequencies between channels 6 and 7 and above channel 13. Amplifiers are used at frequent intervals along the cable to compensate for losses.

11. It is often possible to localize faults in television receivers by looking for the presence, absence, or abnormality of the raster, picture, color, and sound. Once the fault has been localized to one or more stages, there are many types of specialized test equipment that can be used.

12. There are many proposals for high-definition television systems currently being investigated. It seems likely that a simulcast system will be adopted, with NTSC television on one channel and either a complete HDTV signal or the additional information needed to convert NTSC to HDTV on a second channel.

IMPORTANT EQUATIONS

$$N_V = 338 \text{ for NTSC system} \tag{14.1}$$

$$N_H = 449 \text{ for NTSC system} \tag{14.2}$$

$$L_H = B \text{ (MHz)} \times 80 \tag{14.4}$$

$$
\begin{aligned}
Y &= 0.30R + 0.59G + 0.11B \\
I &= 0.60R - 0.28G - 0.32B \\
Q &= 0.21R - 0.52G + 0.31B
\end{aligned}
\tag{14.6}
$$

$$\text{dBmV} = 20 \log \frac{V}{1 \text{ mV}} \tag{14.10}$$

GLOSSARY

aspect ratio   ratio between width and height of a television picture

back porch   portion of horizontal blanking pulse after the sync pulse

black setup   video level corresponding to zero luminance

blanking   period of time when the electron beam in a CRT is cut off

cathode-ray tube (CRT)   a vacuum tube that uses a moving electron beam to produce patterns or images on a phosphorescent screen

chrominance (chroma)   color signal

color burst   several cycles of color subcarrier on the back porch of horizontal sync, for color synchronization

comb filter   filter that can pass (or reject) a fundamental frequency and its harmonics

component color   video system in which color and luminance are sent separately, without frequency interleaving

composite video   video system where luma, sync, and chroma signals (if present) are combined

convergence   alignment of the three electron beams in a color CRT so that they land on the same triad of color phosphor dots

equalizing pulses   pulses in the vertical blanking interval of a video signal that create interlaced scan

field   in an interlaced video system, one-half of a frame, consisting of alternate lines

frame   one complete image in a video system

front porch   portion of horizontal blanking before the sync pulse

intercarrier sound   television receiver design that uses mixing between the picture and sound carriers to generate the sound intermediate frequency

interlace   video scanning system that divides a frame into two fields, to reduce flicker

luminance (luma)   signal that provides brightness information in a video system

NTSC video   North American television standard

peak white   video signal level representing maximum luminance

pixel   picture element

progressive scan   a video system that does not use interlace

purity   in a color CRT, the adjustment of the three electron beams so that each lands on phosphor dots of the appropriate color

raster   the pattern of scanning lines in a video system

resolution   amount of detail produced by a video system

retrace   the return of the electron beam in a CRT from right to left or from bottom to top

RGB color   a color video system where the three primary colors are transmitted separately

ultor   the main accelerating element in a CRT

utilization factor   proportion of scanning lines in a video system that, on average, can be used in determining vertical resolution

vectorscope   specialized oscilloscope designed for the observation of composite color signals

yoke   assembly mounted on the neck of a CRT that contains the deflection coils

QUESTIONS

1. Explain the difference between a frame and a field.

2. Describe interlaced scan and explain its advantages.

3. State the standard for video signals at the inputs and outputs of equipment, and sketch one line of a color signal that complies with these standards. Show an IRE scale on your sketch.

4. How do the vertical blanking intervals differ for odd and even fields?

5. Explain how it is possible to transmit a color signal using the same bandwidth as for monochrome.

6. What is meant by negative picture transmission and why is it used for television broadcasting?

7. Explain briefly how the method used for television stereo sound differs from that used for FM radio broadcasting.

8. Sketch the block diagram of a monochrome television receiver, including only those blocks that process the signal. Above each block note the frequency, or range of frequencies, that it handles.

9. (a) Sketch the frequency response for a typical video IF amplifier, showing the frequencies for picture carrier, sound carrier, and color subcarrier.
   (b) Explain why the sound carrier is at a lower frequency than the picture carrier in the video IF.
   (c) What frequency is used for the sound IF, and why?

10. Give two reasons for the use of the intercarrier sound system in television receivers.

11. Sketch the electron gun in a typical CRT, and explain briefly how the electron beam is focused.

12. What is the aquadag, and what does it do?

13. Describe and compare two methods of color-signal demodulation.

14. Explain briefly the meaning of each of the following:
    (a) color killer
    (b) AFPC
    (c) luminance delay line
    (d) gated AGC

15. Explain briefly what is meant by each of the following, as it refers to a color CRT:
    (a) purity
    (b) static convergence
    (c) dynamic convergence

16. Sometimes a color monitor can be set for "green only" when viewing a monochrome image, as on a computer. Often this gives a sharper image than when white is used. Why?

17. Why is the accelerating voltage much higher in a color CRT than in a monochrome CRT of the same size?

18. Explain the meaning of, and difference between, LNA, LNB, and LNC.

19. Why is it necessary for CATV systems to have isolation between subscribers?

20. Why do CATV amplifiers need automatic gain and slope (frequency response) control?

PROBLEMS

Section 14.2

21. A certain receiver has a vertical retrace time of $0.025V$, where V is the time for one field.

    (a) How much time in seconds is this?
    (b) How many horizontal lines does this represent?

22. For the following pictures, sketch one line of the composite video signal:

    (a) an all-white frame
    (b) two vertical white bars and two black bars equally spaced
    (c) five pairs of equally spaced vertical black and white bars (Why does this signal have more high-frequency content than that in Part (b)?)

23. Suppose the signal from a color camera has $R = 0.8$, $G = 0.4$, and $B = 0.2$, where 1 represents the maximum signal possible.

    (a) Calculate values for Y, I, and Q for the transmitted signal.
    (b) Express the luminance component in IRE units.

24. A certain TV receiver displays 475 scanning lines on the screen, and has a frequency response of 3.2 MHz for luminance and 0.4 MHz for color. Calculate its horizontal and vertical resolution in lines for both monochrome and color components.

Section 14.3

25. Find the frequency of each of the following, for a television signal on channel 47.

    (a) picture carrier
    (b) sound carrier
    (c) color subcarrier

Section 14.4

26. Draw a block diagram for a monochrome television receiver, indicating beside each block the frequency or frequency range at which that block will operate when the receiver is tuned to channel 9.

27. Draw a block diagram for a comb filter that will remove a 2 kHz signal and all its harmonics from a signal.

Section 14.6

28. A CATV system is sketched in Figure 14.39. Find the signal levels at each of the subscriber drops shown.

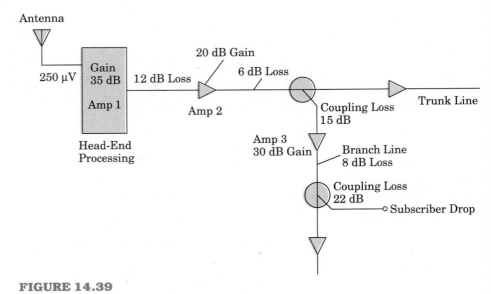

Antenna

250 µV | Gain 35 dB Amp 1 | 12 dB Loss

Head-End Processing

20 dB Gain

6 dB Loss

Amp 2

Coupling Loss 15 dB

Amp 3 30 dB Gain

Trunk Line

Branch Line 8 dB Loss

Coupling Loss 22 dB

Subscriber Drop

FIGURE 14.39

29. Convert 4.5 dBmV in 75 Ω to:

   (a) volts   **(b)** watts

30. Suppose an ordinary TV receiver (without a converter) is connected to a cable system, and that its local oscillator sends some of its signal back into the cable. Find three channels that could be interfered with, and state which channels the TV will be tuned to when it causes the interference.

Section 14.8

31. What would be the effect of each of the following on the performance of a monochrome television receiver?

   (a) mixer stage in VHF tuner not working (assume the set has conventional switch-type tuners)
   (b) gain in video IF amplifier is much too low
   (c) sync separator not working
   (d) horizontal oscillator not functioning

32. What area of a receiver would you suspect first for each of the following symptoms? Give a reason for your choice.

   (a) no sync (neither horizontal nor vertical)
   (b) no sound, picture OK
   (c) no picture, sound OK

33. What area(s) of a color receiver would you suspect for each of the following symptoms?

   (a) no color sync
   (b) no color at all, monochrome picture OK
   (c) no red in the picture

Section 14.9

34. Suppose that a high-resolution television signal used the NTSC standard, except that both vertical and horizontal resolution were doubled. What would be the required bandwidth for the baseband video signal?

35. Suppose that the aspect ratio of a television system were changed from 4:3 to 2:1, with the horizontal resolution keeping the same ratio to vertical resolution as at present. Assume the number of scan lines remains as at present.

    (a) How many details would have to be shown on a horizontal line?

    (b) What would be the bandwidth of the baseband signal?

Comprehensive

36. Suppose a television signal has 10.3 μs of horizontal blanking per line, and 21 lines of vertical blanking per field. Calculate the percentage of total time that is occupied by blanking.

37. Under ideal conditions, the eye can resolve two details subtending 0.5 minutes of arc. Suppose a monitor screen 30 cm wide is located 1.5 m from a viewer. What video bandwidth would be required in order that horizontal details in the center of the screen would be separated by 0.5 minutes of arc, assuming NTSC video?

38. Assume that the horizontal deflection coils in a receiver can be modeled as a pure inductance of 2 mH, and that the current waveform is a sawtooth with a peak current of 2 A. Sketch the voltage waveform, including scales.

39. Express the R signal algebraically in terms of Y, I, and Q.

15

Digital and Data Communications

Objectives

After studying this chapter, you should be able to:

1. Compare analog and digital communication techniques, and discuss the appropriate use of each.
2. Calculate the information capacity of a channel.
3. Calculate the minimum sampling rate for a signal.
4. Describe pulse-code modulation and delta modulation, and perform calculations in PCM systems.
5. Explain the use of time-division multiplexing.
6. Describe and use a variety of character codes and line codes.
7. Compare synchronous and asynchronous communication.
8. Describe the operation of modems, and calculate data rates for a variety of modulation schemes.
9. Use the RS-232C standard to connect a computer to a modem.

15.1 INTRODUCTION

Previous chapters have dealt with the communication of analog signals using analog techniques. While this remains a very important part of communications, an increasing proportion of the signals to be communicated are digital in origin. Digital techniques are also used in the transmission of analog signals.

Figure 15.1 shows several possible types of signal transmission. In Figure 15.1(a), an analog signal is sent over a channel with no modulation. A typical example would be an ordinary public-address system, with a microphone, an amplifier, and a speaker, using twisted-pair wire as a channel.

Figure 15.1(b) shows analog transmission using modulation and demodulation. Much of this book so far has been devoted to this type of communication, of which broadcast radio and television are good examples.

Figures 15.1(c) and (d) start with a digital signal (for example, a data file from a computer). In (c), the link can handle some kind of digital pulse signal directly. In (d), the channel is such that pulses cannot be transmitted directly; a radio channel is one example, and an ordinary telephone connection is another. In that case, the digital signal has to be modulated onto a carrier at one end and demodulated at the other. The **modem** shown is a combination modulator-demodulator.

Lastly, Figures 15.1(e) and (f) show an analog signal that is digitized at the transmitter and converted back to analog form at the receiver. The difference between these two sections is that in Figure 15.1(e), the transmission is digital, while in Figure 15.1(f), the transmission channel cannot carry pulses, so modulation and demodulation are required.

The transmission of analog signals, such as voice, by analog means seems to make sense. It is certainly simpler than converting the signal to digital form and back again. Similarly, it seems obvious that signals that begin digital, such as the contents of computer memories, should be kept in digital form as much as possible. What at first glance seems awkward is the conversion of analog signals to digital form for transmission. Actually, the use of digital techniques with analog signals is one of the fastest growing areas in communications, and there are good reasons for it.

Earlier, we looked at the effects of noise and distortion on analog signals. Once noise and distortion are present, there is usually no way to remove them. In addition, the effects of these impairments are cumulative. Noise will be added in the transmitter, the channel, and the receiver. If the communications system involves several trips through amplifiers and channels, as in a long-distance telephone system, the signal-to-noise ratio will gradually decrease with increasing distance from the source.

Digital systems are not immune from noise and distortion, but it is possible to reduce their effect. Consider the simple digital signal shown in Figure 15.2. Suppose that a transmitter generates 1 V for a binary one, and 0 V for a binary zero. The receiver examines the signal in the middle of the pulse, and has a decision threshold at 0.5 V; that is, it considers any signal with an amplitude greater than 0.5 V to be a one, and any amplitude less than that to represent a zero. Figure 15.2(a) shows the signal as it emerges from the transmitter, and Figure 15.2(b) shows it after its passage through a channel that adds noise and distorts the pulse. In spite of the noise and distortion,

FIGURE 15.1   Analog and Digital Communications

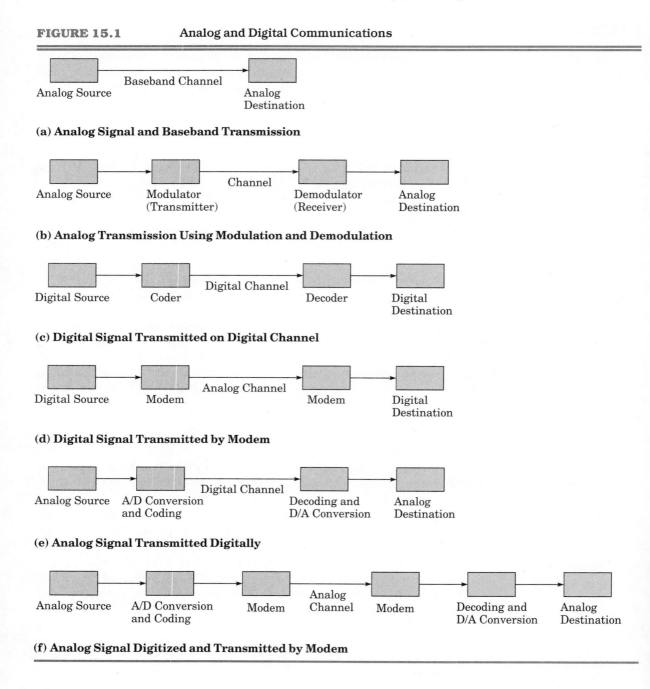

(a) **Analog Signal and Baseband Transmission**

(b) **Analog Transmission Using Modulation and Demodulation**

(c) **Digital Signal Transmitted on Digital Channel**

(d) **Digital Signal Transmitted by Modem**

(e) **Analog Signal Transmitted Digitally**

(f) **Analog Signal Digitized and Transmitted by Modem**

the receiver has no difficulty deciding correctly whether the signal is a zero or a one. Since the binary value of the pulse is the only information in the signal, the distortion has had no effect on the transmission of information.

The perfectly received signal of Figure 15.2(b) could now be used to generate a new pulse train to send further down the channel. This receiver-transmitter combination, called a *repeater* and illustrated in Figure 15.2(c), has not only avoided the addition of any distortion of its own, but has also removed the effects of noise and distortion that were added by the channel

FIGURE 15.2

Removal of Noise and Distortion from Digital Signal

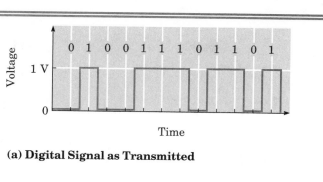

(a) Digital Signal as Transmitted

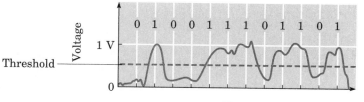

(b) Received Signal with Added Noise and Distortion

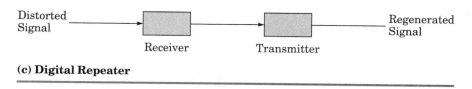

(c) Digital Repeater

preceding the repeater. The elusive goal of distortionless transmission seems to have been achieved!

Unfortunately, as in most areas of life, absolute perfection is elusive. Since noise is random, it is possible for a noise pulse to have any amplitude, including one that will cause a transition to the wrong level. Similarly, extreme distortion of pulses can cause errors. Figure 15.3 demonstrates these problems. Errors can never be eliminated completely, but, by judicious choice of such parameters as signal levels and bit rates, it is possible to reduce the probability of error to a very small value. There are even techniques to detect and correct some of the errors.

The other source of error in the digital transmission of analog signals appears in the conversion of the infinitely variable analog signal to digital form. This will inevitably result in the loss of some information, and the creation of a certain amount of noise and distortion. Once again, however, it is possible to predict quite accurately the amount of error that will be introduced, and to reduce it to any required value.

Other advantages of digital communication include convenience in multiplexing and switching. Time-division multiplexing (TDM) is quite easy with digital signals, and different types of signals (for example, voice and data) can be multiplexed together on the same channel.

The main remaining disadvantages are the greater complexity of digital systems and the greater transmission bandwidth required. Large-scale, low-

FIGURE 15.3

Excessive Noise on a
Digital Signal

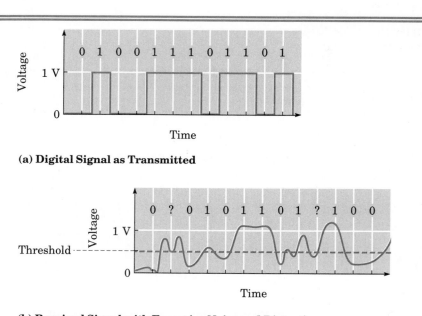

(a) Digital Signal as Transmitted

(b) Received Signal with Excessive Noise and Distortion

cost digital integrated circuits are reducing the difficulty and expense of constructing complex circuitry, and ingenious data-compression techniques are
beginning to decrease the bandwidth penalty. In general, the advantages outweigh the disadvantages.

15.1.1 Channels and Information Capacity

All practical communications channels are band-limited; either they will
pass frequencies from dc to some upper limit, or they have both lower and
upper cutoff frequencies. In addition, all channels are noisy. There is always
thermal noise, and there may be other kinds of noise as well.

There are theoretical limits to the rate at which information can be
sent along a channel with a given bandwidth and signal-to-noise ratio. Of
course, there is no limit to the amount of information that can be sent if
there is sufficient time to send it. The relationship between time, **information capacity,** and channel bandwidth is given by a simple equation called
Hartley's Law:

$$I = ktB \tag{15.1}$$

where I = amount of information to be sent
        k = a constant
        t = time available
        B = channel bandwidth

Hartley's law tells us that the information rate (that is, the amount of
information that can be sent in a given time) is proportional to the bandwidth

FIGURE 15.4

Ideal Bandpass Channel

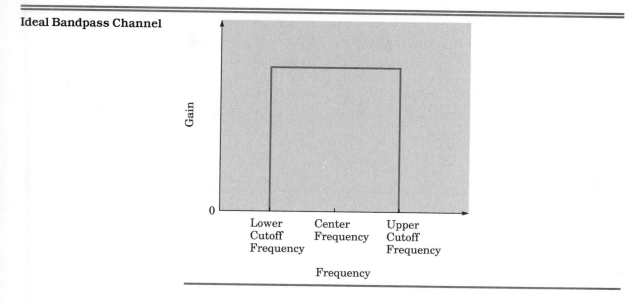

of the channel. It is also proportional to the factor k, which depends on the type of data coding used and the signal-to-noise ratio of the channel.

If the noise is assumed to have the same flat spectrum and random amplitude as thermal noise (the amplitude probability function is of the type called *Gaussian*), and if the bandwidth of the channel has the square-sided shape shown in Figure 15.4, then the maximum theoretical data rate for a given channel can be found using two simple equations.

First, ignoring noise, there is a limit to the amount of data that can be sent in a given bandwidth. This is given by the Shannon-Hartley theorem:

$$C = 2B \log_2 M \tag{15.2}$$

where C = information capacity in bits per second
        B = channel bandwidth in hertz
        M = number of levels transmitted

In the above equation, C and B should be easy to understand. M is a little less obvious. The idea is perhaps best explained by looking at a low-pass, rather than a bandpass, channel. Suppose that the channel can pass all frequencies from zero to some maximum frequency B. Then, of course, the highest frequency that can be transmitted is B. Suppose that a simple binary signal consisting of alternate ones and zeros is transmitted through the channel. This time let a logic 1 be 1 V and a logic 0 be -1 V. The input signal will look like Figure 15.5(a): it will be a square wave with a frequency one-half the bit rate (since there are two bits, a one and a zero, for each cycle). Being a square wave, this signal has harmonics at all odd multiples of its fundamental frequency, with declining amplitude as the frequency increases. At very low bit rates the output signal after passage through the channel will be similar to the input, but as the bit rate, and therefore the frequency, in-

FIGURE 15.5   **Digital Transmission Through a Low-Pass Channel**

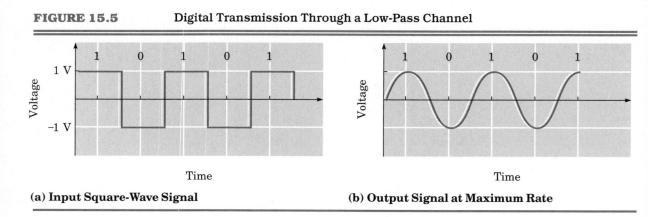

(a) Input Square-Wave Signal   **(b) Output Signal at Maximum Rate**

creases, the output will become more and more distorted. Finally, for a bit rate of $2B$, the frequency of the input signal will be B, and only the fundamental frequency will pass through the channel, as in Figure 15.5(b). Nonetheless, the receiver will still be able to distinguish a one from a zero, and the information will be transmitted. Thus, with binary information, the channel capacity will be

$$C = 2B$$

Next, suppose that instead of only two possible levels, several different levels, each corresponding to a different number, are possible. For instance, each level could represent one of four possibilities. For example, the possible levels could be -1 V, -0.5 V, $+0.5$ V, and $+1$ V. Figure 15.6 shows this four-level code. With this code, measuring the voltage level once at the receiver would actually provide two bits of information, since it takes two bits to express four different possibilities. However, the maximum bandwidth of the signal would not change. We have, it seems, managed to transmit twice as much information in the same bandwidth. This idea can be expanded to any number of levels, in order to give Equation (15.2).

It might seem that any desired amount of information could be transmitted in a given channel by simply increasing the number of levels. This is not

FIGURE 15.6   **Two- and Four-Level Codes**

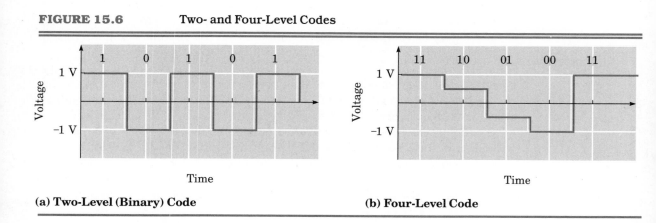

(a) Two-Level (Binary) Code   **(b) Four-Level Code**

possible because of noise. The more levels there are, the closer together they are and the more likely it is that noise will cause the receiver to mistake one level for another. This can also be seen from Figure 15.6. Therefore, there will be, for a given noise level, a maximum data rate that cannot be exceeded no matter how elaborately the data is coded. This maximum rate is called the Shannon limit:

$$C = B \log_2 (1 + S/N) \qquad (15.3)$$

where C = information capacity in bits per second
   B = bandwidth in hertz
   S/N = signal-to-noise ratio (as a power ratio, not in decibels)

There is no contradiction between Equations (15.2) and (15.3). Each represents a maximum rate, so the one that applies in a given situation is the one that gives the lower data rate. Remember also that these rates are theoretical maxima, and real, practical equipment is unlikely ever to reach the limit.

At this point, a small reminder may be in order. Not every calculator is capable of finding logs to the base 2, but they can all find logs to the base 10. A handy way to find logs to the base 2 is to use the following equation:

$$\log_2 N = \frac{\log_{10} N}{\log_{10} 2} \qquad (15.4)$$

EXAMPLE 15.1

A telephone line has a bandwidth of 3.2 kHz and a signal-to-noise ratio of 35 dB. A signal is transmitted down this line using a four-level code. What is the maximum theoretical data rate?

Solution

First, we use the Shannon-Hartley theorem to find the maximum data rate for a four-level code in the available bandwidth, ignoring noise. From Equation (15.2),

$$\begin{aligned} C &= 2B \log_2 M \\ &= 2(3.2 \times 10^3) \times \log_2 4 \\ &= 12.8 \times 10^3 \text{ b/s} \\ &= 12.8 \text{ kb/s} \end{aligned}$$

Next, we use the Shannon limit to find the maximum data rate for *any* code, given the bandwidth and signal-to-noise ratio. Remember that S/N is required as a power ratio.

$$\begin{aligned} S/N &= \text{antilog}_{10} \frac{35}{10} \\ &= 3162 \end{aligned}$$

From Equation (15.3),

$$C = B \log_2 (1 + S/N)$$
$$= (3.2 \times 10^3) \times \log_2 (1 + 3162)$$
$$= 37.2 \text{ kb/s}$$

Since both results are maxima, we take the lesser of the two, 12.8 kb/s. This means that it would be possible to increase the data rate over this channel by using more levels.

15.2 PULSE MODULATION

In order to transmit an analog signal by digital means, it is necessary first to sample the signal at intervals. The amplitude of the samples can then be expressed as a series of binary numbers for transmission. At the receiver, the samples can be reconstituted and used to form a facsimile of the original signal.

15.2.1 Sampling

In 1928, Harry Nyquist showed mathematically that it is possible to reconstruct an analog signal from periodic samples, as long as the sampling rate is at least twice the frequency of the highest frequency component of the signal. This assumes that an ideal low-pass filter prevents higher frequencies from entering the sampler. In practice, the sampling frequency should be considerably greater than twice the maximum frequency to be transmitted. Examples include telephony, where a sample rate of 8 kHz is used for a maximum audio frequency of 3.4 kHz; and the compact disc system, with a 44.1 kHz sampling rate and a maximum audio frequency of 20 kHz.

The penalty for a sampling rate that is too low is called **aliasing**. It is a form of distortion in which frequencies are translated downward. Figure 15.7 shows how it develops. In Figure 15.7(a), the sampling rate is adequate, and the signal can be reconstructed. In Figure 15.7(b), however, the rate is too low, and the attempt to reconstruct the original signal results in a lower-frequency signal. Once aliasing is present it cannot be removed.

15.2.2 Analog Pulse-Modulation Techniques

Sampling alone is not a digital technique. The immediate result of sampling is a **pulse-amplitude modulation** (PAM) signal, like the one in Figure 15.8(b). PAM is an analog scheme, where the amplitude of each pulse is proportional to the amplitude of the signal at the instant at which it is sampled.

For our purpose, the main use of PAM is as an intermediate step; before being transmitted, the PAM signal will have to be digitized. Similarly, at the receiver, the digital signal will be converted back to PAM as part of the demodulation process. The original signal can then be recovered using a low-pass filter.

FIGURE 15.7   **Aliasing**

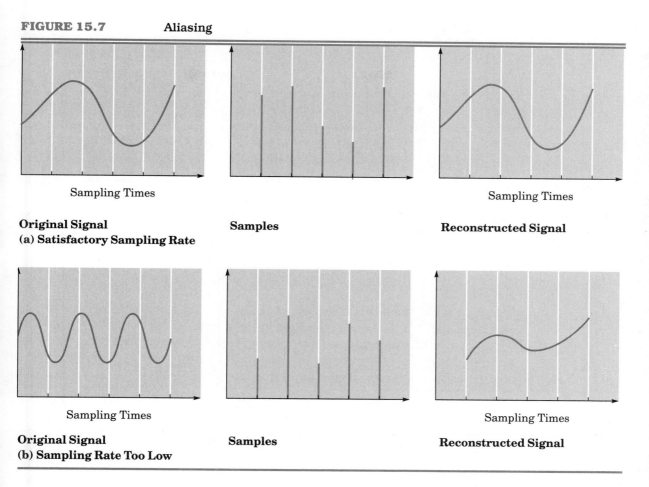

Sampling Times

Original Signal   **Samples**   **Reconstructed Signal**
(a) Satisfactory Sampling Rate

Sampling Times

Original Signal   **Samples**   **Reconstructed Signal**
(b) Sampling Rate Too Low

Before looking at digital schemes, a quick glance at two other analog pulse-modulation techniques is in order. **Pulse-duration modulation** (PDM), shown in Figure 15.8(c), uses pulses which all have the same amplitude. The duration of each pulse depends on the amplitude of the signal at the time it is sampled. PDM has its communications uses; for instance, it is often used in the high-powered audio amplifiers used to modulate AM transmitters. It has also been used for telemetry systems. Though still an analog mode, it is more robust than PAM because it is insensitive to amplitude changes due to noise and distortion. Like PAM, PDM can be demodulated with the aid of a low-pass filter. Sometimes the term **pulse-width modulation** (PWM) is used for this method.

Pulse-position modulation** (PPM), shown in Figure 15.8(d), is closely related to PDM. All pulses have the same amplitude and duration, but their timing varies with the amplitude of the original signal.

15.2.3 Pulse-Code Modulation

Pulse-code modulation (PCM) is the most commonly used digital modulation scheme. In PCM, the available range of signal voltages is divided into levels, and each is assigned a binary number. Each sample is represented by

FIGURE 15.8   Analog Pulse Modulation

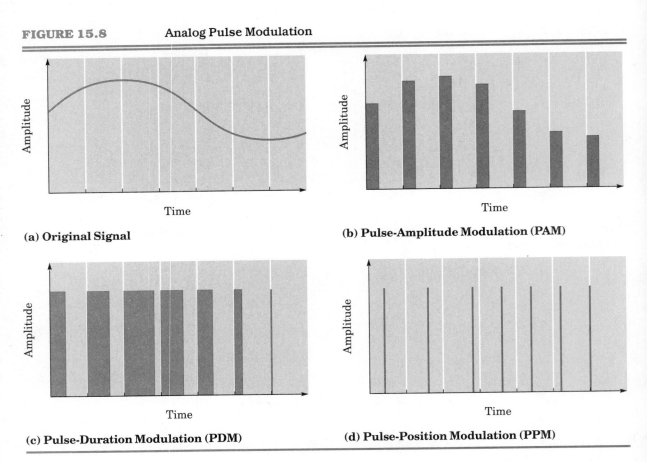

(a) Original Signal

(b) Pulse-Amplitude Modulation (PAM)

(c) Pulse-Duration Modulation (PDM)

(d) Pulse-Position Modulation (PPM)

the binary number representing the level closest to its amplitude. Then this number is transmitted in serial form. In *linear* PCM, levels are separated by equal voltage gradations.

The number of levels available depends on the number of bits used to express the sample value. The number of levels is given by

$$N = 2^m \qquad\qquad (15.5)$$

where N = number of levels
        m = number of bits per sample

EXAMPLE 15.2

Calculate the number of levels if the number of bits per sample is:

(a) eight (as used in telephony)

(b) sixteen (as used in the compact disc audio system)

Solution

(a) The number of levels with eight bits per sample is, from Equation (15.5),

$$N = 2^m$$
$$= 2^8$$
$$= 256$$

(b) The number of levels with sixteen bits per sample is, from the same equation,

$$N = 2^m$$
$$= 2^{16}$$
$$= 65,536$$

This process is called *quantizing*. Since the original analog signal can have an infinite number of signal levels, the quantizing process will produce errors, called *quantizing errors* (or, often, *quantizing noise*).

Figure 15.9 shows how quantizing errors arise. The largest possible error is one-half the difference between levels. Thus, the error is proportionately greater for small signals. This means that the signal-to-noise ratio varies with the signal level and is greatest for large signals. The level of quantizing noise can be decreased by increasing the number of levels, which also increases the number of bits that must be used per sample.

The dynamic range of a system is the ratio between the strongest possible signal that can be transmitted and the weakest discernible signal. For a linear PCM system, the maximum dynamic range in decibels is given approximately by

$$DR = 1.76 + 6.02m \text{ dB} \tag{15.6}$$

where DR = dynamic range in decibels
m = number of bits per sample

This equation ignores any noise contributed by the analog portion of the system.

FIGURE 15.9   Quantizing Error

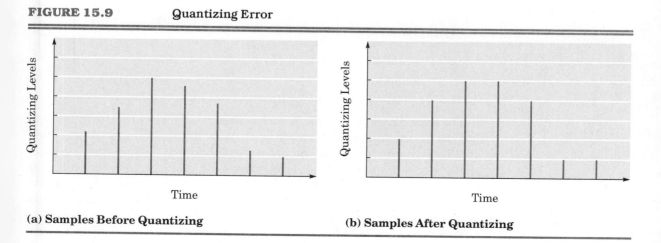

(a) Samples Before Quantizing   (b) Samples After Quantizing

EXAMPLE 15.3

Find the maximum dynamic range for a linear PCM system using sixteen-bit quantizing.

Solution

From Equation (15.6),

$$DR = 1.76 + 6.02m \text{ dB}$$
$$= 1.76 + 6.02 \times 16 \text{ dB}$$
$$= 98.08 \text{ dB}$$

Increasing the number of bits per sample increases the data rate, which is given very approximately by

$$D = f_s m \qquad\qquad (15.7)$$

where D = data rate in bits per second
   f_s = sample rate in samples per second
   m = number of bits per sample

Extra bits are needed to detect and correct errors. A few bits, called *framing bits*, are also needed to ensure that the transmitter and receiver agree on which bits constitute one sample. The actual bit rate will be somewhat higher than calculated above.

EXAMPLE 15.4

Calculate the minimum data rate needed to transmit audio with a sampling rate of 40 kHz and fourteen bits per sample.

Solution

From Equation (15.7),

$$D = f_s m$$
$$= 40 \times 10^3 \times 14$$
$$= 560 \times 10^3 \text{ b/s}$$
$$= 560 \text{ kb/s}$$

The transmission bandwidth varies directly with the bit rate. In order to keep the bit rate and thus the required bandwidth low, companding is often used. This system involves a compressor amplifier at the input, with greater gain for low-level than for high-level signals. The compressor will reduce the quantizing error for small signals. The effect of compression on the signal can be reversed by using an expander at the receiver, with a gain characteristic that is the inverse of that at the transmitter.

It is necessary to follow the same standards at both ends of the circuit, so that the dynamics of the output signal are the same as at the input. The system used in the North American telephone system uses a characteristic known as the μ (*mu*) *law*, which has the following equation for the compressor:

$$V_o = \frac{V_{mo} \ln (1 + \mu V_i / V_{mi})}{\ln (1 + \mu)}$$

(15.8)

where V_o = output voltage from the compressor
V_{mo} = maximum output voltage
V_{mi} = maximum input voltage
V_i = actual input voltage
μ = a parameter that defines the amount of compression
        (Contemporary systems use $\mu = 255$.)

Figure 15.10 shows the μ-255 curve. The curve is a transfer function for the compressor, relating input and output levels. It has been normalized; that is, V_i/V_{mi} and V_o/V_{mo} are plotted, rather than V_i and V_o.

FIGURE 15.10

μ-Law Compression

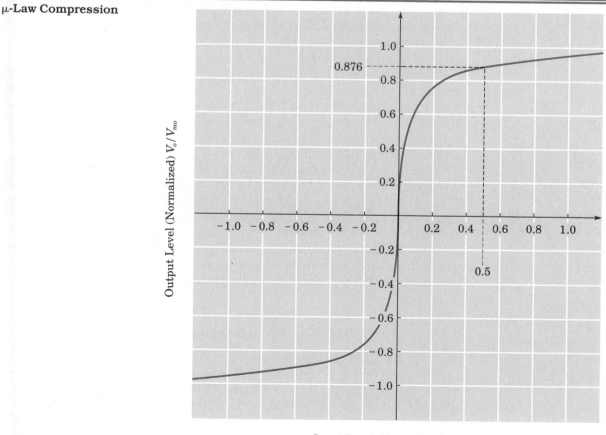

Input Level (Normalized) V_i/V_{mi}

EXAMPLE 15.5

A signal has one-half the maximum input voltage. What proportion of the maximum output voltage is produced?

Solution

From Equation (15.8),

$$V_o = \frac{V_{mo} \ln (1 + \mu V_i / V_{mi})}{\ln (1 + \mu)}$$

$$= \frac{V_{mo} \ln (1 + 255 \times 0.5)}{\ln (1 + 255)}$$

$$= 0.876 \; V_{mo}$$

This problem can also be solved graphically, as shown in Figure 15.10.

15.2.4 Delta Modulation

Although PCM is the most commonly used system of digital transmission, there is one other important technique. **Delta modulation** uses an idea that is, at first glance, breathtakingly simple. Instead of transmitting complete information about the amplitude of every sample, only one bit is transmitted. The bit will be a one if the sample is more positive than the previous sample, zero if it is more negative.

If this seems too good to be true, it is. Since only a small amount of information about each sample is transmitted, delta modulation requires a much higher sampling rate than PCM for equal quality of reproduction. Nyquist did not say that transmitting samples at twice the maximum signal frequency would always give undistorted results, only that it could, provided the samples were transmitted accurately.

Figure 15.11 shows how delta modulation generates errors. In region (i), the signal is not varying at all; the transmitter can only send ones and zeros, however, so the output waveform has a triangular shape, producing a noise signal called granular noise. On the other hand, the signal in region (iii) changes more rapidly than the system can follow, creating an error in the output called *slope overload*. Adaptive delta modulation, where the step size varies according to previous values, is more efficient. Figure 15.12 shows how it works. After a number of steps in the same direction, the step size increases.

15.2.5 Time-Division Multiplexing

Two types of multiplexing were introduced in Chapter 1. Frequency-division multiplexing (FDM) involves the full-time use of part of the available bandwidth for each signal. Though it has not always been pointed out explicitly, most of the systems looked at so far use FDM. Television broadcasting, which we examined in the previous chapter, is a good example. Each television signal occupies a small part (6 MHz) of the available bandwidth for television broadcasting, and does so on a full-time basis.

Time-division multiplexing (TDM) was also mentioned in Chapter 1, but

FIGURE 15.11 **Delta Modulation**

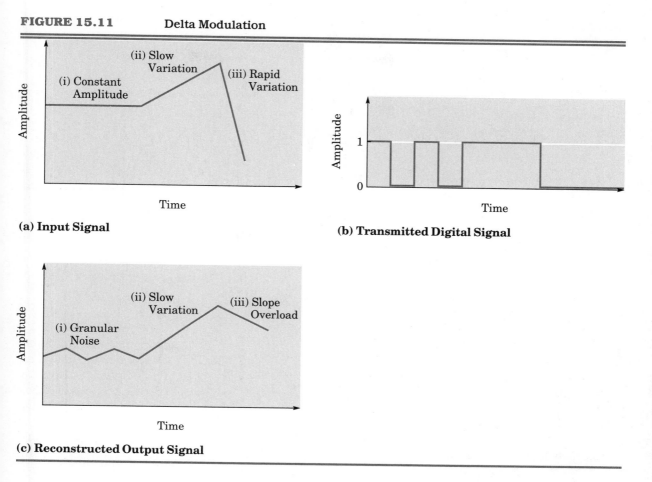

(a) Input Signal

(b) Transmitted Digital Signal

(c) Reconstructed Output Signal

has not been seen since, except for a brief mention in connection with high-definition television in Chapter 14. That is because TDM is used mainly for digital communication. In TDM, each information signal is allowed to use all the available bandwidth, but only for part of the time. From Hartley's Law [Equation (15.1)], it can be seen that the amount of information transmitted is proportional to both bandwidth and time. Therefore, at least in theory, it is equally possible to divide the bandwidth or the time among the users of a channel. Continuously varying amplitude signals, such as analog audio, are not well adapted to TDM, because the signal is present at all times. On the other hand, sampled audio is very suitable for TDM, as it is possible to transmit one sample from each of several sources sequentially, then send the next sample from each source, and so on. As already mentioned, sampling itself does not imply digital transmission, but, in practice, sampling and digitizing usually go together.

Many signals can be sent on one channel by sending a sample from each signal in rotation. Time-division multiplexing requires that the total bit rate be multiplied by the number of channels multiplexed. This means that the bandwidth requirement is also multiplied by the number of signals.

TDM is used extensively in telephony. There are many different standards for TDM. One commonly used arrangement is known as the T1 signal, which consists of 24 PCM voice channels, multiplexed using TDM. Each

FIGURE 15.12   Adaptive Delta Modulation

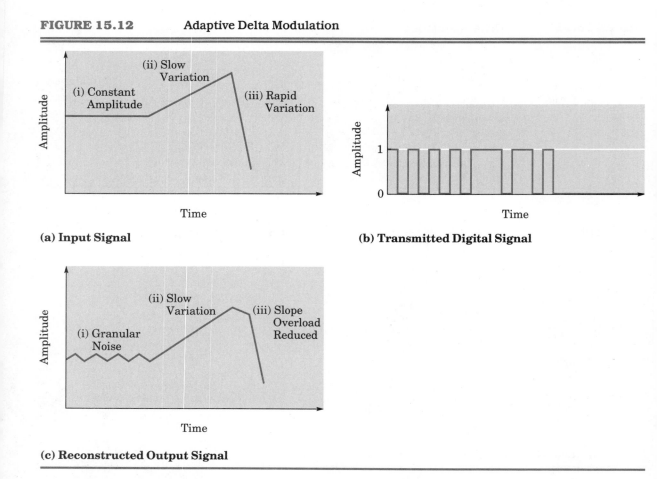

(a) Input Signal

(b) Transmitted Digital Signal

(c) Reconstructed Output Signal

channel is sampled at 8 kHz, with 8 bits per sample, as previously described. This gives a bit rate of 8k × 8 = 64k b/s for each voice channel.

The T1 signal consists of frames, each of which contains the bits representing one sample from each of the 24 channels. One extra bit is added, called the *framing bit*. It helps to synchronize the transmitter and receiver. Each frame contains 24 × 8 + 1 = 193 bits.

The samples must be transmitted at the same rate as they were obtained in order for the signal to be reconstructed at the receiver without delay. This requires the multiplexed signal to be sent at a rate of 8000 frames per second. Thus, the bit rate is 193 × 8000 b/s = 1.544 Mb/s. See Figure 15.13 for an illustration of a frame of a T1 signal.

FIGURE 15.13

**TDM in Telephony
(24 Samples, One from
Each Channel, Eight
Bits Each)**

First Sample Has
Extra Bit for
Framing

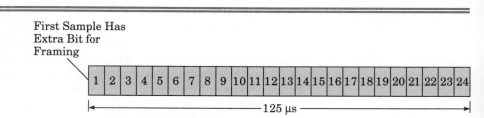

15.3 DATA CODING

PCM and delta modulation serve to convert analog signals, such as voice and video, to binary numbers. The data coding techniques studied in this section will convert alphanumeric characters into binary numbers. When data communication takes place between computers, the code used for communication may or may not be the same as that used internally; of course, data communication also takes place between other devices, such as teletype machines.

A data code is a standardized relationship between signalling elements and characters. Data codes are also called *character sets*, *character languages*, or *character codes*.

Data communication uses three types of characters. Data link control characters facilitate the orderly flow of data; graphic control characters arrange the presentation of data at the receiving terminal; and alphanumeric characters represent the data itself.

15.3.1 Example Character Codes: Baudot and ASCII

Many codes have been used to convert alphanumeric and other characters into electrical signals. In this section, we will look at two examples that are in common use, the Baudot and ASCII codes.

The Baudot code dates from about 1870 and is still in wide use for telex and radioteletype services. It has been designated as International Telegraph Alphabet number two (ITA2) by the International Telegraph and Telephone Consultative Committee (CCITT), a European standards group, some of whose standards are also used in North America. Baudot was originally used with electromechanical teletype machines, but most modern systems use a computer for coding and decoding.

The Baudot code is listed in Table 15.1. Note that it is a five-bit code

TABLE 15.1 Baudot Code (CCITT ITA2)

Code 54321	LTRS	FIGS	Code 54321	LTRS	FIGS
11111		[LTRS]	01111	K	(
11011		[FIGS]	10010	L	)
00000		[BLANK]	11100	M	.
00100		[SPACE]	01100	N	,
01000		[LINE FEED]	11000	O	9
00010		[CARRIAGE RETURN]	10110	P	0
00011	A	-	10111	Q	1
11001	B	?	01010	R	4
01110	C	:	00101	S	’ or [BELL]
01001	D	[WRU] or $	10000	T	5
00001	E	3	00111	U	7
01101	F	[unassigned] or ! or %	11110	V	= or ;
11010	G	[unassigned] or & or @	10011	W	2
10100	H	[unassigned] or # or £	11101	X	/
00110	I	8	10101	Y	6
01011	J	[BELL] or ’	10001	Z	+ or ”

(sometimes called a five-level code). This means that the number of possible combinations is $2^5 = 32$. Additional combinations are created by using two characters called LTRS and FIGS. These act like the shift key on a typewriter, to shift between letters and figures. The FIGS character set also includes control characters. The LTRS set includes uppercase letters only. This is satisfactory for telegraphic communication, but has obvious drawbacks for other forms.

Another problem with Baudot is that an error which is interpreted as a FIGS character will cause all the following information to be interpreted as numerical, until the next LTRS character occurs. To reduce this problem, most modern systems use a technique called *unshift-on-space* (USOS). The receiver will switch back to LTRS whenever a space character is received, so that an error can cause at most one word to be garbled.

Table 15.1 shows the most-significant bit (MSB) first, in standard binary notation. The code is actually transmitted with the least-significant bit (LSB), bit 1, first. A few of the characters for the FIGS case have different interpretations for different variations of the code. The standard CCITT version is given first, followed by the variant(s).

The characters in brackets are nonprinting control characters. Most of them are self-explanatory, but WRU stands for "Who Are You?" and is used to request the identity of the other station. BEL rings the bell on a teleprinter or beeps the speaker in a computer.

A five-bit code has the advantage of faster data transfer for a given bit rate, compared to codes with more bits per character. This makes Baudot useful for low-data-rate channels like HF radio.

The most common code for communication between microcomputers is known as ASCII, which stands for American Standard Code for Information Interchange. This is a seven-bit code, allowing for 128 possibilities without shifting. As with Baudot, bit 1 (LSB) is sent first.

ASCII has certain regularities to make programming easier. For instance, switching from lowercase to uppercase is accomplished by changing bit 6, and bits 4 through 1 of the numbers represent their binary-coded decimal (BCD) values. Many control characters are available; normally, they are generated on a keyboard by pressing the control key and another character simultaneously. Pressing the control key moves a character four columns to the left in Table 15.2, and is the same as subtracting 40 in hexadecimal notation from the character's value. For example, pressing CTRL–G gives the BEL character (the bell rings on a teletype machine or the speaker beeps on a computer).

Besides uppercase and lowercase letters and common punctuation marks, ASCII allows for a number of control characters. Many of these are rather obscure; the ones we need to know about right now are the following:

BEL   Ring the bell (or beep the speaker)

BS   Backspace

LF   Linefeed (advance to the next line)

ESC   Escape

FF   Formfeed (advance to the next page)

CR   Carriage return (move to the beginning of the line)

DEL   Delete

TABLE 15.2

ASCII Code (Binary Notation)	**7**	0	0	0	0	1	1	1	1
	Bit 6	0	0	1	1	0	0	1	1
	5	0	1	0	1	0	1	0	1
	4321								
	0000	NUL	DLE	SP	0	@	P	`	p
	0001	SOH	DC1	!	1	A	Q	a	q
	0010	STX	DC2	"	2	B	R	b	r
	0011	ETX	DC3	#	3	C	S	c	s
	0100	EOT	DC4	$	4	D	T	d	t
	0101	ENQ	NAK	%	5	E	U	e	u
	0110	ACK	SYN	&	6	F	V	f	v
	0111	BEL	ETB	'	7	G	W	g	w
	1000	BS	CAN	(	8	H	X	h	x
	1001	HT	EM	)	9	I	Y	i	y
	1010	LF	SUB	*	:	J	Z	j	z
	1011	VT	ESC	+	;	K	[	k	{
	1100	FF	FS	,	<	L	\	l		
	1101	CR	GS	-	=	M	]	m	}
	1110	SO	RS	.	>	N	^	n	~
	1111	SI	US	/	?	O	_	o	DEL

EXAMPLE 15.6

Write the ASCII codes for the characters B and b.

Solution

Find B in Table 15.2 and proceed upward to find that bits 7, 6, and 5 are 1, 0, and 0, in that order. Then move from B to the left to find bits 4, 3, 2, 1. They are 0010. Therefore the ASCII code for B is 1000010, with the most significant bit first. It would actually be transmitted with the least significant bit first.

The code for b can be found in the same way, as 1100010. Note that the only difference between the codes for B and b is the value of bit 6.

Sometimes ASCII codes are expressed in decimal or hexadecimal notation. The numbers are simply the decimal or hexadecimal translation of the binary numbers given above. Table 15.3 shows the decimal and hexadecimal numbers corresponding to the ASCII codes, along with the full names of all the control characters.

Sometimes an eighth bit is added to the ASCII code to allow for the transmission of graphics characters, mathematical symbols, and foreign-language characters. There is no single standard for this, and some manufacturers use several different character sets depending on the application.

Another bit that may or may not be transmitted after the character is the parity bit. **Parity** is an error-detection system, and will be discussed in Section 15.6.

TABLE 15.3   ASCII Codes (Decimal and Hexadecimal Notation)

Decimal	Hex	ASCII	Full Name	Decimal	Hex	ASCII	Full Name
0	00	NUL	Null	64	40	@	
1	01	SOH	Start of Heading	65	41	A	
2	02	STX	Start of Text	66	42	B	
3	03	ETX	End of Text	67	43	C	
4	04	EOT	End of Transmission	68	44	D	
5	05	ENQ	Enquiry	69	45	E	
6	06	ACK	Acknowledge	70	46	F	
7	07	BEL	Bell	71	47	G	
8	08	BS	Backspace	72	48	H	
9	09	HT	Horizontal Tab	73	49	I	
10	0A	LF	Line Feed	74	4A	J	
11	0B	VT	Vertical Tab	75	4B	K	
12	0C	FF	Form Feed	76	4C	L	
13	0D	CR	Carriage Return	77	4D	M	
14	0E	SO	Shift Out	78	4E	N	
15	0F	SI	Shift In	79	4F	O	
16	10	DLE	Data Link Escape	80	50	P	
17	11	DC1	Device Control 1	81	51	Q	
18	12	DC2	Device Control 2	82	52	R	
19	13	DC3	Device Control 3	83	53	S	
20	14	DC4	Device Control 4	84	54	T	
21	15	NAK	Negative Acknowledge	85	55	U	
22	16	SYN	Synchronous Idle	86	56	V	
23	17	ETB	End of Transmission Block	87	57	W	
24	18	CAN	Cancel	88	58	X	
25	19	EM	End of Medium	89	59	Y	
26	1A	SUB	Substitute	90	5A	Z	
27	1B	ESC	Escape	91	5B	[	
28	1C	FS	File Separator	92	5C	\	
29	1D	GS	Group Separator	93	5D	]	
30	1E	RS	Record Separator	94	5E	^	[Circumflex]
31	1F	US	Unit Separator	95	5F	_	[Underscore]

15.4 TRANSMISSION FORMATS

Either serial or parallel transmission can be used to transmit binary signals. In parallel transmission, multiple channels are used to transmit several bits simultaneously, while in serial transmission, a single channel is used. Data communication over more than very short distances is generally serial to reduce the number of channels needed, so that is the only type that we will study.

In this section, it will be assumed that the transmission channel is capable of carrying voltage pulses. Some of these channels will not be able to support a nonzero average level; that is, they are not dc-coupled. That problem can be dealt with by proper design of the transmission format, and will be considered in this section. Channels such as radio links or **dial-up** telephone lines, which require modulation and demodulation, will be considered later in this chapter.

TABLE 15.3   *Continued*

Decimal	Hex	ASCII	Full Name	Decimal	Hex	ASCII	Full Name
32	20	SP	Space	96	60	`	[Grave Accent]
33	21	!		97	61	a	
34	22	"		98	62	b	
35	23	#		99	63	c	
36	24	$		100	64	d	
37	25	%		101	65	e	
38	26	&		102	66	f	
39	27	'		103	67	g	
40	28	(		104	68	h	
41	29	)		105	69	i	
42	2A	*		106	6A	j	
43	2B	+		107	6B	k	
44	2C	,		108	6C	l	
45	2D	-		109	6D	m	
46	2E	.		110	6E	n	
47	2F	/		111	6F	o	
48	30	0		112	70	p	
49	31	1		113	71	q	
50	32	2		114	72	r	
51	33	3		115	73	s	
52	34	4		116	74	t	
53	35	5		117	75	u	
54	36	6		118	76	v	
55	37	7		119	77	w	
56	38	8		120	78	x	
57	39	9		121	79	y	
58	3A	:		122	7A	z	
59	3B	;		123	7B	{	
60	3C	⟨		124	7C			
61	3D	=		125	7D	}	
62	3E	⟩		126	7E	~	[Tilde]
63	3F	?		127	7F	DEL	Delete

A digital transmission format has to specify standard voltage ranges that will be interpreted as binary 1 and 0. There must be a standard clock speed, and the two clocks must be in phase with each other, so that the receiver will check the level of the line at the correct times. There also needs to be a system of framing; that is, somehow the receiver has to be able to tell which bit represents the first bit of a character or a PCM sample. Of course, the same data code must be in use at both ends of the communications link.

15.4.1 Asynchronous Communication

Data communications schemes are designated as synchronous or asynchronous depending on how the timing and framing information is transmitted. The framing for asynchronous communication is based on a single character, while that for synchronous communication is based on a much longer block of data. This section will deal with asynchronous communication, which was originally intended for use with electromechanical teleprinters, and is now the type of communication most commonly used with microcomputers. Both Baudot and ASCII codes are used, but ASCII is more popular and will be used as our example.

In an asynchronous system, the transmit and receive clocks are free-running, set to approximately the same speed. A **start bit** is transmitted at the beginning of each character, and at least one stop bit is sent at the end of the character. The **stop bit** leaves the line or channel in the **mark** condition, which represents binary 1, and the start bit always switches this to **space** (binary 0). The terms *mark* and *space* date from the early days of telegraphy, when a pen linked to a solenoid marked a moving paper tape with the dots and dashes of Morse code. An energized line caused the pen to mark the tape, and a line with no current caused the pen to lift from the tape, creating a space.

In asynchronous communication, the framing is set by the start bit. The timing will remain accurate enough throughout the limited duration of the character, provided the clocks at the transmitter and receiver are reasonably close to the same speed. This is not a problem with computers using crystal-controlled clocks, but it can be with teleprinters, where the clock rate depends on the speed of a motor. One of the reasons the Baudot code has only five bits is to reduce the length of time during which transmitter and receiver have to remain synchronized.

There is no set length of time between characters with asynchronous transmission. The receiver monitors the line until it receives a start bit. It counts bits, knowing the character length being employed, and, after the stop bit, it begins to monitor the line again, waiting for the next character.

Not every bit that is transmitted contributes directly to the message. The start bit, stop bit(s), and parity bit, if present, are called **overhead**. The **efficiency** of the communication system can be defined as

$$eff = \frac{\text{number of data bits}}{\text{number of total bits}} \tag{15.9}$$

EXAMPLE 15.7

Calculate the maximum efficiency of an asynchronous communication system using ASCII with seven data bits, one start bit, one stop bit, and one parity bit.

Solution

Assuming characters follow immediately one after another, there are ten bits per character, of which seven are data bits. From Equation (15.9),

$$eff = \frac{\text{number of data bits}}{\text{number of total bits}}$$

$$= \frac{7}{10}$$

$$= 0.7 \quad \text{or} \quad 70\%$$

FIGURE 15.14   Simplified Block Diagram of a UART

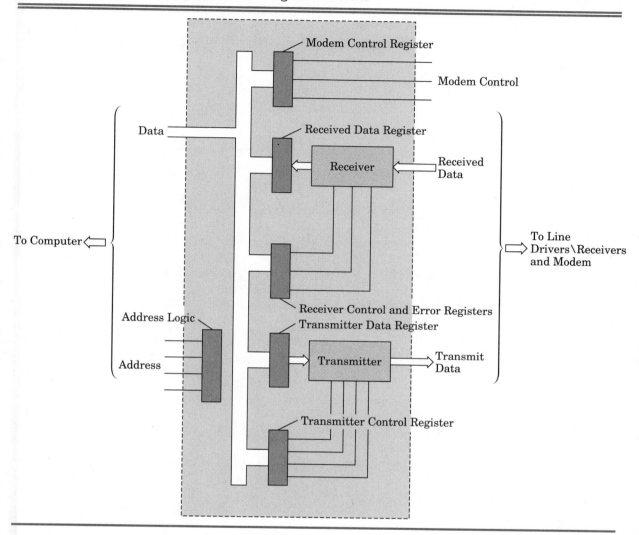

15.4.2 UARTs

The abbreviation UART stands for Universal Asynchronous Receiver-Transmitter. A typical UART is shown in block-diagram form in Figure 15.14. At the transmitting end of an asynchronous communication link, a UART converts parallel data from a computer or terminal into serial form, adds start and stop bits, and clocks it out at the correct rate. It may also add a parity bit, used for error control. Parity will be discussed later in this section.

    The receiver section of a UART must be able to detect the start-bit transition at the beginning of a character. The receiver and transmitter clocks will not be in phase at that point. The receiver must continuously monitor the line until it notes a one-to-zero transition, then must look again at the signal a short time later to make sure it is still low. It then identifies the transition as a start bit rather than a noise pulse. It does this by checking the line at a rate typically sixteen times the bit rate. When a transition is noted, the line

FIGURE 15.15   Asynchronous Reception (Arrows Indicate Sampling Times)

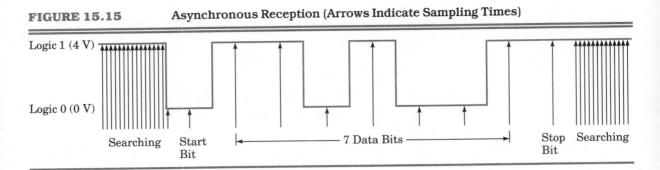

is again observed eight clock cycles (one-half the time for one bit) later. If still low, the pulse is identified as a start bit. Figure 15.15 shows the process.

After the start bit has been identified, the signal will be sampled once per bit time, approximately in the center of the bit, until the end of the character (stop bit). If the stop bit is not a one, an error signal must be sent. An error signal will also occur if the parity bit (if present) is not correct.

Generally the UART will assemble the received character in a shift register, then transfer it to another register for access by the computer while it assembles the next character.

15.4.3 Synchronous Communication

In synchronous communication, the transmitter and receiver are synchronized to the same clock frequency. There are many ways to do this, but the most common method is for the receiver to derive its clock signal from the received data stream. This requires that the transmitted data have a sufficient number of level transitions.

Synchronous communication is more efficient than asynchronous because start and stop bits are not necessary. Instead, data is sent in blocks that are much longer than a single character. Blocks typically begin with an identifying sequence of bits which allows proper framing and often identifies the content of the block as well.

The price that is paid for the increased efficiency of synchronous communication is increased complexity of both hardware and software. In practice, asynchronous systems are commonly used for communication between microcomputers. Synchronous systems are used for the higher-speed communications typical of mainframe computers, and for digitized analog signals such as are found in the telephone system.

15.4.4 Line Codes

So far, our discussion of digital signals has been in terms of zeros and ones. It is now time to convert these binary numbers into voltage or current levels on a line. This can be done in many ways. Probably the simplest **line code** is to use the presence of a current or voltage to represent one logic state, and its absence to indicate the other. For example, TTL voltage levels could be used. Figure 15.15 shows an ASCII character sent using TTL levels. Positive logic, where a high level represents a logical one, was used for the example, but negative logic could be used just as well.

FIGURE 15.16

Current Loop

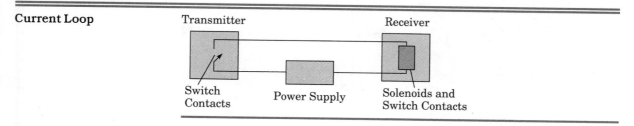

An example of digital transmission using current levels is a teleprinter loop. These machines use a current of either 20 or 60 mA to represent a mark condition (logic 1) and zero current to represent a space (logic 0). Figure 15.16 shows two teleprinters connected in this way. The entire system is electromechanical: signaling is accomplished by opening and closing the loop at the transmitter, and an electromechanical system using a motor and relays interprets the signals at the receiver.

Both of the examples just given require a line with dc continuity, since the current, when present, always flows in the same direction. Codes of this sort are called *unipolar NRZ codes*. The term *unipolar* means that the voltage or current polarity is always the same, and **NRZ** (non-return-to-zero) means that there is no requirement for the signal level to return to zero at the end of each element. For example, the message 111 requires the voltage or current to go to the high state and stay there for three bit periods.

Sometimes, the condition of dc continuity cannot be met (for instance, when there are transformers or ac-coupled amplifiers on the line). Codes that have zero average dc content have been developed for ac-coupled lines. In the long run, the bipolar NRZ code shown in Figure 15.17 will accomplish this, provided the message contains equal numbers of ones and zeros. On the other hand, a long string of ones or zeros will result in a component at very low frequency, and for some systems, such as asynchronous transmission, there is no guarantee that there will be equal numbers of ones and zeros.

Low-frequency ac components and dc components can be eliminated completely with bipolar **RZ** (return-to-zero) codes. Figure 15.18 provides several examples. Shown in Figure 15.18(a) is a system used in telephony; it is called

FIGURE 15.17

Bipolar NRZ Code

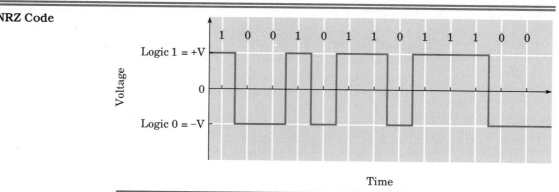

FIGURE 15.18   RZ Codes

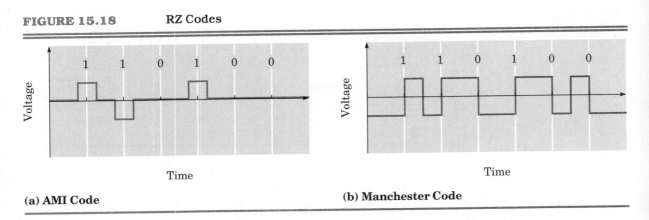

(a) AMI Code   (b) Manchester Code

AMI, for alternate mark inversion. A binary zero is coded as 0 V, and binary ones are recorded alternately by positive and negative voltages. This signal will have no dc or low-frequency ac content, even with long strings of ones or zeros. On the other hand, long strings of zeros must be avoided when synchronous communication takes place with this code. A string of zeros will cause the signal to disappear and can cause timing to be lost. There is some error detection built into this code: any time two consecutive pulses with the same polarity are received, an error must have occurred.

Figure 15.18(b) shows the Manchester code, which is a type of *biphase* code. Every bit has a level transition in the center of the bit period. For a one there is an upward transition; for a zero, a downward transition. There is no dc or low-frequency energy regardless of the proportion of zeros and ones in the signal. The Manchester code also provides strong timing information regardless of the pattern of ones and zeros. Its disadvantage is that it requires more bandwidth than the AMI code.

15.5 TRANSMISSION OF DIGITAL DATA VIA ANALOG CARRIERS

The data referred to in the heading above can be either coded characters or a digitized analog signal. Some communications methods require analog carriers—transmission via radio, for example. Other media, such as coaxial cable and optical fiber, can carry either analog FDM or digital TDM signals. Digital transmission is preferred when practical, but where a system has already been constructed using analog techniques, it is simpler and less expensive to convert data signals to analog form than to rebuild the whole system. This is especially true of the dial-up telephone system, which was set up for analog voice signals. Although many telephone central offices and long-distance trunk lines are now digital, nearly all the individual telephone lines, known as **local loops**, are still analog.

The characteristics of analog channels include limited bandwidth, which prevents the transmission of short pulses. Often, as in the telephone system, neither dc nor high-frequency ac is allowed. Even over the limited frequency range of the channel, the amplitude response may vary with frequency. See

FIGURE 15.19

Voice-Grade Telephone
Channel

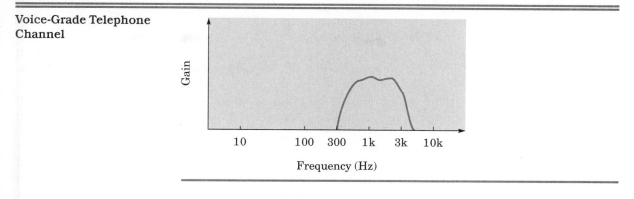

the frequency response of a typical voice-grade telephone channel, shown in
Figure 15.19, for an example. The phase response may also be nonlinear.

All channels are subject to noise of various types. Thermal noise has already been discussed, and noise can also originate from switching, lightning,
and so forth.

Channel characteristics may vary with time. For example, fading due to
rain will reduce the signal-to-noise ratio of a microwave link. Dial-up telephone connections may take different routes each time a call is placed between the same two locations. High-frequency radio links are very susceptible to fading as well as noise and interference.

For transmission via analog channels, the digital information signal is
used to modulate a sine-wave carrier. Demodulation at the other end restores
the original baseband information. A modem is a modulator and demodulator
in one unit.

Figure 15.20 shows a data communication setup using modems where
DTE refers to data terminal equipment and **DCE** to data communications
equipment. Normally, a DTE is a terminal, a computer, or a teleprinter. DCE
is another term for a modem. Confusion sometimes arises when a computer
port that is intended to connect to a DCE is connected directly to another
terminal or a printer. In that case, it may be necessary for the printer, or
terminal, to emulate a DCE as best it can. A device called a **null modem**, to
be described later, is often used for this purpose.

15.5.1 Modulation Techniques

Whether the modulating signal is analog or digital, there are only three carrier parameters that can be varied: amplitude, frequency, and phase. In practice, frequency modulation, known as **frequency-shift keying** (FSK) when

FIGURE 15.20

Data Communication
Using Modems

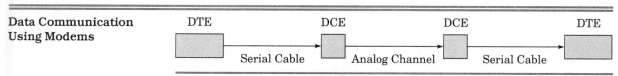

used with digital signals, is generally used at low data rates. Phase modulation (**phase-shift keying** or PSK) is used at moderate data rates, and **quadrature amplitude modulation** (QAM), which is really a combination of amplitude and phase modulation, is used in high-speed modems. (By the way, you may have noticed that we used the abbreviation QUAM for the analog version of quadrature AM. For no logical reason, the analog and digital variations of this technique are usually abbreviated in different ways.) The terms low, moderate, and high are relative, since the possible data rate depends on the channel bandwidth and signal-to-noise ratio as well as the modulation scheme.

Digital signals result in discrete, rather than continuous, changes of the modulated signal. The receiver examines the signal at specified times, and the state of the signal at each such time is called a **symbol**. The eventual output will be binary, but it is certainly possible to use more than two states for the transmitted symbol. As we saw earlier in this chapter, complex schemes using several levels can send more data in a given bandwidth. For example, four bits could be sent at once using sixteen combinations of amplitude and phase.

At this point, and before we begin any discussion of the speed at which information is transmitted by modems, we should distinguish between bit rate and **baud rate**. Theoretically, the bit rate is simply the number of bits transmitted per second, while the baud rate is the number of symbols. Therefore,

$$\text{bit rate} = \text{baud rate} \times \text{bits per symbol} \tag{15.10}$$
$$= \text{baud rate} \times \log_2 (\text{possibilities per symbol})$$

If each symbol has only two possibilities, and therefore carries one bit of information, then the baud rate and the bit rate are the same. Confusion arises, however, when more complex schemes are used that allow several bits to be transmitted at once. With these systems, the bit rate is higher than the baud rate, but the two terms are often used interchangeably to represent the bit rate.

EXAMPLE 15.8

A modem transmits symbols each of which has 64 different possibilities, 10,000 times per second. Calculate the baud rate and bit rate.

Solution

The baud rate is simply the symbol rate, or 10 kbaud. The bit rate is given by Equation (15.10):

$$\text{bit rate} = \text{baud rate} \times \log_2 (\text{possibilities per symbol})$$
$$= (10 \times 10^3) \times \log_2 64$$
$$= (10 \times 10^3) \times 6$$
$$= 60 \times 10^3 \text{ b/s}$$
$$= 60 \text{ kb/s}$$

15.5.2 Frequency-Shift Keying Modems

Frequency-shift keying (FSK) is a simple and reliable modulation scheme and is commonly used for low-data-rate applications. A carrier is switched between two frequencies, one defined as the mark and the other as the space frequency. For full-duplex communication, there will be four frequencies involved, with two transmitted in each direction.

We will look at FSK modems by means of practical examples. First let us consider the Bell 103 standard, intended for use with dial-up telephone lines. The frequencies used are shown in Figure 15.21. This standard allows full duplex operation at 300 bits per second using frequency-division multiplexing. There are two pairs of mark and space frequencies. The modem that places the call is called the *originate modem*, and transmits with a mark frequency of 1270 Hz and a space frequency of 1070 Hz. The other modem is called the *answer modem*, and uses a mark frequency of 2225 Hz and a space frequency of 2025 Hz. All of these frequencies are well inside the telephone passband and sufficiently removed from each other to allow room for the sidebands that are generated by modulation.

Another FSK standard for use with the telephone system is known as the Bell 202 modem. This system, shown in Figure 15.22, allows a higher bit rate, but it is essentially a half-duplex standard. The mark frequency is 1200 Hz and the space frequency is 2200 Hz. The use of only two, more widely separated frequencies allows a data rate of 1200 bits per second. In addition, there is provision for a slow speed (5 bits per second) return channel, using a carrier at 387 Hz that can be keyed on and off to acknowledge transmissions or request a retransmission.

FSK is also used for digital communications via HF radio. Here the carrier frequency is much higher, of course, and depends on the frequency allotted by government regulation to the station. Rather than the actual mark and space frequencies being defined as for telephone modems, radio transmission systems specify the frequency shift between mark and space. There are many possibilities for this, with 170, 425, and 850 Hz being the most common.

Although FSK is actually a form of FM, the most common way of generating it in HF radio communications is to use a single-sideband (SSB) AM

FIGURE 15.21

Bell 103 Modem

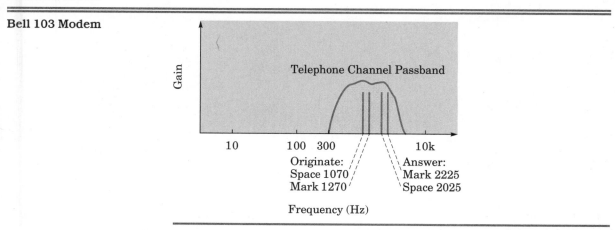

FIGURE 15.22

Bell 202 Modem

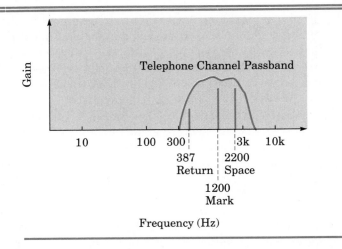

transmitter modulated by audio tones. Recall that when a single-sideband transmitter is modulated by a single audio tone, its output is a sine wave at one frequency. For an upper-sideband signal, the output frequency is the sum of the (suppressed) carrier frequency and the modulating frequency, and for a lower-sideband signal, it is the difference.

Figure 15.23 shows how SSB equipment can be used to produce FSK. At the transmitting station, a modem produces two audio frequencies, one for

FIGURE 15.23   **Radio FSK Transmission Using SSB Equipment**

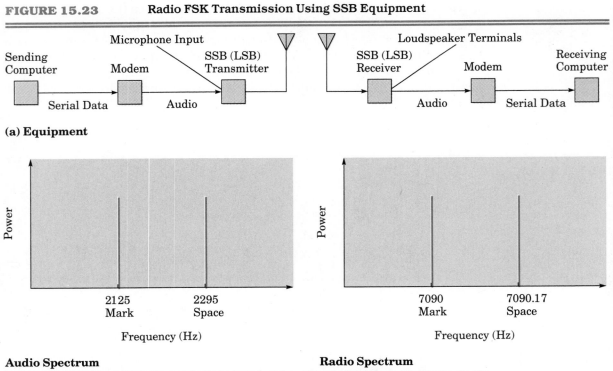

(a) Equipment

Audio Spectrum   Radio Spectrum
(b) Typical Spectra (This Example Is for an Amateur Radio Station with 170 Hz Shift)

FIGURE 15.24   **Radio AFSK Transmission Using FM Equipment**

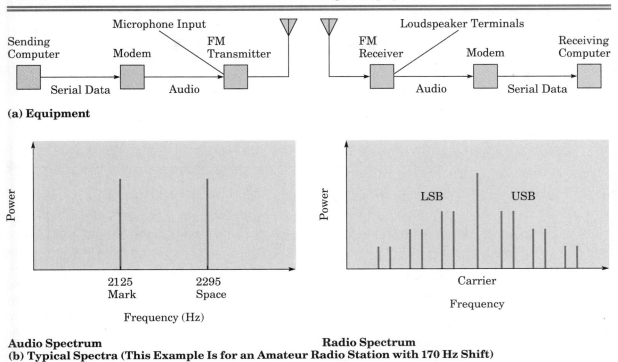

(a) Equipment

Audio Spectrum

Radio Spectrum

(b) Typical Spectra (This Example Is for an Amateur Radio Station with 170 Hz Shift)

mark and one for space, that differ by the required frequency shift. These tones modulate the SSB transmitter, whose output will then consist of a sine wave that shifts between two frequencies. For example, 2125 and 2295 Hz can be used to give a shift of 170 Hz. The receiver will also be set up for SSB operation, and it will produce two tones at its output. They may not be the same two tones as at the transmitter input, depending on the exact tuning of the receiver and its BFO, but they will differ in frequency by the same amount.

A variation called AFSK, for audio frequency-shift keying, is often used on VHF and UHF radio links, which usually employ FM equipment. In this system, the FM transmitter is modulated with an audio tone that shifts frequency from mark to space, just as with the SSB HF system. A conventional FM receiver will demodulate the same two tones as were applied at the transmitter. Figure 15.24 shows the setup.

15.5.3 Phase-Shift Keying Modems

When somewhat faster data rates are required than can be achieved with FSK, phase modulation is often used. Actually, the system used is more completely described as **delta phase-shift keying** (DPSK) because what is measured at the receiver is not the phase of the signal with respect to some permanent reference (which would be awkward to maintain), but rather any shift in phase of the signal with respect to that measured at the previous sampling time. Most DPSK modems use a four-phase system called *quadrature phase-shift keying* (QPSK). In QPSK, each symbol represents two bits, and the bit rate is twice the baud rate. This is often called a *dibit* system.

FIGURE 15.25

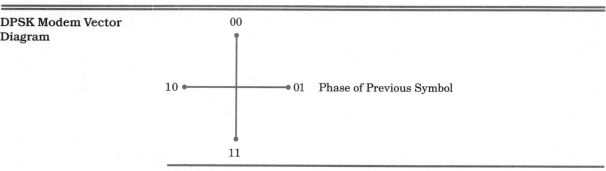

DPSK Modem Vector
Diagram

The Bell 212A modem is an example of a system of this type. Its maximum data rate is 1200 bits per second, full duplex. Since four different phase shifts are used, this is a dibit system and 1200 bits per second is actually 600 baud. Nonetheless, manufacturers that build modems to this standard often call them 1200 baud modems. Modems built to this standard have "fallback" capability; that is, when line conditions are not good enough for reliable communication at 1200 bits per second, they can reduce speed, or "fall back," to 300 bits per second using the Bell 103 standard. Figure 15.25 is a vector diagram that shows how the four possible phase shifts from one symbol to the next are translated into four two-bit combinations.

Another DPSK system is known as the CCITT V.22 bis standard. As mentioned previously, the CCITT is a European standards group, some but not all of whose standards are also observed in North America. The word *bis* is French for *second*, and indicates that this is a revised standard. Modems built to the V.22 bis standard operate at 2400 bits per second, using a dibit system and an actual baud rate of 1200. Modems using this standard and intended for North American use have fallback to the Bell 212A and 103 standards as line quality deteriorates.

15.5.4 Quadrature Amplitude Modulation Modems

The only way to achieve high data rates with a narrowband channel, such as a dial-up telephone line, is to increase the number of bits per symbol, and the most reliable way to do that is to use *quadrature amplitude modulation* (QAM). We have seen analog versions of quadrature AM before, most recently in the color television signal described in Chapter 14. There two amplitude-modulated carriers are used with a 90° phase angle between them. These add to produce a signal with an amplitude and phase angle that can vary continuously.

The digital version of QAM is similar except that there is a finite number of possible amplitude-phase combinations that are allowed for a given system. Figure 15.26 is a "constellation diagram" that shows the possibilities for a hypothetical system with eight phase angles and four amplitudes, for a total of 32 combinations. Thus each transmitted symbol would represent five bits. This diagram is like that of the previous figure except that the vectors are not drawn. Each circle represents a possible amplitude-phase combination. With a noiseless channel, the number of combinations could be in-

FIGURE 15.26

Constellation Diagram

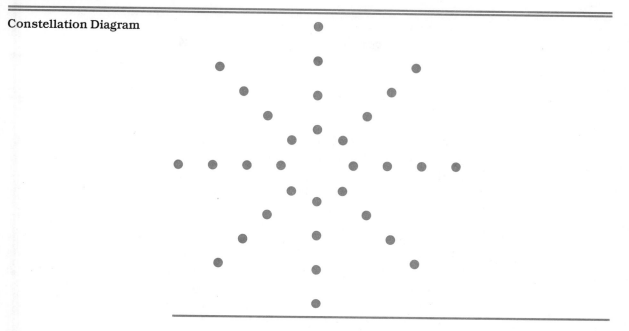

creased indefinitely, but a practical limit is reached when the difference, in amplitude or phase angle, between adjacent combinations becomes too small to be detected reliably in the presence of noise and distortion.

One example of a QAM modem for telephone use is the CCITT V.29 standard. This system operates at 2400 baud, and varies the number of bits per symbol according to the line conditions. There are three possibilities:

9600 b/s using 2 amplitudes and 8 angles, for 16 states or 4 bits per symbol

7200 b/s using 1 amplitude and 8 angles, for 8 states or 3 bits per symbol

4800 b/s using 1 amplitude and 4 angles, for 4 states or 2 bits per symbol

15.5.5 Establishing a Connection

The steps required to establish digital communication over an analog link such as a standard telephone line vary considerably depending on the type of modem and the type of connection. Some high-speed modems are permanently connected to leased lines, for example. In this section, we will look at the procedure for establishing digital communication between two computers over an ordinary **dial-up** telephone line. An example would be connecting to a computer bulletin board. We will assume that asynchronous, full-duplex communication is in use, since that is the most common technique for microcomputer communications.

Full-duplex communication using a single line requires frequency-division multiplexing (FDM); that is, the carrier frequency or frequencies will vary depending on whether the modem is set to originate or answer the call. Formerly, the call would be dialed by hand and answered by a human, and then the modems would be connected, but current practice is to have the

TABLE 15.4

"AT" Modem Command Language: Sample Commands	AT	Attention: starts all command strings
	A	Answer: modem answers call
	D	Dial phone: followed by T (tone) or P (pulse) to indicate type of dialing, then the number to be dialed.
	H	Hang up phone

modem dial the call. The one that does so should be set to the originate mode, of course. Almost all current modems are "smart"; that is, they are programmable to a certain extent, using a language that will be described shortly.

The answer modem will detect the ringing signal on the telephone line and, if it has been programmed to do so, it will answer the phone. Alternatively, it can be programmed to indicate to the computer that the phone is ringing, and then await instructions as to whether to answer. Once it has taken the phone line off-hook, it will send a carrier down the line to indicate to the originating modem that the phone has been answered. The originating modem will also send a carrier, and both modems will indicate to their associated terminals that a connection has been made.

Depending on the modem, the above operations, and others, are usually programmed into the modem by means of commands sent from the computer along the same serial lines used for communications. The language normally used for this was developed by the Hayes company for its Smartmodem,™ but has been adapted by most other modem manufacturers. It is supported by almost all the software that has been written to facilitate modem communications. Sometimes it is called the "AT" language because its command strings begin with those letters. A partial list of some of the more important commands is listed in Table 15.4. There are many others, used for such purposes as instructing the modem whether to answer the phone automatically, turning the modem's internal speaker on and off, and so on.

An example of a command string would be

ATDT 1,855-1155

which instructs the modem to dial the number given (the hyphen is ignored, but the comma is interpreted as a pause). Since commands should be interpreted by the modem rather than being sent down the telephone line, the modem must distinguish between commands and data to be transmitted. When the modem is turned on, it is in the local mode and is receptive to commands; once a connection is made, it goes on line. To return the modem to local mode to receive more commands, a command to disconnect for instance, the computer must send a sequence that is not likely to occur in normal communication. For instance, one commonly used sequence is a pause, followed by three plus signs, then another pause.

15.5.6 The RS-232C Serial Communications Standard

In a previous section, the role of a UART in interfacing between the parallel data of the computer and the serial data necessary for communication was described. Normally, the UART is in the computer or terminal. The modem may also be inside the computer, but often it is a separate device located nearby. Serial communication takes place between the computer and the modem, for data and, in the case of "smart" modems, commands and responses as well. There may also be several other connections between computer and modem, to indicate to each the status of the other and to allow the computer to control the modem.

There are a number of standards for communication between the computer (DTE) and the modem (DCE), but the most popular is the Electronic Industries Association (EIA) standard known as RS-232C. It defines voltage levels and even pin numbers, for both data and control lines. The *serial port* found on most personal computers follows this standard. Not all of the lines are used in all installations, and the port can also be used for non-modem applications, for example, the connection of printers, plotters, and computer mice. The uses (and misuses) of the RS-232C standard could fill a book, so only the basics will be described here.

Table 15.5 shows the lines defined by RS-232C. Two different types of connectors are commonly used. Most computers use the 25-pin DB-25 connector, but some, notably the IBM PC-AT,™ use a nine-pin DB-9 connector to

TABLE 15.5

RS-232C Connections

Connector Pins		Designation		
9-Pin	**25-Pin**	**EIA**	**Common**	**Function**
	1	AA	GND	Protective (Chassis) Ground
3	2	BA	TD	Transmit Data
2	3	BB	RD	Receive Data
7	4	CA	RTS	Request to Send
8	5	CB	CTS	Clear to Send
6	6	CC	DSR	Data Set Ready
5	7	AB	SG	Signal Ground
1	8	CF	DCD	Data Carrier Detect
	9			Reserved
	10			Reserved
	11			Unassigned
	12	SCF		Secondary Received Line Signal Detect
	13	SCB		Secondary Clear to Send
	14	SBA		Secondary Transmit Data
	15	DB		Transmitter Signal Element Timing (DCE)
	16	SBB		Secondary Received Data
	17	DD		Receiver Signal Element Timing
	18			Unassigned
	19	SCA		Secondary Request to Send
4	20	CD	DTR	Data Terminal Ready
	21	CG	SQ	Signal Quality Detector
9	22	CE	RI	Ring Indicator
	23	CH		Data Signal Rate Selector (DCE)
	24	DA		Transmitter Signal Element Timing (DTE)
	25			Unassigned

save space and/or cost. Obviously, the nine-pin connector has room for only a subset of the RS-232C lines, but these are the only ones needed for asynchronous communication with microcomputers.

The EIA has official letter designations for each line, but they are hard to remember, and there are simple mnemonic designations for the most often used lines. These are unofficial but reasonably standardized, and are shown in the "common designation" column in Table 15.5.

The RS-232C interface has lines for data in each direction: Transmit Data (TD) and Receive Data (RD) describe the process as seen by the DTE. That is, the TD line carries data from the computer to the modem, and RD carries it from modem to computer. There are provisions for a second data channel, using the secondary data lines, but these are seldom used.

The RTS and CTS lines provide "handshaking" between the DTE and the DCE; that is, the modem uses CTS (Clear To Send) to signal the DTE that it is ready to transmit data. Normally, this will occur after a connection has been made with a modem at the other end. With an intelligent modem, the function of CTS can be accomplished by a software response along the data lines. RTS (Ready To Send) is a signal from the DTE that it has information to transmit.

The functions of DSR and DTR are related to those of RTS and CTS. DSR (Data Set Ready) indicates the DCE (modem) is turned on, but does not imply that a telephone connection has been made. Similarly, DTR (Data Terminal Ready) indicates that the terminal is on and ready for communications.

The DCD (Data Carrier Detect) line is used for the modem to signal that the analog carrier from another modem is being received, and the RI (Ring Indicator) line is one way the modem can signal that the phone is ringing. Another way, used by intelligent modems, is to send the word "ring" along the data lines in command mode.

It may seem from the above that many of the features of RS-232C are redundant when an intelligent modem is used. This is true, and follows from the fact that the last revision of this standard was in 1966, long before modems with built-in microprocessors and memory were available. To say that confusion reigns in the world of serial interfacing with RS-232C is an understatement.

The RS-232C standard specifies voltage levels and logic type. For all lines, a *high* at the transmitter is in the range of +5 V to +15 V, and a *low* is −5 V to −15 V; the receiver should count anything above +3 V as high, and below −3 V as low. The voltage levels for data are in negative logic; that is, a positive voltage on a data signal line indicates a zero (space), while a negative voltage indicates a one (mark). For the control lines, positive logic is used; that is, a line is *asserted* (active) when it carries a positive voltage.

The official standard allows a maximum data rate of 20 kb per second over lines with not more than 2500 pF of capacitance. This limits its use, with typical cable, to distances of approximately 15 m or less. In practice, the RS-232C standard can be used over longer lines, especially if the data rate is reduced. The problem with line capacitance is that it increases the rise and fall time of pulses on the line, causing them to merge together.

Since one or more RS-232C ports are provided on most personal computers, the interface is often used for other purposes than connection to a modem. For instance, data can be transferred between two computers by connecting their serial ports together directly. In that case, the TD and RD lines must be cross-connected in the connecting cable, as shown in Figure 15.27,

FIGURE 15.27

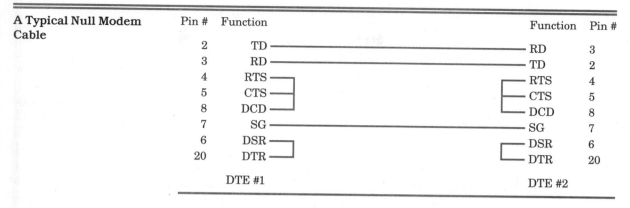

A Typical Null Modem Cable	Pin #	Function		Function	Pin #
	2	TD ———————————— RD			3
	3	RD ———————————— TD			2
	4	RTS		RTS	4
	5	CTS		CTS	5
	8	DCD		DCD	8
	7	SG ———————————— SG			7
	6	DSR		DSR	6
	20	DTR		DTR	20
		DTE #1		DTE #2	

so that information transmitted by one device will be received by the other. Depending on the software in use, it may also be necessary to "fool" each computer into perceiving that a modem is connected and ready, by connecting DTR to DSR and RTS to CTS and DCD. A cable designed for direct connections without a modem is known as a *null modem cable.*

Another common application for an RS-232C port is the connection of a serial printer or plotter. Generally, the manufacturer will provide specifications for the connecting cable, and a program called a *driver* will be needed to ensure operation. Some of the control lines may be used for such functions as reporting whether the printer is ready or out of paper, whether its buffer is full, and so on. This type of application is really outside the scope of our discussion; the point to remember is that an RS-232C connection can be reconfigured by software to perform a variety of functions not envisioned by the original standard.

15.6 ERROR CONTROL

Whether data is sent by digital or analog means, errors will occur. Many of these are due to noise in the channel. The effect of errors varies with their number and with the type of information. For example, a bit error in a digital telephone signal will cause, at worst, a click in the receiver; an error in an exchange of financial data between banks could be much more serious. Usually, errors can be reduced by using a lower data rate, but it may be more efficient to accept a fairly high error rate in the channel and use error correction techniques to remove the errors before they cause problems.

Thermal noise produces random errors, but other types can cause burst errors, where many bits in sequence are in error. Radio systems and tape and disk storage systems are examples of places where burst errors can occur.

There are two stages in the control of errors. The first problem, of course, is to detect the fact that an error has occurred. Then, the error must be corrected, if possible. If the error cannot be corrected, at least the receiver should be notified that the block of data it has just received contains errors. It might seem that with a binary system, detection and correction would be the same thing: if a bit received as a one is in error, then the bit must actually be a zero. However, in many cases an error-detection scheme will only localize an error to a word or a block of data. For instance, we may know that one of the seven bits in an ASCII character is in error, but not which one.

There are many techniques for the detection and correction of errors. All of them involve the addition of some redundancy; that is, bits that do not actually contribute to the information sent. Therefore, error control increases overhead.

Some systems rely on the receiver being able to correct errors from the redundant information. This technique is called **forward error correction** (FEC), and is the only way to correct errors in simplex communication systems. Alternatively, where the communication is half- or full-duplex, a scheme can be used that allows the receiver to request the transmitter to repeat any data blocks that are known to contain errors. Systems of this type are known as **automatic request for retransmission** (ARQ) schemes. In general, ARQ methods do not require the transmission of as much redundant information as do FEC schemes. In this section, we will look at a few systems which are representative of the means that can be employed in both of these situations.

15.6.1 Parity

Parity is a simple means of error detection. It involves the addition of one extra bit to the bits that encode a character. The parity bit is set to make the transmitted character, including the parity bit but not the start and stop bits, have either an even number of ones (even parity) or an odd number of ones (odd parity). The receiver checks the parity. The received parity will be incorrect if one error or an odd number of errors has occurred. It is not possible to use parity to determine which bit is in error; the whole character would have to be retransmitted to correct it. Also, if two errors have occurred, the parity will be correct and the error will not be detected. The same is true for any even number of errors. Sometimes the parity bit is called a *vertical redundancy check* (VRC).

EXAMPLE 15.9

Write the bit pattern for the letter b in ASCII with one start bit, odd parity, and one stop bit, in the order in which the bits would be transmitted.

Solution

From Table 15.2, the ASCII code for b is 1100010. However, this is given in the usual order in which binary numbers are written, with the most significant bit first. It will be transmitted in the reverse order: 0100011. Now we add a parity bit, which will be zero to give odd parity: 01000110. The start bit, sent first, will be 0, and the stop bit, sent last, is 1. This gives the complete character, as sent, as 0010001101.

15.6.2 Longitudinal Redundancy Check

The longitudinal redundancy check (LRC) is an extension of parity that can provide some error correction as well as detection. A block of seven-bit ASCII characters, starting with STX and ending with ETX or ETB, has a parity bit for each character. Another parity bit is made from the first bit of each character, another from the second bits, and so on, so that a *parity character* is made up from the block. It is sent last. By checking for parity for each character, the fact that a single bit error has occurred can be determined. Then, going to the LRC character, the receiver can determine exactly which bit is in error. This allows the correction of single-bit errors.

The amount of redundancy and the power of the system depend on the length of the block. Longer blocks give less redundancy but poorer error correction, as only one bit per block can be corrected with complete reliability.

When there is more than one error per block, the errors may still be detected, and an automatic request for retransmission (ARQ) can be sent. The optimum block length depends on the error rate of the channel. If there are few errors a long block will reduce the amount of redundant information, but with a high channel error rate a long block will result in many ARQ requests, greatly slowing the data rate as long blocks are sent over and over.

EXAMPLE 15.10

Generate the vertical and longitudinal redundancy checks, using odd parity, for the block

STX L R C ETB

Solution

First, look up the code for each character and write it vertically, as below.

Bit	STX	L	R	C	ETB
7	0	1	1	1	0
6	0	0	0	0	0
5	0	0	1	0	1
4	0	1	0	0	0
3	0	1	0	0	1
2	1	0	1	1	1
1	0	0	0	1	1

Next, find the VRC (parity) bit for each character. (Now you can see why parity is called a *vertical* redundancy check.)

Bit	STX	L	R	C	ETB
7	0	1	1	1	0
6	0	0	0	0	0
5	0	0	1	0	1
4	0	1	0	0	0
3	0	1	0	0	1
2	1	0	1	1	1
1	0	0	0	1	1
VRC	0	0	0	0	1

Finally, find the LRC for each row.

Bit	STX	L	R	C	ETB	LRC
7	0	1	1	1	0	0
6	0	0	0	0	0	1
5	0	0	1	0	1	1
4	0	1	0	0	0	0
3	0	1	0	0	1	1
2	1	0	1	1	1	1
1	0	0	0	1	1	1
VRC	0	0	0	0	1	0

EXAMPLE 15.11

The block below contains exactly one error. Find the error, and decode the block. Odd parity is used.

```
0  1  1  1  1  0  1
0  0  1  1  1  0  0
0  0  0  0  0  1  1
0  0  0  0  0  0  1
0  0  0  1  1  1  0
1  1  0  0  0  1  0
0  0  1  1  0  1  0
0  1  0  0  0  1  1
```

Solution

Looking at the VRC bits (the bottom row) reveals one column with incorrect parity. Similarly, looking at the LRC (the right-hand column) shows an error in one row.

```
0  1  1  1  1  0  1
0  0  1  1  1  0  0
0  0  0  0  0  1  1   ←  Error in this row
0  0  0  0  0  0  1
0  0  0  1  1  1  0
1  1  0  0  0  1  0
0  0  1  1  0  1  0
0  1  0  0  0  1  1
```

                    ↑

                Error in
                this column

Next, change the bit at the intersection of that row and column.

```
0 1 1 1 1 0 1
0 0 1 1 1 0 0
0 0 0 1 0 1 1
0 0 0 0 0 0 1
0 0 0 1 1 1 0
1 1 0 0 0 1 0
0 0 1 1 0 1 0
0 1 0 0 0 1 1
```

Now discard the check bits and decode the characters.

```
0 1 1 1 1 0
0 0 1 1 1 0
0 0 0 1 0 1
0 0 0 0 0 0
0 0 0 1 1 1
1 1 0 0 0 1
0 0 1 1 0 1
S B a u d E
T         T
X         B
```

15.6.3 Checksums

Another error-detection method consists of adding together all the data words in a block, for instance by adding their ASCII values, then dividing by some fixed number. The remainder is transmitted at the end of the block. The receiver performs the same division and should get the same remainder. If not, there is an error in the transmitted block. XMODEM is a data-transfer protocol for microcomputers that uses this method. It uses a block of 128 bytes. The checksum is obtained by summing the ASCII values of the transmitted characters, dividing by 128, and using the remainder.

15.6.4 ARQ and FEC Techniques in High-Frequency Radio

Traditionally, radioteletype (RTTY) has used the five-bit Baudot code, which does not have provision for parity. More recently, a special seven-bit code has been adopted, in which each valid character has four zeros and three ones. This allows 35 valid combinations out of 128 possibilities (only uppercase is supported). Any other received combination is rejected as an error. This system can detect two or more errors in a character, provided that they do not occur in complementary pairs; that is, if a one becomes a zero and a zero becomes a one in the same character, it will be interpreted as valid. Table 15.6 lists the valid combinations in the ARQ code. Please note that rather than one and zero, the code is given in terms of B and Y. B is the higher

TABLE 15.6

	Code	LTRS	FIGS
HF Radio ARQ Code	BBBYYYB	A	-
	YBYYBBB	B	?
	BYBBBYY	C	:
	BBYYBYB	D	WRU or $
	YBBYBYB	E	3
	BBYBBYY	F	UNDEFINED or !
	BYBYBBY	G	UNDEFINED or &
	BYYBYBB	H	UNDEFINED or #
	BYBBYYB	I	8
	BBBYBYY	J	BELL or '
	YBBBBYY	K	(
	BYBYYBB	L	)
	BYYBBBY	M	.
	BYYBBYB	N	,
	BYYYBBB	O	9
	BYBBYBY	P	0
	YBBBYBY	Q	1
	BYBYBYB	R	4
	BBYBYYB	S	' or BELL
	YYBYBBB	T	5
	YBBBYYB	U	7
	YYBBBBY	V	= or ;
	BBBYYBY	W	2
	YBYBBBY	X	/
	BBYBYBY	Y	6
	BBYYYBB	Z	+ or "
	YYYBBBB	CARRIAGE RETURN	
	YYBBYBB	LINE FEED	
	YBYBBYB	LTRS	
	YBBYBBY	FIGS	
	YYBBBYB	SPACE	
	YBYBYBB	UNPERFORATED TAPE	

Control and Phasing Signals

	ARQ Mode	FEC Mode
BYBYYBB	Control Signal 1	
YBYBYBB	Control Signal 2	
BYYBBYB	Control Signal 3	
BBYYBBY	Idle Signal Beta	
BBBBYYY	Idle Signal Alpha	Phasing Signal 1
YBBYYBB	Signal Repetition (RQ)	Phasing Signal 2

frequency and Y the lower. The other thing to note is that, unlike the ASCII and Baudot tables, this one specifies the signals in the order they are transmitted. As with Baudot, the ARQ code has variations. Where there are two characters listed for a given bit combination, the first is the CCITT standard, and the second the North American convention.

Unlike Baudot, the ARQ code is used for synchronous communication, so no start and stop bits are needed. When this code is used in half-duplex communication between two stations, such as a ship at sea and a shore station, a sequence of three characters is transmitted, then the transmitting station waits for an acknowledgement from the other station. If the receiving station receives all three characters without detecting any errors, it sends a control character, alternating between control signals 1 and 2. If the receiving station detects an error, it sends the same control signal as last time. In the latter case, the transmitter repeats all three characters and again pauses for an acknowledgement.

The same code can also be used in a forward error correction (FEC) system, when it is desired to broadcast a message to a number of stations. In that case, the transmitter sends each block of three characters twice, and the receiver checks each character for errors. If there are errors in one repetition of a character but not the other, the error-free one is used. If both versions of the character are in error, an error message is displayed. Of course, some errors get through this way, so the ARQ mode is better if it is available.

SUMMARY OF CHAPTER 15

Here are the main points to remember from this chapter.

1. Modern communication systems are often a mixture of analog and digital sources and transmission techniques. The trend is toward digital systems, which tend to have better performance over long distances.

2. The information capacity of a channel is limited by its bandwidth and signal-to-noise ratio. Sometimes the theoretical capacity can be approached more closely by increasing the number of possibilities for each symbol that is transmitted.

3. An analog signal that is to be transmitted digitally must be sampled at least twice per cycle of its highest-frequency component.

4. PCM requires that the amplitude of each sample of a signal be converted to a binary number. The more bits used for the number, the greater the accuracy, but the greater the bit rate required.

5. Delta modulation transmits only one bit per sample, indicating whether the signal level is increasing or decreasing, but it needs a higher sampling rate than PCM for equivalent results.

6. The signal-to-noise ratio for either PCM or delta modulation signals can often be improved by using companding.

7. Asynchronous communication actually involves synchronizing the transmitter and receiver clocks at the start of each character. It is simpler but less efficient than synchronous communication, where the transmitter and receiver clocks are continuously locked together.

8. By appropriate choice of line codes, it is possible to transmit digital signals with no dc content and with sufficient clock information for synchronous communication.

9. When a communications channel is incapable of carrying pulses, a car-

rier is used in conjunction with frequency, phase, and/or amplitude modulation.

10. Baud rate, which is the number of symbols transmitted per second, must be distinguished from bit rate, because in many cases a symbol carries more than one bit of information.

11. The most common way to connect between a microcomputer or terminal and a modem is with a UART connected to a serial port on the computer, which in turn is connected to a modem by a cable using the RS-232C standard. The same standard is commonly used for other computer peripherals.

12. Since noise is present in all communications systems, errors will occur. Errors can be detected and corrected, within limits, by adding redundant information.

IMPORTANT EQUATIONS

$$I = ktB \tag{15.1}$$

$$C = 2B \log_2 M \tag{15.2}$$

$$C = B \log_2 (1 + S/N) \tag{15.3}$$

$$\log_2 N = \frac{\log_{10} N}{\log_{10} 2} \tag{15.4}$$

$$N = 2^m \tag{15.5}$$

$$DR = 1.76 + 6.02m \text{ dB} \tag{15.6}$$

$$D = f_s m \tag{15.7}$$

$$V_o = \frac{V_{mo} \ln (1 + \mu V_i / V_{mi})}{\ln (1 + \mu)} \tag{15.8}$$

$$eff = \frac{\text{number of data bits}}{\text{number of total bits}} \tag{15.9}$$

$$\text{bit rate} = \text{baud rate} \times \text{bits per symbol} \tag{15.10}$$
$$= \text{baud rate} \times \log_2 (\text{possibilities per symbol})$$

GLOSSARY

aliasing   distortion that results when a signal is sampled at too low a rate

automatic request for retransmission (ARQ)   an error-control system based on the repetition of data blocks that contain errors

baud rate   number of symbols per second

data communications equipment (DCE)   a modem

data terminal equipment (DTE)   a computer or terminal

delta modulation   digital communication scheme based on transmitting the direction of amplitude change between samples

delta phase-shift keying (DPSK)   modulation technique based on the change in phase between successive symbols

dial-up line   a temporary telephone connection made by dialing the telephone

efficiency   in data communications, the ratio between information bits and total bits

forward error correction (FEC)   a digital communication system that transmits enough redundant information to allow the receiver to correct errors

frequency-shift keying (FSK)   a communication technique using frequency modulation of a carrier

information capacity   maximum data rate for a channel

line code   means of representing logic levels as electrical signals

local loop   in telephony, the connection to an individual subscriber

mark   the signal corresponding to a logical one

modem   a modulator-demodulator

non-return-to-zero (NRZ)   a line code that does not have the requirement that the voltage return to zero between symbols

null modem   a direct connection between two computers designed to simulate a modem connection

overhead   in data communication, bits that do not carry the message, for example those used for timing and error control

parity   a bit added to a transmitted character to aid in error detection

phase-shift keying (PSK)   data communication system using phase modulation of a carrier

pulse-amplitude modulation (PAM)   analog communication scheme using pulses whose amplitude is proportional to the amplitude of samples

pulse-code modulation (PCM)   digital communication scheme where the amplitude of samples is transmitted as a binary number

pulse-duration modulation (PDM)   analog communication scheme using pulses whose duration is a function of the sample amplitude

pulse-position modulation (PPM)   analog communication scheme using pulses whose timing is a function of the sample amplitude

pulse-width modulation (PWM)   see pulse-duration modulation

quadrature amplitude modulation (QAM or QUAM)   communication scheme that combines amplitude and phase modulation of a carrier

return-to-zero (RZ)   line code that requires the signal level to return to zero once per transmitted symbol

space   the signal corresponding to a logical zero

start bit   a bit transmitted just before a character in asynchronous communication

stop bit   a bit transmitted just after a character in asynchronous communication

symbol   the state of the transmitted signal that carries information

QUESTIONS

1. Give four advantages and one disadvantage of using digital techniques for the transmission of voice telephone signals.
2. What factors limit the theoretical maximum information rate on a channel?
3. What happens when a signal is sampled at less than the Nyquist rate?
4. (a) List three types of analog pulse modulation.
   (b) Which pulse modulation scheme is used as an intermediate step in the creation of PCM?
5. (a) Briefly explain what is meant by companding.
   (b) What advantage does companded PCM have over linear PCM for telephone communications?
6. How are lowercase letters handled with the Baudot code?
7. What advantages does the AMI bipolar code have over unipolar NRZ coding?
8. Compare the bipolar NRZ and Manchester codes:
   (a) Which of these codes is more suitable for synchronous communication, and why?
   (b) Which of these codes has more energy at low frequencies? Illustrate by showing a bit pattern that has a lot of low-frequency energy with this code.
   (c) Which of these codes requires the greater bandwidth? Illustrate by showing a bit pattern which requires the greatest bandwidth with this code.
9. (a) What parameters of a sine-wave carrier can be varied in analog modulation?
   (b) In a data communications system, which parameter(s) are more likely to be varied at (i) low, (ii) moderate, and (iii) high data rates?
10. What is DPSK? What advantage does it have over ordinary PSK?
11. For the Bell 103 standard, explain the difference between the *originate* and *answer* modems in terms of function and frequency.
12. What steps are necessary to set up communications between modems?
13. The Bell 212A modem is often referred to as a 1200 baud modem. Is this correct? Explain.
14. What is meant by fallback, and how does it apply to a V.22 bis modem using North American standards?
15. This question refers to a Hayes-compatible modem.
    (a) What is local mode?
    (b) How can the modem be switched from on-line to local mode?
    (c) What command would be used to get the modem to dial 9 for an outside line, pause, then dial the number 555-1212, using pulse dialing?
16. Even though the nominal maximum length of an RS-232C cable is 15 m, much longer lengths can be used if the data rate is less than maximum. What limits the length, and why does a lower data rate make it less critical?
17. Given that a terminal is a DTE and a modem is a DCE, what would you call a serial printer that plugs into an RS-232C port? Explain briefly why this could be confusing.

18. What is a null modem, and under what circumstances is a null modem required in an RS-232C link?

19. Explain the difference between FEC and ARQ error-control systems. Why is ARQ unsuitable for simplex communications?

20. Explain why the combination of VRC and LRC is more effective than either alone.

PROBLEMS

Section 15.1

21. A broadcast television channel has a bandwidth of 6 MHz.

    (a) Calculate the maximum data rate that could be carried in a television channel, using a sixteen-level code. Ignore noise.

    (b) What would be the minimum permissible signal-to-noise ratio, in decibels, for the data rate calculated above?

Section 15.2

22. A 1 kHz sine wave with a peak value of 1 V and no dc offset is sampled every 250 μs. Assume the first sample is taken as the voltage crosses zero in the upward direction. Sketch the results over 1 ms using:

    (a) PAM with all pulses in the positive direction

    (b) PDM

    (c) PPM

23. The compact disc system of digital audio uses two channels with TDM. Each channel is sampled at 44.1 kHz and coded using linear PCM with sixteen bits per sample. Find:

    (a) the maximum audio frequency that can be recorded (assuming ideal filters)

    (b) the maximum dynamic range in decibels

    (c) the bit rate, ignoring error correction and framing bits

    (d) the number of quantizing levels

24. Suppose an input signal to a μ-law compressor has a positive voltage and an amplitude 25% of the maximum possible. Calculate the output voltage as a percentage of the maximum output.

25. A PCM/TDM system multiplexes 24 voice band channels. Each sample is encoded into seven bits, and a frame consists of one sample from each channel. A framing bit is added to each frame. The sampling rate is 9000 samples per second. What is the line speed in bits per second?

26. Suppose a video signal with a baseband frequency range from 0 to 4 MHz is transmitted by linear PCM, using eight bits per sample and a sampling rate of 10 MHz.

    (a) How many quantization levels are there?

    (b) Calculate the bit rate, ignoring overhead.

    (c) What would be the maximum signal-to-noise ratio, in decibels?

    (d) What type of noise determines the answer to Part (b)?

Section 15.3

27. Suppose that in the TELEX message (using CCITT Baudot Code)

    BE HERE TONITE BY 8

    the space after "HERE" was changed by a noise pulse to a FIGS character. What would the received message read

(a) without USOS?

(b) with USOS?

28. The message RYRYRYRYRYRYRYRYRY is often sent to test and set up radioteletype circuits. Look at the Baudot code in Table 15.1 and suggest a reason why.

29. Decode the message below, assuming it has been sent in ASCII with one start bit,
seven data bits, one stop bit, and even parity. It is sent in the order in which it
reads, from left to right.

1111001000010111111111010010110100010111011111111111111111111

Section 15.4

30. An asynchronous communications system uses ASCII at 300 bits per second,
with one start bit, one stop bit, and no parity bits.

(a) Explain what is meant by efficiency, and calculate the maximum percent efficiency in this system.

(b) Express the data rate in characters per second and in words per minute. (Assume a word has five letters and one space.)

(c) Suppose a synchronous system were used instead, with the same bit rate.
Would you expect the number of characters per second to increase or decrease, and why?

31. Code the following data using each of the codes specified. The data is to be sent in
time as it reads, from left to right.

011000111

(a) Unipolar NRZ

(b) AMI

(c) Manchester code

Section 15.5

32. A constellation diagram for a synchronous modem is shown in Figure 15.28.

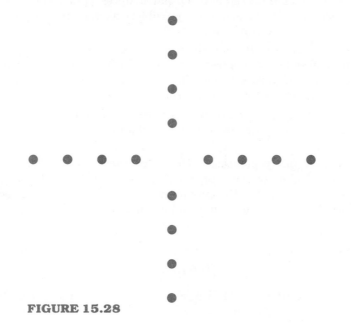

FIGURE 15.28

(a) What type of modulation is this?

(b) If the transmitted bit rate is 9600 b/s, what will be the baud rate using this modem?

(c) Suppose a fallback mode at 4800 b/s, with the same baud rate, were required. Draw a constellation diagram showing how this could be accomplished. (There is more than one right answer.)

(d) When would a fallback mode of operation be useful?

33. Draw a constellation pattern for a modem that uses eight equally spaced phase angles and four equally spaced amplitude levels. If the modem operates at 4800 baud, what is its bit rate?

Section 15.6

34. (a) Calculate the VRC and LRC bits for the block of data below, assuming even parity for both. Indicate which is which.

```
0  1  1  1  0  1  0  1  1  0
1  0  1  0  1  0  1  0  1  0
0  1  0  1  0  1  0  1  0  1
1  1  1  1  0  0  0  0  1  0
1  1  0  0  1  1  0  0  1  1
0  1  0  1  0  1  0  1  0  1
```

35. Assume the LRC uses even parity and VRC uses odd parity in the data block below. There is a one-bit error in the block.

Characters

1	2	3	4	
0	1	0	1	0
0	0	0	0	0
1	0	1	0	0
0	1	0	0	0
0	0	1	0	1
0	0	0	0	0
1	1	1	1	0
1	1	0	1	0

(a) Identify the LRC and VRC.

(b) Circle the erroneous bit.

Comprehensive

36. Calculate the maximum data rate that could be sent, using simplex transmission in a voice-grade telephone channel having a bandwidth of 3 kHz, with a modem that uses QAM modulation with four possible phase angles and two amplitude levels, assuming that the signal-to-noise ratio in the channel is 30 dB.

37. A microwave radio system uses 256-QAM, that is, there are 256 possible amplitude and phase combinations.

(a) How many bits per symbol does it use?

(b) If it has a channel with 40 MHz bandwidth, what would be its maximum data rate, ignoring noise?

16

Fiber Optics

16.1 INTRODUCTION

An optical fiber is essentially a waveguide for light (usually infrared). As shown in Figure 16.1(a), it consists of a **core**, and a **cladding** which surrounds the core. These are both made of transparent material, either glass or plastic, but the **index of refraction** of the cladding is less than that of the core. This causes rays of light leaving the core to be refracted back into the core, so that the light propagates down the fiber.

Figure 16.1(b) shows how an optical fiber can be used for communications by applying a modulated light source to one end, and connecting a photodetector to the other. A light-emitting diode (LED) or a **laser diode** (LD) can be used for the source. Any of the modulation schemes studied so far can be used, but digital transmission is more common than analog. The light source can be turned on and off, or switched between two different power levels, for digital transmission. Analog signals are usually transmitted digitally using PCM, though analog AM and FM schemes are sometimes employed.

Though its operating principle is that of a waveguide, optical fiber is used in practice as a substitute for a copper cable (either coaxial or twisted-pair) or a point-to-point microwave radio link. It is found in many applications; a few random examples include telephone cables, point-to-point transmission of television signals, and computer networks.

Optical fiber has several advantages over electrical cable for communications. In general, fiber has greater bandwidth than coaxial cable, giving it the ability to handle greater data rates. This increased bandwidth allows more signals to be multiplexed. Optical fibers can be built with lower loss than copper cables, increasing the allowable distance between repeaters. The cable itself can be less expensive.

The nonelectrical nature of the signals on optical fiber also confers certain advantages. Since optical cables do not carry electric currents, they are

FIGURE 16.1

Optical Fiber and
Communications
System

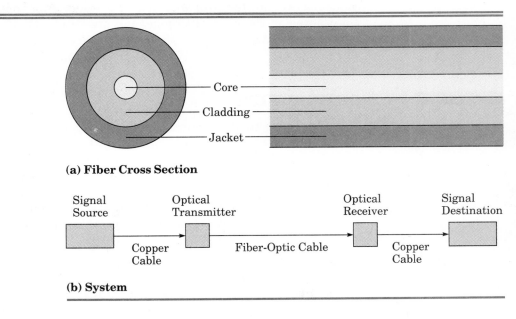

(a) Fiber Cross Section

(b) System

immune to **crosstalk** between cables, and to electromagnetic interference (EMI) from other sources. For the same reason, optical fiber is useful where secure communications are concerned, because it is impossible to "tap" the cable without first breaking it and causing an obvious disruption. Electrical cables can often be monitored by inductive coupling from outside, and of course radio signals can easily be picked up by unauthorized persons.

The lack of electrical connection is also useful in explosive environments where a spark hazard may exist, and in applications where one circuit must be isolated electrically from another. Fiber links are sometimes used to control high-voltage circuits in power stations, for example. In the very different area of medical electronics, fiber optics is often used to avoid shock hazards by isolating the patient from circuits connected to the electrical power line.

Radio links as well as copper cables can be replaced with fiber optics. The two types that are obvious competitors are point-to-point microwave links and geostationary satellite channels. The radio links have the advantage of not needing cable laid. This is especially attractive in the case of satellites, where no access is required to the terrain between source and destination, which can be separated by thousands of kilometers. On the other hand, the delay of about one-half second between a question and the response to it is a considerable nuisance in telephony via satellite. Optical fibers have greater bandwidth than either type of channel, and of course they are much more private.

Physically, optical fibers have size and weight advantages over copper cables. The cost factor depends on the application: optical fiber itself can be less expensive than coaxial cable but is more expensive than twisted-pair. For wide-bandwidth long-distance applications, the cost accounting is in favor of fiber, but for short-distance low-bandwidth applications, copper is still cheaper, especially when the cost of converting an electrical signal to and from optical form is taken into account.

One application where optical fibers cannot substitute for copper cable or microwave waveguide is the transmission of power. They cannot be used to connect a transmitter to an antenna, for instance. Fiber optics are used with power levels in the milliwatt range, and are strictly for the transmission of information, not energy.

16.2 OPTICAL FIBER

The actual fiber used in optical communications is a thin strand of glass or plastic. This fiber has very little mechanical strength, so it is enclosed in a protective jacket, usually made of plastic. Often, two or more fibers are included in one cable, for increased bandwidth, and redundancy in case one fiber breaks. It is also easier to build a full-duplex system using two fibers, one for transmission in each direction, than to send signals in both directions along the same fiber. Figure 16.2 shows several examples of fiber-optic cables.

16.2.1 Total Internal Reflection

Optical fibers work on the principle of **total internal reflection**. This was discussed in connection with radio-wave propagation, but a review seems in

FIGURE 16.2

Fiber-Optic Cables

Cooper Industries, Belden Division.

order at this point. Figure 16.3(a) shows the situation where a wave moves from a medium with lower velocity to one with higher velocity. Rather than specifying the velocity directly, with light it is usual to list the refractive index. This is given very simply by

$$n = \frac{c}{v} \tag{16.1}$$

where n = index of refraction of a medium
c = velocity of light in free space
v = velocity of light in the medium

You may recall that the velocity of propagation of any electromagnetic wave in a medium is closely related to the dielectric constant.

$$\epsilon_r = \left(\frac{c}{v}\right)^2 \tag{16.2}$$

where ϵ_r = relative permittivity, also called the *dielectric constant*, of the medium
c = velocity of light in free space
v = velocity of light in the medium

FIGURE 16.3   Refraction and Total Internal Reflection($n_1 > n_2$)

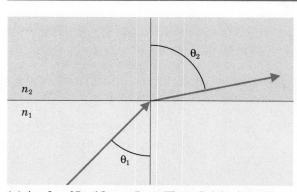

(a) Angle of Incidence Less Than Critical Angle

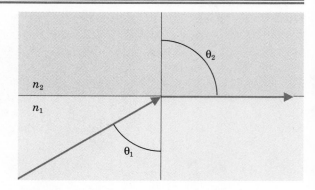

(b) Angle of Incidence Equal to Critical Angle

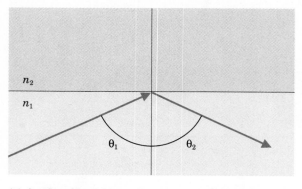

(c) Angle of Incidence Greater Than Critical Angle

Combining these two equations gives a simple relationship between the index of refraction and the dielectric constant.

$$n = \sqrt{\epsilon_r} \tag{16.3}$$

The **angle of refraction** at the interface between two media is governed by Snell's Law:

$$n_1 \sin \theta_1 = n_2 \sin \theta_2 \tag{16.4}$$

where n_1 = index of refraction in the first medium
          n_2 = index of refraction in the second medium
          θ_1 = angle of incidence
          θ_2 = angle of refraction

In each case, the angle is that between a ray of light and the **normal**, which is a line perpendicular to the interface between the media.

By rearranging Equation (16.4), we can get an expression for the angle of refraction.

$$\sin \theta_2 = \frac{n_1}{n_2} \sin \theta_1 \tag{16.5}$$

For an optical fiber, the first medium is the core and the second is the cladding, and n_1 is always greater than n_2. There will be some value of θ_1, less than 90°, for which $\sin \theta_2$ will be 1; that is, θ_2 will be 90° and the refracted ray will lie along the interface between the two media, as shown in Figure 16.3(b). For any greater value of θ_1, the ray will be reflected rather than refracted. In this case, Snell's Law ceases to operate, and the **angle of reflection** becomes equal to the **angle of incidence**, as with any reflection situation. This is shown in Figure 16.3(c). The value of θ_1 for which θ_2 is 90° is called the **critical angle** θ_c, and can be found by rearranging Equation (16.5):

$$\sin \theta_2 = \frac{n_1}{n_2} \sin \theta_1$$

$$\sin \theta_1 = \frac{n_2}{n_1} \sin \theta_2$$

For $\theta_2 = 90°$, $\sin \theta_2 = 1$ and $\theta_1 = \theta_c$:

$$\sin \theta_c = \frac{n_2}{n_1} \sin 90°$$

$$= \frac{n_2}{n_1}$$

$$\theta_c = \arcsin \frac{n_2}{n_1} \tag{16.6}$$

where θ_c = critical angle of incidence
n_1 = index of refraction of the first medium (core)
n_2 = index of refraction of the second medium (cladding)

EXAMPLE 16.1

A fiber has an index of refraction of 1.6 for the core and 1.4 for the cladding. Calculate:

(a) the critical angle
(b) θ_2 for $\theta_1 = 30°$
(c) θ_2 for $\theta_1 = 70°$

Solution

(a) From Equation (16.6),

$$\theta_c = \arcsin \frac{n_2}{n_1}$$

$$= \arcsin \frac{1.4}{1.6}$$

$$= 61°$$

(b) 30° is less than the critical angle, so Equation (16.5) applies.

$$\sin \theta_2 = \frac{n_1}{n_2} \sin \theta_1$$

$$= \frac{1.6}{1.4} \sin 30°$$

$$= 0.571$$

$$\theta_2 = \arcsin 0.571$$

$$= 34.8°$$

(c) 70° is greater than the critical angle, so total internal reflection will take place and the angle of reflection will be equal to the angle of incidence, that is, 70°.

16.2.2 Numerical Aperture

The **numerical aperture** of a fiber is closely related to the critical angle, and is often used in specifications for optical fiber and the components that work with it. The numerical aperture is defined as the sine of the maximum angle a ray entering the fiber can have with the axis of the fiber, and still propagate by internal reflection. Using trigonometry, it can be shown that

$$\text{N.A.} = \sqrt{n_1^2 - n_2^2} \tag{16.7}$$

where   N.A. = numerical aperture

Equation (16.7) assumes that light enters the fiber from free space. (As a practical matter, air can be considered equivalent to free space.) Note that the total **angle of acceptance** is twice that given by the numerical aperture. This is shown in Figure 16.4. Any light entering within the cone of acceptance illustrated will be reflected internally and may propagate down the fi-

FIGURE 16.4   Cone of Acceptance

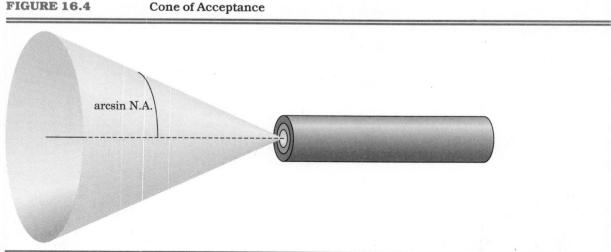

arcsin N.A.

ber. Light entering from outside the cone of acceptance will merely be re-fracted into the cladding and will not propagate. Numerical apertures for optical fiber vary from about 0.1 to 0.5.

EXAMPLE 16.2

Calculate the numerical aperture and the maximum angle of acceptance for the fiber described in Example 16.1.

Solution

From Equation (16.7),

$$
\begin{aligned}
\text{N.A.} &= \sqrt{n_1^2 - n_2^2} \\
&= \sqrt{1.6^2 - 1.4^2} \\
&= 0.775
\end{aligned}
$$

This corresponds to a maximum angle from the fiber axis of

$$\arcsin 0.775 = 50.8°$$

The total width of the cone of acceptance will be twice this, or 101.6°.

16.2.3 Modes and Materials

Since an optical fiber is in reality a waveguide, light can propagate in a num-ber of specific modes. If the diameter of a fiber is relatively large, light enter-ing at different angles will excite different modes. A fiber that is sufficiently narrow, on the other hand, may support only one mode.

Multimode propagation will cause **dispersion**, just as it does in wave-guides. Dispersion results in the spreading of pulses and this limits the us-able bandwidth of the fiber. **Single-mode fiber** has much less dispersion, but it is more expensive to manufacture and its small diameter, coupled with the fact that its numerical aperture is less than that of **multimode fiber**, makes it more difficult to couple light into and out of the fiber.

The maximum allowable diameter for a single-mode fiber varies with the wavelength of the light, and is given by the following equation.

$$r_{max} = \frac{0.383\lambda}{\text{N.A.}} \tag{16.8}$$

where r_{max} = maximum radius of the core
$\quad\quad\quad \lambda$ = wavelength
$\quad\quad\quad$ N.A. = numerical aperture

EXAMPLE 16.3

A single-mode fiber has a numerical aperture of 0.15. What is the maxi-mum core diameter it could have for use with infrared light with a wave-length of 820 nm?

Solution

From Equation (16.8), the maximum radius is

$$
\begin{aligned}
r_{max} &= \frac{0.383\lambda}{\text{N.A.}} \\
&= \frac{0.383 \times 820 \times 10^{-9} \text{ m}}{0.15} \\
&= 2.1 \times 10^{-6} \text{ m} \\
&= 2.1 \text{ } \mu\text{m}
\end{aligned}
$$

Of course, the maximum diameter is twice this radius, or 4.2 μm.

Both types of fiber described so far are known as **step-index fibers** because the index of refraction changes sharply between core and cladding. A compromise is possible using **graded-index fiber**. This is a multimode fiber, but the index of refraction gradually decreases away from the center of the core, resulting in light traveling faster near the outside. Graded-index fiber has reduced dispersion compared with a multimode step-index fiber, though single-mode fibers are still preferred for the most demanding applications. Figure 16.5 shows typical cross sections for all three types of fiber.

As might be expected, most optical fibers are made of high-quality glass chosen for its very great transparency, to reduce losses. Some low-cost multimode fibers designed for short-distance applications (such as optical links in consumer electronics and control-signal lines in automobiles) are made of

FIGURE 16.5   Types of Optical Fiber (Dimensions Are Typical)

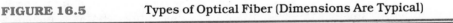

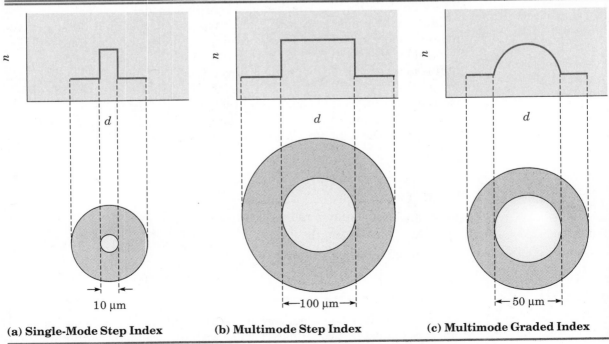

(a) Single-Mode Step Index   **(b) Multimode Step Index**   **(c) Multimode Graded Index**

acrylic plastic. The losses in these fibers are very much greater than in glass fiber, but this is of no importance when the distances involved are a few meters or less.

16.2.4 Losses and Dispersion

Dispersion due to the propagation of more than one mode has already been mentioned. Just as in microwave waveguides, this kind of dispersion results from the fact that in multimode propagation, some modes result in the signal travelling farther than it would in others. Figure 16.6 shows the effect of multimode propagation on the time taken for a signal to get from one end of the fiber to the other.

Single-mode fibers are free of modal dispersion effects. The graded-index multimode fibers described above reduce dispersion by taking advantage of the fact that signals propagating in higher-order modes spend more of their time near the outside of the core than do the low-order modes. This should be clear from Figure 16.6 as well. The reduction of the refractive index toward the outside of the core results in increased velocity in this region. This means that the high-order-mode components of the signal, which have farther to travel, propagate more quickly. The result is that dispersion is greatly reduced, though not eliminated.

Even single-mode fibers exhibit some dispersion, known as *intramodal dispersion* because it takes place within one mode of propagation. It is also called *chromatic dispersion*, because it results from the presence of different wavelengths of light. For visible light, the wavelength corresponds to its color. Infrared light is much more common than visible light in optical communication, but the term "chromatic dispersion" is used anyway.

One form of intramodal dispersion is called *material dispersion*, because it depends on the material of the core. A beam of light, even **laser** light, consists of more than one wavelength, and different wavelengths propagate at different velocities in glass. This is the effect that causes white light to break up into a spectrum as it passes through a prism. The wavelengths corresponding to the various colors of the spectrum have different velocities in the glass, resulting in different indices of refraction. Therefore, they will be refracted at different angles, as demonstrated in Figure 16.7. In an optical fiber, the difference in propagation velocity for different wavelengths causes them to take varying amounts of time to propagate down the fiber.

Another source of dispersion in single-mode fiber is called *waveguide dispersion*. It results from the fact that some of the energy propagates in the cladding rather than the core. This is not obvious from our rather brief study

FIGURE 16.6

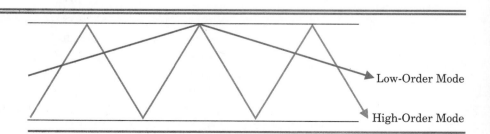

Multimode Propagation

Low-Order Mode

High-Order Mode

FIGURE 16.7

Chromatic Dispersion
in a Prism

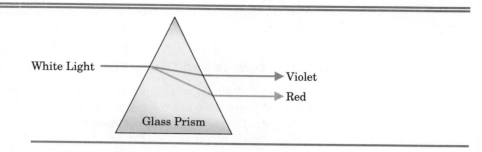

of total internal reflection, but it would be revealed by a more detailed analysis. Since the core and cladding have different indices of refraction, the light propagates at different velocities in the two.

These two sources of dispersion are not necessarily additive. In fact they can, and do, cancel each other out at some wavelength, which for a normal glass fiber is about 1.3 μm, in the infrared region.

Dispersion increases with the bandwidth of the light source. (For LEDs and lasers, the term *linewidth* is generally used rather than *bandwidth*.) Where minimum dispersion is important, the laser diode is greatly preferred over the LED because of its much narrower linewidth.

The effects of dispersion increase with the length of the fiber, since a difference in the velocity of two signals will produce a difference in the time taken to reach the end of the fiber. This difference is proportional to the length of the fiber. The greater the bandwidth of the signal, the greater the effect of a given amount of dispersion. Optical fibers are rated according to the product of distance and bandwidth. Multimode fibers have the lowest bandwidth-distance products, and single-mode fibers have the highest.

EXAMPLE 16.4

An optical fiber has a bandwidth-distance product of 500 MHz-km. If a bandwidth of 85 MHz is required for a particular mode of transmission, what is the maximum distance that can be used between repeaters?

Solution

$$\text{Bandwidth} \times \text{Distance} = 500 \text{ MHz-km}$$
$$\text{Distance} = \frac{500 \text{ MHz-km}}{\text{Bandwidth}}$$
$$= \frac{500 \text{ MHz-km}}{85 \text{ MHz}}$$
$$= 5.88 \text{ km}$$

Losses in optical fibers result from attenuation in the material itself, and from scattering, which causes some of the light to strike the core-cladding boundary at angles less than the critical angle. This light is refracted into the cladding and is lost to the fiber. Figure 16.8 shows approximately how the

FIGURE 16.8

Variance of Loss with Wavelength in Glass Fiber

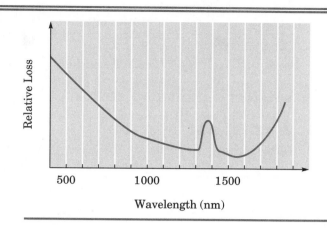

loss varies with wavelength for typical glass fiber. The sharp peak at about 1400 nm is due to the presence of water ions, and operation near this wavelength should be avoided. Loss in glass fibers is lowest at wavelengths around 1550 nm, in the infrared region of the spectrum, while plastic fibers perform best at about 660 nm, which corresponds to visible red light. Losses are an important factor limiting the distance between repeaters, since even with laser diodes the light that can be coupled into the fiber is on the order of a milliwatt.

Optical fibers are not always used at the wavelength of lowest loss. Many short-range systems use a wavelength of about 820 nm, as lasers for this wavelength are less costly than those designed for longer wavelengths. On the other hand, long-distance high-data-rate links may be limited more by dispersion than by power loss. In this case, they can operate at the lowest-dispersion wavelength of 1300 nm. Another possibility is to modify the fiber by changing its composition or the way in which the refractive index varies, to bring its zero-dispersion wavelength closer to the minimum-loss wavelength.

Bending an optical fiber too sharply can also increase losses, by causing some of the light to meet the cladding at less than the critical angle. Figure 16.9 shows how this can happen.

FIGURE 16.9

Losses Due to Bending

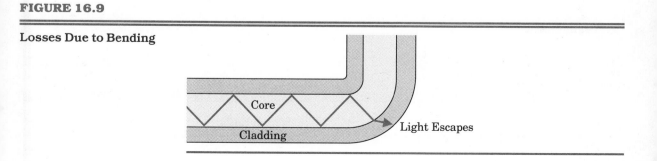

TABLE 16.1

Typical Specifications for Optical Fiber	Multimode	Graded-Index	Single-Mode
Core Diameter	50–100 μm	50–85 μm	8–10 μm
Loss	plastic: 100+ dB/km glass: 8–10 dB/km	1–3 dB/km	0.16–1 dB/km
Bandwidth × Distance	12–20 MHz-km	1 GHz-km	20+ GHz-km
Repeater Spacing	100 m	a few km	40–130 km
Light Source	LED	LED or laser diode	laser diode

Table 16.1 compares the specifications for typical examples of the different types of fiber.

16.3 SPLICES AND CONNECTORS

In an optical communication system, the losses in splices and connectors can easily be more than in the cable itself. Losses result from axial or angular misalignment of the fibers, air gaps between the fibers (which leads to spreading of light), or rough surfaces at the ends of the fibers (which leads to light escaping at various angles). Figure 16.10 shows these problems. For simplicity, only the cores of the fibers are shown in the figure.

Coupling of the fiber to sources and detectors creates losses as well, especially when it involves mismatches in numerical aperture or in the size of optical fibers. For instance, a light source may come with a *pigtail*, a short length of fiber that carries the light away from the source. If this pigtail is joined to a fiber with smaller diameter or lower numerical aperture, some of

FIGURE 16.10   **Coupling Losses**

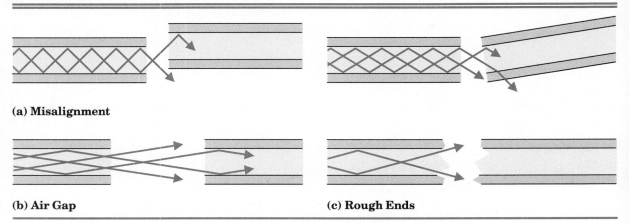

(a) Misalignment

(b) Air Gap   **(c) Rough Ends**

FIGURE 16.11

Loss Due to Coupling to Fiber with Smaller Numerical Aperture

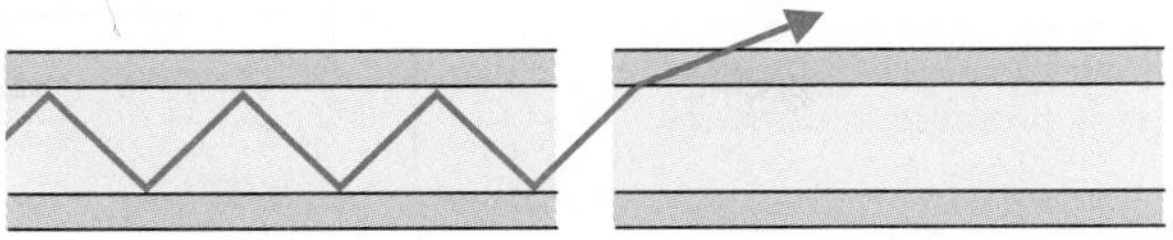

(a) Path of Escaping Ray

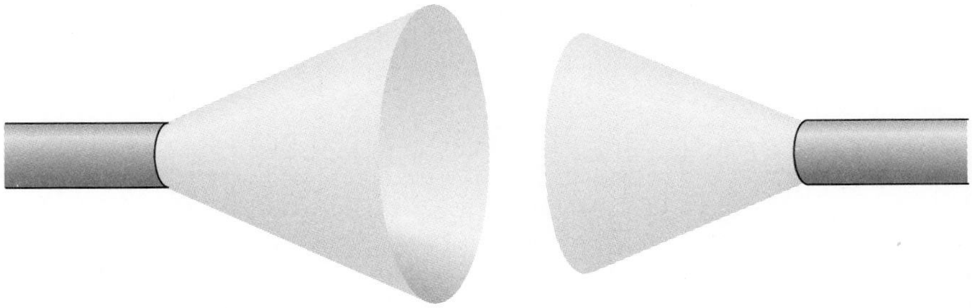

(b) Different Cones of Acceptance

the light will be lost. In the case of coupling to a smaller fiber, some of the light from the first fiber never enters the second. On the other hand, if the second fiber has equal or greater diameter but a smaller numerical aperture, some of the light that enters the second fiber will fail to propagate. Figure 16.11 shows how this happens. Figure 16.11(a) shows a ray entering the second fiber, then escaping from the core. Figure 16.11(b) illustrates the problem by showing the cone of acceptance for each fiber. The first fiber will emit light over a larger angle than the second fiber can accept.

Good connections are more critical with single-mode fiber due to its small diameter and small numerical aperture. The low loss of single-mode fiber also makes connector loss more significant. For example, a connector loss of 1 dB is a more important component in a system where the cable loss is 1 dB/km than where it is 1 dB/m, as it is in some cheap plastic cables.

The terms *splice* and *connector* are related but not equivalent. Generally, a splice is a permanent connection, while connectors are removable. Splices are necessary where sections of cable are joined. For practical reasons, the length of a spool of cable is limited to about 10 km, so longer spans between repeaters require splices to be made in the field. Connectors are needed between sources and detectors and the fiber cable. Generally, the loss in a properly made splice is less than that in a connector. Splices can have losses of 0.2 dB or less, while connector losses are often about 1 dB. One of the reasons for this is that the ends of the fibers touch in a well-made splice, while there is a small air gap in a connector, so that the polished fiber surfaces will not be damaged during the process of connecting or disconnecting.

Figure 16.12 shows some typical commercial connectors. Note the care that is taken to align the fibers. Splices are usually created by cleaving the glass ends and then fusing them together with an electric arc, but some field splices are made using specially formulated adhesives.

FIGURE 16.12   Fiber-Optic Connectors

This fiber optic connector guide presents brief descriptions of popular connectors available for use on any Belden® fiber optic cable assembly. **For assembly ordering information see catalog page 29.**

For further assistance and connector selection for a Belden cable assembly, contact Belden Product Engineering at **1-317-983-5200.**

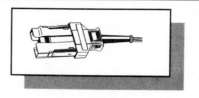

■ **FDDI.** Duplex fiber optic connector system with ceramic ferrule, fully compatible with ANSI FDDI PMD document. For data communcations applications, including FDDI backbone, frontend or backend networks and IEEE 802.4 token bus. Dry connection, with positive latching mechanism. Low insertion loss.

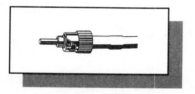

■ **ST Compatible.** Small size connector with keyed bayonet coupling for simple ramp latching or disconnect; dry connection. Available in multimode and single mode versions. Fully compatible with existing ST hardware. For data processing, telecommunications and local area networks, premise installations, instrumentation and other distribution applications. Low insertion and return loss.

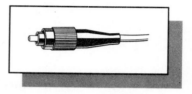

■ **FC.** One piece connector design for easy termination. Compatible with NTT-FC and NTT-D3 hardware. Dry connection with screw type strength member retention. Available in multimode and single mode versions. Applicable for telecommunications and data communications networks, premise installations and instrumentation. Low insertion and return loss.

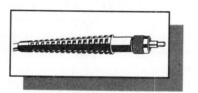

■ **SMA.** Small size connector with SMA coupling nut; dry connection. For use with multimode cables in data communications applications such as local area and data processing networks, premise installations and instrumentation. Low insertion loss. Fully compatible with all existing SMA hardware.

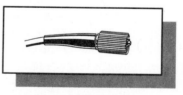

■ **D4.** Compatible with NTT-D4 hardware. Ferrule alignment key for consistent remating. Rugged construction for long life and durability. Low insertion and return loss.

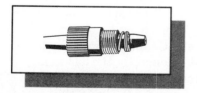

■ **Biconic.** Small size connector with screw thread, cap and spring loaded latching mechanism. Low insertion and return loss. Compatible with all biconic hardware.

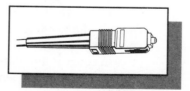

■ **SC.** Square design for high packing density. Push-pull operation simplifies connections. Available in single mode and multimode versions. Low insertion and return loss.

NOTE: Information on connectors is presented for selection assistance only. Belden Wire and Cable does not assume any liability or responsibility for the accuracy of descriptive or performance data herein. Product and performance specifications should be verified with the connector manufacturer.

16.4   OPTICAL EMITTERS AND DETECTORS

Electromagnetic energy can appear only in multiples of a discrete amount known as a **quantum**. These quanta are called **photons** when the energy is radiated. A *photon* is an entity that is somewhere between a wave and a particle. It has a wavelength corresponding to the radiation, and an amount of energy equal to one quantum, but no mass. The details are perhaps best left to physicists, but the existence of photons is important to the operation of the emitters and detectors used with optical fiber.

The amount of energy that corresponds to a quantum, and is found in a photon, is given by a simple equation:

$$E = hf \tag{16.9}$$

where E = energy in one quantum in joules
h = Planck's constant, 6.626×10^{-34} joule-seconds
f = frequency of the radiation in hertz

From Equation (16.9), you can see that the energy in one photon varies directly with frequency. For radio frequencies, the energy per photon is so small, and the number of photons present in even the weakest signals is so large, that it is seldom necessary to consider photons at all. With light, however, the frequency is much higher and the amount of energy in a photon is sometimes sufficient to create a free electron when it collides with an atom. Photons can also trigger the movement of an electron from one valence state to another. Similarly, the movement of an electron from a higher energy level to a lower one can cause the emission of a photon.

When talking about energy levels in atoms, it is usual to use the **electron-volt** (eV), rather than the joule, as the unit of energy. Converting between the two is simple. By definition, the electron-volt is the energy gained or lost by one electron moving through a potential difference of one volt. A volt, of course, is one joule per coulomb, so the energy given to an electron is

$$\begin{aligned} 1 \text{ eV} &= (1 \text{ joule/coulomb})(1.6 \times 10^{-19} \text{ coulomb}) \\ &= 1.6 \times 10^{-19} \text{ joule} \end{aligned} \tag{16.10}$$

EXAMPLE 16.5

Find the energy, in electron-volts, in one photon at a wavelength of 1 μm.

Solution

First find the frequency.

$$c = f\lambda$$

$$f = \frac{c}{\lambda}$$

$$= \frac{3 \times 10^8 \text{ m/s}}{1 \times 10^{-6} \text{ m}}$$

$$= 3 \times 10^{14} \text{ Hz}$$

Next, find the energy in joules (J). From Equation (16.9),

$$E = hf$$
$$= (6.626 \times 10^{-34})(3 \times 10^{14})$$
$$= 1.99 \times 10^{-19} \text{ J}$$

This can easily be converted to electron-volts. From Equation (16.10),

$$1 \text{ eV} = 1.6 \times 10^{-19} \text{ J}$$

so

$$1 \text{ J} = \frac{1}{1.6 \times 10^{-19}} \text{ eV}$$
$$= 6.25 \times 10^{18} \text{ eV}$$

and the energy of our photon, in electron-volts, is

$$E = (1.99 \times 10^{-19})(6.25 \times 10^{18})$$
$$= 1.24 \text{ eV}$$

16.4.1 Light-Emitting Diodes

A light-emitting diode (LED) is a form of junction diode that is operated with forward bias. The recombination of electron-hole pairs in this or any junction diode causes energy to be released. With an ordinary silicon diode, this energy is in the form of heat, but in an LED, a significant proportion of the energy is radiated as photons of either visible or infrared light. Sometimes an infrared LED is called an *infrared-emitting diode* (IRED).

There are several types of structures that can be used for LEDs for use in optical communications. Figure 16.13 is a cross-sectional sketch of one of these, the Burrus etched-well LED. This is a surface-emitting **heterojunction** diode. *Surface-emitting* means that the light is emitted from the

FIGURE 16.13

Simplified Cross
Section of Burrus
Etched-Well LED

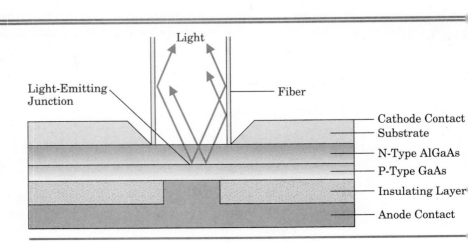

flat surface of the junction, rather than its edge. A *heterojunction* is a junction between two different materials. In this case, the junction is between P-type gallium arsenide (GaAs) and N-type aluminum gallium arsenide (AlGaAs). The diode's geometry is such that a high current density is created in a small region of the junction, producing a small light source that can more easily be coupled to the fiber. The etched well helps to confine the light. The output can be coupled to a fiber *pigtail* as shown, or through a lens inserted into the well.

Since the variation of light output with current is reasonably linear for an LED, it can be used with analog amplitude modulation. It is much more common, however, to use pulse transmission, with the diode simply switched on and off. Analog pulse techniques, like PDM, are sometimes used, though digital techniques are more common.

An idea of the specifications and physical characteristics of a typical LED designed for fiber optics can be gained by studying the data sheet in Figure 16.14. This LED emitter is designed for low-cost fiber-optics systems. Its maximum output is at a wavelength of 820 nm, which is a good match for the silicon photodetectors with which it is usually used. Notice the inefficiency of the device. It typically operates with a forward voltage of 1.5 V and a current of 100 mA, for a total input power of 150 mW. The optical power launched into the fiber is typically 165 μW for an efficiency of about 0.1%. It is easy to see why fiber optics is used to transmit information, not energy.

16.4.2 Laser Diodes

The word *laser* is really an acronym for *light amplification by stimulated emission of radiation*. In operation, a laser acts as an optical oscillator, generating phase-coherent light with a much narrower bandwidth than is possible with other sources. A typical laser diode has an emission linewidth of about 1 nm compared with 40 nm for a typical LED, and some narrow-bandwidth lasers have much better performance still. Laser diodes also produce more light in the narrow cone required for coupling to optical fibers. This advantage is particularly great with single-mode fibers, which are very narrow and also have a small numerical aperture.

The construction of a laser diode is rather similar to that of an LED, and, in fact, at low currents the laser diode will operate as an LED. Figure 16.15 is a simplified cross-sectional diagram of a common type of laser diode known as a *striped heterojunction injection laser*. This diode has a junction surrounded by material that has a lower refractive index than the material used for the junction. Many of the photons produced at the junction will remain in the plane of the junction; if they move away from it, they will be reflected in the same way that light is contained within an optical fiber. The ends of the junction have mirrored surfaces, with one of the mirrors being partly transmissive so that light can escape in one direction only. The whole junction area forms a resonant cavity at the operating frequency.

Laser diodes operate with high current densities of 11–30 kA/cm^2. The actual current is typically only about 100 mA; the high current density is produced by masking off most of the junction so that the area through which the current flows is very small. The high current density causes many electron-hole pairs to be generated. The recombination of these pairs generates photons which stimulate other pairs to combine. Laser operation, some-

FIGURE 16.14   Specifications for a Typical LED

FIBER OPTICS
Low Cost System
FLCS Infrared-Emitting Diode

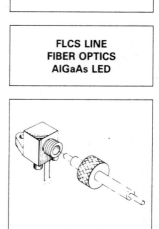

MFOE71

**FLCS LINE
FIBER OPTICS
AlGaAs LED**

CASE 363B-01

. . . designed for low cost, medium frequency, short distance Fiber Optics Systems using 1000 micron core plastic fiber.

Typical applications include: high isolation interconnects, disposable medical electronics, consumer products, and microprocessor controlled systems such as coin operated machines, copy machines, electronic games, industrial clothes dryers, etc.

- Fast Response — > 10 MHz
- Spectral Response Matched to FLCS Detectors: MFOD71, 72, 73
- FLCS Package
  - Low Cost
  - Includes Connector
  - Simple Fiber Termination and Connection
  - Easy Board Mounting
  - Molded Lens for Efficient Coupling
  - Mates with 1000 Micron Core Plastic Fiber (DuPont OE1040, Eska SH4001)

MAXIMUM RATINGS

Rating	Symbol	Value	Unit
Reverse Voltage	V_R	6	Volts
Forward Current	I_F	150	mA
Total Power Dissipation (« T_A = 25 C Derate above 25°C	$P_D(1)$	150 2.5	mW mW/°C
Operating and Storage Junction Temperature Range	T_J, T_{stg}	−40 to +85	°C

(1) Measured with the device soldered into a typical printed circuit board.

ELECTRICAL CHARACTERISTICS (T_A = 25°C unless otherwise noted)

Characteristic	Fig. No.	Symbol	Min	Typ	Max	Unit
Reverse Breakdown Voltage (I_R = 100 μA)	—	$V_{(BR)R}$	2	4	—	Volts
Forward Voltage (I_F = 100 mA)	—	V_F	—	1.5	2	Volts

OPTICAL CHARACTERISTICS (T_A = 25°C unless otherwise noted)

Characteristic	Fig. No.	Symbol	Min	Typ	Max	Unit
Power Launched (I_F = 100 mA)	4, 5	P_L	110	165	—	μW
Optical Rise and Fall Time (I_F = 100 mA)	2	t_r, t_f	—	25	35	ns
Peak Wavelength (I_F = 100 mA)	1	λ_P	—	820	—	nm

For simple fiber termination instructions, see the MFOD71, 72 and 73 data sheet.

Courtesy Motorola Inc.

FIGURE 16.14   *Continued*

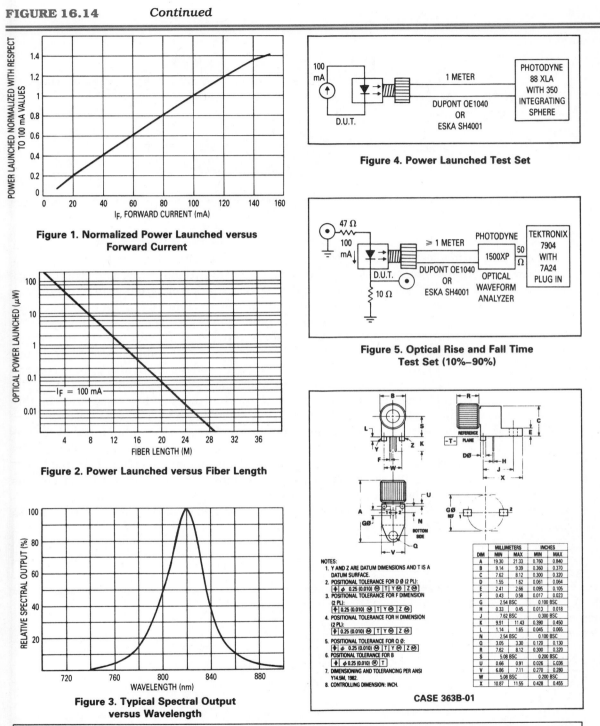

Figure 1. Normalized Power Launched versus Forward Current

Figure 2. Power Launched versus Fiber Length

Figure 3. Typical Spectral Output versus Wavelength

Figure 4. Power Launched Test Set

Figure 5. Optical Rise and Fall Time Test Set (10%–90%)

NOTES:
1. Y AND Z ARE DATUM DIMENSIONS AND T IS A DATUM SURFACE.
2. POSITIONAL TOLERANCE FOR D Ø (2 PL):
3. POSITIONAL TOLERANCE FOR F DIMENSION (2 PL):
4. POSITIONAL TOLERANCE FOR H DIMENSION (2 PL):
5. POSITIONAL TOLERANCE FOR Q Ø:
6. POSITIONAL TOLERANCE FOR B
7. DIMENSIONING AND TOLERANCING PER ANSI Y14.5M, 1982.
8. CONTROLLING DIMENSION: INCH.

DIM	MILLIMETERS		INCHES	
	MIN	MAX	MIN	MAX
A	19.30	21.33	0.760	0.840
B	9.14	9.39	0.360	0.370
C	7.62	8.12	0.300	0.320
D	1.55	1.62	0.061	0.064
E	2.41	2.66	0.095	0.105
F	0.43	0.58	0.017	0.023
G	2.54 BSC		0.100 BSC	
H	0.33	0.45	0.013	0.018
J	7.62 BSC		0.300 BSC	
K	9.91	11.43	0.390	0.450
L	1.14	1.65	0.045	0.065
N	2.54 BSC		0.100 BSC	
Q	3.05	3.30	0.120	0.130
R	7.62	8.12	0.300	0.320
S	5.08 BSC		0.200 BSC	
U	0.66	0.91	0.026	0.036
V	6.86	7.11	0.270	0.280
W	5.08 BSC		0.200 BSC	
X	10.87	11.55	0.428	0.455

CASE 363B-01

FIGURE 16.15

Simplified Cross Section of Diode Laser

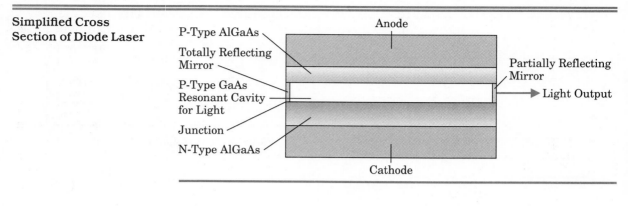

FIGURE 16.16

times called *lasing*, occurs only above a certain threshold current. Below that current, the laser diode acts as a conventional LED. The graph in Figure 16.16 shows the threshold, and the variation of light output with current.

Laser diodes are more difficult to operate than LEDs: their current must be carefully controlled, and the required current varies with temperature. With too little current, lasing stops, but too much will destroy the device. Laser diodes often use thermoelectric coolers for temperature control, and have their light output monitored by a photodetector to allow the current to be adjusted automatically. Lasers are usually modulated by changing the current from a value just below the lasing threshold to some considerably higher value. This can be done at rates exceeding 1 GHz.

A safety note is in order about lasers. Because the light emitted is coherent and intense, even a low-power laser such as a laser diode can damage the eye, which may focus the light onto a small spot on the retina. Infrared lasers are especially dangerous because the light is invisible and it is not obvious that the laser is operating. Anyone working with lasers, even the very low-powered ones used in communications, should be very careful not to look into

FIGURE 16.16

Light Output as a Function of Current for a Laser Diode

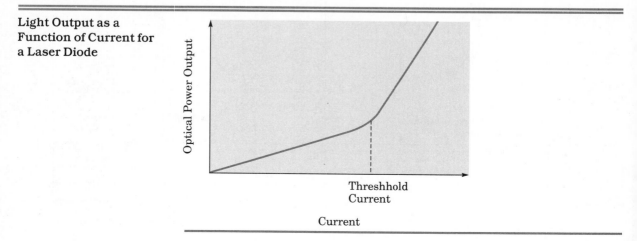

an operating laser. More powerful lasers can cause other damage due to heating effects: they can burn skin, start fires, and so on. These lasers are not found in communications applications, however, and are not within the scope of this book.

16.4.3 Optical Detectors

The most common types of detectors for optical communications are PIN diodes and avalanche photodiodes. The PIN diode, shown in cross section in Figure 16.17, was previously discussed in connection with its uses as a microwave switch and attenuator. As a photodetector, the PIN diode takes advantage of its wide depletion region, in which photons can create electron-hole pairs. If a photon has enough energy to move an electron from the valence to the conduction band, it can, by colliding with an electron, create a free electron. The crystal lattice will contain a hole where the electron used to be. Since the PIN diode is reverse-biased, the free electron and hole will migrate in opposite directions, creating a current that is proportional to the light power. The low junction capacitance of the PIN diode, which results from the wide depletion region, allows for very high-speed operation, into the gigahertz range in some cases.

The avalanche photodiode (APD), shown in Figure 16.18, is also operated with reverse bias. In this case, however, the device is biased just below its reverse-breakdown voltage. The creation of an electron-hole pair due to the absorption of a photon of incoming light may set off avalanche breakdown, creating up to 100 more pairs. This multiplying effect gives the APD high sensitivity, about 5 to 7 dB greater than the PIN diode. On the other hand, the operation of an avalanche photodiode requires a carefully controlled bias voltage, to keep the diode biased on the verge of avalanche breakdown.

Figure 16.19 is a data sheet for the Motorola MFOD71, a low-cost PIN-diode photodetector, which could be used with the LED shown earlier. A brief study of this data sheet will familiarize the reader with photodetector specifications. The most important of these are described below.

Dark current is the leakage current in the absence of light; that is, it is the normal reverse current of the diode. The dark current sets a sensitivity floor. Since the dark current is due to electron-hole pairs created by thermal activity, it increases with temperature, just as for any diode. For this diode, typical dark-current values range from 0.06 nA at 25°C to 10 nA at 85°C.

FIGURE 16.17

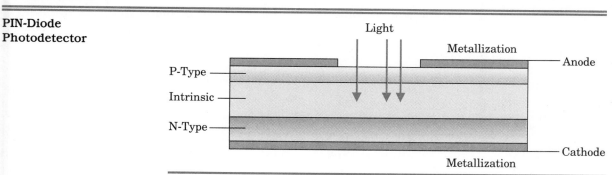

**PIN-Diode
Photodetector**

FIGURE 16.18   Avalanche Photodiode

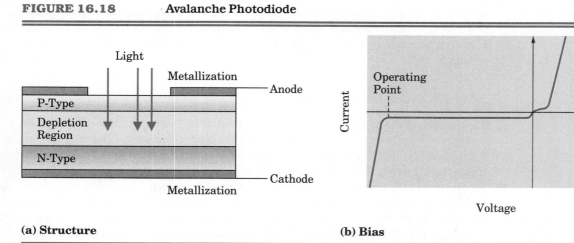

(a) Structure   **(b) Bias**

Responsivity is a measure of the sensitivity of the detector to light. Its basic units are amperes per watt of light; in the case of the MFOD71, a typical value is given as 0.2 μA/μW. (The practical unit is actually the same as the basic one, since the numerator and denominator have both been divided by 10^6.)

The responsivity is given for the wavelength of light to which the detector is most sensitive. If the light source in use has its maximum output at some other wavelength, it is necessary to consider the spectral response of the detector, that is, the variation of detector sensitivity with wavelength. The spectral response curve will provide a percentage or decibel factor that modifies the responsivity according to the source wavelength.

Response time is a very important specification in high-speed systems, as it limits the maximum data rate that can be used.

EXAMPLE 16.6

A typical Motorola MFOD71 photodetector is used with an LED having its maximum output at 600 nm. Calculate the output current for a power input of 50 μW.

Solution

The responsivity of 0.2 μA/μW will have to be multiplied by the relative spectral response at 600 nm. This can be found from the graph in Figure 16.19 as approximately 65% or 0.65. Therefore, the output current will be

$$I_o = 50 \ \mu W \times 0.2 \ \mu A/\mu W \times 0.65$$
$$= 6.5 \ \mu A$$

The data sheet in Figure 16.19 also has suggested circuits for simple optical transmitters and receivers designed to work with TTL-level digital signals.

FIGURE 16.19   Specifications for a PIN-Diode Photodetector

FIBER OPTIC LOW COST SYSTEM
FLCS DETECTORS

... designed for low cost, short distance Fiber Optic Systems using 1000 micron core plastic fiber.

Typical applications include: high isolation interconnects, disposable medical electronics, consumer products, and microprocessor controlled systems such as coin operated machines, copy machines, electronic games, industrial clothes dryers, etc.

- Fast PIN Photodiode: Response Time <5.0 ns
- Standard Phototransistor
- High Sensitivity Photodarlington
- Spectral Response Matched to MFOE71 LED
- Annular Passivated Structure for Stability and Reliability
- FLCS Package
  - Includes Connector
  - Simple Fiber Termination and Connection (Figure 4)
  - Easy Board Mounting
  - Molded Lens for Efficient Coupling
  - Mates with 1000 Micron Core Plastic Fiber (DuPont OE1040, Eska SH4001)

FLCS LINE
FIBER OPTICS
DETECTORS

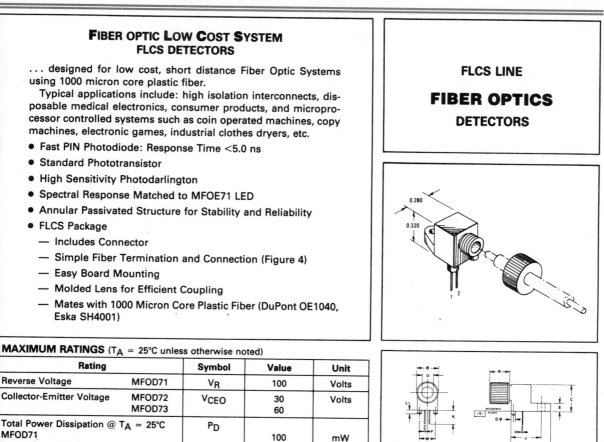

MAXIMUM RATINGS (T_A = 25°C unless otherwise noted)

Rating		Symbol	Value	Unit
Reverse Voltage	MFOD71	V_R	100	Volts
Collector-Emitter Voltage	MFOD72	V_{CEO}	30	Volts
	MFOD73		60	
Total Power Dissipation @ T_A = 25°C		P_D		
MFOD71			100	mW
Derate above 25°C			1.67	mW/°C
MFOD72/73			150	mW
Derate above 25°C			2.5	mW/°C
Operating and Storage Junction Temperature Range		T_J, T_{stg}	−40 to +85	°C

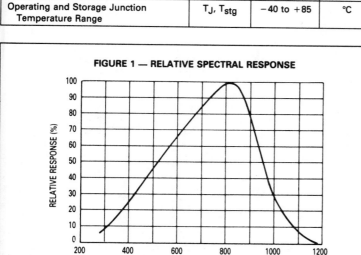

FIGURE 1 — RELATIVE SPECTRAL RESPONSE

NOTES:
1. DIMENSIONS V AND X ARE DATUMS AND [-T-] IS A DATUM SURFACE.
2. POSITIONAL TOLERANCE FOR Lϕ (2 PL):
3. POSITIONAL TOLERANCE FOR F DIMENSION (2 PL):
4. POSITIONAL TOLERANCE FOR H DIMENSION (2 PL):
5. DIMENSIONING AND TOLERANCING PER ANSI Y14.5, 1982.
6. CONTROLLING DIMENSION: INCH.

STYLE 2:
PIN 1. EMITTER
2. COLLECTOR

(MFOD71 ONLY)
STYLE 3:
PIN 1. CATHODE
2. ANODE

DIM	MILLIMETERS		INCHES	
	MIN	MAX	MIN	MAX
A	18.67	19.05	0.735	0.750
B	9.14	9.39	0.360	0.370
C	7.75	8.12	0.305	0.320
D	1.52	1.72	0.060	0.068
E	2.41	2.66	0.095	0.105
F	0.43	0.60	0.017	0.024
G	2.54 BSC		0.100 BSC	
H	0.23	0.55	0.009	0.022
J	7.87	8.25	0.310	0.325
K	10.29	17.14	0.405	0.675
L	1.27	1.52	0.050	0.060
N	3.25	3.35	0.128	0.132
Q	3.05	3.30	0.120	0.130
R	7.49	8.00	0.295	0.315
U	3.56	3.81	0.140	0.150
V	6.86	7.11	0.270	0.280
W	5.33 BSC		0.210 BSC	
X	10.67	11.17	0.420	0.440

CASE 363-01

Courtesy Motorola Inc.

(continued)

FIGURE 16.19   *Continued*

MFOD71

STATIC ELECTRICAL CHARACTERISTICS (T_A = 25°C unless otherwise noted)

Characteristic	Symbol	Min	Typ	Max	Unit
Dark Current (V_R = 20 V, R_L = 1.0 MΩ) T_A = 25°C	I_D	—	0.06	10	nA
$\qquad\qquad\qquad\qquad\qquad\qquad\qquad\qquad T_A$ = 85°C		—	10	—	
Reverse Breakdown Voltage (I_R = 10 μA)	$V_{(BR)R}$	50	100	—	Volts
Forward Voltage (I_F = 50 mA)	V_F	—	—	1.1	Volts
Series Resistance (I_F = 50 mA)	R_S	—	8.0	—	ohms
Total Capacitance (V_R = 20 V; f = 1.0 MHz)	C_T	—	3.0	—	pF

OPTICAL CHARACTERISTICS (T_A = 25°C)

Responsivity (V_R = 5.0 V, Figure 2)	R	0.15	0.2	—	μA/μW
Response Time (V_R = 5.0 V, R_L = 50 Ω)	$t_{(resp)}$	—	5.0	—	ns

MFOD72/MFOD73

STATIC ELECTRICAL CHARACTERISTICS

Collector Dark Current (V_{CE} = 10 V)		I_D	—	—	100	nA
Collector-Emitter Breakdown Voltage (I_C = 10 mA) MFOD72	MFOD72	$V_{(BR)CEO}$	30	—	—	Volts
	MFOD73		60	—	—	

OPTICAL CHARACTERISTICS (T_A = 25°C unless otherwise noted)

Responsivity (V_{CC} = 5.0 V, Figure 2)	MFOD72	R	80	125	—	μA/μW
	MFOD73		1,000	1,500	—	
Saturation Voltage (λ = 820 nm, V_{CC} = 5.0 V)		$V_{CE(sat)}$				Volts
$\qquad\qquad$ (P_{in} = 10 μW, I_C = 1.0 mA)	MFOD72		—	0.25	0.4	
$\qquad\qquad$ (P_{in} = 1.0 μW, I_C = 2.0 mA)	MFOD73		—	0.75	1.0	
Turn-On Time $\quad$ R_L = 2.4 kΩ, P_{in} = 10 μW,	MFOD72	t_{on}	—	10	—	μs
Turn-Off Time $\quad$ λ = 820 nm, V_{CC} = 5.0 V		t_{off}	—	60	—	μs
Turn-On Time $\quad$ R_L = 100 Ω, P_{in} = 1.0 μW,	MFOD73	t_{on}	—	125	—	μs
Turn-Off Time $\quad$ λ = 820 nm, V_{CC} = 5.0 V		t_{off}	—	150	—	μs

TYPICAL COUPLED CHARACTERISTICS

FIGURE 2 — RESPONSIVITY TEST CONFIGURATION   **FIGURE 3 — DETECTOR CURRENT versus FIBER LENGTH**

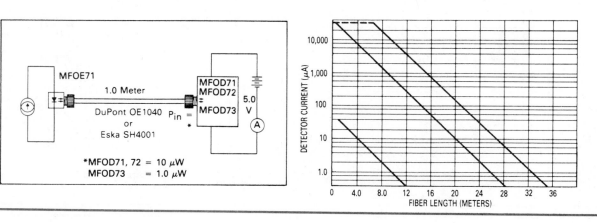

FIGURE 16.19   *Continued*

FLCS WORKING DISTANCES

The system length achieved with a FLCS emitter and detector using the 1000 micron core fiber optic cable depends upon the forward current through the LED and the Responsivity of the detector chosen. Each emitter/detector combination will work at any cable length up to the maximum length shown.

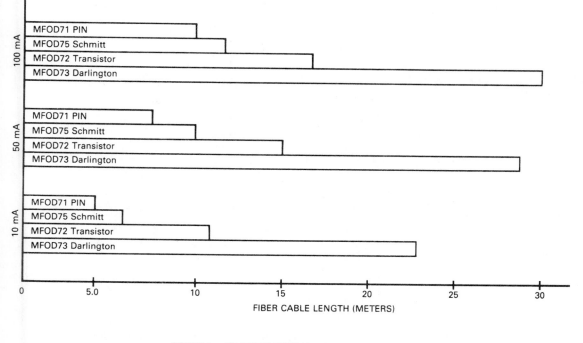

FIGURE 4 — FO CABLE TERMINATION AND ASSEMBLY

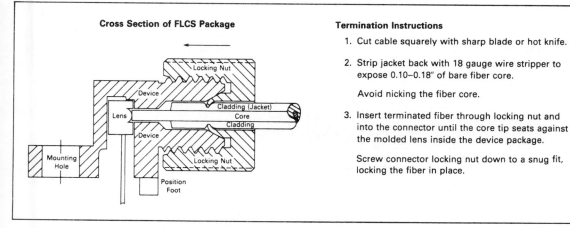

Cross Section of FLCS Package

Termination Instructions

1. Cut cable squarely with sharp blade or hot knife.

2. Strip jacket back with 18 gauge wire stripper to expose 0.10–0.18″ of bare fiber core.

   Avoid nicking the fiber core.

3. Insert terminated fiber through locking nut and into the connector until the core tip seats against the molded lens inside the device package.

   Screw connector locking nut down to a snug fit, locking the fiber in place.

(*continued*)

FIGURE 16.19   *Continued*

Input Signal Conditioning

The following circuits are suggested to provide the desired forward current through the emitter.

TTL TRANSMITTERS

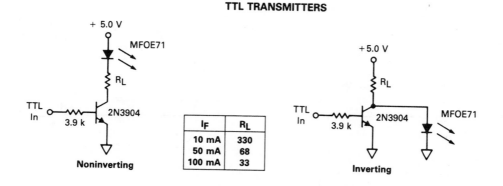

IF	RL
10 mA	330
50 mA	68
100 mA	33

Noninverting   Inverting

Output Signal Conditioning

The following circuits are suggested to take the FLCS detector output and condition it to drive TTL with an acceptable bit error rate.

TTL RECEIVERS

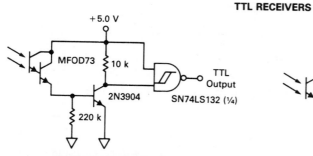

1.0 kHz Darlington Receiver

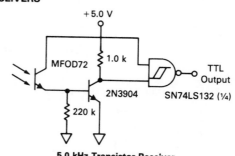

5.0 kHz Transistor Receiver

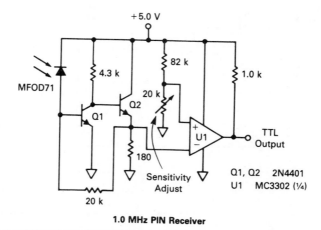

1.0 MHz PIN Receiver

16.5 OPTICAL COMMUNICATIONS SYSTEMS

The principal optical components in a fiber-optic communications system have now been described, and it is time to put them together into a system. As already mentioned, any of the modulation schemes discussed so far can be used, so there is no need to describe them again. This section will describe a generic optical-fiber system. More specific examples will be introduced in the next section.

Depending on its length and data rate and the quality of the components used, a fiber system may consist only of a transmitter, a receiver, a length of fiber, and some connectors and/or splices. If the system is too long for this single-span approach, one or more *repeaters* may be needed. These usually convert the incoming optical signal back to electrical form, regenerate the pulses as required, and then generate a new optical signal. It is actually possible to amplify light directly, using techniques very similar to those used in laser diodes, but as yet this is not a common technique. Figure 16.20 shows optical systems with and without repeaters.

16.5.1 Loss Budget

The maximum possible length of a fiber span between two repeaters is governed by two factors: dispersion and losses. As previously mentioned, dispersion increases with distance, and limits the data rate that can be used at a given distance or, put another way, the maximum length of fiber that can be used for a given data rate. It is also necessary to have sufficient optical power at the receiver, so that the signal-to-noise ratio will be great enough for a specified error rate.

In addition to the loss in the fiber itself, losses occur at the interfaces between the fiber and other components, such as the source and detector, and at splices and connectors.

The received power at the detector (in dBm) can be calculated by subtracting the total losses (in decibels) from the power input (in dBm). A **system margin** of 5 to 10 dB must be allowed. This will account for the deterio-

FIGURE 16.20   **Optical Communication Systems**

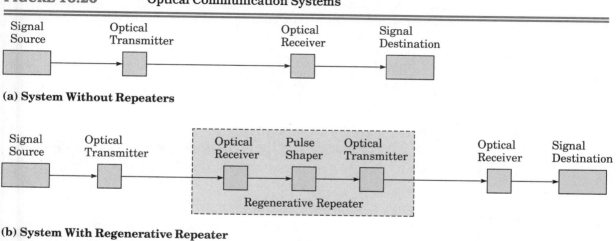

(a) System Without Repeaters

(b) System With Regenerative Repeater

ration of components over time, the possibility that additional splices will be needed (if the cable is accidentally cut, for instance), and so on.

EXAMPLE 16.7

A fiber-optic link extends for 40 km. The laser diode has an output power of 1.5 mW, and the receiver requires a signal strength of -25 dBm for a satisfactory signal-to-noise ratio. The fiber is available in lengths of 2.5 km and can be spliced with a loss of 0.25 dB per splice. The fiber has a loss of 0.3 dB/km. The total of all the connector losses at both ends is 4 dB. Calculate the available system margin. Also, calculate the maximum frequency that can be used with the link if its bandwidth-distance product is 1 GHz-km.

Solution

It is much easier to use decibels throughout to solve this type of problem, so let us begin by converting the input power to the system to dBm.

$$P_{in} \text{ (dBm)} = 10 \log P_{in} \text{ (mW)}$$
$$= 10 \log 1.5$$
$$= 1.76 \text{ dBm}$$

The connector and fiber losses are quite straightforward. The only problem in finding the loss in the splices is determining the number of splices. There will be one less splice than the number of fiber spans. The reason for this is easy to see from the sketch in Figure 16.21.

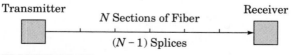

Transmitter   Receiver

N Sections of Fiber

$(N-1)$ Splices

FIGURE 16.21

Consider each length of fiber to be attached to one splice at its output end. Then there is one splice per length, except for the last one. The number of lengths of fiber can be found by dividing the total length by the maximum length for a fiber section, then rounding upward. In this case, the number of lengths is found from

$$\frac{\text{Span Length}}{\text{Fiber Length}} = \frac{40 \text{ km}}{2.5 \text{ km}}$$
$$= 16$$

Therefore, there will have to be 15 splices.

Now it is necessary to find the total losses. These can be calculated as follows:

Connector losses	4 dB
Fiber loss: 40 km $\times$ 0.3 dB/km	12 dB
Splice loss: 15 splices $\times$ 0.25 dB/splice	3.75 dB
Total loss	19.75 dB

Next, find the output level:

$$P_{out} = 1.76 \text{ dBm} - 19.75 \text{ dB}$$
$$= -17.99 \text{ dBm}$$

The receiver requires a power of -25 dBm, so the available system margin is

$$\text{System Margin} = -17.99 \text{ dBm} - (-25 \text{ dBm})$$
$$= 7.01 \text{ dB}$$

This system should be satisfactory in terms of signal strength. To find the maximum frequency, we divide the bandwidth-distance product by the distance.

$$f_{max} = \frac{1 \text{ GHz-km}}{40 \text{ km}}$$
$$= 25 \text{ MHz}$$

16.5.2 Repeaters

One of the advantages of the use of digital techniques in communications is the fact that regenerative repeaters can be used; that is, a misshapen pulse that is received can be decoded into ones and zeros from which a new, perfect pulse train can be reconstructed. Provided that repeaters are spaced closely enough that the error rate is very low, digital systems can be spared the accumulation of noise and distortion that plagues analog systems.

Repeaters need electrical power, of course, and it is preferable for them to be accessible for maintenance. Consequently, the trend in optical system design is to minimize the number of repeaters by increasing the bandwidth-distance product and reducing the losses of the fiber.

16.5.3 Wavelength-Division Multiplexing

Most optical-fiber systems use time-division multiplexing to take advantage of the available bandwidth using one LED or laser diode. This bandwidth, which is usually limited by dispersion, is really only a small fraction of the actual bandwidth available on a fiber, however. There is no theoretical reason why there cannot be several light sources, each operating at a different wavelength and coupled into the same fiber. At the receiving end, the different wavelengths would be sent to separate receivers.

A diagram of this scheme, called **wavelength-division multiplexing** (WDM), is given in Figure 16.22. Practical implementation requires lasers with narrow bandwidth. WDM is really a form of frequency-division multiplexing (FDM). Think of the laser diodes as carrier oscillators, each with a different carrier frequency and separately modulated. One difference between FDM on radio-frequency carriers and WDM on optical fibers is that for FDM the separation between carriers is limited by the sidebands created by modulation, while with even the best lasers, the width of the carrier sig-

FIGURE 16.22

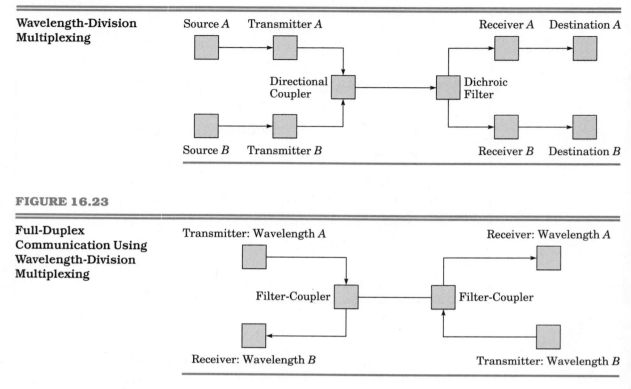

Wavelength-Division
Multiplexing

FIGURE 16.23

Full-Duplex
Communication Using
Wavelength-Division
Multiplexing

nal itself is such that this—rather than the modulation rate and type—determines the signal bandwidth. That is, the laser can be considered as an optical oscillator, but by radio-frequency standards it is a rather unstable oscillator with plenty of spurious signals. This is changing, however, with improvements in laser technology.

One obvious use for WDM is to implement full-duplex communications on a single fiber. Figure 16.23 shows how this can be done. At present, however, it is still usually more cost-effective to use two fibers and dedicate each to service in one direction.

16.6 PRESENT AND FUTURE APPLICATIONS

In this section, we will look at some current uses of optical fiber, and speculate a little as to future uses. Fiber optics will never supplant other modes completely (it is useless for portable and mobile communications, for instance), but it appears that, in the future, fiber will replace copper wire as the most common way of transmitting signals.

Perhaps the most obvious, and some of the earliest, uses of fiber were for long-distance, high-traffic lines such as telephone trunk lines. Many telephone companies have stopped using copper for long-distance trunks, and the use of optical fiber for such demanding applications as undersea cables is increasing. For example, the first transatlantic undersea fiber cable was completed in December of 1988. The repeater spacing is 70 km, and 109 repeaters are required altogether. Its primary data rate is 295.6 Mb/s, which is equivalent to 40,000 simultaneous telephone calls. Of course, the system can handle other types of communications, for example video and data, as well.

Two single-mode fibers are used, one for communication in each direction. Laser-diode emitters and PIN-diode detectors operating at a wavelength of 1.3 μm are used.

By comparison, the first transatlantic telephone cable, installed in 1956, could handle 36 telephone calls, and the most recent copper cable (1983) has a capacity of 8500 calls.

In contrast to its wide acceptance in long-distance applications, the use of fiber for telephone service to individual subscribers is in its infancy. The reason is that local loops are still two to three times as expensive with fiber as with copper. The existing copper twisted-pair loops have more bandwidth than required for telephone service, so there is no overwhelming advantage to fiber. Nonetheless, many telephone companies are experimenting with *fiber-in-the-loop* (FITL) systems, especially for areas where many businesses and homes have multi-line services. Fiber lends itself very well to the new digital telephone service called the Integrated Services Digital Network (ISDN) in which signals can be converted into digital form in the telephone itself, for transmission throughout the system.

It is quite likely that at some future date optical fiber will be in wide use for the delivery of both telephone and cable-television services to homes. One connection could easily provide enough bandwidth for both, and, in fact, additional services could be added, still using one fiber. Possibilities include, among others, the ability to transmit full-motion video from one home to another, and on-demand pay television, where the viewer could request a specific program. This would require a great change in the way cable television is organized, of course. It would have to change from the broadcast-oriented tree structure described in Chapter 14 to a switched network like that of the telephone system, and closely associated with it. There are no technical reasons why this could not be done. The economic and political problems are quite intimidating, however: economic in terms of the amount of new plant that would have to be built and old plant that would become obsolete; and political in the broad sense, since at present telephone and cable-television systems are operated by different companies, under different regulations, and with different outlooks and priorities.

Since most optical-fiber communications are digital, the medium seems naturally suited to data communications. It is widely used for high-speed long-distance data links, and has made some inroads into short-distant systems called *local-area networks* (LANs). As with the local loop, however, the advantages of fiber over coaxial or twisted-pair cable are not as great with short distances and low data rates, so this is another idea whose time is coming, but is perhaps not quite here.

Optical communications are still in their infancy. The LED switched on and off as a transmitter has a certain resemblance to early spark transmitters in radio. They too had wide bandwidth. The laser diode represents the same kind of improvement that the vacuum tube made in radio transmitters. In terms of receivers, the PIN photodiode resembles very much the primitive

FIGURE 16.24

Optical Version of
Superheterodyne
Receiver

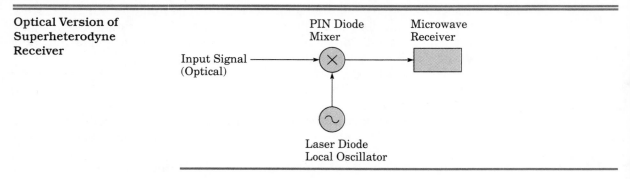

"crystal" receiver from the early days of radio. One might wonder whether an
optical equivalent to the superheterodyne receiver could be built. The answer
is Yes, and it has been done. Figure 16.24 shows how: a laser diode is used as
a local oscillator, a photodiode as a mixer, and the intermediate frequency is
a microwave signal. This kind of thing is still largely experimental, but it
shows the direction in which technology is moving. If the twentieth century
has been dominated by radio and its variations, it appears that the twenty-
first may be the century of light-wave communication.

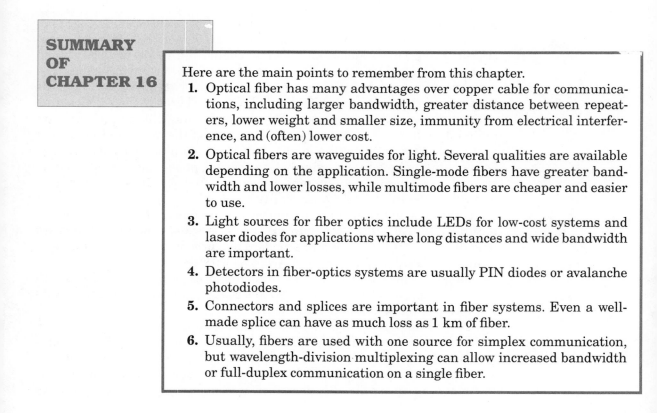

SUMMARY OF CHAPTER 16

Here are the main points to remember from this chapter.

1. Optical fiber has many advantages over copper cable for communications, including larger bandwidth, greater distance between repeaters, lower weight and smaller size, immunity from electrical interference, and (often) lower cost.

2. Optical fibers are waveguides for light. Several qualities are available
   depending on the application. Single-mode fibers have greater bandwidth and lower losses, while multimode fibers are cheaper and easier
   to use.

3. Light sources for fiber optics include LEDs for low-cost systems and
   laser diodes for applications where long distances and wide bandwidth
   are important.

4. Detectors in fiber-optics systems are usually PIN diodes or avalanche
   photodiodes.

5. Connectors and splices are important in fiber systems. Even a well-
   made splice can have as much loss as 1 km of fiber.

6. Usually, fibers are used with one source for simplex communication,
   but wavelength-division multiplexing can allow increased bandwidth
   or full-duplex communication on a single fiber.

IMPORTANT EQUATIONS

$$n = \frac{c}{v} \tag{16.1}$$

$$n = \sqrt{\epsilon_r} \tag{16.3}$$

$$n_1 \sin \theta_1 = n_2 \sin \theta_2 \tag{16.4}$$

$$\theta_c = \arcsin \frac{n_2}{n_1} \tag{16.6}$$

$$\text{N.A.} = \sqrt{n_1^2 - n_2^2} \tag{16.7}$$

$$r_{max} = \frac{0.383\lambda}{\text{N.A.}} \tag{16.8}$$

$$E = hf \tag{16.9}$$

$$1 \text{ eV} = 1.6 \times 10^{-19} \text{ joule} \tag{16.10}$$

GLOSSARY

angle of acceptance maximum angle between the axis of an optical fiber and a ray of light entering the fiber

angle of incidence angle an incident ray makes with the normal to a reflecting or refracting surface

angle of reflection angle a reflected ray makes with the normal to a reflecting surface

angle of refraction angle a refracted ray makes with the normal to a refracting surface

cladding in an optical fiber, the material of lower refractive index that surrounds the core

core in an optical fiber, the central part of the fiber in which the light propagates

critical angle the maximum angle of incidence for which refraction takes place

crosstalk interference between signals on separate cables in close proximity

dark current in a photodetector, the current that flows in the absence of light

dispersion variation of propagation velocity with wavelength

electron-volt (eV) the energy given to or absorbed by an electron that moves through a potential difference of one volt ($1 \text{ eV} = 1.6 \times 10^{-19}$ joule)

graded-index fiber optical fiber in which the index of refraction of the core decreases gradually with increasing distance from the center

heterojunction A P–N junction in which the two sides of the junction are made of different materials

index of refraction ratio between the velocity of light in free space and that in a given medium

laser   acronym for Light Amplification by Stimulated Emission of Radiation. A laser is an optical oscillator that produces phase-coherent light with narrow bandwidth.

laser diode (LD)   a low-power laser resembling an LED in its construction

multimode fiber   a fiber that allows light to travel along it in more than one waveguide mode, at different velocities

normal   a line perpendicular to a reflecting or refracting surface

numerical aperture   for an optical fiber, the sine of the angle of acceptance

photon   a quantum of electromagnetic radiation

quantum   the smallest amount in which energy can exist. The size of a quantum depends on the wavelength of the energy.

responsivity   the relationship between output current and input light power for a photodetector

single-mode fiber   an optical fiber whose core is sufficiently narrow that only one waveguide mode can propagate

step-index fiber   an optical fiber that has one index of refraction for the core and a second, lower index for the cladding, with a sharp transition between them

system margin   excess, in decibels, of the power at the receiver of a communication system over that required for a satisfactory signal-to-noise ratio

total internal reflection   reflection at the boundary between two media when the angle of incidence is greater than the critical angle

wavelength-division multiplexing (WDM)   in an optical communication system, the use of two or more light sources at different wavelengths to increase the communications bandwidth or to implement full-duplex communication over a single fiber

QUESTIONS

1. Give four possible advantages of optical fiber over wire cables, and explain each.

2. Give one advantage and one disadvantage of fiber compared with a geostationary-satellite radio link.

3. Explain briefly the difference between single-mode and multimode fiber. Which gives better performance, and why?

4. What are the advantages of using a laser diode with single-mode fiber, instead of an LED?

5. State Snell's Law, and illustrate it with a diagram.

6. What is total internal reflection, and under what circumstances does it occur?

7. What is meant by the numerical aperture of an optical fiber? What happens if light moves from a fiber to another with a lower numerical aperture?

8. List the three types of optical fiber discussed in this chapter, and order them in terms of dispersion and loss.

9. Describe the mechanisms by which dispersion takes place in optical fibers. Which of these mechanisms apply to single-mode fiber?

10. Which type of fiber has the highest bandwidth-distance product, and why?

11. What is a photon?

12. Draw a cross section of an LED, and describe its operation.

13. Draw a cross section of a laser diode, and describe its operation.

14. For what is the term *laser* an acronym?

15. What safety precautions are necessary with infrared laser diodes, and why?

16. Name and compare two types of photodetectors that are used in optical communications.

17. What is meant by a loss budget for a fiber-optic system?

18. How can full-duplex communication be implemented on a single optical fiber?

19. Why are fiber-optic cables more popular in telephone trunk applications than for local loops?

20. How can the equivalent of a superheterodyne receiver be implemented in an optical communication system?

PROBLEMS

Section 16.1

21. Calculate the frequency that corresponds to each of the following wavelengths of light. Assume free-space propagation.
    (a) 400 nm (violet)
    (b) 700 nm (red)
    (c) 900 nm (infrared)

Section 16.2

22. Suppose the indices of refraction of the core and cladding of an optical fiber are 1.5 and 1.45, respectively.
    (a) Calculate the dielectric constants for the core and the cladding.
    (b) Calculate the speed of light in the core and the cladding.

23. For the fiber in Problem 22:
    (a) What is the critical angle for a ray moving from the core to the cladding?
    (b) What is the numerical aperture?
    (c) What is the maximum angle from the axis of the fiber, at which light will be accepted?

24. For a cable with indices of refraction of 1.46 and 1.41 for the core and cladding, respectively, calculate the maximum diameter the core could have for single-mode propagation at a wavelength of 1.5 μm.

25. A commonly used wavelength for optical communication is 1.55 μm.
    (a) What type of light is this?
    (b) Calculate the frequency corresponding to this wavelength, assuming free-space propagation.

26. A fiber-optic cable has a bandwidth-distance product of 500 MHz-km. What bandwidth can be used with a cable that runs 30 km between repeaters?

27. A fiber is installed over a distance of 15 km and it is found experimentally that

the maximum operating bandwidth is 700 MHz. Calculate the bandwidth-distance product for the fiber.

28. From the chart labeled "Power Launched versus Fiber Length" on the MFOE71 data sheet in Figure 16.14, find the loss of the fiber used, in dB/km.

Section 16.4

29. Calculate the energy in one photon of a light wave at a wavelength of 400 nm. Express the result in both joules and electron-volts.

30. Figure 16.25 is a graph relating light output to input current for a laser diode. Find the threshold current level and the power output for a current of 120 mA.

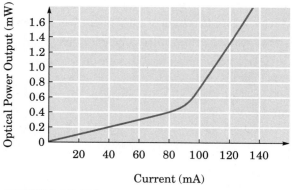

FIGURE 16.25

31. A Motorola MFOD71 photodiode is connected in series with a resistor of 10 kΩ and reverse biased. Light with a wavelength of 700 nm and a power of 75 μW is applied. Calculate the current through the diode and the output voltage across the resistor.

Section 16.5

32. A communications link runs 50 km. The transmitter power output is 3 mW, and the losses are as follows:

    Connector loss (total):   5 dB

    Splice loss: 0.3 dB per slice, splices are 2 km apart.

    Fiber loss: 1.5 dB/km

Calculate the power level at the receiver, in dBm.

33. A laser diode emits a power of 1 mW. It is to be used in a fiber-optic system with a receiver that requires a power of at least 1 μW for the required bit error rate. The losses in the system are as follows:

    Coupling and connector losses, transmitter to cable: 10 dB

    Cable loss: 0.5 dB per km

    Splice loss: 0.2 dB per splice

    Connector loss between cable and receiver: 2 dB

Determine whether the system will work over a 10 km distance. Assume that it will be necessary to have a splice every 2 km of cable.

34. An optical communications link is to be built using fiber rated for 500 MHz-km. The light source is a laser diode producing 0.25 mW. The fiber has losses of

0.4 dB/km, and is available in 2 km lengths. It can be spliced with a loss of 0.2 dB per splice, and there will be 5 dB loss in the connectors throughout the system. The receiver used has a sensitivity of -30 dBm.

(a) Prepare a loss budget for a link 20 km long, and calculate the system margin in decibels.

(b) What would be the maximum bandwidth that could be used with the system in Part (a)?

Review of Basic AC Theory

This section is intended as a review of ac signals. Therefore the coverage will be brief, with no derivations. The reader is referred to any standard electric circuits text for greater detail.

The simplest, most common form of ac signal is known as the *sinusoid*, because its defining equation is that of either a sine or a cosine wave. This appendix will deal only with sinusoids. Other periodic waveforms can be constructed from sine and cosine waves by using Fourier series, as explained in Chapter 1.

A.1 EQUATION FOR SINUSOIDAL WAVES

Consider a sinusoidal voltage waveform. The instantaneous voltage is given by an equation of the form

$$v = V \sin (\omega t + \theta) \tag{A.1}$$

or

$$v = V \cos (\omega t + \theta) \tag{A.2}$$

where v = instantaneous voltage at time t
V = maximum (peak) voltage
ω = radian frequency in radians per second
t = time in seconds
θ = phase angle, with respect to some reference phase, in radians

Actually, the two forms are equivalent because the sine and cosine waves differ only by an angle of 90° or $\pi/2$ rad; that is,

$$\cos \omega t = \sin \left(\omega t + \frac{\pi}{2} \right) \tag{A.3}$$

Equations (A.1) and (A.2) can be used for a sinusoidal current wave by simply changing v to i and V to I.

Often the phase angle θ in Equations (A.1) and (A.2) can be set to zero. If there is only one waveform under consideration, the time that is chosen for $t = 0$ is arbitrary so it can be chosen to make $\theta = 0$. Figure A.1 shows the effect of changing the zero point for time.

FIGURE A.1   Two Sinusoids with Different Phase Angles

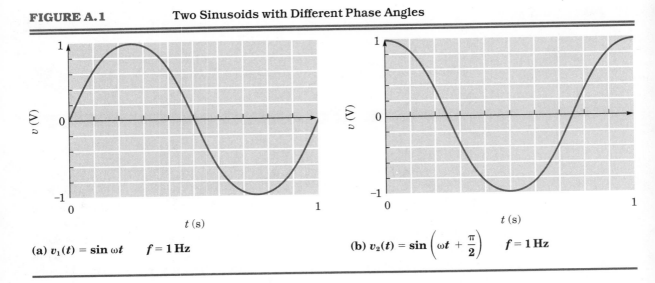

(a) $v_1(t) = \sin \omega t$ $f = 1\,\text{Hz}$ (b) $v_2(t) = \sin\left(\omega t + \dfrac{\pi}{2}\right)$ $f = 1\,\text{Hz}$

Radian measurement of angle is generally used in the mathematics dealing with sine waves. Degrees are more common in practical situations, however. The conversion between radians and degrees is quite simple. From basic geometry, one complete revolution is 2π rad or $360°$. Therefore,

$$2\pi \text{ rad} = 360°$$
$$1 \text{ rad} = \frac{180°}{\pi}$$
$$= 57.3° \tag{A.4}$$

When there are two or more waveforms with the same frequency ω, it is usual to use one as the reference, and calculate the phase of the other(s) with respect to it. The phase angle can be expressed in two different ways, as illustrated in Figure A.2. For instance, one signal can be said to *lead* another by $90°$ or *lag* it by $270°$.

The period T of any repetitive waveform is the time taken for one complete cycle. For a sinusoid, it is the time taken for the angle to go through $360°$ or 2π rad. The frequency in hertz is the number of cycles per second, and is equal to $1/T$. The angular velocity ω is given by

$$\omega = 2\pi f \tag{A.5}$$

There are several ways of expressing the amplitude of an ac voltage or current waveform. Some of them are illustrated in Figure A.3. Equations (A.1) and (A.2) give the instantaneous value. The amplitude V in those equations is the peak value, which occurs once per cycle. There will also be one instant per cycle when the voltage (or current) is equal in magnitude to the peak value but with the opposite sign. Physically, this means that the voltage or current has reversed polarity.

FIGURE A.2

Representation of Phase
Angles

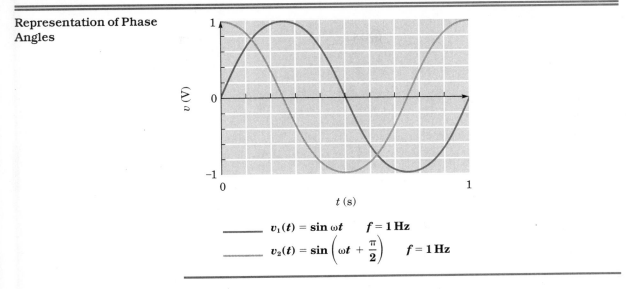

$$v_1(t) = \sin \omega t \qquad f = 1\,\text{Hz}$$

$$v_2(t) = \sin\left(\omega t + \frac{\pi}{2}\right) \qquad f = 1\,\text{Hz}$$

When making measurements with an oscilloscope, it is easier to find the peak-to-peak than the peak value. For a simple sine wave with no dc offset, it should be obvious that the peak-to-peak value is twice the peak value.

Sometimes people get confused when "average" values are used. For a sine wave with no dc offset, the average over one or more complete cycles is zero. If the wave is rectified, as shown in Figure A.4, there is an average value. Using calculus, it can be shown that for half-wave rectification the average is

$$V_{avg} \text{ (half-wave)} = \frac{V_p}{\pi} \tag{A.6}$$

where V_p = peak voltage

FIGURE A.3

Peak, Peak-to-Peak,
and RMS Values
of a Sinusoid

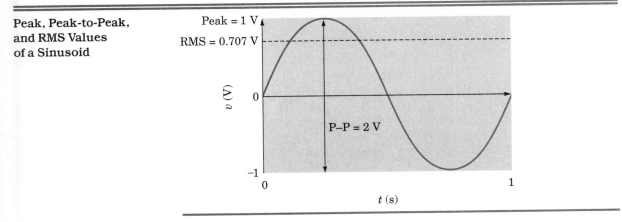

FIGURE A.4

Peak and Average
Values of a Rectified
Sinusoid

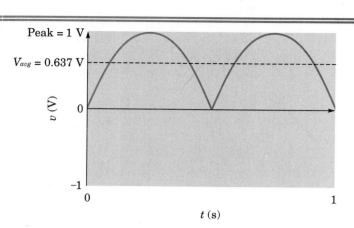

(a) Full-Wave Rectification

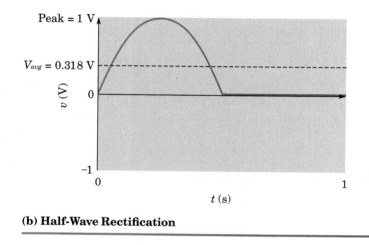

(b) Half-Wave Rectification

With full-wave rectification, it should be obvious that the average level will be twice that given above:

$$V_{avg} \text{ (full-wave)} = \frac{2V_p}{\pi} \tag{A.7}$$

A very common way of expressing the level of an ac waveform is the RMS (Root-Mean-Square) method. The idea behind RMS quantities is that the heating effect produced in a resistor by applying a given RMS voltage or current should be the same as that obtained by using a dc voltage or current with the same magnitude. For example, connecting a 120 V RMS ac supply to a toaster should produce toast just as quickly as using a dc voltage of 120 V.

The relationship between peak and RMS voltage or current depends on the waveshape. For any wave, the relationship can be found using calculus; the expression is

$$V_{RMS} = \sqrt{\frac{1}{T} \int_0^T v^2 \, dt} \tag{A.8}$$

For a sinusoidal wave, the relation is quite simple:

$$V_{RMS} = \frac{V_p}{\sqrt{2}} \tag{A.9}$$

It is quite important to remember, however, that Equation (A.9) is not valid for other waveshapes.

RMS values of voltage and current are very commonly used. In fact, in common usage an ac voltage or current level is assumed to be RMS unless otherwise stated.

Phasor Notation. Very often, ac calculations involve voltages or currents with the same frequency but different phase angles. Complex numbers can be used to simplify these calculations. A voltage can be expressed in phasor form as

$$\mathbf{V} = V \angle \theta \tag{A.10}$$

where $\mathbf{V}$ = phasor voltage
V = RMS magnitude
θ = phase angle in degrees

Current can be expressed in a similar way. Of course, these expressions can be converted to rectangular form whenever that method is more convenient. A voltage could be expressed as

$$\mathbf{V} = a + jb \tag{A.11}$$

where $a = V \cos \theta$
$b = V \sin \theta$
$j = \sqrt{-1}$

Similarly, a voltage given in rectangular form can be converted to polar (magnitude and phase) form by using the equations

$$V = \sqrt{a^2 + b^2} \tag{A.12}$$

$$\theta = \arctan \frac{b}{a} \tag{A.13}$$

Figure A.5 shows the relationship between the polar and rectangular representations.

FIGURE A.5

Phasor Representation
of a Signal

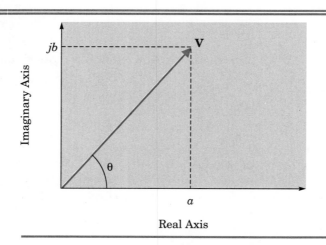

A.2 CAPACITANCE AND INDUCTANCE

A capacitor operates by storing electrical charge. There is no current through a capacitor, but when a capacitor is used with an ac signal, the continual charging and discharging can give the appearance of current flow through the device.

When an ac signal is applied to a capacitor, the current will lead the voltage by 90°. The magnitude of the current can be found by making use of the capacitive reactance X_C. This is given by the equation

$$X_C = \frac{1}{\omega C} \tag{A.14}$$

where X_C = capacitive reactance in ohms
       ω = frequency in radians per second
       C = capacitance in farads

A similar relation holds for inductors, except that here the voltage leads the current by 90°. The inductive reactance is given by

$$X_L = \omega L \tag{A.15}$$

where X_L = inductive reactance in ohms
       ω = frequency in radians per second
       L = inductance in henrys

A.2.1 Series and Parallel Connections

Inductors in series and parallel combine in the same way as resistances. For inductors in series, add the individual inductances. Inductors in parallel combine in the following way:

$$L_T = \cfrac{1}{\cfrac{1}{L_1} + \cfrac{1}{L_2} + \cfrac{1}{L_3} + \cdots} \qquad (A.16)$$

Capacitors in parallel add, just as do resistors or inductors in series. To find the total value of several capacitors in parallel, add the individual values. For capacitors in series, use the same method as for resistors or inductors in parallel; that is,

$$C_T = \cfrac{1}{\cfrac{1}{C_1} + \cfrac{1}{C_2} + \cfrac{1}{C_3} + \cdots} \qquad (A.17)$$

A.2.2 Complex Impedance

Although inductive and capacitive reactance are expressed in ohms as is resistance, their effects in a circuit are different. The use of complex impedance allows voltage, current, and impedance calculations to be made with resistance, inductance, and capacitance, alone or in any combination, using the same equations as for dc circuits, provided that the ac voltages and currents are expressed in phasor form.

The impedance of a capacitance is

$$\mathbf{Z}_C = -jX_C \qquad (A.18)$$
$$= X_C \angle -90°$$

and that of an inductance is

$$\mathbf{Z}_L = jX_L \qquad (A.19)$$
$$= X_L \angle 90°$$

Using this system, resistance is expressed as a real number.

The total impedance of a number of impedances connected together can be found in the same way as for resistive circuits, except that complex arithmetic must be used. For the series RLC combination shown in Figure A.6, for instance,

$$Z_T = R + jX_L - jX_C$$
$$= R + j\omega L - \frac{j}{\omega C}$$

A.2.3 Simple RC and RL Filters

Filters are designed to pass a frequency range called the *passband* while attenuating another range called the *stop-band*. They are characterized as low-pass, high-pass, bandpass, or band-reject, as illustrated in Figure A.7. An ideal filter, like those in the figure, has a sharp transition between stop-band

FIGURE A.6

Series *RLC* Circuit

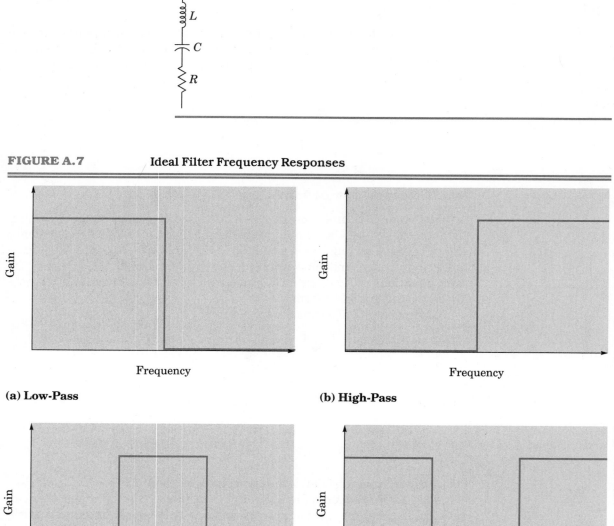

FIGURE A.7   **Ideal Filter Frequency Responses**

(a) Low-Pass

(b) High-Pass

(c) Bandpass

(d) Band-Reject

FIGURE A.8

First-Order *RC*
Low-Pass Filter

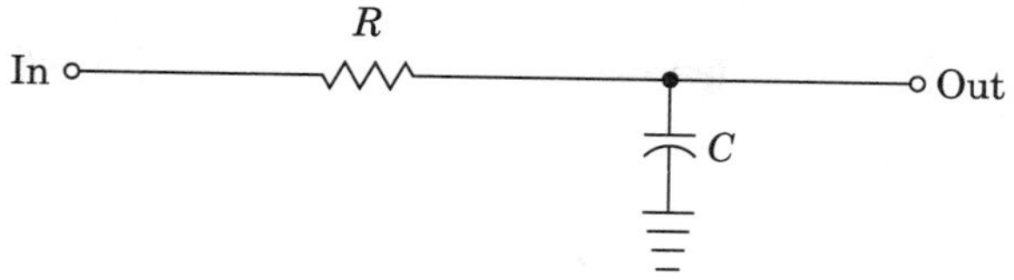

and passband at a frequency called the *cutoff frequency*. In a real filter, the transition is not perfectly sharp, and the cutoff frequency is the frequency at which the filter output is 3 dB below its maximum amplitude.

Figure A.8 shows a simple low-pass filter using a resistor and a capacitor. Using the voltage-divider rule from dc electricity, it can be seen that

$$
\frac{\mathbf{V}_o}{\mathbf{V}_i} = \frac{-jX_C}{R - jX_C}
$$

$$
= \frac{-j/(\omega C)}{R - j/(\omega C)}
$$

$$
= \frac{1}{1 + j\omega RC} \tag{A.20}
$$

The cutoff frequency ω_c is given by

$$
\omega_c = \frac{1}{RC} \tag{A.21}
$$

When Equation (A.21) is substituted into Equation (A.20), the latter becomes

$$
\frac{\mathbf{V}_o}{\mathbf{V}_i} = \frac{1}{1 + j\omega/\omega_c} \tag{A.22}
$$

This *transfer function* is a complex quantity whose magnitude will be

$$
\left|\frac{V_o}{V_i}\right| = \frac{1}{\sqrt{1 + (\omega/\omega_c)^2}} \tag{A.23}
$$

When $\omega = \omega_c$, the magnitude of the transfer function will be $1/\sqrt{2}$. The phase of the transfer function will be

$$
\phi = -\tan^{-1}\frac{\omega}{\omega_c} \tag{A.24}
$$

Figure A.9 shows some other simple *RC* and *RL* filters. All of these circuits are called *first-order filters* because their transfer functions contain only first-order terms. They are characterized by curves of voltage with respect to

First-Order Filters

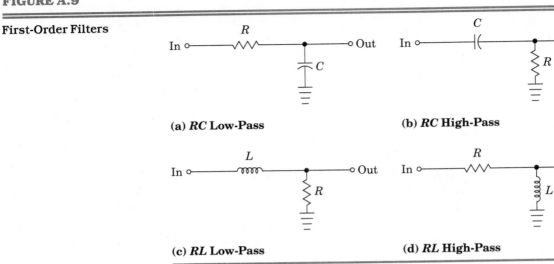

(a) *RC* Low-Pass

(b) *RC* High-Pass

(c) *RL* Low-Pass

(d) *RL* High-Pass

frequency that change at a rate of 20 dB per decade for frequencies far from cutoff. For the *RL* versions, the cutoff frequency is

$$\omega_c = \frac{R}{L} \qquad\qquad (A.25)$$

Figure A.10 shows the amplitude response for the circuits of Figure A.9. The graphs are a straight-line approximation known as a *Bode plot*. The actual response at the cutoff frequency is 3 dB below that in the passband.

FIGURE A.10   **Bode Plots for First-Order Filters**

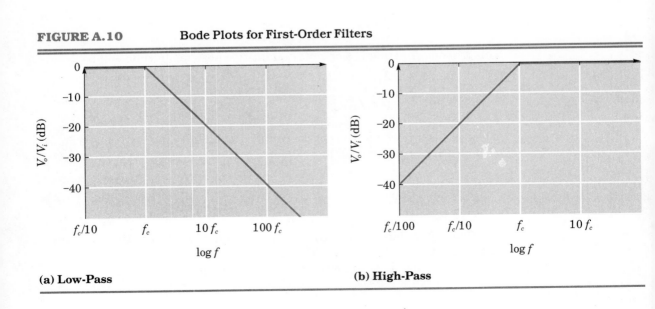

(a) **Low-Pass**

(b) **High-Pass**

A.3 *LC* RESONANT CIRCUITS

Consider the series *RLC* circuit of Figure A.6. Its impedance is

$$\mathbf{Z}_T = R + j\omega L - \frac{j}{\omega C}$$

There will be a frequency, called the *resonant frequency* ω_0, where $j\omega_0 L = j/(\omega_0 C)$ and $\mathbf{Z}_T = R$. The resonant frequency can be found by solving for ω_0.

$$j\omega_0 L = \frac{j}{\omega_0 C}$$

$$\omega_0 = \frac{1}{\sqrt{LC}} \tag{A.26}$$

At ω_0, the impedance will be minimum and resistive. At frequencies above ω_0, the circuit will be inductive, as $X_L > X_C$. Below resonance, $X_C > X_L$ and the circuit will be capacitive. Figure A.11 shows the variation of impedance with frequency. In Figure A.11, the horizontal scale is labeled f rather than ω, because frequency in practical situations is nearly always measured in hertz rather than in radians per second. Recalling from Equation (A.5) that

$$\omega = 2\pi f$$

and therefore

$$f = \frac{\omega}{2\pi}$$

it is easy to see that

$$f_0 = \frac{1}{2\pi\sqrt{LC}}$$

where f_0 = resonant frequency in hertz.

FIGURE A.11

Variation of Impedance in a Series-Resonant Circuit

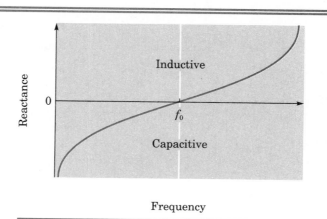

The resonant circuit will have a bandwidth, defined as the difference in frequency between the two half-power points. At these frequencies, the response will be 3 dB down from that at resonance. For a series-resonant circuit, the bandwidth will be

$$B = \frac{f_0}{Q}$$

where

$$Q = \frac{X_L}{R}$$

and X_L is measured at the resonant frequency.

Parallel RLC circuits like that of Figure A.12(a) can also be resonant. In this case, the impedance will be maximum and equal to R at resonance, which will occur at the same frequency as for the series circuit. Above resonance, the circuit will appear capacitive, and below resonance, inductive. In a practical circuit, most of the resistance is likely to be that of the coil, and a *parallel resonant* circuit will actually take the form of the series-parallel circuit of Figure A.12(b). For this circuit, it can be shown that

$$\omega_0 = \sqrt{\frac{1}{LC} - \left(\frac{R}{L}\right)^2}$$

For a circuit with Q greater than about 10, the resonant frequency of this circuit is approximately the same as for the series-resonant circuit. The impedance at resonance is no longer equal to the resistance of the coil, however. It is given by

$$Z_0 = R(Q^2 + 1) \tag{A.27}$$

where Z_0 = impedance of the circuit at resonance
R = coil resistance
Q = X_L/R, at the resonant frequency

FIGURE A.12

Parallel-Resonant Circuits

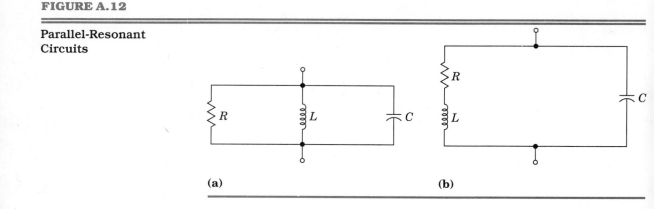

(a)   (b)

FIGURE A.13

Simple Filters Using
Resonant Circuits

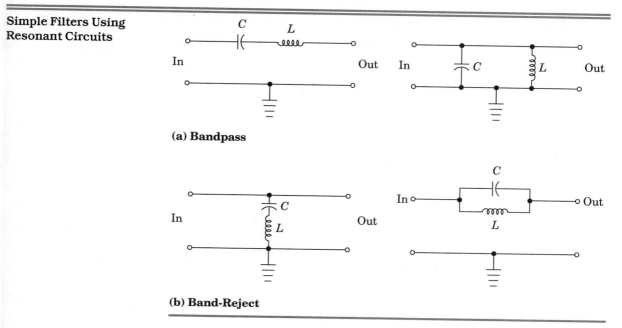

(a) Bandpass

(b) Band-Reject

Observe that Equation (A.27) indicates that the resonant impedance of the circuit increases as the coil resistance decreases.

If an external impedance loads the circuit, the Q will be reduced and the bandwidth of the filter will increase.

Figure A.13 shows how series- and parallel-resonant circuits can be used as simple bandpass and band-reject filters. More complex filters with greater attenuation and sharper transitions between passband and stopband can be created by combining the circuits shown here. The design of complex filters is a subject for a book in itself, and will not be covered here.

A.4 TRANSFORMERS

Transformers use magnetic coupling to transfer a signal from one winding to another or, in the case of autotransformers, within a single winding. The coefficient coupling k expresses the ratio of flux linking the secondary to flux linking the primary winding. An ideal transformer has complete coupling between windings; that is, $k = 1$. It also has no losses. For an ideal transformer, the power entering the primary must equal that leaving the secondary winding(s) and, for each secondary winding, the voltage is given by the simple equation

$$\frac{V_2}{V_1} = \frac{N_2}{N_1}$$

(A.28)

where V_1 = primary voltage
V_2 = secondary voltage

N_1 = number of turns on primary winding

N_2 = number of turns on secondary winding

Assuming there is only one secondary winding, the current ratio must be the reciprocal of that for voltage, since the power in each winding must be equal. Therefore,

$$\frac{I_2}{I_1} = \frac{N_1}{N_2}$$
(A.29)

where I_1 = primary current

I_2 = secondary current

Transformers are often used for impedance matching. By combining Equations (A.28) and (A.29), it is easy to show that the impedance looking into the primary is related to the impedance of the load connected to the secondary by

$$\frac{Z_1}{Z_2} = \frac{N_1^2}{N_2^2}$$
(A.30)

where Z_1 = impedance looking into the primary winding

Z_2 = impedance connected to the secondary winding

Real transformers deviate from the ideal. The magnetic coupling between primary and secondary windings is not complete, and there are losses in the windings. When iron or ferrite cores are used, there are losses in the core as well. The above equations are reasonably accurate at power-line and audio frequencies, but radio-frequency transformers often depart from the ideal.

APPENDIX
B
Fourier Series

In Chapter 1, the concept of the Fourier series was introduced. Some examples were shown using "cookbook" equations for simple waveforms. A table of formulas for other common signals will be found in Table B.1.

It is now time to consider what to do if presented with a waveform that is not in the table. One possibility is to find a longer list. They are certainly available, but you might wonder how these equations were arrived at in the first place.

Any periodic function has a Fourier series, but certain conditions must be satisfied for it to converge to the function. These conditions, known as the *Dirichlet conditions*, are as follows:

1. The function is defined and single-valued, except possibly at a finite number of points within one period.
2. The function is periodic.
3. The function and its first derivative are piecewise continuous.

Any electrical signal, for instance a voltage, will satisfy conditions 1 and 3. It would not be possible for an actual voltage to have two values at the same time, or an undefined value. Nor is a discontinuity, where a signal jumps in zero time from one voltage to another, actually possible. Electrical signals may have a very short rise time, but not zero!

The second condition must be viewed carefully. Many useful signals in electronics are nonperiodic. Some of them are close enough to being periodic that a Fourier series can be a useful approximation. Others are not. Consider, for instance, a single voltage pulse. This does not satisfy Dirichlet's conditions, and cannot be represented—even approximately—by a Fourier series. Such signals can still be represented in the frequency domain, however, by means of Fourier transforms, rather than series.

Note that the function does not have to be a function of time. Fourier series are more general than that. At the moment, however, we are only interested in functions of time.

If Dirichlet's conditions are satisfied, and assuming the function is a function of time t, the series will have the form given in Equation (1.5), which is repeated here for convenience:

$$f(t) = \frac{A_0}{2} + A_1 \cos \omega t + B_1 \sin \omega t + A_2 \cos 2\omega t$$

$$+ B_2 \sin 2\omega t + A_3 \cos 3\omega t + B_3 \sin 3\omega t + \cdots \quad \text{(B.1)}$$

where A_n, B_n = real-number coefficients (that is, they can be positive, negative, or zero)

ω = radian frequency of the fundamental

B-1

TABLE B.1   Fourier Series for Common Repetitive Waveforms

1. Half-Wave Rectified Sine Wave

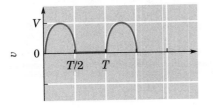

$$v(t) = \frac{V}{\pi} + \frac{V}{2} \sin \omega t - \frac{2V}{\pi} \left(\frac{\cos 2\omega t}{1 \times 3} + \frac{\cos 4\omega t}{3 \times 5} + \frac{\cos 6\omega t}{5 \times 7} + \cdots \right)$$

2. Full-Wave Rectified Sine Wave
(a) With time zero at voltage zero

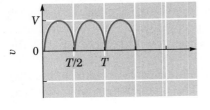

$$v(t) = \frac{2V}{\pi} - \frac{4V}{\pi} \left(\frac{\cos 2\omega t}{1 \times 3} + \frac{\cos 4\omega t}{3 \times 5} + \frac{\cos 6\omega t}{5 \times 7} + \cdots \right)$$

(b) With time zero at voltage peak

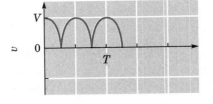

$$v(t) = \frac{2V}{\pi} \left(1 + \frac{2 \cos 2\omega t}{1 \times 3} - \frac{2 \cos 4\omega t}{3 \times 5} + \frac{2 \cos 6\omega t}{5 \times 7} - \cdots \right)$$

3. Square Wave
(a) Odd function

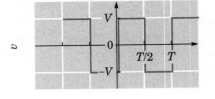

$$v(t) = \frac{4V}{\pi} \left(\sin \omega t + \frac{1}{3} \sin 3\omega t + \frac{1}{5} \sin 5\omega t + \cdots \right)$$

(b) Even function

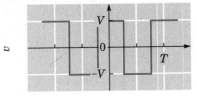

$$v(t) = \frac{4V}{\pi} \left(\cos \omega t - \frac{1}{3} \cos 3\omega t + \frac{1}{5} \cos 5\omega t - \cdots \right)$$

TABLE B.1 *Continued*

4. Pulse Train

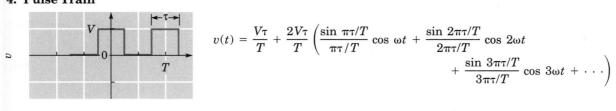

$$v(t) = \frac{V\tau}{T} + \frac{2V\tau}{T}\left(\frac{\sin \pi\tau/T}{\pi\tau/T} \cos \omega t + \frac{\sin 2\pi\tau/T}{2\pi\tau/T} \cos 2\omega t \right.$$
$$\left. + \frac{\sin 3\pi\tau/T}{3\pi\tau/T} \cos 3\omega t + \cdots\right)$$

5. Triangle Wave

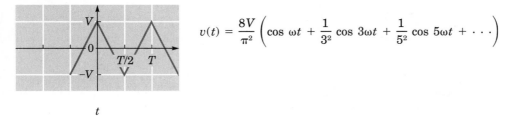

$$v(t) = \frac{8V}{\pi^2}\left(\cos \omega t + \frac{1}{3^2} \cos 3\omega t + \frac{1}{5^2} \cos 5\omega t + \cdots\right)$$

6. Sawtooth Wave
(a) With no dc offset

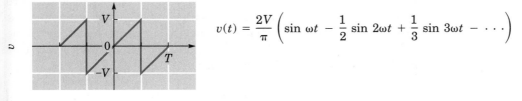

$$v(t) = \frac{2V}{\pi}\left(\sin \omega t - \frac{1}{2} \sin 2\omega t + \frac{1}{3} \sin 3\omega t - \cdots\right)$$

(b) Positive-going

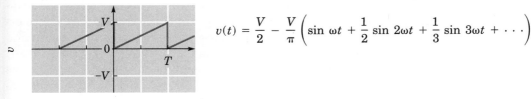

$$v(t) = \frac{V}{2} - \frac{V}{\pi}\left(\sin \omega t + \frac{1}{2} \sin 2\omega t + \frac{1}{3} \sin 3\omega t + \cdots\right)$$

There is a fairly simple set of equations, involving integral calculus, for finding the coefficients:

$$A_n = \frac{2}{T} \int_0^T f(t) \cos \frac{2n\pi t}{T} \, dt \tag{B.2}$$

$$B_n = \frac{2}{T} \int_0^T f(t) \sin \frac{2n\pi t}{T} \, dt \tag{B.3}$$

where T = period of $f(t)$.

Notice that there is an A_0 term, which will be a constant equal to twice the average value of the function. This can be seen from Equation (B.2). When $n = 0$, this equation contains cos 0, which is equal to 1. Therefore, the integral reduces to:

$$A_0 = \frac{2}{T} \int_0^T f(t) \, dt \tag{B.4}$$

A_0 will sometimes be zero, of course. There is no B_0 term, because for $n = 0$, the integral would be that of the function multiplied by sin 0, which is zero.

For arbitrary signals it may not be possible to perform the above integrals analytically. In that case, numerical integration may be used.

B.1 ODD AND EVEN FUNCTIONS

In the examples in Chapter 1, we found that many of the coefficients were zero. If we can predict in advance which coefficients are zero, we can save a lot of work in solving the above integrals. There are several rules that can be used for this. They will be presented with explanations, not proofs. The proofs can be found in any standard text on Fourier analysis.

1. A function $f(t)$ is called an **odd** function if $f(-t) = -f(t)$. Odd functions contain only sine terms. A simple example of an odd function is a sine wave. Figure B.1 shows some other examples. Since such functions are symmetrical about the x axis, they have an average value of zero. Therefore, they cannot have a constant term.

   Sometimes students are bothered by the concept of negative time. This will not pose a problem if you remember that time zero is an arbitrary point that can be set anywhere in time. All other times are referenced to it. Negative time is simply time before that reference point, and is no more mysterious than negative voltage.

2. A function $f(t)$ is called an **even** function if $f(-t) = f(t)$. A cosine wave is an even function. Such functions have cosine terms in the Fourier series, and sometimes a constant term, but they do not have sine terms. Figure B.2 has examples of even functions with and without constant terms. It is generally easy to see which ones have such a term by noting whether the average amplitude over a complete cycle is zero. It is for the cosine wave, for instance, but not for the full- or half-wave rectified signals.

FIGURE B.1   Odd Functions: $f(-t) = -f(t)$

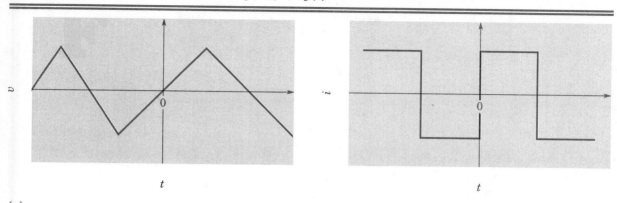

(a)   **(b)**

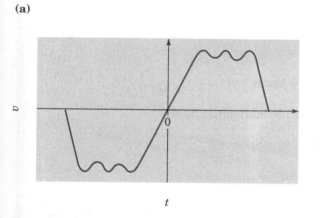

(c)

FIGURE B.2   Even Functions: $f(-t) = f(t)$

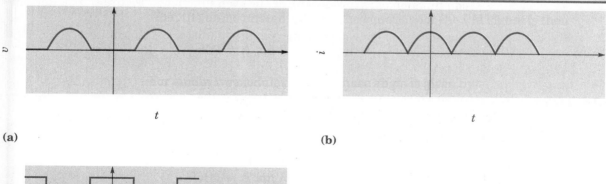

(a)   **(b)**

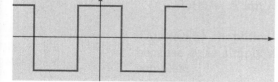

(c)

FIGURE B.3   **Changing Functions to Odd or Even**

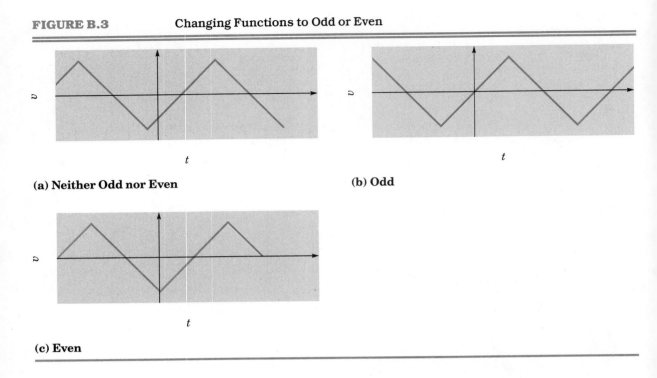

(a) Neither Odd nor Even   **(b) Odd**

(c) Even

It may seem that this division of functions is rather arbitrary. After all, we can turn a sine wave into a cosine wave by shifting it 90°, or by redefining the time reference. This is true, and just as the time-domain representation changes when we do this, so does that in the frequency domain. It is certainly allowed to change the zero reference to that which makes a particular problem easiest to solve. For instance, the waveform in Figure B.3(a) is neither odd nor even. By simply moving the zero reference, we can convert it into the odd function of Figure B.3(b), or the even function of Figure B.3(c). Remember, however, that zero time can be set only once for a given problem. If two signals add at a given point in a circuit, for instance, they must both have the same zero reference.

EXAMPLE B.1

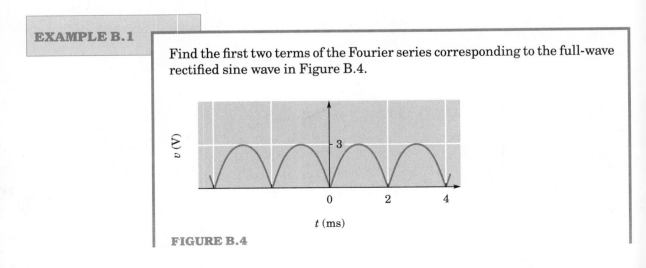

Find the first two terms of the Fourier series corresponding to the full-wave rectified sine wave in Figure B.4.

FIGURE B.4

Solution

This wave has a period of 2 ms. As shown, this wave is an even function of time. Over its period, the equation of this wave is that of a sine wave, but the period of the sine wave is twice that of the function. This is easy to see when you consider that this 500 Hz rectified waveform could be produced by full-wave rectification of a 250 Hz sine wave.

The equation in the time domain can be found from the general equation of a sine wave:

$$v(t) = A \sin \omega t \tag{B.5}$$

where v = instantaneous voltage in volts
A = peak amplitude in volts
ω = frequency in radians per second
t = time in seconds

In this case, the sine wave has a frequency of 250 Hz, as explained above.

$$A = 3 \text{ V} \qquad \omega = 2\pi f = 2\pi \times 250 = 500\pi \text{ rad/s}$$

We can then write the equation over the range of time t from 0 to 2 ms as:

$$v = 3 \sin (500\pi t)$$

Since the function is even, it will not have any sine-wave components in the frequency domain. Its average voltage is obviously greater than zero, so it will have a constant component. This can be found from Equation (B.4):

$$A_0 = \frac{2}{T} \int_0^T f(t) \, dt$$

$$= \frac{2}{T} \int_0^T A \sin \omega t \, dt$$

$$= \frac{2A}{T} \int_0^T \sin \omega t \, dt$$

$$= \frac{2A}{T} \left(-\frac{1}{\omega} \cos \omega t \right)_0^T$$

$$= \frac{2A}{\omega T} (-\cos \omega t)_0^T$$

$$= \frac{2 \times 3}{500\pi \times 2 \times 10^{-3}} [-\cos 500\pi(2 \times 10^{-3}) + \cos 0]$$

$$= \frac{6}{\pi} (-\cos \pi + \cos 0)$$

$$= \frac{6}{\pi} (1 + 1)$$

$$= \frac{12}{\pi}$$

Remembering that the dc component is $A_0/2$, we find that this dc component has a voltage of:

$$\frac{A_0}{2} = \frac{12}{2\pi} = \frac{6}{\pi} = 1.91 \text{ V}$$

This may remind you of the average-voltage calculation for an unfiltered full-wave power supply, which is usually given as $0.637V_p$. The exact value is $2V_p/\pi$, which agrees with the above result.

By a similar process, all the other components can be found by integration. It can be done more quickly by doing the integration for all components at once:

$$A_n = \frac{2}{T} \int_0^T f(t) \cos \frac{2n\pi t}{T} \, dt \tag{B.6}$$

$$= \frac{2}{T} \int_0^T 3 \sin 500\pi t \cos \frac{2n\pi t}{T} \, dt$$

$$= \frac{6}{T} \int_0^T \sin 500\pi t \cos \frac{2n\pi t}{T} \, dt$$

This integral is of the form

$$\int \sin mx \cos nx \, dx$$

whose solution can be found in any table of integrals, as

$$\int \sin mx \cos nx \, dx = -\frac{\cos (m - n)x}{2(m - n)} - \frac{\cos (m + n)x}{2(m + n)}$$

So, setting $T = 2$ ms,

$$A_n = \frac{6}{0.002} \left[\frac{-\cos (500\pi - 2n\pi/0.002)t}{2(500\pi - 2n\pi/0.002)} \right.$$

$$\left. - \frac{\cos (500\pi + 2n\pi/0.002)t}{2(500\pi + 2n\pi/0.002)} \right]_0^{0.002}$$

$$= 3000 \left[\frac{-\cos (500\pi - 1000n\pi)t}{2(500\pi - 2n\pi/0.002)} - \frac{\cos (500\pi + 1000n\pi)t}{2(500\pi + 2n\pi/0.002)} \right]_0^{0.002}$$

$$= 3000 \left[\frac{-\cos (500\pi - 1000n\pi)0.002}{2(500\pi - 2n\pi/0.002)} - \frac{\cos (500\pi + 1000n\pi)0.002}{2(500\pi + 2n\pi/0.002)} \right.$$

$$\left. + \frac{\cos 0}{2(500\pi - 2n\pi/0.002)} + \frac{\cos 0}{2(500\pi + 2n\pi/0.002)} \right]$$

$$= 3000 \left[\frac{-\cos (\pi - 2n\pi)}{1000(\pi - 2n\pi)} - \frac{\cos (\pi + 2n\pi)}{1000(\pi + 2n\pi)} \right.$$

$$\left. + \frac{1}{1000(\pi - 2n\pi)} - \frac{1}{1000(\pi + 2n\pi)} \right]$$

For $n = 1$,

$$A_1 = 3000\left(\frac{1}{-1000\pi} + \frac{1}{3000\pi} + \frac{1}{-1000\pi} + \frac{1}{3000\pi}\right)$$

$$= 3000\left(\frac{-3 + 1 - 3 + 1}{3000\pi}\right)$$

$$= \frac{-4}{\pi}$$

$$= -1.27 \text{ V}$$

The other coefficients can be found in the same way.

The result agrees with the formula, but getting it was a considerable amount of work. In practice, Fourier series (and transforms) are usually either looked up in a table, or calculated numerically using a computer.

APPENDIX C

Tables of Modulation Types

These designations are used as abbreviations for the various modulation schemes and are often found in regulations, etc. They are often preceded by a number which represents the bandwidth in kilohertz.

Amplitude Modulation

Designation	Description
A0	Unmodulated carrier.
A1	Telegraphy by switched carrier. Also commonly referred to as on-off keying (OOK) or continuous waves (CW). Usually employs Morse code.
A2	Telegraphy by switched modulating tone. The carrier is on continuously during a transmission, and an audio modulating tone is switched on and off to convey the information. Sometimes called *modulated continuous waves* (MCW). Usually employs Morse code.
A3	Telephony by double-sideband full-carrier (DSBFC) AM. AM broadcasting is an example of this method.
A3A	Telephony by single-sideband reduced-carrier (SSBRC) AM.
A3B	Telephony by independent-sideband (ISB) AM.
A3H	Telephony by single-sideband full-carrier (SSBFC) AM.
A3J	Telephony by single-sideband suppressed-carrier (SSBSC) AM.
A3Y	Digital voice modulation using AM.
A4	Facsimile or slow-scan television using AM.
A5C	Television using vestigial sideband (VSB) AM. The picture portion of ordinary television broadcasting is an example.
A9B	Telephony or telegraphy with independent sidebands.
A9Y	Nonvoice digital modulation.

Frequency Modulation

Designation	Description
F1	Telegraphy using frequency-shift-keying (FSK) of the carrier. This is the method commonly employed for radioteletype (RTTY) transmissions on HF radio.
F2	Telegraphy using on-off keying of a tone that frequency-modulates the carrier. The carrier is sent continuously during the transmission.
F3	Telephony using frequency or phase modulation.
F3Y	Digital voice modulation.
F9Y	Nonvoice digital modulation.
F4	Facsimile using FM.
F5	Television using FM. Satellite TV uses this method.
F6	Telegraphy using a four-frequency diplex system.

Radio-Frequency Spectrum

Frequency Band	Abbrev.	Frequency	Wavelength	Typical Uses
Very Low	VLF	3 to 30 kHz	10 to 100 km	Submarine communications
Low	LF	30 to 300 kHz	1 to 10 km	Navigation (LORAN) Standard time/ frequency
Medium	MF	300 kHz to 3 MHz	100 m to 1 km	AM broadcast Marine safety
High	HF	3 to 30 MHz	10 to 100 m	SW broadcast Amateur radio CB radio Military News services Marine Aircraft Standard time/ frequency
Very High	VHF	30 to 300 MHz	1 to 10 m	FM radio broadcast TV broadcast Utilities Aircraft Amateur
Ultra High	UHF	300 MHz to 3 GHz	10 cm to 1 m	TV broadcast Cellular phone Utilities Amateur
Super High	SHF	3 to 30 GHz	1 to 10 cm	Satellite TV Radar Point-to-point microwave
Extremely High	EHF	30 to 300 GHz	1 to 10 mm	

Note: The bands above about 1 GHz are also called *microwaves*. Frequencies above about 20 GHz are also referred to as *millimeter waves*.

E

Decibels

The decibel is a logarithmic way of expressing the ratio of two power levels or, sometimes, voltage levels. Decibels are used in almost every part of electronic communications. Since the decibel is based on logarithms, a brief review of logarithms is in order before beginning to discuss it.

E.1 REVIEW OF LOGARITHMS

The *logarithm* (log) of a number to a given base is the power to which the base must be raised to give the number. Although any number can be used as the base, the base 10 is used in decibel calculations. Logarithms to the base 10 are called *common logarithms* and are abbreviated $\log_{10}$ or just log. For example, the log of 100 to the base 10 is 2, because 10 raised to the power of 2 equals 100. Expressed in mathematical notation,

$$\log_{10} 100 = 2$$

Similarly,

$$\log_{10} 0.01 = -2$$

Negative numbers do not have logs, because no matter what power 10 is raised to, the result is always positive.

The *antilog* of a number to a given base is simply the base raised to that number. For example, using base 10,

$$\text{antilog } 2 = 10^2$$
$$= 100$$

The antilog is the inverse operation to the log of a number; that is, the antilog of the log of a number is the number itself. The antilog of x can also be written as inverse log x, $\log^{-1} x$, or 10^x (assuming the base 10 is being used).

E.1.1 Simple Operations with Logarithms

The following results can easily be proved by going back to the definition of a logarithm. It is assumed that all logs are to the same base.

$$\log ab = \log a + \log b \qquad \text{(E.1)}$$

E-2 Appendix E Decibels

$$\log \frac{a}{b} = \log a - \log b \tag{E.2}$$

$$\log a^b = b \log a \tag{E.3}$$

E.2 DECIBELS

The decibel, in its simplest form, expresses the ratio of two power levels, logarithmically. If P_1 and P_2 are two power levels, P_2 can be said to be greater than P_1 by a number of decibels given by

$$\frac{P_2}{P_1} \text{ (dB)} = 10 \log \frac{P_2}{P_1} \tag{E.4}$$

From now on, to reduce clutter in the equations, the base 10 will be assumed. Note that, if P_1 happens to be greater than P_2, the result will be negative.

EXAMPLE E.1

Find the ratio between P_2 and P_1, in decibels, if

(a) $P_1 = 2$ W, $P_2 = 3$ W   (b) $P_1 = 3$ W, $P_2 = 2$ W

Solution

(a) $\dfrac{P_2}{P_1}$ (dB) $= 10 \log \dfrac{P_2}{P_1}$

$\qquad\qquad = 10 \log \dfrac{3}{2}$

$\qquad\qquad = 1.76$ dB

(b) $\dfrac{P_2}{P_1}$ (dB) $= 10 \log \dfrac{P_2}{P_1}$

$\qquad\qquad = 10 \log \dfrac{2}{3}$

$\qquad\qquad = -1.76$ dB

E.2.1 Decibel Gain and Loss

If P_o is the output power of a device, and P_i is the input power, then the power gain in decibels is

$$A_P \text{ (dB)} = 10 \log \frac{P_o}{P_i} \tag{E.5}$$

A negative result simply means that the output power is less than the input and there is a power loss in the device. In decibels, a gain of $-x$ dB is equivalent to a loss of x dB.

EXAMPLE E.2

An amplifier has an input of 100 mW and an output of 4 W. Find its gain in decibels.

Solution

$$A_P \text{ (dB)} = 10 \log \frac{P_o}{P_i}$$

$$= 10 \log \frac{4 \text{ W}}{0.1 \text{ W}}$$

$$= 16 \text{ dB}$$

EXAMPLE E.3

An attenuator has a loss of 26 dB. If a power of 2 W is applied to the attenuator, find the output power.

Solution

Equation (E.5) can be rearranged quite easily to solve this.

$$A_P \text{ (dB)} = 10 \log \frac{P_o}{P_i}$$

$$\frac{P_o}{P_i} = \text{antilog} \frac{A_P \text{ (dB)}}{10}$$

$$P_o = P_i \text{ antilog} \frac{A_P \text{ (dB)}}{10}$$

Since the 26 dB is given as a loss, its sign must be changed before it can be used in this equation.

$$P_2 = 2 \text{ W antilog} \frac{-26}{10}$$

$$= 5.02 \times 10^{-3} \text{ W}$$

$$= 5.02 \text{ mW}$$

E.2.2 Use of Reference Power Levels: dBm, dBW, dBf

The decibel expresses the ratio between two power levels, but there is no requirement for both of these two signals to exist physically. For instance, we could ask, by how many decibels is the power in the circuit greater than one milliwatt? This does not imply that we actually have a power of 1 mW somewhere in the circuit. Power levels expressed in this way are said to be in **dBm**.

$$P \text{ (dBm)} = 10 \log \frac{P}{1 \text{ mW}} \qquad\qquad (E.6)$$

Power is often measured in dBm in both radio- and audio-frequency applications.

EXAMPLE E.4

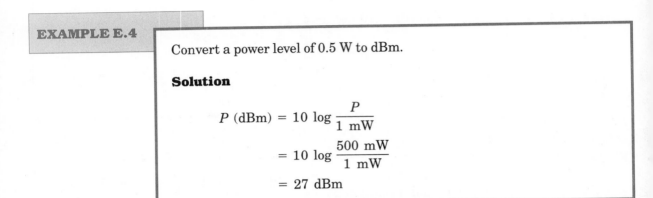

Convert a power level of 0.5 W to dBm.

Solution

$$P \text{ (dBm)} = 10 \log \frac{P}{1 \text{ mW}}$$

$$= 10 \log \frac{500 \text{ mW}}{1 \text{ mW}}$$

$$= 27 \text{ dBm}$$

Note that the ratio in Equation (E.6) must be dimensionless; that is, a power level given in some unit other than milliwatts must be converted to milliwatts before the log is found.

Other reference power levels can be used whenever it is more convenient. Two common references are the femtowatt and the watt.

$$P \text{ (dBf)} = 10 \log \frac{P}{1 \text{ fW}} \qquad\qquad (E.7)$$

$$P \text{ (dBW)} = 10 \log \frac{P}{1 \text{ W}} \qquad\qquad (E.8)$$

The femtowatt reference is useful where power levels are very low, as in the calculation of receiver sensitivity. For relatively large power levels, such as the output power of transmitters, the watt is a more useful reference.

E.2.3 Operations with Decibels

These follow directly from the properties of logarithms and the fact that decibels are logarithmic.

1. Decibel gains add.

FIGURE E.1

Decibel Gains and Losses

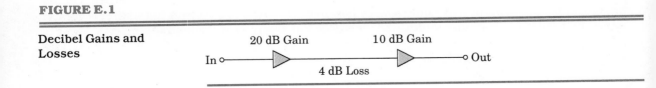

Example: An amplifier with 20 dB gain is connected to another with 10 dB gain by means of a transmission line with a loss of 4 dB. (See Figure E.1.) The total system gain is

$$A_P \text{ (dB)} = 20 - 4 + 10$$
$$= 26 \text{ dB}$$

Note that a loss of 4 dB is equivalent to a gain of -4 dB, as explained above.

2. Gains in decibels can be added directly to powers expressed in dBm, dBf, dBW, etc., giving an output in the same unit as the input.

Example: If a signal with a power level of -12 dBm were applied to the system of Figure E.1, the output would be

$$P_o = -12 \text{ dBm} + 26 \text{ dB}$$
$$= 14 \text{ dBm}$$

It may seem that we are adding different quantities in the above example, but in fact we are not. Both the quantities are actually the logarithms of power ratios, and thus dimensionless. Decibels and dBm are not units like amperes or volts: the "dB" indicates the operation that has been performed on a ratio, and the "m" keeps track of a reference level.

3. To change reference levels, express the old reference level in terms of the new one, then add this amount to every value.

Example: To convert dBW to dBm, we note that

$$0 \text{ dBW} = 1 \text{ W} = 30 \text{ dBm}$$

so any power given in dBW can be converted to dBm by adding 30 dB. For example, 2 dBW = 32 dBm, and -10 dBW = 20 dBm.

E.2.4   Decibels and Voltages

Until now, we have dealt only with power ratios. If signal voltages are compared instead, the ratio of the voltages will be the square of the power ratio, provided that the voltages are developed across identical impedances. This is very often the case in radio-frequency amplifiers like that shown in Figure E.2. In this case, both the input impedance and the load impedance are 50 Ω, so

$$\frac{P_o}{P_i} = \left(\frac{V_o}{V_i}\right)^2$$

Using the fact that

$$\log a^b = b \log a$$

FIGURE E.2

RF Amplifier

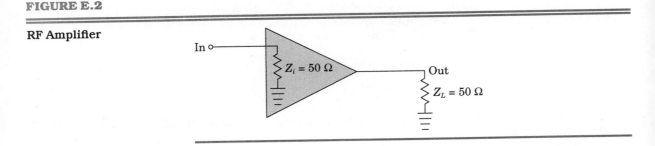

it is easy to show that

$$A_P \text{ (dB)} = 20 \log \frac{V_o}{V_i}$$

Please note carefully that this is only true when both voltages are measured across the same impedance. It would not be true for a typical audio power amplifier, for instance. Here the input impedance Z_i might be 10 kΩ, while the load resistance Z_L is commonly about 8 Ω. It is possible to define a decibel voltage gain as follows:

$$A_V \text{ (dB)} = 20 \log \frac{V_o}{V_i} \tag{E.9}$$

A_V (dB) and A_P (dB) will be equal only when $Z_i = Z_L$. In practice, this seldom causes problems, because it is usually clear from the context which type of decibel gain is being referred to.

Just as for power, it is possible to use decibels to compare a voltage with a reference level. For example, the term dBmV is used in cable-television systems. It is a measure of a voltage level compared to 1 mV; that is,

$$V \text{ (dBmV)} = 20 \log \frac{V}{1 \text{ mV}} \tag{E.10}$$

EXAMPLE E.5

A signal in a cable-television system has an amplitude of 3 mV in 75 Ω (resistive). Calculate its level in

(a) dBmV   (b) dBm

Solution

(a) From Equation (E.10),

$$V \text{ (dBmV)} = 20 \log \frac{V}{1 \text{ mV}}$$

$$= 20 \log \frac{3 \text{ mV}}{1 \text{ mV}}$$

$$= 9.54 \text{ dBmV}$$

(b) Since dBm is a measure of power, we first find the power in the signal.

$$P = \frac{V^2}{R}$$

$$= \frac{(3 \times 10^{-3} \text{ V})^2}{75 \ \Omega}$$

$$= 120 \times 10^{-9} \text{ W}$$

Now, Equation (E.6) can be used to calculate the power in dBm:

$$P \text{ (dBm)} = 10 \log \frac{P}{1 \text{ mW}}$$

$$= 10 \log \left(\frac{120 \times 10^{-9} \text{ W}}{1 \times 10^{-3} \text{ W}} \right)$$

$$= -39.2 \text{ dBm}$$

E.2.5 Decibel Measurements

Since decibels simply express power or voltage ratios in a logarithmic way, it is not necessary to have specialized equipment to measure decibels. Power and voltage levels can be measured in any of the usual ways, and then decibel quantities can be calculated as explained above. Many meters, however, have scales calibrated directly in decibels. These include, among others, output-level meters on RF generators, audio voltmeters, and, sometimes, the ac ranges on multimeters.

As always, of course, the decibel measurements provided by these meters must be with respect to some reference, either stated or implied. A typical audio voltmeter, for instance, will have two scales, one labeled dBV and the other dBm. The dBV scale is self-explanatory, but one might ask how a volt-meter measures power in order to find a level in dBm. It does not actually measure power, of course: it measures voltage and calculates power, based on an assumption about the impedance across which the voltage appears. The most common impedance value used for audio voltmeters is 600 Ω, but some multimeters allow this to be changed. RF meters, on the other hand, use 50 Ω, or sometimes 75 Ω, as the reference impedance.

Relationship Between Frequency Modulation and Phase Modulation

As mentioned in Chapter 8, either FM or PM will result in changes in both the frequency and phase of the modulated waveform. It was also pointed out that frequency (in radians per second) is the derivative of phase (in radians). This leads to a relatively simply relationship between FM and PM that can make it easier to understand both.

Any FM or PM signal can be described by an equation of the form

$$v(t) = A \sin \theta(t) \tag{F.1}$$

For an unmodulated carrier, assuming an angle of zero at $t = 0$,

$$\theta(t) = \omega_c t$$

and we have the well-known sine-wave equation

$$v(t) = A \sin \omega_c t$$

If the signal is phase-modulated,

$$\theta(t) = \omega_c t + \phi(t)$$

because the phase deviation simply adds on to the angle that would result from the unmodulated carrier. The equation for the PM signal is then

$$v(t) = A \sin [\omega_c t + \phi(t)] \tag{F.2}$$

For sine-wave modulation, the phase angle is given by

$$\phi(t) = \phi_c + m_p \sin \omega_m t \tag{F.3}$$

We can combine this equation with Equation (F.2) to get

$$v(t) = A \sin (\omega_c t + \phi_c + m_p \sin \omega_m t)$$

This can be simplified a little by letting ϕ_c be equal to zero. That is allowed because ϕ_c is simply a reference phase angle. The result is still relatively complex:

$$v(t) = A \sin (\omega_c t + m_p \sin \omega_m t) \tag{F.4}$$

Now let us consider FM. In order to fit this modulation scheme into the form of Equation (F.2), it is necessary to have an expression for the phase angle of an FM signal as a function of time. Since radian frequency is the derivative of phase angle,

$$\omega(t) = \frac{d}{dt}\,\theta(t) \tag{F.5}$$

The frequency of an FM signal with sine-wave modulation as a function of time is

$$f_{sig}(t) = f_c + m_f f_m \sin \omega_m t$$

Of course, this frequency is in hertz, but we can convert to radians per second by simply multiplying by 2π.

$$\begin{aligned}
\omega(t) &= 2\pi f_{sig}(t) \\
&= 2\pi f_c + 2\pi m_f f_m \sin \omega_m t
\end{aligned} \tag{F.6}$$

Combining Equation (F.5) with Equation (F.6), we get

$$\frac{d}{dt}\,\theta(t) = 2\pi f_c + 2\pi m_f f_m \sin \omega_m t \tag{F.7}$$

This is a differential equation, but a simple one. The two terms on the right can be integrated separately, giving

$$\theta(t) = \int 2\pi f_c\, dt + \int 2\pi m_f f_m \sin \omega_m t\, dt \tag{F.8}$$

Since f_c is a constant, the integral of the first term in Equation (F.8) is easily found.

$$\int 2\pi f_c\, dt = 2\pi f_c t + C_1 \tag{F.9}$$

C_1 is a constant of integration, which will be dealt with later. As for the second term in Equation (F.8), everything there is a constant except for $\sin \omega_m t$, which is easy to integrate.

$$\int 2\pi m_f f_m \sin \omega_m t\, dt = \frac{2\pi m_f f_m \cos \omega_m t}{\omega_m} + C_2 \tag{F.10}$$

C_2 is another integration constant which will be taken care of later. Remembering that

$$\omega_m = 2\pi f_m$$

allows us to simplify Equation (F.10) to

$$\int 2\pi m_f f_m \sin \omega_m t\, dt = m_f \cos \omega_m t + C_2 \tag{F.11}$$

Now we can combine Equations (F.9) and (F.11) to get

$$\theta(t) = 2\pi f_c t + C_1 + m_f \cos \omega_m t + C_2$$

C_1 and C_2 can be combined into a single constant, which can be set to zero if we simply decide that the signal has an angle of 0 at time zero. Now we have

$$\theta(t) = 2\pi f_c t + m_f \cos \omega_m t$$

A further slight simplification is possible by noticing that the first term is just the phase angle of the carrier, which we earlier called θ_c.

$$\theta(t) = \theta_c + m_f \cos \omega_m t \qquad\qquad\qquad\text{(F.12)}$$

This equation is worth the effort expended to derive it. A comparison with Equation (F.3), given earlier for PM:

$$\theta(t) = \theta_c + m_p \sin \omega_m t$$

shows a remarkable similarity: m_f, like m_p, is the peak phase deviation of the signal (in radians). The difference between them is that m_f varies with the modulating frequency, while m_p does not. Recall the two definitions given in Chapter 8.

$$m_f = \frac{\delta}{f_m} \qquad \text{and} \qquad m_p = \phi_{max}$$

The fact that phase and frequency modulation are so close mathematically allows the same functions to be used to find bandwidth and power relationships, and is also of practical use in the design of transmitters and receivers. FM is easily converted to PM and vice versa as the need arises.

G

CB Channel Frequencies

Channel	Frequency (MHz)	Channel	Frequency (MHz)
1	26.965	21	27.215
2	26.975	22	27.225
3	26.985	23	27.255
4	27.005	24	27.235
5	27.015	25	27.245
6	27.025	26	27.265
7	27.035	27	27.275
8	27.055	28	27.285
9	27.065	29	27.295
10	27.075	30	27.305
11	27.085	31	27.315
12	27.105	32	27.325
13	27.115	33	27.335
14	27.125	34	27.345
15	27.135	35	27.355
16	27.155	36	27.365
17	27.165	37	27.375
18	27.175	38	27.385
19	27.185	39	27.395
20	27.205	40	27.405

APPENDIX

H

Television Channel Frequencies

VHF: Low Band

Channel	Frequency (MHz)
2	54–60
3	60–66
4	66–72
5	76–82
6	82–88

VHF: High Band

Channel	Frequency (MHz)
7	174–180
8	180–186
9	186–192
10	192–198
11	198–204
12	204–210
13	210–216

UHF

Channel	Frequency (MHz)	Channel	Frequency (MHz)	Channel	Frequency (MHz)
14	470–476	34	590–596	54	710–716
15	476–482	35	596–602	55	716–722
16	482–488	36	602–608	56	722–728
17	488–494	37	608–614	57	728–734
18	494–500	38	614–620	58	734–740
19	500–506	39	620–626	59	740–746
20	506–512	40	626–632	60	746–752
21	512–518	41	632–638	61	752–758
22	518–524	42	638–644	62	758–764
23	524–530	43	644–650	63	764–770
24	530–536	44	650–656	64	770–776
25	536–542	45	656–662	65	776–782
26	542–548	46	662–668	66	782–788
27	548–554	47	668–674	67	788–794
28	554–560	48	674–680	68	794–800
29	560–566	49	680–686	69	800–806
30	566–572	50	686–692	70	806–812
31	572–578	51	692–698	71	812–818
32	578–584	52	698–704	72	818–824
33	584–590	53	704–710		

Cable-Television Frequencies

Note: There are several schemes in use with different designations. The list below includes the most popular variations. Channels are sometimes designated by numbers and sometimes by letters, and there is more than one system for each! Not all systems carry all these channels. In addition to the channels listed, cable-television systems also use the normal VHF channels 2 through 13, at the same frequencies as for broadcast television. Some systems also have a *hyperband* consisting of channels above 64, at 6 MHz intervals.

Midband Channels (Between Broadcast Channels 6 and 7)	Letter	Number	Frequency Range (MHz)	Letter	Number	Frequency Range (MHz)
	A-5	95	90–96 *	C	16	132–138
	A-4	96	96–102 *	D	17	138–144
	A-3	97	102–108 *	E	18	144–150
	A-2	98	108–114	F	19	150–156
	A-1	99	114–120	G	20	156–162
	A	14	120–126	H	21	162–168
	B	15	126–132	I	22	168–174

*These channels cannot be used when the cable carries FM broadcasting.

Superband Channels (Above Broadcast Channel 13)	Letter	Number	Frequency Range (MHz)	Letter	Number	Frequency Range (MHz)
	J	23	216–222	Q	30	258–264
	K	24	222–228	R	31	264–270
	L	25	228–234	S	32	270–276
	M	26	234–240	T	33	276–282
	N	27	240–246	U	34	282–288
	O	28	246–252	V	35	288–294
	P	29	252–258	W	36	294–300

Hyperband Channels

Letter	Letter + Number	Number	Frequency Range (MHz)
AA	W+1	37	300–306
BB	W+2	38	306–312
CC	W+3	39	312–318
DD	W+4	40	318–324
EE	W+5	41	324–330
FF	W+6	42	330–336
GG	W+7	43	336–342
HH	W+8	44	342–348
II	W+9	45	348–354
JJ	W+10	46	354–360
KK	W+11	47	360–366
LL	W+12	48	366–372
MM	W+13	49	372–378
NN	W+14	50	378–384
OO	W+15	51	384–390
PP	W+16	52	390–396
QQ	W+17	53	396–402
RR	W+18	54	402–408
SS	W+19	55	408–414
TT	W+20	56	414–420
UU	W+21	57	420–426
VV	W+22	58	426–432
WW	W+23	59	432–438
XX	W+24	60	438–444
YY	W+25	61	444–450
	W+26	62	450–456
	W+27	63	456–462
	W+28	64	462–468

Waveguide Table

The following table consists of selected rectangular waveguides. Dimensions are internal. The useful ranges and cutoff frequencies (f_c) given are for single-mode propagation using the TE_{10} mode.

| U.S. RG-Number | | European | | | | |
Brass	Alumi-num	IEC Number	Width (mm)	Height (mm)	f_c (GHz)	Useful Range (GHz)
69/U	103/U	R 14	165	82.6	0.91	1.14–1.73
104/U	105/U	R 22	109	54.6	1.38	1.72–2.61
112/U	113/U	R 26	86.4	43.2	1.74	2.17–3.30
48/U	75/U	R 32	72.1	34.0	2.08	2.60–3.95
49/U	95/U	R 48	47.6	22.1	3.16	3.94–5.99
50/U	106/U	R 70	34.9	15.8	4.29	5.38–8.17
51/U	68/U	R 84	28.5	12.6	5.26	6.57–9.99
52/U	67/U	R 100	22.9	10.2	6.56	8.20–12.5
91/U	107/U	R 140	15.8	7.9	9.49	11.9–18.0
53/U	121/U	R 220	10.7	4.3	14.08	17.6–26.7

CHAPTER 1

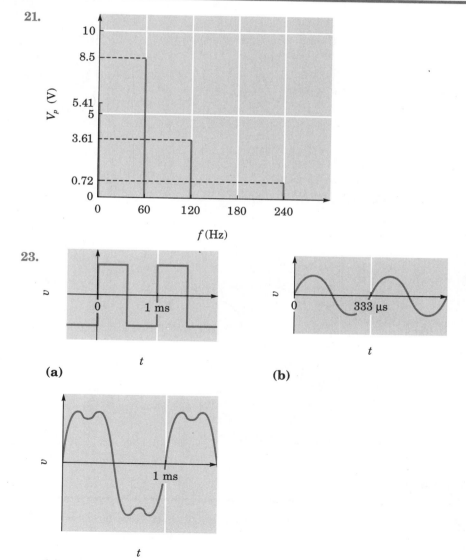

21.

23.

(a)

(b)

(c)

25. **(a)** 90.1 nV   **(b)** 2.2 μV

27. **(a)** 2.7 dB   **(b)** 249 K
29. **(a)** 1.29 dB
     (b) Better: lower NF means less noise contribution and better S/N
31. $A_T = 6000 = 37.8$ dB, $NF = 3.32 = 5.2$ dB
33. **(a)** $f = 867$ MHz, $P = -50$ dBm = 10 nW, $V = 707$ μV
     (b) $f = 80$ MHz, $P = -19$ dBm = 12.5 μW, $V = 25$ mV
     (c) $f = 260$ kHz, $P = 12$ dBm = 15.8 mW, $V = 890$ mV
35. 2.27 dB
37. 42.5 dB

CHAPTER 2

21. **(a)** 17.6 pF   **(b)** 1357   **(c)** 3.39 V
23. **(a)** Class C: zero bias with no input   **(b)** 0

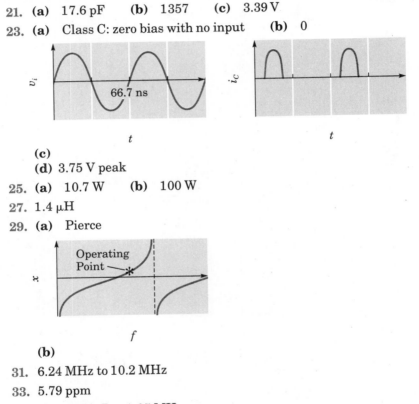

     (c)
     (d) 3.75 V peak
25. **(a)** 10.7 W   **(b)** 100 W
27. 1.4 μH
29. **(a)** Pierce

     (b)
31. 6.24 MHz to 10.2 MHz
33. 5.79 ppm
35. $A_V = 6.64$, $B = 1.05$ MHz
37. 120 MHz to 220 MHz

CHAPTER 3

13. $v(t) = [5 + 2 \sin (3.14 \times 10^3 t)] \sin (18.8 \times 10^6 t)$, $m = 0.4$
15. 0.56
17. $m = 0.75$, $V_c = 283$ V RMS
19. 7.2 MHz, 7.2015 MHz, 7.203 MHz, 7.1985 MHz, 7.197 MHz

21. (a) 0.8 **(b)** 6 kHz **(c)** 20 V

23.

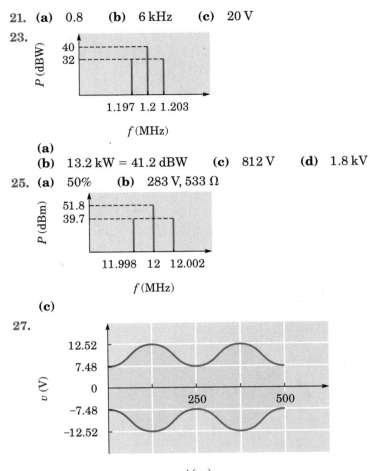

(a)
(b) 13.2 kW = 41.2 dBW **(c)** 812 V **(d)** 1.8 kV

25. (a) 50% **(b)** 283 V, 533 Ω

(c)

27.

CHAPTER 4

21. 27.2064 MHz, 27.2036 MHz

23. 63.7%

25.

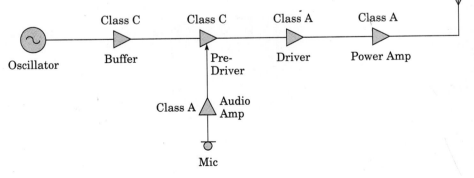

27. 17.9 W
29. **(a)** 857 mA   **(b)** 58.3 Ω   **(c)** 21.4 W   **(d)** 200 V
31. 0.38 dB
33. 20 W
35. **(a)** LED glows when collector voltage goes negative.
     (b) LED will not glow until collector voltage exceeds about − 2 V.
37. 500 μS
39.

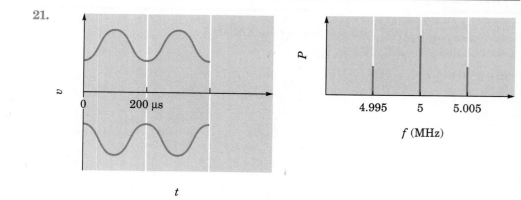

            (a)   **(b)**

CHAPTER 5

23. **(a)** 1.65 MHz   **(b)** High-side injection
25. 108 dB
27. **(a)** 17.5 MHz   **(b)** 14 MHz
29. 15 kHz
31. **(a)** 58.6 dB   **(b)** 9.08 dB
33. 359 kHz
35. **(a)** S-7   **(b)** S-9 + 6 dB   **(c)** S-9 + 32 dB
37. **(a)** Audio sine wave   **(b)** AM at IF
     (c) AM at IF   **(d)** AM at signal frequency

CHAPTER 6

21.

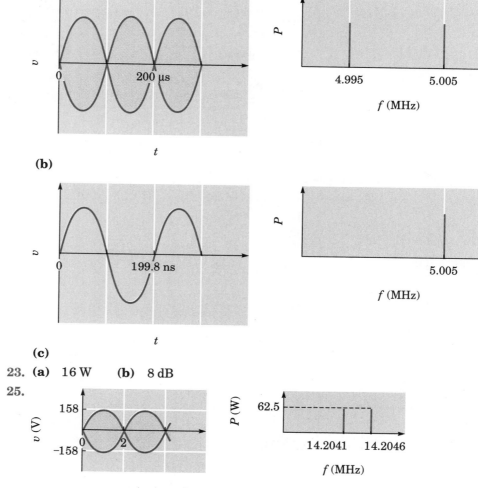

(b)

(c)

23. (a) 16 W   **(b)** 8 dB

25.

t (ms)

27. 4.99837 MHz

29. (a) 14.9985 MHz to 15.4985 MHz and 4.4985 MHz to 4.9985 MHz
  (b) USB for both   **(c)** 10.0015 MHz

33. 3 dB reduction

35.

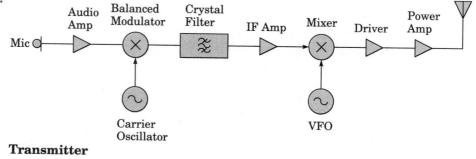

Transmitter

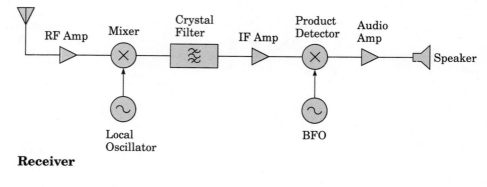

Receiver

37. Transmitter 2, by 8.8 dB
39. **(a)** 0.9 kHz, 1.9 kHz, 2.9 kHz   **(b)** 1.1 kHz, 2.1 kHz, 3.1 kHz

CHAPTER 7

21.

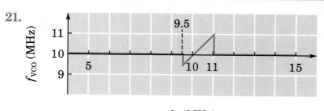

23. **(a)** 3.025 MHz   **(b)** 3.1 MHz
25. N ranges from 2 to 20.

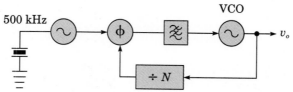

27. 10.2 MHz
29. 8.5 MHz
31. 562 nV
33. 0.01%, 1%, no count
35. **(a)** 300 kHz   **(b)** 266 kHz   **(c)** 480 kHz   **(d)** 0.83
37. One solution is shown below. Others are possible.

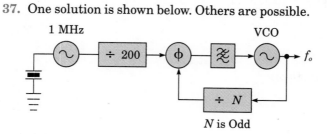

39. One solution is shown below. Others are possible.

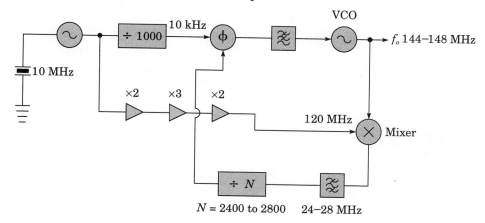

CHAPTER 8

21. 5

23.

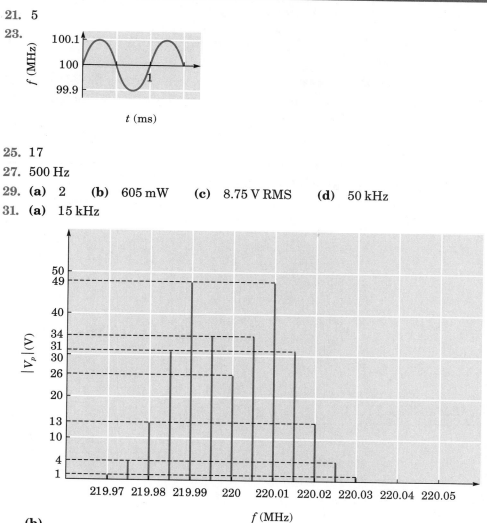

25. 17

27. 500 Hz

29. **(a)** 2   **(b)** 605 mW   **(c)** 8.75 V RMS   **(d)** 50 kHz

31. **(a)** 15 kHz

(b)

     (c)   70 kHz   (d)   40 kHz

33. (a)   16 kHz   (b)   156 kHz

35. (a)   De-emphasis, $R = 7.5$ kΩ   (b)   Pre-emphasis, $C = 75$ nF

37. Use $f_m = 31.25$ kHz for first null or 13.6 kHz for second null.

39. For full-carrier AM, sideband power is in addition to carrier power, which remains constant. For SSB, the carrier power is zero.

CHAPTER 9

23. $f_c = 648$ MHz, $\delta = 162$ kHz

25. (a)   16.7 MHz   (b)   0.21 rad   (c)   Converts PM to FM

27. (a)   911   (b)   26.5 kHz/V

29. (a)   1.5 V (peak)   (b)   60 kHz

31. (a) 563×10^3 or 115 dB   (b)   Yes, but the noise advantage of FM would be lost.

33. (a)   156.7 MHz, 139.8 MHz
    (b)   Transmitter is direct-FM using PLL.
    (c)   16 kHz (from Carson's rule)

35. 4 μV

37. Increase signal level slightly and measure SINAD again.

39. Approach (b), because modulator output is proportional to deviation.

CHAPTER 10

21. (a)   6 V   (b)   155 Ω   (c)   0.567

23. 359 Ω

25. (a) -0.371 (b)   43.1 W

27. 5 m

29. (a)   1.5   (b) -0.2 (c)   400 μW

31. 0.594 μV

33. (a)   0.082λ   (b)   52.5 Ω

35. Length 2.48 m, position 0.81 m from load

37. (a)   1.5 m   (b)   190 MHz   (c)   3.33   (d)   167 Ω

39. 112 Ω

CHAPTER 11

21. 107×10^6 m/s

23. 135 Ω

25. (a)   0.897 m²   (b)   5.95 pW

27. (a)   49 dBW   (b)   178 pW

29. 21°

31. 4.4 MHz approx.

33. **(a)** 34.2 km **(b)** 43.4 km

35. 10 μs

37. Satellite 253 ms; cable 5 ms; terrestrial microwave 3.3 ms

39. **(a)** Ionospheric propagation is more likely at these frequencies at night.
    (b) Ionospheric propagation is common at 27 MHz, especially near peaks of the solar cycle.
    (c) This phenomenon causes disturbances in the ionosphere.
    (d) The likely cause is tropospheric ducting.
    (e) There is refraction in the troposphere.

CHAPTER 12

21. 9.5 m

23. 2.71 mV/m

25. Gain = 5 dBi, beamwidth = 20°

27. 71.3 m

29. Draw lines at right angles to each antenna's axis; find point of intersection.

31. **(a)** Add base or center inductive loading.
    (b) No, because of losses due to low radiation resistance and presence of coil.

33. $L_2 = 2.14$ m, $L_3 = 3.06$ m, $D_1 = 2.8$ m, $D_2 = 4$ m, $D_3 = 5.7$ m

35. **(a)** 3 dB increase in gain   **(b)** Cancellation

37. **(a)** 39 dBi   **(b)** 39 dBi

39. 16.7λ

41. **(a)** 415 μW   **(b)** 13.1 μW

43. 7.68 dBi

CHAPTER 13

21. **(a)** 397×10^6 m/s   **(b)** 226×10^6 m/s
    (c) 39.7 mm   **(d)** 500 Ω

23. 5 μm

25. **(a)** 0.001   **(b)** 1.67 MW   **(c)** 1 kW   **(d)** 1.67 kW   **(e)** 670 W

27. Max. 750 km, min. 7.5 km

29. **(a)** 500 m/s   **(b)** Toward

31. $f_D = 1.85 v_r f_i$

33. **(a)** 8.4 m   **(b)** 58.5 dBW   **(c)** 40.1 dBi   **(d)** 53 dBf

35. **(a)** 638 mW   **(b)** 535 μW

CHAPTER 14

21. **(a)** 417 μs   **(b)** 6.6

23. **(a)** $Y = 0.498$, $I = 0.304$, $Q = 0.022$ **(b)** 53.6 IRE

25. **(a)** 669.25 MHz   **(b)** 672.83 MHz   **(c)** 673.75 MHz

27.

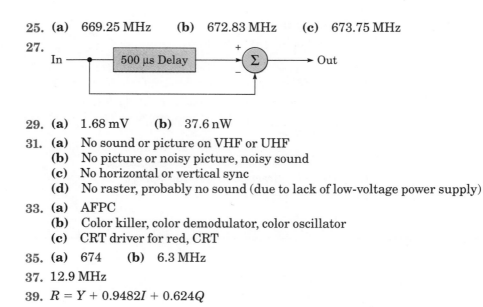

29. **(a)** 1.68 mV   **(b)** 37.6 nW

31. **(a)** No sound or picture on VHF or UHF
 (b) No picture or noisy picture, noisy sound
 (c) No horizontal or vertical sync
 (d) No raster, probably no sound (due to lack of low-voltage power supply)

33. **(a)** AFPC
 (b) Color killer, color demodulator, color oscillator
 (c) CRT driver for red, CRT

35. **(a)** 674   **(b)** 6.3 MHz

37. 12.9 MHz

39. $R = Y + 0.9482I + 0.624Q$

CHAPTER 15

21. **(a)** 48 Mb/s   **(b)** 24 dB

23. **(a)** 22.05 kHz   **(b)** 98.08 dB   **(c)** 1.41 Mb/s   **(d)** 65536

25. 1.521 Mb/s

27. **(a)** BE HERE59,853 ?6 8   **(b)** BE HERE59,853 BY 8

29. Bit

31.

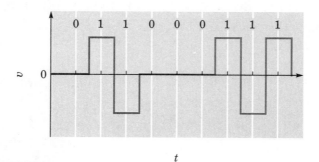

(a) **Unipolar NRZ**

(b) **AMI**

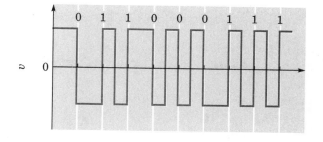

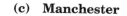

(c)   Manchester

33. Bit rate is 24 kb/s.

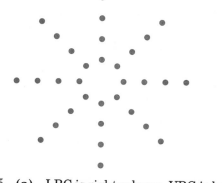

35. **(a)**   LRC is right column, VRC is bottom row.
    (b)   Error is 4th bit down in second character.
37. **(a)**   8   **(b)**   640 Mb/s

CHAPTER 16

21. **(a)**   750 THz (1 THz = 1×10^{12} Hz)
    (b)   429 THz
    (c)   333 THz
23. **(a)**   75°   **(b)**   0.384   **(c)**   22.6°
25. **(a)**   Infrared   **(b)**   194 THz
27. 10.5 GHz-km
29. 497×10^{-21} J, 3.11 eV
31. 12.75 μA, 127.5 mV
33. Yes

Index